Adsorption of Metals by Geomedia

Adsorption of Metals by Geomedia

Variables, Mechanisms, and Model Applications

Edited by

Everett A. Jenne
Senior Staff Scientist (Retired)
Battelle, Pacific Northwest National Laboratory
Richland, Washington

Academic Press

San Diego London Boston New York Sydney Tokyo Toronto

Cover photograph: SEM micrograph of hydroxylapatite from BioRad.
(See Figure 5 in chapter 6.)

This book is printed on acid-free paper. ∞

Academic Press
a division of Harcourt Brace & Company
525 B Street, Suite 1900, San Diego, California 92101-4495, USA
http://www.apnet.com

Academic Press Limited
24-28 Oval Road, London NW1 7DX, UK
http://www.hbuk.co.uk/ap/

Library of Congress Card Catalog Number: 98-84493

International Standard Book Number: 0-12-384245-X

PRINTED IN THE UNITED STATES OF AMERICA
98 99 00 01 02 03 MM 9 8 7 6 5 4 3 2 1

Contents

Chapter 2

Uranium^{VI} Adsorption on Model Minerals: Controlling Factors and Surface Complexation Modeling

T. E. Payne, G. R. Lumpkin, and T. D. Waite

Chapter 3
Uranium^{VI} Sorption onto Selected Mineral Surfaces: Key Geochemical Parameters

Roberto T. Pabalan, David R. Turner, F. Paul Bertetti, and James D. Prikryl

Chapter 4
Studies of Neptunium^{V} Sorption on Quartz, Clinoptilolite, Montmorillonite, and α-Alumina

F. Paul Bertetti, Roberto T. Pabalan, and Michael G. Almendarez

Chapter 5

Factors Affecting Trivalent *f*-element Adsorption to an Acidic Sandy Soil

S. B. Clark, A. L. Bryce, A. D. Lueking, J. Gariboldi, and S. M. Serkiz

Chapter 6

Lead Sorption, Chemically Enhanced Desorption, and Equilibrium Modeling in an Iron-Oxide-Coated Sand and Synthetic Groundwater System

Mohammad F. Azizian and Peter O. Nelson

Chapter 7

Uranium Sorption onto Natural Sands as a Function of Sediment Characteristics and Solution pH

Jeffrey J. Rosentreter, H. Swantje Quarder, Robert W. Smith, and Travis McLing

Chapter 8

Intraparticle Diffusion of Metal Contaminants in Amorphous Oxide Minerals

Lisa Axe and Paul Anderson

Chapter 9

Copper Sorption Kinetics and Sorption Hysteresis in Two Oxide-Rich Soils (Oxisols): Effect of Phosphate Pretreatment

Luiz Roberto G. Guilherme and Sharon J. Anderson

Chapter 10

Influence of pH, Metal Concentration, and Soil Component Removal on Retention of Pb and Cu by an Illitic Soil

Raymond N. Yong and Elaine M. MacDonald

Chapter 11

Immobilization of Pb by Hydroxylapatite

Valérie Laperche and Samuel J. Traina

Chapter 14

Surface and Solution Speciation of Ag^{I} in a Heterogeneous Ferrihydrite-Solution System with Thiosulfate

Colin G. Ong and James O. Leckie

Chapter 15

Extended X-Ray Absorption Fine Structure (EXAFS) Analysis of Aqueous Sr^{II} Ion Sorption at Clay–Water Interfaces

Chia-Chen Chen, Charalambos Papelis, and Kim F. Hayes

Chapter 16

Structure and Composition of Uranium^{VI} Sorption Complexes at the Kaolinite–Water Interface

H. A. Thompson, G. A. Parks, and G. E. Brown, Jr.

Chapter 17

Surface Charge and Metal Sorption to Kaolinite

Patrick V. Brady, Randall T. Cygan, and Kathryn L. Nagy

Chapter 18

Molecular Models of Cesium Sorption on Kaolinite

Randall T. Cygan, Kathryn L. Nagy, and Patrick V. Brady

Chapter 21
Fluorescence Quenching and Aluminum Adsorption to Organic Substances

D. Scott Smith and James R. Kramer

Chapter 22
Modeling of Competitive Ion Binding to Heterogeneous Materials with Affinity Distributions

Michal Borkovec, Ursula Rusch, and John C. Westall

Chapter 23

Ion Binding to Humic Substances: Measurements, Models, and Mechanisms

D. G. Kinniburgh, W. H. van Riemsdijk, L. K. Koopal, and M. F. Benedetti

Chapter 24

Predictive Double-Layer Modeling of Metal Sorption in Mine-Drainage Systems

Kathleen S. Smith, James F. Ranville, Geoffrey S. Plumlee, and Donald L. Macalady

Epilogue
Priorities for Future Metal Adsorption Research

Everett A. Jenne

Contributors

The numbers in parentheses indicate the pages on which the authors' contributions begin.

Michael G. Almendarez (131), Center for Nuclear Waste Regulatory Analyses, Southwest Research Institute, San Antonio, Texas 78238

Paul Anderson (193), Department of Chemical and Environmental Engineering, Illinois Institute of Technology, Chicago, Illinois 60616

Sharon J. Anderson (209), Department of Crop and Soil Sciences, Michigan State University, East Lansing, Michigan 48824[1]

Janick F. Artiola (427), Soil, Water and Environmental Science, University of Arizona, Tucson, Arizona 85721

Lisa Axe (193), Department of Civil Environmental Engineering, New Jersey Institute of Technology, Newark, New Jersey 07102

Mohammad F. Azizian (165), Department of Civil, Construction, and Environmental Engineering, Oregon State University, Corvallis, Oregon 97331

M. F. Benedetti (483), URA CNRS-1762, Laboratoire de Géochimie et Métallogénie, Université Curie, Paris, France

F. Paul Bertetti (99, 131), Cambrian Systems, Inc., San Antonio, Texas 78238

Michal Borkovec (467), Institute of Terrestrial Ecology, Swiss Federal Institute of Technology, Schlieren, Switzerland

[1]Present address: Institute for Earth Systems Science and Policy, California State University/Monterey Bay, Seaside, California 93955

Patrick V. Brady (371, 383), Sandia National Laboratories, Albuquerque, New Mexico 87123

Gordon E. Brown, Jr. (349), Department of Geological and Environmental Sciences, Stanford University, Stanford, California 94305

Mark L. Brusseau (427), Soil, Water and Environmental Science, University of Arizona, Tucson, Arizona 85721

A. L. Bryce (149), University of Georgia, Savannah River Ecology Laboratory, Aiken, South Carolina[2]

Chia-Chen Chen (333), Environmental and Water Resources Engineering, Department of Civil and Environmental Engineering, University of Michigan, Ann Arbor, Michigan 48109

Sue B. Clark (149), University of Georgia, Savannah River Ecology Laboratory, Aiken, South Carolina[3]

Randall T. Cygan (371, 383), Sandia National Laboratories, Albuquerque, New Mexico 87123

Harold S. Forster (401), USDA-ARS, U.S. Salinity Laboratory, Riverside, California 92507

J. Gariboldi (149), University of Georgia, Savannah River Ecology Laboratory, Aiken, South Carolina

Sabine Goldberg (401), USDA-ARS, U.S. Salinity Laboratory, Riverside, California 92507

Luiz Roberto G. Guilherme (209), Department of Crop and Soil Sciences, Michigan State University, East Lansing, Michigan 48824[4]

Kim F. Hayes (333), Environmental and Water Resources Engineering, Department of Civil and Environmental Engineering, University of Michigan, Ann Arbor, Michigan 48109

Harold B. Hume (277), AECL, Whiteshell Laboratories, Pinawa, Manitoba R0E 1L0, Canada

Everett A. Jenne (1, 549), Battelle, Pacific Northwest National Laboratory, Richland, Washington 99352[5]

D. G. Kinniburgh (483), Department of Soil Science and Plant Nutrition, Wageningen Agricultural University, The Netherlands[6]

[2]Present address: United States Nuclear Regulatory Commission, Washington, DC 20555

[3]Present address: Department of Chemistry, Washington State University, Pullman, Washington 99164

[4]Present address: Departamento di Ciencia de Solo, Universidade Federal de Lavras, Lavras, Brazil

[5]Retired

[6]On leave from the British Geological Survey, Wallingford, United Kingdom

L. K. Koopal (483), Department of Physical and Colloid Chemistry, Wageningen Agricultural University, The Netherlands

James R. Kramer (445), Department of Geology, McMaster University, Hamilton, Ontario, Canada L8S 4M1

Valérie Laperche (255), School of Natural Resources, Ohio State University, Columbus, Ohio 43210

James O. Leckie (291, 317), Department of Civil and Environmental Engineering, Stanford University, Stanford, California 94305

Jinhe Li (291), Department of Civil and Environmental Engineering, Stanford University, Stanford, California 94305

A. D. Lueking (149), University of Georgia, Savannah River Ecology Laboratory, Aiken, South Carolina[7]

G. R. Lumpkin (75), Australian Nuclear Science and Technology Organisation, Menai, NSW 2234, Australia

Donald L. Macalady (521), Department of Chemistry and Geochemistry, Colorado School of Mines, Denver, Colorado

Elaine M. MacDonald (229), Department of Civil Engineering, McGill University, Montreal, Canada

Travis McLing (181), Idaho National Engineering and Environmental Laboratory, Idaho Falls, Idaho 83415

Kathryn L. Nagy (371, 383), Department of Geological Sciences, University of Colorado, Boulder, Colorado 80309

Peter O. Nelson (165), Department of Civil, Construction, and Environmental Engineering, Oregon State University, Corvallis, Oregon 97331

Colin G. Ong (317), Department of Civil and Environmental Engineering, Stanford University, Stanford, California 94305

Dennis W. Oscarson (277), AECL, Whiteshell Laboratories, Pinawa, Manitoba R0E 1L0, Canada

Roberto T. Pabalan (99, 131), Center for Nuclear Waste Regulatory Analyses, Southwest Research Institute, San Antonio, Texas 78238

Charalambos Papelis (333), Desert Research Institute, Water Resources Center, University and Community College System of Nevada, Las Vegas, Nevada 89119

George A. Parks (349), Department of Geological and Environmental Sciences, Stanford University, Stanford, California 94305

[7]Present address: Department of Civil Engineering, University of Michigan, Ann Arbor, Michigan 48109

T. E. Payne (75), Australian Nuclear Science and Technology Organisation, Menai, NSW 2234, Australia

Geoffrey S. Plumlee (521), U.S. Geological Survey, Denver, Colorado 80225

James D. Prikryl (99), Center for Nuclear Waste Regulatory Analyses, Southwest Research Institute, San Antonio, Texas, 78238

H. Swantje Quarder (181), Department of Chemistry, Idaho State University, Pocatello, Idaho 83209

James F. Ranville (521), Department of Chemistry and Geochemistry, Colorado School of Mines, Denver, Colorado

George Redden (291), Department of Civil and Environmental Engineering, Stanford University, Stanford, California 94305

Jeffrey Rosentreter (181), Department of Chemistry, Idaho State University, Pocatello, Idaho 83209

Ursula Rusch (467), Institute of Terrestrial Ecology, Swiss Federal Institute of Technology, Schlieren, Switzerland

S. M. Serkiz (149), Westinghouse Savannah River Company, Savannah River Technology Center, Aiken, South Carolina

D. Scott Smith (445), Department of Geology, McMaster University, Hamilton, Ontario, Canada L8S 4M1

Kathleen S. Smith (521), U.S. Geological Survey, Denver, Colorado 80225

Robert W. Smith (181), Idaho National Engineering and Environmental Laboratory, Idaho Falls, Idaho 83415

Chunming Su (401), USEPA, National Risk Management Research Laboratory, Ada, Oklahoma 74820

Hillary A. Thompson (349), Department of Geological and Environmental Sciences, Stanford University, Stanford, California 94305

Samuel J. Traina (255), School of Natural Resources, Ohio State University, Columbus, Ohio 43210

David R. Turner (99), Center for Nuclear Waste Regulatory Analyses, Southwest Research Institute, San Antonio, Texas 78238

W. H. van Riemsdijk (483), Department of Soil Science and Plant Nutrition, Wageningen Agricultural University, The Netherlands

T. D. Waite (75), Department of Water Engineering, University of New South Wales, Sydney NSW 2052, Australia

Wei-Zi Wang (427), Soil, Water and Environmental Science, University of Arizona, Tucson, Arizona 85721

John C. Westall (467), Clarkson University, Potsdam, New York

Raymond N. Yong (229), Geoenvironmental Engineering Research Centre, Cardiff School of Engineering, University of Wales Cardiff, United Kingdom

Preface

Metal adsorption and desorption are crucial processes in the remediation of contaminated surficial geomedia (e.g., soils) and aquifers. Contaminant transport modeling, which necessarily includes some adsorption parameter, has important uses in anticipating the direction, speed, and extent of contaminant migration in ground water. Adsorption processes are of vital importance in many industrial processes in the beneficiation of various ores (some of which were formed largely as the result of adsorption processes).

Metal adsorption is a rapidly expanding, very dynamic area of earth science research. The advent of surface analysis instrumentation and methodologies, which provide information on the nature of adsorbed complexes, has opened new vistas for the interpretation of metal adsorption by geomedia. Surface complexation models, which utilize information on the nature of metal complexes adsorbed at solid surfaces, have largely displaced the many variants of Freundlich, Langmuir, and mass action equilibrium constant relationships that have been in use since the early days of adsorption research. Concurrently, surface complexation-based adsorption models are being increasingly used in field applications, most commonly in an interpretive mode, but increasingly in a predictive mode. In field applications of surface complexation adsorption models, there is a subtle change from refining adsorption constants for a geomedia or site-specific model to attempting to use preexisting constants to predict *de facto* the results of experiments. Both surface analysis research and the application of models to heterogeneous geomedia in the laboratory and in the field have focused renewed interest in important variables. The detailed analysis of variables shown in this book and elsewhere indicates a high degree of interrelationships among variables. This points to the acute need for future studies that include numerous variables on a relatviely few, well-characterized samples.

Metal and radioisotope adsorption modeling in the earth sciences is currently in a state of tension. On one side are those investigators carrying out detailed analyses of variables or producing mechanistic information at the molecular level. On the other side is the traditional ground water and soil contaminant modeling sector using generic distribution coefficients. Each generic distribution coefficient generally represents a single point in a multidimensional space (i.e., on a multivariable response surface) and contains little or no information on the slope along any variable axis other than perhaps pH and metal concentration. The determination of distribution coefficients has been covered in many prior conference proceedings and is not specifically addressed in this book.

In April of 1996, numerous researchers from five or more nations gathered at the American Chemical Society Meeting for $2\frac{1}{2}$days of presentations and discussions. Of the presentations made at that meeting, 24 have made it into this book, surviving two or more peer reviews plus editorial reviews. Along with an introductory chapter to acquaint students and investigators who have recently entered this research area with the terminology of the field, the breadth and depth of coverage of metal adsorption by single-phase and heterogenous geomedia will make this book of value to all levels of investigators and students in the metal adsorption, metal precipitation, and remediation fields.

Chapter 1

Adsorption of Metals by Geomedia: Data Analysis, Modeling, Controlling Factors, and Related Issues

Everett A. Jenne (Retired)
Battelle, Pacific Northwest Laboratory, Richland, Washington

An overview of common adsorption models is presented, indicating that they are all variants of the mass action equation, either with simplifications or added terms. The basis of isotherm linearization by log–log plots is presented. The success of Freundlich versus Langmuir equations depends on the concentration range of the adsorbent, the capacity of the adsorbent, and the geomedia concentration. Artifacts common to adsorption data are noted (e.g., precipitation). The principal variables effecting the extent of metal adsorption are reviewed and a high degree of interaction among these variables is identified. This extensive interaction of variables (pH, $P\text{co}_2$, metal concentration, loading of sites, aqueous speciation, surface complexes, stoichiometry, site density, time dependency, and others) raises questions concerning the value of traditional single- and bivariable adsorption experiments. The nature of surface complexes is reviewed and the insight provided by surface analysis is noted. The importance of terminology and the need for adequate adsorbent characterization are noted. This chapter is written with the intent that it can be understood by upper level students.

Adsorption of Metals by Geomedia

I. INTRODUCTION

Major progress in understanding adsorption mechanisms on oxide and silicate surfaces, as well as in modeling adsorption, has occurred in the past two decades (e.g., Davis and Hayes, 1986; Davis and Kent, 1990). These studies have led to the development of models based on the conceptualization of adsorption[1] as a system of complexation reactions at the adsorbent–water interface. However, in the adsorption field, terminology remains somewhat confused and, in the absence of adequate system characterization and multivariable experiments, controversy has continued as to the importance of certain variables and their effect on the extent of surface complexation reactions of added adsorbate.[2] The objectives of this chapter are to assist in the clarification of terminology, to overview and evaluate the important factors affecting the extent of adsorption, to show some relationships among available models, to stress the importance of the interaction of variables and the mode of presenting adsorption data, and to note the important insight provided by surface analysis. This chapter draws heavily on the U literature as this was an item of focus at the time this chapter was started. This chapter is written such that upper level students should be able to understand the material with little difficulty. With students in mind, cited literature includes some of the older references rather than only those of the past few years. An effort has been made to cite original sources when feasible and appropriate.

A. Importance of Adsorption and Literature Citation

Adsorption is critical in many geochemical processes such as the removal of contaminants from ground and surface waters used for drinking water, in many chemical engineering processes, and in the formation of uranium ore deposits although precipitation is often the primary process (Langmuir, 1978; Giblin, 1980). Adsorption data are critical to radionuclide retardation calculations which are in turn integral to performance assessments for the siting of both low- and high-level nuclear waste repositories and the decommissioning of sites contaminated with radioactive materials. An increased understanding of adsorption processes and the factors affecting them can improve the transferability of adsorption measurements and the reliability of models used in interpreting past, and estimating future, migration of contaminants.

[1]Adsorption may be defined as "the process by which atoms, molecules, or ions are taken up from the aqueous or gas phase and retained on the surfaces of solids by chemical or physical binding" (modified from Soil Science Society of America, 1996, p. 2).

[2]Adsorbate is a dissolved specie, or multiple species, that becomes concentrated at the surface of some solid, which is the adsorbent.

One of the deficiencies in the current adsorption literature is inadequate citation of extant literature. As Roy (1996) has pointed out, this omission of appropriate citations is often not noticed by peer reviewers, who may not be personally aware of relevant papers, and is countenanced by editors, as a limited review of the literature tends to shorten papers. Inadequate literature review and summarization leads to rediscovery of the past and slows progress in the adsorption field. As Roy also noted, an author's omission of appropriate references tends to give the illusion that the work is "original" rather than simply redoing, or elaborating on, earlier work. For example, Horowitz *et al.* (1996) claim that during the development of the U.S. Geological Survey's new "clean" protocol (Horowitz *et al.*, 1994), it became apparent that based on aqueous concentrations of Fe and Al 0.45-μm membranes allowed sufficient colloidal matter to pass that significant amounts of colloid-associated trace elements were include in the filtrate (Horowitz *et al.*, 1992). In fact, these results were unequivocally demonstrated in the 1974 paper of Kennedy *et al.* (see also Jones *et al.*, 1974; Kennedy *et al.*, 1976).

B. Importance of Terminology

Clarity of terminology is a factor that either facilitates or hinders progress in a field of research by impacting the thought processes of investigators. The importance of clarity of nomenclature to progress in a field is easily underestimated. This may be particularly important for those new to a field. See for example, the clarity and conciseness provided by the use of Roman superscripts to indicate formal valence and Arabic numbers for specie charge provided in the chapter by Ong and Leckie (1998, this volume). The several pages of acronyms and abbreviations (Appendix A) required for the 24 Chapters of this book is a testimony to the complexities of the field and the difficulties in understanding adsorption papers in areas other than one's own subdiscipline. In spite of considerable editorial effort, the redundance in the use of individual symbols is considerable (m for instance). The use of a system of intuitive abbreviations, with extensive use of subscripts and superscripts as used in this chapter would greatly reduce the number of symbols required in the field.

In his 1918 paper, Langmuir viewed "adsorption" as being a chemical process with similarities to the chemical bonding between atoms in a crystal with unsatisfied bonds in contact with solvent water. However, in a 1977 (p. 235) review, Thomas concluded that "The question raised over 100 years ago, as to whether cation exchange is chemical or an adsorption reaction, is still not answered." Although this issue has since been resolved in favor of adsorption being a chemical reaction it appears that the generally used term "adsorption" is somewhat of a misnomer in that it apparently conveys the impression of a one-way process to some investigators.

The term adsorption is a carryover from early studies of the adsorption of N_2, CH_4, CO, Ar, O_2, and CO_2 onto mica, glass, and Pt surfaces (Langmuir, 1918). Adsorption onto such surfaces may or may not involve exchange with a previously adsorbed molecule. Adsorbates, such as those used by Langmuir, may or may not be charged depending on the individual gases and media involved. In contrast, adsorption of a cationic metal by geomedia (i.e., rock, sediment, soil, mineral) involves the displacement of H or other previously adsorbed cations,[3] i.e., cation exchange (Sposito, 1980a). The general absence of information on desorbed cations, other than the H, suggests that not all investigators appreciate that the adsorption of metal cations involves the simultaneous desorption of previously adsorbed cations. However, when current investigators specify cation exchange they commonly mean adsorption at pH-independent sites, i.e., permanent charge sites (Azizian and Nelson, 1998, this volume; Clark *et al.,* 1998, this volume). Chen *et al.* (1998, this volume) point out that most Sr^{II} adsorption studies assume that the only Important process is ion exchange. Indeed, they concluded that Sr^{II} maintained the same hydration geometry upon adsorption (following removal of carbonates, reactive particulate organic carbon {RPOC}, and oxides of Fe and Mn) as it had in the aqueous phase, which was taken as evidence that it sorbed as a mononuclear outer-sphere complex irrespective of clay type (i.e., kaolinite, illite, hectorite, and montmorillonite) or geochemical conditions. This is the type of surface complex is generally the one found on permenant charge sites.

An unfortunate result of the frequently inadequate conceptualization of the adsorption process is that the composition of equilibrated solutions is generally unreported although exceptions occur (e.g., Anderson *et al.,* 1973; Lieser *et al.,* 1986; Thomas, 1987; Turner *et al.,* 1996). Similarly, only rarely are desorbed cations, other than the adsorbate of interest, determined (e.g., Harter, 1991; Turner *et al.,* 1996).

Contributing to the confusion regarding the extent to which adsorption involves exchange with previously adsorbed cations is the lack of apparent stoichiometric exchange between adsorbing and desorbing components (i.e., hydrogen and metals) that is often observed. As will be discussed subsequently, the failure to distinguish between the effect of ionic strength ($\boldsymbol{I}$) and the concentration of competing cations ($C_{\text{competing cations}}$) is the probable cause of large differences in the reported importance of $\boldsymbol{I}$ on the adsorption of metals.

The terms "rocks," "soils," and "sediments" are individually too exclusive for use as a general term in referring collectively to earth materials that include individual minerals. Inclusive terms include earth materials, geomedia, and solids.

[3]The charge of ionic species is generally omitted in the text but usually included in equations. The ensemble of dissolved species of a metal is indicated by either the element symbol or a superscripted valence state since it is incorrect to designate the multiple dissolved species of a metal using the notation for the single uncomplexed specie.

Geomedia is generally used in this chapter following ASTM practice (ASTM, 1993, 1994). The term solids is preferentially used when referring to synthetic or essentially single-phase materials.

C. Importance of System Characterization

Available adsorption data are commonly obtained in one of three types of investigations, i.e., (1) a causal characterization of adsorption properties of a geomedia sample; (2) systematic study of one or, less frequently, two or more variables; and (3) mechanistic investigations. Current modeling studies are commonly done in conjunction with type 2 or type 3 studies. The first category is probably the most numerous because of the large numbers of such studies funded by Federal agencies (e.g., the former Atomic Energy Commission, the former Energy Research and Development Administration, and the current Department of Energy). Much of the type 1 and 2 data are published in nonpeer reviewed conference proceedings and national laboratory reports. In many cases, these experiments were intended to mimic specific systems (Benjamin, 1979, p. 30) and/or were of a survey nature funded by Federal agencies with particular missions. Although these considerations presumably account for the minimal range of the variable(s) studied and system characterizations reported, there is no justification for the frequently encountered situation where the specification of experimental details is inadequate to allow the reader to recalculate published data in units other than that of the author(s) for comparison with other data.

The adsorption literature indicates that reaction (equilibration) time (t), pH, partial pressure of CO_2 (P_{CO_2}), final aqueous metal concentration ($C_{M,aq,f}$), solids concentration ($C_{geomedia}$), surface area (A_S), the identity and concentration of competing cations (e.g., C_H, $C_{Ca,}$ C_{Na}), the extent of filling of the capacity of an absorbent (loading) with a particular adsorbate ($C_{M,ads}/C_{M,ads,max}$), the valence state (e.g., U^{VI}, U^{IV}) of redox sensitive metal, prior or simultaneous adsorption of multivalent anions, and temperature can each have an important impact on the adsorption curve. There is, of course, some redundancy in the preceding list of variables. Inadequate attention to the effects of sample preservation and preparation on adsorption data is another important issue. The report of Roy *et al.* (1991) provides a relatively thorough treatment of this aspect of adsorption research.

There are infrequent cautions in the adsorption literature concerning the importance of characterization of the adsorbent and the aqueous solution (e.g., Townsend, 1986; Pabalan, 1994). A recent survey of U adsorption data (E. A. Jenne, unpublished report) suggests that in a surprising number of instances, variables not specifically examined in these studies were neither controlled nor measured (e.g., final pH, final solution composition). As Cremers and Maes (1986) have pointed out, too little attention in metal adsorption investigations is given to the sometimes critical changes that may occur in the composition of the liquid phase and the ensuing changes in "speciation of the nuclide after equilibration." As a consequence of inadequate definition of variables in adsorption experiments, much of the extant adsorption data are of marginal value for purposes of extending our understanding of adsorption process or the development and the testing of adsorption models. In particular, the concentration of displaced cations, which are generally ignored, need to be determined (Townsend, 1986; Pabalan *et al.,* 1998, this volume). Harter (1991), for example, found that the molar quantities of Ca, H, and Mg displaced were equivalent to the amount of Ni adsorbed by two soils for the first several minutes. Beyond 10 to 12 min, the amount of the three cations displaced slightly exceeded the amount of Ni adsorbed. The limitations of system characterization in the extant data have occasionally been noted. For example, Turner (1993, p. 2-9) remarks that he was unable to model the data of Breeuwsma and Lyklema (1973) because of inadequate description of their experimental conditions. Adsorbent characterization is also important. For example, Bradbury and Baeyens (1991) note that data sets which have both surface area measurements and defined water chemistry are relatively rare.

II. DATA TREATMENT AND PRESENTATION

Pabalan *et al.* (1998, this volume) note that "Plotting data in terms of K_d (distribution coefficient, see Eq. [5A]) instead of percent U sorbed makes comparison of different sets of experimental data easier" since in this form of presentation the data show little effect of $C_{geomedia}$. Indeed, the traditional sigmoidal plots of percentage or fraction absorbed versus pH or versus the logarithm (log) of $C_{M,aq,f}$ make meaningful comparison among treatments and samples difficult. Furthermore, sigmoidal plots do not lend themselves to interpolation or extrapolation since an increase in $C_{M,aq,i}$ or $C_{geomedia}$ (and certain other variables) generally causes an increase in the pH and a decrease in the slope of the adsorption edge[4] on sigmoidal curves. The fraction-adsorbed versus pH format has been recommended by Benjamin and Leckie (1982) as a "convenient format for presenting adsorption data." This approach emphasizes the importance of pH or $C_{M,aq,f}$ on $C_{M,ads}$ but neither Benjamin and Leckie nor others using such plots provide a rationale for using semilog plots to depict the pH dependence of adsorption. Such semilog plots (e.g., Fig. 1A) implicitly test the hypothesis that the $C_{M,ads}$ varies

[4]The pH adsorption edge is defined by Dent *et al.* (1992) as "For each polyvalent ion there is also a critical pH range, usually less than 1 pH unit wide, over which the percentage of total metal sorbed onto an oxide surface increases from nearly zero to about 100%," Ong and Leckie (1998, this volume) defined it as the "surface concentration of a sorbing species as a function of pH," whereas Axe and Anderson (1997) describe the adsorption edge for Cd on $Fe(OH)_3$(a) as "a steep S shaped curve." The pH adsorption edge was earlier defined by Benjamin and Leckie (1981b, p. 414) as "where percentage adsorption increases from low to high values as the pH increases by a unit or so."

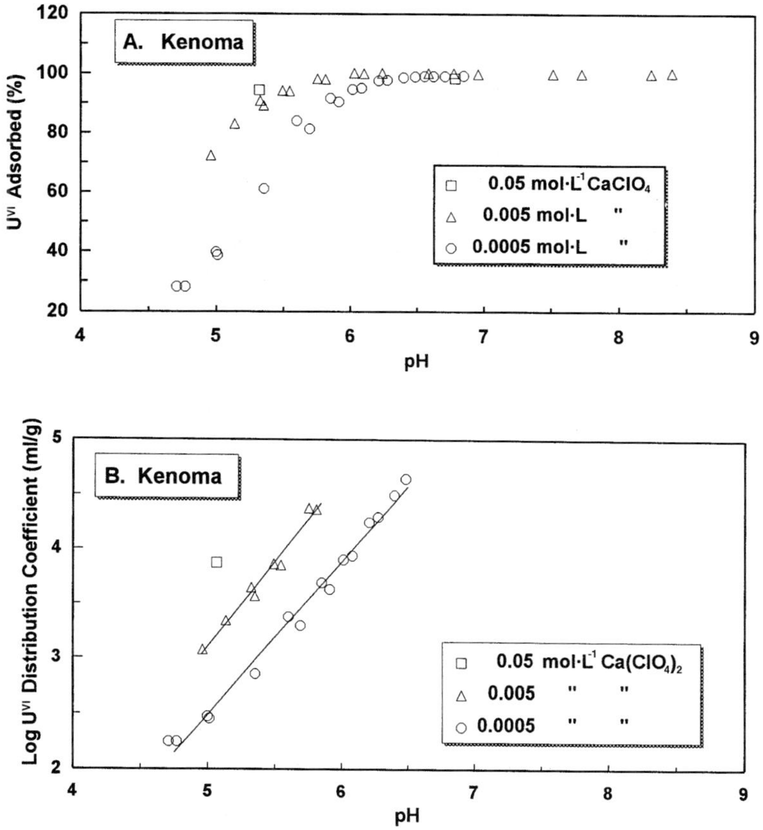

Figure 1 Increase in U^{VI} adsorption with pH for a smectite-rich separate from the Kenoma subsoil (after Zachara *et al.*, 1992, Fig. 6). (Lines on this and subsequent figures are plots of a simple linear regression of the experimental values falling within the pH range spanned by the regression line.)

linearly with the log of the H activity ($\mathbf{a}_H$), which it rarely does; hence, it is an inappropriate hypothesis. Since K_d values are commonly assumed to be log-normally distributed (Kap-lan *et al.,* 1994), semilog plots could be rationalized if log K_d were plotted against C_H, although the wide range in $\mathbf{a}_H$ generally employed makes this impractable. The coherence of data sets displayed as sigmoidal plots where, for example, the adsorption edge is shifted to higher pH value as $C_{M,aq,i}$ is increased or $C_{geomedia}$ is decreased is not intuitive using fractional adsorption plots, whereas visual comparisons are readily made on log–log plots (e.g., compare Figs. 1A and 1B). The data shown in Figure 1 are anomalous in that the $K_{d,U}$ increases as the concentration of Ca increases, whereas one normally finds that the K_d decreases as the concentration of a competing cation increases (see discussion in Section V.C).

Several authors have addressed the visualization problem of semilog plots by approximately linearizing their data with plots of log K_d versus log $\mathbf{a}_H$ (e.g., Allen *et al.,* 1995), log $C_{M,ads}$ ($mol \cdot g^{-1}$ or $mol \cdot mol^{-1}$ of adsorbate:adsorbent) versus log $\mathbf{a}_H$ (Murray 1975), log $C_{M,des}$ versus log $\mathbf{a}_H$ (Benjamin and Felmy, 1981), log $C_{M,ads}$ versus log $C_{M,aq,f}$ (Avotins, 1975; Murray, 1975; Benjamin and Leckie, 1980, 1981; Eylem *et al.,* 1989; Zachara *et al.,* 1989; Amacher *et al.,* 1990; Axe and Anderson, 1998, this volume; Clark *et al.,* 1998, this volume; Kinniburgh *et al.,* 1998, this volume), log $C_{M,ads}$ versus log $C_{M,aq,i}$ (Kurbatov *et al.,* 1951), or log $C_{Cd/Zn,ads}$ versus log $C_{Cd/Zn,aq,f}$ (Tiller *et al.,* 1979). Adsorption density (log $C_{M,ads,area}$ or log $C_{M,ads,mol}$) ($mol \cdot nm^{-2}$ or $mol \cdot mol^{-2}$) is sometimes plotted versus log $C_{Me,aq,f}$ (MacNaughton 1973; Benjamin and Leckie, 1980).

Plots of log K_d versus pH or the fraction adsorbed or desorbed {log ($C_{M,ads}/C_{M,ads,max}$), log ($C_{M,des}/C_{M,des,max}$)} versus log $C_{M,aq,f}$ commonly approximately linearize pH adsorption curves so long as variables other than $C_{M,aq,f}$ and the activity of H^+ ($\mathbf{a}_H$) remain constant (see Jenne, 1995) (Section III.I). Log–log plots fail to linearize K_d versus $\mathbf{a}_H$ adsorption curves when (1) one or more additional variables become important as a result of, for example, a hitherto unimportant competing cation becoming quantitatively important as a result of further desorption or increased competition at a sufficiently low pH (Figs. 2 and 3); (2) the solubilization of the adsorbent at a sufficiently low or high pH produces competing cations (Figs. 1, 2, and 3); (3) the activity of positively charged hydroxy species decreases due to the increased formation of carbonate species (Figs. 2 and 3); (4) at a sufficiently high $C_{M,aq,f}$ the value of $C_{M,ads}$ must approach some

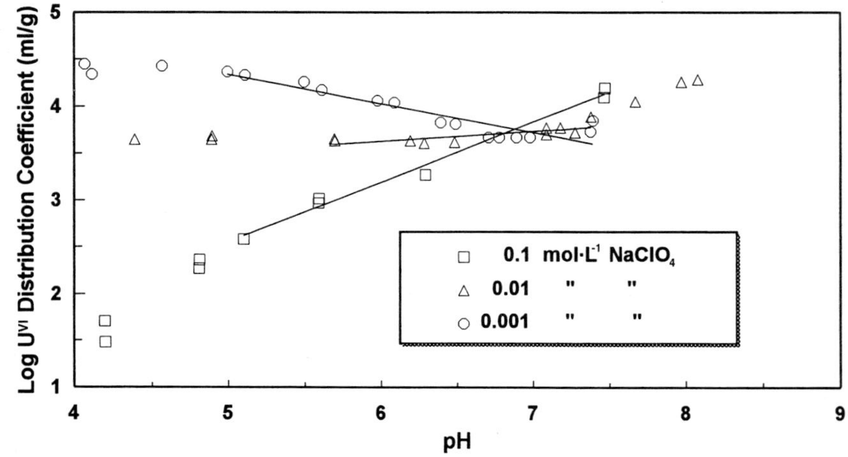

Figure 2 Variation of log K_d values for U^{VI} adsorption on a smectite, with pH and sodium concentration (after McKinley *et al.,* 1995, Fig. 4).

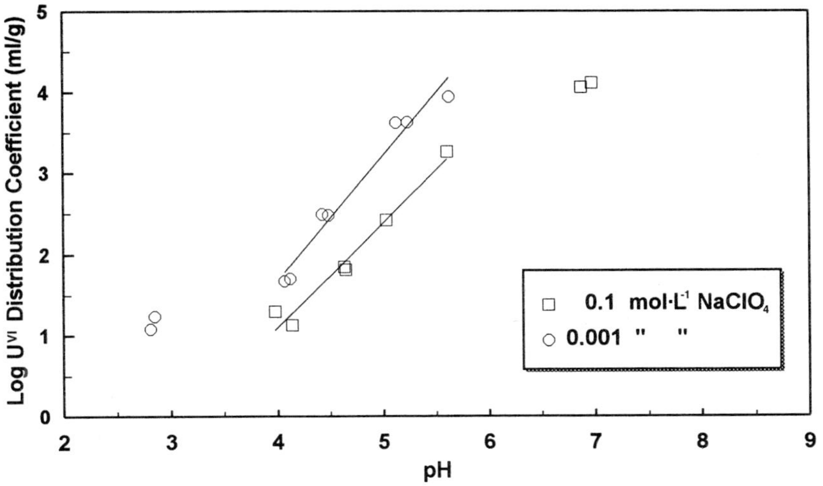

Figure 3 Increase in log K_d values for U^{VI} adsorption on Gibbsite as a function of pH and sodium concentration (after Zachara and McKinley, 1993, Fig. 2).

asymmtopic value as Kinniburgh *et al.* (1998, this volume) illustrate in their Figure 8, or (5) a significant decrease in the size of the adsorbate ions as a consequence of decarbonization and dehydroxylation as pH values decrease (Turner, 1993). The change in adsorbate ion size is clearly difficult to separate from the effect of increased H competition for adsorption sites. As pH is raised from an initially low value, the aqueous concentration of $Al(OH)_x^{3-x}$ becomes significant relative to Al^{3+} in clay or zeolite systems. The pH-dependent ratio of $Al(OH)_x^{3-x}$ to Al^{3+} may change sufficiently to cause the linearity of log $C_{U,ads}$ versus log $C_{U,aq,f}$ to be lost due, presumably, to the competition of $Al(OH)_x^{3-x}$ with U species (Pabalan and Turner, 1993a; Turner *et al.,* 1996). In log–log plots where an additional variable becomes important in some pH range, experimental K_d values may be equal, higher, or lower than values obtained by linear extrapolation, depending on the identity and concentration of the competing cation. In the pH region above 6.5 to 7.5, CO_3^{2-} complexation of metals, such as U, that form strong carbonate complexes causes a decrease in the amount adsorbed (Fig. 5).

The linearity of the increase in log $C_{M,ads}$ with log $C_{M,aq,f}$ as $C_{Maq,f}$ increases, i.e., an essentially constant log $(\Delta C_{M,ads}/\Delta C_{M,aq,f})$, may decrease as the surface coverage increases. At low surface coverage, where the availability of sites not occupied by the adsorbate of interest is not a limiting factor, incremental adsorption is expected to be independent of adsorption density. However, log $(\Delta C_{M,ads}/\Delta C_{M,aq,f})$ generally decreases as log $C_{M,aq,f}$ increases after only

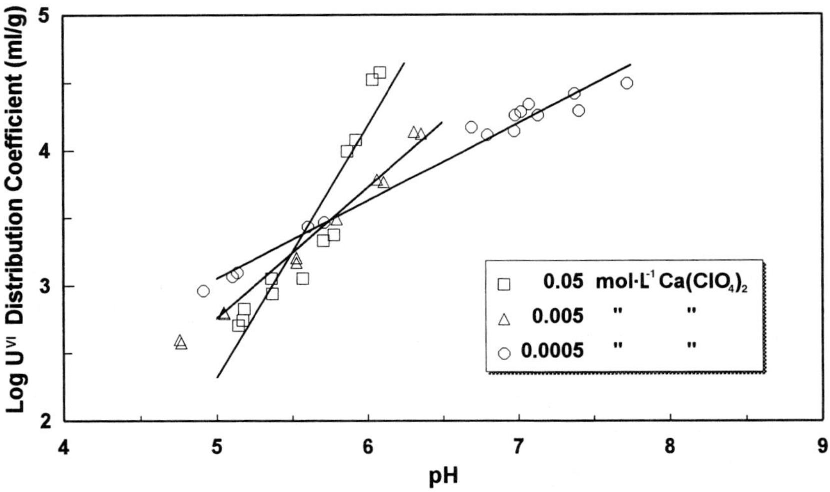

Figure 4 Increase in K_d values for U^{VI} with pH for a smectite-rich separate from the Ringold subsoil (after Zachara *et al.,* 1992, Fig. 7).

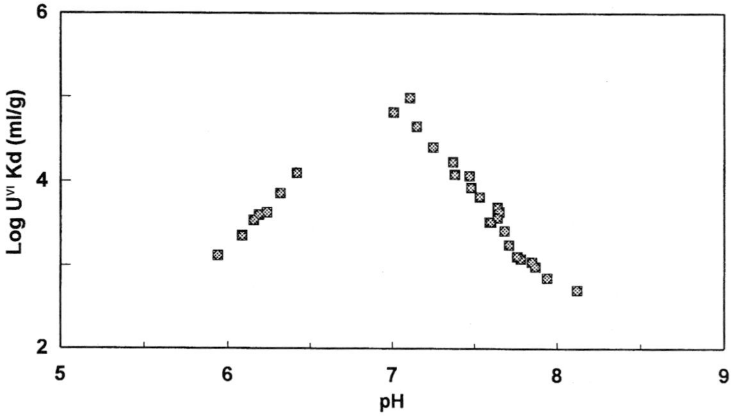

Figure 5 Variation in the ratio of U^{VI} adsorbed to that remaining in solution as a function of pH (after Tripathy, 1984); $P\text{CO}_2$ of $10^{-3.5}$ atm.

a small percentage of the adsorption capacity has been utilized. This is presumably because only a fraction of the total number of sites have high specificity for the adsorbate (Benjamin and Leckie, 1981a; McKinley and Jenne, 1991). Of course, the slope of log–log plots varies significantly depending upon the adsorbent and other system variables. In those instances where the log–log treatment does not linearize the data, this visualization indicates the point at which other variables (e.g., competing cation, carbonate complexation) become important.

Comparability between data sets for the same variables is greatly increased by normalizing the incremental amount adsorbed ($\Delta C_{M,ads}$) to the amount of metal already adsorbed or not yet desorbed ($C_{M,ads}$), as this partially reduces the effects of the differing values of $C_{M,ads}$, $C_{M,aq,i}$, or $C_{geomedia}$ (Jenne, 1996).

The data reduction process used in preparing the plots presented in this chapter has been given in detail in recent papers (Jenne 1995, 1996). Briefly, it consists of digitizing experimental values from published figures and calculating $C_{M,ads}$ ($\mu mol \cdot g^{-1}$ or ($\mu mol \cdot mol^{-1}$) as a function of time, $C_{M,aq,f}$, $C_{geomedia}$, etc. The slopes of log–log plots for a given set of variables depend significantly upon the values of system variables.

III. ADSORPTION MODELS

Models are used to interpret and generalize experimental adsorption data. However, in using adsorption models for interpretative purposes, it is well to heed the admonition that the adherence of experimental adsorption data to a particular adsorption equation provides no evidence of the actual mechanism of an adsorption process (Sposito, 1986). Schulthess and Dey (1996, p. 441) extend this advice by saying

> Adherence of data to a proposed equation does not prove that the assumed (or corresponding) adsorption mechanism is actually occurring. It does, however, establish a platform (albeit often temporary) that allows the process of scientific inquiry to proceed.

Selected aspects of the mass action equation, Freundlich and Langmuir isotherms, and the more recent surface complexation (SC) models are overviewed in this section. The Freundlich and Langmuir equations have historically been used, and are still being used, by investigators in many fields. Surface complexation models have recently become widely used in many scientific disciplines. All these models either explicitly or implicitly invoke the mass action law; hence, all of these models are inherently related.

For additional detail on these models, the reader is referred to earlier reviews (Kent *et al.,* 1988; Serne, 1992; Goldberg, 1995; Appelo and Postma, 1993; other chapters in this volume).

A. Mass Action

The mass action (MA) equation, in one form or another, has historically been used by soil scientists to model cation exchange. A difficulty in the use of this relationship is that an *a priori* assumption must be made as to the effective valence of the sites (see Sposito, 1977; Appelo and Postma, 1993). For monovalent exchange the sites are normally assumed to be monodentate, and in divalent exchange the sites are commonly assumed to be bidentate (i.e., two adjacent monovalent sites giving an effective charge of +2). Thus,

$$\mathrm{Na}\mathbf{X} + \mathrm{K}^+ = \mathrm{K}\mathbf{X} + \mathrm{Na}^+ \qquad [1A]$$

and

$$\mathrm{Ca}\mathbf{X} + \mathrm{Cd}^{2+} = \mathrm{C}\mathbf{X} + \mathrm{Ca}^{2+}, \qquad [1B]$$

respectively, where **X** is the conventional symbol for exchange sites. The exchange sites, to which the MA equation has traditionally been conceptualized as applying, are pH-independent, fixed charge adsorption sites. However, metals that readily form inorganic aqueous complexes compete more effectively for pH-dependent sites (e.g., $S{:}O^-$, $S{:}OH^0$, $S{:}OH_2^+$ in the nomenclature of SC models) than for pH-independent, fixed charge adsorption sites as indicated by the minimal effect on adsorption that results from the addition of supporting electrolytes. The pH-independent sites result from deficiencies in positive charges within the octahedral and tetrahedral layers of layer silicates. These charge deficiencies are expressed at basal surfaces whereas pH-dependent sites occur on the edges of layer silicate minerals and are the predominate sites on hydrous oxides. The pH, the nature of the adsorbate metal, the nature and concentrations of competing cations, and the extent of loading determine the distribution of a metal between pH-dependent and -independent sites. Therefore, in formulating mass action equilibrium constant relationships, S′:, rather than S:, is used here to indicate the inclusion of both pH-dependent and pH-independent sites. Since the MA equation has traditionally been used to describe cation exchange on pH-independent sites, the exchanger component is written omitting the screening[5] oxygen, i.e., $S'{:}Na^0$ rather than $S'{:}ONa^0$ or S′:(OH)Na as is written in SC models for pH-dependent sites.

The mass action equilibrium constant (K_{MA}) is preferably calculated using the activity of one or more aqueous species of the dissolved metal and the estimated activity of the adsorbed metal by use of either equivalent fractions or mole fractions, yielding a $^{a}K_{MA}$. This constant is quite often calculated using dissolved metal concentrations, $^{c}K_{MA}$, rather than activities. An obvious reason for the contin-

[5]Because of the reactivity of metal ions, such as Fe and Al, at the edges of crystal structures or amorphic precipitates they are not exposed directly to the liquid media in which they are suspended but are always "screened" by oxygen or hydroxyl ions.

ued use of concentrations rather than activities, which is not mentioned in recent texts (e.g., Appelo and Postma, 1993; Langmuir, 1997) is the problem of selecting one versus a subset of the multiple aqueous species of a given metal. As alkali and alkaline earth metals have minimal tendencies to form aqueous complexes, the uncomplexed species, i.e., Na^+ or Ca^+, are obvious choices. However, for a metal such as Co^{II} or Hg^{II}, which may occur only to a minor extent as the uncomplexed metal specie, the choice is more difficult. For the metals which tend to be hydrolyzed in aqueous media, one or more of the hydrolysis products may be a good choice. For trivalent metals, the choice is still more difficult and may depend upon the pH range being investigated.

In writing the activity-based, mass action equilibrium constant (${}^{a}K_{MA}^{Na\backslash K}$) for the monovalent exchange as depicted in Eq. [1A], it makes no difference whether one uses equivalent fractions or molar fractions to estimate the activity of adsorbed metals,

$$ {}^{a}K_{MA}^{Na\backslash K} = \{(S'{:}Na)\,(\mathbf{a}_{K})\} \,/\, \{(S'{:}K)\,(\mathbf{a}_{Na})\}; \quad [2A] $$

nor does it make any difference for the exchange of divalent Ca by Cd (Eq. [1B])

$$ {}^{a}K_{MA}^{Ca\backslash Cd} = \{(S'{:}Ca)\,(\mathbf{a}_{Cd})\} \,/\, \{(S'{:}Cd)\,(\mathbf{a}_{Ca})\}. \quad [2B] $$

In Eq. [2A], the site is clearly assumed to be monovalent, and in Eq. [2B] the site is assumed to be divalent.

For heterovalent exchange, i.e., the exchange of adsorbed Ca for Na, following the Gapon (G) convention we can write

$$ S'{:}Cd_{1/2} + Na = S'{:}Na + \tfrac{1}{2}Cd, \quad [3A] $$

which yields a ${}^{a}K_{MA,G}$ of

$$ {}^{a}K_{MA,G}^{Na\backslash Cd} = \{(S'{:}Na)\,(\mathbf{a}_{Cd})^{1/2} \,/\, (S'{:}Cd_{1/2})(\mathbf{a}_{Na})\}. \quad [4A] $$

In this form, it makes no difference whether molar or equivalent fractions are used as estimates of the activity of adsorbed metals (Appelo and Postma, 1993). However, if we write

$$ \tfrac{1}{2}S'_{2}{:}Cd + \mathbf{a}_{H} = S'{:}H + \tfrac{1}{2}\mathbf{a}_{Cd}, \quad [3B] $$

with

$$ {}^{a}K_{MA}^{H\backslash Cd} = \{(S'{:}H)\,(\mathbf{a}_{Cd})^{1/2}\} \,/\, \{(S'_{2}{:}Cd)^{1/2}(\mathbf{a}_{H})\}, \quad [4B] $$

it makes a difference in the numerical value of ${}^{a}K_{MA}$ whether we use molar or equivalent fractions. If we use equivalents (commonly in units of meq per 100 g), we follow the Gaines–Thomas convention. If we use molar units we follow the Kerr or Vanslow convention (Sposito 1980b; Appelo and Postma, 1993) and have

$$ S'_{2}{:}Cd + 2H = 2S'{:}H + Cd \quad [3C] $$

with

$$ {}^{a}K_{MA}^{H\backslash Cd} = \{(S':H)^2 (\mathbf{a}_{Cd})\} / \{(S'_2:Ca)(\mathbf{a}_H)\}. \quad [4C] $$

Considering only pH-dependent, monodentate sites for Cd displacement by H, Eq. [3B] becomes

$$ S:OCd^{+} + H^{+} = S:OH^{0} + Cd^{2+} \quad [3D] $$

and

$$ {}^{a}K_{MA}^{Cd\backslash H} = \{(S:OH^{0}) (\mathbf{a}_{Cd^{2+}})\} / \{(S:OCd^{+})(\mathbf{a}_{H^{+}})\} \quad [4D] $$

$$ = \{(S:OH^{0}) / (S:OCd^{+})\} \{(\mathbf{a}_{Cd^{2+}}) / (\mathbf{a}_{H^{+}})\}. \quad [4E] $$

The mass action relationship can commonly be linearized by taking logarithms to yield

$$ \log {}^{a}K_{MA}^{Cd\backslash H} = \log\{(S:OH^{0}) / (S:OCd^{+})\} + \log(Cd^{2+}) + pH. \quad [4F] $$

Plotting $\log\{(S:OH^{0}) / (S:OCd^{+})\} + \log(\mathbf{a}_{Cd^{2+}})$ versus pH yields $\log {}^{a}K_{MA}^{Cd\backslash H}$ as the intercept and the average net proton coefficient[6] as the slope of the line. In evaluating stoichiometry, it must be recognized that geomedia adsorbents are rarely, if ever, homoionic. This is especially true of H-saturated clays because of the rapidity with which H breaks structural O:Al bonds, leading to mixed H^{+}-$Al(OH)_y^{(3-x)}$ saturation of both fixed charge exchange and $S:O^{-}$ or $S:OH^{0}$ adsorption sites. Thus, to evaluate stoichiometry, additional cations beyond those whose exchange is being studied may need to be analyzed.

Mass action equilibrium constants are used less frequently to describe metal adsorption since the advent of SC models although the MA continues to be used occasionally. For example, Swallow *et al.* (1980) calculated a ${}^{c}K_{MA}$ for the adsorption of Cu^{II} onto $Fe(OH)_3$(a) and found that the data scattered about a straight line with a slope of about 1.5 above pH 5.75 but fell above the extrapolated line at lower pH values. More recently, Erikson *et al.* (1990, p. 4.11) modeled the U^{VI} adsorption data of Hsi and Langmuir (1985) by parameterizing the mass action equation as outlined by Krupka *et al.* (1988).

B. Distribution Coefficients

In various fields of applied science such as predicting metal uptake by plant roots (Kirk and Staunton, 1989) but particularly in the nuclear waster disposal, adsorption data are presented as the distribution coefficient, K_d, where

$$ K_d = \frac{\text{(fraction sorbed/kg geomedia)}}{\text{(fraction in solution/L system volume)}}. \quad [5A] $$

[6]Proton stoichiometry may also be obtained by the alternate technique of Perona and Leckie (1985).

The common units of milliliter per gram and of liter per kilograms are of course interchangeable but conversion is required from the SI unit of kilograms per cubic meter which is often used in European countries.

In the conventions of this chapter, Eq. [5A] becomes

$$K_d = (C_{M,ads}/C_{M,aq,f}) \quad [5B]$$

where the units of $C_{M,ads}$ are moles of metal (M) per gram (or per mole) of adsorbent and of $C_{M,aq,f}$ is in moles per liter. Rearranging terms yields

$$C_{M,ads} = K_d\, C_{M,aq,f}. \quad [5C]$$

From this equation it is obvious that for every unit increase in $C_{M,aq,f}$ there must be a corresponding increase in $C_{M,ads}$, that is, a linear concentration isotherm; the term isotherm simply denotes that the adsorption experiments were at a constant temperature.

Distribution coefficient tabulations are useful in that they present adsorption data in the form generally used in contaminant transport models. Indeed, a number of countries that utilize nuclear power have developed national data bases of K_d values. Radionuclide K_d tabulations have been compiled for Canada, Finland, Germany, and Sweden (Vandergraaf, 1982; Allard *et al.,* 1991; Albinsson, 1991; Hakanen and Hölttä, 1992; Vandergraaf *et al.,* 1992, 1993; Stenhouse, 1994; Vandergraaf and Ticknor, 1994). The compilation of K_d geometric means by Thibault *et al.* (1990) supersedes those of Baes and Sharp (1983) and Shepard *et al.* (1984). The compilation of Thibault *et al.* (1990) for United States and Canadian soils gives mean K_d values by soil texture, i.e., sand, silt, clay, or organic. The K_d compilation of Vandergraaf (1982) has been replaced by a data base of polynomial coefficients (Vandergraaf and Ticknor, 1994). These tabulations have been prepared in national for national nuclear waste disposal or site decommissioning programs with the prime objective of being conservative, i.e., providing minimum K_d values as opposed to best estimates. These compilations generally give little attention to specifying the values of significant factors that affect the tabulated K_d values other than pH and sometimes $C_{M,aq,i}$ values. The data included in these compilations have generally not been subjected to critical peer review. The development of national data bases is carried to the extent of estimating K_d values from the radionuclide content of plants and selected properties of the soils in which the plants grew (see Section III.I).

Extensive K_d measurements on geomedia (e.g., sediments, soils) from U.S. Atomic Energy Commission sites (particularly the Nevada Test Site) were initiated in the United States in the early 1960s (e.g., Wahlberg and Fishman, 1962; Baker *et al.,* 1964; Wahlberg *et al.* 1965). After such studies had been underway for some years, these investigations were criticized for not being carried to reaction completion. One expedient response to this criticism was to define an explicitly nonequilibrium distribution parameter, termed the retardation coefficient or simply R_d (see, for example, the discussion of Serne, 1992; ASTM, 1993, 1994). However, as shown elsewhere (Jenne, 1995, 1996), adsorption studies rarely achieve the equilibrium state. Thus, the artifice of defining the K_d as an equilibrium value and the R_d as a nonequilibrium value has little merit. The subscript "f" for final point in time series measurements is used in conjunction with aqueous concentrations in this chapter to emphasize the fact that relatively few adsorption data may have been actually collected at adsorption equilibrium (see Section V.G; Axe and Anderson, 1998, this volume).

Some authors implicitly assume that metals are sorbed reversibly onto geomedia and others report that a portion of the adsorption metal is held irreversibly. The amount of metal adsorbed irreversiblely is commonly implicitly defined as the amount adsorbed that was not desorbed within some time frame which is commonly that of the adsorption experiment. Frequently, reversibility is measured by the addition of a competing cation which is often H^+. For example, Millward (1980) used a pH of one and found that all the Cd adsorbed by $Fe(OH)_3$(a) was immediately released at this pH. However, since there tends to be an inverse relationship between the amount not desorbed in some time period and the length of the adsorption period (which some investigators refer to as aging), it is not surprising that given an adsorption period of only approximately 10 min and the dissolution of 40% of the solid that all the Cd was desorbed. It follows that the thermodynamic requirement that adsorption be completely reversible (e.g., Turner and Millward 1994) is meaningful only if the time frames and other experimental conditions of the adsorption and desorption experiments are specified. Furthermore, it would seem that reversibility does not need to be instantaneous as is sometimes specified in the contaminant transport literature (e.g., Turner, 1993, p. 1-1). Rather, in transport modeling, it is necessary only that the reaction is reversible within the time frame of the movement of a particle of water through a specified volume of geomedia. This is sometimes referred to as local equilibria. Attainment of local equilibria will then be dependent upon infiltration or groundwater flow rates as well as on desorption rates. It is important that individual adsorption studies should report time series data for the experimental conditions expected to give the highest and lowest loadings of adsorbate on the adsorbent(s) investigated to indicate the probable extent to which the equilibrium state was approached in a given study. Such time series should extend for a time period of perhaps 10-fold longer than the expected reaction period needed (see also Section V.G).

Distribution coefficients are generally calculated using the total dissolved concentration of the metal without regard to the aqueous speciation although a few investigators have used particular aqueous species in calculating K_d values (see Section V.D). Kaplan *et al.* (1994, p. 2.4) suggest using a $K_{d,\Sigma sp}$ calculated using the sum of the adsorbed surface species and the sum of the aqueous species of *j*th metal; i.e.,

$$K_{d,\Sigma sp} = (\Sigma C_{Mj,ads}/\Sigma C_{Mj,aq,f}). \qquad [5D]$$

Retardation factors (R_f) are used widely in contaminant transport models. The R_f can be defined as,

$$R_f = V_w/V_M, \qquad [6A]$$

where V_w and V_M are the velocity of groundwater and of metal, respectively. In practice, adsorption data in the form of K_d values are used to calculate R_f via a relationship that accounts for the porosity and the bulk density of geomedia; i.e.,

$$R_f = 1 + K_d\,(\rho/\epsilon_b), \qquad [6B]$$

where ρ is the bulk density ($g{\cdot}cm^{-3}$) and ϵ_b is the bulk porosity ($cm^3{\cdot}cm^{-3}$) of the geomedia. Vandergraaf *et al.* (1993) compile literature values for porosity of 0.003 to 0.005, 0.10 to 0.20, and > 0.5 for intact rock, fracture zones, and crushed rock in packed columns, respectively.

Vandergraaf *et al.* (1993) use regression equations (see Section III.I) to obtain "retardation coefficients" (i.e., K_d values) for particular minerals in a given flow path segment which they then weight using the average fraction of that mineral present in the flow segment and they then sum the K_{d,M_R} values for the *j*th radioisotope over all minerals yielding K^{Σ}_{d,M_R}. Since adsorption is a surface phenomenon, they normalized the K^{Σ}_{d,M_R} to the surface area of the geomedia in that flow segment to yield K^{Σ}_{d,M_R,A_S}, which they term the equivalent bulk sorption coefficient; thus,

$$K^{\Sigma}_{d,M_R,A_S} = K^{\Sigma}_{d,M_R}/A_S, \qquad [7]$$

where A_S = (specific) surface area. Vandergraff *et al.* (1993) compile A_S values of 0.4 to 4, 10^2 to 10^4, and 10^2 to 10^4 $m^2{\cdot}kg^{-1}$, for intact rock, fracture zones, and curshed rock in packed columns, respectively. They note that the surface area can be estimated from porosity; i.e.,

$$A_S = \epsilon_b/(r_h\rho_S\{1-\epsilon_b\}), \qquad [8]$$

where ϵ_b = bulk porosity of adsorbing medium, and r_h = mean hydraulic radius of pores (which Vandergraaf *et al.* {1993} give as 0.14 to 0.18 μm for granite and sandstone), and ρ_S = bulk density of adsorbing medium.

Equation [6B] is valid for saturated moisture conditions in porous media and for linear adsorption curves. For unsaturated moisture conditions, the volumetric moisture content replaces the porosity term. The disagreements among investigators and theoretical considerations as to whether K_d values remains constant as the volumetric water content changes were discussed by Kaplan *et al.* (1994, pp. 2.9–2.10). For fractured systems, as opposed to porous systems, estimates of the fracture surface area, opening width, and matrix diffusion attributes and other equations are used to define the retardation factor (Holtta *et al.,* 1991; Abelin *et al.,* 1986; Rasmuson and Neretnieks, 1986). For transport modeling involving a nonlinear isotherm, see Wang *et al.* (1998, this volume).

According to Vandergraaf *et al.* (1993), metal transport velocities in porous media can be approximated with the relationship

$$V_M = V_w/(1 + \{\rho_S[1-\epsilon_b]/\epsilon_b\}K^{\Sigma}_{d,M_R}). \qquad [9A]$$

Substituting Eqs. [7] and [8] into Eq. [9A] yields

$$V_M = V_w/(1 + K^{\Sigma}_{d,M_R,A_S}), \qquad [9B]$$

since the porosity-dependent factors cancel out. They point out that if both the K_d and the mean hydraulic radius remain constant with distance, then the retardation and the retardation factor, R_f, can be obtained from the relationship

$$R_f = 1 + (\rho_S\{[1-\epsilon_b]/\epsilon_b\}K^{\Sigma}_{d,M_R}). \qquad [10]$$

The limitations of a single-valued distribution coefficients have been discussed in detail elsewhere (Reardon, 1981; Kent *et al.,* 1988; Turner, 1991, 1993). As they are commonly determined, K_d values generally represent the effect of a single variable (e.g., pH, $C_{M,aq,f}$) in multidimensional space. Bradbury and Baeyens (1991) comment that as a consequence of this [K_d] methodology, adsorption data, however good, are only strictly valid under the experimental conditions at which they were measured, and extrapolation to other conditions and systems may be questionable. They note that implicit in the K_d approach is the need to generate large amounts of data for a simple K_d description of adsorption under the conditions which are likely to arise in space and time along a migration pathway from contaminated site to Man. Clearly, reliable modeling of adsorption with one-dimensional K_d values requires either a matrix of values or equations such that the effect of changing $C_{Me,aq,}$ pH, $C_{competing\ cation}$, ionic strength (***I***), or salinity (S) on K_d values can be predicted.

In general, prediction of radioactive metal transport using Eq. [6B] requires that the adsorption be linear, i.e., a constant ($\Delta C_{M,ads}/\Delta C_{M,aq,f}$) ratio. Alternatively, a set of equations yielding log K_d or a matrix of K_d values as a function of key variables are needed.

Under the appropriate conditions, the K_d is a simplification of the classical mass action equilibrium constant (Wahlberg *et al.,* 1965). For divalent metal exchange on bivalent sites, the fraction adsorbed, normalized to mass and volume, equals $\{(Cd_{ads})/(Cd_{aq,i})\}$ (V/W) and the fraction in solution equals $\{(C_{Cd,aq,f})/(C_{Cd,aq,i})\}$ where Cd_{ads} is the number of moles of Cd adsorbed, and $Cd_{aq,i}$ and $C_{Cd,aq,f}$ are the moles of Cd initially added to the system and present in the solution at the end of the experiment, respectively; W is the mass of adsorbent in grams, and V is the volume in liters of the aqueous phase. Therefore,

$$K_d = (S':Cd)/(C_{Cd,aq,f}), \qquad [5E]$$

where S′:Cd is the concentration of adsorbed Cd. Clearly, an implicit assumption of the K_d simplification of the mass action relationship is that the increase in the aqueous concentration of the desorbed cation (unspecified in the case of Eq. [5C]) is unimportant. In practice, it is also assumed that the correction of the metal from a concentration to a thermodynamic activity basis is unnecessary. Obviously, the validity of the first assumption depends on the composition of the solution, the cation being desorbed, and the geomedia:solution ratio. The validity of the second assumption depends largely on the change in aqueous speciation that occurs.

C. Power Function

Langmuir (1981, 1997) espoused a power function approach to adsorption modeling using the activity of the predominate aqueous specie and mole fraction of the adsorbed metals. Rewriting the mass action relationship of Eq. [1B] for the general case of metal $M_{k,aq}$ displacing metal $M_{j,ads}$, we have

$$S'{:}M_j + M_{k,aq,f} = S'{:}M_k + M_{j,aq,f}. \qquad [1C]$$

Recognizing that both the ratio of adsorbed metals and the power function equilibrium constant (K_p) are likely to be nonintegers, we have

$$K_p = \{(M_{j,aq,f}) / (M_{k,aq,f})\}\{S'{:}M_k) / (S'{:}M_j)\}^{n_p}, \qquad [11A]$$

where the adsorbed concentrations are mole fractions raised to the *n*th power. Langmuir (1981) noted that this relationship is identical to the Rothmund–Kornfield cation exchange equation. Clearly, the power function is a variant of the mass action equation where the sites include surface complexion (pH-dependent) and fixed charge (pH-independent) sites, the mono- versus bidentate nature of sites is unspecified, and the stoichiometry of exchange is unspecified. These variables are subsumed in the value of log K_p and n_p.

Equation [11] is linearized (Langmuir, 1997) by taking logarithms and rearranging to give

$$\log\{(M_{j,aq,f})/(M_{k,aq,f})\} = \log K_p + n_p \log\{(S'{:}M_j)/S'{:}M_k)\}. \qquad [11B]$$

Plotting $\log\{(M_{j,aq,f})/(M_{k,aq,f})\}$ versus $\log\{(S'{:}M_k)/(S'{:}M_j)\}$ yields a slope of n_p and an intercept of log K_p. Power function plots of Langmuir (1997) show slopes (at low $M_{j,aq,f}$ and $M_{k,aq,f}$ values on log–log plots) ranging from 0.95 to 2.00. Langmuir (1981) reported that binary exchange on clays and whole soils at both constant and variable surface charge of sorbent, such plots are highly linear, with correlation coefficients (r values) usually between 0.98 and 1.00. Langmuir (1981) also noted that binary exchange between Na or K and H on clay minerals, and between the alkaline earth metals and alkali metals often fited power exchange functions with *n* equal to 1.00 corresponding to simple ion exchange behavior and that n_p values were also near unity for exchange between the alkaline earths or alkali metals and transition metals including Cd^{2+}, Co^{2+}, Cu^{2+} and Zn^{2+} when the competing cations were approximately equal in aqueous concentrations.

However, it is important to note that, as a rule, Langmuir's data show that when the concentration of one of the competing metals is orders of magnitude less than that of the other, adsorption of the lower concentration metal is preferred over the major adsorbate with the result that n_p is not equal to 1 or 2. Further, the value of n_p generally increases as the ratio of trace metal to the competing macrocation decreases; i.e., on average sites of greater specificity are occupied with a lower loading by the trace metal.

This power function approach is expanded in Section III.I to deal with multiple variables, which is essential for any general interpretive or predictive adsorption modeling.

D. Freundlich Equation

Freundlich and Langmuir equations were early adsorption models and are still frequently used. The Freundlich and Langmuir models differ from each other with respect to how the free energy of adsorption is assumed to vary with $C_{M,ads}$ and in their assumption concerning the existence of a finite adsorption capacity.

The current use of the Freundlich equation makes it desirable to consider its conceptual basis, its relation to the mass action equation, variants of the equation, and selected literature comments. According to Sposito (1980a) and Harter (1986), it was van Bemmelen (1878, 1888) who showed that the mathematical relationship

$$x/W = KC^{1/n} \qquad [12A]$$

provided a useful description of adsorption as a function of dissolved metal concentration, but it was Freundlich (1909) who demonstrated its general applicability. In this equation, x/W is the quantity of metal adsorbed normalized to unit mass of geomedia, K is a constant, C is the aqueous concentration of the metal, and 1/n is a constant. In terms of the notation of this chapter, we write

$$C_{M,ads} = K_F\, C_{M,aq,f}^{1/n_F}. \qquad [12B]$$

where $C_{M,ads}$ and $C_{M,aq,f}$ have their usual meaning; K_F and n_F are positive-valued, empirical constants (where $< 0 < n_F < 1$) that depend upon the nature of the adsorbate, and adsorbent, and the values of system variables. Values of n_F other than 1 specify a nonlinear change in the $\Delta C_{M,ads}$ to $\Delta C_{M,aq,f}$ ratio. This equation differs from the distribution equation (Eq. [5C]) only in the presence of an exponent on $C_{M,aq,f}$ which allows the relationship to be fitted to nonlinear isotherm.

The linearized form is

$$\log C_{M,ads} = \log K_F + 1/n_F \log C_{M,aq,f}. \qquad [12C]$$

The constants are determined by plotting log $C_{M,ads}$ versus log $C_{M,aq,f}$, log K_F being the intercept and $1/n_F$ the slope of the isotherm. The equation obviously does not predict an adsorption maximum; i.e., $C_{M,ads}$ would be expected to increase so long as $C_{M,aq,f}$ increases.

From time to time it has been demonstrated that adherence to a Freundlich or Langmuir equation is not proof that adsorption, as opposed to precipitation, has occurred (Fisher, 1922; Sposito, 1982). This observation has been widely ignored. Since the Freundlich equation does not include an adsorption maximum, it is obviously not appropriate to apply it to data sets wherein a sizable portion of sites are filled with the adsorbate. Similarly, little attention appears to have been given to the possibility of obtaining information on surface heterogeneity via the Freundlich equation as was suggested by earlier authors (Benjamin and Leckie, 1980; Sposito, 1980a).

A variant of the Freundlich equation that includes a time variable has been used in phosphate adsorption studies, but so far as this author is aware, it has not been used in metal adsorption modeling. The equation given by Torrent (1987) is

$$X = a\,C^{b1}\,t^{b2}, \qquad [13A]$$

which in the terminology of this chapter is

$$C_{M,ads} = K_{F_t}\,C_{M,aq,f}^{1/n_{F_t}}\,t^{F} \qquad [13B]$$

where K_{F_t} is the Freundlich constant (the t subscripted K_F indicates that this constant is from the time variable form of the equation) and t^F is time to the F power.

It may be noted that the log form of the power function (Eq. [11B]) and the log form of the Freundlich equation (Eq. [12C]) are similar except that the Eq. [11B] includes terms for the competing cation. The relationship between K_d and the Freundlich equation is considered in Section III.I.

E. Langmuir Equation

In systems that obey a Langmuir equation (Langmuir, 1918), $C_{M,ads}$ approximates a linear function of $C_{M,aq,f}$ at sufficiently low fractional occupancy and is largely independent of $C_{M,eq,f}$ at sufficiently high fractional occupancy.[7] The Langmiur equation includes a term for the maximum amount of metal that can be adsorbed; hence it predicts a decrease in incremental adsorption, i.e., in the $\Delta C_{M,ads}/\Delta C_{M,aq,f}$ ratio at sufficiently high $C_{M,ads}$. The original derivation of the Langmuir equation (Langmuir, 1918) was for gas adsorption onto solids.

[7]Additional aspects of the Langmuir equations are discussed in Benjamin (1979, Chap. 4), Sposito (1980a), and Langmuir (1997).

Langmuir's derivation assumed the increased adsorption of gas molecules in response to an increasing partial pressure's of the gas until a monolayer of molecules was formed on a planar surface, i.e.,

$$q = (a\,bP)/(1 + aP), \qquad [14A]$$

where q in the amount adsorbed, a and b are constants, and P is the partial pressure of the gas. In the nomenclature of this chapter, Eq. [14A] becomes

$$C_{M,ads} = (K_{L,aff}\,K_{L,max}\,C_{M,aq,f})/(1 + K_{L,aff}\,C_{M,aq,f}), \qquad [14B]$$

where a = $K_{L,aff}$, b = $K_{L,max}$, and P (gas pressure at equilibrium) is replaced by the activity of the dissolved metal at equilibrium (or more commonly, the final concentration measured). In the application of this equation to the adsorption of solutes it is assumed that the $C_{M,ads}$ can be represented by the mole fraction of the adsorbing metal. The linearized forms of this equation (Veith and Sposito, 1977; Campbell and Davies, 1995) include

$$C_{M,aq,f}/C_{M,ads} = (K_{L,aff}\,K_{L,max}\,C_{M,aq,f})/(1 + K_{L,aff}\,C_{M,aq,f}) \qquad [14C]$$

$$C_{M,aq,f}/C_{M,ads} = \{1/(K_{L,aff}\,K_{L,max})\} + C_{M,aq,f}/K_{L,max}) \qquad [14D]$$

$$C_{M,ads} = (1/K_{L,max}) + (1/K_{L,aff}\,K_{L,max}\,C_{M,aq,f}), \qquad [14E]$$

where $C_{M,ads}$ and $C_{M,aq,f}$ have their usual meaning, $K_{L,aff}$ is the Langmuir affinity constant ($mg{\cdot}kg^{-1}$ or $mol{\cdot}L^{-1}$) which is sometimes said to represent the energy of adsorption, and $K_{L,max}$ is the adsorption maximum. Equations [14A] and [14B] yield a straight line upon plotting $C_{M,aq,f}/C_{M,ads}$ versus $C_{M,aq,f}$ if the data fit the Langmuir equation. The values of $K_{L,max}$ and $K_{L,aff}$ are obtained as the reciprocal of the slope and as the intercept, respectively, where the plot is linear. Dowd and Riggs (1965) discuss the merits of these different formulations.

MacNaughton (1973) described Hg adsorption by an area-normalized Langmuir equation where A_S is in square centimeters per gram; i.e.,

$$C_{M,ads}/A_S = \{K_{L,aff}\,K_{L,max}\,C_{M,aq,f})\,/\,(1 + K_{L,aff}\,C_{M,aq,f})\}. \qquad [14F]$$

Borkovec (1981) described U adsorption by the Langmiur equation using a fractional occupancy form

$$C_{M,ads}/C_{M,ads,max} = K_{L,aff}\,C_{M,aq,f}/(1 + K_{L,aff}\,C_{M,aq,f}). \qquad [14G]$$

He gives the linearized form as

$$C_{M,aq,f}/C_{M,ads} = (1/K_{L,aff}\,C_{M,ads,max}) + (C_{M,aq,f}/C_{M,ads,max}). \qquad [14H]$$

Borkovec (1981) plotted $C_{M,aq,f}/C_{M,ads}$ versus $C_{M,aq,f}$ which yields a slope of $(1/C_{M,ads,max})$. The affinity constant, $K_{L,aff}$, was obtained from the shift of the slope on the $(C_{M,aq}/C_{M,ads})$ axis for $C_{M,aq} = 0$.

F. Selected Freundlich and Langmuir Equation Considerations

In the range where high affinity (i.e., strong) sites are only beginning to be filled by M, the relationship between K_d and $C_{M,aq}$ may be linear with a near zero slope, often called a "constant K_d" (see plots of Wahlberg *et al.* 1965). In the next range, where only a small fraction of the strong sites are filled with M_j, the relationship may be linear but with a decrease in slope (on a K_d versus $C_{M,aq,f}$ plot). When a sufficient portion of the total available sites are filled with the adsorbate, the K_d falls off more rapidly due to a reduced number of available sites that are unfilled with M, increased competition from the desorbed cations, and a decreasing adsorbate concentration where $C_{Mj,aq,i}$ is constant.

One frequently finds discussions in the literature of the relative merit of various equations to describe particular data sets, i.e., whether or not a particular data set is linearized more effectively by the Freundlich or the Langmuir equation. Fisher (1922) pointed out that linearization of adsorption data by the Freundlich equation was often dependent upon the concentration range over which the experiments were conducted and that a sufficiently large concentration range was not likely to yield a single linear relationship. Thus, the apparent success of Freundlich versus Langmuir equations may depend on the concentration range of the adsorbent, the adsorption capacity of the absorbent, and the $C_{geomedia}$. Several investigators report the $C_{M,aq,i}$ range over which an adsorption isotherm obeys the Freundlich or Langmuir isotherm. However, it is important to note that such an aqueous concentration range is meaningless unless the $C_{geomedia}$ is also specified.

Some investigators have broken their adsorption isotherm into segments to obtain an adequate linear fit (e.g., Stollenwerk and Grove, 1985). For example, Kinniburgh and Jackson (1982) broke their adsorption isotherms into from one to three regions and fit each region separately with the Freundlich equation. This segmentation of adsorption isotherms is simply a device for approximating a curve with multiple straight line segments. This approach has little merit. Such segmented isotherms for single-metal-adsorbent systems have been used to infer two or more types of bonding. These include multisite Langmuir bonding (Loganathan and Burau, 1973; Zasoski and Burau, 1988) and specific and nonspecific bonding depending on pH (Healy *et al.,* 1968). However, given the largely unevaluated effects of competing cations, loading, reaction time, and other variables on adsorption isotherm linearity, it unlikely that these inferences relative to multiple sites and bond types are meaningful.

Both Freundlich and Langmiur equations have been reported for many combinations of adsorbents and metals (Benjamin, 1979, Ch. 4); however, it is significant that in the literature reviewed for this chapter in no instance have Freundlich or Langmurian constants calculated by one investigator been used by another investigator for comparison with their own data or to predict adsorption in another system. For example, Eylem *et al.* (1989) analyze their adsorption data using both the Freundlich and the Dubinin–Radushkevich models but they failed to compare their constants with a single literature value. Similarly, Clark *et al.* (1998, this volume) calculate Freundlich equation constants but do not compare their values to those previously determined. Thus, it is an unescapable inference that these constants, which generally are evaluated for only one, or occasionally two, variables, have little or no transfer value, which suggests that there is little or no merit in calculating and publishing these constants. This lack of transferability presumably arises from the multidimensional nature of adsorption; the Freundlich or Langmuir constants for pH or $C_{M,aq}$ adsorption curves implicitly include the effect of other variables such as $C_{competing\ cation}$, loading, and the speciation consequences of differing water chemistries. A further complication not generally recognized in the many literature discussions of the relative merits of describing a set of data with the Freundlich or Langmuir equations is that the time required for both the fast and the slow adsorption processes to approach completion is quite variable (Jenne, 1995, 1996). For example, Sharpley *et al.* (1981) cite literature showing that the Langmuir constants depend upon the length of the aging period between the addition of P to the soil and the start of desorption. Thus, the dependence of $C_{M,ads}$ upon time may confound comparisons of calculated $K_{L,aff}$ and $K_{L,max}$ values.

G. Surface Complexation

Recently, surface complexation models (Davis and Leckie, 1978a,b) has been proposed for use in contaminant transport models (Turner *et al.,* 1993; Pabalan *et al.,* 1994; McKinley *et al.,* 1995), particularly for performance assessment of nuclear waste disposal sites. There is basic agreement among investigators that the conceptual model of surface complexation is valid, although appreciable data and experience are needed to exploit effectively SC models. Regrettably, no clear consensus has developed regarding which of the three principal implementations of the electrostatic aspects of the SC model is the preferred form; hence, there are three partial sets of SC model constants in the adsorption literature.

The SC model conceptualizes adsorption reactions of pH-dependent surface sites as analogous to those of aqueous-phase ligands. The structural metal, S:, acts as a Lewis acid and exchanges cations or anions such as H^+, OH^-, M^{2+} and MOH^{z-1}. The protonation and deprotonation reactions of neutral sites, respectively, are

$$S{:}OH^0 + H^+ = S{:}OH_2^+ \qquad [15A]$$

and

$$S{:}OH^0 = S{:}O^- + H^+. \qquad [16A]$$

Metal cation adsorption reactions are typically of the type

$$S{:}OH^0 + M^{2+} = S{:}OM^+ + H^+ \qquad [15B]$$

or

$$S{:}OH^0 + MOH^+ = S{:}OM^+ + HOH^0. \qquad [16B]$$

Surface complexation models simultaneously solve equations defining the mass balance of sites, mass action, and charge balance, and where the total number of adsorption sites is calculated from $C_{geomedia}$, A_S, and $C_{sites,area}$ (i.e., site density). The work of bringing an ion from the bulk solution to the surface is given by the Boltzmann factor, $e^{-\Psi_i F/RT}$, which describes the electrostatic portion of the change in activity between an ion in bulk solution and at the surface; i.e.,

$$(S{:}OH^0) = \mathbf{a}_{H,aq}\, e^{-\Psi_i F/RT}, \qquad [17A]$$

and for the general case,

$$\mathbf{a}_{M,ads} = \mathbf{a}_{M,aq}[e^{-\Psi_i F/RT}]^z, \qquad [17B]$$

where $S{:}OH^0$ is the singly protonated uncharged adsorption site, $\mathbf{a}_{M,ads}$ and $\mathbf{a}_{M,aq}$ are the activity of the M_{ads} cation and in bulk solution, respectively, $e^{-\Psi_i F/RT}$ is the Boltzmann factor, Ψ_i is the electrostatic potential (V), z is the charge of the cation, F is the Faraday constant (J V^{-1} eq^{-1}), R is the ideal gas constant (J $K^{-1}\cdot mol^{-1}$), and T is the absolute temperature (K).

Mechanistically, it is difficult to determine if only a H^+, or both the O^{2-} and H^+, of a $S{:}OH^0$ complex are replaced when a MOH^0 complex is adsorbed. The total dissolved concentration of a metal is sometimes used on SC models rather than ion activities. It is generally assumed that the free energy difference between Eqs. [15A] and [15B] is small. The replacement of a surface proton by metal or ligand exchange for a hydroxyl is considered to be equivalent to the formation of an inner-sphere complex (Schindler *et al.* 1976; Stumm *et al.,* 1980; Sposito, 1984), an inner-sphere surface complex being one which contains no water molecule between the surface site and adsorbed cation (Sposito, 1984). Guilherme and Anderson (1998, this volume) state that the high affinity of RPOC for Cu is due to the formation of inner-sphere complexes which are often referred to as chemisorption or specific adsorption.

Current adsorption research generally uses one of three different implementations of the conceptual SC model. The primary disagreement among modelers is where within the interfacial region the shear plane is located and where the potential determining ions are located. Yates *et al.* (1974) proposed the existence of ion pairs formed between charged surface sites and ions located at the inner Helmholtz plane, which corresponds to the constant capacitance (CC) SC model (Schindler and Kamber, 1968). Assumptions of the CC SC model are that the background electrolytes do not participate to a significant extent in the ligand exchange reactions and that the net surface charge and the surface potential are unchanged by the ligand exchange reactions. In the CC SC model, the specifically adsorbed metals (i.e., those held by noncoulombic forces) are located in the surface plane as are adsorbed H and OH. The CC model uses a single potential plane at the surface; hence, all surface complexation reactions are considered to form inner-sphere complexes.

The diffuse double layer (DL) SC model (Stumm *et al.,* 1970; Dzombak and Morel, 1990) requires acidity constants (K_{-1} K_{+}) for the protonation and deprotonation of neutral surface sites, metal adsorption constants, and adsorbent site concentration ($C_{site,area}$ in mol·nm^{-2} but sometimes $C_{site,volume}$ in mol·L^{-1}). Dzombak and Morel (1990) modeled Cd with the DL SC model. Subsequently, Waite *et al.* (1994) and Payne *et al.* (1998, this volume) modeled U^{VI} adsorption with the TL SC model, using both strong and weak bidentate sites, 0.0018 and 0.873 ($C_{sites,tot,mol\ Fe}$ = 0.875) moles of sites per mole of Fe, respectively. McKinley *et al.* (1995) specify face and edge sites. Various investigators specify mono- and/or bidentate sites. The strong sites were presumed to account for the initial linear portion of the $C_{Cd,ads}$ adsorption isotherm at low fractional occupancy and adsorption to remaining unfilled strong plus weak sites to explain the nonlinear portion of their adsorption isotherm.

Davis (1977) showed that background electrolyte ions are involved in the development of surface charge and extended the single-potential-determining single-plane CC SC model to include triple layers (TL). The change in calculated acidity constants with varying ionic strength was interpreted as ion pair formation between a surface site and background electrolyte cation. Thus, the K^{avg} values calculated by the two models will be significantly different since $S{:}O^-$ and $S{:}OH_2^+$ vary with pH. The TL SC model is the most general SC model implementation in that it can operate over a range of ionic strengths and can accommodate both coordinated surface complexes (inner-sphere) and ion pair (outer-sphere) complexes. The SC model parameters needed are A_S, $C_{site,area}$, surface acidity constants (${}^{int}K_{a1}$, ${}^{int}K_{a2}$), and adsorption constants for background electrolytes and other competing cations (K_{i-j}), and inner- and outer-layer capacitances. Some version of the FITEQL model (Westall 1982a,b; Herbelin and Westall, 1994) is nearly always used to calculate intrinsic equilibrium constants, ${}^{int}K$.

The mass action equation, including the work of bringing a cation to the adsorbent surface, can be written for the triple layer (TL) SC model as (Pabalan and Turner, 1993b)

$$K_{MA,H/M} = \left\{ \frac{(S{:}OM^{(z-1)})(H^+)(e^{-\Psi_i F/RT})}{(S{:}OH^0)(M^{z+})(e^{-\Psi_i F/RT})^z} \right\}, \quad [18]$$

where $K_{MA,H/M}$ is the mass action equilibrium constant for the exchange of M^{z+} for H^+. The log of the intrinsic surface complexation constant can be obtained by plotting the log of the conditional equilibrium constant versus $C_{M,ads}$, but preferably $\mathbf{a}_{M,aq}$, and extrapolating to a zero value of the net surface charge (Stumm *et al.*, 1980). This extrapolation is based on the assumption that the net surface charge is equal to $S{:}OH_2^+$ below the point of zero charge and is equal to $S{:}O^-$ above the point of zero charge. The point of zero charge, pH_{PZC}, is the pH of no net charge on the particle; i.e., $C^+_{S:H2,area} = C^-_{S:O,area}$.

A significant difficulty in applying SC models to published data for natural systems is that K_a and C_{site} values are rarely reported (Pabalan and Turner, 1993a,b). Indeed, K_{a1} and pK_{a2} are normally measured for individual minerals or amorphic solids rather than for mixtures such as soils. Intrinsic adsorption constants for one or more dominant individual minerals may be used or these constants may be determined for the particular geomedia. Pabalan and Turner (1993c) took the uniform parameter concept of Davis and Hayes (1986) a step further and used the ${}^{int}K_U^{TL}$ value derived for U^{VI} adsorption onto goethite to model the adsorption of U^{VI} on clinoptilolite. Only $C_{U,aq,i}$, A_S, and C_{solid} were specific to the clinoptilolite system. They obtained acceptable fits to their experimental data for the adsorption of U^{VI} to clinoptilite using only the ${}^{int}K_U^{TL}$ constants for $S{:}OH(UO_2)(OH)_4^2$ and $S{:}OH_2(UO_2)(CO_3)_3^{3-}$, respectively, for carbonate free and carbonate-containing goethite systems. Recently, other simplifications of SC model applications have been made. These approaches may be viewed as a combination of the SC model and the activity K_d model, $K_d^{\mathbf{a}}$, introduced some years ago as an adsorption option in the MINTEQ geochemical model (Felmy *et al.*, 1984). In these recent TL SC model applications (e.g., Koß, 1988; Koß and Kim, 1990; Bradbury and Baeyens, 1993), a small subset (one or more) of the aqueous species of a metal is used rather than only the activity of the uncomplexed metal itself in conjunction with ${}^{int}K$ values to calculate the extent of adsorption. The subset of aqueous species used is specific to an individual metal or radionuclide. Thus, a way is available to incorporate a state-of-the-art description of adsorption in transport models without a significant increase in computational requirements by using previously calculated $K_d^{\mathbf{a}}$ and indexed lookup tables available to the transport code.

Although SC models are commonly considered to be thermodynamic models, their application frequently becomes a data-fitting exercise to determine the set of adsorption reactions which best describe the data set. For example, Bradbury and Baeyens (1991) acknowledge that the ${}^{int}K$ values that they calculated from the data of Girvin *et al.* (1991) represent an adjustable parameter. This is partially a result of the variation of acidity constants, and hence of ${}^{int}K$ value because of the $C_{site,area}$ used in its calculation (Davis and Kent, 1990; Bradbury and Baeyens, 1991). The data fitting aspect includes identifying the number of the adsorption reactions needed to quantify adequately the adsorption reactions (aqueous adsorbate and competing cation species), including identifying which reactions are best described as inner-sphere versus outer-sphere adsorption reactions. Indeed, a given surface complex may change from predominantly an inner-sphere to predominantly an outer-sphere surface complex with pH, as with Ag adsorption on $Fe(OH)_3$(a) in the presence of thiosulfate (Ong and Leckie, 1998, this volume).

H. Historical Developments of Surface Complexation Modeling

Modeling the partitioning of metals between solid and aqueous phases on geomedia systems has traditionally involved selecting one or more, but frequently three, single-phase adsorbents from among clay, reactive particulate organic carbon (RPOC), $Fe(OH)_3$(a), MnO_x, SiO_2, carbonate, amorphic aluminosilicate, and Fe sulfides to represent the adsorption sites (Guy *et al.,* 1975; Vuceta and Morgan, 1978; Oakley *et al.,* 1981; Luoma and Davis, 1983; Davis-Colley *et al.,* 1984; Tessier *et al,.* 1985, 1989; Jenne *et al.,* 1986).

An early modeling effort used bentonite, MnO_x, and solid humate as adsorbents for Cu with tannic acid as an aqueous complexent (Guy *et al.,* 1975). They modeled the adsorption of Cu on MnO_x and solid humic acid with Langmuir equation coefficients and onto bentonite with distribution coefficients. They state that "the model was at least in qualitative agreement with the reported behaviors of natural waters" (p. 669), but did not show any comparison of model predictions and their experimental results. Subsequently, the elegant modeling effort of Vucenta and Morgan (1978) speciated eight trace metals, four macroconcentration cations, eight inorganic ligands, and five organic ligands and calculated the relative uptake by three adsorbents (SiO_2, $Fe(OH)_3$(a), and MnO_2).

Oakley *et al.* (1981) chose carbonates, sulfides, and clay as the absorbents for aerobic marine sediments. They used the linear portion of an adsorption isotherm at seawater pH to develop conditional adsorption constants. They concluded that clays were the dominant absorbent in oxidized sediments. This conclusion was disputed by Luoma and Davis (1983) and Laxen (1983). A subsequent paper by the Oregon State University group, of which Oakley had been a member, found that crystalline aluminosilicates were of little importance relative to amorphic Fe oxides and organic substances (Davies–Colley *et al.,* 1984). This change in view came about because the earlier paper (Oakley *et al.,* 1981) did not consider the presence of Mn oxides and disregarded the Fe oxide plus organic substance coating commonly present on natural clay mineral surfaces. Davies-Colley *et al.* (1984) compared predictions made using mass action selectivity coefficients determined for each of five adsorbents ($Fe(OH)_3$(a), MnO_x, Wyoming bentonite, amorphic aluminosilicate, and humic substances) to the measured adsorption onto an estuarine sediment. Their measured Cd adsorption on three of four estuarine samples was somewhat higher than that forecast using their five-adsorbent model. The authors suggested that sulfides were present in the samples and were responsible for the measured adsorption being higher than that predicted by their five-component model. The underprediction of adsorption by the model would presumably have been appreciably greater if they had made a more realistic estimate of the Fe component, e.g., had estimated the amorphic Fe oxide fraction (Tessier *et al.,* 1985; Jenne and Crecelius, 1988) rather than total Fe in the sediment.[8]

Subsequently Luoma and Davis (1983) presented a conceptual model of metal adsorption in marine waters which requires knowing, for each important adsorbent present in the sediment, (1) the adsorption capacity, $C_{M,ads,maxi}$; (2) the adsorption constant (which they envisioned as a conditional surface complexation constant formulated using thermodynamic activity of the uncomplexed metal specie {without considering the displaced cations} and the moles of sites); (3) the quantity of each absorbent, $C_{solid,j-k}$; and (4) the system variables, e.g., pH, C_{ligand}, and $C_{competing\ cation}$. In this model, the $C_{competing\ cation}$ displaced from the surface is not treated as a variable because of the high and essentially constant cation concentration of marine waters. They consider that is advantageous in that the details of pH dependence (e.g., proton release from the surface complexes) as well as surface charge and potential are not required. However, as they note, this limits the application of their approach to conditions of essentially constant water chemistry, e.g., marine waters.

The historical information presented so far in this section illustrates a considerable diversity among investigators as to what are considered the important adsorbents. The same diversity occurs today even among the chapters of this volume. Azizian and Nelson (1998, this volume) consider RPOC[9], Fe oxides, and clay to be the principal adsorbents, citing King (1988) for the importance of clay as an adsorbent. Similarly, Wang *et al.* (1998, this volume) attribute the adsorption of metals (Cd, Ni, and Sr) principally to the clay minerals present in the soils that they studied, rather than oxide coatings, citing literature of the 1960s (Keay and Wild, 1961; Hsu and Bates, 1964). Yong and MacDonald (1998, this volume) use a selective extraction technique to asses the partitioning of added Pb and Cu among carbonates, oxyhydroxides, and particulate organic matter adsorbents. In contrast, Smith *et al.* (1998, this volume) and Payne *et al.* (1998, this volume) consider Fe oxides to be the principal adsorbent. Indeed, since the comprehensive review of Jenne (1968) there has been widespread acceptance of Fe oxides (and Mn or Al oxides where they are quantitatively important) as the major adsorbent in oxic soils, sediments, and clays (Jones, 1957; Theng, 1971; Gonzalez *et al.,* 1974; Jenne, 1977; Jarrell and Dawson, 1978; Karimian and Cox, 1978; Vuceta and Morgan, 1978; Cavallaro and McBride, 1984a,b; Davies-Colley *et al.,* 1984; Stollenwerk and Grove, 1985; McLaren *et al.,* 1986; Jenne and Zachara, 1987; Zachara

[8]The method of determination of oxidic Fe was not specified.

[9]Jenne and Crecelius (1988) proposed the use of 0.5 mol·L^{-1} KOH to estimate the RPOC content of geomedia. However, no consensus has developed as to the appropriate means of estimating the RPOC parameter.

et al., 1989; Harter, 1991; Warren and Zimmerman, 1994; Goldberg *et al.,* 1996; Axe and Anderson, 1997; Payne *et al.,* this volume). Other references attest to the importance of Fe (and Mn) oxides in metal adsorption include Balistrieri *et al.* (1981); Lion *et al.* (1982); McLaren *et al.* (1986); Manceau and Charlet (1992a,b); Waychunas *et al.* (1993); Coughlin and Stone (1995). Thus, Smith *et al.* (1998, this volume) report that using only $Fe(OH)_3$(a) as the adsorbent was adequate to model the adsorption of Pb, Cu, and Zn in an acid-mine drainage system. They concluded that metal partitioning in iron-rich mine-drainage systems occurs primarily in the suspended phase as a consequence of the precipitation of dissolved Fe. Payne *et al.* (1998, this volume) measured similar amounts of U adsorbed on Fe coated kaolinite (1.4% Fe) as on ferrihydrite, which suggests that the coatings were about twice as effective as the bulk ferrihydrite. (Generally, 89 $mg{\cdot}L^{-1}$ of ferrihydrite—as $Fe_2O_3{\cdot}H_2O$—and 4 $g{\cdot}L^{-1}$ of kaolinite were used in each of these adsorption experiments, hence, the Fe in the coating apparently had a higher adsorption capacity than the ferrihydrite since $\{4\ g{\cdot}L^{-1}\}\ \{0.014\} = 56\ g{\cdot}L^{-1}$ of Fe oxide in the coated kaolinite suspension.)

The preponderance of evidence is that for transition metals such as Cd and Ni, iron oxides are generally the primary adsorbent in oxic environments unless unusual amounts of MnO_x, Al oxides, POC, and rarely carbonates, etc., are present relative to the amount of oxidic Fe surfaces available. It is presently incumbent upon investigators to establish by experimental means that adsorbents other than Fe oxides are important adsorbents in their system(s). This was exactly the approach of Payne *et al.* (1998, this volume). They found that although Fe oxides were the dominant adsorbent, the presence of a limited number of Ti-rich particles (anatase, and rutile) with a high affinity for U^{IV} made it difficult to model the adsorption of U^{IV} by two reference kaolinites.

An approach sometimes used in evaluating the importance of Fe and Mn oxides are so-called selective extractions. However, selective extractions yield results that generally can be interpreted only in a speculative manner as found by Yong and MacDonald (1998, this volume). They report that the removal of amorphic oxides of Fe and Al increased the amount of Pb and Cu "retained" (they do not distinguish between adsorption and precipitation) by two to three times that found initially for pulverized illitic shale. Their characterization of the pulverized illitic shale indicated a sizable decrease in cation exchange capacity ($C_{sites,cation\ exchange}$) in spite of a large increase in S_A (determined with ethylene glycol monoethyl ether) following NaOAc–HOAc treatment (intended to remove carbonates but which also removes Mn oxides according to Jackson, 1956). Yong and MacDonald (1998, this volume) do not speculate on the cause of the decrease in $C_{sites,cation\ exchage}$.

Where Fe and Mn oxides are minor components and reactive organic carbon is present in substantial quantities, the organic substances may dominate metal adsorption (Langston, 1982). Lion *et al.* (1982) found that, after removal of reactive organic substances from a sediment with 0.1 mol·L^{-1} NaOH, the preextraction adsorption capacity was restored by adding back to this estuarine sediment an amount of "sediment derived humic substances" equal to that which was removed by extraction. The sediment derived humic substances were presumably obtained by 0.1 mol·L^{-1} NaOH extraction of another aliquot of the sediment. Amorphous aluminosilicates (Davies-Colley *et al.,* 1984) and/or carbonates (Jenne and Wahlberg, 1965) may be dominate adsorbents in sediments containing minimal amounts of oxidic Fe and Mn or organic matter. In anaerobic soils and sediments, Fe sulfides are presumed to be the important absorbents along with RPOC (Di Toro *et al.,* 1990; Allen *et al.,* 1993). Dissolved sulfide (S^{-II}) limits the $C_{M,aq}$ in oxygen-limited sediments if S^{-II}_{aq} exceeds the $C_{M,aq}$, generally by precipitation rather than by adsorption (Di Toro *et al.,* 1990; Allen *et al.,* 1993). However, the relationship between toxicity and the S^{-II}_{aq} to $C_{M,aq}$ ratio is not always straightforward (Besser *et al.,* 1996).

The extent to which adsorption measurements on laboratory oxide or humic acid preparations represent the real world is of long standing concern. Therefore, it is significant that Tessier *et al.* (1989) found that the apparent equilibrium adsorption constants calculated from measured metal concentrations in pore water and amorphic iron extracted from the sediment generally fell within the range of similar constants recalculated from literature data. All multiadsorbent models assume (1) linear additivity among absorbents (i.e., they act independent of each other), (2) no effect of $C_{geomedia}$ given that $C_{M,aq,i}$ and $C_{geomedia}$ are such that an approximately constant $C_{M,aq,f}$ results, and (3) an adequate state of dispersion. Honeyman (1984), carrying out studies with mixed adsorbents, concluded that an error sometimes resulted from the additivity assumption, especially in mixtures containing TiO_2. However, Davies-Cooley *et al.* (1984) found no evidence of a reduction in adsorption effectiveness as a result of mixing of selected solids (e.g., $Fe(OH)_3$(a)–Wyoming bentonite–humic substance mixture). E. A. Jenne and R. J. Serne (unpublished data) showed that the $C_{geomedia}$ effects reported in the literature are primarily a consequence of differing loadings and of different $C_{M,aq,f}$ among the comparisons. However, the mixing of highly unstable adsorbents can result in significant changes in surface chemistry that can strongly impact adsorption. Thus, Anderson and Benjamin (1990) found that mixing fresh precipitates of Al and Fe at pH 8 resulted in physical changes in the size distribution of $Fe(OH)_3$(a) particles and the presence of a significant amount of Al on the $Fe(OH)_3$(a) surfaces which changed its sorptive properties. Smith *et al.* (1998, this volume), following the approach used by Loux *et al.* (1989), modeled their acid-mine drainage system using chemical analyses of the water and DL SC model constants from Dzombak and Morel (1990) without any fitting of parameters to site data. Both Loux *et al.* (1989) and Smith *et al.* (1998, this volume) use $Fe(OH)_3$(a) as the sole adsorbent and simplified the modeling by neglecting the competition of other cations.

There is little information on the effect of the extent of aggregate dispersion on adsorption. However, the apparent effect of aggregate size, for at least one sediment, as shown to be due to differing adsorption capacities among the various size fractions of sediment from a freshwater stream (Jenne, 1995). It seems likely that the effect of particle and aggregate size reported by Turner and Millward (1994) was a result of mineralogy differences in the transect through the estuary.

I. Regression Models

The nature of adsorption experiments is such that many adsorption-determining variables change in the course of a given experiment and across experiments. A suite of experiments where $C_{M,aq,i}$ is varied is likely to result in a systematic variation in pH unless it is adjusted. Other variables, e.g., $C_{competing\ cation}$, may also vary as a result of the adsorption process. Therefore, the useful modeling of adsorption data and reliable prediction of future adsorption require either a set of equations containing important variables or a matrix of K_d, K_F, $K_{L,aff}$, $K_{L,max}$, K_{MA}, or K_P values such that the effects of changing $C_{Me,aq}$, pH, $C_{competng\ cation}$, loading, ***I***, etc., on $C_{M,ads}$ are adequately accounted for. The use of these adsorption constants may require the generation of large amounts of data necessary for these empirical descriptions of adsorption. The interpretative and modeling problem presented by the simultaneous variation of multiple variables can be dealt with by use of the semitheoretical SC models discussed previously. Alternatively, regression analysis can be used to treat explicitly the multidimensional nature of adsorption. Multivariable regression analysis of log-normalized $C_{M,ads}$ data allows the development of linear multivariable regression models (MVRM) that explicitly treat pH, $C_{M,aq,f}$, $C_{competing\ cation}$, ***I***, etc.

Linear regression models have also been used to evaluate the dependence of metal adsorption on measured soil and rock characteristics. For example, Borkovec (1981) found a relationship between the $C_{U,ads,max}$ and the $C_{sites,cation\ exchange}$ of silicate minerals at pH 7 and 20°C as

$$C_{U,ads,max} = 1.56 + 0.90(C_{sites,cation\ exchange}), \quad [19]$$

where both variables are in milliequivalents per 100 g. The equation fit the observed data with an r^2 value of 0.945. He noted that the average surface area for the clay minerals used was 2.73 nm^2 per adsorbed UO_2^{2+} ion. Torrent (1987) carried out regression analysis using the log form of his time variant Freundlich equation (Eq. [13B]), i.e.,

$$\log C_{ads} = \log K_{F_t} + 1/n_F \log C_{M,aq,f} + F \log t, \quad [20]$$

to obtain the coefficients describing his PO_4^{3-} adsorption data.

Thibault *et al.* (1990) used soil-to-plant concentration ratios ($C_{soil/plant}$), taken from Baes *et al.* (1984) and certain soil properties in a nonlinear MVRM to estimate K_d values for soils on which K_d values had not been measured. Their equation is

$$\ln K_d = c_0 = c_1 + c_3 \ln C_{soil/plant}, \qquad [21]$$

where $C_0 = 4.62$; c_1 for sand, loam, clay, and reactive particulate organic matter (> 30% RPOM) is −2.51, −1.26, −0.84, and 0, respectively; and $c_3 = 0.56$. The "texture" categories were defined as sand ≥ 70% sand-sized particles; clay as ≥ 30% clay-sized material; and loam represents roughly equal amounts of sand, clay, and silt or materials that contains up to 80% silt-sized particles. The Canadian nuclear waste storage program is making extensive use of a nonlinear MVRM to explicitly account for dissolved radionuclide concentration (M_R), total dissolved solids (TDS, $g \cdot L^{-1}$), and interaction effects,

$$K_d = c_0 + c_1 (\log C_{TDS}) + c_2 (\log C_{M_R,aq}) + c_{11} (\log C_{TDS})^2 + c_{22} (\log C_{M_R,aq})^2 + c_{12} (\log C_{TDS}) (\log C_{M,aq}), \qquad [22]$$

where $C_{M_R,aq}$ is presumably the final concentration, and $C_{j \to k}$ are regression coefficients (Vandergraaf and Ticknor, 1994). They found that this model over constrained the system; hence, one of the C_{TDS}, $C_{M_R,aq}$, or $(\log C_{TDS}) (\log C_{M_R,aq})$ terms was unnecessary and therefore set to zero. The use of nonlinear regression may provide a better fit to a set of data than linear regression; however, it is not assured that the improved fit is worth the increased effort generally required to optimize the fit of a nonlinear equation to a data set.

According to Vandergraaf *et al.* (1993), their use of logarithmic values of their variables stems in part from the common practice of varying the $C_{M,aq,i}$, in batch adsorption experiments, over several orders of magnitude. The K_{d,M_R} is calculated for individual minerals present and then weighted by the estimated fractional amount of each mineral present, and these individual K_{d,M_R} values are summed to obtain the K_d^{Σ} for a given radionuclide for that segment of GEONET. In some instances, the K_{d,M_R} values are for a rock type, e.g., gray granite, gabbro. The Canadian radionuclide transport model (GEONET) requires K_d values that are independent of the radionuclide concentration, C_{M_R}. This is accomplished by treating the concentration dependence as an additional uncertainty on the K_{d,M_R} (Vandergraaf *et al.*, 1993).

The approximately constant relationship between ***I*** and the concentration of divalent cations, in transects from fresh water through estuaries into marine waters, allows freshwater K_d values to be used to estimate K_d values for associated marine water through a regression equation that includes salinity (Turner and Millward, 1994). Their equation is

$$\ln K_{d,S_x} = c_o \ln (S_x + 1) + \ln K_{d,S_x=0}, \qquad [23]$$

where S_x is the salinity > 0 (parts per thousand) of a particular sample, $K_{d,S_x=0}$ is the K_d for the fresh water end member, and c_o is the slope of a plot of ln K_{d,S_X} versus ln $(S_x + 1)$. This equation indicates that the ln $K_{d,Sr}$, at least for selected individual metals, is a linear function of a constant times ln $(S_x + 1)$ plus the fresh water $K_{d,S=0}$ value.

It is instructive to consider how a linear single variable regression model relates to other adsorption models. All types of adsorption models are either simplifications or elaborations of the mass action relationship. Recalling the Freundlich equation,

$$C_{M,ads} = K_F\, C_{M,aq,f}^{1/n_F}, \qquad [12B]$$

where $C_{M,ads}$ is the adsorbed metal concentration $(mol \cdot g^{-1})$, K_F is the isotherm coefficient, $C_{M,aq}$ is the aqueous concentration of the metal at some designated time, and $1/n_F$ is a dimensionless constant. Taking logarithims yields

$$\log C_{M,ads} = \log K_F + 1/n_F \log C_{M,aq,f}, \qquad [12C]$$

where log K_F is the *Y*-axis intercept and $1/n_F$ the slope of a plot of log $C_{M,ads}$ versus log $C_{M,aq,f}$. Thus, linear regression on single variables are seen to be the linearized form of the Freundlich equation and with two variables, it is similar to the Torrent relationship (Eq. [20]). Note also that where $1/n_F$ equals one, Eq. [12B] reduces to

$$K_F = C_{M,ads}/C_{M,aq,f}, \qquad [19]$$

indicating that the K_d conceptual model is a special case of the Freundlich equation where it is assumed that the stoichiometry of exchange is equal to one regardless of the valence of the exchanging cations, all adsorption sites are homo-dentate, there is no competition from desorbed cations, and there are no loading effects.

It follows from the preceding discussion that MVRMs of log transformed data are a multidimensional Freundlich equation where one of the variables may or may not be metal concentration. Adding additional variables is sometimes called extending the Freundlich equation (Streck and Richter, 1997). Zachara *et al.* (1992) reported the percentage of U^{VI} adsorbed by two smectite-rich, < 2-μm soil clay separates over a range of pH values. Background electrolyte concentrations of $Ca(ClO_4)_2$ were 0.05, 0.005, and 0.0005 $mol \cdot L^{-1}$, and the minimum H was 4.7; all data points with less than 99% of the U^{VI} adsorbed were used in the MVRM developed from these data. The MVRM for the Ringold and Kenoma separates are, respectively,

$$\log K_d = 3.42 - 1.15(\log C_{U,aq,f}) + 0.00875(pH) - 0.00220(\log \boldsymbol{I}), \qquad [24]$$

$$\log K_d = 0.327 - 0.614(\log C_{U,aq,f}) + 0.737(pH) + 0.373(\log \boldsymbol{I}). \qquad [25]$$

The agreement between experimental, (filled symbols) and simulated open symbols values for the Kenoma subsoil (Fig. 6) is excellent; many of the simulated

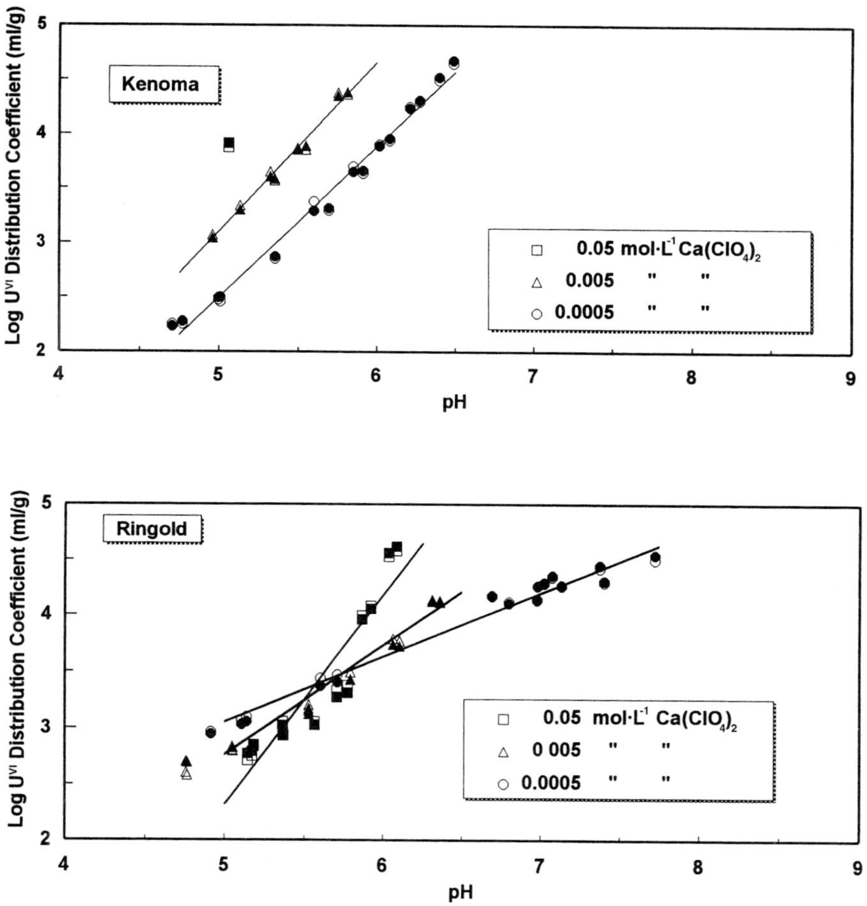

Figure 6 Simulation of experimental log K_d values (open symbols) for U^{VI} for smectite-rich separates from the Kenoma and Ringold subsoils using multiple variable linear regression models (filled symbols) developed from their respective data sets (after Zachara *et al.*, 1992, Figs. 6 and 7, respectively).

values virtually overlay the experimental values in spite of the single adsorption value for the 0.05 $mol \cdot L^{-1}$ $Ca(ClO_4)_2$ concentration. The model fit for the Ringold subsoil is very good, notwithstanding the markedly different slopes of the K_d values at each of the three electrolyte concentrations (Fig. 6). These relatively large changes in slope due to different $C_{competing\ Ca}$ and $\boldsymbol{I}$ illustrate why Freundlich and Langmuir coefficients have virtually no transfer value. These slope differences also indicate why adequate system characterization is vital.

One of the important questions in model development is evaluation of the trans-

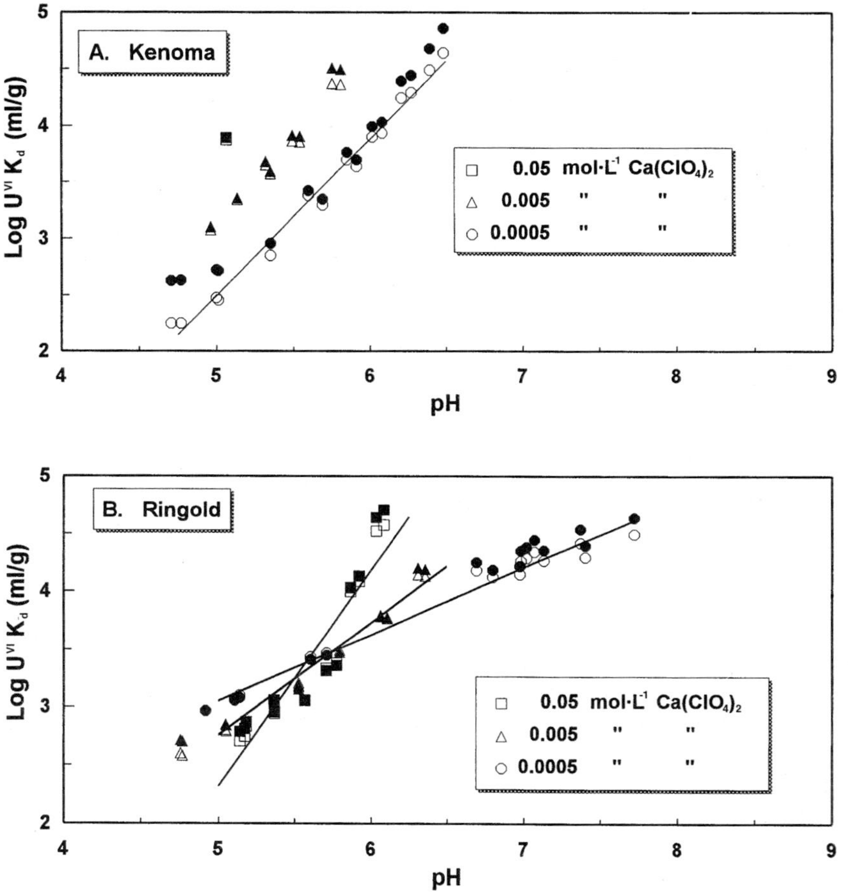

Figure 7 Simulation of experimental (open symbols) U^{VI} K_d values for the Kenoma and Ringold subsoil data sets with a single multiple variable linear regression model (filled symbols) developed by combining the data sets of both subsoils (after Zachara *et al.,* 1992, Figs. 6 and 7, respectively).

fer potential of the calibrated models. Because the Ringold and Kenoma sorbents are both smectite-rich, <2-μm soil clay fractions from which free iron[10] had been removed and the structural iron reoxidized, the adsorption data collected on these sorbents were combined, yielding

[10]'Free' oxide (terminology from the field of soil science) signifies that the iron is present exclusively in oxidic form, rather than being all or partially present in the structure of minerals other than iron oxides, and is operationally defined as the Fe removed by a dithionite-citrate extraction (Jackson, 1956).

$$\log K_d = 3.41 - 1.19(\log C_{U,aq,f}) + 0.0183(pH) - 0.0247(\log \boldsymbol{I}). \quad [26]$$

Equation [26] was used to simulate separately the Kenoma and Ringold experimental results. As shown in Figure 7, the linear MVRM values (open symbols) simulate the experimental K_d values (filled symbols) adequately for the Kenoma subsoil, the larger divergence occurring at the higher and lower ends of the pH range. The fit is even better for the Ringold subsoil, which has several times as many experimental values as does the Kenoma subsoil.

A comparison of Figures 2 and 7 suggests that there is a sizable difference in the pH at which $C_{competing\ Ca}$, as compared to $C_{competing\ Na}$, results in increased $K_{d,U}$ for the higher competing cation concentrations. Thus, for field applications, it would appear desirable to develop such models using C_{Ca} and C_{Na} rather than $\boldsymbol{I}$ or TDS and to include all monovalent cations with Na and other divalent cations with Ca.

IV. ARTIFACT EFFECTS

Artifact effects are relatively common in metal adsorption investigations (Jenne and Zachara, 1987; Jenne, 1996). These effects include: (1) pretreatment and purification procedures, (2) removal by precipitation instead of adsorption, (3) water composition changes during experiments (e.g., mineral dissolution, increase in $C_{competing\ cation}$ by desorption); (4) uptake of metal by bacterial growth, (5) bulk diffusion limitation on the adsorption rate, (6) blank corrections errors due to widely differing pH and $C_{M,aq,i}$, (7) valence reduction of adsorbate at the surface of the adsorbent (Griffin *et al.*, 1985), and (8) inadequate phase separation (Koskinen *et al.*, 1985).

Bertetti *et al.* (1998, this volume) note that in the case of adsorption by individual minerals, the amount and nature of impurities present may be a significant variable. Payne *et al.* (1998, this volume) found that Ti oxide (anatase and rutile), an impurity in their reference kaolinites, adsorbed sufficient U that they were unable to satisfactorily model U adsorption on kaolinite. Pretreatments can have a marked effect on metal adsorption; for example, Clark *et al.* (1998, this volume) report that following aging of a low $C_{site,cation\ exchange}$ sandy soil at pH 2.0 the adsorption capacity for Eu increased significantly. Acidic conditions commonly leach cations from minerals, e.g., feldspars, and form microcrystalline to amorphic oxides of Al and/or Si which can increase the adsorption capacity, especially of low adsorptive capacity geomedia. Pretreatment to remove oxidic Fe from pulverized shale (Yong and MacDonald, 1998, this volume) with 0.1 mol·L^{-1} oxalic acid resulted in increased adsorption capacity. Oxalate removes readily soluble Fe and Al by acid dissolution and complexation and would not be expected to reduce structural Fe^{III} to Fe^{II} and thereby increase permanent charge adsorption sites as strong reducing agents can do. When strong reducing agents are used, it is advisable to treat clays with dilute hydrogen peroxide to reoxidize structural Fe and minimize this source of error.

Some investigators who have not attempted to separate adsorption and precipitation effects have called upon precipitation to explain higher than expected metal uptake. It may be difficult to determine whether or not heterogeneous precipitation is responsible for a greater than expected uptake. The occurrence of precipitation during adsorption experiments (1) is preferably identified by a surface analysis that indicates that new solids have formed, but (2) is sometimes inferred from a break in a plot of the K_d versus some appropriate parameter, such as $C_{geomedia}$ or saturation index (SI) (Fig. 8, discussed further on), or (3) is inferred from an amount adsorbed well in excess of the calculated number of

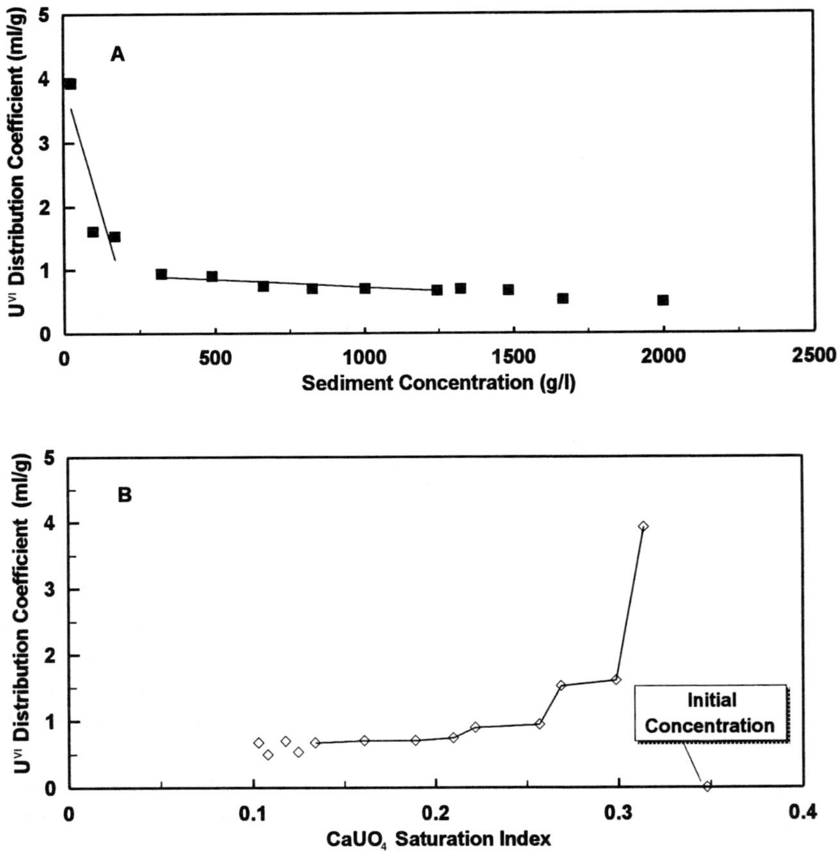

Figure 8 Uranium^{VI} K_d values as a function of (A) sediment concentration and (B) the $CaUO_4$ saturation index (after Koß, 1988, Fig. 5).

sites (McBride, 1979). Because of the general occurrence of fast and slow adsorption processes (Jenne, 1995, 1996; Axe and Anderson, 1998, this volume), the presence of both fast and slow adsorption processes provides no evidence of precipitation.

The precipitation of metal compounds requires that the product of the thermodynamic activity of the aqueous uncomplexed metal specie and other component ions that form the solid, each raised to the power of their stoichiometry in the solid, exceeds the solubility of one or more plausible compounds of the metal. Payne *et al.* (1998, this volume) state that at U concentrations of 10^{-4} mol·L^{-1} and greater, polynuclear species become important and may result in precipitation if the aqueous concentration is not lowered rapidly enough by adsorption. For precipitation to actually affect the aqueous concentration of a metal at time t ($C_{M,aq,t}$), precipitation must occur at a rate that is significant relative to the adsorption rate of the metal. This appears possible in the data shown in Figure 8A. The rapid decrease in $K_{d,U}$ in conjunction with a large increase in $C_{sediment}$ is accompanied by a decrease in the saturation indices of $CaUO_4$, a plausible U solid (Fig. 8B). (Note that the smallest $C_{sediment}$ values are at the left side of Fig. 8A and the right side of Fig. 8B). The need for surface analyses to resolve possible precipitation of a new solid phase is clearly indicated in this case.

Inferring precipitation by loss of a reactant from solution may be complicated by the occurrence of multiple reactions. For example, the uptake of phosphate on layer silicate surfaces has been frequently studied by soil scientists. Phosphate uptake can often be resolved into multiple separate apparent rates which are generally interpreted as (1) the initial adsorption at the surface and (2) the reaction of PO_4 with surface structural Al to form Al–PO_4 compounds, which is followed by (3) diffusion of PO_4 through the surface Al–PO_4 layer to form a precipitate with the underlying structural Al. Thus, adsorption may be a necessary first step for precipitation where the solid provides one of the components of the precipitation reaction or where the solid provides a nucleation surface (McBride, 1979; Zachara *et al.,* 1989).

The sometimes important role of biological uptake and release, and the turnover rates involved, has been identified as the major factor causing differences, said to be of 1 to 3 orders of magnitude, between K_d values calculated from laboratory data as compared to field measurements in the marine environment (Coale and Bruland, 1985, 1987; Morel and Hudson, 1985; Jannasch *et al.,* 1988). In analyzing laboratory data for a marine system, Jannasch *et al.* (1988) treated the combined fast and slow adsorption processes as a "simple overall first order reaction." Such treatment appears to obscure the actual processes. It is questionable if these calculated rates have any relevance.

Redden *et al.* (1998, this volume) note that container blanks need to be determined as a function of $C_{M,aq,f}$ and the appropriate value from this curve used in making blank corrections. Otherwise, blank corrections from a single $C_{M,aq,f}$ can increase the error in the results as compared to no blank correction. This follows since containers display low-capacity adsorption isotherm in the manner of other adsorbents (Redden *et al.* 1998, this volume).

Inadequate phase separation frequently produces artifact effects of varying magnitude. For example, modest increases in K_d values were found when centrifuged supernatants (60 min at 8000 rpm; the relative centrifugal force was unspecified) were filtered (Koskinen *et al.,* 1985).

Investigators have generally disregarded valence changes during the adsorption of redox sensitive elements although, for example, the adsorption of U onto Fe oxides may involve valence reduction in nominally oxic systems, especially during adsorption periods of weeks to months (e.g., Hsi and Langmuir, 1985). Only rarely have the actual valence states of the adsorbing specie(s) been determined. Studies involving Co (e.g., Murray *et al.,* 1984) and As (Scott and Morgan, 1995) are exceptions to the foregoing as there are several studies of Co and of As oxidation following adsorption onto oxide surfaces. Oxidation reactions may be particularly important where they result in significantly increased K_d values for the metal and more so at sites were colloidal transport of contaminants occurs. For metals with high affinity for colloidal organic substances, higher valence states may result in increased amounts of the metal passing through filter membrane and biasing experimental results.

V. VARIABLES

Important variables and some of their interactions are reviewed in this section. The variables reviewed include pH, metal concentration and loading, competing cations versus ionic strength, aqueous speciation, surface complexation and stoickiometry, surface area and site density, surface area normalization, and time dependency of reaction.

A. pH AND P_{CO_2}

The most studied variable in adsorption investigations is, without doubt, pH. The reasons for this include the (1) the often dramatic effect of pH on the fraction of added metal that is adsorbed; (2) the availability of reliable pH control instrumentation; (3) the likelihood of significant change of pH in the system of interest during the reaction period, particularly if the imposed pH is outside the normal pH of that geomedia-water system; (4) the possibility of obtaining mechanistic information, and (5) the ease of achieving multiple orders of magnitude change in H activity.

Distribution coefficient values for U adsorption from 14 sources are shown in Figure 9. These $K_{d,U}$ range over six orders of magnitude for relatively narrow pH

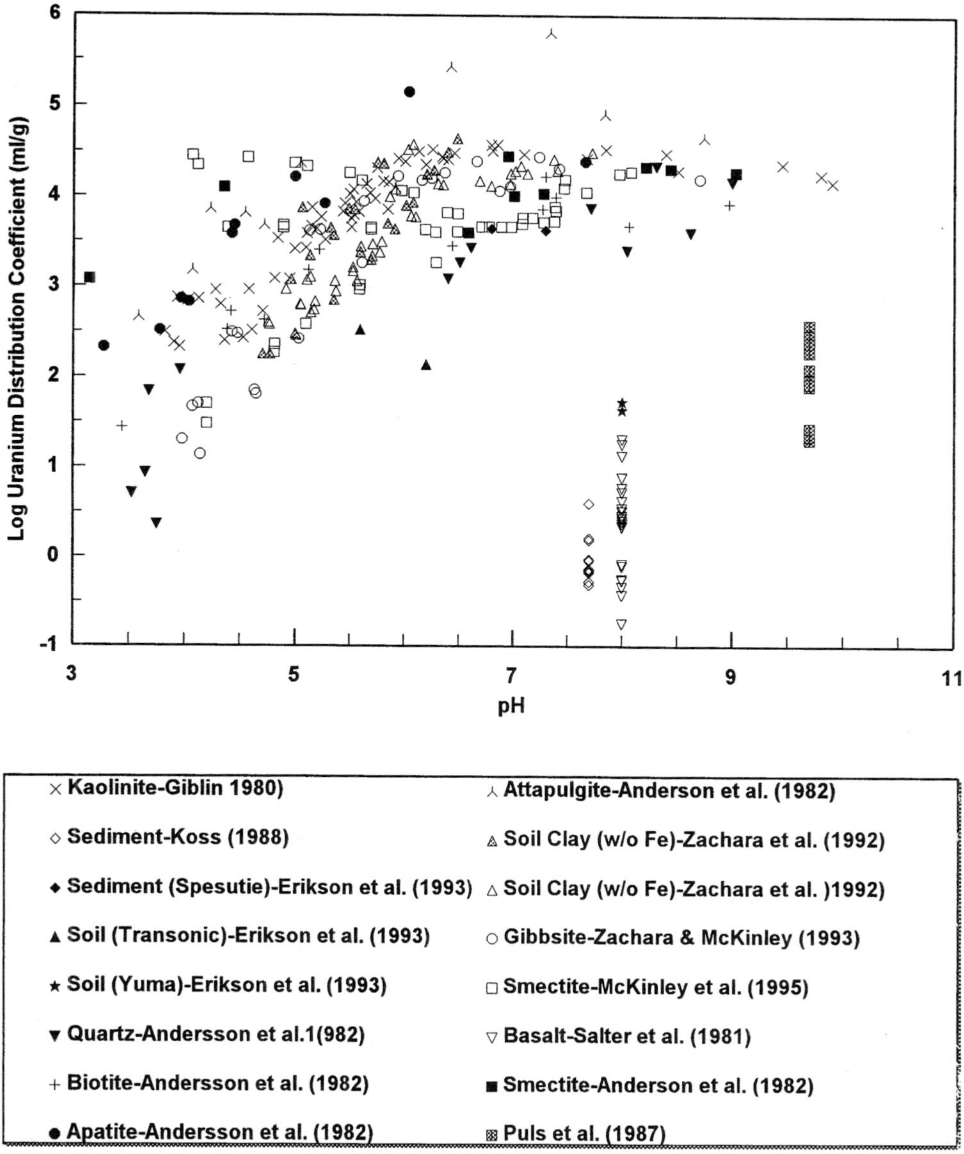

Figure 9 Composite plot of log U^{VI} K_d versus pH for numerous solids from several sources.

ranges, particularly below pH 5. The range of $K_{d,U}$ values at a given pH can not be attributed solely or even primarily to the different adsorbents used. It is apparent that a major portion of the variation in K_d must be attributed to variables other than pH and the adsorbent itself. For example, the K_d values for the smectitic clay separate of Zachara *et al.* (1992) cover three orders of magnitude as a result of the interplay of pH with competing cations (Na or Ca) as shown in Figures 2 and 6B. Note also that if the metal to adsorbent ratio is such that adsorption approaches 100% in the near neutral pH region, the isotherm must plateau even if the suspension is free of CO_2 as is the case for the Zachara and McKinley (1993) data in Figure 9. All other data were presumably in equilibrium with atmospheric CO_2. The geomedia whose K_d values fall well below those of the main grouping of samples are adsorbents with particularly low adsorption capacity (e.g., basalt, aquifer sands).

Numerous investigators have found that the slope of a log $C_{M,ads}$ versus pH plot decreases significantly somewhere between pH 6 and 8.5, in the presence of CO_2, with geomedia as diverse as quartz, secondary layer silicates, and amorphic Fe oxides (e.g., Allard *et al.* 1980), whereas this decrease is not observed in CO_2-free systems (Tripathi, 1984; Hsi and Langmuir, 1985; Zachara and McKinley, 1993; McKinley *et al.,* 1995; Bertetti *et al.,* 1998, this volume; Pabalan *et al.,* 1998, this volume). This plateau or maximum in the adsorption curve in the pH 6 to 8.5 region normally occurs only with metals that form aqueous complexes with carbonate to a significant degree. Bertetti *et al.* (1998, this volume) found that when CO_2 was excluded, the adsorption of Np^{V} increased with pH over the entire range examined.

Desorption of actinides, induced by dissolved CO_3^{2-}, occurs at some pH above 7 (Hsi and Langmuir, 1985; Tripathi, 1984; LaFlamme and Murray 1987; Payne and Waite, 1991; Kohler *et al.,* 1992; Turner, 1993). The pH at which net desorption becomes apparent varies with the particular actinide, $C_{actinide}$, $C_{geomedia}$, competing cation, and $C_{competing\ cation}$. Carbonate-induced desorption at pH values above neutrality has been attributed to the failure of negatively charged metal–CO_3 complexes to adsorb and to the effect of competition between the anionic species of the metal and adsorbed CO_3^{2-} and/or HCO_3^- on negatively charged sites. Zachara *et al.* (1987) found that the fit of the TL SC model for CrO_4^{2-} was improved by considering the competitive adsorption of CO_3^{2-} at a CO_2 partial pressure of $10^{-2.46}$ atm, whereas Payne *et al.* (1992) model U^{VI} adsorption on ferrihydrite with the TL SC model in the presence of CO_2 without invoking competitive CO_3^{2-} adsorption. In accord with the finding of Payne *et al.* (1992), Pabalan and Turner (1993c) concluded from the literature that little CO_3^{2-} was adsorbed except when P_{CO_2} were higher than the normal atmospheric value of $10^{-3.48}$ atm. Pabalan and Turner (1993a) note that the P_{CO_2} in Yucca Mountain (Nevada) groundwaters varied from $10^{-3.8}$ to $10^{-0.8}$ atm. Payne *et al.* (1998, this volume; Fig. 7) show that an increase in P_{CO_2} from $10^{-3.5}$ to 10^{-2} caused the high pH adsorption edge to be lowered by one pH unit.

The pH at which significant metal adsorption begins is dependent upon the strength of interaction of the metal and surface adsorption sites, the concentration of $S:OH^0$ sites, and $C_{geomedia}$ (Schindler, 1981; Honeyman, 1984). The pH at which individual metals exhibit maximum adsorption is expected to vary with the first hydrolysis constant of cationic metals (e.g., Nyffeler *et al.,* 1984). Both Bertetti *et al.* (1998, this volume) and Pabalan *et al.* (1998, this volume) assume that the decreased adsorption at pH values below the maximum adsorption of Np^V and U^{VI}, respectively, is due to a change in the aqueous speciation of the metal. They do not appear to have evaluated the likelihood that the decreased adsorption with decreasing pH is a direct result of competition of H with the actinide for adsorption sites, i.e., the increasing proportion of $S:OH_2^+$ sites as the C_H increased. The increase in adsorption capacity with increasing pH is presumably due to the conversion of positively charged diprontonated sites, $S:OH_2^+$, to monoprontonated sites, $S:OH^0$, which in turn convert to negatively charged $S:O^-$. This increases the number of available adsorption sites and reduces the competition of H^+_{aq} with the adsorbate.

The site protonation diagram of Krupka *et al.* (1988, Fig. 7-1), based on the experimental data of Davis (1977) plus that of Hsi and Langmuir (1985), shows that the fraction of $S:OH_2^+$ increases slightly as the pH drops from 7 to 6 and increases rapidly below pH 6. The $S:O^-$ noticeably increases above zero at pH 9 and increases to about 10% at pH 10.5.

Given the wide range in $K_{d,U}$ values at a given pH for similar geomedia (Fig. 9), it is not surprising that the pH of $C_{M,ads,max}$ occurs over a range of pH values for different systems. For example, Lieser *et al.* (1992) observed that the maximum adsorption of U in systems open to the air occurred at a pH between 6.5 and 7.5 for $SiO_2 \cdot H_2O(a)$, $SiO_2(c)$, $TiO_2 \cdot H_2O(a)$, and $Al_2O_3(c)$. It should be borne in mind that individual geomedia tend to approach some particular pH when suspended in a small volume of water (Keller *et al.,* 1963). For example suspensions of kaolonite and montmorillonite clays were found to converge toward a pH of 6.5 to 7.5 when subjected to shear stress in a cylindrical viscometer (Langston and Jenne, 1964). This is of consequence in two regards. The pH of the aqueous media used to suspend the geomedia often changes significantly during adsorption studies unless it is adjusted. Unfortunately, in many adsorption studies, particularly where the objective is to calculate K_d values, only the *initial* pH value has been reported. As a consequence of the significant alteration in the pH values that frequently occurs as a result of metal desorption, adsorbent dissolution, or solute precipitation during sorption experiments, the absence of final pH values noticeably degrades the reliability of the determined adsorption values.

B. Metal Concentration and Loading

A decrease in the amount of metal adsorbed per increment of increased $C_{M,aq,f}$, after an initial nearly linear relationship (if the $C_{M,aq,i}$ is sufficiently low relative to the mass of adsorbent present), is the well known "loading effect." Loading describes the extent to which the total number of adsorption sites are occupied by the adsorbate(s) of interest. The loading effect results in an equilibrium constant that decreases as the loading increases beyond some small fraction of the total number of sites. Benjamin and Leckie (1981a) found that their equilibrium constant for Cd ($^{avg}K_{TL,Cd}$) varied only slightly below an occupancy of less than $1 \cdot 10^{-5}$ $mol \cdot mol^{-1}$ of Cd adsorbed on $Fe(OH)_3$(a) but decreased 100-fold at higher loadings. Smith and Jenne (1991) show that the $K^{SC}_{TL,CdOH^+}$ decreased by an order of magnitude as the loading of Cd on $Fe(OH)_3$(a) increased. The $K_{d,U^{VI}}$ values of Pabalan *et al.* (1998, this volume) decrease with increasing $C_{U,aq,f}$, particularly in the intermediate pH range. Similarly, Koß and Kim (1990) found $K_{d,Sr}$ values to vary by a factor of 35 with modest variations in $C_{Sr,aq,i}$ (some properties of the aquifer sediment–water samples also varied). Being aware of the loading effect, Borkovec (1981) showed that U^{VI} adsorption by clay minerals from concentrations greater than approximately $1 \cdot 10^{-4}$ $mol \cdot L^{-1}$ (presumably using 2.4 $g \cdot L^{-1}$ of clay) could be appropriately described by a fractional occupancy form of the Langmuir equation (Eq. [14G]).

The conventional interpretation of the loading effect is that the number of sites highly preferred by a given metal are a small portion of the total number of sites (Benjamin and Leckie, 1981a; Payne *et al.,* 1998, this volume). Dzombak and Morel (1986) calculated that a small number of strong sites and a proportionally larger number of weak sites (0.0018 and 0.2 $mol \cdot mol^{-1}$ of sites:Fe, respectively) gave the best DL SC model fit. Benjamin and Leckie (1981a) inferred from competition experiments that $Fe(OH)_3$(a) contains multiple binding sites; hence, the preferred binding sites for one metal are not necessarily preferred by other metals. This conclusion does not meld smoothly with the Dzombak and Morel (1986) hypothesis of a set of strong and a set of weak sites that pertain to all cationic metals.

A complication arises in interpreting loading effects where the pH varies because the number of neutral sites is a function of pH. One frequently finds statements in the literature to the effect that since "trace" elements are present at low concentrations whereas binding sites on surfaces of sedimentary phases are in great excess, adsorption is expected to be independent of metal concentration" (e.g., Davies-Colley *et al.,* 1984). The inference being that adsorption sites are vacant, whereas, these sites are always occupied by other metals or H. Of the papers reviewed for this chapter, none have investigated the possibility that conventional pH isotherms are biased because longer equilibration time are required at higher pH values (Axe and Anderson, 1998, this volume).

The determination of a loading effect may also be complicated by the lack of apparent stoichiometric exchange due to the formation of bidentate surface complexes (Benjamin and Leckie, 1981a); formation of multiple surface complexes (see Table 1); a change from mononuclear to multinuclear surfaces species with increasing pH (Thomson *et al.,* 1998, this volume); and aqueous speciation, especially the hydrolysis of displaced metal ions (Benjamin and Leckie, 1981a). Given the number of papers reporting bidentate complexes, one wonders if bidentate sites are preferentially filled over monodentate sites.

Benjamin and Leckie (1982) noted that the pH–adsorption edge for metals on $Fe(OH)_3$(a) often shifts to higher pH with increasing loading even though an excess of surface sites remain available and the electrical properties of the surface are approximately constant. Pabalan and Turner (1993c) also observed a shift of the adsorption edge toward higher pH values with a 10-fold increase in $C_{U,aq,i}$ and note that Tripathi (1984) and Payne *et al.* (1992) found similar effects on goethite and ferrihydrite, respectively. This phenomenon has been attributed to nonuniformity of surface sites and may be more a characteristic of trace metal adsorption than ligand adsorption (Benjamin and Leckie, 1981a).

Axe and Anderson (1998, this volume) found linear isotherms over six orders of magnitude of $C_{Sr,aq,f}$ where $C_{Sr,aq}$ was held constant at each of a series of concentrations throughout the equilibrium period. In contrast, the majority of studies reporting a decrease in $\Delta C_{M,ads}/\Delta C_{M,aq,f}$ with increasing $C_{M,aq,f}$ are from systems where $C_{M,aq,f}$ decreased during the experiments. Another difference in the Axe and Anderson (1998, this volume) experiment is that they chose a reaction time just sufficient to yield equilibrium with external sites. In contrast, most other studies used a time period 50 to 100% in excess of the time to saturate external sites. That is, as $C_{S:OH^{-0}}$ becomes large relative to the $C_{M,aq}$, progressively more of the adsorbing metal must diffuse to interior sites to reach equilibrium which requires longer and longer times for equilibrium to be reached. Given that Axe and Anderson (1998, this volume) found that periods of 1 month to 5 yr was required to reach equilibrium in the Sr–$Fe(OH)_3$(a) system for pH 6 and 7, respectively, the failure to reach equilibrium may be responsible for many of the reports of loading effects. Axe and Anderson (1998, this volume) found that roughly equal amounts of Cd were adsorbed on external and on internal adsorption sites of ferrihydrite. It may be that the metal versus H selectivity of internal sites differs from that on external sites. It appears that progressively longer times may be required to reach equilibrium as the loading, hence the diffusivity, increases.

C. Competing Cations and Ionic Strength

It is generally accepted that Cd, Cu, Pb, and Zn compete with one another for adsorption sites (Benjamin and Leckie, 1980, 1981b). However, the literature is ambivalent regarding the importance of competing macroelement cations. Some authors report significant competition effects (Kurbotov *et al.*, 1951; Schindler, 1975; Barney, 1981; Balistrieri and Murray, 1982; Harter, 1991; Zachara *et al.*, 1992), others find little effect (e.g., O'Connor and Renn, 1964), and still other investigators simply disregard possible effects of macrocations (which are often added in variable amounts as a background electrolyte). The competition of macrocations has been systematically examined in a number of papers (Borkovec, 1981;

O'Connor and Renn, 1964; Posselt *et al.,* 1968; Gadde and Laitenen, 1974; Guy and Chakrabarti, 1976; Zasoski and Burau, 1988). The controversy over the importance of competing cations appears to be a result of: (1) the failure to test the importance of competition of macrocations in many studies, (2) the confounding of ionic strength effects by covarying macrocation concentrations, (3) the dependence of competing macrocation effects on pH and loading, and (4) differences among macrocations in their competitive ability and concentration effects.

Disregard for macrocation effects may, in part, derive from the general use of the term "adsorption," which denotes a one-way process, rather than use of the term "exchange," which is clearly a two-way process. Another reason that competitive cations are given little attention is that it is generally agreed that monovalent cations such as Na and K do not form inner-sphere surface complexes and are located exclusively in the diffuse layer and bulk aqueous phase. For this reason, Bradbury and Baeyens (1991) did not include the competitive effect of these cations in their simplified SC model of Np adsorption.

Qualitative indications of significant cation competition include the success of Kurbatov *et al.* (1951) in linearizing their adsorption data for the adsorption of Co onto $Fe(OH)_3(a)$ by adding a term for the competing cation (NH^+) in their mass action equation, the success of the power function in linearizing adsorption isotherms by explicit use of competing cation concentrations (Eq. [11A]), and the observations of Schindler (1975) and Balistrieri and Murray (1982) that Mg and Ca compete with trace metals for adsorption sites in marine waters. Quantitatively, Barney (1981) found that $\ln K_{d,Sr}$ decreased approximately linearly with increasing $\ln C_{Na,aq,f}$ and Harter (1991) reported near equality between the quantities of cations displaced (Ca, H, and Mg) from and Ni adsorbed by a soil.

The pH region used may determine if a competing cation effect is observed. Azizian and Nelson (1998, this volume) report that Ca and Na compete with Pb, but only below pH 5, on Fe(OH)3(a) coated sand. Pabalan *et al.* (1998, this volume) state that an increase in $\boldsymbol{I}$ (0.1 to 1.0 $mol \cdot L^{-1}$ of $NaNO_3$) suppressed U^{VI} adsorption when cation exchange (i.e., adsorption at pH-independent sites) was important (as with montmorillonite) and to a smaller extent with clinoptilolite at low pH, recognizing that there may have been a significant effect of $\boldsymbol{I}$ on the activity of aqueous species of the adsorbate (e.g., increase in concentration of $UO_2NO_3^+$ specie). Payne *et al.* (1998, this volume) state that $\boldsymbol{I}$ affected U adsorption more at high pH values, where anionic carbonate complexes of U are more abundant, than a low pH. Unfortunately, competition of Ca, Na, and other macrocations with metal adsorption with increasing $\boldsymbol{I}$ was not evaluated in these studies. Brady *et al.* (1998, this volume) found that the pH-dependent charge of kaolinite resides on edge Al and Si sites and pH-independent charge sites are located on basal planes. McKinley *et al.* (1995) estimated pH-independent sites as the cation exchange capacity and found that competition of U^{VI} with Na was primarily at pH-independent charge sites. Thus, the ratio of pH-dependent to pH-independent charge sites and the extent of loading may impact the competitive effect of macrocations. The $K_{d,U}$ values for the two smectitic clay separates of Zachara *et al.* (1992) cover three orders of magnitude as a result of the interplay of competing cations (Na or Ca) with pH. These authors report that the adsorption of U^{VI} to clay mineral-rich separates of two subsoils increased as the $Ca(ClO_4)_2$ background electrolyte concentration increased from 0.0005 to 0.05 mol·L^{-1}, postulating that this effect was due to a dependence of the hydrolysis of UO_2^{2+} to $(UO_2)_x(OH)_y^{2x-y}$ and noting that the log formation constants of UO_2^{2+} and $UO_2(OH)_2^{2+}$ have both been reported to become 0.02 more negative as ***I*** increased fro 0.0 to 0.1. However, it may be that the observed effect on U^{VI} adsorption was a result of competition from the added cations. Differences in $C_{U,ads}$ between the two soil clays were hypothesized to result from different ratios of the area of exposed basal planes, $A_{S,basal}$, to exposed edge area of platelets, $A_{S,edge}$ of the expansible clays.

Clark *et al.* (1998, this volume) found that Ca was more competitive than Al^{III} with Eu adsorption and compare crystal radii of Ca^{II} and Al^{III} as an interpretative assist in evaluating their ability to compete with Eu for adsorption sites. However, at pH values above 4 (which encompasses nearly all of their data) very little of the aqueous Al exists as the Al^{3+} specie, the majority of the Al^{III} being present as $Al(OH)^{2+}$, $Al(OH)_2^+$, etc., species. For this reason, the size comparison between uncomplexed Ca and Al may not be meaningful in terms of Eu adsorption above pH 4.

Decreased adsorbate adsorption due to increases in the concentration of a background electrolyte are typically attributed to ***I*** effects and interpreted as evidence of nonspecific adsorption, i.e., formation of outer-sphere complexes as occurs on pH-independent charge site (Hayes and Leckie, 1986; Zachara *et al.,* 1992). Separation of the competing cation and ionic strength effects may also be complicated by changes occurring in the aqueous-phase composition during adsorption experiments. Only rarely have measurements been made of the concentration of macrocations desorbed from sediment (Lieser *et al.,* 1986) or desorbed plus leached from laboratory oxide preparations (Anderson *et al.,* 1973; Benjamin, 1979). Evidence that the adsorption of added metal is accompanied by the desorption of both structural metal ions (K, Mn) and sorbed metal cations residual from oxide preparation was provided by Anderson *et al.* (1973) and Loganathan and Burau (1973). Anderson *et al.* (1990) suspected that impurities in the Si oxide that they used was a factor in the observed pH change in the aqueous phase. Pabalan and Turner (1993c) concluded that the pH increase in suspensions of a zeolite was largely a result of exchange of H_{aq} for Na_{ads}.

The presumptive importance of ***I*** is implied by the Swedish nuclear program which segregates measured K_d values into "low" and "high" ***I*** systems. Andersson *et al.* (1982) attributed the 3.1 to 31.2% reduction in Cs adsorption, on rocks ranging from granite to gneiss, to ***I*** and/or competing cation effects resulting from increasing the NaCl from 0.1 to 4 mol·L^{-1}. Koß (1988) reports a significant effect of ***I*** on U adsorption onto largely quartz aquifer sediments. The percentage adsorption decreased from about 90% to about 80% as ***I*** increased from 0 to 0.6. Given the result of Zachara *et al.* (1992) shown in Figures 4 and 6, it seems equally likely that the effects of ***I*** reported by Andersson *et al.* (1992), Koß (1988), and Payne *et al.* (1998, this volume) are due more to competing cation effects than to ***I***. Similarly, it is unclear whether the ability to estimate estuarine $K_{d,M}$ values through a regression equation (Eq. [23]; Turner and Millward, 1994) based on salinity, and freshwater K_d values, is really due to an ***I*** effect or more probably to the covarying concentration of divalent cations.

In evaluating ***I*** effects, the use of ***I*** data is preferable to the use of total dissolved solids (TDS) as is done in the Canadian nuclear program (Eq. [22]) because dissolved silica (silicic acid) may contribute an important portion of the TDS but contributes little to ***I*** below neutrality. The contribution of silicic acid to ***I*** becomes progressively more important at pH values well above neutrality, particularly in volcanic rocks (tuff, breccia, basalt) which contain relatively large amounts of amorphic silica. A further advantage of using ***I*** is that it is computed by geochemical models such as MINTEQ (Felmy *et al.,* 1984).

D. Aqueous Speciation

The important role of complexation in the transport of U, presumably largely as U^{VI}, has been known at least since the 1950s when Barker and Scott (1958) showed that the U content of water of the extensive Ogallala Formation in Texas and New Mexico increased significantly with increasing ***I*** and C_{CO_3} in the water. Presumably, the increased aqueous complexation of U with increasing C_{CO_3}, in addition to competition from increased cation concentrations associated with the higher C_{CO_3}, reduced the extent of U^{VI} adsorption. Unfortunately, the effect of C_{CO_3} complexation on U adsorption cannot be separated from the effects of competing cations and ***I*** since Barker and Scott (1958) did not give analyses of individual water samples. Similarly, Wolfsberg (1978) noted that $K_{d,U^{VI}}$ values for alluvium from Dutchman Flat at the Nevada Test site were about 7 to 9 $ml{\cdot}g^{-1}$ from groundwater but about 60 $ml{\cdot}g^{-1}$ from distilled water which he speculated was a consequence of the complexation of added U^{VI} by CO_3^{2-} present in the groundwater. Unfortunately, the analytical information presented on individual water samples by Wolfsberg (1978) is insufficient to separate the pH and CO_3^{2-} effects in these data.

The value of aqueous speciation calculations lies primarily in increasing the accuracy of interpretation and the reliability predictive modeling of adsorption isotherms but also allow inferences concerning the nature of surface species. Phelan and Mattigod (1984) used speciation calculations in planning their experiments to determine the pH range over which the concentration of MnO_4^{2+} was maximized. Smith *et al.* (1998, this volume) note that agreement between model and experimental results was improved when the activity of the uncomplexed M^{2+}, rather than the total dissolved metal, was used in their model. As Payne *et al.* (1998, this volume) point out, the effect of aqueous complexing ligands may either enhance or reduce metal adsorption. Thomson *et al.* (1998, this volume) note that the change from mononuclear to multinuclear surface complexes of U^{VI} with increasing pH parallels these changes in the aqueous phase. Thus, speciation computations can be a valuable aid in planning experiments as well as interpreting the effect of organic ligands on metal adsorption.

Pabalan *et al.* (1998, this volume) note that there is a close correspondence between the curves defining the adsorption of U^{VI} on several minerals and of the sum of the aqueous hydroxy species. Lieser *et al.* (1992) inferred from their species distribution diagrams and adsorption data that the adsorption-determining aqueous species of U^{VI} are UO_2^{2+}, $UO_2(OH)^+$, $UO_2(OH)_3^-$. However, it appears questionable to draw conclusions such as those of Lieser *et al.* (1992) without appropriate spectroscopic data or at least confirming mass balance calculations.

E. Surface Complexation and Stoichiometry

Identification of surface species has progressed from inferences drawn from aqueous speciation through (1) identification of probable surface species through SC modeling and (2) determination of the number of, and distances to, near neighbors of the adsorbate ions through surface spectroscopy.

Model fitting is used to make distinctions among possible surface complexes and between adsorption to pH-dependent and pH-independent sites. Koß (1988) reasoned that the surface complex that came the closest to yielding a zero slope on plots of log ${}^{avg}K$ (which Koß calls log ${}^{app}K$ since a single aqueous species, a single surface complex, and a single type of site was used to calculate the equilibrium constant) versus pH was the most probable dominant surface complex, which he identified as $S{:}O(UO_2)(OH)^0$. Koß and Kim (1990) used a similar approach for Sr, plotting log ${}^{app}K$ versus the log C_{sites} for a set of geomedia samples for multiple postulated surface complexes and concluded that $(S{:}OSr)^+$, rather than the bidentate $(\{S{:}O\}_2Sr)^0$ or hydroxy $(S{:}OSrOH)^0$ complex, was the most probable surface specie. Of course, the $(S{:}OSrOH)^0$ surface complex is improbable because of the weakness of the aqueous $SrOH^+$ and $Sr(OH)_2^0$ complexes. Pabalan and Turner (1993b) calculated TL SC model constants from the data of Tripathy (1984) for U^{VI} adsorption on $Fe(OH)_3(a)$. They concluded that the single $S{:}OH(UO_2)(OH)_4^{2-}$ complex was able to reproduce the general aspects of the observed U^{VI} behavior over the bulk-phase pH range of 4 to 10 although the adsorption edge was lightly overpredicted, whereas the maximum adsorption in the pH 6.25 to 7.25 region was underpredicted. Various proposed surface active complexes for U^{VI} are given in Table I. Other chapters in this volume provide addi-

Table I

Surface Complexes of U^{vi} Inferred from Surface Complexation Models

Surface species	CO_3 present	Substrate	No. sites	Protonation	Dentate	Coordination	Species rejected	Reference
$S{:}O(UO_2)(OH)^0$	Yes	Aq sed[a]	One	$S{:}O^-$	Mono	Outer	$S{:}O(UO_2)^+$, $S{:}O(UO_2)(OH)_2^+$	Koß, 1988
$S{:}O(UO_2)(OH)^{0}$ [b] $S{:}O(UO_2)_3(OH)_5^{0}$ [b] $S{:}OH_2(UO_2)(CO_3)_2^{-}$ [b] $S{:}OH_2(UO_2)(CO_3)_3^{3-}$ [b]	Yes	Goethite	NE[c]	$S{:}O^-$ $S{:}O^-$ $S{:}OH_2^+$ $S{:}OH_2^+$	NE	NE	NE	Kohler *et al.*, 1992
$S{:}OH_2(UO_2)(OH)_4^{-}$ [d]	No	Goethite	One	$S{:}OH_2^+$	Mono	NE	[e]	Pabalan and Turner, 1993b
$S{:}OH_2(UO_2)(OH)_4^{2-}$ [d] $S{:}OH_2(UO_2)(CO_3)_3^{3-}$ [d]	No Yes	Goethite	One	$S{:}OH^0$ $S{:}OH_2^+$	Mono	NE	[e]	Pabalan and Turner, 1993c
$Si{:}O(UO_2)^+$ $Si{:}O(UO_2)(OH)^0$ $Si{:}OH(UO_2)_3(OH)_5^{0-}$	Yes	Quartz	NE	$S{:}O^-$ $S{:}O^-$ $S{:}O^-$	NE	Inner	None	Pabalan *et al.*, 1998
$S{:}OH_2(UO_2)(OH)_4^{-}$ [d]	No	Goethite	One	$S{:}OH_2^+$ $S{:}OH^0$ $S{:}OH^0$	Mono	Outer	$S{:}OH(UO_2^{2+})$, $S{:}OH(UO_2)(OH)^+$, $S{:}OH(UO_2)(OH)^+$	Turner *et al.*, 1993

$S{:}OH(UO_2)^{2+}$ $Si{:}O(UO_2)_3(OH)_5^0$	No	Mont	Three	$S{:}OH^0$	Mono	NE	$S{:}OH(UO_2)(OH)^+$	McKinley *et al.*, 1995
$S{:}O(UO_2)^0$ $(S{:}O)_2(UO_2)(CO_3)^{2-}$ [f]	Yes	$Fe(OH)_3$(a)	Two	$S{:}O^-$	Mono Bi	Inner	None	Waite *et al.*, 1994

Note. CO_3, presence or absence of CO_2; No. sites, number of types of sites for which adsorption reactions written; Protonation, No. of H^+ on adsorption site; Dentate, inferred number of (monovalent) sites participating in reaction; Coordination, inner or outer sphere; Species rejected, surface species tested and rejected on basis of fit to experimental data.

[a] Aquifer sediment.

[b] Four species were used simultaneously.

[c] NE, not evaluated.

[d] Data from Tripathy (1984).

[e] Sought single complex with best fit to data.

[f] This surface species was needed at high pH and P_{CO_2}:CO_3 adsorption was included in model.

tional information on surface complexation species. It is important to bear in mind that there are significant differences in the thermodynamic data used by the various authors, which is likely to account for much of the differences in identified surface species. Recently, McKinley *et al.* (1995) concluded that the most significant U adsorption reactions were adsorption of UO_2^{2+} to AlOH edge sites and $(UO_2)_3(OH)_5^+$ to SiOH edge sites. Brady *et al.* (1998, this volume) report that edge Al and Si provide the pH-dependent sites on kaolinite, whereas and pH-independent charge sites are located on basal planes.

Evaluating the presence of adsorbed metal as inner- or outer-sphere complexes has been an important aspect of studying surface complexes. In the absence of surface analysis techniques, this has been done by testing various reactions via adsorption models. For example, Zhang and Sparks (1989) report that MoO_4^{2-} adsorption to goethite yielded the best fit model, assuming inner-sphere complexation. Wehrli *et al.* (1990) considered that V^{IV} and Cr^{III} adsorption onto δ-Al_2O_3 must involve inner-sphere complexes because these metals were adsorbing against a positive charge on the oxide surface. However, this is likely to be an inadequate basis as positively charged oxides commonly adsorb metals.

Surface spectroscopy is the method of choice for characterization of surface species. Manceau and Charlet (1991) reported the formation of bidentate mononuclear U^{VI}-hydroxy surface complexes on $Fe(OH)_3$(a) from millimolar aqueous concentrations of U^{VI} at a single pH value. Thompson *et al.* (1998, this volume) concluded that in the pH range of 6 to 7 only monomeric inner-sphere surface complexes of U^{VI} were found, whereas in the pH range of 7.0 to 7.9 the dominate species were multinuclear inner-sphere complexes, all of these species being identified in the presence of air (i.e., dissolved carbonate was present).

The generally reported apparent nonstoichiometric exchange (e.g., Benjamin, 1979) may result from (1) failure to measure displaced cations other than H, (2) varying ratio of monodentate to bidentate sites becoming occupied by the adsorbate as loading increases (see loading discussion in Section V.B), (3) changes in both surface and aqueous speciation as a result of ligand and cation (particularly of H) adsorption or desorption, and (4) changes is the net surface charge.

It is likely that H, being a small cation, may exchange more rapidly than metal cations, causing the stoichiometry between an adsorbing metal and H to change with reaction time although the total stoichiometry may not necessarily change. As noted in Section V.C, desorbed cations other than H are only rarely determined (Anderson *et al.,* 1973; Loganathan and Burau, 1974; Benjamin, 1979; Lieser *et al.,* 1986).

Payne *et al.* (1998, this volume) suggest that the steepness of the adsorption edge may reflect the proton stoichiometry involved. Kinniburgh *et al.* (1998, this volume) show that the H^+:M^{2+} exchange ratio is dependent upon the pH although they do not appear to evaluate the extent to which this is a function of variations in the S:OH_2^+ to SOH^0 ratio. They state that the measured H^+:M^{2+} exchange ratio for oxides were approximately equal to the horizontal shift of the log–log adsorption isotherm.

It is not clear if there is a progression of bidentate to monodentate surface complexes with increased pH and/or loading. Kinniburgh *et al.* (1998, this volume) indicate that the presence of bidentate sites allows a maximum $H^+:M^{2+}$ exchange ratio of 2 for oxides but note that adsorption of partially hydrolyzed metal species may also increase the $H^+:M^{2+}$ exchange ratio. However, for adsorbents that contain adsorbed cations other than H, the amount of displaced metals must also be determined in calculate exchange stoichiometry.

F. Surface Area, Site Density, and Surface Area Normalization

Notwithstanding four decades of research on methods of surface area (A_s) and adsorption site concentration measurement, (C_{site}) A_s measurements on similar materials differ by up to an order of magnitude and *a priori* determinations of $C_{site,tot}$ on oxides and clay minerals by titration or tritium exchange are generally higher than $C_{M,ads,max}$ by up to an order of magnitude (Brady *et al.*, 1998, this volume). There appear to be similar uncertainties in acidity constants as Smith and Jenne (1991) found that there was considerably less variation in surface complexation constants (K^{SC}_{TL}), in which the acidity constant had been subtracted from published ${}^{int}K_{TL,M}$ constants, than in the original ${}^{int}K_{TL,M}$ values. Their interpretation was that acidity constants were the largest general source of error in ${}^{int}K_{TL,M}$ constants. Perhaps it was for this reason that Bradbury and Baeyens (1991) used identical K_a values (for the protonation and deprotonation of $S{:}OH^0$) for high- and low-affinity sites, although they used different ${}^{int}K_{Ca}$ values for the formation of $S{:}OCa^+$ high- and low-affinity sites.

Site density is expressed both as sites per nanometer ($C_{site,area}$) and as mole per mole Fe ($C_{site,mol\ Fe}$). Measurement remains a problem as several variables affect A_S and C_{site} values. According to Dzombak and Morel (1990), the concentration of high-affinity sites (those filled in the range where $C_{M,ads}$ is proportional to $C_{M,aq,f}$) for $Fe(OH)_3$(a) varies from 0.001 to 0.01 $mol{\cdot}mol^{-1}$, with a mean of 0.005 $mol{\cdot}mol^{-1}$ of metal to Fe. For low-affinity sites (all sites other than high-affinity sites), the maximum adsorption values range from about 0.1 to 0.3 $mol{\cdot}mol^{-1}$ of metal to Fe with a mean value of 0.2 $mol{\cdot}mol^{-1}$ (Dzombak and Morel, 1990).

Surface area is commonly measured by the adsorption of a polar liquid (typically ethylene glycol monoethyl either) on soils, sediments, and clay minerals, while for oxides and primary minerals, the BET inert gas adsorption method is generally used. Ethylene glycol hydrates the surface area of expansible clay; hence, it tends to measure the total external plus internal area of expansible clays. The BET method is carried out under a relatively high vacuum and measures primarily the external area of particles and aggregates. However, Bradbury and Baeyens (1991) argue that since the pH-dependent sites are on the edges of clay minerals, the BET method provides a better estimate of the area of pH-dependent sites (see normalization discussion at end of this section). Neither the glycol nor the BET method includes the area of smaller pores that may be accessible to metals by diffusion. As pointed out by Turner (1993), the use of a surface area measured on dry samples in a vacuum to represent the surface area of a mineral in suspension is uncertain. Axe and Anderson (1998, this volume) suggest that the BET method probably underestimates the hydrated A_s of oxide precipitates. In some cases, A_S has been used as a fitting parameter (Davis, 1977; Hsi and Langmuir, 1985). Bradbury and Baeyens (1991) believed that the external A_S was the key component of the reliability of $C_{site,area}$ for their aquifer samples since the permanent charge of planar and interlayer charge due to isomorphic substitution does not exhibit significant pH-dependent charge densities. Davis and Kent (1990) suggested that the variation of site density of $Fe(OH)_3$(a) may reflect differing microporosities which could also be true of other oxides and aggregates as well. To the extent that the microporosity hypothesis of Davis and Kent (1990) is true, use of a uniform A_S may lead to apparent differences in calculated intrinsic constants. For clay minerals, it is desirable to estimate separately the pH-dependent and -independent sites since competition of U^{VI} with Na was found to be primarily at the pH-independent sites (Pabalan *et al.,* 1998, this volume). McKinley *et al.* (1995) estimated pH-independent sites as the conventional cation exchange capacity.

Because of the uncertainties in A_S and $C_{site,area}$ measurements, Davis and Kent (1990) recommended that the value of 2.31 sites·mm^2 (3.84 μmol m^{-2}) previously selected by Dzombak and Morel (1986) be used because "it is more important that a uniform value be accepted for modeling than it is that accurate site densities be used for each mineral surface in a composite sample." Subsequently, numerous modeling studies in the past few years have used uniform A_S values rather than experimentally measured $C_{site,area}$ values (Davis and Kent, 1990; Dzombak and Morel, 1990; Hayes *et al.,* 1991; Bradbury and Baeyens, 1991, 1993; Pabalan and Turner, 1993c; Turner *et al.,* 1993; Smith *et al.,* this volume). Several have used a uniform A_S of 600 m^2·g^{-1} (e.g., Kent *et al.,* 1988; Clark *et al.,* 1998, this volume), although Turner *et al.* (1993) used 50 m^2·g^{-1}. In accepting the $C_{site,area}$ of 3.84 μmol·m^{-2} used by Dzombak and Morel (1986) as recommended by Davis and Kent (1990), Bradbury and Baeyens (1991) note that the fit of their model simulations to the experimental data was relatively insensitive to the $C_{site,area}$ value used. Although no data was found showing that the internal sites have different $^{int}K_{TL,M}$ values that external sites, such results may be forthcoming in the future. If such differences exist, they could be important in view of the finding of Axe and Anderson (1995) that for their $Fe(OH)_3$(a), 40% of the total sites were internal sites.

For each of the past three decades, there has been at least one paper relating metal adsorption to surface area. MacNaughton (1973) wrote the Langmuir equation on an area basis (Eq. [14F]), whereas Borkovec (1981) simply regressed $C_{M,ads}$ on ethylene glycol monoethyl ether surface area and obtained the equation

$$C_{M,ads} = 0.11\ A_S + 2.05\ (r = 0.97). \qquad [27]$$

Borkovec (1981) calculated that on average each UO_2^{2+} occupied 2.73 ± 0.24 nm^2 at the $C_{M,ads,mas}$ of the individual clay minerals.

There is currently increased interest in surface area normalization. As discussed in section III.B, Vandergraaf *et al.* (1993) normalize their K_{d,M_R} values to surface area, which they estimate from porosity, for use in the Canadian radionuclide transport model. O'Day *et al.* (1994) also reported their data for Co^{II} adsorption by kaolinite on an area basis. Zachara *et al.* (1995) reported that area normalization of their Co^{II} adsorption onto oxides and naturally Fe coated subsurface sands did not bring the isotherms into coherence. The K_{d,Np^V} values obtained by Bertetti *et al.* (1998, this volume) for a number of minerals ranged over two orders of magnitude but when the K_{d,Np^V} values were normalized to S_A the differences among minerals was greatly reduced although α-alumina and quartz had noticeably greater K_{d,Np^V} values than montmorillonite or clinoptilite on a BET-area normalized basis. Reasoning that Np adsorption occurred primarily at the edges of the montmorillonite and clinoptilite and using literature information that about 10% of the BET area represents edge area, they normalized the area of these two silicates to their presumed edge area and obtained a much improved fit of all four minerals to a single pH isotherm. Pabalan *et al.* (1998, this volume) obtained successful area normalization results for U^{IV} adsorption by the same suite of minerals. Similarly, Rosentretter *et al.* (1998, this volume) were largely succesful in area normalizing their K_d values for U^{IV} adsorption to sands coated naturally with Fe oxides.

G. Time Dependency

The importance of reaction time beyond an initial period of 1 to 8 hr is controversial since some investigators find a relatively large time dependence, whereas other investigators report little or no time dependence of adsorption beyond an initial period. An adequate review of the rates and kinetics of adsorption is beyond the scope of this chapter; however, the ramifications of time as a variable are so important and poorly dealt with in the literature that a number of points (reaction order, data plotting, pH, decreasing $C_{M,aq}$, and Pco_2 diffusion into samples) need to be made and some salient aspects of diffusion need to be mentioned. Hopefully, this discussion will promote the evolution of an understanding of the currently disparagate time-dependency results.

Kinetic studies with geomedia tend to be limited to some initial "straight line" segment of $C_{M,ads}$ versus time isotherm (e.g., Harter, 1991; Guilherme and Anderson, 1998, this volume). If the order of reaction is to be evaluated, it needs to be recognized that the kinetics of adsorption must be a minimum of second order since the rate necessarily depends upon the concentration of adsorbent and adsorbate. In a particular kinetic adsorption study, the initial concentration of adsorbent is commonly held constant; hence, changing a variable such as adsorbate concentration may yield a pseudo-first-order reaction. Harter (1991) comments that, for visualizing time-dependent data, it is convenient to plot the fraction of adsorbate remaining as a function of time. The fraction remaining is essentially the inverse of plotting K_d (but without normalization for mass of geomedia and volume of solution). The plotting of the K_d (normalized fraction adsorbed) minimizes the effects of differences in sample mass and volume; thus it is preferable to plotting the fraction of metal remaining in solution.

Guilherme and Anderson (1998, this volume) contend that it is often experimentally impossible to separate the effects of reaction kinetics from those of transport phenomena, such as diffusion. They reason that the initial rate of adsorption onto surface sites in contact with the bulk liquid should be neither transport limited or site limited; hence, the initial net reaction rate should closely approximate the forward reaction rate. However, a number of investigators show that the main part of their adsorption isotherms is linearized on $time^{1/2}$ plots which implies a diffusion limited process (e.g., Duursma and Bosch, 1970; Bérubé, 1967; Davis, 1977; Bruemmer *et al.,* 1988; Wehrlic *et al.,* 1990). Jenne (1995, 1996) showed that metal uptake ($mol \cdot g^{-1}$) in virtually all of the adsorption data sets examined was composed of a fast and a slow process (presumably adsorption to external and internal sites) and that the initial rapid reaction process in nearly all of these data sets was linearized on $time^{1/2}$ plots. This suggests that the initial rapid rate is also transport (diffusion) limited. Thus, the initial rates of Harter (1991) and Guilherme and Anderson (1998, this volume) may be (coating?) diffusion limited rather that pseudo-first-order reaction rates. Axe and Anderson (1997) were able to separately estimate the number of external (60%) and internal (40%) sites by an appropriate choice of time for completion of the fast reaction. However, the fast and slow reactions are not always so readily separated as in the system they used. A more generally reliable approach may be to extrapolate the slow reaction to zero time on plots of $C_{M,ads}$ versus $time^{1/2}$ and take the intercept as the amount of adsorption onto external sites (Jenne, 1995). It seems likely that the concept of multiple site domains (commonly two) widely used in transport modeling is actually a proxy for the two dominate rate processes, a rapid rate on surfaces in direct contact with the solution and a slow diffusion-limited process (Jenne, 1995, 1996).

Interpretation of the time dependency of metal adsorption is confounded by a continuously decreasing adsorbate concentration in traditional isotherms (Axe and Anderson, 1995). This decrease in the driving force of the reaction may account for the reduction in the rate of adsorption, which occurs with time, but which is attributed to the filling of lower energy sites. There are two approaches to dealing with this decreasing driving force. One is to maintain $C_{M,aq,f}$ constant during the experiment (Axe and Anderson, 1995). The other is to plot the adsorption rate $\Delta C_{M,ads,tx}$, normalized to the amount of metal already adsorbed

($C_{M,ads,tx}/C_{geomedia}$), for the time interval Δt_x, at successive sampling times t_x, i.e., $\{(C_{M,ads,tx}/C_{geomedia})/\Delta t_x\}$. The $C_{M,ads,tx}$ is in moles per liter, $C_{geomedia}$ is in moles or grams per liter, and t is in hr. Plots of the normalized rate $\log[\{(\Delta C_{M,ads,tx}/C_{geomedia})/\Delta t_x\}/(C_{M,ads,tx}/C_{geomedia})]$ versus log (time) generally yield linear relationships (Jenne, 1995) unless precipitation (or dissolution) occurs or analytical uncertainties become large relative to the rates. In the only constant $C_{M,aq,f}$ experiments encountered in this review, Axe and Anderson (1995, 1997) found that equilibrium was reached in about 1 hr between Sr in the aqueous phase and the external surface of $Fe(OH)_3$(a).

Smith *et al.* (1998, this volume) suggest that adsorption rates are likely to vary with pH. Indeed, Axe and Anderson (1997, 1998, this volume), report that the equilibrium adsorption in the Cd-$Fe(OH)_3$(a) system was reached in approximately 1 month at pH 6 but diffusion model calculations indicate that at pH 7 about 5 yr would be required to reach equilibrium. Their explanation for this marked difference in time to equilibrium is that the concentration of $S{:}OH^0$ sites is lower at pH 6 than at pH 7; therefore, the jump distance between active sites not filled with Cd is greater at pH 6 which results in a greater diffusivity. However, the surface speciation plot of Krupka *et al.* (1988) indicates only about a 5% increase in $S{:}OH^0$ on $Fe(OH)_3$(a) between pH 6 and 7. This seems an insufficient increase in $S{:}OH^0$ to account for the nearly 5 yr difference in time to equilibrium. Bertetti *et al.* (1998, this volume) suggest that a slow mass transfer of CO_2 from laboratory air into experimental solutions that were undersaturated with CO_2 may be one cause of slow equilibrium. It is unclear to what extent differences in P_{CO_2} and pH are responsible for the wide variation in reported times to adsorption equilibrium for solids such as $Fe(OH)_3$(a).

Diffusion is a molecular process, i.e., random molecular motion, which allows a metal to be transported in the absence of bulk flow in the direction of a concentration gradient. Adsorption rates may be limited by diffusion in the bulk aqueous phase, through the film of adsorbed water at external sites, and through pores. Bulk diffusion can be rendered negligible by appropriate mixing, film diffusion is generally considered to be rapid, and pore diffusion may vary markedly in its speed. Diffusion in micropores can occur in both the aqueous phase and along pore surfaces by the ion jumping from site to site. However, Axe and Anderson (1998, this volume) argue that where adsorption is significant, molecular diffusion as a transport process through the aqueous phase in micropores is insignificant, hence, diffusion along surfaces is the dominate diffusion process in small pores. They visualize that the dominant transport process in pores is one in which the adsorbate ions vibrate at a local site until they acquire sufficient vibrational energy to jump to a neighboring site along the pore wall. Axe and Anderson (1995) visualize this jump occurring "only if the site is vacant." The conceptualization of movement by ions jumping from site to site along the pore wall is analogous to the "contact exchange" concept of Jenny and Overstreet (1939) for cation transfer from mineral surfaces to root surfaces. This diffusion-limited slow adsorption process is no doubt responsible for the "so-called" aging effects and for adsorption/desorption

Table II

Experimental Diffusivities of Sr and Cd

Effective Diffusivity ($cm^2 \cdot sec^{-1}$)	Adsorbate(s)	Adsorbent	Source
$1.7{\cdot}10^{-19}$ to $1.7{\cdot}10^{20}$	Ni,Zn,Cd at pH 4.67–5.47	Geothite	Bruemmer *et al.*, 1988
$1.7{\cdot}10^{-12}$ to $1.7{\cdot}10^{-11}$	Cd, Se^{III}	$Fe(OH)_3$(a)	Papelis *et al.*, 1995
$4.0{\cdot}10^{-13}$ [a]	Sr at pH 7	$Fe(OH)_3$(a)	Axe and Anderson, 1995
$3.3{\cdot}10^{-11}$ [b]	Co at pH 6.5	$Fe(OH)_3$(a)	Misak *et al.*, 1966
$2.0{\cdot}10^{-11}$ [b]	Zn at pH 6.5	$Fe(OH)_3$(a)	Misak *et al.*, 1966
$1.0{\cdot}10^{-14}$ [a,c]	Cd at pH 7	$Fe(OH)_3$(a)	Axe and Anderson, 1997
$6.0{\cdot}10^{-13}$ [a]	Cd at pH 6	$Fe(OH)_3$(a)	Axe and Anderson, 1997

[a]Surface diffusivities.
[b]This diffusivity would require 5 yr for adsorption equilibrium.
[c]From $1{\cdot}10^{-3}$ $mol{\cdot}L^{-1}$ solutions; oxide previously dried at 50°C.

hysteresis in general (Barrow, 1986; Bruemmer *et al.,* 1988; Ghilherme and Anderson, 1998, this volume).

The diffusion coefficient, D, measures the net flux in some direction, that direction generally being taken as along the maximum concentration gradient. Where the metal concentration is dilute, i.e., $C_{M,aq}$ is small in comparison with the number of adsorptions sites, the likelihood is small that the metal will have to compete with itself. Where D is constant, it is referred to as the self-diffusion coefficient (D_s). The surface diffusion coefficient can be calculated as (Axe and Anderson, 1997)

$$D_s = \lambda\, (E_A/2\,W)^{1/2} \exp\{(-E_A/R)\,T\}, \qquad [28]$$

where D_s is the surface diffusion coefficient, λ is the mean distance between sites, E_A is the energy barrier between adjacent sites, W is the molecular mass (molecular weight at dilute concentrations), $\exp\{(-E_A/R)\,T\}$ is the Boltzmann factor representing the probability that the vibrating atom will have the required energy to jump to the next site, *R* is the gas constant, and T the temperature. Some effective diffusivities are collected in Table 2. It is not clear if the 10^7 range in effective diffusivities are real or if part of the range is due to differences in experimental methods.

H. Other Variables

Temperature

The extent of adsorption commonly increases with temperature (Avotins, 1975; Bruemmer *et al.,* 1988; Axe and Anderson, 1998, this volume). However, as Harter (1991) pointed out, adsorption data obtained at different temperatures are likely to be confounded by physical and chemical changes that occur in sorbent surfaces of amorphic and organic substances and of crystalline minerals whose degree of hydration may change within the experimental temperature range (Avotins, 1975). Adsorption experiments carried out over an extended time necessitates obtaining experimental data establishing that the amorphic solids and organic substances have not undergone significant changes in properties (Avotins, 1975) such as crystallinity, pore size, and A_S.

Obviously, the temperature range used will impact the extent to which temperature-dependent changes in adsorbents will occur. Axe and Anderson (1998, this volume), using temperatures in the range of 4 to 25°C, evaluated the enthalpy of Cd and Sr adsorption onto $Fe(OH)_3$(a) and found K_d value increased from 1.28 to 38.8 $L \cdot g^{-1}$ as the temperature increased from 4 to 25°C.

Although temperature is obviously an important variable, one rarely finds a study where adsorption constants have been corrected for the difference between the common laboratory temperature (20 to 25°C) and the lower soil, groundwater, or stream temperatures that are normally encountered at northern hemisphere field sites.

Anion Effect

Polyvalent anions (e.g., PO_4) readily form precipitates. Thus, (Sposito 1986) noted that the increased metal adsorption sometimes observed after a strongly adsorbing anion has been reacted with a hydrous oxide adsorbent may be modeled either as a metal-anion surface precipitate effect or as metal-anion surface complex effect. He concludes that sorption data themselves do not provide for a choice of model, unless the ion-activity product (i.e., saturation index) for a proposed surface precipitate exceeds the corresponding solubility product constant. It appears that much of the enhancement of metal adsorption attributed to polyvalent anions may be a result of the failure to distinguish between adsorption and precipitation. Guilherme and Anderson (1998, this volume) show that the effect of PO_4 sorption on Cu adsorption by oxide-rich soils is quite complex.

Additivity

One of the unresolved controversies in the application of adsorption data is that of "additivity;" i.e., are the adsorptive properties of individual adsorbents the same in mixtures as when no other adsorbents are present. Honeyman (1984) contends that he found significant changes in adsorptive properties when certain adsorbents were present in mixtures, whereas Davies-Colley *et al.* (1984) state that they found no evidence there was a reduction in adsorption effectiveness as a result of mixing selected solids (e.g., a $Fe(OH)_3$(a)–Wyoming bentonite–humic substance mixture). Anderson and Benjamin (1990) found that decreased adsorption in $Fe(OH)_3$(a)–$Al(OH)_3$ mixtures was due to $Al(OH)_y^{x+}$ adsorption by the

$Fe(OH)_3(a)$. The largest effect that Honeyman (1984) reported was with thallium oxide, a quadravalent metal.

VI. CONCLUSIONS

• All adsorption equations discussed in this chapter are variants of the mass action equation, either simplifications (e.g., a distribution coefficient) or with an added electrostatic term (e.g., triple-layer implementation of the surface complexation model).

• Log–log plots are a linearized form of the Freundlich equation, and as such, are linear in certain pH regions and in certain metal to adsorbent concentration ratios.

• Multivariable linear regression models are n-dimensional versions of the linearized Freundlich equation.

• The numerous discussions in the literature of the respective "fit" of Freundlich versus Langmuir equations are of little merit in the absence of an evaluation of the effect of increasing site occupancy (loading) and of competing cation effects on adsorption isotherms.

• Unevaluated loading and competing cation (including hydrogen) effects make suspect the published inferences derived from Freundlich and Langmuir equations, concerning differing site energies, bond types, and heterogeneity.

• Log–log plots of the concentration of adsorbed metal versus its aqueous concentration or versus the hydrogen activity have significantly more information content than do sigmoidal semilog plots.

• Effectiveness of macrocation competition depends on the pH, the nature of the cation, its concentration, and the portion of permanent charge charge sites versus pH-dependent sites the adsorbate occupies.

• The loss of isotherm linearity in log–log plots at low pH values may be due to the increased concentration of positively charged multivalent competing cations (e.g., Al, Fe) from adsorbate dissolution and to the increase in positively charged sites (i.e., $S{:}OH_2^+$). At pH values near or above neutrality, the loss of linearity can be attributed to the formation of anionic aqueous complexes (i.e., complexes of OH^-, CO_3^{2-}, etc.) of the metal resulting in a decrease in the activity of hydroxylated and uncomplexed species of the adsorbate.

• Time to equilibrium is impacted by pH, differing microporosities of amorphic oxides, aggregate formation, differing degrees of floc formation, the extent of required equilibration with P_{CO_2}, and the absorbate to adsorbent ratio.

• Different time periods considered adequate by investigators for equilibration of the metal–$Fe(OH)_3(a)$ systems (e.g., 1 hr, 3 weeks) generally result in equilibration with external sites but presumably result in great variations in the extent to which internal sites reach are equilibrated with the metal adsorbate.

• Apparent nonstoichiometric exchange results from (1) failure to measure all cations displaced, (2) reaction of MOH^+ with two $S{:}OH^0$ sites to form a bidentate complex releasing one HOH^0 and one H^+, (3) varying ratio of monodentate to bidentate sites occupied with increased loading, and (4) changes in aqueous speciation as a result of ligand and cation adsorption or desorption.

• Metal reaction rates with both external and internal sites are permissibly diffusion limited.

• Use of a uniform surface area and a uniform site density for $Fe(OH)_3(a)$ preparations in surface complexation models increases the comparability of calculated intrinsic equilibrium constants but may obscure differences among preparations that have variable microporosity.

• Characterization of both aqueous and solid phases is frequently insufficient to adequately interpret the experimental data collected.

ACKNOWLEDGMENTS

The careful and constructive review of this chapter by Jeff Serne and by Daniel Kaplan are much appreciated.

REFERENCES

Abelin, H., Birgerson, L., Gidlund, J., Moreno, L., Neretnieks, I., and Tunbrant, S. 1986. Flow and tracer experiments in crystalline rocks: Results from several Swedish in-situ experiments. *In* "Scientific Basis for Nuclear Waste Management IX," Vol. 50, pp. 627–639. Materials Research Society, Pittsburg, PA.

Albinsson, Y. 1991. "Sorption of Radionuclides in Granitic rock," SKB Working Report AR 91-07. Svensk Kärnbränslehanterin SKB Arbetsrapport, Stockholm, Sweden.

Allard, B., Beall, G. W., and Krajewski, T. 1980. The sorption of actinides in igneous rocks. *Nuclear Technol.* 49:474–480.

Allard, B., Karlsson, F., and Neretnieks, I. 1991. "Concentrations of Particulate Matter and Humic Substances in Deep Groundwaters and Estimated Effects on the Adsorption and Transport of Radionuclides," Technical Report 91-50. Svensk Kärnbränslehanterint SKB Arbetsrapport, Stockholm, Sweden.

Allen, H. E., Chen, Y.-T., Li, Y., and Huang, C. P. 1993. The significance of trace metal speciation for water, sediment and soil quality criteria and standards. *Sci. Total Environ.* Supplement 1993, pp. 23–45.

Allen, H. E., Chen, Y.-T., Li, Y., Huang, C. P., and Sanders, P. F. 1995. Soil partition coefficients for Cd by column desorption and comparison to batch adsorption measurements. *Environ. Sci. Technol.* 29:1887–1891.

Amacher, M. C., Selim, H. M., and Iskandar, I. K. 1990. Kinetics of mercuric chloride retention by soils. *J. Environ. Qual.* 19:382–388.

American Soil Science Society of America 1966. "Glossary of Soil Science Terms 1996." Madison, Wisconsin.

Anderson, B. J., Jenne, E. A., and Chao, T. T. 1973. The sorption of silver by poorly crystallized manganese oxides. *Geochim. Cosmochim. Acta* 37:611–622.

Anderson, M. A., Palm-Gennen, M. H., Renard, P. N., Defosse, C., and Rouxhet, P. G. 1984. Chemical and XPS study of the adsorption of iron(III) onto porous silica. *J. Colloid Interface Sci.* 102:328–336.

Anderson, P. R., and Benjamin, M. M. 1990. Surface and bulk characteristics of binary oxide suspensions. *Environ. Sci. Technol.* 24:692–698.

Andersson, K., Torstenfelt, B., and Allard, B. 1982. "Sorption Behaviour of Long-Lived Radionuclides in Igneous Rock," IAEA-SM-2507/20. International Atomic Energy Agency, Vienna.

Appelo, C. A. J., and Postma, D. 1993. "Geochemistry, Groundwater and Pollution." A. A. Balkema, Rotterdam.

ASTM (American Society of Testing Materials) 1993. D 4319-83: Standard test method for distribution ratios by the short-term batch method. *In* "Annual Book of ASTM Standards," Vol. 04.08, pp. 693–698. ASTM, Philadelphia, PA.

ASTM, (American Society of Testing Materials) 1994. D 4646-87: Standard test method for 24-h batch-type measurement of containment sorption by soil and sediments. *In* "Annual Book of ASTM Standards," Vol. 11.04, pp. 125–128. ASTM, Philadelphia, PA.

Avotins, P. V. 1975. "Adsorption and Coprecipitation Studies of Mercury on Hydrous Iron Oxide." Ph.D. Thesis, Stanford Univ., Stanford, CA.

Axe, L., and Anderson, P. R. 1995. Sr diffusion and reaction within Fe oxides: Evaluation of the rate-limiting mechanism for sorption. *J. Coll. Interface Sci.* 175:157–165.

Axe, L., and Anderson, P. R. 1997. "Experimental and theoretical diffusivities of Cd and Sr in hydrous ferric oxide. *J. Colloid Sci.* 185:436–448.

Axe, L., and Anderson, P. 1998. Intraparticle diffusion of metal contaminants in amorphous oxide minerals. *In* "Adsorption of Metals by Geomedia" (E. A. Jenne, Ed.). Academic Press, San Diego.

Azizian, M. F., and Nelson, P. O. 1998. Lead sorption, chemically enhanced desorption, and equilibrium modeling In iron-oxide coated sand, syntheric groundwater system. *In* "Adsorption of Metals by Geomedia" (E. A. Jenne, Ed.). Academic Press, San Diego.

Baes, C. F., III, and Sharp, R. D. 1983. A proposal for estimation of soil leaching and leaching constants for use in assessment models. *J. Environ. Qual.* 12:17–28.

Baes, C. F., III, Sharp, R. D., Sjoreen, A. L., and Shor, R. W. 1984. "A Review and Analysis of Parameters for Assessing Transport of Environmentally Released Radionuclides through Agriculture," ORNL-5786. Oak Ridge National Laboratory, Oak Ridge, TN.

Baker, J. H., Beetem, W. A., and Wahlberg, J. S. 1964. "Adsorption Equilibria between Earth Materials and Radionuclides, Cape Thompson, Alaska," Open-File Report. U.S. Geological Survey, Reston, VA.

Balistrieri, L. S., Brewer, P. G., and Murray, J. W. 1981. Scavanging residence times of trace metals and surface chemistry of sinking particles in the deep ocean. *Deep-Sea Res.* 28A: 103–121.

Balistrieri, L. S., and Murray, J. W. 1982. The adsorption of Cu, Pb, Zn, and Cd on goethite from major ion seawater. *Geochim. Cosmochim. Acta* 46:1253–1267.

Barker, F. B., and Scott, R. C. 1958. Uranium and radium in the ground water of the Liano Estascado, Texas and New Mexico. *Trans. Am. Geophys. Un.* 39:459–466.

Barney, G. S. 1981. "Radionuclide Reactions with Groundwater and Basalts from Columbia River Basalt Formations," RHO-SA-217. Rockwell Hanford Operations, Richland, WA.

Barrow, G. M. 1970. Comparison of the adsorption of molybdate, sulfate and phosphate by soils. *Soil Sci.* 109:282–288.

Barrow, N. J. 1986. Testing a mechanistic model. II. The effects of time and temperature on the reaction of zinc with a soil. *J. Soil Sci.* 37:277–286.

Benjamin, M. M. 1979. "Effects of Competing Metals and Complexing Ligands on Trace Metal Adsorption at the Oxide/Solution Interface." Ph.D. Thesis, Stanford Univ., Stanford, CA.

Benjamin, M. M., and Felmy, A. 1981. Trace metal exchange between ferromanganese nodules and artificial seawater. *Mar. Min.* 3:151–183.

Benjamin, M. M., and Leckie, J. O. 1980. Adsorption of metals at oxide interfaces: Effects of the concentrations of adsorbate and competing metals. *In* "Contaminants and Sediments" (R. A. Baker, Ed.), Vol. 2, pp. 305–322. Ann Arbor Science, Ann Arbor, MI.

Benjamin, M. M., and Leckie, J. O. 1981a. Multiple-site adsorption of Cd, Cu, Zn, and Pb on amorphous iron oxyhydroxide. *J. Colloid Interface Sci.* 79:209–221.

Benjamin, M. M., and Leckie, J. O. 1981b. Competitive adsorption of Cd, Cu, Zn, and Pb on amorphous iron oxyhydroxide. *J. Colloid Interface Sci.* 83:410–419.

Benjamin, M. M., and Leckie, J. O. 1982. Effects of complexation by Cl, SO_4, and S_2O_3 on adsorption behavior of Cd on oxide surfaces. *Environ. Sci. Technol.* 16:162–170.

Bertetti, F. P., Pabalan, R. T., Turner, D. R., and Almendarez, M. G. 1998. Studies of neptunium(V) sorption on montmorillonite, clinoptilolite, quartz and α-alumina. *In* "Adsorption of Metals by Geomedia" (E. A. Jenne, Ed.). Academic Press, San Diego.

Bérubé, Y. G., Onoda, Jr., G. Y., and De Bruyn, P. L. 1967. Proton adsorption at the ferric oxide:aqueous solution interface. *Surf. Sci.* 8:448–461.

Besser, J. M., Ingersoll, C. G., and Giesy, J. P. 1996. Effects of spatial and temporal variability of acid–volatile sulfide on the bioavailability of copper and zinc in freshwater sediments. *Environ. Tox. Chem.* 15:286–293.

Borkovec, Z. 1981. The adsorption of uranyl species by fine clay. *Chem. Geol.* 32:45–58.

Borkovec, M., Rusch, U., and Westall, J. C. 1998. Modeling of Competitive ion binding to heterogeneous materials with affinity distributions. *In* "Adsorption of Metals by Geomedia" (E. A. Jenne, Ed.). Academic Press, San Diego.

Bradbury, M. H., and Baeyens, B. 1991. A mechanistic approach to the generation of sorption databases. *In* "Proceedings NEA Workshop, Interlaken, Switzerland, October 1991," pp. 121–162. OECD/OCDE, Paris, France.

Bradbury, M. H., and Baeyens, B. 1993. A general application of surface complexation to modeling radionuclide sorption in natural systems. *J. Colloid Interface Sci.* 158:364–371.

Brady, P. V., Cygan, R. T., and Nagy, K. L. Surface charge and metal sorption on kaolinite. *In* "Adsorption of Metals by Geomedia" (E. A. Jenne, Ed.). Academic Press, San Diego.

Breeuwsma, A., and Lyklema, J. 1973. Physical and chemical adsorption of ions in the electrical double layer on hematite (α-Fe_2O_3). *J. Colloid Interface Sci.* 43:437–448.

Bruemmer, G. W., J. Gerth, and K. G. Tiller. 1988. Reaction kinetics of the adsorption and desorption of nickel, zinc, and cadmium by goethite. I. Adsorption and diffusion of metals. *J. Soil Sci.* 39:37–52.

Campbell, L. S., and Davies, B. E. 1995. Soil sorption of caesium modeled by the Langmuir and Freundlich isotherm equations. *Appl. Geochem.* 10:715–723.

Cavallaro, N., and McBride, M. B. 1984a. Zinc and copper sorption and fixation by an acid soil clay: Effect of selective dissolution. *Am. Soil Sci. Soc. J.* 48:1050–1054.

Cavallaro, N., and McBride, M. B. 1984b. Effect of selective dissolution on charge and surface properties of an acid soil clay. *Clays Clay Miner.* 32:283–290.

Chen, C.-C., Papelis, C., and Hayes, K. F. 1998. Extended X-ray absorption fine structure (EXAFS) analysis of aqueous Sr^{II} Ion sorption at clay–water interfaces. *In* "Adsorption of Metals by Geomedia" (E. A. Jenne, Ed.). Academic Press, San Diego.

Clark, S. B., Amy L. Bryce, A. L., A. D. Leuking, A. D., J. Gariboldt, J., and S. M. Serkiz, S. M. Factors affecting tri-valent f-element adsorption to an acidic sandy soil. *In* "Adsorption of Metals by Geomedia" (E. A. Jenne, Ed.). Academic Press, San Diego.

Coale, K. H., and Bruland, K. W. 1985. ^{234}Th:^{238}U disequilibria within the California current. *Limnol. Oceanogr.* 30:22–33.

Coale, K. H., and Bruland, K. W. 1987. Oceanic stratified euphotic zone as elucidated by ^{234}Th:^{238}U disequilibria. *Limnol. Oceanogr.* 32:189–200.

Coughlin, B. R., and Stone, A. T. 1995. Nonreversible adsorption of divalent metal ions (Mn, Co, Ni,

Cu, Pb) onto goethite: Effects of acidification, Fe addition, and picolinic acid addition. *Environ. Sci. Technol.* 29:2445–2455.

Cremers, A., and A. Maes, B. E. 1986. Radionuclide partitioning in environmental systems: A critical analysis. *In* "Applications of Distribution Coefficients to Radiological Assessment Models" (T. H. Sibley and C. Myttenaere, Eds.), pp. 4–14. Elsevier Science, Amsterdam.

Davies-Colley, R. J., Nelson, P. O., and Williamson, K. J. 1984. Copper and cadmium uptake by estuarine sedimentary phases. *Environ. Sci. Technol.* 18:491–499.

Davis, J. A. 1977. Adsorption of Trace Metals and Complexing Ligands at the Oxide/Water Interface." Ph.D. Thesis, Stanford Univ., Stanford, CA.

Davis, J. A., and Hayes, K. F. 1986. Geochemical processes at mineral surfaces: An overview. *In* "Geochemical Processes at Mineral Surfaces" (J. A. Davis and K. F. Hayes, Eds.), Am. Chem. Soc. Sym. Ser. 323, pp. 2–18. American Chemical Society, Washington, DC.

Davis, J. A., and Kent, D. B. 1990. Surface complexation modeling in aqueous geochemistry. *In* "Reviews in Mineralogy" (M. F. Hochella and A. F. White, Eds.), pp. 177–260. Mineralogist Society of America, Washington, DC.

Davis, J. A., and Leckie, J. O. 1978a. Surface ionization and complexation at the oxide/water interface. I. Computation of electrical double layer properties in simple electrolytes. *J. Colloid Interface Sci.* 63:480–499.

Davis, J. A., and Leckie, J. O. 1978b. Surface ionization and complexation at the oxide/water interface. II. Surface properties of amorphous iron oxyhydroxide and adsorption of metal ions. *J. Colloid Interface Sci.* 67:90–106.

Dent, A. J., Ramsay, J. D. F., and Swanton, S. W. 1992. An EXAFS study of uranyl ion in solution and sorbed onto silica and montmorillonite clay colloids. *J. Colloid Interface Sci.* 150:45–60.

Di Toro, D. M., Mahony, J. D., Hansen, D. J., Scott, K. J., Hicks, M. B., Mayr, S. M., and Redmond, M. S. 1990. Toxicity of cadmium in sediments: The role of acid volatile sulfide. *Environ. Toxicol. Chem.* 9:1487–1502.

Dowd, J. E., and Riggs, S. 1965. A comparison of estimates of Michaelis–Menten kinetic constants from various linear transformations. *J. Biol. Chem.* 240:863–869.

Duursma, E. K., and Bosch, C. J. 1970. Theoretical, experimental and field studies concerning diffusion of ratioisotopes in sediments and suspended particles of the sea. B. Methods and experiments. *Neth. J. Sea Res.* 4:395–469.

Dzombak, D. A., and Morel, F. M. M. 1986. Sorption of cadmium on hydrous ferric oxide at high sorbate/sorbent ratios: Equilibrium, kinetics, and modeling. *J. Colloid Interface Sci.* 112:588–598.

Dzombak, D. A., and Morel, F. M. M. 1990. "Surface Complexation Modeling: Hydrous Ferric Oxide." Wiley, New York.

Erikson, R. L., Hostetler, C. J., and Kemmer, M. L. 1990. "Mobilization and Transport of Uranium at Uranium Mill Tailings Disposal Sites: Application of a Chemical Transport Model," NUREG/CR-5169. U.S. Nuclear Regulatory Commission, Washington, DC.

Erikson, R. L., Hostetler, C. J., Serne, R. J., Divine, J. R., and Parkhurst, M. A. 1993. "Geochemical Factors Affecting Degradation and Environmental Fate of Depleted Uranium Pentrators in Soil and Water," PNL-8527. Pacific Northwest Laboratory, Richland, WA. [limited distribution]

Eylem, C., Erten, H. N., and Göktürk, H. 1989. Sorption of barium on kaolinite, montmorillonite and chlorite. *Analyst* 114:351–353.

Felmy, A. R., Girvin, D. C., and Jenne, E. A. 1984. "MINTEQ-A Computer Program for Calculating Aqueous Geochemical Equilibria," NTIS PB 84157148. National Technical Information Service, Washington, DC.

Fisher, E. A. 1922. The phenomena of adsorption in soils: A critical discussion of the hypotheses put forward. *Trans. Faraday Soc.* 17:305–316.

Freundlich, H. 1909. Kapillarchemie, *Akademische Verlagsgesellschaft,* Leipziz.

Gadde, R. R., and Laitinen, H. A. 1974. Studies of heavy metal adsorption by hydrous iron and manganese oxides. *Anal. Chem.* 46:2022–2026.

Giblin, A. M. 1980. The role of clay adsorption in genesis of uranium ores. *In* "Uranium in the Pine Creek Geosyncline" (J. Ferguson and A. B. Goleby, Eds.), pp. 521–529. IAEA, Vienna.

Girvin, D. C., Ames, L. L., Schwab, A. P., and McGarrah, J. E. 1991. Neptunium adsorption on synthetic amorphous iron oxyhydroxide. *J. Colloid Interface Sci.* 141:67–78.

Goldberg, S. 1985. Chemical modeling of anion competition on goethite using the constant capacitance model. *Soil Sci. Soc. Am. J.* 49:851–856.

Goldberg, S. 1995. Adsorption models incorporated into chemical equilibrium models, *In* "Chemical Equilibrium and Reaction Models," pp. 75–95. Soil Science Society of America, Madison, WI.

Goldberg, S., Forster, H. S., and Godfrey, C. L. 1996. Molybdenum adsorption on oxides, clay minerals, and soils. *Soil Sci. Soc. Am. J.* 60:425–432.

Gonzalez, B. R., Appelt, H., Schalscha, E. B., and Bingham, F. T. 1974. Molybdate adsorption characteristics of volcanic-ash-derived soils in Chile. *Soil Sci. Soc. Am. J.* 38:903–906.

Griffin, R. A., Sack, W. A., Roy, W. R., Ainsworth, C. C., and Krapac, I. G. 1985. Batch-type 24-hour distribution ratio for contaminant adsorption by soil materials. *In* "Hazardous and industrial Solid Waste Testing and Disposal" (D. Lorenzen *et al.,* Eds.), Vol. 6, pp. 390–408, STP 933. American Society for Testing and Materials, Philadelphia, PA.

Guilherme, L. R. G., and Anderson, S. J. 1998. Copper sorption kinetics and sorption hysteresis in two oxide-rich soils (oxisols): Effect of phosphate pretreatment. *In* "Adsorption of Metals by Geomedia" (E. A. Jenne, Ed.). Academic Press, San Diego.

Guy, R., and Chakrabarti, C. 1976. Studies of metal–organic interactions in model systems pertaining to natural waters. *Can. J. Chem.* 54:2600–2611.

Guy, R. D., Chakrabarti, C. L., and Schramm, L. L. 1975. The application of a simple chemical model of natural waters to metal fixation in particulate matter. *Can. J. Chem.* 53:661–669.

Hakanen, M., and Hölttä, P. 1992. "Review of Sorption and Diffusivity Parameters for TVO-92," Technical Report YJT-92-14. Nuclear Waste Commission of Finnish Power Companies, Helsinki, Finland.

Harter, R. D. 1986. Editor's comments on papers 2 through 5. *In* "Adsorption Phenomena" (R. D. Harter, Ed.), pp. 12–16. Van Nostrand Reinhold, New York.

Harter, R. D. 1991. Kinetics of sorption/desorption processes in soil. *In* "Rates of Soil Chemical Processes" (D. L. Sparks and D. L. Suarez, Eds.), Special publ. 27, pp. 135–149. Soil Science Society of America, Madison, WI.

Hayes, K. F., and Leckie, J. O. 1986. Mechanism of lead ion adsorption at the goethite-water interface. *In* "Geochemical Processes at Mineral Surfaces" (J. A. Davis and K. F. Hayes, Eds.). Am. Chem. Soc. Sym. Ser. 323, pp. 114–141. American Chemical Society, Washington, DC.

Hayes, K. F., Redden, G., Ela, W., and Leckie, J. O. 1991. Surface complexation models: An evaluation of model parameter estimation using FITEQL and oxide mineral titration data. *J. Colloid Interface Sci.* 142:442–469.

Healy, T. W., James, R. O., and Cooper, R. 1968. "The Adsorption of Aqueous Co(II) at the Silica–Water Interface," Am. Chem. Soc. Adv. Chem. Ser. 79, pp. 62–73. American Chemical Society, Washington, DC.

Herbelin, A. L., and Westall, J. C. 1994. "A Computer Program for Determination of Chemical Equilibrium Constants," version 3.1, report 94-01. Department of Chemistry, Oregon State Univ., Corvallis, OR.

Holtta, P., Hakanen, M., and Hautojarrvi, A. 1991. Migration of radionuclides in fracture columns, *In* "Scientific Basis for Nuclear Waste Management XIV" (T. Abrajano, Jr., and L. H. Johnson, Eds.), Materials Res. Soc. Sym. Proc. Vol. 212, pp. 669–676. Materials Research Society, Pittsburgh, PA.

Honeyman, B. D. 1984. "Cation and Anion Adsorption at the Oxide/Solution Interface in Systems Containing Binary Mixtures of Adsorbents: An Investigation of the Concept of Adsorptive Additivity." Ph.D. thesis, Stanford Univ. Stanford, CA.

Horowitz, A. J., Demas, C. R., Fitzgerald, K. K., Miller, T. L., and Rickert, D. A. 1994. "U.S. Geological Survey Protocol for the Collection and Processing of Surface-Water Samples for the Subsequent Determination of Inorganic Constituents in Filtered Water" Open-File Report, No. 94-539. U.S. Geological Survey. Reston, VA.

Horowitz, A. J., Elrick, K. A., and Colberg, M. R. 1992. Effect of membrane filtration artifacts on dissolved trace element concentrations. *Water Res.* 26:753–763.

Horowitz, A. J., Lum, K. R., Garbarino, J. R., Hall, G. E. M., Lemieux, C., and Demas, C. R. 1996. Problems associated with using filtration to define dissolved trace element concentrations in natural water samples. *Environ. Sci. Technol.* 30:954–963.

Hsi, C.-K. D., and Langmuir, D. 1985. Adsorption of uranyl onto ferric oxyhydroxides: Application of the surface complexation site-binding model. *Geochim. Cosmochim. Acta* 49:1931–1941.

Hsu, P. H., and Bates, T. F. 1964. Fixation of hydroxy-aluminum polymers by vermiculite. *Soil Sci. Soc. Am. J.* 28:763–766.

Jackson, M. L. 1956. "Soil Chemical Analysis—Advanced Course." Univ. of Wisconsin, M. L. Jackson, Madison, WI.

Jannasch, H. W., Honeyman, B. D., Balistrieri, L. S., and Murray, J. W. 1988. Kinetics of trace element uptake by marine particles. *Geochim. Cosmochim. Acta.* 52:567–577.

Jarrell, W. M., and Dawson, M. D. 1978. Sorption and availability of molybdenum in soils of western Oregon. *Soil Sci. Soc. Am J.* 42:412–415.

Jenne, E. A. 1968. "Controls on Mn, Fe, Co, Ni, Cu and Zn Concentrations in Soils and Water—The Significant Role of Hydrous Mn and Fe Oxides," Am. Chem. Soc. Adv. Chem. Vol. 73, pp. 337–387. American Chemical Society; Washington, DC.

Jenne, E. A. 1977. Trace element sorption by sediments and soil—Sites and processes. *In* "Symposium on Molybdenum in the Environment" (W. Chappell and K. Petersen, Eds.), Vol. 2, pp. 425–553. Dekker, New York.

Jenne, E. A. 1995. Metal adsorption onto and desorption from sediments. I. Rates. *In* "Metal Speciation and Contamination of Aquatic Sediments" (A. E. Allen, Ed.), pp. 81–112. Ann Arbor Press, Ann Arbor, MI.

Jenne, E. A. 1996. Metal adsorption onto and desorption from sediments 2. Artifact effects. *Mar. Freshwater Res.* 64:1–18.

Jenne, E. A., and Crecelius, E. A. 1988. Determination of sorbed metals, amorphic Fe, oxidic Mn, and reactive particulate organic carbon in sediments and soils. *In* "Environmental Contamination 3rd International Conference, September 26–29, Venice, Italy" (A. A. Orio, Ed.), pp. 88–93.

Jenne, E. A., Di Toro, D. M., Allen, H. E., and Zarba, C. S. 1986. An activity-based model for developing sediment criteria for metals. *In* "Proceedings, International Conference on Chemical in the Environment, Lisbon, 1–3 July 1986." Selper Ltd., London.

Jenne, E. A., and Wahlberg, J. S. 1965. "Role of Certain Stream-Sediment Components in Radio-Ion Adsorption," Prof. Paper 433-F. U.S. Geological Survey, Reston, VA.

Jenne, E. A., and Zachara, J. M. 1987. Factors influencing the sorption of metals. *In* "Fate and Effects of Sediment Bound Chemicals in Aquatic Systems," (K. L. Dickson, A. W. Maki, and W. Brungs, Eds.), pp. 83–98. Pergamon, New York.

Jenny, H., and Overstreet, R. 1939. Cation exchange between plant roots and soil colloids. *Soil Sci.* 47:257–272.

Jones, B. F., Kennedy, V. C., and Zellweger, G. W. 1974. Comparison of observed and calculated concentrations of dissolved al and Fe in stream water. *Water Resources Res.* 10:791–793.

Jones, L. H. P. 1957. The solubility of molybdenum in simplified systems and aqueous soil suspensions. *J. Soil Sci.* 8:313–327.

Kaplan, D. I., Serne, R. J., and Piepho, M. G. 1994. "Geochemical Factors Affecting Radionuclide

Transport through Near and Far Fields at a Low-Level Waste Disposal Site: Available Sorption Constants and Recommendations for Future Studies," PNL-10379. Pacific Northwest National Laboratory, Richland, WA.

Karimian, N., and Cox, F. R. 1978. Adsorption and extractability of molybdenum in relation to some chemical properties of soil. *Soil Sci. Soc. Am. J.* 42:757–761.

Keay, J., and Wild, A. 1961. The kinetics of cation exchange in vermiculite. *Soil Sci.* 92:54–60.

Keller, W. D., Balgord, W. D., and Reesman, A. L. 1963. Dissolved products of artificially pulverized silicate minerals and rocks: Part I. *J. Sed. Pet.* 35:191–204.

Kennedy, V. C., Jenne, E. A., and Burchard, J. M. 1976. "Backflushing Filters for Field Processing of Water Samples prior to Trace-Element Analyses," Open-File Report 76-126. U.S. Geological Survey, Reston, VA.

Kennedy, V. C., Zellweger, G. W., and Jones, B. F. 1974. Filter pore-size effects on the analysis of Al, Fe, Mn, and Ti in water. *Water Resources Res.* 10:785–790.

Kent, D. B., Tripathi, V. S., Ball, N. B., Leckie, J. O., and Siegel, M. D. 1988. "Surface-Complexation Modeling of Radionuclide Adsorption in Surface Environments," Final Report, NUREG/CR-4807. U.S. Nuclear Regulatory Commission, Washington, DC.

King, L. D. 1988. Retention of metals by several soils of the southeastern United States. J. *Environ. Qual.* 17:239–246.

Kinniburgh, D. G., and Jackson, M. L. 1982. Concentration and pH dependence of calcium and zinc adsorption by iron hydrous oxide gel. *Soil Sci. Soc. Am. J.* 46:56–61.

Kinniburgh, D. G., van Riemsdijk, W. H., Koopal, L. K., and Benedetti, M. F. 1998. Ion binding to humic substances: Measurements, models, mechanisms. *In* "Adsorption of Metals by Geomedia" (E. A. Jenne, Ed.). Academic Press, San Diego.

Kirk, G. J. D., and Staunton, S. 1989. On predicting the fate of radioactive caesium in soil beneath grassland. J. *Soil Sci.* 40:71–84.

Kohler, M., Wieland, E., and Leckie, J. O. 1992. Metal-ligand-surface interactions during sorption of uranyl and neptunyl on oxides and silicates. *In* "Proceedings of the 7th International Symposium on Water–Rock Interaction-WRI-7" (Y. K. Kharaka and A. S. Maest, Ed.), Vol. I, Low Temperature Environments. A. A. Balkema, Roterdam.

Koskinen, A., Alaluusua, M., Pinnioja, S., Jaakkola, T., and Lindberg, A. 1985. "Sorption of Iodine, Neptunium, Technetium, Thorium and Uranium on Rocks and Minerals," YJT-85-36. Geological Survey of Findland, Helsinki,

Koß V. 1988. Modeling of uranium(VI) sorption and speciation in a natural sediment–groundwater system. *Radiochim. Acta* 44/45:403–406.

Koß, V., and Kim, J. I. 1990. Modeling of strontium sorption and speciation in a natural sediment-groundwater system. *J. Contam. Hydrol.* 6:267–280.

Krupka, K. M., Erikson, R. L., Mattigod, S. V., Schramke, J. A., and Cowan, C. E. 1988. "Thermochemical Data Used by the FASTCHEM™ Package," EPRI EA-5872. Electric Power Research Institute, Palo Alto, CA.

Kurbatov, M. H., Wood, G. B., and Kurbatov, J. D. 1951. Isothermal adsorption of cobalt from dilute solutions. *J. Phys. Chem.* 55:1170–1182.

LaFlamme, B. D., and Murray, J. W. 1987. Solid/solution interaction: The effect of carbonate alkalinity on adsorbed thorium. *Geochim. Cosmochim. Acta* 51:243–250.

Langmuir, D. 1978. Uranium solution–mineral equilibria at low temperatures with application to sedimentary ore deposits. *Geochim. Cosmochim. Acta* 42:547–569.

Langmuir, D. 1981. The power exchange function; A general model for metal adsorption onto geological materials. *In* "Adsorption from Aqueous Solutions" (P. H. Tewari, Ed.), pp. 1–18. Plenum, New York.

Langmuir, D. 1997. "Aqueous Environmental Geochemistry." Prentice Hall, Toronto.

Langmuir, I. 1918. The adsorption of gases on plane surfaces of glass, mica, and platinum. *J. Am. Chem. Soc.* 40:1361–1403.

Langston, R. B., and Jenne, E. A. 1964. NaOH dissolution of some oxide impurities from kaolins. *Clays Clay Miner.* 12:633–647.

Langston, W. J. 1982. The distribution of mercury in British estuarine sediments and its availability to deposit-feeding bivalves. *J. Mar. Bio. Assoc. U.K.* 62:667–684.

Laxen, D. P. H. 1983. Cadmium adsorption in fresh waters—A quantitative appraisal of the literature. *Sci. Total Environ.* 30:129–146.

Lieser, K. H., Gleitsmann, B., and Steinkopff, Th. 1986. Sorption of trace elements or radionuclides in natural systems containing groundwater and sediments. *Radiochim. Acta* 40:33–37.

Lieser, K. H., Quandt-Klenk, S., and Thybusch, B. 1992. Sorption of uranyl ions on hydrous silicon dioxide. *Radiochim. Acta* 57:45–50.

Lion, L. W., Altmann, R. S., and Leckie, J. O. 1982. Trace-metal adsorption characteristics of estuarine particulate matter: Evaluation of contributions of Fe/Mn oxide and organic surface coatings. *Environ. Sci. Technol.* 16:660–666.

Loganathan, P., and Burau, R. 1973. Sorption of heavy metal ions, by a hydrous manganese oxide. *Geochim. Cosmochim. Acta* 37:1277–1293.

Loux, N. T., Brown, D. S., Chafin, C. R., Allison, J. D., and Hassan, S. M. 1989. Chemical speciation and competitive cationic partitioning on a sandy aquifer material. *Chem. Spec. Bioavail.* 1:111–125.

Luoma, S. N., and Davis, J. A. 1983. Requirements for modeling trace metal partitioning in oxidized estuarine sediments. *Marine Chem.* 12:159–181.

MacNaughton, M. D. 1973. "Adsorption of Mercury(II) at the Solid–Water Interface." Ph.D. Thesis. Stanford Univ. Stanford, CA.

Manceau, A., and Charlet, L. 1991. Sorption and speciation of heavy metals at the oxide/water interface: From microscopic to macroscopic. *In* "Environ. Pollution 1-ICEP. 1," pp. 401–408. Interscience Enterprises Ltd., Jersey, Channel Islands.

Manceau, A., and Charlet, L. 1992a. X-ray absorption spectroscopic study of the sorption of Cr(III) at the oxide–water interface. I. Molecular mechanism of Cr(III) oxidation on Mn oxides." *J. Colloid Interface Sci.* 148:425–442.

Manceau, A., and Charlet, L. 1992b. X-Ray absorption spectroscopic study of the sorption of Cr(III) at the oxide-water interface. II. Adsorption, coprecipitation, and surface precipitation on hydrous ferric oxide. *J. Colloid Interface Sci.* 148:443–458.

Manceau, A., Charlet, L., Boisset, M. C., Didier, B., and Spadini, L. 1992. Sorption and speciation of heavy metals on hydrous Fe and Mn oxides. From microscopic to macroscopic. *Appl. Clay Sci.* 7:201–223.

McBride, M. B. 1979. Chemisorption and precipitation of Mn^{2+} at $CaCO_3$ surfaces. *Soil Sci. Soc. Am. J.* 43:693–698.

McKinley, J. P., and Jenne, E. A. 1991. An experimental investigation and review of the "solids concentration" effect in adsorption studies. *Environ. Sci. Technol.* 25:2082–2087.

McKinley, J. P., Zachara, J. M., Smith, S. C., and Turner, G. D. 1995. The influence of hydrolysis and multiple site-binding reactions on adsorption of U(VI) to montmorillonite. *Clays Clay Miner.* 43:586–598.

McLaren, R. G., Lawson, D. M., and Swift, R. S. 1986. Sorption and desorption of cobalt by soils and soil components. *J. Soil Sci.* 37:413–426.

Millward, G. E. 1980. The adsorption of cadmium by iron(III) precipitates in model estuarine solutions. *Environ. Tech. Lett.* 1:394–399.

Misak, N. Z., Ghoneimy, H. F., and Morcos, T. N. 1996. Adsorption of Co^{2+} and Zn^{2+} ions on hydrous Fe(III), Sn(IV) and Fe(III)/Sn(IV) oxides. II. Thermal behavior of loaded oxides, isotopic exchange equilibria, and percentage adsorption-pH curves. *J. Colloid Interface Sci.* 184:31–43.

Morel, F. M. M. and Hudson, R. J. M. 1985. The geobiological cycle of trace elements in aqueous systems: Redfield revisited. *In* "Chemical Processes in Lakes" (W. Stumm, Ed.), pp. 389–426. Wiley, New York.

Murray, J. W. 1975. The interaction of cobalt with hydrous manganese dioxide. *Geochim. Cosmochim. Acta* 39:635–647.

Murray, J. W., Balistrieri, L. S., and Paul, B. 1984. The oxidation state of manganese in marine sediments and ferromanganese nodules. *Geochim. Cosmochim. Acta* 48:1237–1247.

Nyffeler, U. P., Li, Y.-H., and Santschi, P. H. 1984. A kinetic approach to describe trace-element distribution between particles and solution in natural aquatic systems. *Geochim. Cosmochim. Acta* 48:1513–1522.

Oakley, S. M., Nelson, P. O., and Williamson, K. J. 1981. Model of trace-metal partitioning in marine sediments. *Environ. Sci. Technol.* 15:474–480.

O'Conner, J. T., and Renn, C. E. 1964. Soluble-adsorbed zinc equilibrium in natural waters. *J. Am. Water Works Assoc.* 56:1055–1061.

O'Day, P. A., Brown, G. E., Jr., and Parks, G. A. 1994. X-Ray Absorption Spectroscopy of Cobalt(II) Multinuclear Surface Complexes and Surface Precipitates on Kaolinite. *J. Coll. Interface Sci.* 165: 269–289.

Ong, C. G., and J. O. Leckie. 1998. Surface and solution speciation of Ag^{I} in a heterogeneous ferrihydrite-solution system with thiosulfate. *In* "Adsorption of Metals by Geomedia" (E. A. Jenne, Ed.). Academic Press, San Diego.

Pabalan, R. T. 1994. Thermodynamics of ion exchange between clinoptilolite and aqueous solutions of Na^+/K^+ and Na^+/Ca^{2+}. *Geochim. Cosmochim. Acta* 58:4573–4590.

Pabalan, R. T., Turner, D. R., Bertetti, F. P. and Prikryl, J. D. 1998. $Uranium^{VI}$ sorption onto selected mineral surfaces: Key geochemical parameters. *In* "Adsorption of Metals by Geomedia" (E. A. Jenne, Ed.). Academic Press, San Diego.

Pabalan, R. T., and Turner, D. R. 1993a. "Sorption Modeling for HLW Performance Assessment (1991)," NRC high-level radioactive waste research at CNWRA, NUREG/CR-5817, CNWRA 91-01A, Vol. 2, pp. 1–8.62.

Pabalan, R. T., and Turner, D. R. 1993b. "Sorption Modeling for HLW Performance Assessment (January–June 1992)," NRC high-level radioactive waste research at CNWRA, NUREG/CR-5817, CNWRA 92-01S, Vol. 3(1), pp. 8.1–8.24.

Pabalan, R. T., and Turner, D. R. 1993c. "Sorption Modeling for HLW Performance Assessment (July–December 1992)," NCR high-level radioactive waste research at CNWRA, NUREG/CR-5817, CNWRA, 92-02S, Vol. 3(2), pp. 8.1–8.18.

Pabalan, R. T., Turner, D. R., and Bertetti, F. P. 1994. "Sorption Modeling for HLW Performance Assessment, (January–June 1994), NRC high-level radioactive waste research at CNWRA, CNWRA 94-01S, pp. 77–95.

Papelis, C., Roberts, P. V., and Leckie, J. O. 1995. Modeling the rate of cadmium and selenite adsorption on micro-and mesoporous transition aluminas. *Environ. Sci. Technol.* 29:1099–1108.

Payne, T. E., Lumpkin, G. R., Waite, T. D. 1998. Uranium (VI) adsorption on model minerals: Controlling factors and surface complexation modeling. *In* "Adsorption of Metals by Geomedia" (E. A. Jenne, Ed.). Academic Press, San Diego.

Payne, T. E., Sekine, K., Davis, J. A., and Waite, T. D. 1992. Modeling of radionuclide sorption processes in the weathered zone of the Koongarra ore body. *In* "Alligator Rivers Analogue Project Annual Report, 1990–1991" (P. Duerden, Ed.), pp. 57–85. Austalian Nuclear Science and Technology Organization (ANSTO), Menai, Australia.

Payne, T. E., and Waite, T. D. 1991. Surface complexation modeling of uranium sorption data obtained by isotopic techniques. *Radiochim. Acta* 52/53:487–493.

Perona, M. J., and Leckie, J. O. 1985. Proton stoichiometry for the adsorption of cations on oxide surfaces. *J. Colloid Interface Sci.* 106:64–69.

Phelan, P. J., and Mattigod, S. V. 1984. Adsorption of molybdate anion (MoO_4^{2-}) by sodium-saturated kaolinite." *Clays Clay Miner.* 32:45–48.

Posselt, H. S., Anderson, F. J., and Weber, W. J., Jr. 1968. Cation sorption on colloidal hydrous dioxide. *Environ. Sci. Technol.* 2:1087–1093.

Puls, R. W., Ames, L. L., and McGarrah, J. E. 1987. "Sorption and Desorption of Uranium, Selenium, and Radium in a Basalt Geochemical Environment," WHC-SA-0003-FP. Westinghouse Hanford, Richland, WA.

Rasmuson, A., and Neretnieks, I. 1986. Radionuclide transport modeling in fissured zone and channels. *In* "Scientific Basis for Nuclear Waste Management IX." *Mater. Res. Soc. Sym. Proc.* 50:641–653.

Reardon, E. J. 1981. Kd's—Can they be used to describe reversible ion sorption reactions in contaminant migration? *Ground Water* 19:279–2285.

Redden, G. D., Li, J., and Leckie, J. O. 1998. Adsorption of U^{VI} and citric acid on goethite, gibbsite, and kaolinite: Comparing results for binary and ternary systems." *In* "Adsorption of Metals by Geomedia" (E. A. Jenne, Ed.). Academic Press, San Diego.

Roy, R. 1996. Citing scientific literature. *Chem. Eng. News.* April 8, pp. 5, 66.

Roy, W. R., Krapac, I. G., Chou, S. F. J., and Griffin, R. A., 1991. "Batch-Type Procedures for Estimating Soil Adsorption of Chemicals," EPA/530-SW-87-006-F. U.S. Environmental Protection Agency, Cincinnati, OH.

Salter, P. F., Ames, L. L., and McGarrah, J. E. 1981. "The Sorption Behavior of Selected Radionuclides on Columbia River Basalts," RHO-BWI-LD-48. Rockwell, Hanford, Richland, WA.

Schindler, P. W. 1975. Removal of trace metals from the oceans: A zero order model. *Thal. Jugoslav.* 11:101–111.

Schindler, P. W. 1981. Surface complexes at oxide–water interfaces. *In* "Adsorption of Inorganics" (M. A. Anderson and A. J. Rubin, Eds.), pp. 1–49. Ann Arbor Science, Ann Arbor, MI.

Schindler, P. W., Fürst, B., Dick, R., and Wolf, P. U. 1976. Ligand properties of surface silanol groups. I. Surface complex formation with Fe^{3+}, Cu^{2+}, Cd^{2+}, and Pb^{2+}. *J. Colloid Interface, Sci.* 55:469–475.

Schindler, P. W., and Kamber, H. R. 1968. The acidity of silanol groups. *Helv. Chim. Acta* 51:1781–1786.

Scott, M. J., and Morgan, J. J. 1995. Reactions at oxide surfaces. 1. Oxidation of As(III) by synthetic birnessite. *Environ. Sci. Technol.* 29:1898–1905.

Schulthess, C. P., and Dey, D. K. 1996. Estimation of Langmuir constants using linear and nonlinear least squares regression analyses. *Environ. Sci. Technol.* 60:433–442.

Serne, R. J. 1992. Current adsorption models and open issues pertaining to performance assessment. *In* "Proceedings of the DOE/Yucca Mountain Site Characterization Project Radionuclide Adsorption Workshop at Los Alamos National Laboratory" (J. A. Canepa, Ed.), LA-12325-C, pp. 43–74. Los Alamos National Laboratory, Los Alamos, NM.

Sharpley, A. N., Ahuja, L. R., Yamamoto, M., and Menzel, R. G. 1981. The kinetics of phosphorus desorption from soil. *Soil Sci. Soc. Am. J.* 45:493–496.

Sheppard, M. I., Beals, D. I., Thibault, D. H., and O'Connor, P. 1984. "Soil Nuclide Distribution Coefficients and Their Statistical Distribution," AECL-8364, Atomic Energy Canada Limited, Pinawa, Manitoba, Canada.

Smith, K. S., Ranville, J. F., Plumlee, G. S., and Macalady, D. L. 1998. Predictive Double-Layer Modeling of Metal Sorption in Mine-Drainage Systems. *In* "Adsorption of Metals by Geomedia" (E. A. Jenne, Ed.). Academic Press, San Diego.

Smith, R. W., and Jenne, E. A. 1991. Compilation, evaluation, and prediction of triple-layer model constants for ions on Fe(III) and Mn(IV) hydrous oxides. *Environ. Sci. Tech.* 25:525–531.

Sposito, G. 1977. The Gapon and the Vanselow selectivity coefficients. *Soil Sci. Soc. Am. J.* 41:1205–1206.

Sposito, G. 1980a. Derivation of the Freundlich equation for ion exchange reactions in soils. *Soil Sci. Soc. Am. J.* 44:652–654.

Sposito, G. 1980b. Cation exchange in soils: An historical and theoretical perspective. *In Chemistry in the Soil Environment* (R. H. Dowdy, D. Baker, V. Volk, and J. Ryan, Eds.), Soil Sci. Soc. Am. Special Publ. 40, pp. 13–30. Soil Science Society of America, Madison, WI.

Sposito, G. 1982. Soil chemistry. *In* "McGraw-Hill Encyclopedia of Science and Technology" (S. P. Parker, Ed.), pp. 548–555. McGraw-Hill, New York.

Sposito, G. 1984. "The Surface Chemistry of Soils." Oxford Univ. Press, New York.

Sposito, G. 1986. Distinguishing adsorption from surface precipitation. *In* ;"Geochemical Processes at Mineral Surfaces" (J. A. Davis and K. F. Hayes, Eds.), Am. Chem. Soc. Sym. Series, pp. 217–228. American Chemical Society, Washington, DC.

Stenhouse, M. J. 1994. "Sorption Databases for Crystalline, Marl and Bentonite for Performance Assessment," NAGRA Technical Report NTB 93-06. Baden, Switzerland.

Stollenwerk, K. G., and Grove, D. B. 1985. Adsorption and desorption of hexavalent chronium in an alluvial aquifer near Telluride, Colorado. *J. Environ. Qual.* 14:150–155.

Streck, T., and Richter, J. 1997. Heavy Metal Displacement in a Sandy Soil at the Field Scale: I. Measurements and Parameterization of Sorption. *J. Environ. Qual.* 26:49–56.

Stumm, W., Huang, C. P., and Jenkins, S. R. 1970. Specific chemical interactions affecting the stability of dispersed systems. *Croat. Chem. Acta* 42:223–244.

Stumm, W., Kummert, R., and Sigg, L. 1980. A ligand exchange model for the adsorption of inorganic and organic ligands at hydrous oxide interfaces. *Croat. Chem. Acta* 532:291–312.

Swallow, K. C., Hume, D. N., and Morel, F. M. M. 1980. Sorption of copper and lead by hydrous ferric oxide. *Environ. Sci. Technol.* 14:1326–1331.

Tessier, A., Carignan, R., Dubreuil, B., and Rapin, F. 1989. Partitioning of zinc between the water column and the oxic sediments in lakes. *Geochim. Cosmochim. Acta* 53:1511–1522.

Tessier, A., Rapin, F., and Carignan, R. 1985. Trace metals in oxic lake sediments: Possible adsorption onto iron oxyhydroxides. *Geochim. Cosmochim. Acta* 49:183–194.

Theng, B. K. G. 1971. Adsorption of molybdate by some crystalline and amorphous soil clays. *New Zealand J. Sci.* 14:1040–1056.

Thibault, D. H., Shepard, M. I., and Smith, P. A. 1990. "A Critical Compilation and Review of Default Soil Solid/Liquid Partition Coefficients, K_d, for Use in Environmental Assessments," AECL-10125. Atomic Energy Canada Ltd., Pinawa, Manitoba, Canada.

Thomas, G. W. 1977. Historical developments in soil chemistry: Ion exchange. *Soil Sci. Soc. Am. J.* 41:230–238.

Thomas, K. W., 1987. "Summary of Sorption Measurements Performed with Yucca Mountain, Nevada, Tuff Samples and Water from Well J-13," LA-10960-MS. Los Alamos National Laboratory, Los Alamos, NM.

Thompson, H. A., Parks, G. A., and Brown, G. E., Jr. 1998. Structure and composition of uranium^{VI} sorption complexes at the kaolinite-water interface. *In* "Adsorption of Metals by Geomedia" (E. A. Jenne, Ed.). Academic Press, San Diego.

Tiller, K. G., Nayyar, M. K., and Clayton, P. M. 1979. Specific and non-specific sorption of cadmium by soil clays as influenced by zinc and calcium, *Aust. J. Soil Res.* 17:17–28.

Torrent, J. 1987. Rapid and slow phosphate sorption by mediterranean soils: Effect of iron oxides. *Soil Sci. Soc. Am. J.* 51:78–82.

Townsend, R. P. 1986. Ion exchange in zeolites: Some recent developments in theory and practice. *Pure Appl. Chem.* 58:1359–1366.

Tripathi, V. S. 1984. Uranium (VI) Transport Modeling: Geochemical data and Submodels." Ph.D. Thesis, Stanford Univ. Stanford, CA.

Turner, A., and Millward, G. E. 1994. Partitioning of trace metals in a macrotidal estuary, implications for contaminant transport models. *Estuarine Coastal Shelf Sci.* 39:45–58.

Turner, D. R. 1991. "Sorption Modeling for High-Level Waste Performance Assessment: A Literature Review," CNWRA 91011. Center for Nuclear Waste Regulatory Analyses, San Antonio, TX.

Turner, D. R. 1993. "Mechanistic Approaches to Radionuclide Sorption Modeling," CNWRA 93-019. Center for Nuclear Waste Regulatory Analyses, San Antonio, TX.

Turner, D. R., Griffin, T., and Dietrich, T. B. 1993. Radionuclide sorption modeling using the MINTEQA2 speciation code. *Mater. Res. Soc. Sym. Proc.* 294:783–789.

Turner, G. D., Zachara, J. M., McKinley, J. P., and Smith S. 1996. Surface-charge properties and UO_2^{2+} adsorption of a subsurface smectite. *Geochim. Cosmochim. Acta* 60:3399–3414.

van Bemmelen, J. M. 1878. Das Adsorptionsvermogen der Ackererde. *Landwirtsch. Vers. Stat.* 21:135–191.

van Bemmelen, J. M. 1888. Die absorptions Verbindungen und das Absorptions Vermögen der Ackererde. *Landwirtsch. Vers. Stat.* 35:69–136.

Vandergraaf, T. T. 1982. "A Compilation of Sorption Coefficients for Radionuclides on Granites and Granitic Rocks," AECL, TR-120. Atomic Energy of Canada Ltd., Pinawa, Manitoba.

Vandergraaf, T. T., and Ticknor, K. V. 1994. "A Compilation and Evaluation of Sorption Coefficients Used in the Geosphere Model of SYVAC for the 1990 Assessment of the Whiteshell Research Area," AECL-10546. Atomic Energy Canada Ltd., Pinawa, Manitoba, Canada.

Vandergraaf, T. T., Ticknor, K. V., and Melnyk, T. W. 1992. The selection and use of a sorption database for the Geosphere model in the Canadian Nuclear Fuel Waste Management Program. *In* "Radionuclide Sorption from the Safety Evaluation Perspective," pp. 81–120. Proceedings, NEA Workshop, Interlaken, Switzerland, 1991.

Vandergraaf, T. T., Ticknor, K. V., and Melnyk, T. W. 1993. The selection of a sorption database for the Geosphere model in the Canadian Nuclear Fuel Waste Management Program. *J. Contam. Hydrol.* 1:327–345.

Veith, J. A., and Sposito, G. 1977. On the use of the Langmuir equation in the interpretation of "adsorption" phenomena. *Soil Sci. Soc. Am. J.* 41:469–702.

Vuceta, J., and Morgan, J. J. 1978. Chemical modeling of trace metals in fresh waters: Role of compexation and adsorption. *Environ. Sci. Tech.* 12:1302–1309.

Wahlberg, J. S., Baker, J. H., Vernon, R. W., and Dewar, R. S. 1965. "Exchange Adsorption of Strontium on Clay Minerals," Bull. 1140-C. U.S. Geological Survey, Reston, VA.

Wahlberg, J. S., and Fishman, M. J. 1962. "Adsorption of Cesium on Clay Minerals," Bull. 1140-A. U.S. Geological Survey, Reston, VA.

Waite, T. D., Davis, J. A., Payne, T. E., Waychunas, G. A., and Xu, N. 1994. Uranium(VI) adsorption to ferrihydrite: Application of a surface complexation model. *Geochim. Cosmochim. Act* 58:5465–5478.

Wang, W.-Z., Brusseau, M. L., and Artiola, J. F. 1998. Nonequilibrium and nonlinear transport of cadmium, nickel, and strontium through subsurface soils. *In* "Adsorption of Metals by Geomedia" (E. A. Jenne, Ed.). Academic Press, San Diego.

Warren, L. A., and Zimmerman, A. P. 1994. The importance of surface area in metal sorption by oxides and organic mater in a heterogeneous natural sediment. *Appl. Geochem.* 9:245–254.

Waychunas, G. A., Rea, B. A., Fuller, C. C., and David, J. A. 1993. Surface chemistry of ferrihydrite. 1. EXAFS studies of the geometry of coprecipitated and adsorped arsenate. *Geochim. Cosmochim. Acta.* 57:2251–2269.

Wehrlic, B., Ibric, B., and Stumm, W. 1990. Adsorption kinetics of vanadyl(IV) and chromium(III) to aluminum oxide: Evidence for a two-step process. *Colloids Surf.* 51:77–88.

Westall, J. C. 1982a. "FITEQL: A Computer Program for Determination of Chemical Equilibrium Constants from Experimental Data," version 1.2, Report 82-01. Department of Chemistry, Oregon State Univ., Corvallis, OR.

Westall, J. C. 1982b. "FITEQL: A Computer Program for Determination of Chemical Equilibrium Constants from Experimental Data," Version 2.1, Report 82-02. Department of Chemistry, Oregon State Univ., Corvallis, OR.

Wolfsberg, K. 1978. Sorption–Desorption Studies of Nevada Test Site Alluvium and Leaching Studies of Nuclear Test Debris," LA-7216-MS (Informal Report). Los Alamos National Laboratory, Los Alamos, NM.

Yates, D. E., Levine, S., and Healy, T. W. 1974. Site-binding model of the electrical double layer at the oxide/water interface. *J. Chem. Soc. Faraday Trans.* 70:1807–1818.

Yong, R. N., and MacDonald, E. M. 1998. Influence of pH, metal concentration, and soil component removal on retention of Pb and Cu by an illitic soil. *In* "Adsorption of Metals by Geomedia" (E. A. Jenne, Ed.). Academic Press, San Diego.

Zachara, J. M., Ainsworth, C. C., Cowan, C. E., and Resch, C. T. 1989. Adsorption of chromate by subsurface soil horizons. *Soil Soc. Sci. Am. J.* 53:418–428.

Zachara, J. M., Ainsworth, C. C., Cowan, C. E., and Thomas, B. L. 1987. Sorption of binary mixtures of aromatic heterocyclic compounds on subsurface materials. *Environ. Sci. Technol.* 21:397–402.

Zachara, J. M., Ainsworth, C. C., McKinley, J. P., Murphy, E. M., Westall, J. C., and Rao, P. X. C. 1992. Subsurface chemistry of organic ligand–radionuclide mixtures. *In* "Pacific Northwest Laboratory Annual Report for 1991 to the DOE Office of Energy Research. 2. Environmental Science, " PNL-8000 Pt. 2, pp. 1–12. Pacific Northwest Laboratory, Richland, WA.

Zachara, J. M., and McKinley, J. P. 1993. Influence of hydrolysis on the sorption of metal cations by smectites: Importance of edge coordination reactions. *Aquatic Sci.* 55:250–261.

Zachara, J. M., Smith, S. C., Kuzel, L. S. 1995. "Adsorption and Dissociation of Co-EDTA Complexes in Iron Oxide-Containing Subsurface Sands." *Geochim. Cosmochim. Acta.* 59:4825–4844.

Zasoski, R. J., and Burau, R. G. 1988. Sorption and sorptive interactions of cadmium and zinc on hydrous manganese-oxide. *Soil Sci. Soc. Am. J.* 52:81–87.

Zhang, P. C., and Sparks, D. L. 1989. Kinetics and mechanism of molybdate adsorption/desorption at the goethite/water interface using pressure-jump relaxation. *Soil Sci. Soc. Am. J.* 53:1028–1034.

Chapter 2

Uranium^{VI} Adsorption on Model Minerals

Controlling Factors and Surface Complexation Modeling

T. E. Payne,[1] G. R. Lumpkin,[1] and T. D. Waite[2]
[1]Australian Nuclear Science and Technology Organisation, Menai, New South Wales, Australia; [2]Department of Water Engineering, University of New South Wales, Sydney, New South Wales, Australia

Uranium^{VI} sorption on ferrihydrite and kaolinite is influenced by a large number of factors, including pH, ionic strength, partial pressure of CO_2, adsorbent loading, total amount of U present, and the presence of ligands such as phosphate and humic acid. The effect of complexing ligands may be to enhance or reduce U uptake. The adsorption model being used for ferrihydrite is a surface complexation model with a diffuse double layer, and both strong and weak sites for U sorption. In terms of the amount of U sorbed per gram of adsorbent, U uptake on kaolinites KGa-1 and KGa-1B is much weaker than U uptake on ferrihydrite under similar experimental conditions. Titanium-rich impurity phases play a major role in the sorption of U by these standard kaolinites. Trace impurities and mineral coatings such as ferrihydrite can play a dominant role in determining U sorption in both environmental and model systems.

I. INTRODUCTION

In order to predict the environmental impact of human activities such as uranium mining and radioactive waste disposal, it is necessary to understand the mi-

Adsorption of Metals by Geomedia

gration of actinides in the environment. Adsorption on mineral surfaces is an important mechanism which reduces the mobility of actinides in natural systems. Consequently, we have been undertaking a detailed study of U^{VI} adsorption on model minerals, including kaolinite and ferrihydrite, as a prelude to developing models for actinide sorption in the natural environment.

A number of approaches to modeling sorption have been advocated, ranging from simple models in which uptake is represented by a single parameter (e.g., a K_d value), to detailed mechanistic models such as the surface complexation approach (Davis and Kent, 1990). In the surface complexation model (SCM), interactions between dissolved species and surface functional groups are modeled in a similar way to complex formation with ligands in solution. A significant component of the model is the inclusion of electrostatic terms to allow for changes in surface charge due to adsorption reactions. The SCM has been applied to modeling the adsorption of a wide range of inorganic cations and anions on the surface of hydrous ferric oxide (Dzombak and Morel, 1990).

A mechanistic model such as the SCM offers the attraction of being generally applicable to metal sorption on geological materials. The increasing use of the SCM in the past decade can also be attributed to the widespread availability of computers and chemical speciation codes. Software such as FITEQL (Herbelin and Westall, 1994) can be used to derive formation constants for surface complexes, and speciation programs such as MINTEQA2 (Allison *et al.,* 1990) can then be applied to compute the equilibrium distribution of all species in the system, including dissolved, precipitated, and adsorbed forms. The modeling of U sorption data with the SCM has been advanced in recent years by the availability of reliable thermodynamic data for aqueous species and compounds of U, particularly a review carried out by the OECD/NEA (Grenthe *et al.,* 1992).

The application of SCMs is now becoming widespread, but there are two areas in which information is lacking. Firstly, the identity of surface species has been difficult to establish. In the case of U, numerous surface species have been postulated and used in modeling experimental sorption data. Although the models can usually be made to fit the data, this may simply be a consequence of including a sufficient number of adjustable fitting parameters. In recent years, surface spectroscopic techniques such as EXAFS (extended X-ray absorption fine structure) have provided an independent means of identifying surface species (Waychunas *et al.,* 1993; Waite *et al.,* 1994), and these techniques have given considerable support to modeling studies in some instances.

The second major requirement of the SCM approach is reliable experimental sorption data for metal uptake on various substrates, with parameters such as pH, metal loading, and ionic strength varied over wide ranges. The work described in this chapter is being carried out for the purpose of providing a data base for U sorption on model materials, with the ultimate goal of developing a general model of U sorption in environmental systems.

II. AQUEOUS SPECIATION AND SORPTION MODELING

A. Uranium Speciation

Most of our experiments were undertaken in systems which were open to the atmosphere, with a background electrolyte of 0.1 mol·L^{-1} $NaNO_3$. Under these conditions, U is present in its +6 oxidation state, and equilibration with atmospheric CO_2 plays a major role in determining U^{VI} speciation. The speciation of U^{VI} as a function of pH in systems exposed to CO_2 of $10^{-3.5}$ atm is shown in Figure 1A. The speciation was calculated for a typical experiment with a total U (Σ U) of 10^{-6} mol·L^{-1} in 0.1 mol·L^{-1} $NaNO_3$. Uranium speciation is dominated by a series of hydrolyzed U species, with uranyl carbonate complexes becoming dominant at high pH values.

The computations were carried out using MINTEQA2 (Allison *et al.,* 1990), with a thermodynamic data base for aqueous species which contained the same reactions and constants given in a previous paper (Waite *et al.,* 1994). This data set was based on the NEA compilation (Grenthe *et al.,* 1992), supplemented by some thermodynamic data selected by Tripathi (1983). The only species added for the computations described in the present chapter was $UO_2NO_3^+$ with a log K of 0.3 (Grenthe *et al.,* 1992), which is of some importance in $NaNO_3$ solutions below pH 7 (Fig. 1A).

At higher U concentrations, polynuclear U complexes become more important (Fig. 1B). In this illustration (Fig. 1B), possible precipitation of solid U phases was excluded, although this was considered in the subsequent modeling of experimental results.

Due to the importance of carbonate speciation, some experiments were carried out in a glove box with an atmosphere containing an elevated P_{CO_2} of 10^{-2} atm, which is similar to many shallow groundwater environments. Under these conditions, carbonate complexes begin to dominate U speciation at lower pH values than shown in Figures 1A and 1B.

B. Previous U-Sorption Work

The adsorption of U^{VI} onto geological materials has been studied for several decades, and since about 1980 quantitative data suitable for mathematical modeling have been obtained in several studies. Notable among these are investigations of U uptake on goethite (Tripathi, 1983), and on various iron oxide minerals (Hsi and Langmuir, 1985). In both cases the surface complexation approach was applied to describe the experimental results.

As with many other metals, the uptake of U on geological materials increases with increasing pH, up to approximately pH 6.0. In this chapter, this is referred to

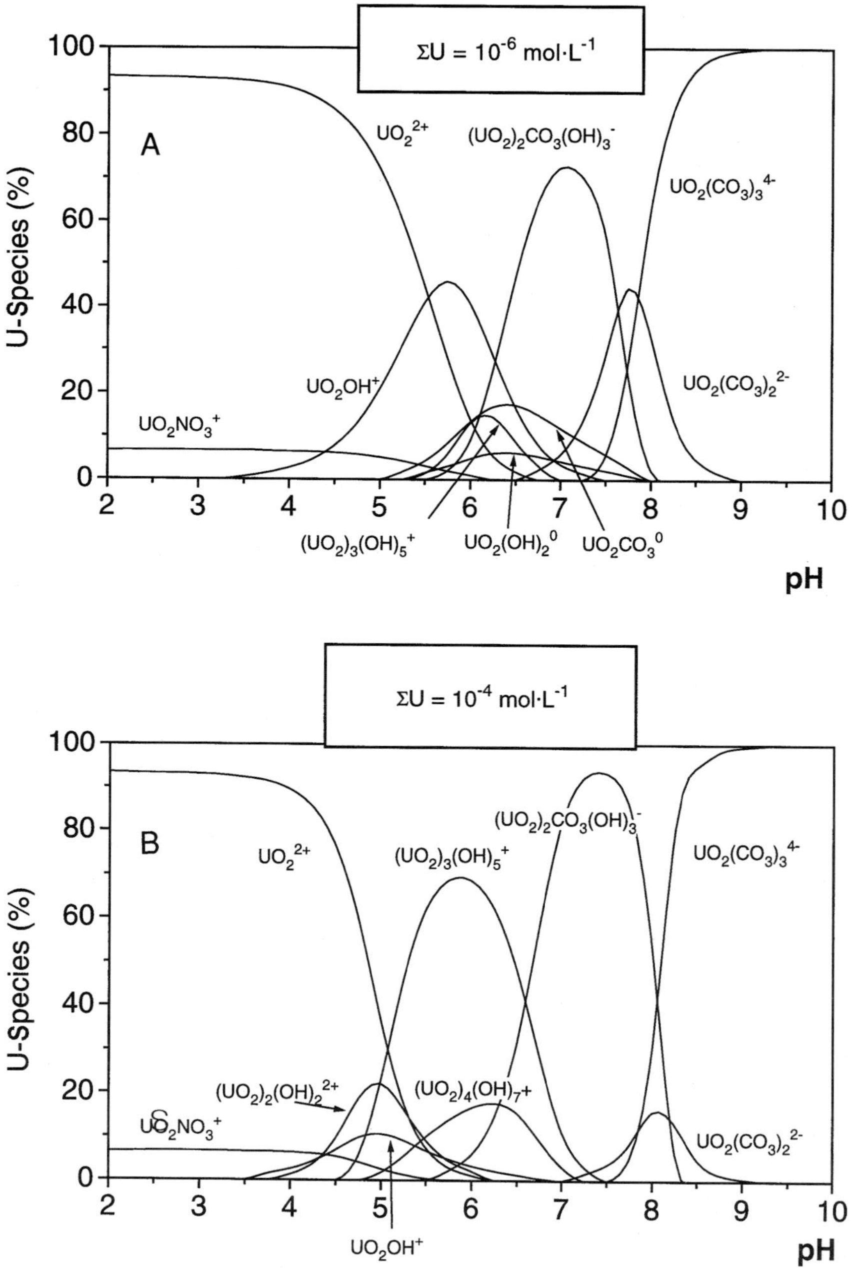

Figure 1 Uranium speciation in 0.1 $mol \cdot L^{-1}$ $NaNO_3$, at equilibrium with air ($P_{CO_2} = 10^{-3.5}$ atm): (A) Σ U = 10^{-6} $mol \cdot L^{-1}$; (B) Σ U = 10^{-4} $mol \cdot L^{-1}$ (precipitation excluded).

as the "low-pH adsorption edge." In the case of ferrihydrite, this occurs at pH 4 to 5, with the exact position depending on the experimental conditions (Hsi and Langmuir, 1985; Waite *et al.,* 1994). In systems containing carbonate, there is a marked decrease in sorption at high pH values, referred to as the "high-pH edge." The positions and shapes of these edges under a variety of experimental conditions are useful constraints on U-sorption models.

C. Surface Complexation Modeling

A surface complexation model for U adsorption on ferrihydrite was developed in a previous paper (Waite *et al.,* 1994). It was shown that in air-equilibrated systems at pH values below 7, U^{VI} uptake on ferrihydrite could be modeled adequately as the uptake of UO_2^{2+} on weak and strong bidentate surface sites (denoted as Fe_w and Fe_s, respectively), with the surface species being $(\equiv FeO_2)UO_2^0$, an inner-sphere, bidentate complex (reactions 5 and 6 in Table I). The existence of this species was supported by the results of EXAFS spectroscopy.

The suite of surface reactions includes protonation and deprotonation of surface sites (reactions 1 and 2 in Table I) together with adsorption of carbonate ions (reactions 3 and 4). For these reactions, strong and weak surface sites are assumed to have the same reaction constants.

A ternary uranyl–carbonate surface complex was included in the suite of surface reactions on both site types (reactions 7 and 8), in order to provide a satisfactory fit to the experimental sorption data under conditions of elevated P_{CO_2} and high pH values (Waite *et al.,* 1994). Similar uranyl–carbonate surface complexes

Table I

Ferrihydrite Surface Reactions

	Reaction	log K[a]
1	$\equiv FeOH^0 + H^+ = \equiv FeOH_2^+$	6.51
2	$\equiv FeOH^0 = \equiv FeO^- + H^+$	−9.13
3	$\equiv FeOH^0 + H_2CO_3^0 = \equiv FeCO_3H^0 + H_20$	2.90
4	$\equiv FeOH^0 + H_2CO_3^0 = \equiv FeCO_3^- + H_20 + H^+$	−5.09
5	$(\equiv Fe_s(OH)_2) + UO_2^{2+} = (\equiv Fe_sO_2)UO_2^0 + 2H^+$	−2.57
6	$(\equiv Fe_w(OH)_2) + UO_2^{2+} = (\equiv Fe_wO_2)UO_2^0 + 2H^+$	−6.28
7	$(\equiv Fe_s(OH)_2) + UO_2^{2+} + CO_3^{2-} = (\equiv Fe_sO_2)UO_2CO_3^{2-} + 2H^+$	3.67
8	$(\equiv Fe_w(OH)_2) + UO_2^{2+} + CO_3^{2-} = (\equiv Fe_wO_2)UO_2CO_3^{2-} + 2H^+$	−0.42

Note. From Waite *et al.* (1994). The total site density of ferrihydrite is taken to be 0.875 moles of sites per mole of Fe, with a strong site (Fe_s) density of 0.0018 moles sites per mole of Fe. The surface area is taken to be 600 $m^2 \cdot g^{-1}$.

[a]log K values are for $\boldsymbol{I} = 0.1$ (most experiments were carried out in 0.1 $mol \cdot L^{-1}$ $NaNO_3$).

have also been postulated in order to model the adsorption of U^{VI} on goethite (Hsi and Langmuir, 1985; Kohler *et al.*, 1992).

In this chapter, we will test the ability of the model containing the full suite of surface reactions in Table I to describe uranyl adsorption under a wide range of conditions. Calculations were carried out using MINTEQA2, which allows different stoichiometric coefficients in mass balance and mass action equations for surface complexes (mass balance equations with bidentate surface complexes require a stoichiometric coefficient of two for $\equiv$FeOH, although the corresponding mass action equation has a coefficient of one). The adsorption of U on kaolinite is not modeled, for reasons to be discussed.

III. EXPERIMENTAL

A. Materials

Ferrihydrite was precipitated by raising the pH of a Fe^{III}/HNO_3 solution to 6.0 and then aging for 65 hr at 25°C in a stirred vessel. The resulting precipitate is amorphous hydrous Fe oxide with a stoichiometry near $Fe_2O_3 \cdot H_2O$ and a surface area of 600 $m^2 \cdot g^{-1}$ (Waite *et al.*, 1994). Aliquots of the aged ferrihydrite were transferred to open polypropylene centrifuge tubes for sorption experiments. For most experiments, the amount of solid corresponded to a total Fe content (Σ Fe) of 1 $mmol \cdot L^{-1}$ (89 $mg \cdot L^{-1}$ as $Fe_2O_3 \cdot H_2O$). The sorption properties of ferrihydrite may change with time as it transforms to more crystalline phases such as goethite and hematite (Payne *et al.*, 1994). For this reason, fresh ferrihydrite was prepared for each experiment.

Kaolinite standards KGa-1 and KGa-1B were obtained from the Clay Minerals Society Source Clays Repository (Columbia, USA). KGa-1 has been extensively characterized (Van Olphen and Fripiat, 1979), and KGa-1B is a mineralogically similar sample which was selected to replace the exhausted stock of KGa-1 (Pruett and Webb, 1993). The mass loading of kaolinite in the experiments was usually 4 $g \cdot L^{-1}$. Prior to sorption experiments we established that leaching of natural U from these clay materials was insufficient to significantly increase the amount of U in any of the experiments.

Uranium sorption on mixed substrates containing both kaolinite and ferrihydrite was also studied. These substrates were prepared by raising the pH of an acidic Fe^{III} solution in the presence of a weighed amount of kaolinite (KGa-1). After an aging time of 65 hr, the mixed slurry was vigorously agitated and aliquots were transferred to centrifuge tubes for sorption experiments. The amounts of ferrihydrite and kaolinite were chosen so that the final suspension had the same mass loadings of these minerals as were present in other experiments (89 $mg \cdot L^{-1}$ and 4 $g \cdot L^{-1}$, respectively).

B. Procedure for Sorption Experiments

Batch experiments were carried out in polypropylene centrifuge tubes which had holes drilled in their lids in order to ensure equilibration with the atmosphere. An aliquot of the ferrihydrite slurry or a weighed amount of kaolinite was suspended in $NaNO_3$ (usually 0.1 mol·L^{-1}). For studies at pH > 7.0, sufficient $NaHCO_3$ was added to the system to achieve equilibrium with air at the desired pH. This ensured pH stability and fixed the carbonate content of the system. For some experiments, phosphate (as NaH_2PO_4) or humic acid (a filtered solution of Aldrich sodium humate) was added at this stage. The amounts of added phosphate and humic acid were 10^{-4} mol·L^{-1} and 9 mg·L^{-1}, respectively.

The U^{VI} was added after 24 hr "preequilibration" during which the system was allowed to come to equilibrium under the appropriate chemical conditions (pH, ionic strength, etc.). The contact time for U was 48 hr, with the pH being checked and adjusted if necessary after 24 hr. The equilibration time was selected after carrying out preliminary experiments in which the uptake of U was studied as a function of time (Waite *et al.,* 1994). These experiments showed that uptake of U was rapid during the first 24 hr, and was followed by a slower process, which probably involves diffusion into aggregates of particles.

The experimental tubes were gently shaken in a waterbath at 25°C. Immediately prior to sampling the pH was again determined and the aqueous phase was then separated by high-speed centrifugation (RCF $> 6500g$) for 30 min. Uranium concentrations in the supernate were determined using the Kinetic Phosphorescence Analyser (model KPA-10, Chemchek Instruments, Richland, USA), or by alpha spectrometry. Experiments at high P_{CO_2} were carried out in a glove box with a gas mixture of CO_2 (1%), N_2 (89%), and O_2 (10%). Further details of the techniques are available elsewhere (Waite *et al.,* 1994).

C. Container Materials

Trace metals and radionuclides tend to adsorb on any available surface. As a result, blank experiments carried out without a mineral sorbing phase present often indicate substantial uptake of metal on the container walls. Recent work with several plastic materials (polypropylene, Teflon, and polycarbonate) has shown that these materials can remove 50% or more of U^{VI} from solution, particularly at low total U concentrations (Pabalan *et al.,* 1994). Teflon vessels were not used for our work, because fluoride has been detected in experimental solutions contained in Teflon, even in nonirradiated Teflon at low temperatures (Solomah, 1983; Crine, 1987). This presumably results from leaching of F from the Teflon. If this occurred, complexation of U^{VI} with F would influence experimental results. Polypropylene centrifuge tubes were selected for the sorption experiments.

Blank experiments with polypropylene tubes in the absence of a sorbing solid showed that uptake on the vessel walls reached almost 25% at pH values between approximately 5 and 7 (Fig. 2). In comparison, U sorption on ferrihydrite (10^{-3} mol·L^{-1} as Fe) was close to 100% across this pH range. In order to determine the quantity of U adsorbed by the vessel when ferrihydrite was present, we carried out some experiments in which the tube contents were vigorously shaken and then decanted, removing both the solid and the liquid. The vessel walls were then leached with HNO_3 and the U content was determined. Iron was also measured in order to quantify the amount of U bound to ferrihydrite particles which were themselves attached to the vessel walls. It was concluded that direct uptake of U on the vessel walls was greatly reduced by the presence of ferrihydrite, and was probably less than 5% under the conditions of these experiments (Fig. 2).

Similar experiments to directly measure tube sorption with kaolinite present were unsuccessful because the kaolinite particles tended to aggregate and adhere to the vessel walls. However, wall sorption was expected to be a potential problem in experiments with kaolinite, which has a smaller specific surface area, and is a weaker U adsorber than ferrihydrite. Consequently, much higher mass loadings were used for kaolinite sorption experiments.

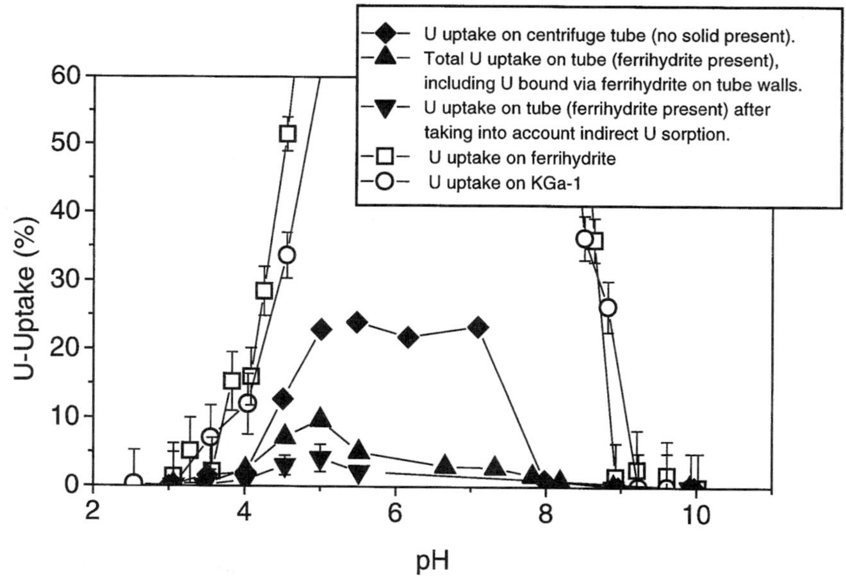

Figure 2 Uranium uptake on polypropylene centrifuge tubes (0.1 mol·L^{-1} $NaNO_3$, Σ U = 10^{-6} mol·L^{-1}) in the presence and absence of ferrihydrite (Σ Fe = 1 mmol·L^{-1}). Uptake curves of U on ferrihydrite (Σ Fe = 1 mmol·L^{-1}) and kaolinite KGa-1 (4 g·L^{-1}) are shown for comparison.

D. Analytical Electron Microscopy

The association of U with the mineral surfaces in the sorption experiments was studied using a JEOL 2000FX transmission electron microscope (TEM) equipped with a Tracor-Northern Si(Li) energy dispersive X-ray (EDX) spectrometer and a Link Isis analyzer (Oxford Instruments, High Wycombe, United Kingdom). Samples were prepared by evaporating a small droplet of clay suspension onto a holey carbon TEM grid. Samples which had been suspended in $NaNO_3$ were centrifuged and resuspended in Milli-Q water prior to the evaporation step, to reduce the amount of $NaNO_3$ on the grid. Analytical electron microscopy (AEM) was carried out with an operating voltage of 200 kV and a beam current of 110 μA. Analyses were acquired for 400 to 600 sec live time and reduced to elemental weight percentages using the Link Isis software package, TEMQuant. Experimental k-factors for the data reduction were obtained from a large suite of silicate mineral standards.

IV. URANIUM SORPTION ON MODEL MINERALS IN $NaNO_3/CO_2$ SYSTEMS

In this section, we discuss the effects of pH, Σ U, ionic strength, partial pressure of CO_2 (P_{CO_2}), and mass loading on U sorption by ferrihydrite and kaolinite. In terms of surface complexation studies, this means varying the aqueous components H^+, UO_2^{2+}, Na^+, NO_3^-, and CO_3^{2-}, and the quantity of sorbing minerals.

A. Effect of Total U

Figure 3 shows sorption curves for U on ferrihydrite at a range of Σ U from 10^{-6} to 10^{-4} mol·L^{-1}. The uptake of U in air-equilibrated systems is greatest at near-neutral pH values. The strong complexation of U^{VI} with carbonate in the solution phase (Fig. 1) greatly reduces the adsorption of U^{VI} at pH values above 8. Tripathi (1983) showed that there is no decrease in the extent of U^{VI} sorption at high pH values in systems which are free of CO_2.

The positions of the sorption edges in Figure 3 are dependent on Σ U. For a given pH, the percentage of U adsorbed decreases with increasing Σ U. This can be explained on the basis of the existence of a relatively small number of "strong" surface sites, and a larger number of "weak" sites. Such two-site models have been found to be applicable to the sorption of a range of metals on ferrihydrite (Dzombak and Morel, 1990). Based on the previous EXAFS study (Waite *et al.,* 1994), these sites were modeled as bidentate surface sites. The model curves shown in Figure 3 were calculated using MINTEQA2 and the reactions in Table I. There is generally good agreement between the model predictions and the experimental data, which cover a wide range of Σ U and pH.

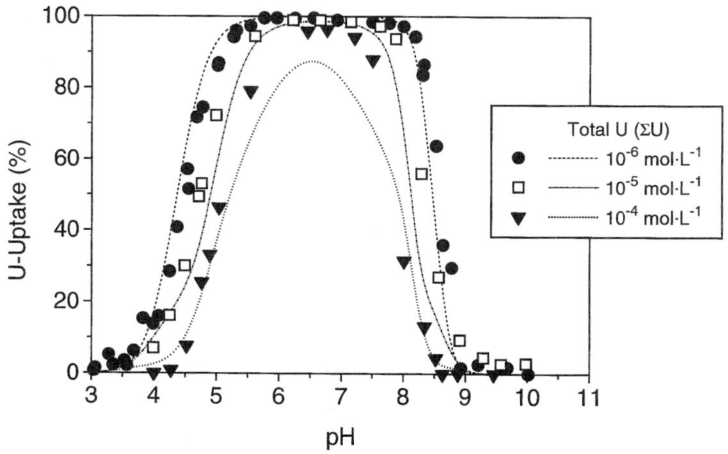

Figure 3 U^{VI} sorption on ferrihydrite for a range of Σ U. Curves were calculated using MINTEQA2, with the suite of surface reactions in Table I. (Experimental conditions: 0.1 mol·L^{-1} $NaNO_3$ in equilibrium with air, Σ Fe = 1 mmol·L^{-1}.)

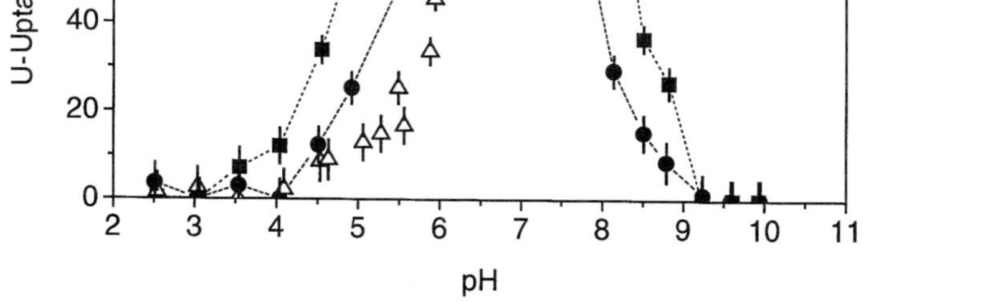

Figure 4 Uranium uptake on kaolinite KGa-1 (mass loading, 4 g·L^{-1}) for a range of Σ U, showing possible precipitation in experiments with Σ U = 10^{-4} mol·L^{-1}. Experiments were carried out in 0.1 mol·L^{-1} $NaNO_3$ in equilibrium with air.

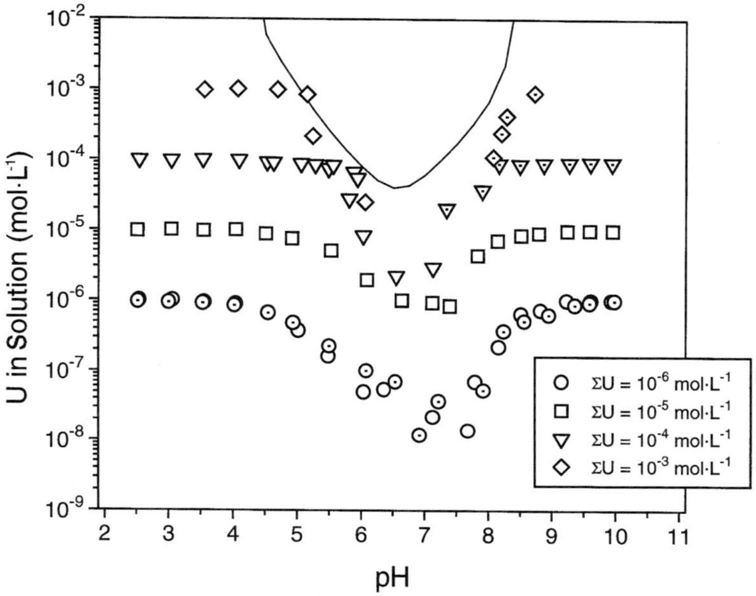

Figure 5 Total U in solution phase in experiments with kaolinite for Σ U ranging from 10^{-6} to 10^{-3} mol·L^{-1}. The aqueous phase was 0.1 mol·L^{-1} $NaNO_3$ in equilibrium with air. The curve was calculated from the solubility of uranyl hydroxide with log K_{so} ($UO_2(OH)_2$) = −22.43 (Tripathi, 1983). Data are for experiments with kaolinites KGa-1 (open symbols) and KGa-1B (dot center) with a mass loading of 4 g·L^{-1}.

Corresponding experimental data on the KGa-1 kaolinite are shown in Figure 4. The pH dependence (steepness of the edges) is not as great as for ferrihydrite, possibly indicating a different proton stoichiometry of adsorption, and there is a substantial shift in the position of the low-pH edge between the Σ U of 10^{-6} and that of 10^{-5} mol·L^{-1}. Uptake curves for 10^{-4} mol·L^{-1} Σ U have an entirely different pH dependence, with the pH edge being much steeper, suggesting precipitation may be important.

The precipitating solid in this type of experiment is unlikely to be a crystalline phase such as β-$UO_2(OH)_2$ or schoepite, and has previously been modeled as being an unspecified "uranyl hydroxide" (Tripathi, 1983). Tripathi defined a solubility product for this phase as $\{UO_2^{2+}\}\cdot\{OH^-\}^2$, with a value of log K_{so} = −22.43. When the experimental results are plotted as U in solution as a function of pH, it appears that the data are consistent with the solubility of this phase placing an upper limit on dissolved U (Fig. 5). Data for the higher Σ U tend to lie below (rather than on) the calculated solubility curve, suggesting that U concentrations are further reduced by adsorption.

B. Mass Loading, P_{CO_2} and Ionic Strength

As might be expected, the uptake of U on both kaolinite and ferrihydrite is a function of the mass loading of the solid in the system (Fig. 6). In the case of ferrihydrite, the model predictions provide a reasonable fit to the data, although slightly underpredicting adsorption at high pH values.

As already noted, carbonate is important in uranyl speciation (Fig. 1), so addi-

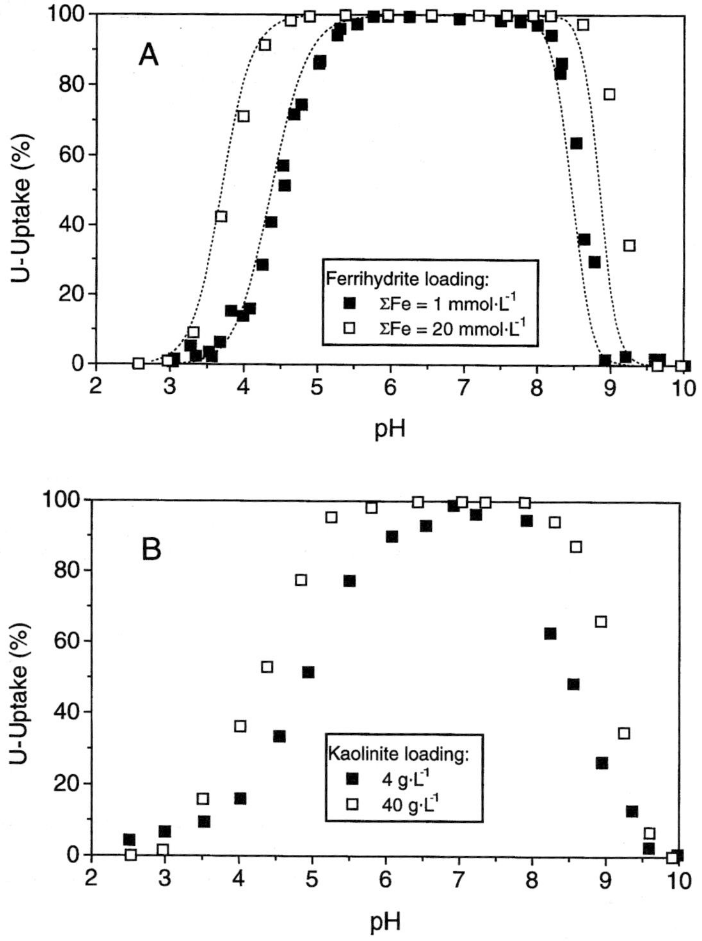

Figure 6 Effect of solid mass loading on U sorption: (A) ferrihydrite (curves calculated using reactions in Table I); (B) kaolinite (KGa-1B). The background electrolyte was 0.1 mol·L^{-1} $NaNO_3$ in equilibrium with air with $\Sigma U = 10^{-6}$ mol·L^{-1}.

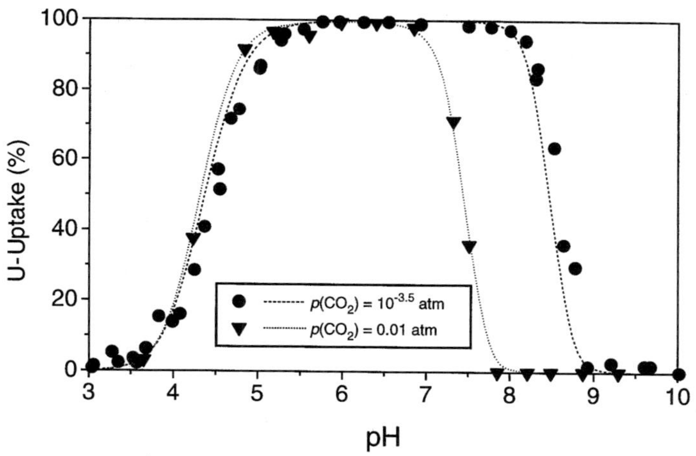

Figure 7 Uranium adsorption by ferrihydrite (Σ Fe = 1 mmol·L^{-1}) at two partial pressures of CO_2. Curves were calculated with MINTEQA2 using the reactions in Table I. Experiments were carried out in 0.1 mol·L^{-1} $NaNO_3$ with a Σ U of 10^{-6} mol·L^{-1}.

tional experiments were carried out in a glove box, with an elevated P_{CO_2} of 10^{-2} atm. The sorption of U on ferrihydrite in the high-pH range was significantly reduced at the higher partial pressure of CO_2, due to the dominance of weakly sorbing anionic uranyl–carbonate species. The model provides an excellent fit to the high-pH data (Fig. 7), and also predicts a slight increase in U uptake on the low-

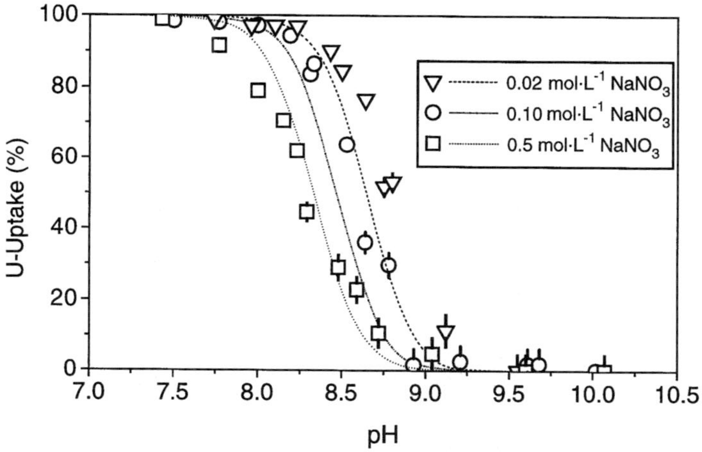

Figure 8 Effect of ionic strength on the high-pH U-sorption edge for ferrihydrite (Σ Fe = 1 mmol·L^{-1}, Σ U = 10^{-6} mol·L^{-1}). Curves were calculated using the reactions in Table I.

pH edge (due to ternary uranyl–carbonate surface complexes), which is consistent with the experimental results.

Ionic strength has a small effect on the low-pH edge, but at high pH values increasing ionic strength reduces U sorption on ferrihydrite (Fig. 8). Although an ionic strength dependence can be attributed to outer-sphere surface complexes, the diffuse double-layer model does not accommodate outer-sphere surface complex formation. However, even with only inner-sphere complexes included, the model correctly predicts a decrease in sorption with increasing ionic strength. This is because the activity of the highly charged carbonate species, which prevail in the high-pH range, is significantly affected by ionic strength. This influences the distribution of aqueous species, and in turn the uptake of U.

V. EFFECT OF TRACE IMPURITIES ON URANIUM SORPTION BY KAOLINITE

A. U Sorption in Mixed Ferrihydrite–Kaolinite Systems

Uranium sorption was studied on a mixed substrate containing kaolinite and ferrihydrite, with mass loadings of 4 $g \cdot L^{-1}$ and 89 $mg \cdot L^{-1}$, respectively. This was prepared following the procedure previously described. Under the experimental conditions, the ferrihydrite probably coats the surface of kaolinite, as has been observed in nature. The resulting solid had a slight pink tinge, whereas pure kaolinite is off-white in color.

The sorption results (Fig. 9) showed that the sorption properties of the mixed system were very similar to those of ferrihydrite, although it was the minor phase on a mass basis (about 2.2 wt %). This supports the hypothesis that strongly sorbing trace phases may dominate U sorption in the natural environment.

B. Analytical Electron Microscopy Results

The association of sorbed U with the various mineral surfaces was studied using AEM. A sample of kaolinite KGa-1, which had been used in an experiment with a Σ U of 10^{-4} $mol \cdot L^{-1}$, was found to consist of abundant kaolinite particles and a smaller proportion of Ti-rich particles. The AEM data for 25 kaolinite grains in this sample are summarized in Table II. The elemental results for Al and Si are similar to that expected for a pure kaolinite, and are close to reported values for the bulk composition of KGa-1. The average U content of the kaolinite particles examined was only 0.06%, which was much less than was expected on the basis

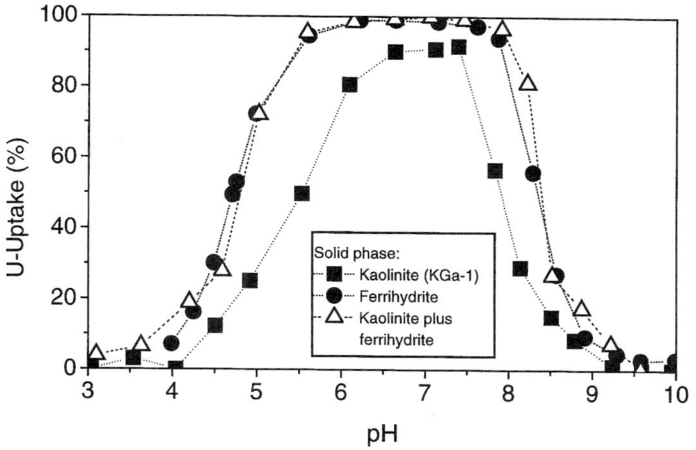

Figure 9 Adsorption of U on kaolinite (KGa-1, 4 g·L^{-1}) and ferrihydrite (Σ Fe = 1 mmol·L^{-1}) and a kaolinite–ferrihydrite mixture (Σ U = 10^{-5} mol·L^{-1}, 0.1 mol·L^{-1} $NaNO_3$ in equilibrium with air).

of the amount of U taken up by the solid in this experiment. Therefore, the kaolinite particles in this sample were only responsible for a small proportion of the U uptake.

The AEM results showed that separate Ti-rich particles were present, and these were identified by electron diffraction as being anatase and rutile. Anatase is the more abundant TiO_2 polymorph, and occurs as clusters consisting of grains generally less than 0.2 to 0.3 μm in size, and also as single crystals up to 0.5 μm in size. Rutile is less common, but several single crystals up to approximately 1 μm in length were observed. The presence of these phases is consistent with previous work on Georgia kaolinite (Weaver, 1976). The amount of U associated with Ti-rich particles was about 0.7%, more than 10 times higher than that associated with the kaolinite particles (Table II). Therefore, the Ti-rich particles, despite being less abundant than the kaolinite particles, were responsible for a substantial proportion of the total U adsorbed by the KGa-1 sample. We observed a significant negative correlation between the amount of U present and the grain size of the Ti-rich particles, indicating that U is preferentially adsorbed by the fine-grained anatase clusters.

Mass balance calculations showed that the anatase and kaolinite particles could not be responsible for all the loss of U from solution. This discrepancy was resolved when a grain containing predominantly U in the EDX spectrum was observed in the sample (Table II). This indicated that some precipitation had occurred in experiments with high Σ U on the KGa-1 sample, and was consistent with the solubility calculations (Section IV.A). Based on the presence of diffuse rings in its

Table II

Summary of AEM Data for Samples from U-Sorption Experiments (All Data Expressed as wt %)

		Kaolinite (KGa-1) with added U of 0.6 wt %			KGa-1 ferrihydrite (Fe of 1.4 wt %, added U of 0.06 wt %)	
Element	KGa-1 bulk composition[a]	Kaolinite grains (n = 25)	Ti-rich grains (n = 12)	U-rich grain (n = 1)	Kaolinite grains (n = 25)	Ti-rich grains (n = 6)
Al	21.0	20.7	1.23	1.35	20.6	0.85
Si	20.7	21.6	0.75	0.69	21.4	0.36
Ti	0.83	0.14	56.0	0.00	0.12	56.9
U	—	0.06	0.71	76.4	0.05	0.08
Fe	0.15	0.15	0.62	0.15	0.76	0.94

[a]From Van Olphen and Fripiat (1979).

diffraction pattern, the U-rich particle was predominantly amorphous. However, it had some crystallinity which may have been acquired by evaporation during the sample preparation, or by electron beam heating while under observation in the TEM. Crystalline spots in the diffraction pattern were consistent with a partially dehydrated form of schoepite.

We also studied mixed KGa-1–ferrihydrite samples with adsorbed U. Kaolinite and anatase particles were both examined, and the results showed that Fe was fairly uniformly distributed over the clay and anatase particles (Table II). The Fe content of the kaolinite grains averaged 0.76% compared to 0.15% in pure KGa-1. The Fe associated with the anatase increased slightly from 0.62 to 0.94%. Presumably the U was adsorbed on Fe coatings, since separate particles rich in Fe or U (or both) were not observed.

C. Implications

The influence of Ti-rich particles on the U-sorption behavior of KGa-1 may be important in any study in which KGa-1 or a similar kaolinite is used, and suggests that the uptake of U (and possibly other species) by KGa-1 should not be modeled on the basis of surface coordination by Al or Si surface sites alone. The much higher sorption on anatase should be taken into account. The greater Fe content of the anatase particles (relative to the kaolinite particles) may also be of significance (Table II). Clearly, minor phases such as anatase or ferrihydrite potentially play a major role in systems where much greater amounts of kaolinite are present.

In previous work, extractions of weathered natural samples with the DCB (dithionite–citrate–bicarbonate) reagent, which dissolves amorphous and crystalline iron minerals, removed less than 2% of the contained titanium (Payne *et al.*, 1994). Therefore, extraction techniques which aim to dissolve trace iron impurities may well be ineffective in dissolving anatase. Indeed, it was previously shown that extraction with the DCB reagent had no effect on U uptake by the KGa-1 kaolinite (Payne *et al.*, 1992). For the present study, we did not carry out chemical treatments to remove amorphous or trace mineral phases. The presence of chemically resistant anatase in the KGa-1 sample (and possibly other kaolinites) may well influence the results of sorption experiments, even after prior chemical extraction.

VI. EFFECT OF COMPLEXING LIGANDS

The experimental work just described shows that the sorption of U^{VI} in simple laboratory systems is influenced by a number of experimental parameters, including pH, ionic strength, Σ U, mass loading, and P_{CO_2}. Considering the complexity of natural systems, it would be expected that U adsorption in the environment would be influenced by additional factors, including the presence of inorganic and organic ligands in the aqueous phase.

In our initial studies of the importance of complexing ligands, we have investigated the effect of phosphate and humic acid on U uptake by ferrihydrite. These ligands were chosen because they may be important in the migration of U at the Koongarra uranium deposit in Northern Australia (Yanase *et al.*, 1995). Koongarra is a field site at which we are studying U migration in the natural environment.

A. Phosphate

Experiments with added phosphate (Σ PO_4 of 10^{-4} mol·L^{-1}) showed that the presence of phosphate substantially increased U uptake on ferrihydrite at low pH values (Fig. 10A). The adsorption of PO_4^{3-} was almost 100% for pH values up to about 6, with a gradual decrease above this pH (Fig. 10A).

In other experiments involving both U and phosphate, in which the amount of U present (Σ U) was varied, we found that adsorption of phosphate on ferrihydrite between pH values of 6 and 8 increased if a sufficient amount of U was present in the system (Payne *et al.*, 1996). Similar results have been found in studies of zinc and phosphate adsorption on goethite (Bolland *et al.*, 1977). These authors reported that "the presence of phosphate or zinc on the goethite surface was found to increase slightly the adsorption of zinc and phosphate respectively," and suggested the formation of a surface complex involving both Zn and phosphate ions. The formation of a ternary surface complex involving both U and phosphate on the ferrihydrite surface provides a possible explanation for our experimental results on ferrihydrite. However, we also considered the possibility that precipitation may be occurring in these experiments.

Calculations using MINTEQA2, together with the uranyl–phosphate complexes and solids from the NEA data base (Grenthe *et al.*, 1992), showed that precipitation of $(UO_2)_3(PO_4)_2 \cdot 4H_2O$ was possible between pH values of approximately 4.75 and 7.25 for systems with a Σ U of 10^{-6} mol·L^{-1} and a Σ PO_4 of 10^{-4} mol·L^{-1}, if no sorbing solids were present. When ferrihydrite was present (and the equilibrium concentrations of U and phosphate were consequently greatly reduced by sorption), no supersaturation was predicted. Figure 10A shows that the presence of phosphate increased U sorption in the pH range below 4.5, in which no precipitation is expected, even in the absence of ferrihydrite. Further evidence that precipitation of U^{VI}–phosphate phases is not responsible for removing U from solution in these experiments comes from the experiments with kaolinite (described below) in which much higher equilibrium concentrations of both U and phosphate were measured for pH values below 6. This would not be expected if precipitation was occurring in the experiments with ferrihydrite.

A ternary surface complex involving both U and phosphate on the ferrihydrite

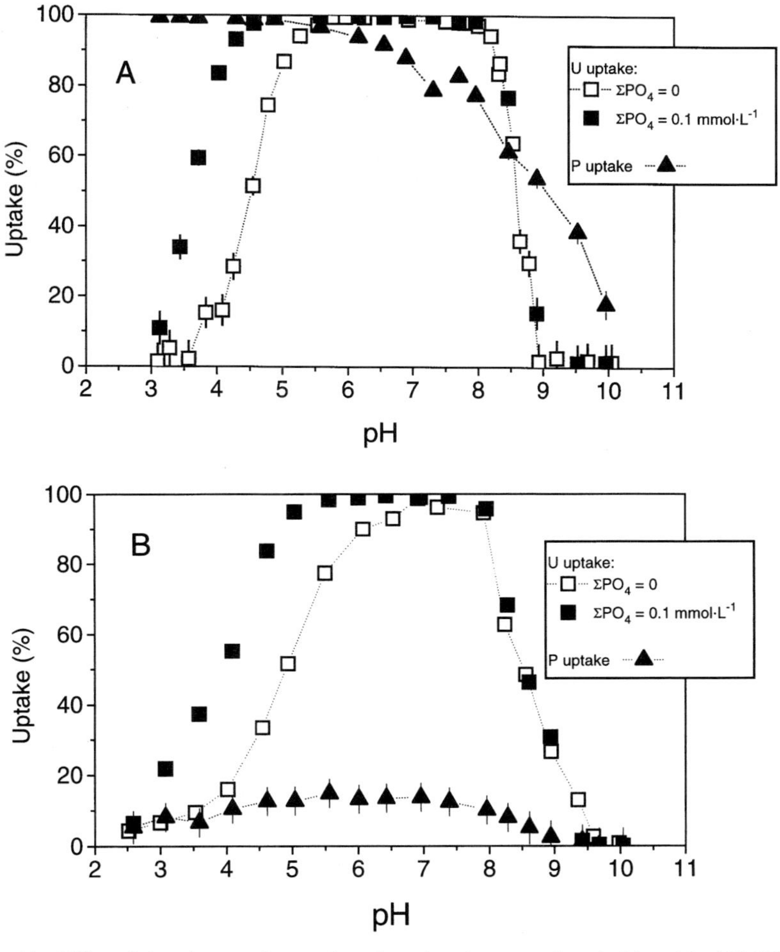

Figure 10 Effect of phosphate on the sorption of uranium (aqueous phase is 0.1 mol·L^{-1} $NaNO_3$ in equilibrium with air; total U = 10^{-6} mol·L^{-1}): (A) ferrihydrite (Σ Fe = 1 mmol·L^{-1}); (B) kaolinite KGa-1B (4 g·L^{-1}). The uptake of phosphate by the solid is also shown.

surface is therefore a reasonable explanation of our experimental results. However, another possible mechanism has been suggested by Benjamin and Bloom (1981), who studied adsorption of Zn on ferrihydrite in the presence of phosphate. Their experimental system contained equal amounts of phosphate and Fe (10^{-3} mol·L^{-1}), and it was proposed that a secondary surface phase formed in the presence of strongly sorbing anions (such as phosphate), and this phase adsorbed trace metals more strongly than the original ferrihydrite. Benjamin and Bloom (1981) hypothesized that adsorption was the dominant mechanism at low surface coverage and there was some surface density at which an iron–anion surface precipitate starts forming. The systems studied in our work contain an order of magnitude more Fe than phosphate and it appears that the uptake of phosphate could be explained by sorption alone. Nonetheless, surface precipitation of another solid phase (such as ferric phosphate) could influence the experimental results, particularly if it bound uranyl very strongly. The possibility of surface precipitation is discussed by Dzombak and Morel (1990), who included it in their overall SCM for adsorption at the ferrihydrite surface.

In the case of kaolinite, the low-pH U-sorption edge was displaced to more acidic pH values when phosphate was present, although only a relatively small proportion of phosphate was adsorbed (Fig. 10B). The presence of phosphate increased U sorption across a wide pH range from 3 to 7.5. In this case, a surface precipitate of Fe and PO_4^{3-} appears unlikely, and the increased U sorption may be attributable to the formation of a ternary surface complex. However, as already noted, the important role of trace impurities in the standard kaolinites complicates the interpretation of the kaolinite data.

B. Humic Acid

The effect of natural organic materials on the mobility of various pollutants, including actinides, has been studied by several workers. It has been suggested that the sorption behavior of oxide surfaces is modified when adsorption sites for trace metal ions become dominated by the functional groups of adsorbed humic compounds (Davis and Leckie, 1978). Uranium uptake on hematite was increased by the presence of humic acid (Ho and Miller, 1985). The formation of an organic coating increased the uptake of americium on silica and alumina at low pH values, whereas at higher pH values, humic substances were predominately in solution and prevented Am sorption through aqueous complex formation (Moulin and Ouzounian, 1992).

In our experiments, the addition of humic acid (9 $mg \cdot L^{-1}$) moved the low-pH U-sorption edge on ferrihydrite to the left (increased sorption), and had little effect on the high-pH edge (Fig. 11). These results are similar to the reported behavior of americium, and suggest that humic materials have a substantial influence on U^{VI} sorption by ferrihydrite. During the sorption experiments, over 80% of the humic acid was adsorbed to the ferrihydrite (Payne *et al.,* 1996), and the final concentrations of humic acid in solution (1 to 2 $mg \cdot L^{-1}$) were comparable to the amount of total organic carbon (TOC) reported for Koongarra groundwaters by Yanase *et al.* (1995). Consequently, it seems probable that organic matter has a significant affect on U sorption at the Koongarra field site, although only present at low concentrations in the groundwater. However, natural humic materials are dis-

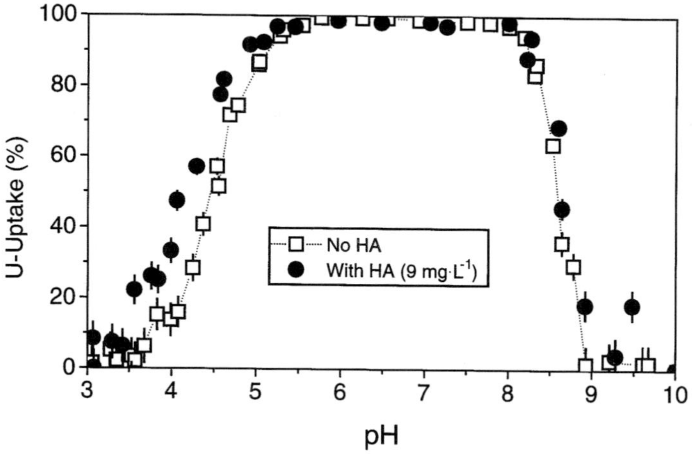

Figure 11 Effect of humic acid (HA) on the sorption of uranium by ferrihydrite (Σ Fe = 1 mmol·L^{-1}). Aqueous phase was 0.1 mol·L^{-1} $NaNO_3$ in equilibrium with air, with a total U of 10^{-6} mol·L^{-1}.

similar in many respects to commercial humic substances, such as the Aldrich humic acid used in this work (Malcolm and MacCarthy, 1986), and the natural organic materials present in Koongarra groundwaters have not been characterized in detail. Consequently, the effect of natural organic matter on U sorption in the Koongarra system is yet to be fully understood.

VII. SUMMARY

We have studied U sorption on ferrihydrite under a wide range of conditions. With increasing total U, the proportion of U sorbed became progressively less, indicating the presence of strong and weak sorption sites. At low total U the stronger sites predominate, leading to strong uptake. Ionic strength affected sorption, particularly at high pH values when highly charged carbonate complexes dominate the aqueous chemistry of U^{VI}. Adsorption of U was strongly dependent on the carbonate content of the system, and was influenced by the presence of humic acid and phosphate. The surface complexation model generally provides an excellent simulation of the U uptake data on ferrihydrite.

For a given set of experimental conditions, U uptake by kaolinite (expressed as U sorbed per gram of sorbent) is much less than for ferrihydrite. However, the amount of U adsorbed by kaolinite is similarly dependent on experimental variables such as pH, total U, and mass loading. Uranium sorption on model kaolinites appears to be strongly influenced by the presence of Ti-rich impurity phases (predominantly anatase), which occur in many natural kaolinites. Consequently, we have not yet modeled the U-sorption data for the kaolinite samples. Due to the weak sorption on kaolinite, precipitation of a U-rich phase occurred in some experiments. The removal of impurities from model kaolinites by chemical extractions is not straightforward, as the anatase is chemically resistant. A small amount of a ferrihydrite phase added to the model kaolinite completely modified its behavior, further demonstrating the importance of trace impurity phases in the natural environment.

ACKNOWLEDGMENTS

We thank Dr. J. A. Davis (United States Geological Survey) for generously providing advice and encouragement throughout the uranium sorption study. We also thank the United States Nuclear Regulatory Commission for financially supporting much of this work. This study was carried out as part of the ASARR (Analogue Studies in the Alligator Rivers Region) project.

REFERENCES

Allison, J. D., Brown, D. S., and Novo-Gradec, K. J. 1990. "MINTEQA2, a Geochemical Assessment Model for Environmental Systems." USEPA, Athens.

Benjamin, M. M., and Bloom, N. S. 1981. Effects of Strong Binding of Anionic Adsorbates on Adsorption of Trace Metals on Amorphous Iron Oxyhydroxide. *In* "Adsorption from Aqueous Solution" (P. H. Tewari, Ed.), pp. 41–60. Plenum Press, New York.

Bolland, M. D. A., Posner, A. M., and Quirk, J. P. 1977. Zinc Adsorption by Goethite in the Absence and Presence of Phosphate. *Aust. J. Soil Res.* 15:279–286.

Crine, J-P. 1987. Ionic Impurities in Water Contained in Plastic and Pyrex Bottles. *Polym. Eng. Sci.* 27:611–614.

Davis, J. A., and Kent, D. B. 1990. Surface Complexation Modeling in Aqueous Geochemistry. *Rev. Mineral.* 23:177–260.

Davis, J. A., and Leckie, J. O. 1978. Effect of Adsorbed Complexing Ligands on Trace Metal Uptake by Hydrous Oxides. *Environ. Sci. Technol.* 12:1309–1315.

Dzombak, D. A., and Morel, F. M. M. 1990. "Surface Complexation Modeling, Hydrous Ferric Oxide." John Wiley, New York.

Grenthe, I., Fuger, J., Konings, R. J. M., Lemire, R. J., Muller, A. B., Nguyen-Trung, C., and Wanner, H. 1992. "Chemical Thermodynamics of Uranium." Elsevier, Amsterdam.

Herbelin, A., and Westall, J. 1994. "FITEQL—A Computer Program for Determination of Chemical Equilibrium Constants from Experimental Data." Oregon State University, Corvallis.

Ho, C. H., and Miller, N. H. 1985. Effect of Humic Acid on Uranium Uptake by Hematite Particles. *J. Colloid Interface Sci.* 106:281–288.

Hsi, C-K. D., and Langmuir, D. 1985. Adsorption of Uranyl onto Ferric Oxyhydroxides: Application of the Surface Complexation Site-Binding Model. *Geochim. Cosmochim. Acta* 49:1931–1941.

Kohler, M., Wieland, E., and Leckie, J. O. 1992. Metal-Ligand-Surface Interactions during Sorption of Uranyl and Neptunyl on Oxides and Silicates. *In* "Water-Rock Interaction" (Y. K. Kharaka and A. S. Maest, Eds.), pp. 51–54. Balkema, Rotterdam.

Malcolm, R. L., and MacCarthy, P. 1986. Limitations in the Use of Commercial Humic Acids in Water and Soil Research. *Environ. Sci. Technol.* 20:904–911.

Moulin, V., and Ouzounian, G. 1992. Role of Colloids and Humic Substances in the Transport of Radio-elements through the Geosphere. *Appl. Geochem. Suppl.* 1:179–186.

Pabalan, R. T., Turner, D. R., and Bertetti, F. P. 1994. Sorption Modelling for HLW Performance Assessment. *In* "NRC High-Level Radioactive Waste Research at CNWRA, January–June 1994" (B. Sagar, Ed.), pp. 77–96. Center for Nuclear Waste Regulatory Analyses, San Antonio.

Payne, T. E., Davis, J. A., and Waite, T. D. 1994. Uranium Retention by Weathered Schists—The Role of Iron Minerals. *Radiochim. Acta* 66/67:297–303.

Payne, T. E., Davis, J. A., and Waite, T. D. 1996. Uranium Adsorption on Ferrihydrite—Effects of Phosphate and Humic Acid. *Radiochim. Acta* 74:239–243.

Payne, T. E., Sekine, K., Davis, J. A., and Waite, T. D. 1992. Modelling of Uranium Sorption Processes in the Weathered Zone of the Koongarra Ore Body. *In* "Alligator Rivers Analogue Project, 3rd Annual Report" (P. Duerden, Ed.), pp. 57–86. Australian Nuclear Science and Technology Organisation, Menai, Australia.

Pruett, R. J., and Webb, H. L. 1993. Sampling and Analysis of KGa-1B Well Crystallized Kaolin Source Clay. *Clays Clay Miner.* 41:514–519.

Solomah, A. G. 1983. Effects of Temperature on the Leaching Behavior of Sintered Modified Synroc-B Waste Form. *Nucl. Technol.* 62:311–316.

Tripathi, V. S. 1983. Uranium (VI) Transport Modelling: Geochemical Data and Sub-models. Ph.D. Thesis, Stanford University.

Van Olphen, H., and Fripiat, J. J. 1979. "Data Handbook for Clay Materials and Other Non-metallic Minerals." Pergamon, Oxford.

Waite, T. D., Davis, J. A., Payne, T. E., Waychunas, G. A., and Xu, N. 1994. Uranium(VI) Adsorption to Ferrihydrite: Application of a Surface Complexation Model. *Geochim. Cosmochim. Acta* 58:5465–5478.

Waychunas, G. A., Rea, B. A., Fuller, C. C., and Davis, J. A. 1993. Surface Chemistry of Ferrihydrite: Part I. EXAFS Studies of the Geometry of Co-precipitated and Adsorbed Arsenate. *Geochim. Cosmochim. Acta* 57:2251–2269.

Weaver, C. E. 1976. The Nature of TiO_2 in Kaolinite. *Clays Clay Miner.* 24:215–218.

Yanase, N., Payne, T. E., and Sekine, K. 1995. Groundwater Geochemistry in the Koongarra Ore Deposit, Australia (I): Implications for Uranium Migration. *Geochem. J.* 29:1–29.

Chapter 3

Uranium^{VI} Sorption onto Selected Mineral Surfaces

Key Geochemical Parameters

Roberto T. Pabalan, David R. Turner, F. Paul Bertetti, and James D. Prikryl
Center for Nuclear Waste Regulatory Analyses, San Antonio, Texas

The description and prediction of actinide retardation and transport in groundwater systems are complicated by the dependence of sorption processes on various geochemical parameters, including aqueous solution properties and sorptive phase characteristics, as well as temperature and Eh condition. To determine which parameters are key to understanding actinide sorption behavior, experiments were conducted to study U^{VI} sorption on quartz, clinoptilolite, montmorillonite, and α-alumina, which are sorbents of distinct mineralogic and surface characteristics, as a function of pH, U^{VI} concentration, ionic strength, P_{CO_2}, and the solid-mass to solution-volume ratio (M/V). The results from this study and from the literature show that U^{VI} sorption is important under conditions in which U^{VI} hydroxy complexes are favored to form in the aqueous phase. The similarity in the pH dependence of U^{VI} sorption on quartz, α-alumina, clinoptilolite, montmorillonite, amorphous silica, kaolinite, and hydrous titanium oxide suggests that U^{VI} sorption is not sensitive to the surface charge characteristics of the sorbent as compared to the effect of changing the number of available sorption sites. The magnitudes of U^{VI} sorption (at a specific pH, initial U concentration, and P_{CO_2}) are essentially the same for the different minerals if normalized to an "effective" surface area, and changing M/V has little influence on U^{VI} K_d, except at very low values. Ionic strength effects are limited for U^{VI} surface complexation reactions, but can be important if ion exchange is the predominant sorption mecha-

Adsorption of Metals by Geomedia

nism and can indirectly influence U^{VI} sorption behavior by affecting the activity of aqueous U^{VI} complexes.

Sorption modeling based on a diffuse layer model (DLM) and parameterized using subsets of the experimental data effectively simulates the complex sorption behavior observed in the U^{VI}–H_2O–CO_2–sorbent system. U^{VI} sorption modeling results are presented for quartz and montmorillonite which show that DLM-predicted values agree well with experimental data. The good agreement indicates that the surface complexation approach holds promise for developing models which can be used to extrapolate radionuclide sorption behavior beyond laboratory conditions and for quantitative assessment of the effects of various geochemical parameters on radionuclide retardation and transport.

I. INTRODUCTION

A fundamental concern in safety assessments of nuclear waste repositories and in nuclear waste management is the potential release of radionuclides, particularly actinides such as U, Np, and Pu, to the accessible environment as dissolved constituents in groundwater. An important mechanism for retarding radionuclide migration is sorption onto minerals present along groundwater flow paths. Thus, a quantitative understanding of actinide sorption behavior is important in evaluating the suitability of proposed geologic repositories for nuclear wastes. However, this understanding is complicated by the possible dependence of sorption processes on various geochemical parameters, including aqueous solution properties (e.g., pH, ionic strength, radionuclide concentration, and complexing ligands) and sorptive phase characteristics (e.g., composition, surface area, sorption site density, and surface charge), as well as temperature and Eh conditions. The dependence of sorption on various parameters makes it difficult to describe and predict actinide retardation and transport in geochemical systems of variable and composite mineralogic composition and changing aqueous speciation.

In this study, batch experiments were conducted to investigate the sorption of U^{VI} on the minerals quartz, clinoptilolite, montmorillonite, and α-alumina over wide ranges of experimental conditions. These minerals were selected because their mineralogic and surface characteristics, which could potentially influence the sorption behavior of U^{VI}, are very distinct from each other. For example, the reported point-of-zero-charge (pH_{pzc}) for quartz, clinoptilolite, and α-alumina are 2.9, 3.0, and 9.1, respectively (Davis and Kent, 1990; Gainer, 1990), and pH_{pzc} values reported for montmorillonite are in the range 6.5 to 8, depending on the ionic strength of the background electrolyte (Wanner *et al.*, 1994). In addition, quartz, clinoptilolite, and montmorillonite are important mineral phases at the proposed high-level nuclear waste repository at Yucca Mountain, Nevada (e.g., Bish and Chipera, 1989). The experiments were designed to evaluate the possible effect on U^{VI} sorption of pH, U concentration, P_{CO_2}, ionic strength, and solid-mass to solution-volume ratio (M/V). Results of these experiments, as well as data from published literature, were used in determining which of the various geochemical parameters are key to understanding U^{VI} sorption behavior. The data were also used to develop and parameterize surface complexation models for describing and predicting U^{VI} sorption onto mineral substrates.

II. EXPERIMENTAL PROCEDURE

A. Mineral Preparation

Minerals were acquired from commercial sources and, in most cases, were physically and chemically pretreated to isolate particular size fractions and to remove unwanted mineral impurities. Clinoptilolite was obtained by purification of a clinoptilolite-rich tuff from Death Valley Junction, California, purchased from Minerals Research Co. (Clarkson, New York). Powdered material was prepared by first using a rock hammer to break the as-received tuff samples into pieces less than one-third of an inch in diameter. These small pieces were then ground in a Spex #8000 Mixer/Mill using a tungsten carbide vial and sieved using a Ro-Tap sieve shaker and 8-in.-diameter stainless-steel sieves to obtain material in the 100 to 200 mesh (150 to 75 μm) size range. This material was first treated with buffered (pH 5.0) acetic acid solution to dissolve any carbonate minerals. Other mineral impurities were then separated from clinoptilolite by density separation using heavy liquid mixtures of tetrabromoethane and *N,N*-dimethyl formamide (Hutchison, 1974). The clinoptilolite was washed several times with acetone with the aid of an ultrasonic cleaner, and was rinsed with deionized water. Subsequently, iron oxide minerals were dissolved using a sodium citrate–dithionite–bicarbonate mixture (Mehra and Jackson, 1960). The purified clinoptilolite was converted to Na-form by ion exchange with 3 $mol \cdot L^{-1}$ NaCl solutions at 90°C for about 2 weeks. Details of the pretreatment of clinoptilolite can be found in Pabalan (1994).

Quartz was obtained as quartz sand (Wedron #510) from Wedron Silica Co. (Wedron, Illinois). The quartz sand was sieved to isolate the 60 to 100 mesh fraction (250 to 150 μm). Although the sand's powder X-ray diffraction (XRD) pattern exhibited no nonquartz peaks, transmitted and reflected light microscopy showed that minor mineral impurities ($<1\%$), predominantly Fe-oxide grain coatings or pyrite inclusions, are present. Soluble salts and carbonates were removed by washing the sand in deionized water and in buffered (pH 5.0) acetic acid solution. The sand was washed in sodium citrate–dithionite–bicarbonate solution to remove iron oxides. Other minerals and grains (e.g., those with pyrite inclusions) with densities above that of quartz (2.65 g·cm^{-3}) were then removed by density separation using heavy liquid (Na-polytungstate, Geoliquids). Details of preparation of the quartz material are described in Bertetti *et al.* (in preparation).

Montmorillonite (SAz-1, Apache County, Arizona) was obtained from the Source Clays Mineral Repository (Columbia, Missouri). The less than 2-μm size fraction, which was used in the experiments, was separated and collected by centrifugation of several batches of clay suspensions. The originally Ca-form montmorillonite was converted into Na-form by ion exchange with 2 mol·L^{-1} NaCl solutions. The clay material was subsequently lyophilized for storage prior to use in the experiments. Preparation of the montmorillonite material is described in Pabalan and Turner (1997).

The α-alumina (α-Al_2O_3) material was obtained from the National Institute of Standards and Technology (NIST), which issues the material as certified reference materials for measurements of specific surface area of powders. Reference materials (RM) 8006 and 8007, with reported surface areas of 0.229 and 0.0686 mg^2·g^{-1}, respectively, were used in the sorption experiments. To minimize altering the surfaces of the α-alumina, the powders were not washed or pretreated in any way prior to their use.

The external surface areas of the sorbent materials were determined using a multipoint N_2-BET isotherm measured with a Coulter SA3100 surface area analyzer. The surface area samples were prepared by outgassing at 350°C (100°C for montmorillonite) for 24 hr. Replicate analyses gave values of 0.03 ± 0.01, 97 ± 2, and 10.1 ± 0.3 mg^2·g^{-1}, respectively, for the quartz, montmorillonite, and clinoptilolite. Measured values for the α-alumina RM 8006 and 8007 are 0.23 ± 0.01 and 0.083 ± 0.005 mg^2·g^{-1}, respectively; the latter value is 21% greater than the NIST certified value. The purity of the mineral sorbents was checked by powder XRD analysis, scanning electron microscopy/energy dispersive spectrometry (SEM/EDS), and inductively coupled plasma (ICP) emission spectrometry.

B. Uranium Solutions

Uranium^{VI} experimental solutions were prepared by dilution of a purchased ^{233}U standard solution (Isotope Products, Inc., Burbank, California) consisting of 99.5% ^{233}U by mass of U. A 2.15·10^{-6} mol·L^{-1} U^{VI} stock solution was prepared initially, and stock solutions of lower concentrations were prepared by dilution. The stock solutions were sampled immediately before initiation of the experiments to determine the initial U concentration of the experimental solutions.

Uranium was analyzed by measurement of ^{233}U α-decay using liquid scintillation counting in a Packard 1900TR or 2505 TR/AB liquid scintillation analyzer

(LSA). Prior to counting, duplicate 0.5-ml aliquots taken from experimental solutions were mixed with 0.5 ml of 0.02 m HNO_3 solution and 5 ml of Ultima-Gold (Packard) scintillation cocktail in 7-ml glass vials. The samples were acidified in order to minimize U sorption onto the glass LSA vials, which could affect counting results. The counting efficiency for the LSA procedure used is at or very near to 100% for α-particles, although the energy for the counting region of interest is quenched to 100 to 350 keV. Because the original standard solution is radiochemically pure, the contribution to the total activity of the sample from other α- or β-emitting U isotopes and decay daughters is less than 0.1% within the counting region of interest and was neglected in the calculations.

Each sample was counted for a period of time such that the 2σ error of the reported sample activity in counts per minute (cpm) was ±3% for experimental solutions with initial U concentrations of $2 \cdot 10^{-7}$ mol·L^{-1} or greater, or ±5% for solutions which initially had less than $2 \cdot 10^{-7}$ mol·L^{-1} U. For calculation purposes, raw data in counts per minute, which in this case are equivalent to decays per minute, were converted into concentration units. Uranium concentrations were subsequently converted into mass (g) of U using the measured weight of solution. Uncertainties in sorption data presented in later sections were propagated based on the 2σ counting errors.

The pH of the U solutions was measured using a Ross combination pH electrode and an Orion 920A pH/ISE/mV/°C meter.

C. Batch Experiments

Equilibrium batch experiments were conducted at room temperature (20 ± 2°C) by reacting weighted amounts of the sorbent minerals with weighed quantities of U^{VI} solutions in Teflon–FEP (fluorinated ethylene propylene) centrifuge tubes or in polycarbonate bottles. All experimental mixtures were agitated using gyratory shakers. Most of the experiments were conducted under conditions in equilibrium with atmospheric CO_2. One set of experiments involving clinoptilolite was done in a controlled-atmosphere glove box at an elevated P_{CO_2} (10^{-2} atm) to investigate the effect of carbonate complexation on U^{VI} sorption. Most of the U^{VI} solutions had a matrix of 0.1 mol·L^{-1} $NaNO_3$ to keep the ionic strength relatively constant. One set of experiments with clinoptilolite used a 1.0 mol·L^{-1} $NaNO_3$ solution to determine the possible effect of ionic strength on U^{VI} sorption. The initial conditions of the various experiments are summarized in Table I.

The sorption experiments were started by adding ^{233}U solution to each Teflon or polycarbonate container. The initial pH of each solution was adjusted to a value in the range of 2.0 to 9.0 at approximately 0.25 pH intervals by addition of HNO_3 or $NaHCO_3$ solution. The amount of reagent needed to achieve the desired initial pH of the U^{VI} solutions was estimated using the EQ3NR geochemical code

Table I

Summary of U^{VI} Sorption Experimental Conditions[a]

Experiment	Mineral	Initial U Concentration ($mol \cdot L^{-1}$)	M/V ($g \cdot L^{-1}$)
A1	α-Alumina (NIST RM 8007)	$5.00 \cdot 10^{-7}$	2.88
A2	α-Alumina (NIST RM 8006)	$4.84 \cdot 10^{-7}$	2.79
C1	Na-clinoptilolite	$2.17 \cdot 10^{-7}$	2.04
C2	Na-clinoptilolite	$2.22 \cdot 10^{-6}$	2.09
C3	Na-clinoptilolite	$1.90 \cdot 10^{-8}$	2.05
C4	Na-clinoptilolite	$2.17 \cdot 10^{-7}$	2.07
C5	Na-clinoptilolite	$2.10 \cdot 10^{-7}$	20.28
C6	Na-clinoptilolite	$2.10 \cdot 10^{-7}$	2.43
M1	Na-montmorillonite	$2.45 \cdot 10^{-7}$	3.2
M2	Na-montmorillonite	$2.06 \cdot 10^{-7}$	0.27
M3	Na-montmorillonite	$2.10 \cdot 10^{-7}$	0.028
M4	Na-montmorillonite	$2.16 \cdot 10^{-6}$	0.28
Q1	Quartz	$2.14 \cdot 10^{-7}$	2
Q2	Quartz	$2.11 \cdot 10^{-7}$	20
Q3	Quartz	$2.06 \cdot 10^{-7}$	50
Q4	Quartz	$2.00 \cdot 10^{-8}$	20
Q5	Quartz	$2.15 \cdot 10^{-6}$	50

[a]The experiments were conducted over a pH range of 2 to 9 and with a 0.1 $mol \cdot L^{-1}$ $NaNO_3$ matrix, except for C4, which used a 1.0 $mol \cdot L^{-1}$ $NaNO_3$ matrix. All the U solutions were maintained in equilibrium with atmospheric P_{CO_2} ($10^{-3.5}$atm), except for the C6 solutions, which were equilibrated with $10^{-2.0}$ atm P_{CO_2}. The amounts of U sorbed in the A, C, and M experiments were calculated from initial and final U solution concentrations (corrected for container sorption), whereas the amounts of U sorbed in the Q experiments were calculated from the amount desorbed from the solid with HNO_3 solution. Additional data were also acquired from a desorption step for experiments C1, C3, C5, M2, and M3.

(version 7) with the database Data0.com.R12 (Wolery, 1992). The mineral sorbents were then added to each solution and the mixtures allowed to equilibrate for at least 10 or 14 days. Kinetics experiments indicated that 10 days were sufficient to reach sorption equilibrium in mixtures with quartz, α-alumina, and montmorillonite (e.g., Prikryl *et al.,* 1994; Bertetti *et al.,* in preparation), and 14 days were sufficient to reach equilibrium in mixtures with clinoptilolite (Pabalan *et al.,* 1993). After equilibrium has been achieved, the final pH of each solution was measured and the amount of U^{VI} sorbed on the mineral was determined either from the difference in initial and final U concentration in solution or by desorbing the U from the solid phase with 0.1 $mol \cdot L^{-1}$ HNO_3 (Table I). In many cases the former method required corrections to be made due to container losses, which are strongly dependent on pH, on the amount of solid added, and on the relative affinity of the sorbent for U^{VI}. Detailed descriptions of the experimental procedures can be found in Pabalan and Turner (1997) and in Bertetti *et al.* (in preparation).

III. EXPERIMENTAL RESULTS AND DISCUSSION

The results of selected sets of experiments are plotted in the figures as a function of pH. Although U^{VI} sorption data are typically plotted in terms of percent U sorbed, a useful representation is in terms of a distribution coefficient, K_d, which may be defined as

$$K_d\,(\text{ml}\cdot\text{g}^{-1}) = \frac{\text{equilibrium amount of U sorbed}}{\text{equilibrium amount of U in solution}} \times \left(\frac{V}{M}\right), \qquad [1]$$

where V is the volume of experimental solution (ml) and M is the mass of solid (g). The use of K_d provides a means of normalizing sorption results with respect to the sorbent concentration (or M/V ratio) and of taking into account the decrease in solution concentration of the radionuclide due to sorption.

Figure 1 shows some of the U^{VI} sorption data on quartz, α-alumina, clinoptilolite, and montmorillonite as a function of pH. The data in the figure demonstrate that U^{VI} sorption on these minerals is strongly affected by solution pH. Although the minerals used in the experiments have different mineralogic and surface properties, U^{VI} sorption on these minerals is similar with respect to dependence on pH. In all cases, U sorption is at a maximum at near-neutral pH (~6.0 to ~6.8) and decreases sharply toward more acidic or more alkaline conditions.

To aid in evaluating possible surface species, it is useful to compare U^{VI} sorption behavior with U^{VI} aqueous speciation. Figure 2 shows the relative stabilities of the aqueous U^{VI} species as a function of pH for a $2.1\cdot10^{-7}$ mol·L^{-1} U^{VI} solution (0.1 mol·L^{-1} $NaNO_3$ matrix) in equilibrium with atmospheric CO_2(g) ($P\text{CO}_2 = 10^{-3.5}$ atm). The U^{VI} speciation was calculated using the MINTEQA2 geochemical code (Allison *et al.,* 1991) and stability constants of U^{VI} aqueous species taken from the Nuclear Energy Agency (NEA) database (Wanner and Forest, 1992), with the exception of the value for $UO_2(OH)_2^0$(aq) species. The log K value used for $UO_2(OH)_2^0$(aq) and taken from Fuger (1992) is −13.0, compared with an upper limit of −10.3 recommended in Wanner and Forest (1992). The thick curve in Figure 2 represents the sum of the molalities of the monomeric U^{VI} hydroxy complexes.

A comparison of Figures 1 and 2 indicates a close correspondence between the pH dependence of U^{VI} sorption and the predominance field of the U^{VI} hydroxy complexes. Uranium^{VI} sorption occurs in the pH range where the U^{VI} hydroxy

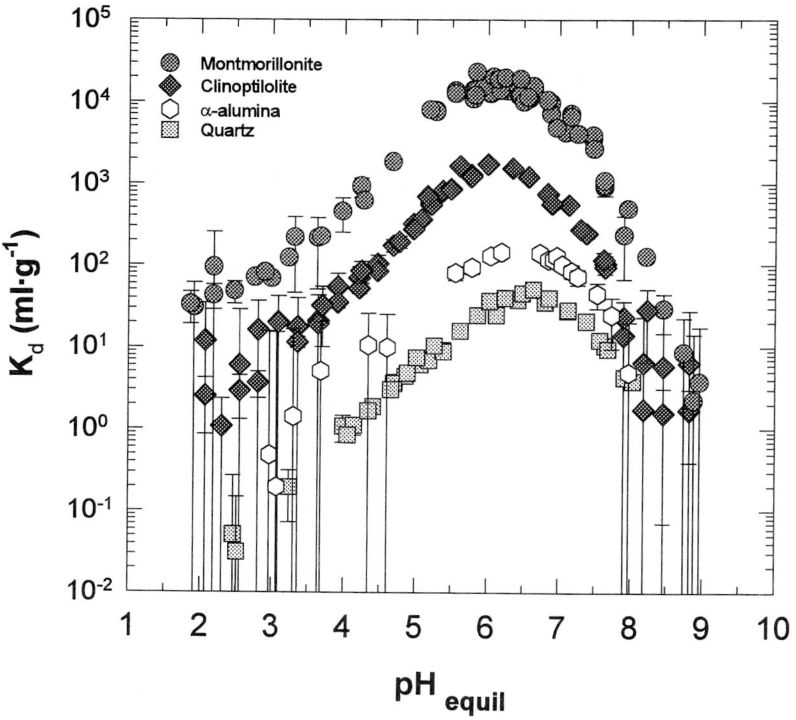

Figure 1 Uranium^{VI} sorption on montmorillonite, clinoptilolite, α-alumina, and quartz as a function of pH. Data are from experiments M1, M2, C1, A2, Q2, and Q3, which were all conducted under conditions in equilibrium with atmospheric P_{CO_2}. Error bars indicate 2σ uncertainties associated with the sorption data.

complexes are important. The decrease in the amount of U^{VI} sorbed at alkaline pH can be related to the increasing importance of aqueous U^{VI} carbonate complexes with increasing pH. In carbonate-free systems, decreasing U^{VI} sorption with increasing pH is generally not observed (e.g., Tripathi, 1984; Allard *et al.,* 1980; Hsi and Langmuir, 1985; McKinley *et al.,* 1995; Turner *et al.,* 1996).

At low pH values where the uranyl cation UO_2^{2+} is predominant, U^{VI} sorption is weak for the minerals and experimental conditions used in this study. However, other studies have shown that U^{VI} sorption through an ion exchange mechanism can be important for cation exchangers such as montmorillonite and, to a lesser degree, for clinoptilolite at low pH and low ionic strength (e.g., Zachara and McKinley, 1993; McKinley *et al.,* 1995; Andreeva and Chernyavskaya, 1982; Pabalan *et al.,* 1993). Thus, it appears that the relatively high Na^+ concentration of the U solutions (0.1 mol·L^{-1} $NaNO_3$ matrix) used in this study suppressed ion exchange between the UO_2^{2+} species in the aqueous phase and the interlayer cations of montmorillonite and the intracrystalline cations of clinoptilolite.

The importance of aqueous carbonate complexation in reducing U^{VI} sorption onto mineral sorbents is emphasized by data in Figure 3. The figure compares results of experiments on U^{VI} sorption on clinoptilolite conducted under atmospheric P_{CO_2} conditions ($10^{-3.5}$ atm) and at an elevated P_{CO_2} (10^{-2} atm) typical of groundwater systems. The data show that an increase in P_{CO_2} from $10^{-3.5}$ to 10^{-2} atm reduces the maximum K_d by about an order of magnitude and shifts the high-pH side to lower pH. The low-pH side, however, does not shift. The effect of increased P_{CO_2} is directly related to its effect on the aqueous speciation of U^{VI}. Figure 4 shows the aqueous speciation of U^{VI} as a function of pH for a $2.1 \cdot 10^{-7}$ $mol \cdot L^{-1}$ U^{VI} solution (0.1 $mol \cdot L^{-1}$ $NaNO_3$ matrix) at a P_{CO_2} of $10^{-2.0}$ atm. A comparison of Figures 2 and 4 shows that at higher P_{CO_2} the predominance field of the monomeric U^{VI} hydroxy species (thick curve in Figure 4) becomes reduced

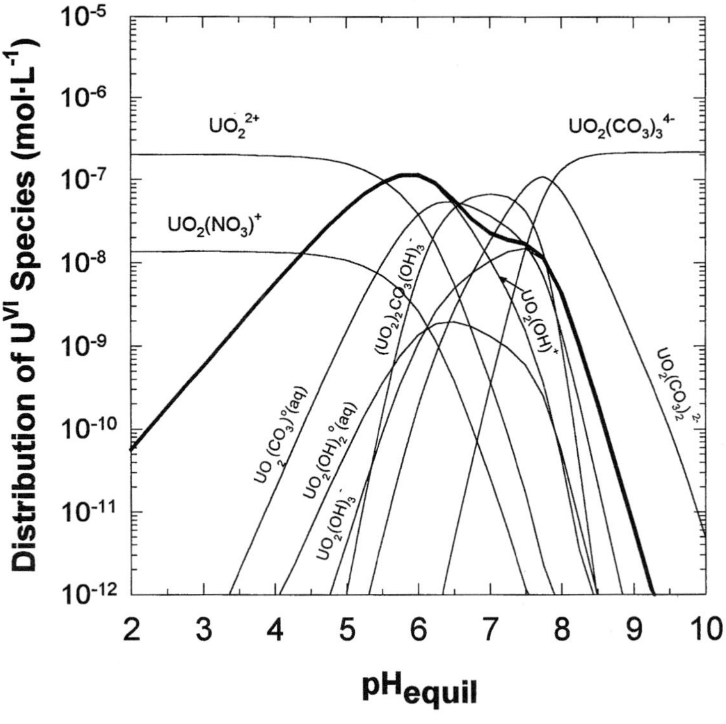

Figure 2 Aqueous U^{VI} speciation at 25°C as a function of pH for a $2.1 \cdot 10^{-7}$ $mol \cdot L^{-1}$ U^{VI} solution (0.1 $mol \cdot L^{-1}$ $NaNO_3$ matrix) in equilibrium with $P_{CO_2} = 10^{-3.5}$ atm. Some minor species are not shown. The heavy curve represents the total concentration of U^{VI} hydroxy complexes.

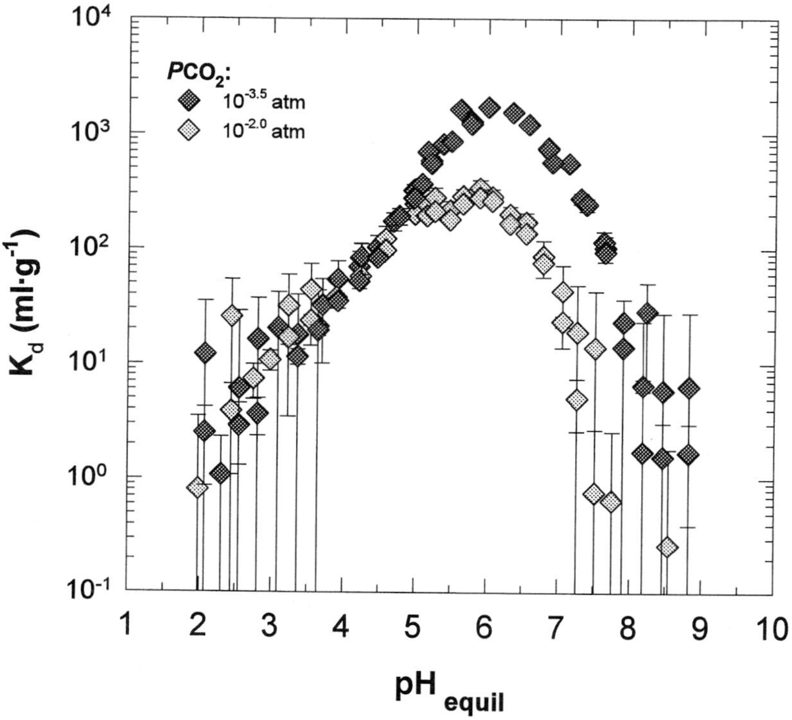

Figure 3 Data on sorption of U^{VI} on clinoptilolite from experiments C1 and C6, which had initial U concentrations of $2 \cdot 10^{-7}$ mol·L^{-1} and a $P\text{CO}_2$ of $10^{-3.5}$ or $10^{-2.0}$ atm, respectively. Error bars indicate 2σ uncertainties associated with the sorption data.

in a manner similar to the reduction in U^{VI} sorption; i.e., the maximum is lowered and the high-pH side is shifted to lower pH. This is due to the increased importance of the U^{VI} carbonate complexes at higher $P\text{CO}_2$. There is also the possibility that some of the mineral surface sites are occupied by sorbed carbonate species, particularly at higher $P\text{CO}_2$ (Van Geen *et al.,* 1994). However, no data on carbonate sorption are currently available to allow an evaluation of its relative importance in reducing U^{VI} sorption on the minerals used in this study.

The initial concentration of U in solution also influences U^{VI} sorption. Examples are shown in Figures 5A and 5B, which plot K_d versus pH for quartz and montmorillonite, respectively, at different initial U concentrations. These figures show that K_d decreases with increasing initial U concentration particularly in the intermediate pH range. The effect is greater at higher initial U concentrations due to the nonlinearity of the sorption isotherm. In Figure 6, for example, plots of the amount of U^{VI} sorbed on quartz at a fixed pH versus equilibrium concentration in solution are fit well by nonlinear Freundlich isotherms. These data indicate that sorption of U^{VI} is not linearly proportional to dissolved U concentration except at very low U concentrations. The K_d, which is the slope of the isotherm at a fixed U solution concentration, is larger at lower solution concentrations and approaches a constant value at low enough solution concentrations where the isotherm is likely to be linear. Thus, although experimental data in this study are outside the concentration range where the isotherm is linear, K_d values at initial U concentrations lower than those plotted in Figure 5 ($\sim 2 \cdot 10^{-8}$ mol·L^{-1}) will likely approach a constant value with decreasing initial U concentration.

In addition to the results presented here, decreasing sorption with increasing initial U concentration has also been observed for the U^{VI}–ferrihydrite system (Waite *et al.,* 1994). This behavior may be due to increasing coverage of available sites at higher sorbate concentrations (Dzombak and Morel, 1990). In the presence of excess sorption sites, the nonlinear trend in sorption with increasing U concen-

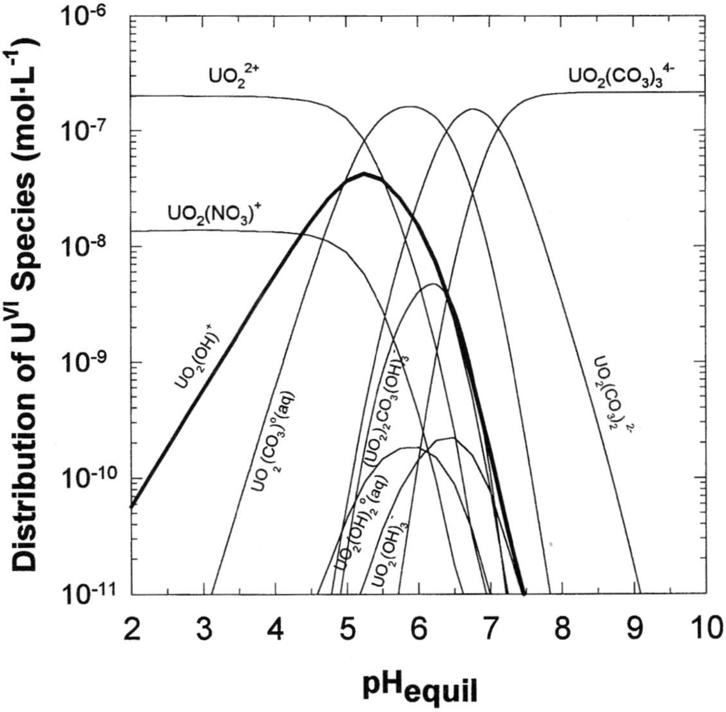

Figure 4 Aqueous U^{VI} speciation at 25°C as a function of pH for a $2.1 \cdot 10^{-7}$ mol·L^{-1} U^{VI} solution (0.1 mol·L^{-1} $NaNO_3$ matrix) in equilibrium with $P_{CO_2} = 10^{-2.0}$ atm. Some minor species are not shown. The heavy curve represents the total concentration of U^{VI} hydroxy complexes.

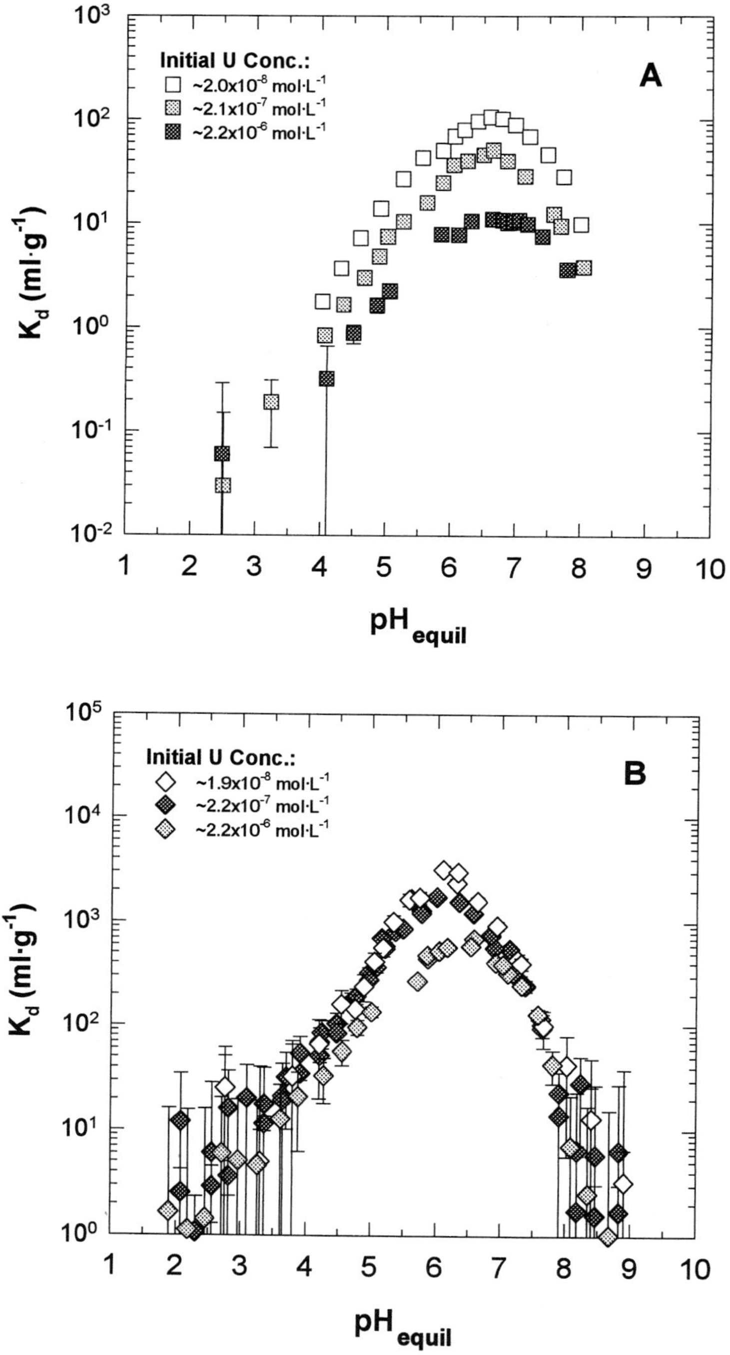

Figure 5 Uranium^{VI} sorption as a function of initial U concentration and pH. Data in Figure 5A are from quartz experiments (Q3, Q4, Q5), and data in Figure 5B are from clinoptilolite experiments (C1, C2, C3). Error bars indicate 2σ uncertainties associated with the sorption data.

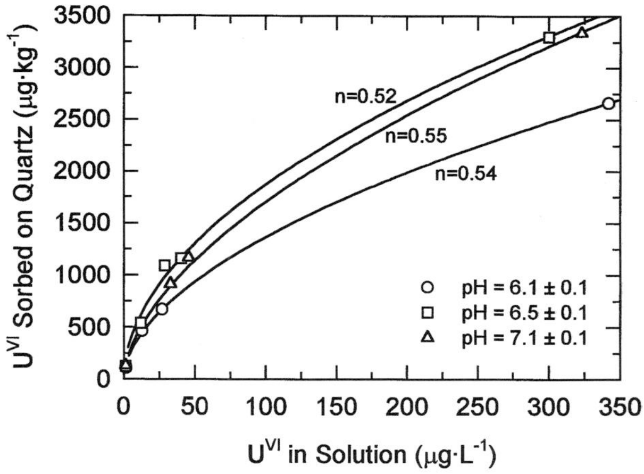

Figure 6 Freundlich isotherms of U^{VI} sorption on quartz at pH 6, 6.5, and 7. Data were fitted by nonlinear least squares regression using the equation $S = KC^n$, where S and C are the equilibrium concentrations of U^{VI} on the quartz and in solution, respectively, and K and n are constants. Values of n other than 1 indicate a nonlinear trend in the increase of sorbed concentration with increased solution equilibrium concentration.

tration may also be influenced by the formation of polynuclear aqueous complexes (O'Day, 1994). Likewise, the decrease in the proportion of sorbed U^{VI} associated with an increase in initial U concentration has been used as evidence against formation of polynuclear U^{VI} surface complexes (Waite *et al.*, 1994).

Uranium^{VI} sorption onto clinoptilolite as a function of ionic strength and pH is shown in Figure 7. The results indicate that ionic strength of the solution has little influence on U^{VI} sorption on clinoptilolite, at least for the ionic strengths used in this study (0.1 and 1.0 mol·L^{-1} $NaNO_3$). There is a subtle indication of ionic strength effect at pH values between 7 and 8 in Figure 7. Waite *et al.* (1994), who used $NaNO_3$ concentrations from 0.004 to 0.5 mol·L^{-1}, observed a relatively more pronounced ionic strength dependence of U^{VI} sorption on ferrihydrite in alkaline solutions. These apparent ionic strength effects, however, can be accounted for by an increase in the activity or concentration of the U^{VI}–carbonate species in solution at higher $NaNO_3$ concentration (at a fixed pH and $P\text{CO}_2$). At low pH, the concentration of the $UO_2NO_3^+$ aqueous species also increases with increasing $NaNO_3$ molality, which could explain the somewhat lower U^{VI} sorption in acidic solutions with higher $NaNO_3$ concentrations observed in this study and in Waite *et al.* (1994). The effect of U^{VI}–nitrate complex formation on U^{VI} sorption is not pronounced; however, a stronger complexing ligand such as SO_4^{2-} could substantially reduce U^{VI} sorption at low pH, as shown by data from Venkataramani and Gupta (1991), either by forming uranyl sulfate complexes or by competing for available sites (Dzombak and Morel, 1990; Davis and Kent, 1990).

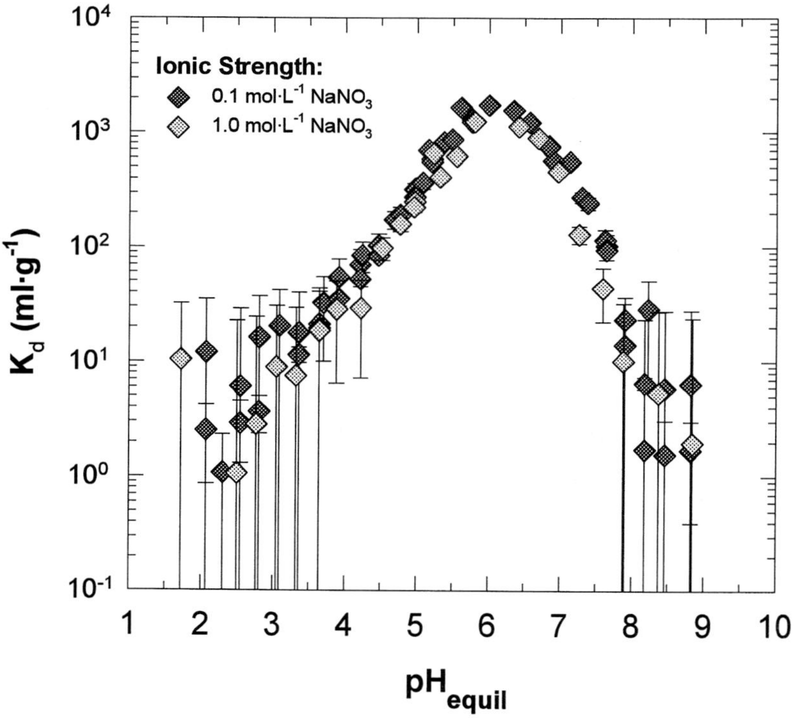

Figure 7 Uranium^{VI} sorption on clinoptilolite as functions of ionic strength and pH. Data are from experiments C1 and C4, which had initial U concentrations of $2 \cdot 10^{-7}$ mol·L^{-1} and a $P\text{CO}_2$ of $10^{-3.5}$ atm. Error bars indicate 2σ uncertainties associated with the sorption data.

The lack of ionic strength dependence of sorption is consistent with the formation of inner-sphere surface complexes (Davis and Kent, 1990). However, sorption reactions involving an ion exchange mechanism, e.g., UO_2^{2+} in solution exchanging with $2Na^+$ in montmorillonite, will be affected by changes in ionic strength due to changes in the concentration of the competing cation (e.g., Na^+). This effect is more pronounced in dilute solutions and low pH (e.g., Zachara and McKinley, 1993; McKinley *et al.,* 1995) and would be dependent on the competing cation present (e.g., Turner *et al.,* 1996). Under the conditions of the experiments in this study (0.1 and 1.0 mol·L^{-1} $NaNO_3$), ion exchange between UO_2^{2+} in solution and Na^+ in montmorillonite and clinoptilolite was mostly suppressed.

Results of experiments conducted at different M/V ratios are shown in Figure 8 in the typical manner of percent U sorbed versus pH. The sorption curves are broader at higher M/V; i.e., the amount of U^{VI} sorbed relative to the initial amount of U^{VI} in solution increases with an increasing M/V ratio. The apparent M/V effect, however, is mostly eliminated if the results are plotted in terms of K_d, a ratio of equilibrium concentrations in the solid versus in solution. For example, data in Figure 8 replotted in terms of K_d in Figure 9 show that, for the M/V range studied, changes in M/V ratio have little or no effect on U^{VI} sorption. Plotting data in terms of K_d instead of percent U sorbed makes comparison of different sets of experimental data easier.

As shown previously, U^{VI} sorption on quartz, α-alumina, clinoptilolite, and montmorillonite is similar with respect to pH dependence. However, the K_d values for the different minerals vary over three orders of magnitude. This variation is an artifact of representing sorption data in terms of K_d, which normalizes the amount of U^{VI} sorbed to the sorbent mass and not to the number of available sorption sites. Surface area measured by gas adsorption (e.g., N_2-BET) methods is a relative index of the number of sorption sites on the mineral surface. Thus, it is more useful to represent sorption data normalized to the specific surface area of the mineral sorbent. Figure 10 presents the results of U^{VI} sorption on quartz, α-alumina, clinoptilolite, and montmorillonite shown previously in Figure 1 and replotted in terms of K_a ($ml \cdot m^{-2}$), where K_a is K_d normalized to the mineral's N_2-BET specific surface area (S_A, $m^2 \cdot g^{-1}$) (i.e., $K_a = K_d/S_A$). As shown in Figure 10, surface area normalized sorption data for clinoptilolite and montmorillonite are indistinguishable, whereas surface area normalized sorption data for quartz and α-alumina are almost coincident. The α-alumina K_a values in the intermediate pH range are lower than those of quartz due to the higher initial U concentration of the α-alumina experiments.

The results plotted in Figure 10 appear to indicate that quartz and α-alumina sorb more U^{VI} per unit area than either clinoptilolite or montmorillonite. However, surface areas determined by N_2-BET methods most likely overestimate the amount of sorption sites on layered silicates such as montmorillonite and zeolitic minerals such as clinoptilolite. For example, it is believed that surface complex formation of U^{VI} on montmorillonite occurs on the hydroxylated edge sites of the mineral (Zachara and McKinley, 1993). Wanner *et al.* (1994) estimated that only 10% of the N_2-BET specific surface area is accounted for by the crystallite edges of montmorillonite. Assuming that the "effective" surface area (S_{EA}) for montmorillonite and clinoptilolite is equivalent to 10% of the measured S_A, sorption data for motmorillonite and clinoptilolite can be recast in terms of $K_{a'}$, where $K_{a'}$ is K_d normalized to the mineral's S_{EA} (i.e., $K_{a'} = K_d/S_{EA}$). For nonlayered and nonporous minerals such as quartz and α-alumina, $K_a = K_{a'}$. Figure 11 plots $K_{a'}$ values for quartz, clinoptilolite, and montmorillonite. As shown in the figure, U^{VI} sorption on these minerals, which have distinct mineralogic and surface properties, are essentially equivalent when recast in terms of $K_{a'}$.

The very good correspondence between the $K_{a'}$ for the different minerals indicates that S_{EA} may be a useful parameter for comparing and estimating U^{VI} sorption on various sorbents. To test this usefulness, literature data on U^{VI} sorp-

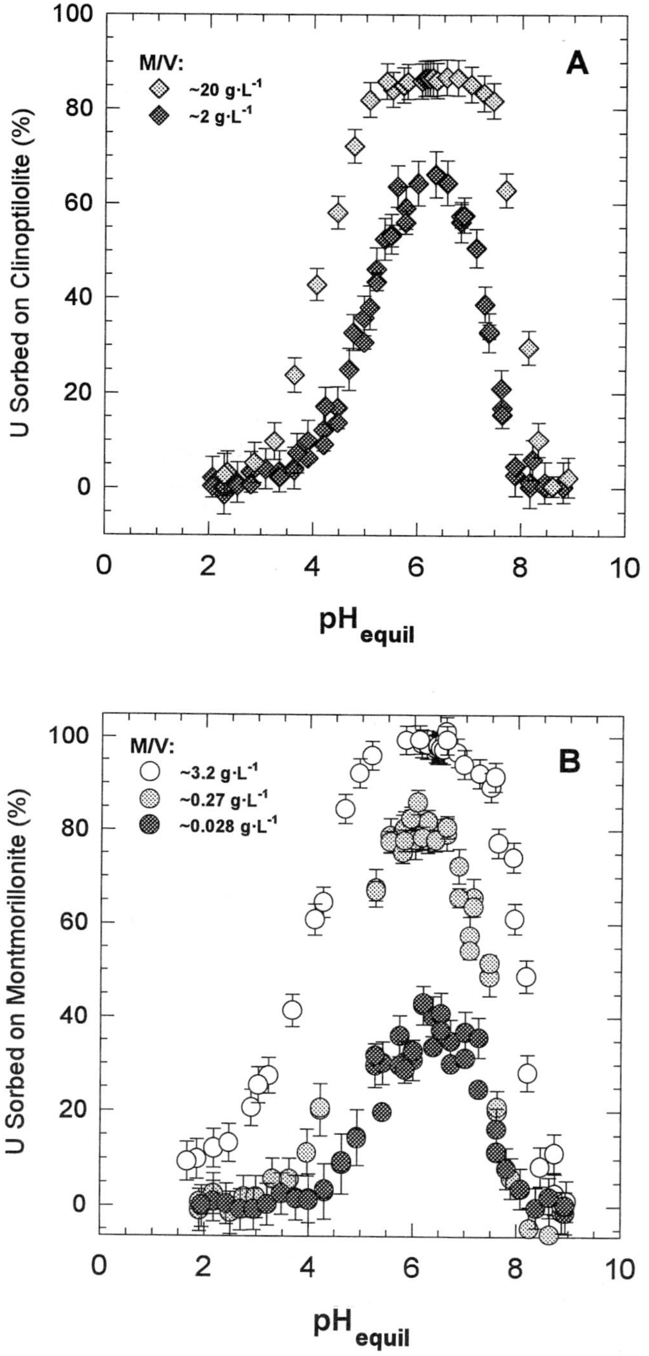

Figure 8 Uranium^{VI} sorption as functions of M/V and pH plotted in terms of percent U sorbed. Data in Figure 8A are from clinoptilolite experiments (C1, C5), and data in Figure 8B are from montmorillonite experiments (M1, M2, M3). Error bars indicate 2σ uncertainties associated with the sorption data.

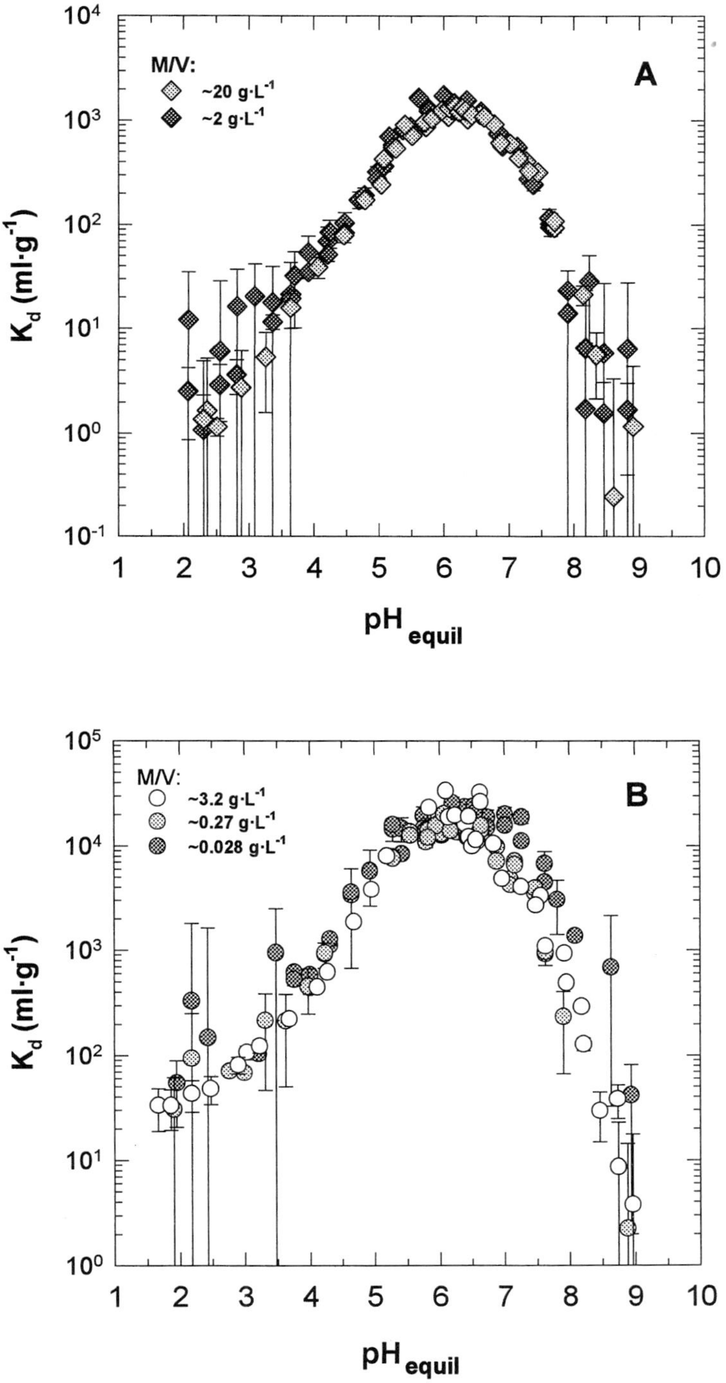

Figure 9 Uranium^{VI} sorption data on (A) clinoptilolite and (B) montmorillonite taken from Figure 8 and replotted in terms of K_d.

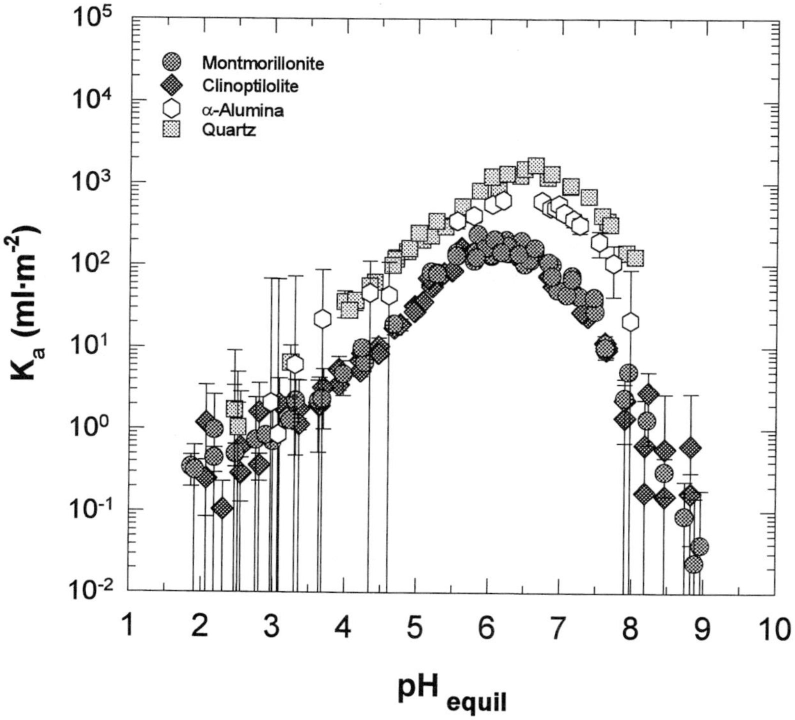

Figure 10 Data on U^{VI} sorption on montmorillonite, clinoptilolite, α-alumina, and quartz taken from Figure 1 and replotted in terms of K_a, which is K_d normalized to the mineral's measured N_2-BET specific surface area ($K_a = K_d/S_A$). The data for montmorillonite, clinoptilolite, and quartz are from experiments with initial U concentrations of $2 \cdot 10^{-7}$ mol·L^{-1}, whereas the α-alumina data are from experiments with initial U concentrations of $5 \cdot 10^{-7}$ mol·L^{-1}.

tion on different solids were compiled and recast in terms of K_a, versus pH. To facilitate comparison of literature data with those from this study, a reference line was derived from a weighted nonlinear regression of K_a, versus pH data. This reference line, which is shown in Figures 11 and 12, is with respect to initial U solution concentration of $2.1 \cdot 10^{-7}$ mol·L^{-1} and equilibrium with atmospheric P_{CO_2}. Literature data on U^{VI} sorption on various minerals (Lieser and Thybusch, 1988; McKinley *et al.*, 1995; Payne *et al.*, 1992) are plotted in Figure 12 in terms of K_a, versus pH. The reported N_2-BET specific surface areas of $TiO_2 \cdot xH_2O$, Swy-1 montmorillonite, amorphous SiO_2, and KGa-1 kaolinite are 190, 31, 182, and 10.1 mg^2·g^{-1}, respectively (Lieser *et al.*, 1992; McKinley *et al.*, 1995; Van Olphen and Fripiat, 1979). Effective surface areas for montmorillonite and kaolinite are assumed equal to 10% of their N_2-BET specific surface areas. The literature data in Figure 12 show good agreement with the results from this study. The $K_{a'}$ values derived from literature experiments are slightly lower than those from this study due to the higher initial U concentrations of the former. The literature data are from experiments which were reportedly done under carbonate-free conditions. Thus, except for the $TiO_2 \cdot xH_2O$ results, those data do not exhibit the decreasing trend at alkaline pH observed in results from this study. $CO_2(g)$ contamination is possible in the $TiO_2 \cdot xH_2O$ experiments, which could explain the decreasing trend of $K_{a'}$ at high pH. Other than capping the bottles, no special precautions (e.g., glove box) were taken to exclude $CO_2(g)$ in those experiments.

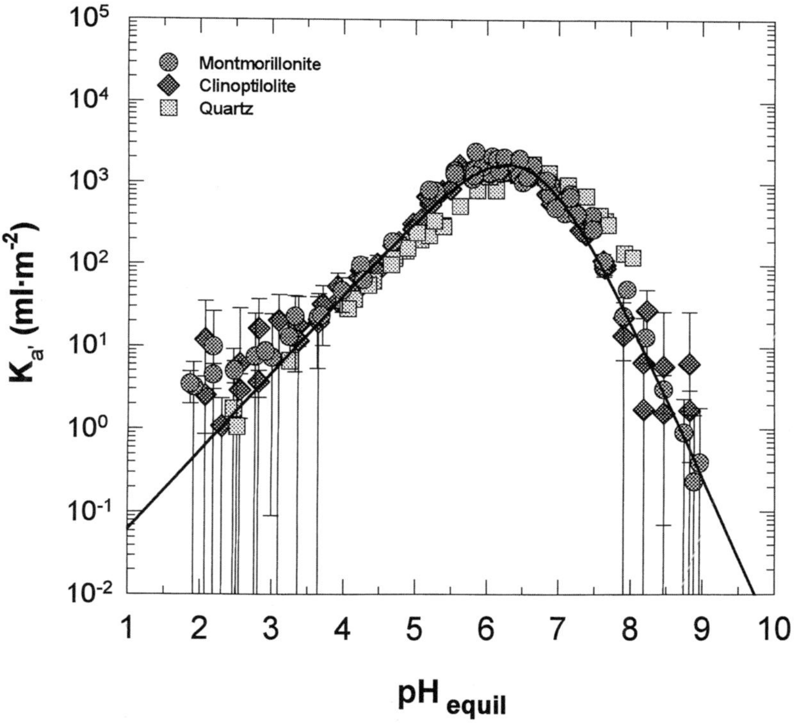

Figure 11 Data on U^{VI} sorption on montmorillonite, clinoptilolite, and quartz taken from Figure 1 and replotted in terms of $K_{a'}$, which is K_d normalized to the mineral's "effective" surface area ($K_{a'} = K_d/S_{EA}$). For montmorillonite and clinoptilolite, S_{EA} is assumed equal to 10% of N_2-BET S_A, whereas for quartz and α-alumina S_{EA} is equal to N_2-BET S_A. The curve is a reference line derived from a weighted nonlinear regression of the montmorillonite, clinoptilolite, and quartz $K_{a'}$ values. The regression curve is defined by the equation $K_{a'} = a(1 + n)^{-(d+1)/d}n(d + 1)^{(d+1)/d}/d$, where $n = \exp((pH + d \ln(d) - b)/c)$, a = 1679.0744, b = 6.2683501, c = 0.45176211, and d = 0.48419989.

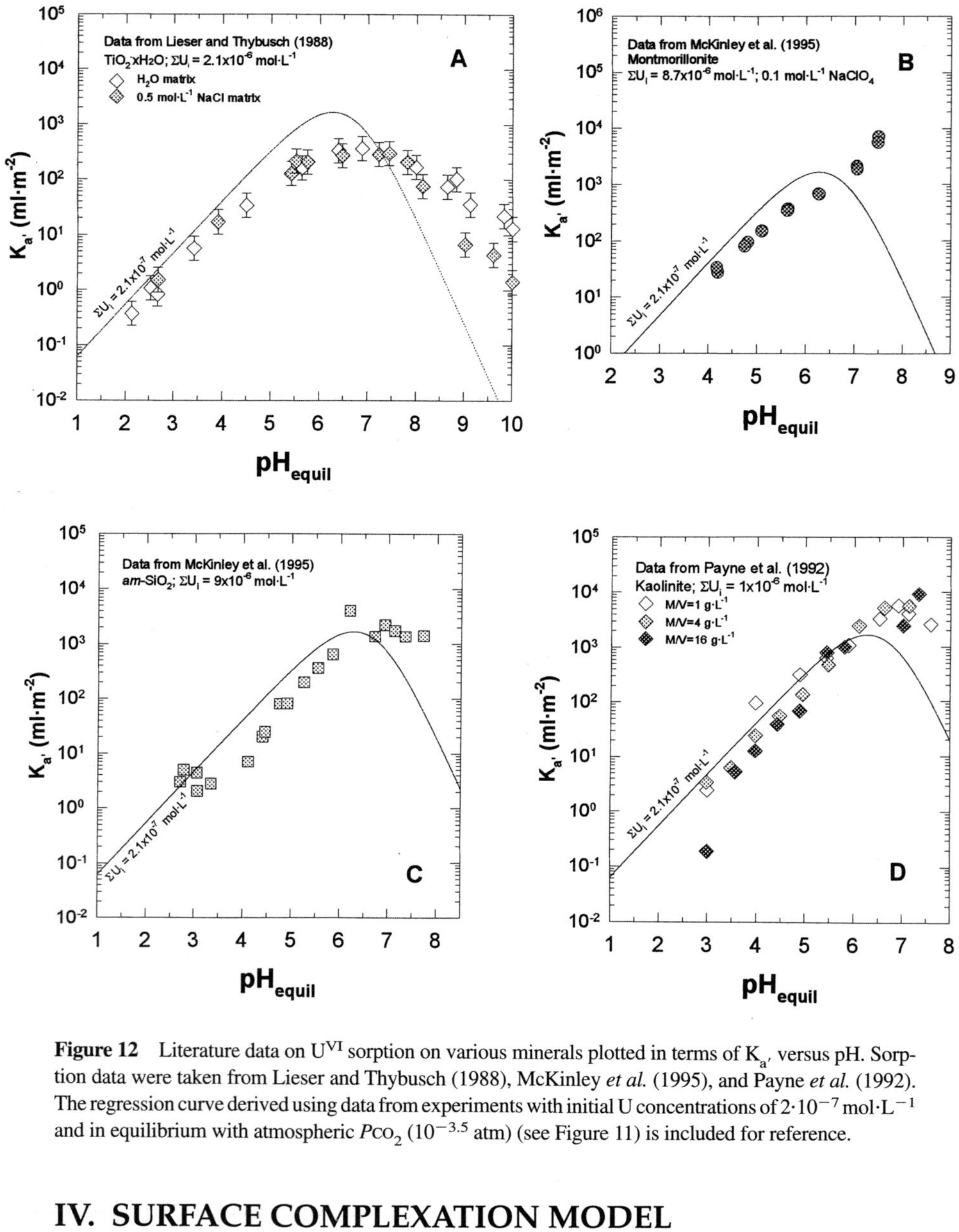

Figure 12 Literature data on U^{VI} sorption on various minerals plotted in terms of $K_{a'}$ versus pH. Sorption data were taken from Lieser and Thybusch (1988), McKinley *et al.* (1995), and Payne *et al.* (1992). The regression curve derived using data from experiments with initial U concentrations of $2 \cdot 10^{-7}$ $mol \cdot L^{-1}$ and in equilibrium with atmospheric P_{CO_2} ($10^{-3.5}$ atm) (see Figure 11) is included for reference.

IV. SURFACE COMPLEXATION MODEL

The strong pH dependence observed in the experiments indicates that predictive sorption modeling requires geochemical models capable of simulating the influence of changing physicochemical conditions on U^{VI} aqueous speciation and sorption. Although U^{VI} sorption appears relatively insensitive to surface charge, a class of models that has been used with success in modeling pH-dependent sorption is the electrostatic surface complexation model (SCM) (Davis and Leckie, 1978; Westall and Hohl, 1980; Davis and Kent, 1990; Hayes *et al.,* 1991). The model is now briefly described, including applications to modeling of U^{VI} sorption onto quartz and montmorillonite.

A. Model Description

SCMs are based on the assumption of analogous behavior between aqueous complex formation in the bulk solution and formation of surface complexes with functional binding sites at the mineral–water interface. Surface reactions are written for sorbing species, and mass action and mass balance relations are used to determine sorption at the mineral surface as a function of system chemistry. Of the different SCMs, the diffuse-layer model (DLM) is perhaps the simplest, using a one-layer representation of the mineral–water interface. The details of the DLM and the simplified uniform approach used here are described elsewhere (Dzombak and Morel, 1990; Davis and Kent, 1990; Turner and Sassman, 1996), and only a brief overview is presented here.

In SCMs, neutral amphoteric surface sites ($>XOH^0$) are assumed to form charged surface sites represented as $>XOH_2^+$ and $>XO^-$ through the addition or removal of protons, respectively. In setting up the chemical equilibrium model, protonation and deprotonation reactions are expressed as

$$>XOH^0 + H^+ = >XOH_2^+ \quad \text{(protonation, } K_+^{app}\text{)} \qquad [2]$$

$$>XOH^0 = >XO^- + H^+ \quad \text{(deprotonation, } K_-^{app}\text{).} \qquad [3]$$

The equilibrium constants K_+^{app} and K_-^{app} are referred to as apparent surface acidity constants, which include surface effects and are, therefore, dependent on the extent of surface protonation and deprotonation (Dzombak and Morel, 1990).

For U^{VI}–mineral sorption, a generalized surface reaction can be written as

$$>XOH^0 + pUO_2^{2+} + nH_2O \leftrightharpoons [>XOH_q\text{–}(UO_2)_p(OH)_n]^{2p+q-n-1} + (1 + n - q)H^+, \qquad [4]$$

where q is the protonation state of the sorption site (q = 0, 1, or 2), and UO_2^{2+} and $[>XOH_q\text{–}(UO_2)_p(OH)_n]^{2p+q-n-1}$ represent the uranyl aqueous species and the U^{VI} surface complex, respectively. Apparent equilibrium constants ($K^{app}_{[>XOH_q-(UO_2)_p(OH)_n]}$) for reactions of the type given in Eq. [4] are similar to K_+^{app} and K_-^{app} for Eqs. [2] and [3] in that they include the effects of surface charge. The reactions in Eqs. [2]–[4] illustrate the pH dependence of surface charge development and radionuclide sorption.

In the SCM approach, the change in activity of sorbed surface complexes due to electrostatic forces is assumed to be governed by the Boltzmann relation (Dzombak and Morel, 1990). This coulombic correction is incorporated into the mass action expressions for surface reactions (Eqs. [2]–[4] to extract the intrinsic equilibrium constants (e.g., K_{+}^{int}, K_{-}^{int}, and $K^{int}_{[>XOH_q-(UO_2)_p(OH)_n]}$) that are independent of surface charge. In the SCM approach, K_i^{int} is related to K_i^{app} according to the relationship $K_i^{int} = K_i^{app}$ (exp($\Delta zF\Psi/RT$), where Δz is the change in charge at the surface, Ψ is the surface potential (V), F is the Faraday constant (96,485 C·mol^{-1}), R is the gas constant (8.314 J·mol^{-1}·K^{-1}), and T is temperature (K). For sorption reactions of the type given in Eq. [4], $K^{int}_{[>XOH_q-(UO_2)_p(OH)_n]}$ is commonly referred to as the binding constant.

Turner and Sassman (1996) used the numerical, nonlinear parameter optimization code FITEQL, version 2.0 (Westall, 1982), to interpret potentiometric titration data and determine the DLM acidity constants (K_{+}^{int} and K_{-}^{int}) necessary to define the acid–base behavior of a number of different mineral surfaces. The values are model specific, but once defined, the acid–base behavior of the surface is characterized and these values become fixed in the geochemical model. To determine the binding constants for U^{VI} sorption reactions of the general type given in Eq. [4], FITEQL was used in this study to regress subsets of the sorption data using equations for mass action equilibria among aqueous and surface species. The equilibrium constants used in the chemical equilibrium model for the U^{VI}–H_2O–CO_2 system were taken from the NEA thermodynamic database (Wanner and Forest, 1992), with the exception of the value for the neutral hydroxy complex $UO_2(OH)_2^0$(aq), as discussed previously.

In the presence of CO_2, U^{VI} sorption typically decreases at alkaline conditions. In the absence of detailed information on what surface complexes form, the conceptual model developed here attributes the trend of decreasing sorption at higher pH to the increased carbonate concentration and the more effective complexation of U^{VI} in the bulk solution by the carbonate ligand relative to that by the surface sites.

B. Modeling of U^{VI} Sorption on Quartz

An additional concern in modeling sorption on minerals with low surface area such as quartz was how to take into account the competing effects of sorption loss to the container walls (Bertetti *et al.,* in preparation). To develop the necessary pH dependence, two reactions were postulated that bind the positively charged species UO_2OH^+ and $(UO_2)_3(OH)_5^+$ to a hypothetical, negatively charged container surface site represented as X^-:

$$X^- + UO_2^{2+} + H_2O \rightleftarrows XUO_2OH^0 + H^+ \quad (K_{XUO_2OH}) \quad [5]$$

$$X^- + 3UO_2^{2+} + 5H_2O \rightleftarrows X(UO_2)_3OH_5^0 + 5H^+ \quad (K_{X(UO_2)_3(OH)_5}). \quad [6]$$

FITEQL was used to interpret U^{VI} container sorption using the control data (no quartz present) from experiment Q2 ($\Sigma U_i = 2.11 \cdot 10^{-7}$ mol·L^{-1}, M/V = 20 g·L^{-1}). No characterization of the container surface was done to yield information on the sorption site density. Instead, the total site concentration (T_{X^-}) on the container surface was treated as a fitting parameter and was derived using the Q2 container sorption data. No electrostatic terms were included in the mass action equations for container sorption. The resulting value of T_{X^-} was $2.3 \cdot 10^{-8}$ mol sites·L^{-1}, and the derived binding constants (log K^{int}) for Eqs. [5] and [6] were 2.85 and −5.25, respectively. The parameters derived from the Q2 data were used to predict container sorption for experiments Q3 and Q4. Despite the number of assumptions made, the model was successful in reproducing the observed container sorption (Figure 13A).

For the purposes of modeling U^{VI} sorption on quartz, a separate analysis with FITEQL was conducted to derive binding constants for U^{VI} sorption on quartz alone using data from experiment Q2. Based on the U^{VI} aqueous speciation over the pH range of interest, a combination of three reactions forming the surface complexes $>SiO{-}UO_2^+$, $>SiO{-}UO_2OH^0$, and $>SiO{-}(UO_2)_3(OH)_5^0$ was considered and found to be the most capable of reproducing the Q2 quartz sorption results. The full reactions and the estimated binding constants are given in Table II. Modeling results for sorption on quartz alone are shown in Figure 13B, including values predicted for other experimental conditions. As shown in the figure, the model is able to reproduce the observed sorption behavior quite well. Using the independently derived binding constants for container sorption and quartz sorption in a single chemical equilibrium model, it is possible to simulate sorption competition between the container wall and the quartz surface. Calculated values are shown in Figure 13C, which indicate that the model is generally successful in reproducing the observed pH-dependent total sorption (container + quartz).

C. Modeling of U^{VI} Sorption on Montmorillonite

Previous studies have established that sorption on a clay such as montmorillonite consists of ion exchange at interlayer sites and pH-dependent sorption at crystallite edge sites (Davis and Kent, 1990; Zachara and McKinley, 1993; Zachara and Smith, 1994; Wanner *et al.*, 1994; Degueldre *et al.*, 1994). In modeling U^{VI} sorption on montmorillonite, it was assumed that edge sites dominate sorption behavior at the relatively high ionic strength (0.1 mol·L^{-1} $NaNO_3$) used in this study, and ion exchange was not explicitly incorporated in the model construction (Pabalan and Turner, 1997).

Recent modeling efforts suggest that it is possible to simulate the pH-dependent sorption behavior of aluminosilicates such as montmorillonite by assuming the edge sites to comprise silanol ($>SiOH^0$) and aluminol ($>AlOH^0$) sites which

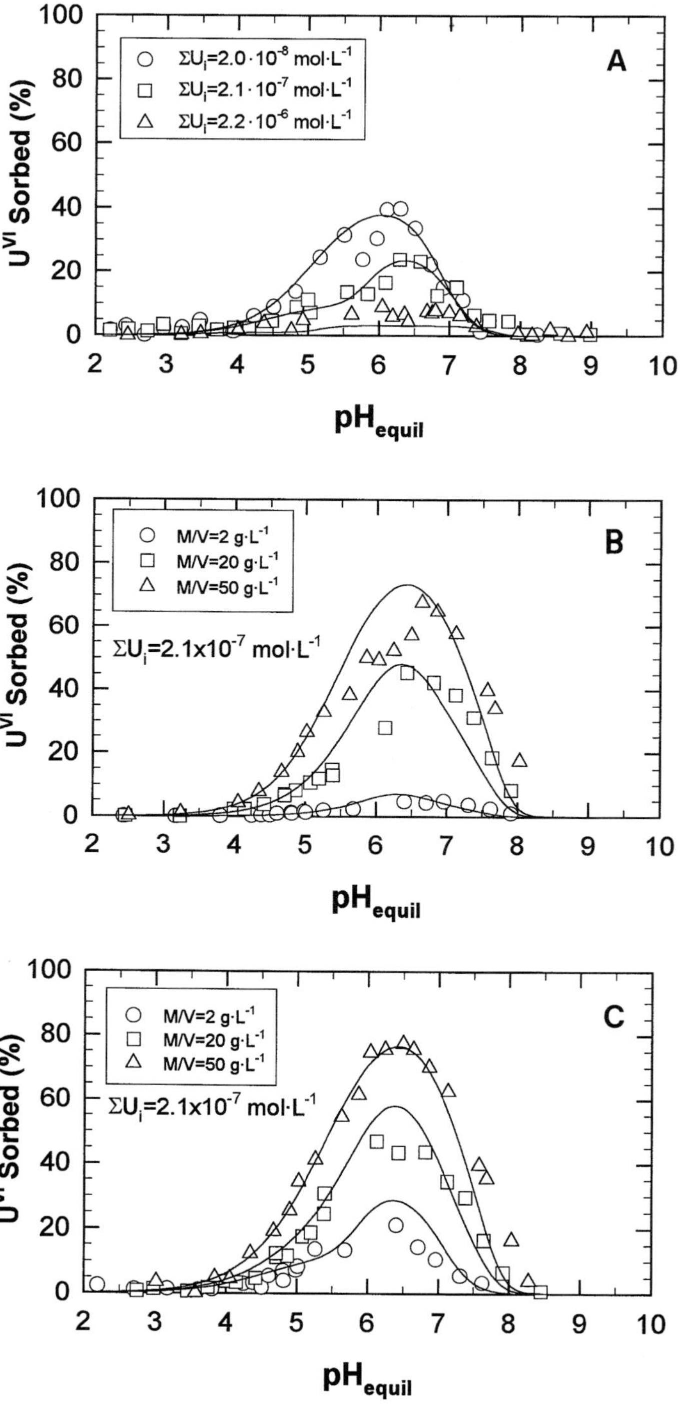
A
ΣU_i=2.0·10^{-8} mol·L^{-1}
ΣU_i=2.1·10^{-7} mol·L^{-1}
ΣU_i=2.2·10^{-6} mol·L^{-1}
U^{VI} Sorbed (%)
pH_{equil}
B
M/V=2 g·L^{-1}
M/V=20 g·L^{-1}
M/V=50 g·L^{-1}
ΣU_i=2.1x10^{-7} mol·L^{-1}
U^{VI} Sorbed (%)
pH_{equil}
C
M/V=2 g·L^{-1}
M/V=20 g·L^{-1}
M/V=50 g·L^{-1}
ΣU_i=2.1x10^{-7} mol·L^{-1}
U^{VI} Sorbed (%)
pH_{equil}

compete for available adsorbent, but otherwise do not interact with each other (Zachara and Smith, 1994; Turner, 1995). Calculation of total sorption site concentration requires some estimate of an effective edge site surface area and an assumed $>AlOH^0/>SiOH^0$ ratio. The montmorillonite used in this study has a measured (N_2-BET) external surface area of 97 $m^2 \cdot g^{-1}$. For modeling pH-dependent U^{VI}–montmorillonite sorption, the crystallite edges were assumed to compose 10% of the N_2-BET specific surface area (9.7 $m^2 \cdot g^{-1}$), consistent with observations of Wanner *et al.* (1994) and with the previous discussion of $K_{a'}$. It was also assumed that the ratio of $>AlOH^0$ to $>SiOH^0$ sites is 0.83, as proposed for montmorillonite by White and Zelazny (1988). The acidity constants (K_+^{int} and K_-^{int}) used for the $>SiOH^0$ and $>AlOH^0$ edge sites were based on potentiometric titration data for SiO_2 and α-Al_2O_3, as described in Turner and Sassman (1996). The pH_{pzc} calculated using this approach is 7.4. This value is within the experimental range of 6.5 to 8.0 observed by Wanner *et al.* (1994), although it is higher than their model value of 6.1 for the edge sites.

Sorption data from experiment M4 ($\Sigma U_i = 2.16 \cdot 10^{-6}$ mol·L^{-1}, M/V = 0.28 g·L^{-1}) were used to derive binding constants for possible U^{VI} sorption reactions with the aid of FITEQL. A combination of four surface complexes, $>AlO–UO_2^+$, $>AlO–(UO_2)_3(OH)_5^0$, $>SiO–UO_2^+$, and $>SiO–(UO_2)_3(OH)_5^0$, was found to best reproduce the M4 sorption results. The binding constants determined for these complexes are given in Table II. The model developed here reproduces the M4 sorption data very well as a function of pH (Figure 14A). Using the same model parameters, U^{VI} sorption on montmorillonite at other experimental conditions was predicted and compared with measured values. For $\Sigma U_i = 2 \cdot 10^{-7}$ mol·L^{-1}, the sorption predicted by the DLM for an M/V of 3.2 and 0.27 g·L^{-1} agree well with data from experiments M1 and M2, respectively. The underprediction at low pH (<3.5) is likely due to the contribution of ion exchange to the measured sorption which is not accounted for in the model. The agreement between predicted and measured (M3 experiment) sorption at M/V = 0.028 g·L^{-1} is acceptable, although there is some underprediction of sorption at pH > 6.

For the purposes of transport calculations, it is useful to express sorption modeling results in terms of K_d. Comparisons between model and experimental results are shown in Figure 14B. It is clear from these figures that the model adequately represents the pH dependence of U^{VI} sorption on montmorillonite, except at very low pH (<3.5) where ion exchange is expected to occur. Figure 14B also shows

Figure 13 Comparison of experimental and DLM-predicted sorption for the system U^{VI}–quartz–H_2O–CO_2–container. The sorption models are described in the text. (A) U^{VI}–container sorption, (B) U^{VI} sorption on quartz alone, and (C) U^{VI} sorption for quartz + container. Predicted U^{VI} sorption in (C) was derived by combining reactions and model parameters used in (A) and (B). Model parameters were determined through regression of data from experiment Q2 using FITEQL, version 2.0. Parameter values are given in Table II.

Table II

FITEQL Model Conditions Used to Determine DLM Binding Constants for U^{VI} Sorption on Quartz and Montmorillonite

DLM parameters and chemical model reactions	Quartz	Montmorillonite
Site density[a]	2.3 sites·nm^{-2}	2.3 sites·nm^{-2}
Surface area[b]	0.03 $m^2 \cdot g^{-1}$	9.7 $m^2 \cdot g^{-1}$
Ionic strength	0.1 mol·L^{-1} $NaNO_3$	0.1 mol·L^{-1} $NaNO_3$
Pco_2	$10^{-3.5}$atm	$10^{-3.5}$atm

	log K_{int}	
Sorption reactions	Quartz	Montmorillonite
$>AlOH^0 + H^+ \Leftrightarrow >AlOH_2^+$	—	8.33[c]
$>AlOH^0 \Leftrightarrow >AlO^- + H^+$	—	−9.73[c]
$>AlOH^0 + UO_2^{2+} \Leftrightarrow >AlO\text{–}UO_2^+ + H^+$	—	2.70[d]
$>AlOH^0 + 3UO_2^{2+} + 5H_2O \Leftrightarrow >AlO\text{–}(UO_2)_3(OH)_5^0 + 6H^+$	—	−14.95[d]
$>SiOH^0 \Leftrightarrow SiO^- + H^+$	−7.20[c]	−7.20[c]
$>SiOH^0 + UO_2^{2+} \Leftrightarrow >SiO\text{–}UO_2^+ + H^+$	0.30[d]	2.60[d]
$>SiOH^0 + UO_2^{2+} + H_2O \Leftrightarrow SiO\text{–}UO_2OH^0 + 2H^+$	−5.65[d]	—
$>SiOH^0 + 3UO_2^{2+} + 5H_2O \Leftrightarrow >SiO\text{–}(UO_2)_3(OH)_5^0 + 6H^+$	−16.75[d]	−15.29[d]

	log K	
Aqueous speciation reactions	Quartz	Montmorillonite
$UO_2^{2+} + H_2O \Leftrightarrow UO_2OH^+ + H^+$	−5.20	−5.20
$UO_2^{2+} + 2H_2O \Leftrightarrow UO_2(OH)_2^0(aq) + 2H^+$	−13.0[e]	−13.0[e]
$UO_2^{2+} + 3H_2O \Leftrightarrow UO_2(OH)_3^- + 3H^+$	−19.20	−19.20
$2UO_2^{2+} + 2H_2O \Leftrightarrow (UO_2)_3(OH)_2^{2+} + 2H^+$	−5.62	−5.62
$3UO_2^{2+} + 5H_2O \Leftrightarrow (UO_2)_2(OH)_5^+ + 5H^+$	−15.55	−15.55
$UO_2^{2+} + CO_3^{2-} \Leftrightarrow UO_2CO_3^0$	9.68	9.68
$UO_2^{2+} + 2CO_3^{2-} \Leftrightarrow UO_2(CO_3)_2^{2-}$	16.94	16.94
$UO_2^{2+} + 3CO_3^{2-} \Leftrightarrow UO_2(CO_3)_3^{4-}$	21.60	21.60
$2UO_2^{2+} + CO_3^{2-} + 3H_2O \Leftrightarrow (UO_2)_2CO_3(OH)_3^- + 3H^+$	−0.86	−0.86
$UO_2^{2+} + NO_3^- \Leftrightarrow UO_2NO_3^+$	0.30	0.30

[a]Site density recommended for all minerals by Davis and Kent (1990), based on ferrihydrite work of Dzombak and Morel (1990).
[b]Effective edge site surface area for montmorillonite assumed to be 10% of total N_2-BET surface area (97 $m^2 \cdot g^{-1}$). See text for detailed discussion.
[c]Intrinsic acidity constants for am$-SiO_2$ and $\alpha-Al_2O_3$ from Turner and Sassman (1996).
[d]This study. Binding constants log K_i^{int} determined using FITEQL, version 2.0 (Westall, 1982).
[e]All aqueous speciation log K values are from the NEA Uranium Thermodynamic Database (Wanner and Forest, 1992), except for $UO_2(OH)_2^0$(aq), which is taken from Fuger (1992).

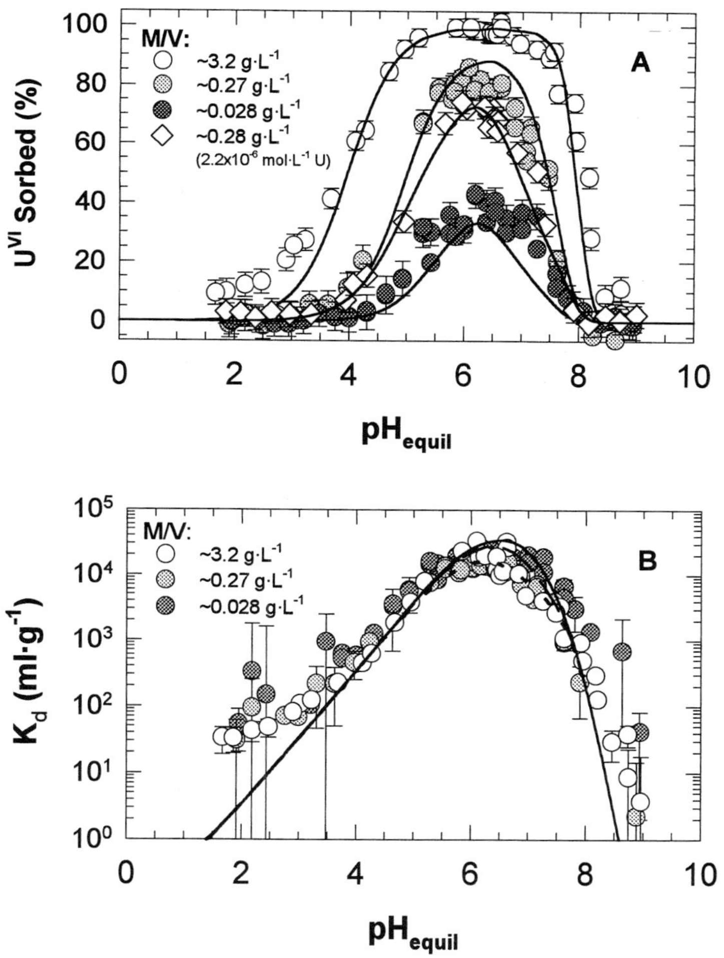

Figure 14 (A) Comparison of experimental and DLM-predicted U^{VI} sorption on montmorillonite expressed as percent U sorbed. (B) Data in (A) replotted in terms of K_d ($ml \cdot g^{-1}$). Solid, dashed, and dotted curves in (B) are DLM-calculated values for M/V of 3.2, 0.27, and 0.028 $g \cdot L^{-1}$, respectively. Model parameters were determined through regression of data from experiment M4 using FITEQL, version 2.0. Parameter values are given in Table II.

that for the range of M/V used in this study, both modeling and experimental results indicate that changes in M/V have relatively little effect on the equilibrium partitioning of U^{VI} between the aqueous and the solid phases.

An objective in developing mechanistic sorption models is to have a modeling approach that is flexible and robust enough for use in predicting changes in sorp-

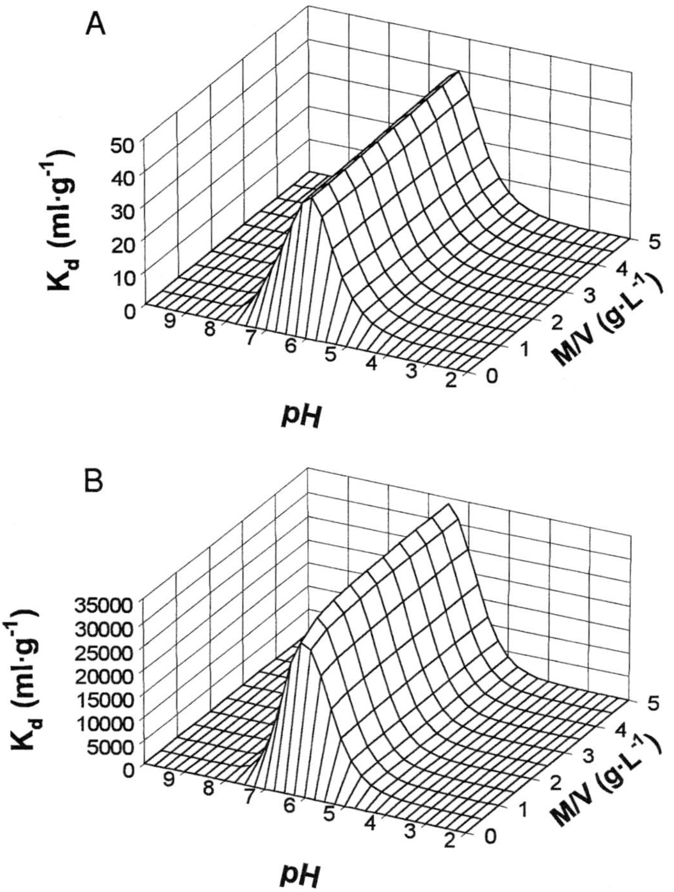

Figure 15 Predicted U^{VI} sorption as a function of pH and M/V for (A) quartz–CO_2–H_2O and (B) montmorillonite–CO_2–H_2O. Model parameters are given in Table II.

tion behavior beyond experimental conditions to those likely to be encountered in natural systems and for evaluating the sensitivity of sorption to various geochemical parameters. For example, the total number of available sorption sites relative to the number of metal ions is directly proportional to the M/V ratio. Intuitively, the greater the number of available sites, the more effective the solid will be at removing trace concentrations of radionuclides from solution. As is shown in Figure 15, the model predicts that although U^{VI} concentration on the solid phase is strongly affected by pH, it is relatively insensitive to changes in M/V, except at very low values (<0.5 g·L^{-1}). This prediction holds true for both quartz and montmorillonite, even though the maximum partitioning to the two solids differs by three orders of magnitude. This suggests that for most mineral–radionuclide–water systems, above a certain threshold value of M/V, the total number of available sites is in such excess relative to the radionuclide concentration that further increases in M/V have relatively little effect on the total amount of radionuclide removed from solution. A similar relationship was observed by Rogers and Meijer (1993), where measured K_d values appeared to be relatively insensitive to specific surface area, another parameter affecting the number of available sites.

V. CONCLUSIONS

The experimental results show an unambiguous link between the aqueous speciation of U^{VI} and its sorption behavior. Uranium^{VI} sorption is important under conditions in which U^{VI} hydroxy complexes are favored to form in the aqueous phase. Geochemical conditions which inhibit the formation of U^{VI} hydroxy complexes, e.g., low pH and carbonate complex formation, suppress U^{VI} sorption. The similarity in the pH dependence of U^{VI} sorption on quartz, α-alumina, clinoptilolite, montmorillonite, amorphous silica, kaolinite, and hydrous titanium oxide, which are substrates of distinct surface charge properties, suggest that U^{VI} sorption is not sensitive to the surface charge characteristics of the sorbent as compared to the effect of changing the total number of available sites. The experimental and modeling results demonstrate that changing M/V has little influence on U^{VI} K_d, except at very low values. Ionic strength effects are limited for surface complexation reactions, although these effects can be important if ion exchange is the predominant sorption mechanism and ionic strength effects on the activity of aqueous complexes can indirectly influence sorption behavior. The data derived in this study and from the literature suggest that the magnitudes of U^{VI} sorption (at a specific pH, initial U concentration, and P_{CO_2}) are essentially the same for different minerals if normalized to an "effective" surface area.

Sorption modeling based on a DLM approach effectively simulates the complex sorption behavior observed in the U^{VI}–H_2O–CO_2 system. This includes uranium sorption onto simple (hydr)oxides such as quartz and more complex layered silicates such as montmorillonite. It should be noted that in the absence of detailed information on site characteristics and surface speciation, several model parameter values have been assumed *a priori.* Where possible, a parsimonious approach has been adopted to minimize the number of adjustable parameters and limit the amount of curve-fitting. Although the validity of the model results is dependent on the assumptions made, the good agreement between measured and DLM-predicted sorption values presented here suggests that conceptual models based on an SCM approach will be useful in sensitivity analyses, i.e., in exploring which of the various geochemical parameters control actinide sorption. Model assumptions can be modified as more detailed information on sorption site characteristics and surface speciation become available. With additional constraints, the models could ultimately be used to extrapolate radionuclide sorption behavior over a range of geochemical conditions based on parameters derived from a limited set of experimental data. This type of flexible modeling approach provides a better alternative to the constant K_d model currently used in most transport calculations.

ACKNOWLEDGMENTS

We gratefully acknowledge the help of M. Almendarez, T. Dietrich, and P. Muller in conducting the experiments. Reviews by W. Murphy, E. Jenne, and two anonymous reviewers are greatly appreciated. This work was funded by the U.S. Nuclear Regulatory Commission (NRC), Office of Nuclear Regulatory Research, Division of Regulatory Applications, and by the NRC Office of Nuclear Material Safety and Safeguards, Division of Waste Management, under Contract No. NRC-02-93-005. This paper is an independent product of the Center for Nuclear Waste Regulatory Analyses (CNWRA) and does not necessarily reflect the views or regulatory position of the NRC.

REFERENCES

Allard, B., Beall, G. W., and Krajewski, T. 1980. The sorption of actinides in igneous rocks. *Nucl. Technol.* 49:474–480.

Allison, J. D., Brown, D. S., and Novo-Gradac, K. J. 1991. MINTEQA2/PRODEFA2, a Geochemical Assessment Model for Environmental Systems: Version 3.0 User's Manual. EPA/600/3-91/921, Environmental Protection Agency, Athens, GA.

Andreeva, N. R., and Chernyavskaya, N. B. 1982. Uranyl ion sorption by mordenite and clinoptilolite. *Radiokhimiya* 24:9–13.

Bertetti, F. P., Pabalan, R. T., Turner, D. T., and Almendarez, M. G. Experimental and modeling study of uranium(6+) sorption on quartz. In preparation.

Bish, D. L., and Chipera, S. J. 1989. Revised Mineralogic Summary of Yucca Mountain, Nevada. LA-11497-MS, Los Alamos National Laboratory, Los Alamos, NM.

Davis, J. A., and Kent, D. B. 1990. Surface complexation modeling in aqueous geochemistry. *In* "Reviews in Mineralogy," Vol. 23, pp. 177–260, "Mineral–Water Interface Geochemistry" (M. F. Hochella, Jr., and A. F. White, Eds.). Mineralogical Society of America, Washington, DC.

Davis, J. A., and Leckie, J. O. 1978. Surface ionization and complexation at the oxide/water interface. II. Surface properties of amorphous iron oxyhydroxide and adsorption of metal ions. *J. Colloid Interface Sci.* 67:90–107.

Degueldre, C., Ulrich, J. J., and Silby, H. 1994. Sorption of ^{241}Am onto montmorillonite, illite and hematite colloids. *Radiochim. Acta* 65:173–179.

Dzombak, D. A., and Morel, F. M. M. 1990. "Surface Complexation Modeling: Hydrous Ferric Oxide." John Wiley, New York.

Fuger, J. 1992. Thermodynamic properties of actinide species relevant to geochemical problems. *Radiochim. Acta* 58/59:81–91.

Gainer, G. M. 1990. Boron adsorption on hematite and clinoptilolite. M.S. Thesis, Univ. of Texas at El Paso, TX.

Hayes, K. F., Redden, G. G., Ela, W., and Leckie, J. O. 1991. Surface complexation models: An evaluation of model parameter estimation using FITEQL and oxide mineral titration data. *J. Colloid Interface Sci.* 142:448–469.

Hsi, C-K. D., and Langmuir, D. 1985. Adsorption of uranyl onto ferric oxyhydroxides: Application of the surface complexation site-binding model. *Geochim. Cosmochim. Acta* 49:1931–1941.

Hutchison, C. S. 1974. "Laboratory Handbook of Petrographic Techniques." John Wiley, New York.

Lieser, K. H., Quandt-Klenk, S., and Thybusch, B. 1992. Sorption of uranyl ions on hydrous silicon dioxide. *Radiochim. Acta* 57:45–50.

Lieser, K. H., and Thybusch, B. 1988. Sorption of uranyl ions on hydrous titanium dioxide. *Fresenius Z. Anal. Chem.* 332:351–357.

McKinley, J. P., Zachara, J. M., Smith, S. C., and Turner, G. D. 1995. The influence of hydrolysis and multiple site-binding reactions on adsorption of U(VI) to montmorillonite. *Clays Clay Miner.* 43:586–598.

Mehra, O. P., and Jackson, M. L. 1960. Iron oxide removal from soils and clays by a dithionite-citrate system with sodium bicarbonate. *Clays Cay Miner. Proc. 7th Conf.*, 317–327.

O'Day, P. A. 1994. Free energies of adsorption of divalent metal ions on quartz. *GSA Abstracts Programs* 26(7):A-111.

Pabalan, R. T. 1994. Thermodynamics of ion-exchange between clinoptilolite and aqueous solutions of Na^+/K^+ and Na^+/Ca^{2+}. *Geochim. Cosmochim. Acta* 58:4573–4590.

Pabalan, R. T., Prikryl, J. D., Muller, P. M., and Dietrich, T. B. 1993. Experimental study of uranium(6+) sorption on the zeolite mineral clinoptilolite. *In* "Materials Research Society Symposium Proceedings," Vol. 294, pp. 777–782, "Scientific Basis for Nuclear Waste Management XVI" (C. G. Interrante and R. T. Pabalan, Eds.). Materials Research Society, Boston, MA.

Pabalan, R. T., and Turner, D. T. 1997. Uranium(6+) sorption on montmorillonite: Experimental and surface complexation modeling study. *Aquatic Geochem.* 2:203–226.

Payne, T. E., Sekine, K., Davis, J. A., and Waite, T. D. 1992. Modeling of radionuclide sorption processes in the weathered zone of the Koongarra ore body. *In* "Alligator Rivers Analogue Project Annual Report, 1990–1991" (P. Duerden, Ed.), pp. 57–85. Australian Nuclear Science and Technology Organization (ANSTO), Menai NSW, Australia.

Prikryl, J. D., Pabalan, R. T., Turner, D. R., and Leslie, B. W. 1994. Uranium sorption on α-alumina: Effects of pH and surface-area/solution-volume ratio. *Radiochim. Acta* 66:291–296.

Rogers, P. S. Z., and Meijer, A. 1993. Dependence of radionuclide sorption on sample grinding, surface area, and water composition. *In* "Proceedings of the Fourth Annual International Conference on High Level Radioactive Waste Management," pp. 1509–1516. American Nuclear Society, La Grange Park, IL.

Tripathi, V. S. 1984. Uranium(VI) transport modeling: Geochemical data and submodels. Unpublished Ph.D. Thesis, Stanford Univ., Stanford, CA.

Turner, D. 1995. A Uniform Approach to Surface Complexation Modeling of Radionuclide Sorption. CNWRA 95-001, Center for Nuclear Waste Regulatory Analyses, San Antonio, Texas.

Turner, D. R., and Sassman, S. A. 1996. Approaches to sorption modeling for high-level waste performance assessment. *J. Contam. Hydrol.* 21:311–332.

Turner, G. D., Zachara, J. M., McKinley, J. P., and Smith, S. C. 1996. Surface-charge properties and UO_2^{2+} adsorption of a subsurface smectite. *Geochim. Cosmochim. Acta* 60:3399–3414.

Van Geen, A., Robertson, A. P., and Leckie, J. O. 1994. Complexation of carbonate species at the goethite surface: Implications for adsorption of metal ions in natural waters. *Geochim. Cosmochim. Acta* 58:2073–2086.

Van Olphen, H., and Fripiat, J. J. 1979. "Data Handbook for Clay Minerals and Other Non-metallic materials." Pergamon Press, Oxford, England.

Venkataramani, B., and Gupta, A. R. 1991. Effect of anions on the sorption of uranyl ions on hydrous oxides—Application of the surface hydrolyis model. *Colloids Surf.* 53:1–19.

Waite, T. D., Davis, J. A., Payne, T. E., Waychunas, G. A., and Xu, N. 1994. Uranium(VI) adsorption to ferrihydrite: Application of a surface complexation model. *Geochim. Cosmochim. Acta* 58: 5465–5478.

Wanner, H., Albinsson, Y., Karnl, O., Wieland, E., Wersin, P., and Charlet, L. 1994. The acid/base chemistry of montmorillonite. *Radiochim. Acta* 66/67:733–738.

Wanner, H., and Forest, I. (Eds.). 1992. "Chemical Thermodynamics of Uranium." Elsevier, New York.

Westall, J. C. 1982. FITEQL: A Computer Program for Determination of Chemical Equilibrium Constants from Experimental Data, Version 2.0. Rpt. 82-02, Department of Chemistry, Oregon State University, Corvallis, OR.

Westall, J. C., and Hohl, H. 1980. A comparison of electrostatic models for the oxide/solution interface. *Adv. Colloid Interface Sci.* 12:265–294.

White, G. N., and Zelazny, L. W. 1988. Analysis and implications of the edge structure of dioctahedral phyllosilicates. *Clays Clay Miner.* 36:141–146.

Wolery, T. J. 1992. EQ3NR, a Computer Program for Geochemical Aqueous Speciation-Solubility Calculations: Theoretical Manual, User's Guide, and Related Documentation (Version 7.0). UCRL-MA-110662, Pt. 3, Lawrence Livermore National Laboratory, Livermore, CA.

Zachara, J. M., and McKinley, J. P. 1993. Influence of hydrolysis on the sorption of metal cations by smectites: Importance of edge coordination reactions. *Aquatic Sci.* 55:250–261.

Zachara, J. M., and Smith, S. C. 1994. Edge complexation reactions of cadmium on specimen and soil-derived smectite. *Soil Sci. Soc. Am. J.* 58:762–769.

Chapter 4

Studies of Neptunium^{V} Sorption on Quartz, Clinoptilolite, Montmorillonite, and α-Alumina

F. Paul Bertetti,* Roberto T. Pabalan, and Michael G. Almendarez
Center for Nuclear Waste Regulatory Analyses, San Antonio, Texas

Because ^{237}Np is a radionuclide of concern in safety assessments of high-level nuclear waste repositories, sorption of Np^{V} onto quartz, clinoptilolite, montmorillonite, and α-alumina was studied by conducting batch experiments in which solution pH, Np^{V} concentration, ionic strength, P_{CO_2}, sorbent surface area, and the solid-mass to solution-volume ratio were varied. The results show that Np^{V} sorption on the four minerals studied have similar pH-dependent trends, even though the minerals have different mineralogic and surface charge characteristics. For experiments in equilibrium with atmospheric P_{CO_2}, a distinct sorption maximum is observed for all four minerals at a pH ~8 to 8.5, coincident with the pH of maximum aqueous concentration of $NpO_2OH^0(aq)$ complex, and there is reduction in sorption toward more alkaline conditions due to increased concentration of Np^{V} carbonate complexes and toward more acidic conditions due to increased stability of the NpO_2^+ species. For low-P_{CO_2} or CO_2-free conditions, Np^{V} sorption increases with increasing pH over the entire pH range studied, consistent with the increasing stability of Np^{V} hydroxy com-

*Present address: Cambrian Systems, Inc., 6502 Bandera Rd., San Antonio, TX 78238.

Adsorption of Metals by Geomedia

plexes in the aqueous phase at higher pH. These observations are compatible with those of other studies regarding the similarity in the tendency of actinides to form hydroxy complexes in solution and their tendency to interact with surface hydroxyl groups. The data also suggest that the magnitude of Np^{V} sorption (at a specific pH, initial Np^{V} concentration, and Pco_2) is essentially the same for different minerals if normalized to the number of available sites using an "effective" surface area. In addition, for the experimental conditions in this study, changes in M/V and ionic strength had little influence on Np^{V} sorption.

I. INTRODUCTION

Sorption onto minerals present along groundwater flow paths may be an important mechanism for retarding the migration of radionuclides from nuclear waste repositories. ^{237}Np has been identified as a radionuclide of concern with respect to the disposal of high-level nuclear waste (HLW), especially at longer time frames ($\sim$10,000 yr) (Wilson *et al.,* 1994; TRW, 1995; Wescott *et al.,* 1995), due to its long half-life ($2.14 \cdot 10^6$ yr), suspected high radiotoxicity (Thompson, 1982), and reportedly low sorption characteristics (e.g., Triay *et al.,* 1993). Because of worldwide interest in geologic disposal of HLW, the number of investigations on the sorption behavior of Np have increased over the past several years (e.g., Allard *et al.,* 1984; Bidoglio *et al.,* 1988; Lieser and Mühlenweg, 1988; Nakayama and Sakamoto, 1991). However, numerous uncertainties remain with respect to the magnitude of Np sorption under the oxidizing conditions and in bicarbonate-rich groundwaters relevant to the proposed HLW repository at Yucca Mountain, Nevada. For example, reported distribution coefficients (K_d) for Np sorption on quartz or quartz-bearing rocks vary significantly and are dependent on the initial Np concentration, mineral impurities present, and activity of CO_2 (e.g., Allard *et al.,* 1984; Nakayama *et al.,* 1988; Triay *et al.,* 1993).

In this study, quartz, clinoptilolite, montmorillonite, and α-alumina were reacted with ^{237}Np-bearing solutions to characterize the sorption behavior of Np^{V} on these minerals. These sorbents were selected because their mineralogic and surface charge characteristics are distinct from each other. For example, the reported points-of-zero-charge (pH_{pzc}) for quartz, clinoptilolite, and α-alumina are 2.9, 3.0, and 9.1, respectively (Davis and Kent, 1990; Gainer, 1990), and the pH_{pzc} for montmorillonite is in the range 6.5 to 8, depending on the ionic strength of the background electrolyte (Wanner *et al.,* 1994). In addition, quartz, clinoptilolite, and montmorillonite are representative of mineral phases that occur both in the rock matrix and as fracture coatings at Yucca Mountain (Bish and Chipera, 1989).

II. EXPERIMENTAL PROCEDURE

A. Mineral Preparation and Characterization

Minerals were acquired from readily available commercial sources. Clinoptilolite was obtained from a clinoptilolite-rich tuff from Death Valley Junction, California (Minerals Research Co., Clarkson, New York). The tuff was crushed and sieved to isolate the 100 to 200 mesh (0.150 to 0.074 mm) fraction and chemically pretreated to remove soluble salts, carbonates, Fe-oxides, and other mineral impurities. The clinoptilolite was then converted to its Na-form by ion exchange with 3 mol·L^{-1} NaCl solutions. Details of the clinoptilolite pretreatment can be found in Pabalan (1994). Quartz was obtained as quartz sand (Wedron #510) from Wedron Silica Co. (Wedron, Illinois). The quartz sand was sieved to isolate the 60 to 100 mesh (0.250 to 0.149 mm) fraction and was pretreated to remove impurities in a fashion similar to that used for clinoptilolite. A portion of the 60 to 100 mesh fraction was crushed using an agate mortar and pestle to create a fine-grained fraction (0.044 to 0.004 mm) for use in evaluating the effects of changes in specific surface area on sorption. Montmorillonite (SAz-1, Apache Co., Arizona) was obtained from the Source Clays Mineral Repository (Columbia, Missouri). Following ion exchange with 3 mol·L^{-1} NaCl and isolation of the less than 2-μm size fraction, the clay was lyophilized for storage prior to use in experiments. The α-alumina (α-Al_2O_3) was obtained as a powder from the National Institute of Standards and Technology (NIST), which issues the material as a certified reference material for measurements of specific surface area of powders. Reference material (RM) 8006, with a reported surface area of 0.229 m^2·g^{-1}, was used in the sorption experiments. To minimize altering the surface of the α-alumina, the powder was not washed or pretreated prior to its use.

The external surface areas of the sorbent materials were determined using a multipoint N_2-BET isotherm measured using a Coulter SA3100 surface area analyzer. The surface area samples were prepared by outgassing for 24 hr at 350°C (100°C in the case of montmorillonite). Surface areas measured for quartz, fine-grained quartz, clinoptilolite, montmorillonite, and α-alumina were 0.03 ± 0.01, 0.5 ± 0.05, 10.1 ± 0.3, 97 ± 2, and 0.23 ± 0.01 mg^2·g^{-1}, respectively. The purity of the mineral sorbents was verified by XRD, SEM/EDS, and ICP analyses.

B. ^{237}Np Solutions

Neptunium-bearing solutions were prepared by diluting a purchased ^{237}Np standard (Isotope Products, Burbank, California) in a 0.01 or 0.1 mol·L^{-1} $NaNO_3$ matrix to make a stock solution of $1 \cdot 10^{-6}$ mol·L^{-1} ^{237}Np. Lower-concentration solutions were made by dilution of the stock solution with 0.01 or 0.1 mol·L^{-1} $NaNO_3$. Initial Np^V concentrations used in the experiments were either $1 \cdot 10^{-7}$ or

$1 \cdot 10^{-6}$ mol·L^{-1}. The oxidation state of Np^V was confirmed by (i) observing the 981-nm peak of NpO_2^+ using near-infrared (NIR) spectroscopy (Perkin Elmer Lambda 9) in stock solutions of 10^{-4} and 10^{-5} mol·L^{-1} Np, and (ii) solvent exchange and subsequent activity counting (Bertrand and Choppin, 1982) in solutions of $1 \cdot 10^{-6}$ mol·L^{-1} Np. For all solutions, the activity of ^{237}Np was determined by counting its α-activity in an α–β discriminating liquid scintillation analyzer (Packard 2505 TR/AB) for a period of time such that the 2σ error of the reported sample activity in counts per minute (cpm) was ±3%. Neptunium activity was subsequently converted to mass of Np^V in solution (or on solid) for calculation of all sorption results. Uncertainties in sorption data presented in later sections were propagated based on the 2σ counting errors.

C. Batch Experiments

The initial conditions of the Np^V sorption experiments are summarized in Table I. Equilibrium batch experiments were conducted at room temperature (20 ± 2°C) and over a pH range of about 4 to 11 for all four minerals. Solution pH was adjusted using HNO_3, $NaHCO_3$, or NaOH; pH was measured using a glass combination electrode (Orion-Ross). Generally, the solids were immersed in 25 ml of Np^V solution

Table I

Summary of Np^V Sorption Experiment Initial Conditions[a]

Experiment	Mineral	M/V (g·L^{-1})	Initial Np concentration (mol·L^{-1})	Concentration of $NaNO_3$ matrix (mol·L^{-1})	P_{CO_2} conditions
NpQ1	Quartz	40	$1 \cdot 10^{-7}$	0.1	Atmosphere
NpQ2	Quartz	80	$1 \cdot 10^{-7}$	0.1	Atmosphere
NpQ3	Quartz	40	$1 \cdot 10^{-6}$	0.1	Atmosphere
NpQ4	Quartz	40	$1 \cdot 10^{-6}$	0.1	Capped
NpQ5	Quartz (fine-grained)	4	$1 \cdot 10^{-6}$	0.1	Capped
NpC1	Na-clinoptilolite	4	$1 \cdot 10^{-6}$	0.1	Atmosphere
NpC2	Na-clinoptilolite	4	$1 \cdot 10^{-6}$	0.1	Capped
NpC3	Na-clinoptilolite	8	$1 \cdot 10^{-6}$	0.01	Atmosphere
NpC4	Na-clinoptilolite	8	$1 \cdot 10^{-6}$	0.01	Glove box
NpM1	Na-montmorillonite	4	$1 \cdot 10^{-6}$	0.1	Capped
NpM2	Na-montmorillonite	4	$1 \cdot 10^{-6}$	0.1	Atmosphere
NpM3	Na-montmorillonite	4	$1 \cdot 10^{-6}$	0.1	Glove box
NpA1	α-Alumina	4	$1 \cdot 10^{-6}$	0.01	Capped
NpA2	α-Alumina	4	$1 \cdot 10^{-6}$	0.01	Glove box

[a]All experiments were conducted with a solution volume of 25 ml.

contained in 50-ml polycarbonate Oak Ridge-type centrifuge tubes. Appropriate amounts of acid or base, estimated using the EQ3NR geochemical code (version 7.2a with database data0.com.R22a) (Wolery, 1992), were then added to the experimental mixtures. The containers were left open for experiments in which solutions were allowed to equilibrate with atmospheric P_{CO_2}. Experiments at low P_{CO_2} were conducted by (i) immediately capping the tubes after initial pH adjustment, or (ii) conducting the experiment in a controlled-atmosphere glove box using certified CO_2-free (less than 1 ppm CO_2) simulated air (21% O_2 and 79% N_2 by volume).

The experimental mixtures were agitated using gyratory shakers for at least 14 days to allow for equilibrium of both pH and the sorption reaction. Kinetics experiments conducted with clinoptilolite and $1 \cdot 10^{-6}$ mol·L^{-1} Np^{V} showed that the sorption reactions were ~98% complete within 48 hr, but some mixtures required up to 10 days for equilibration of pH due to somewhat slow mass transfer of CO_2(g) into the aqueous phase during equilibration to atmospheric P_{CO_2}. Reversibility of the sorption reactions on some of the minerals was confirmed by changing the pH of experimental mixtures, reequilibrating for about 10 days, and resampling the experimental solutions.

To determine the amount of Np^{V} sorbed on the minerals after sorption equilibrium had been reached, Np^{V} was desorbed off the solid surface with HNO_3 solution. The procedure involved centrifuging the experimental mixtures to separate the solid from the aqueous phase, taking samples from the solution for Np analysis, decanting most of the aqueous phase into a polypropylene tube, and adding ~8 ml of ~0.04 mol·L^{-1} HNO_3 to the polycarbonate tube, which contained the solid and some amount of solution, to desorb the Np. The solids were subsequently redispersed and a second set of samples for Np analysis were taken from the acidic solutions after about 10 days. Additional tests were also conducted to evaluate potential losses of Np^{V} to container walls. These tests indicated that under the conditions of the experiments in this study, sorption of Np^{V} onto the polycarbonate containers was negligible.

III. RESULTS AND DISCUSSION

The distribution coefficient (K_d) is a convenient empirical ratio for representation of sorption data and is commonly used to represent retardation in transport models. The K_d (ml·g^{-1}) can be defined as

$$K_d\ (\text{ml}\cdot\text{g}^{-1}) = \frac{\text{equilibrium mass of Np sorbed on solid}}{\text{equilibrium mass of Np in solution}}\left(\frac{V}{M}\right),$$

where V is the volume of experimental solution in milliliters, and M is the mass of solid in grams. The use of K_d has the effect of normalizing sorption results to the solid-mass to solution-volume (M/V) ratio used in the experiments and provides a means of accounting for the change in solution concentration that occurs during the course of the experiment. Results are presented in the following sections in terms of K_d versus pH or in terms of K_a ($ml \cdot m^{-2}$) versus pH, where K_a is K_d normalized to the mineral's measured specific surface area (S_A) (i.e., $K_a = K_d/S_A$).

A. Quartz

The results of experiments investigating sorption of Np^V onto coarse-grained quartz are shown in Figure 1. The data show that Np^V sorption onto quartz is strongly influenced by solution pH. Although the magnitude of sorption is low over the entire pH range studied, Np sorption onto quartz at pH values below about 8 increases with increasing pH regardless of the P_{CO_2} conditions. For experiments in equilibrium with atmospheric P_{CO_2}, sorption reaches a maximum at about pH 8 to 8.5 and decreases again at more alkaline conditions. On the other hand, for low-P_{CO_2} conditions sorption continues to increase throughout the pH range stud-

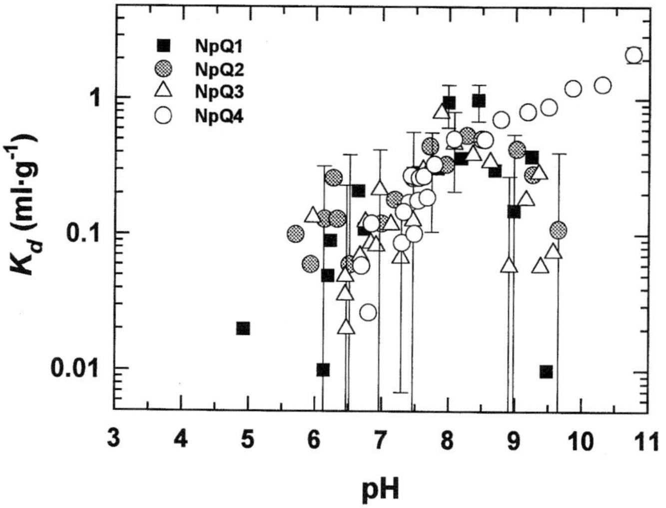

Figure 1 Sorption of Np^V on quartz. Data from experiments NpQ1, NpQ2, and NpQ3 are from solutions in equilibrium with atmospheric P_{CO_2}, whereas NpQ4 was conducted at low P_{CO_2} (capped vials). NpQ1 and NpQ2 had similar initial Np concentrations ($1 \cdot 10^{-7}$ mol·L^{-1}), but different M/V ratios (40 and 80 g·L^{-1}, respectively). Initial experimental conditions are given in Table I. Error bars for selected points show propagated uncertainties based on 2σ counting errors.

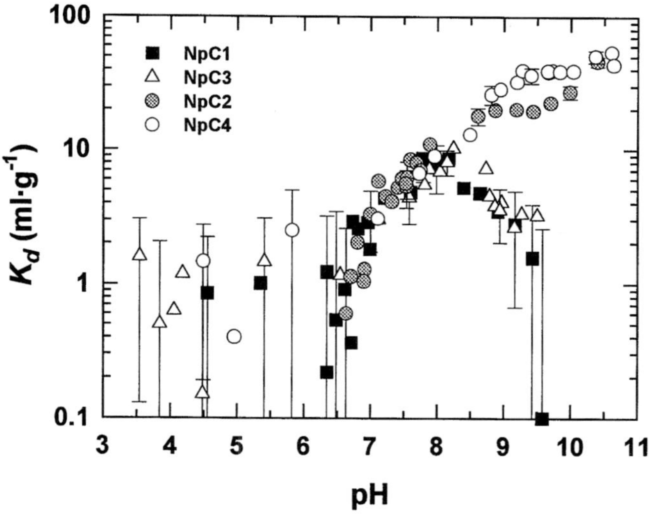

Figure 2 Sorption of Np^{V} on clinoptilolite. Data from NpC1 and NpC3 are from solutions in equilibrium with atmospheric P_{CO_2}, whereas experiments NpC2 and NpC4 were conducted under low-P_{CO_2} (capped vials) or CO_2-free (glove box) conditions. Initial experimental conditions are given in Table I Error bars for selected points show propagated uncertainties based on 2σ counting errors.

ied. A change in the initial concentration of Np^{V} from $1 \cdot 10^{-6}$ to $1 \cdot 10^{-7}$ mol·L^{-1} resulted in no observable effect on the magnitude of Np^{V} solution. Likewise, K_d data plotted in Figure 1 show that changing the M/V ratio from 40 to 80 g·L^{-1} made no difference in the magnitude of Np^{V} sorption.

The trend of increasing sorption with increasing pH throughout the pH range studied under low-P_{CO_2} conditions has also been observed in column-based studies of Np sorption onto quartz (Nakayama *et al.*, 1988). Although other studies indicated much larger K_d for Np^{V} sorption onto SiO_2 (Allard *et al.*, 1984; Nakayama *et al.*, 1988; Righetto *et al.*, 1991), these other studies used much lower initial Np^{V} concentrations and high-surface-area colloidal silica or untreated quartz that could have contained highly sorptive mineral impurities.

B. Clinoptilolite

Similar to the data for quartz, results of experiments on Np^{V} sorption onto clinoptilolite also show strong pH dependence (Fig. 2). Sorption increases with increasing pH for conditions with and without CO_2 present. For solutions in equilibrium with atmospheric P_{CO_2}, Np sorption onto clinoptilolite reaches a maximum K_d of about 10 near pH 8 and decreases again with further increases in pH.

In contrast, Np sorption under low-$P\text{CO}_2$ or CO_2-free conditions exhibits continued increase in sorption as the solution becomes more alkaline, with maximum sorption measured at the highest pH (~10) studied. The difference in K_d observed for the capped vials compared to the glove box experiments between pH 9 and 10 is indicative of CO_2 contamination in some of the capped vial experiments which resulted in formation of Np^V aqueous carbonate complexes.

Variation in ionic strength of the $NaNO_3$ matrix from 0.1 to 0.01 mol·L^{-1} had no discernible effect on sorption for the pH range studied. Clinoptilolite has been inferred to sorb UO_2^{2+} at low pH and low ionic strength through an ion exchange mechanism (Andreeva and Chernyavskaya, 1982; Pabalan *et al.*, 1993), although UO_2^{2+} ion exchange is apparently suppressed in Na^+ concentrations of 0.01 mol·L^{-1} or higher (Pabalan *et al.*, this volume). Thus, it is possible that the relatively high Na^+ solution concentrations used in this study also suppressed ion exchange between NpO_2^+ in solution and intracrystalline Na^+ in clinoptilolite.

C. Montmorillonite

Over the entire pH range considered, the magnitude of Np^V sorption onto Na–montmorillonite (Fig. 3) is greater than that observed for the other minerals studied. However, the sorption trends as a function of pH are very similar to those observed for the other minerals. For solutions in equilibrium with atmospheric $P\text{CO}_2$, maximum sorption occurs near pH 8 to 8.5 and decreases toward more acidic or alkaline values. Solutions that remained under low-$P\text{CO}_2$ or CO_2-free conditions show a continuous increase in sorption with increasing pH over the pH range examined. There is good agreement between data from forward and reverse experiments (open and closed circles, respectively), which indicates that sorption of Np^V onto montmorillonite is reversible.

A comparison of results from this study with data on Np^V sorption on Na-smectite (Kozai, 1994) shows poor agreement between the two data sets, except at pH values above 6 (Fig. 3). The Na–smectite data show an increase in Np^V sorption with decreasing pH below a pH of about 5.5, whereas sorption results from this study decrease with decreasing pH. The experiments reported by Kozai (1994) were conducted using solutions with Na^+ concentrations lower than those used in this study [0.01 mol·L^{-1} $NaClO_4$ and $6 \cdot 10^{-7}$ mol·L^{-1} ^{237}Np (T. Banba, personal communication)]. The higher Np^V sorption at lower pH has been attributed to ion exchange of NpO_2^+ in solution and interlayer Na^+ in smectite (T. Banba, personal communication) which is analogous to ion exchange observed between UO_2^{2+} solutions and montmorillonite at low pH and low ionic strength (Zachara and McKinley, 1993; McKinley *et al.*, 1995). The lower Np^V sorption observed in this study at pH values of less than 5.5 com-

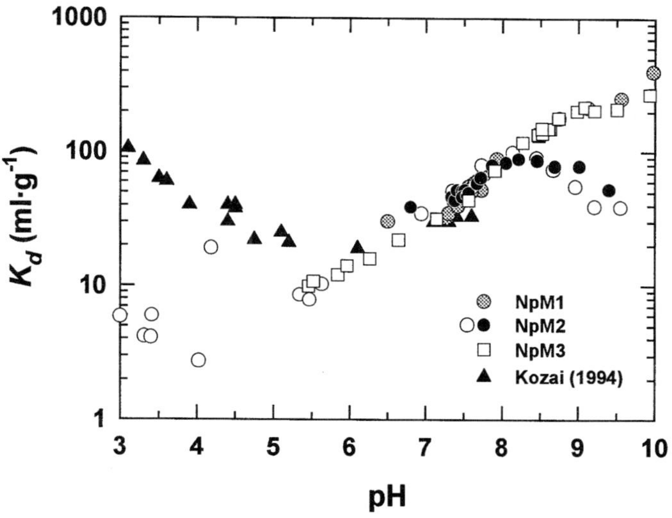

Figure 3 Sorption of Np^V on montmorillonite. Experiment NpM2 was conducted in equilibrium with atmospheric P_{CO_2}, whereas experiments NpM1 and NpM3 were conducted under low-P_{CO_2} (capped vials) or CO_2-free (glove box) conditions. Initial experimental conditions are given in Table I. The open and closed circles represent data from the forward and reverse phase, respectively, of experiment NpM2. Also shown are data on Np^V sorption on Na-smectite ($\Sigma Np_i \sim 6 \cdot 10^{-7}$ M, ionic strength ~0.01 mol·L^{-1} $NaClO_4$; M/V ~10 g·L^{-1}) from Kozai (1994).

pared to data from Kozai (1994) is likely a result of the higher Na^+ solution concentration (0.1 mol·L^{-1} $NaNO_3$) which suppressed the ion exchange reaction. Note that in contrast to the clay data, Np^V sorption on clinoptilolite through an ion exchange mechanism is negligible in a 0.01 mol·L^{-1} $NaNO_3$ matrix, probably due to steric hindrance arising from the more rigid crystal structure of the zeolite mineral.

D. α-ALUMINA

Experiments using α-alumina (RM 8006) were conducted under low-P_{CO_2} (capped vials) or CO_2-free (glove box) conditions. The results in Figure 4 show that Np^V sorption onto α-alumina has a similar pH dependence as Np^V sorption onto quartz, clinoptilolite, and montmorillonite, and has a magnitude which is about the same as that observed for clinoptilolite. Neptunium^V sorption data from Nakayama and Sakamoto (1991) plotted in Figure 4 also reveal a similar pH trend, although their results show much higher sorption values probably due to the larger surface area of α-alumina (2.5 m^2·g^{-1}) used in their experiments.

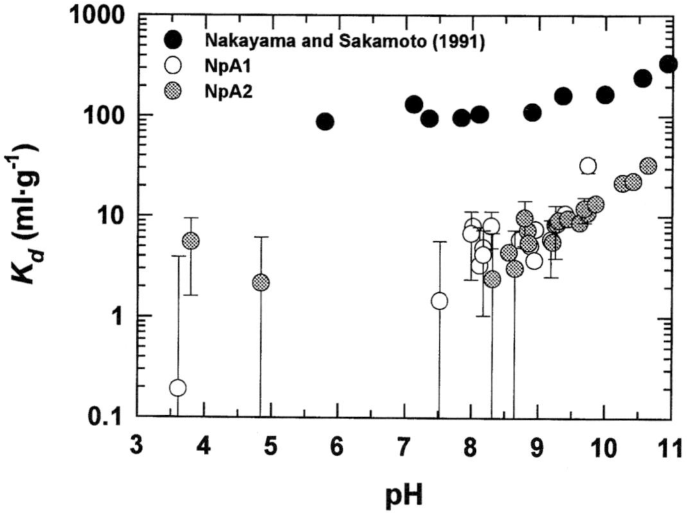

Figure 4 Sorption of Np^V on α-alumina. Experiments NpA1 and NpA2 were conducted under low-P_{CO_2} (capped vials) or CO_2-free (glove box) conditions. Initial experimental conditions are given in Table I. Also shown are data from Nakayama and Sakamoto (1991) who used α-alumina with a reported surface area of 2.5 $m^2 \cdot g^{-1}$ ($\Sigma Np_i = 6 \cdot 10^{-6}$ M, ionic strength = 0.1 mol·L^{-1} $NaNO_3$; M/V = 1 g·L^{-1}; capped bottles). Error bars for selected points show propagated uncertainties based on 2σ counting errors.

E. Surface Area Effects

As shown previously, Np^V sorption on quartz, clinoptilolite, montmorillonite, and α-alumina is similar with respect to pH dependence. However, at a given pH, the K_d for the different minerals varies by over two orders of magnitude. This variation is an artifact of representing data in terms of K_d, which normalizes the amount of Np^V sorbed to the sorbent mass and not to the number of available sorption sites on the mineral surface. Surface areas measured by gas adsorption (e.g., N_2-BET) methods can be used as a relative measure of the number of sorption sites on the mineral surface. The effect of increasing quartz surface area on Np^V sorption was studied by using two types of quartz powder of different grain size and surface area. As shown by the data in Figure 5A, the fine-grained quartz sorbed significantly more Np^V compared to the coarse-grained quartz. However, when normalized with respect to surface area, Np^V sorption (K_a) onto the two types of quartz material is nearly indistinguishable (Fig. 5B). Moreover, when α-alumina data from this study and from Nakayama and Sakamoto (1991) are surface area normalized, the apparent differences in the magnitude of Np^V solution between the different data sets are mostly eliminated (Figs. 5A and 5B). The reason for the apparent divergence from the general trend at low pH of some of Nakayama and Sakamoto's (1991) results is not clear.

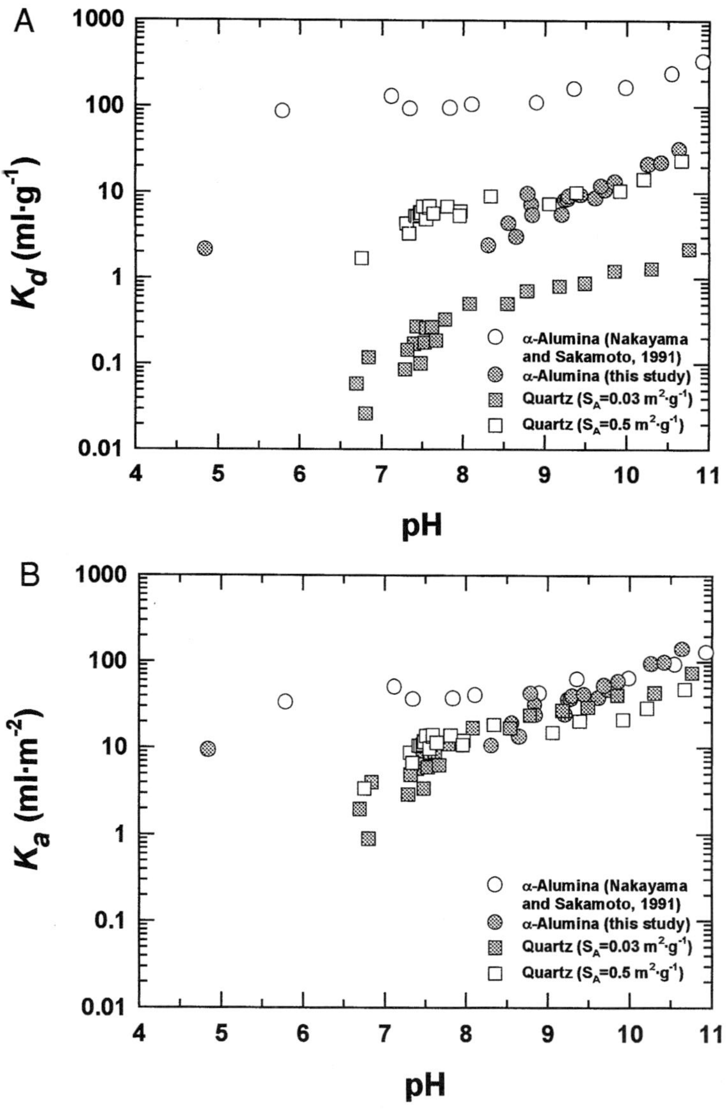

Figure 5 Effect of surface area and normalization on Np^{V} sorption. (A) K_d data from experiments using fine-grained quartz (0.5 $m^2 \cdot g^{-1}$; NpQ5 data), coarse-grained quartz (0.03 $m^2 \cdot g^{-1}$; NpQ4 data), and α-alumina (0.23 $m^2 \cdot g^{-1}$; NpA2 data). Initial experimental conditions are given in Table I. Also plotted are K_d data from Nakayama and Sakamoto's (1991) experiments which used α-alumina with a reported specific surface area of 2.5 $m^2 \cdot g^{-1}$. (B) Sorption data in Figure 5A normalized to the sorbent's specific surface area ($K_a = K_d/S_A$).

A comparison of surface-area-normalized sorption data for quartz, clinoptilolite, montmorillonite, and α-alumina (Fig. 6) appears to indicate that quartz and α-alumina sorb more Np^V per unit area than clinoptilolite and montmorillonite. This "excess" Np^V sorption by quartz relative to clinoptilolite and montmorillonite has also been observed by others (e.g., Triay *et al.,* 1993). However, surface complex formation of actinides on montmorillonite has been inferred to occur primarily on the hydroxylated edge sites of the clay mineral (Zachara and McKinley, 1993; McKinley *et al.,* 1995). Wanner *et al.* (1994) estimated that only 10% of the N_2-BET specific surface area is accounted for by the crystallite edges of montmorillonite. Thus, an "effective" surface area (S_{EA}) may be defined for layered silicates which, in the case of montmorillonite, is equivalent to 10% of its measured N_2-BET specific surface area. Assuming that the zeolite mineral clinoptilolite, because of its porous, open-framework crystal lattice, can be treated in a similar fashion, sorption data for montmorillonite and clinoptilolite can be recast in terms of $K_{a'}$, where $K_{a'}$ is K_d normalized to the mineral's S_{EA} (i.e., $K_{a'} = K_d/S_{EA}$). For nonlayered and nonporous minerals such as quartz and α-alumina, $K_a = K_{a'}$. Figure 7 plots Np^V sorption $K_{a'}$ values versus pH for quartz, clinoptilolite, montmorillonite, and α-alumina. As shown in the figure, Np^V sorption on these minerals is essentially equivalent when recast in terms of $K_{a'}$. The apparent scatter of data points at low pH is a result of larger uncertainties in the experimental data at low sorption values.

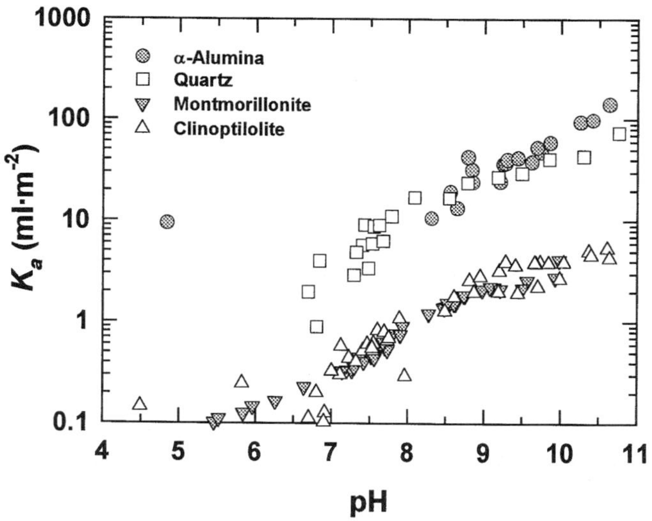

Figure 6 Comparison of K_a versus pH for α-alumina (NpA2), quartz (NpQ4), montmorillonite (NpM1, NpM3), and clinoptilolite (NpC2, NpC4) for data sets determined under similar experimental conditions ($\Sigma Np_i = 1.0 \cdot 10^{-6}$ mol·L^{-1}; low-P_{CO_2} or CO_2-free conditions).

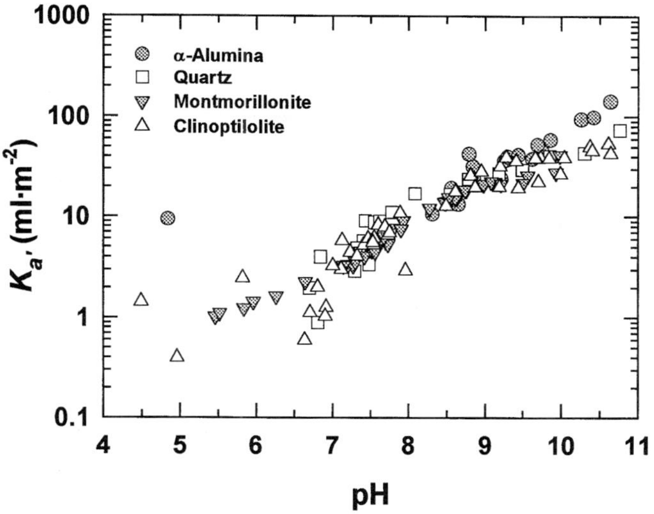

Figure 7 Comparison of $K_{a'}$ (where $K_{a'} = K_d/S_{EA}$) calculated from the results shown in Figure 6. For clinoptilolite and montmorillonite, S_{EA} is assumed equal to 10% of the mineral's N_2-BET specific surface area, S_A. For α-alumina and quartz, $S_{EA} = S_A$.

F. Neptunium^{V} Speciation

Comparisons of experimental conditions used in this study with solubility data for NpO_2OH(s), Np_2O_5, and Na–neptunyl-carbonates (Itagaki *et al.,* 1992; Lemire *et al.,* 1993; Nitsche *et al.,* 1993, 1995; Neck *et al.,* 1995) indicate that the experimental solutions remained undersaturated with respect to these phases under all experimental conditions; therefore, the possibility of Np precipitation is unlikely. The results presented here compare favorably with other studies that have correlated a decrease in Np^{V} sorption with an increase in Np–carbonate complexation in solution (Bidoglio *et al.,* 1985, 1987). Also of note are published data on Np^{V} sorption on other minerals such as biotite, hematite, and feldspar. Though reported magnitudes of sorption on these minerals vary, the continuous increase in sorption with increasing pH when solutions are under low-P_{CO_2} conditions were also observed (e.g., Allard *et al.,* 1984; Nakayama and Sakamoto, 1991; Righetto *et al.,* 1991).

To aid in evaluating possible surface species, it is useful to compare Np^{V} sorption behavior with Np^{V} aqueous speciation. Speciation calculations were performed with the equilibrium geochemical code MINTEQA2 (Allison *et al.,* 1991) using a database modified to include relevant actinide species (Turner, 1993). Equilibrium constants for Np species used in the calculations and the pertinent references are given in Table II. The calculated relative stabilities of Np^{V} species in

Table II
Equilibrium Constants Used in Np^V Aqueous Speciation Calculations

Np^V species reaction	log K	References
$NpO_2^+ + H_2O \rightleftharpoons NpO_2(OH)^0(aq) + H^+$	−10.0	Lemire and Garisto (1989)
$NpO_2^+ + 2H_2O \rightleftharpoons NpO_2(OH)_2^- + 2H^+$	−22.4	Lemire and Garisto (1989), Fuger (1992)
$NpO_2^+ + CO_3^{2-} \rightleftharpoons NpO_2(CO_3)^-$	4.6	Lemire (1984), Fuger (1992), Lemire *et al.* (1993)
$NpO_2^+ + 2(CO_3^{2-}) \rightleftharpoons NpO_2(CO_3)_2^{3-}$	7.0	Lemire (1984), Fuger (1992), Lemire *et al.* (1993)
$NpO_2^+ + 3(CO_3^{2-}) \rightleftharpoons NpO_2(CO_3)_3^{5-}$	8.5	Lemire (1984), Lemire *et al.* (1993)
$NpO_2^+ + NO_3^- \rightleftharpoons NpO_2(NO_3)^0(aq)$	−0.5	Danesi *et al.* (1971), Patil *et al.* (1978)

a $1 \cdot 10^{-6}$ mol·L^{-1} Np solution (0.1 mol·L^{-1} $NaNO_3$ matrix) are shown in Figures 8 and 9 for conditions in equilibrium with atmospheric P_{CO_2} and without CO_2 present, respectively.

As shown in these figures, Np^V aqueous speciation is dominated by NpO_2^+ at pH values below 7 whether CO_2 is present or not. However, near pH ~7, Np hydrolysis becomes significant (~0.1 mole percent of Np in solution) and the amount of the Np^V hydroxy species, $NpO_2OH^0(aq)$, increases with increasing pH. Under atmospheric P_{CO_2} conditions (Fig. 8), the stability of the neutral hydroxy species

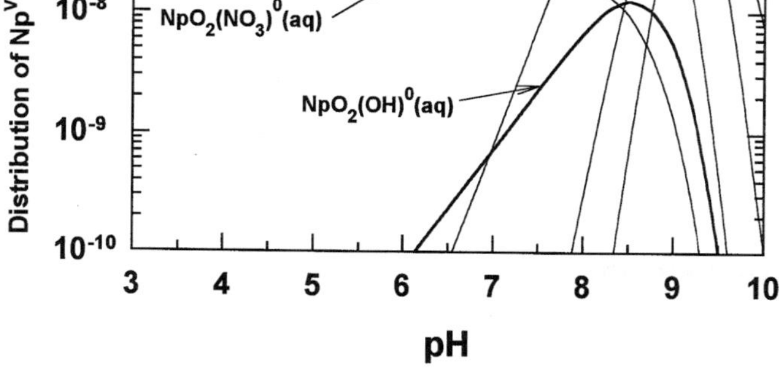

Figure 8 Calculated speciation of a $1 \cdot 10^{-6}$ mol·L^{-1} Np^V solution (0.1 mol·L^{-1} $NaNO_3$ matrix) in equilibrium with atmospheric P_{CO_2}. Log K values for the species shown are listed in Table II.

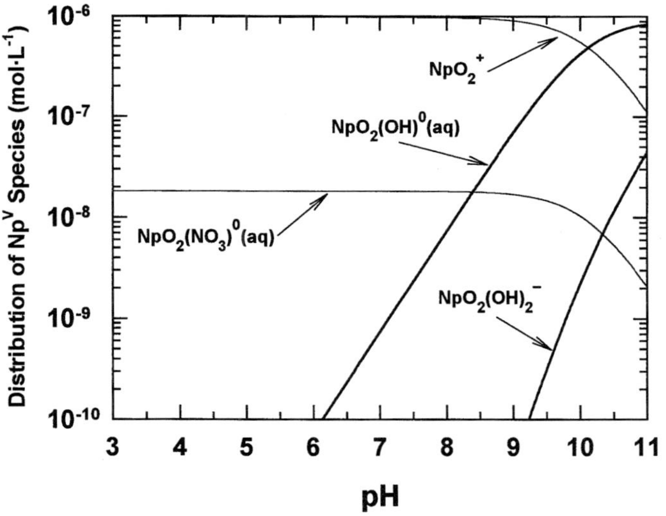

Figure 9 Calculated speciation of a $1 \cdot 10^{-6}$ mol·L^{-1} Np^{V} solution (0.1 mol·L^{-1} $NaNO_3$ matrix) in the absence of CO_2. log K values for the species shown are listed in Table II.

reaches a maximum near pH 8.5 and decreases with further increases in pH. Although the neutral hydroxy complex does not become a predominant species in this solution, a comparison of Np^{V} sorption data and aqueous speciation indicates that the pH dependence of the stability of the NpO_2OH^0(aq) species is distinctly similar to the pH dependence of Np^{V} sorption from solutions in equilibrium with atmospheric P_{CO_2}. In the absence of CO_2, the relative stability of Np^{V} hydroxy complexes increases continuously with increasing pH and eventually dominates Np^{V} aqueous speciation (Fig. 9). Again, the pH dependence of the relative stability of the Np^{V} hydroxy species mimics the observed pH-dependent sorption behavior of Np^{V} under similar conditions. These observations are consistent with those of other studies (e.g., Beall and Allard, 1981; Allard *et al.,* 1984; Righetto *et al.,* 1991; Pabalan *et al.,* this volume) regarding the similarity in the tendency of actinides to form hydroxy complexes in solution and their tendency to interact with surface hydroxyl groups.

IV. CONCLUSIONS

The pH-dependent trends in observed Np^{V} sorption behavior are similar for clinoptilolite, quartz, α-alumina, and montmorillonite, even though these minerals have different mineralogic and surface charge characteristics. This similarity suggests that Np^{V} sorption is not sensitive to the surface charge properties of the sorbent as compared to the effect of changes in the surface area or number of sorption sites. On the other hand, Np^{V} sorption is strongly dependent on pH. For experiments with solutions in equilibrium with atmospheric P_{CO_2}, a distinct sorption maximum is observed for all four minerals. The pH of maximum sorption in these solutions is ~8 to 8.5, which is coincident with the pH of maximum aqueous concentration of the $NpO_2OH^0(aq)$ species, and there is a distinct reduction in sorption toward more alkaline conditions due to increased concentration of Np^{V}carbonate complexes and toward more acidic conditions due to increased concentration of the NpO_2^+ [and $NpO_2(NO_3)^0(aq)$] species. For conditions in which Np solutions are under low-P_{CO_2} or CO_2-free conditions, Np^{V} sorption increases with increasing pH over the entire pH range studied, which is consistent with the increasing stability of Np^{V} hydroxy complexes in the aqueous phase at higher pH. The data also suggest that the magnitude of Np^{V} sorption (at a specific pH, initial Np^{V} concentration, and P_{CO_2}) is essentially the same for different minerals if normalized to the number of available sites using an "effective" surface area.

Observations of data from this study and those derived from similar studies on sorption of U^{VI} and other actinides (e.g., Beall and Allard, 1981; Allard *et al.,* 1984; Righetto *et al.,* 1991; Pabalan *et al.,* this volume) indicate a common pattern for actinide sorption that is related to the formation of hydroxy complexes in solution. This pattern suggests that modeling approaches that are capable of accounting for changes in solution chemistry (e.g., surface complexation models) are required for successful description and prediction of sorption of Np and other actinides on mineral surfaces over wide ranges of geochemical conditions.

ACKNOWLEDGMENTS

The reviews by D. Turner, W. Patrick, E. Jenne, and two anonymous reviewers are gratefully acknowledged. This work was funded by the U.S. Nuclear Regualtory Commission (NRC), Office of Nuclear Regulatory Research, Division of Regulatory Applications, and by the NRC Office of Nuclear Material Safety and Safeguards, Division of Waste Management, under Contract No. NRC-02-93-005. This paper is an independent product of the CNWRA and does not necessarily reflect the views or regulatory position of the NRC.

REFERENCES

Allard, B., Olofsson, U., and Torstenfelt, B. 1984. Environmental actinide chemistry. *Inorg. Chim. Acta* 94:205–221.

Allison, J. D., Brown, D. S., and Novo-Gradac, K. J. 1991. MINTEQA2/PRODEFA2, a Geochemical

Assessment Model for Environmental Systems: Version 3.0 User's Manual. EPA/600/3-91/021, Environmental Protection Agency, Athens, GA.
Andreeva, N. R., and Chernyavskaya, N. B. 1982. Uranyl ion sorption by mordenite and clinoptilolite. *Radiokhimiya* 24:9–13.
Beall, G. W., and Allard, B. 1981. Sorption of actinides from aqueous solutions under environmental conditions. "Adsorption from Aqueous Solutions" (P. H. Tewari, Ed.), pp. 193–212. Plenum Press, New York.
Bertrand, P. A., and Choppin, G. R. 1982. Separation of actinides in different oxidation states by solvent extraction. *Radiochim. Acta* 31:135–137.
Bidoglio, G., Avogadro, A., DePlano, A., and Lazarri, G. P. 1988. Reaction pathways of Pu and Np in selected natural water environments. *Radiochim. Acta* 44/45:29–32.
Bidoglio, G., Offermann, P., and Saltelli, A. 1987. Neptunium migration in oxidizing clayey sand. *Appl. Geochem* 2:275–284.
Bidoglio, G., Tanet, G., and Chatt, A. 1985. Studies on neptunium(V) carbonate complexes under geologic repository conditions. *Radiochim. Acta* 38:21–26.
Bish, D. L., and Chipera, S. J. 1989. Revised Mineralogic Summary of Yucca Mountain, Nevada. LA-11497-MS, Los Alamos National Laboratory, Los Alamos, NM.
Danesi, P. R., Chiarizia, R., Scibona, G., and D'Alessandro, G. J. 1971. Stability constants of nitrate and chloride complexes of Np(IV), Np(V) and Np(VI) ions. *Inorg. Nucl. Chem.* 33:3503–3510.
Davis, J. A., and Kent, D. B. 1990. Surface complexation modeling in aqueous geochemistry. *In* "Reviews in Mineralogy," Vol. 23, pp. 177–260, "Mineral-Water Interface Geochemistry" (M. F. Hochella, Jr., and A. P. F. White, Eds.). Mineralogical Society of America, Washington, DC.
Fuger, J. 1992. Thermodynamic properties of actinide species relevant to geochemical problems. *Radiochim. Acta* 59:81–91.
Gainer, G. M. 1990. Boron adsorption on hematite and clinoptilolite. Unpublished M.S. Thesis, Univ. of Texas at El Paso, TX.
Itagaki, H., Nakayama, S., Tanaka, S., and Yamawaki, M. 1992. Effect of ionic strength on the solubility of neptunium(V) hydroxide. *Radiochim. Acta* 58/59:61–66.
Kozai, N. 1994. Sorption characteristics of neptunium by smectite. *In* Progress Report on Safety Research on Radioactive Waste Management for the Period April 1992 to March 1993, pp. 39–41. JAERI-M94-027, Japan Atomic Energy Research Institute, Tokai, Japan.
Lemire, R. J. 1984. An Assessment of the Thermodynamic Behavior of Neptunium in Water and Model Groundwaters from 25 to 150°C. AECL-7817, Atomic Energy of Canada Limited, Pinawa, Manitoba.
Lemire, R. J., Boyer, G. D., and Campbell, A. B. 1993. The solubilities of sodium and potassium dioxoneptunium(V) carbonate hydrates at 30, 50 and 75°C. *Radiochim. Acta* 61:57–63.
Lemire, R. J., and Garisto, F. 1989. The Solubility of U, Np, Pu, Th, and Tc in a Geologic Disposal Vault for Used Nuclear Fuel. AECL-10009, Atomic Energy of Canada Limited, Pinawa, Manitoba.
Lieser, K. H., and Mühlenweg, U. 1988. Neptunium in the hydrosphere and geosphere. I. Chemistry of neptunium in the hydrosphere and sorption of neptunium from groundwaters on sediments under aerobic and anaerobic conditions. *Radiochim. Acta* 43:27–35.
McKinley, J. P., Zachara, J. M., Smith, S. C., and Turner, G. D. 1995. The influence of hydrolysis and multiple site-binding reactions on adsorption of U(VI) to montmorillonite. *Clays Clay Miner.* 43:586–598.
Nakayama, S., Arimoto, H., Yamada, N., Moriyama, H., and Higashi, K. 1988. Column experiments on migration behaviour of neptunium(V). *Radiochim. Acta* 44/45:179–182.
Nakayama, S., and Sakamato, Y. 1991. Sorption of neptunium on naturally-occurring iron-containing minerals. *Radiochim. Acta* 52/53:153–157.
Neck, V., Runde, W., and Kim, J. I. 1995. Solid-liquid equilibria of neptunium(V) in carbonate solutions of different ionic strengths: Solubility of the solid phases. *J. Alloys Compounds* 225:295–302.

Nitsche, H., Gatti, R. C., Standifer, E. M., Lee, S. C., Müller, A., Prussin, T., Deinhammer, R. S., Maurer, H., Becraft, K., Leung, S., and Carpenter, S. 1993. Measured Solubilities and Speciations of Neptunium, Plutonium, and Americium in a Typical Groundwater (J-13) from the Yucca Mountain Region Milestone. Report 3010-WBS 1.2.3.4.1.3.1. LA-12562-MS, Los Alamos National Laboratory, Los Alamos, NM.

Nitsche, H., Roberts, K., Becraft, K., Prussin, T., Keeney, D., Carpenter, S., and Hobart, D. 1995. Solubility and Speciation Results from Over- and Undersaturation Experiments on Neptunium, Plutonium, and Americium in Water from Yucca Mountain Region Well, UE-25p#1. LA-13017-MS, Los Alamos National Laboratory, Los Alamos, NM.

Pabalan, R. T. 1994. Thermodynamics of ion-exchange between clinoptilolite and aqueous solutions of NA^+/K^+ and Na^+/Ca^{2+}. *Geochim. Cosmochim. Acta* 58:45–73.

Pabalan, R. T., Prikryl, J. D., Muller, P. M., and Dietrich, T. B. 1993. Experimental study of uranium(6+) sorption on the zeolite mineral clinoptilolite. *In* "Materials Research Society Symposium Proceedings," Vol. 294, pp. 777–782, "Scientific Basis for Nuclear Waste Management XVI" (C. G. Interrante and R. T. Pabalan, Eds.). Materials Research Society, Pittsburgh, PA.

Pabalan, R. T., Turner, D. R., Bertetti, F. P., and Prikryl, J. P. 1998. Uranium(VI) sorption onto selected mineral surfaces: Key geochemical parameters. *In* "Absorption of Metals by Geomedia" (E. A. Jenne, Ed.). Academic Press, San Diego.

Patil, S. K., Ramakrishna, V. V., and Ramaniah, M. V. 1978. Aqueous coordination complexes of neptunium. *Coordination Chem. Rev.* 25:133–171.

Righetto, L., Bidoglio, G., Azimonti, G., and Bellobono, I. R. 1991. Competitive actinide interactions in colloidal humic acid-mineral oxide systems. *Environ. Sci. Technol.* 25:1913–1919.

Thompson, R. C. 1982. Neptunium—the neglected actinide: A review of the biological and environmental literature. *Radiat. Res.* 90:1–32.

Triay, I. R., Robinson, B. A., Lopez, R. M., Mitchell, A. J., and Overly, C. M. 1993. Neptunium retardation with tuffs and groundwaters from Yucca Mountain. *In* "Proceedings of the Fourth Annual International Conference on High-Level Radioactive Waste Management," pp. 1504–1508. American Nuclear Society and American Society of Civil Engineers, La Grange Park, IL.

TRW. 1995. Total System Performance Assessment—1995: An Evaluation of the Potential Yucca Mountain Repository. B00000000-01717-2200-00136, Rev. 01, TRW Environmental Safety Systems, Inc., Las Vegas, NV.

Turner, D. 1993. Mechanistic Approaches to Radionuclide Sorption Modeling. CNWRA 93-001, Center for Nuclear Waste Regulatory Analyses, San Antonio, TX.

Wanner, H., Albinsson, Y., Karnl, O., Wieland, E., Wersin, P., and Charlet, L. 1994. The acid/base chemistry of montmorillonite. *Radiochim. Acta* 66/67:733–738.

Wescott, R. G., Lee, M. P., McCartin, T. J., Eisenberg, N. A., and Baca, R. G. (Eds.). 1995. NRC Iterative Performance Assessment Phase 2: Development of Capabilities for Review of a Performance Assessment for a High-Level Waste Repository. NUREG-1464, Nuclear Regulatory Commission, Washington, DC.

Wilson, M. L., Gauthier, J. H., Barnard, R. W., Barr, G. E., Dockery, H. A., Dunn, E., Eaton, R. R., Guerin, D. C., Lu, N., Martinez, M. J., Nilson, R., Rautman, C. A., Robey, T. H., Ross, B., Ryder, E. E., Schenker, A. R., Shannon, S. A., Skinner, L. H., Halsey, W. G., Gansemer, J. D., Lewis, L. C., Lamont, A. D., Triay, I. R., Meijer, A., and Morris, D. E. 1994. Total-System Performance Assessment for Yucca Mountain—SNL Second Iteration (TSPA-1993), Vols. 1 and 2. SAND93-2675, Sandia National Laboratories, Albuquerque, NM.

Wolery, T. 1992. EQ3/6, a Software Package for Geochemical Modeling of Aqueous Systems, Vols. 1–4. URCL-MA-110662, Lawrence Livermore National Laboratory, Livermore, CA.

Zachara, J. M., and McKinley, J. P. 1993. Influence of hydrolysis on the sorption of metal cations by smectites: Importance of edge coordination reactions. *Aquatic Sci.* 55:250–261.

Chapter 5

Factors Affecting Trivalent *f*-Element Adsorption to an Acidic Sandy Soil

S. B. Clark,[1,*] A. L. Bryce,[1,†] A. D. Lueking,[1,‡] J. Gariboldi,[1] and S. M. Serkiz[2]

[1]University of Georgia, Savannah River Ecology Laboratory, Aiken, South Carolina; [2]Westinghouse Savannah River Company, Savannah River Technology Center, Aiken, South Carolina

Europium adsorption to a well-characterized acidic sandy soil has been studied under a range of relevant geochemical conditions. As expected, Eu adsorption increased with increasing pH; 100% retention of Eu by the soil was obtained at pH values of 5.5 and greater, or approximately one order of magnitude above the natural soil pH of 4.4. Increasing the ionic strength tended to reduce adsorption for most conditions studied, suggesting a nonspecific interaction such as ion exchange as a plausible adsorption mechanism. Increasing the total Eu concentration resulted in different adsorption behavior as a function of pH, and may indicate more specific interactions (e.g., possibly surface complexation) in addition to the nonspecific processes that occur at a higher metal loading. The effect of the presence of Ca and Al as competing cations was also investigated, and Ca appears to be more effective than Al in affecting Eu sorption. Finally, aging the soil under acidic conditions increased its adsorption capacity with respect to Eu, but did not interfere with dissolution of Al phases from the soil, sug-

*Current address: Washington State University, Department of Chemistry, Pullman, WA 99164.
†Current address: United States Nuclear Regulatory Commission, Washington, DC 20555.
‡Current address: University of Michigan, Department of Civil Engineering, Ann Arbor, MI 48109.

Adsorption of Metals by Geomedia

gesting that Eu adsorption occurred independently of Al dissolution. These results provide useful insight into factors that control trivalent *f*-element transport in acidic sandy soils.

I. INTRODUCTION

The ability to predict the long-term fate and transport of transuranic or actinide elements (Np, Pu, Am, Cm, etc.) from their sources is required to address the risk resulting from past releases of these contaminants as well as to assess the performance of planned nuclear waste disposal facilities. The transport of trivalent actinides in the subsurface has been documented under various geochemical conditions (Kaplan *et al.,* 1995; Marley *et al.,* 1993; Penrose *et al.,* 1990). Although different transport mechanisms were proposed in each case, it is generally agreed that the mobility of the contaminants is related, in part, to their distributions between the solid and the solution phases. Consequently, improving our understanding of the processes that control sorption of transuranic contaminants will enhance our ability to predict the transport of these constituents in natural systems.

The lanthanides and actinides compose the 4*f* and 5*f* series of elements in the periodic table. The use of the naturally occurring lanthanides or rare earth elements (REE) as chemical analogues for the trivalent actinides (Pu^{3+}, Am^{3+}, and Cm^{3+}) is well established (Choppin, 1989; Krauskopf, 1986). In groundwater, dissolved REE are believed to originate from weathering processes affecting the host rock formation, and are typically present at very low concentrations (Hanson, 1980). Thus, the REE can serve as tracers to address questions concerning groundwater sources, flow, and mixing (Smedley, 1991), thereby providing information on far-field transport. For example, observed REE and major ion concentrations in groundwater from the vicinity of underground nuclear tests and a proposed high-level waste repository have been used to predict the solution speciation of trivalent actinides in these systems (Johannesson *et al.,* 1995).

Metal interactions at the solid–solution interface have been studied extensively, where many experimental studies have focused on metal sorption to well-characterized clay minerals. For example, Righetto *et al.* (1988) reported the observation of reduced adsorption of 0.5 $nmol{\cdot}L^{-1}$ Am^{3+} onto alumina at fixed pH as ionic strength was increased. While no interpretation of this observation was discussed, similar results have been reported by others (e.g., Shiao *et al.,* 1981). Ledin *et al.* (1994) reported that 10 $nmol{\cdot}L^{-1}$ Eu adsorbed onto iron oxyhydroxides by both electrostatic and specific chemical interactions, because adsorption occurred at pH values where the surface was positively charged, e.g., below the pH_{pzc} (Stumm and Morgan, 1981), and the amount of Eu adsorbed was sensitive to the ionic strength. Fairhurst *et al.* (1995) have reported that 1 $nmol{\cdot}L^{-1}$ Eu adsorption onto sand as a function of pH exhibits two distinct sorption processes. At low to intermediate pH, simple electrostatic processes such as ion exchange between the trivalent lanthanide and the surface were proposed, whereas metal binding by specific sites was suggested at higher pH values. On the other hand, Marmier *et al.* (1995) were able to adequately describe Yb adsorption to kaolinite over a broad range of pH by applying unmodified surface complexation constants determined for the specific adsorption of Yb to alumina and silica.

Identifying mechanisms of metal sorption to soils is more difficult due to the heterogeneity of the solid phase. Although naturally occurring REE are generally considered to be bound within the mineral phases of most soils and sediments, recent work has demonstrated that in certain highly weathered native soils, a significant fraction of the natural REE are "exchangeable," or "sorbed, but labile" (Clark *et al.*, 1996). Furthermore, data published by Johanesson *et al.* (1994) suggest that dissolved REE concentrations increase at low pH and decrease as salinity increases. Kaplan *et al.* (1994) recently reported enhanced transport of Am^{3+} and Cm^{3+} in a subsurface area composed of primarily highly weathered, acidic sandy soils where groundwater pH was low and salinity was high. Colloidal transport was deemed insignificant, and the rapid movement of the trivalent actinides was attributed to decreased contaminant sorption by the matrix under such geochemical conditions.

Here we report on the factors affecting Eu adsorption to an acidic sandy soil. Factors that were varied include pH, ionic strength, Eu concentration, competing cations, and soil aging under acidic conditions. The results of this study indicate that simple electrostatic interactions cannot completely explain Eu adsorption for all conditions, particularly at higher Eu concentrations and at increased ionic strength. Laboratory results suggest several geochemical factors that may have enhanced REE transport observed in the field with this type of soil.

II. MATERIALS AND METHODS

All solutions were prepared from reagent grade chemicals using Milli-Q water in acid washed, double rinsed glassware, unless otherwise specified. Adsorption experiments were completed using low-carbonate water prepared by boiling Milli-Q water, followed by cooling in air stripped of carbonate using 1 $mol \cdot L^{-1}$ NaOH. Hydrogen ion activity was adjusted with standardized NaOH prepared from a saturated solution; HNO_3 (Fisher Scientific) served as the standardized acid. Stock Eu solutions were prepared from a 100 $mg \cdot L^{-1}$ standard (GFS Chemicals).

A. Soil Collection

A single soil sample was collected from an unimpacted location in the General Separations Area of the Department of Energy's Savannah River Site. The sample was collected at the 0.5- to 3-ft (0.15- to 0.91-m) depth interval from the Barnwell Aquifer unit using a grab technique. The soil was stored in a sealed container at room temperature, and sieved to less than 2 mm prior to use.

B. Soil Characterization

Soils were characterized according to standard methods published by the American Society of Agronomy, as described below. Soil moisture content was determined by drying a 10-g soil aliquot at 105°C overnight. The ratio of dry to wet soil was used to estimate the soil mass corresponding to given soil aliquots. Soil pH was determined in water and 1.0 mol·L^{-1} KCl by suspending a soil aliquot in solution for 30 min, followed by measurement with a standardized Fisher Scientific Accument 50 pH meter. Cation exchange capacity (CEC) was measured using 10 g soil in 30 ml of 1.0 eq·L^{-1} $BaCl_2$. Tubes were shaken for 30 min, and the supernatant solution was then separated by centrifugation at 7.02 relative centrifugal force (RCF) for 30 min. An aliquot of the supernatant was withdrawn and recentrifuged at 7796 RCF for 15 min. The supernatant was analyzed for Na, Ca, K, Mg, and Al using flame atomic absorption spectrometry (AAS), as described below. Organic matter was determined by the Walkley–Black procedure (Nelson and Sommers, 1982).

The particle size distribution of the soil was measured by the pipette method (Gee and Bauder, 1986). The soil was not treated with peroxide prior to the measurement, as the organic matter content was negligible. The quantity of noncrystalline aluminosilicates and hydrous Fe and Al oxides was determined by extraction with ammonium oxalate under darkness (AOD), as described by Jackson *et al.* (1986). Free Fe and Al oxide content (crystalline and noncrystalline) was determined by extraction with dithionite bicarbonate citrate (DCB), where the extractant was filtered gravimetrically prior to analysis (Jackson *et al.,* 1986). Iron and Al were analyzed by AAS as described below. The clay fraction was characterized by X-ray diffraction using Cu, K_α radiation and a Phillips Norelco diffractometer equipped with a graphite monochromator. The clay fraction was isolated from the bulk soil after dispersion using a saturated solution of Na_2CO_3 (pH 10), followed by centrifugation (Jackson, 1979). Clay concentrates were deposited on petrographic slides using the Drever method (Drever, 1973). The samples were treated with Mg, Mg–ethylene glycol (Mg-EG), and K saturation. Potassium-saturated samples were heat treated at 110, 300, and 550°C prior to X-ray analysis.

C. Adsorption Experiments

The conditions affecting Eu adsorption to the characterized soil were studied by systematically varying the total Eu concentration ($[Eu]_T$), pH, and ionic strength. Ten grams of soil was combined with water and the desired amount of electrolyte solution (either $NaNO_3$ or $Ca(NO_3)_2$) in polyallomer centrifuge tubes. To monitor Eu loss to container walls during the course of an experiment, controls containing no soil were included. Container loss was negligible under the conditions of the study. Soil suspensions were equilibrated on an orbital shaker at 300 rpm for 1 hr. Adsorption experiments were initiated by adjusting the pH of the suspension with acid or base and adding the desired amount of Eu to produce a final volume of 30 ml. Suspensions were then returned to the shaker for overnight equilibration. After equilibration, the supernatant was separated by centrifugation for 15 min at 15,542 RCF, followed by filtration through 0.1-μm polycarbonate filters. The first few milliliters of filtrate was discarded, and metal loss to the filters was not observed. A small fraction of the collected sample was preserved with HNO_3 for metal analysis; the remainder was used for pH determination with a Radiometer electrode.

In cases where $[Eu]_T$ was varied over a wide range at constant pH and ionic strength (i.e., "Eu isotherms," or simply "isotherms" for this chapter), the pH of the stock Eu solution was adjusted to pH 4.0 ± 0.5 prior to addition to the soil suspension to prevent large changes in pH. When adjusting the pH of the Eu stock solution, care was taken not to induce Eu precipitation. Controls for each experiment were prepared as already described. The dissolution of the soil matrix over the pH and ionic strength ranges of this study were determined in the absence of Eu. Here, overnight equilibration began after adding the electrolyte and adjusting the pH.

Dissolution of the soil matrix has been suggested in field systems experiencing high acid concentrations and where transport of trivalent actinides has been reported (Kaplan *et al.,* 1994). To study the effect of low-pH conditions on the subsequent adsorption of trivalent *f*-elements, an aliquot of the soil in this study was aged at pH 2.0 (HNO_3) prior to adding Eu. After the aging period (24 hr), the samples were centrifuged as previously described. Approximately 90% of the supernatant solution was removed, where a portion was filtered for analysis, and the remainder discarded. Supernatant that was not removed from the centrifuge tubes was accounted for in subsequent calculations. Fresh water, stock Eu, sodium nitrate, and base were then added to the samples. The soil was resuspended by vortexing, and then allowed to shake overnight. The supernatant was then separated and analyzed as previously described.

D. Metal Analysis

Metals determined during soil characterization (Na, K, Mg, Ca, Fe, Al) were analyzed by flame AAS (Perkin Elmer). For the adsorption experiments, dissolved Eu and Al solution were determined using graphite furnace AAS with Zeeman corrected background (Perkin Elmer). Furnace and flame conditions were those recommended by the manufacturer. All measurements were made in triplicate.

III. RESULTS AND DISCUSSION

A. Soil Characteristics

The soil used in this study had a natural pH of 4.4, was composed of 94.3% sand, 3.3% silt, and 2.4% clay, and had a low organic carbon content (<0.0015 $g \cdot g^{-1}$ of C to soil). Additional physical and chemical properties of the soil are summarized in Table I. The total CEC (sum of exchangeable cations) of the soil was low (0.346 $cmol_+ \cdot kg^{-1}$) when compared to other southeastern acidic sandy soils, which typically have CEC values ranging from 0.5 to 10 $cmol_+ \cdot kg^{-1}$ (King, 1988). X-ray diffraction of the clay fraction indicated primarily kaolinite, with

Table I

Physical and Chemical Soil Characteristics

Soil pH	
Milli-Q water	4.92 ± 0.03
KCl (1 $eq \cdot L^{-1}$)	4.34 ± 0.01
CEC[a]	($cmol_+ \cdot kg^{-1}$)
Na	0.001 ± 0.007
Ca	0.017 ± 0.0072
Mg	0.0035 ± 0.0007
K	0.028 ± 0.026
Al	0.296 ± 0.0095
Total CEC	0.346 ± 0.029
Oxides	($g \cdot kg^{-1}$)
Fe-DCB[b]	1.9 ± 0.2
Fe-AOD[c]	0.3 ± 0.2
Al-DCB[b]	8 ± 3.5
Al-AOD[c]	1.4 ± 0.2

[a]CEC, cation exchange capacity.
[b]DCB, extraction of free Fe and Al oxides using dithionite-citrate-bicarbonate.
[c]AOD, extraction of noncrystalline aluminosilicates and hydrous Fe and Al oxides using ammonium oxalate in the dark.

some contribution from hydroxy-interlayered vermiculite, gibbsite, and quartz; goethite was not present. This soil had a high exchangeable Al content, where Al contributed approximately 80% of the total CEC; furthermore, the amount of measurable Al in solution after AOD and DCB extractions surpassed that of Fe.

B. Europium Adsorption

1. Effect of pH and Total Eu Concentration

Typical for metal sorption to soils and clays, the fraction of Eu sorbed increased as a function of pH for all conditions studied (Fig. 1). Eu adsorption was significant at all pH values, increasing from approximately 30% adsorbed at pH 2.5 to 100% adsorbed at pH 5.5. In this pH region, and in the absence of the soil, Eu is expected to be in solution as hydrolysis only becomes important above pH 7 (log $K_{Eu(OH)^{2+}} = 7.8$ at 25°C and zero ionic strength (Baes and Mesmer, 1986)) and the solubility of the $Eu(OH)_3$ solid is not limiting (log $K_{sp} = 16.5$ (Baes and Mesmer, 1986)). The shape and position of the adsorption edge depended on $[Eu]_T$, where decreasing $[Eu]_T$ from $1 \cdot 10^{-5}$ to $1 \cdot 10^{-6}$ mol·L^{-1} increased the fraction of Eu adsorbed. At pH 3.5, for example, a 10-fold decrease in $[Eu]_T$ corresponded to a 12% increase in the fraction of Eu adsorbed.

At the higher $[Eu]_T$ concentration (Fig. 1A), two distinct adsorption regions were apparent, which suggested the occurrence of more than one adsorption mode. In the first region, the fraction of Eu adsorbed increased moderately between pH 3 and 4.5. In the second region, the fraction of Eu adsorbed increased sharply at pH > 4.5. At the lower $[Eu]_T$ concentration (Fig. 1B), the distinction between two adsorption regions was not as readily apparent. Other studies have shown that such observations can be sensitive to the experimental conditions, including the adsorbing metal used and the metal:sorbent ratio (Schulthless and Huang, 1990; Spark *et al.*, 1995). For example, Spark *et al.* (1995) showed in a study of Co^{II} and Zn^{II} adsorption to kaolinite that two adsorption modes were less apparent when the metal:sorbent ratio was decreased.

At low to moderate pH, ion exchange in which the metal may compete with protons and other cations during sorption can be a more important uptake mechanism on soils than specific adsorption (Stahl and James, 1991; Zachara *et al.*, 1992). On the other hand, specific adsorption to oxides may become the more predominant mechanism at higher pH. The effects of pH and the metal:sorbent ratio can also be evaluated by varying the metal:sorbent ratio over a wide range of fixed pH, as done in the isotherms shown in Figure 2. At each pH, the relationship between adsorbed and dissolved Eu can be described by a Freundlich isotherm over the range of $[Eu]_T$ studied ($5 \cdot 10^{-7}$ to $5 \cdot 10^{-4}$ mol·L^{-1}). The empirical parameters were obtained by linear regression to the experimental data using a linearized form

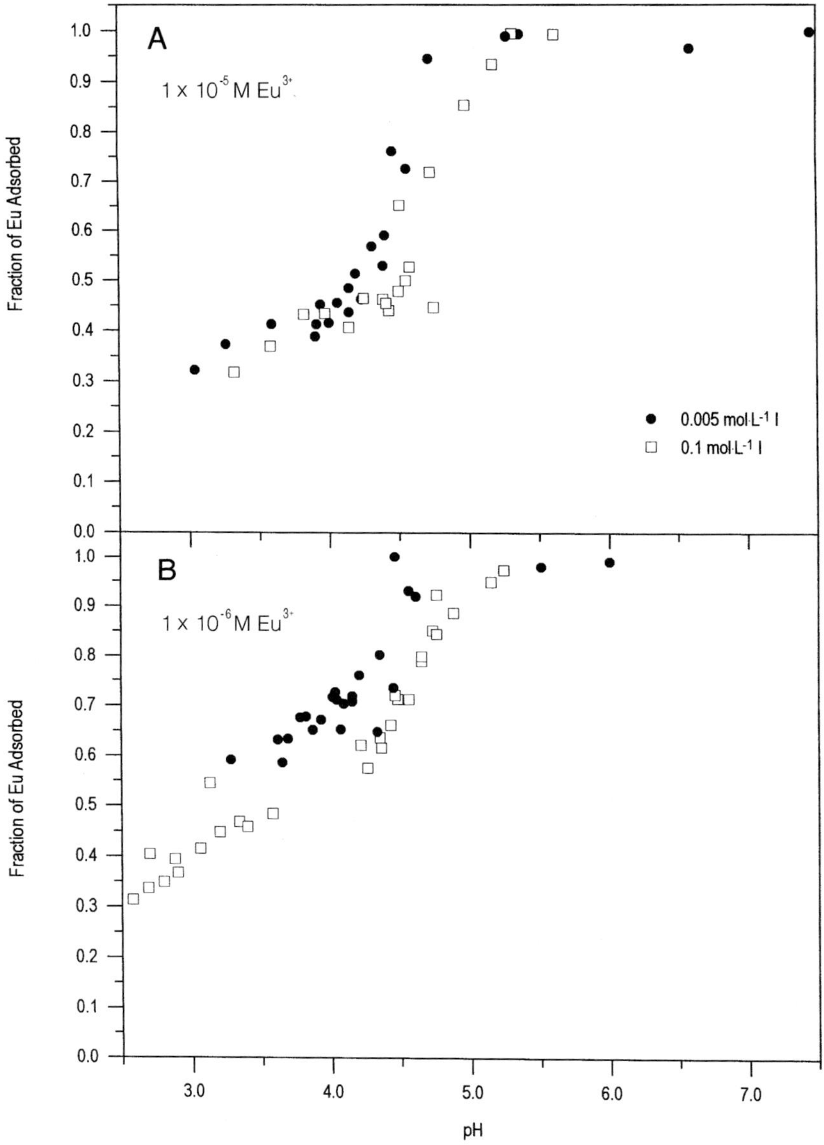

Figure 1 Europium adsorption to Savannah River site soil. Total europium as specified, with 10 g of soil and 30 ml solution. Ionic strength was adjusted with sodium nitrate.

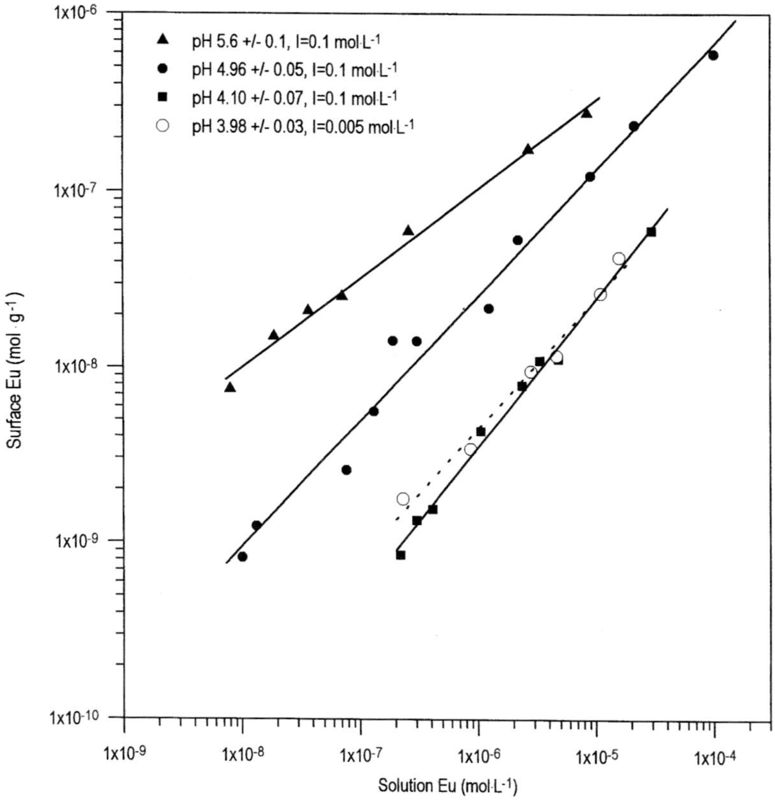

Figure 2 Europium isotherms on Savannah River site soil. The ionic strength is 0.1 mol·L^{-1} sodium nitrate unless otherwise specified. In all cases 10 g soil and 30 ml solution were used.

of the Freundlich equation, and are given in Table II. At a fixed ionic strength, increasing pH resulted in a decrease in slope, providing additional evidence for a change in adsorption mode (Zachara *et al.*, 1992).

2. Effect of Ionic Strength

Increasing the ionic strength appeared to decrease Eu adsorption at pH values less than 5.5 for both $[Eu]_T$ concentrations studied (Fig. 1). However, the ionic strength effect on Eu adsorption for $[Eu]_T = 1 \cdot 10^{-5}$ mol·L^{-1} was less significant when compared to $[Eu]_T = 1 \cdot 10^{-6}$ mol·L^{-1}. For both cases, the change in the fraction Eu adsorbed due to changes in the ionic strength is summarized using the pH at which 50% adsorption occurs (pH_{50}) in Table III.

Changes in ionic strength that lead to changes in sorbate adsorption are typi-

Table II
Freundlich Iostherm Parameters[a]

pH	Intercept, K	Slope, n
4.0	−3.4 ± 0.2	0.85 ± 0.03
5.0	−3.3 ± 0.7	0.71 ± 0.03
5.6	−4.0 ± 0.5	0.51 ± 0.02
4.0, $I = 0.005$ mol·L^{-1}	−3.9 ± 0.3	0.74 ± 0.05

Note. Ionic strength is 0.1 mol·L^{-1} sodium nitrate unless otherwise specified.
[a]The linearized form of the Freundlich equation is given as log (Γ) = log (K) + n log (C), where Γ is defined as the surface concentration of Eu (mol·g^{-1}), C is the solution concentration of Eu (mol·L^{-1}), and K and n are empirical constants.

cally interpreted as evidence of nonspecific adsorption, including ion exchange (Hayes and Leckie, 1987; Naidu *et al.,* 1994; Zachara *et al.,* 1992). In Figure 2, comparing the isotherms for ionic strengths of 0.005 and 0.1 mol·L^{-1} at pH 4 appears to suggest negligible differences over the entire range of $[Eu]_T$ studied, although the isotherms diverge at lower $[Eu]_T$, similar to the trend observed in Figure 1. A measurable difference was calculated for the isotherm slopes (Table II). Decreases in the fraction of adsorbed metal at the higher ionic strengths may result from competition between the Na and the Eu and/or a change in the surface charge of the soil. The dependence of an ionic strength effect on $[Eu]_T$ suggested that Na could compete with Eu, but only at low $[Eu]_T$.

3. Competing Cations

Although Na does not appear to compete with Eu over most of the conditions studied, cations with higher charge may. We chose to study the affect of the two pri-

Table III
Summary of Eu(III) Adsorption Characteristics

Total Eu(III)		Ionic Strength	pH_{50}
1×10^{-5} M		0.005 M $NaNO_3$	4.20
		0.1 M $NaNO_3$	4.53
		0.03 M $Ca(NO_3)_2$	4.69
	Aged	0.1 M $NaNO_3$	<4.01
1×10^{-6} M		0.005 M $NaNO_3$	<3.3
		0.1 M $NaNO_3$	3.63

Note. "Aged" refers to soil treated at pH 2.0 (HNO_3) for 24 hr prior to adsorption experiments.

mary exchangeable cations, Ca^{2+} and Al^{3+}, on Eu adsorption. In addition to charge, cation size may be an important factor, where a large cation may be sterically hindered from reaching some available adsorption sites. Although Ca^{2+} is less charged than Eu^{3+}, it is smaller, whereas Al^{3+} has equivalent charge and is significantly smaller. The Pauling crystal radii for Na^{+}, Ca^{2+}, Al^{3+}, La^{3+}, and Lu^{3+}, which span Eu^{3+}, are 0.95, 0.99, 0.5, 1.18, and 0.97 Å, respectively (Huheey, 1978).

Dissolution of the Al oxides present in this soil occurred to some degree under all conditions of this study, as shown in Figure 3. The solubility of gibbsite is indicated by the solid and dashed lines at ionic strengths of 0.1 and 0.005 mol·L^{-1}, respectively. Although gibbsite was identified in the clay fraction of the soil by XRD, the solution Al concentration at 0.1 mol·L^{-1} ionic strength after 24 hr was supersaturated with respect to gibbsite at pH > 4.5. Instead, the change in the so-

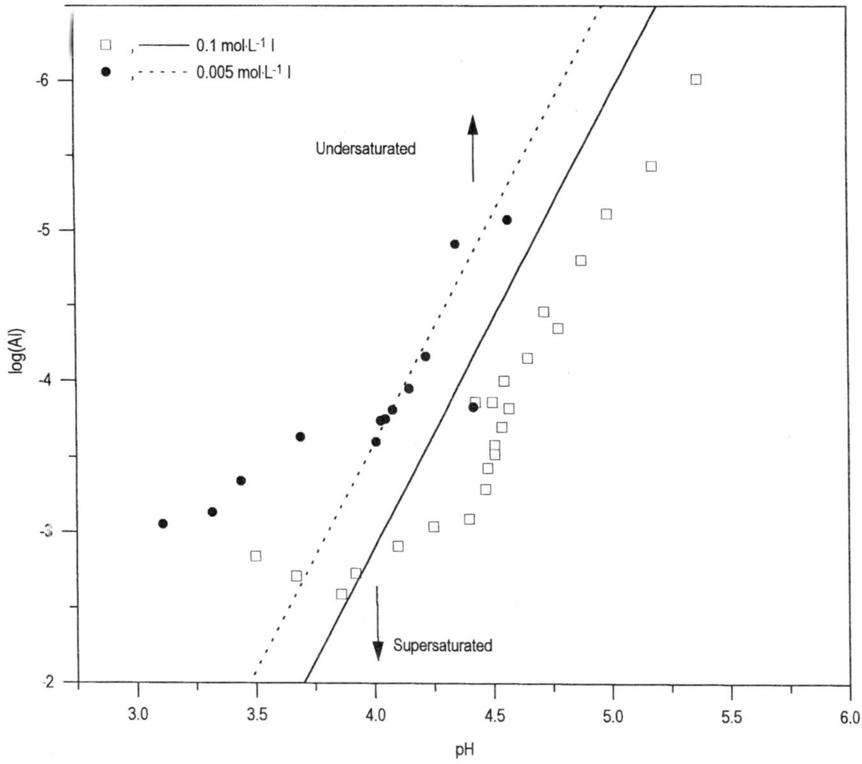

Figure 3 Aluminum dissolution from Savannah River site soil. The lines indicate the gibbsite solubility limit assuming log K_{sp} = 8.11 at zero ionic strength, and adjusted to the appropriate ionic strength using the Davies equation. In all cases 10 g soil and 30 ml solution were used.

lution Al concentration with pH at 0.1 mol·L^{-1} ionic strength was consistent with the dissolution of an amorphous Al phase at pH > 4.5 (Mulder and Stein, 1994). At pH < 4.5 and 0.1 mol·L^{-1} $NaNO_3$, the solution Al concentration began to level off at approximately 1.5·10^{-3} mol·L^{-1}, representing roughly 10% of the noncrystalline Al oxide content measured by AOD extraction. When the ionic strength was decreased to 0.005 mol·L^{-1}, the solution Al concentration was also decreased. In this case, the observed Al concentrations are consistent with the calculated dissolved Al based on gibbsite solubility, suggesting that gibbsite may be controlling the solution Al concentration for pH > 4.0 (Fig. 3). For both ionic strengths at pH < 4.5, the solution Al concentration is several orders of magnitude larger than the total Eu concentration, and hydrolysis products of Al are expected (Baes and Mesmer, 1986). The large concentrations of dissolved Al and the marked effect of ionic strength on dissolved Al concentrations did not appear to cause similar trends in Eu adsorption (Figs. 1A and 1B), suggesting that Eu sorption is relatively insensitive to dissolved Al concentrations, Al speciation, and the solid phase controlling Al dissolution at pH values less than 4.5.

The amount of adsorbed Eu decreased over the pH range studied when the electrolyte used to maintain the ionic strength was changed from $NaNO_3$ to $Ca(NO_3)_2$ (Fig. 4A). At a constant ionic strength of 0.1 mol·L^{-1}, changing the electrolyte salt resulted in a total Ca^{2+} concentration of 0.03 mol·L^{-1}, compared to 0.1 mol·L^{-1} Na^+. Eu adsorption decreased by as much as 15% (pH < 4.5), again suggesting ion exchange as a plausible mechanism. Al dissolution in the presence of Eu and 0.03 mol·L^{-1} Ca (Fig. 4B) was equivalent to Al dissolution at 0.1 mol·L^{-1} $NaNO_3$ (Fig. 4B), indicating that Al dissolution is occurring independently of the Eu adsorption mechanism.

4. Effect of Acid Aging

At pH 2.0, the dissolution of Al and Fe oxides is important, although dissolution may be kinetically slow. The dissolved Al concentration observed in the supernatant after aging was consistent with the amounts of Al measured at 0.1 mol·L^{-1} ionic strength below pH 4.5, e.g., on the order of 10^{-3} mol·L^{-1}. Using the aged soil and a fresh solution of sodium nitrate, Eu adsorption after 24 hr equilibration was greater than the Eu adsorption measured on the unaged soil by approximately 20% at pH values below the sorption edge (Fig. 4A). The solution Al concentration was not significantly different when compared to that of the unaged soil (Fig. 4B). This indicated that replacing the supernatant with a fresh electrolyte solution allowed Al dissolution to resume, and at pH < 4.5 after both equilibration periods approximately 30% of the amorphous Al had dissolved. Although the aging treatment used in this study was not a selective extraction, it does mimic conditions that can exist downfield of nuclear waste disposal sites, albeit for much longer time periods. However, because aging was not selective it was difficult to

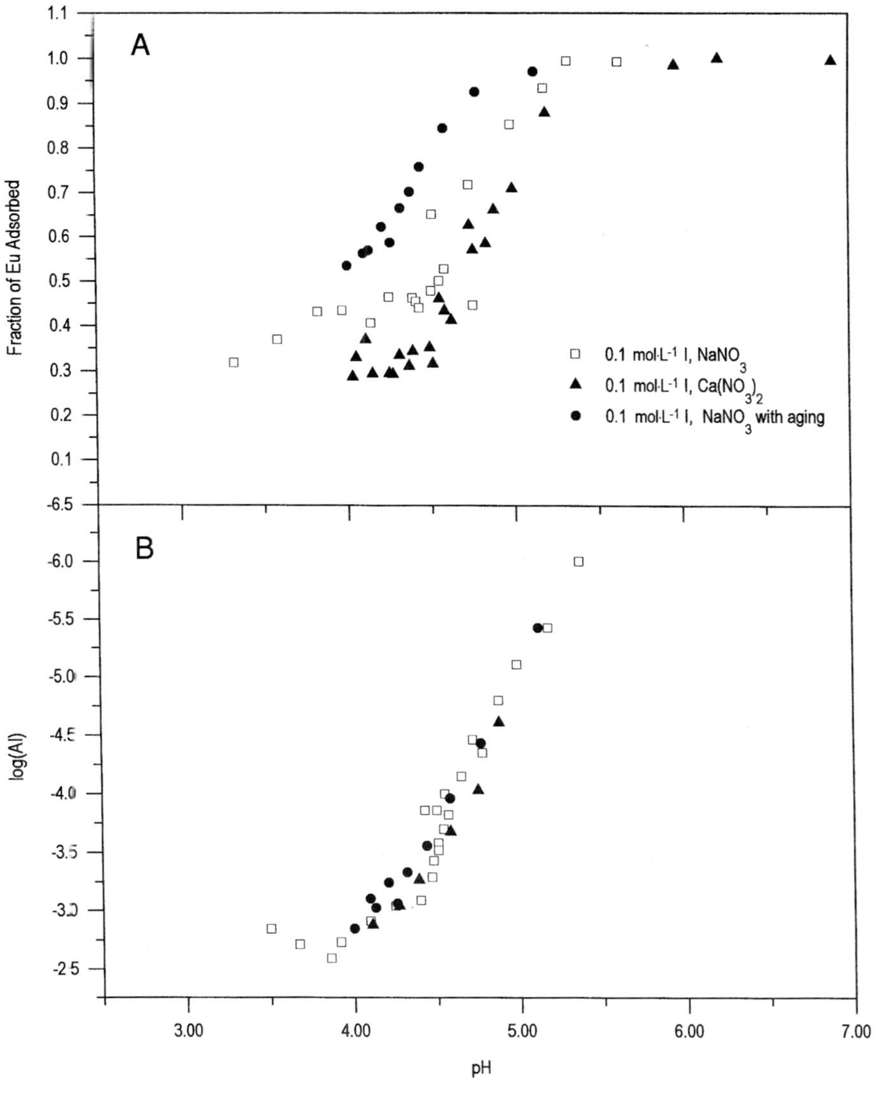

Figure 4 Effect of index cation and aging: (A) europium adsorption and (B) aluminum dissolution. $[Eu]_T = 1 \times 10^{-5}$ M. Aging period for soil prior to Eu adsorption was 24 hr at pH 2.0 (HNO_3). In all cases 10 g soil and 30 ml solution were used. In the following two cases Eu measurements and Al measurements correspond to the same samples: (▲) 0.1 mol·L^{-1} ionic strength as 0.03 mol·L^{-1} calcium nitrate, and (●) 0.1 mol·L^{-1} ionic strength after acid aging at pH 2.0.

specify the cause of the increased Eu adsorption. The addition of Fe and Al oxide coatings has been shown to inhibit Cd^{2+} adsorption on clays (Zachara *et al.*, 1992). In this study, aging the soil at pH 2.0 may have removed coatings on the clays, thereby exposing previously unavailable adsorption sites. Subsequent increases in pH then exhibited increased Eu adsorption. However, additional study is required to elucidate these processes.

IV. CONCLUSIONS

Europium adsorption to an acidic sandy soil from the Savannah River Site was significant at the natural pH of the soil, but the fraction adsorbed decreased with increasing $[Eu]_T$. Increasing the pH increased Eu adsorption, although pH values in excess of 5.5 were required for complete retention of Eu at the highest $[Eu]_T$ studied. Europium adsorption was sensitive to the ionic strength at low total Eu concentrations, and Ca^{2+} at higher total Eu concentrations. This suggested that ion exchange may be an important uptake mechanism for Eu, particularly at pH < 4.5. Furthermore, these results support field observations that suggest conditions of low to moderate pH and high ionic strength will lead to increased mobility of Eu in the environment. Aging the soil at a low pH resulted in increased Eu retention on these soils, possibly due to the removal of amorphous Al oxide coatings. If the uptake capacity of the soil does increase with acid aging, trivalent actinide adsorption should be enhanced *in situ* and its transport reduced if the pH of the system is returned to natural conditions.

ACKNOWLEDGMENTS

This work was funded by the U.S. Department of Energy under Contracts AA46420T/C00066ST with the University of Georgia's Savannah River Ecology Laboratory via the Education, Research, and Development Association of Georgia Universities, Contract DE-AC09-76SR00-819 with the University of Georgia, and Contract DE-AC09-90SR-18035 with Westinghouse Savannah River Company.

REFERENCES

Baes, C. F., Jr. and Mesmer, R. E. 1986. "The Hydrolysis of Cations," 2nd ed. Robert E. Krieger Publishing, Malabar, FL.

Choppin, G. R. 1989. Soluble rare earth and actinide species in seawater. *Mar. Chem.* 28:19–26.

Clark, S. B., Johnson, W. H., Malek, M. A., Serkiz, S. M., and Hinton, T. G. 1996. A comparison of sequential extraction techniques to estimate geochemical controls on the mobility of fission product, actinide, and heavy metal contaminants in soil. *Radiochim. Acta* 74:173–179.

Drever, J. I. 1973. The preparation of oriented clay mineral specimens for X-ray diffraction analysis by a filter membrane peel technique. *Am. Mineral.* 58:553–554.

Fairhurst, A. J., Warwick, P., and Richardson, S. 1995. The effect of pH on europium-mineral interactions in the presence of humic acid. *Radiochim. Acta* 69:103–111.

Gee, G. W., and Bauder, J. W. 1986. Particle-size analysis. *In* "Methods of Soil Analysis Part I. Physical and Mineralogical Methods" (A. Klute, Ed.), Vol. 1, pp. 383–412. American Society of Agronomy, Madison, WI.

Hanson, G. B. 1980. Rare earth elements in petrogenetic studies of igneous systems. *Ann. Rev. Earth Plant. Sci.* 8:371–406.

Hayes, K. F., and Leckie, J. O. 1987. Modeling ionic strength effects on cation adsorption at hydrous oxide/solution interfaces. *J. Colloid Interface Sci.* 115:564–572.

Huheey, J. E. 1978. "Inorganic Chemistry: Principles of Structure and Reactivity," 2nd ed. Harper & Row, New York, NY.

Jackson, M. L. 1979. "Soil Chemical Analysis: Advanced Course," 2nd ed. Department of Soil Science, University of Wisconsin, Madison, WI.

Jackson, M. L., Lim, C. H., and Zelazny, L. W. 1986. Oxides, hydroxides, and aluminosilicates. *In* "Methods of Soil Analysis Part 1. Physical and Mineralogical Methods" (A. Klute, Ed.), Vol. 1, pp. 101–150. American Society of Agronomy, Madison, WI.

Johannesson, K. H., Lyons, W. B., Fee, J. H. Gaudetter, H. E., and McArthur, J. M. 1994. Geochemical processes affecting the acidic groundwaters of Lake Gilmore, Yilgarn Block, Western Australia: A preliminary study using neodymium, samarium, and dysprosium. *J. Hydrol.* 154:271–289.

Johannesson, K. H., Stetzenbach, K. J., and Hodge, V. F. 1995. Speciation of the rare earth element neodymium in groundwaters of the Nevada test site and Yucca Mountain and implications for actinide solubility. *Appl. Geochem.* 10:565–572.

Kaplan, D. I., Bertsch, P. M., and Adriano, D. C. 1995. Facilitated transport of contaminant metals through an acidified aquifer. *Ground Water* 33:708–717.

Kaplan, D. I., Bertsch, P. M., Adriano, D. C., and Orlandini, K. A. 1994. Actinide association with groundwater colloids in a coastal plain aquifer. *Radiochim. Acta* 66/67:181–187.

King, L. D. 1988. Retention of metals by several soils of the southeastern United States. *J. Environ. Qual.* 17:239–246.

Krauskopf, K. B. 1986. Thorium and rare-earth metals as analogs for actinide elements. *Chem. Geol.* 55:323–335.

Ledin, A., Karlsson, S., Duker, A., and Allard, B. 1994. The adsorption of europium to colloidal iron oxyhydroxides and quartz—The impact of pH and an aquatic fulvic acid. *Radiochim. Acta* 66/67:213–220.

Marley, N. A., Gaffney, J. S., Orlandini, K. A., and Cunningham, M. M. 1993. Evidence for radionuclide transport and mobilization in a shallow, sandy aquifer. *Environ. Sci. Technol.* 27:2456–2461.

Marmier, N., Dumonceau, J., Chupeau, J., and Fromage, F. 1995. Modeling of Yb(III) sorption on dalinite by using single oxide surface complexation models. *Mater. Res. Symp. Proc.* 353:1085–1092.

Mulder, J., and Stein, A. 1994. The solubility of aluminum in acidic forest soils: Long-term changes due to acid deposition. *Geochim. Cosmochim. Acta* 58:85–94.

Naidu, R., Bolan, N. S., Kookana, R. S., and Tiller, K. G. 1994. Ionic-strength and pH effects on the sorption of cadmium and the surface charge of soils. *Eur. J. Soil Sci.* 45:419–429.

Nelson, D. W., and Sommers, L. E. 1982. Total carbon, organic carbon, and organic matter. *In* "Methods of Soil Analysis Part 2. Chemical and Microbiological Properties" (A. L. Page, R. H. Miller, and D. R. Keeney, Eds.), Vol. 2, pp. 539–580. American Society of Agronomy, Madison, WI.

Penrose, W. R., Polzer, W. L., Essington, E. H., Nelson, D. M., and Orlandini, K. A. 1990. Mobility of plutonium and americium through a shallow aquifer in a semiarid region. *Environ. Sci. Technol.* 24:228–234.

Righetto, L., Bidoglio, G., Marcandalli, B., and Bellobono, I. R. 1988. Surface interactions of actinides with alumina colloids. *Radiochim. Acta* 44/45:73–75.

Schulthless, C. P., and Huang, C. P. 1990. Adsorption of heavy metals by silicon and aluminum oxide surfaces on clay minerals. *Soil Sci. Soc. Am. J.* 54:679–688.

Shiao, S.-Y., Egozy, Y., and Meyer, R. E. 1981. Adsorption of Cs(I), Sr(II), Eu(III), Co(II) by Al_2O_3. *J. Inorg. Nucl. Chem.* 43:3309–3315.

Smedley, P. L. 1991. The geochemistry of rare earth elements in groundwater from the Carnmenellis area, southwest England. *Geochim. Cosmochim. Acta* 55:2767–2779.

Spark, K. M., Johnsnon, B. B., and Wells, J. D. 1995. Characterizing heavy-metal adsorption on oxides and oxyhydroxides. *Eur. J. Soil Sci.* 46:621–631.

Stahl, R. S., and James, B. R. 1991. Zinc sorption by iron-oxide-coated sand as a function of pH. *Soil Sci. Soc. Am. J.* 55:1287–1290.

Stumm, W., and Morgan, J. J. 1981. "Aquatic Chemistry," 2nd ed. John Wiley & Sons, Inc., New York, NY.

Zachara, J. M., Smith, S. C., Resch, C. T., and Cowan, C. E. 1992. Cadmium sorption to soil separates containing layer silicates and iron and aluminum oxides. *Soil Sci. Soc. Am. J.* 56:1074–1084.

Chapter 6

Lead Sorption, Chemically Enhanced Desorption, and Equilibrium Modeling in an Iron-Oxide-Coated Sand and Synthetic Groundwater System

Mohammad F. Azizian and Peter O. Nelson
Civil, Construction, and Environmental Engineering Department, Oregon State University, Corvallis, Oregon

Oxides of iron are ubiquitous subsurface mineral constituents and control the mobility of metal ions in soils and groundwater by adsorptive retardation. In this laboratory study, iron-oxide-coated sand (IOCS) was used to model Pb chemical behavior in soil using batch reactor experiments. Several parameters were varied in order to ascertain their effects: pH, ionic strength, complexation by organic ligands, competing cations, and reaction time. Results indicated that equilibrium was reached in less than 24 hr in completely mixed batch systems. In equilibrium experiments, pH was the major factor that controlled the adsorption process, and increasing Ca^{2+} and Na^{+} electrolytes somewhat decreased Pb adsorption below pH values of 5, but had no influence at higher pH values. Using a surface complexation, triple-layer model, inner-sphere surface binding was successful in describing the effect of pH and ionic strength on Pb adsorption on IOCS over a range of Pb concentrations. EDTA and NTA greatly decreased Pb adsorption at equimolar and greater concentrations over the pH range of 3 to 10.

Adsorption of Metals by Geomedia

$PbEDTA^{2-}$ was adsorbed on the IOCS by ligand-like adsorption, increasing with decreasing pH, while $PbNTA^{-}$ adsorption on IOCS was metal-like, increasing with increasing pH.

I. INTRODUCTION

Lead is a ubiquitous trace constituent of the environment that has been known for centuries to be a cumulative metabolic poison. The groups most susceptible to Pb poisoning are fetuses and young children. Their developing central nervous systems may be affected, leading to hyperactivity, irritability, headaches, and learning and concentration difficulties (Bjorklund *et al.,* 1980; Putka, 1992). Lead contamination can occur in groundwater, surface waters, and soils due to mining and smelting activities, battery plant emissions, battery reprocessing plant wastes, automotive exhaust emissions, leaded fuel spills, incinerator ash disposal, and municipal or industrial land fill leachates. Almost one-third of the sites on the EPA's National Priority List (NPL) for superfund sites have Pb concentrations significantly higher than normal background levels (Anonymous, 1988).

Lead forms a number of hydroxide and polynuclear hydroxy species. The polynuclear species become important only at concentrations greater than 10^{-3} $mol{\cdot}L^{-1}$ (Nraigu, 1978). Lead also forms a number of chloride, sulfate, sulfide, and phosphate complexes. However, in the vast majority of freshwater systems in contact with the atmosphere, lead carbonate ($PbCO_3$) dominates the inorganic chemistry of dissolved Pb. Santillan-Medrano and Jurinak (1975) found that in calcareous soils, $PbCO_3(s)$ appeared to control the solubility, whereas in noncalcareous soils the solubility of Pb appeared to be controlled by $Pb(OH)_2(s)$, $Pb_3(PO_4)_2(s)$, $Pb_4O(PO_4)_2(s)$, and $Pb_5(PO_4)_3OH(s)$, depending on the pH.

Lead adsorption and desorption have been extensively studied on soils and mineral phases. The major finding was that Pb may be rapidly immobilized by sorption and precipitation mechanisms (Scokart *et al.,* 1983; Hem, 1976). However, changing conditions, such as lowered pH in acidic soils, and the presence of complexing ligands in leachates, can remobilize Pb concentrations well above drinking water standards.

In the absence of organic ligands or chelators, other soil parameters affecting Pb affinity are oxides of iron and clay content (King, 1988). Adsorption of Pb onto hydrous oxides and hydroxides has been reported (Huang and Stumm, 1973; Hohl and Stumm, 1976; Hayes and Leckie, 1987; Dzombak and Morel 1990). The mechanism of Pb bonding with hydrous oxide is essentially an ion exchange process in which Pb adsorption results in replacing a bound proton (Hayes and Leckie, 1987). The adsorption reactions that occur between metallic ions and charged surfaces may involve either the formation of relatively weak outer-sphere complexes (ion pairs) through cation exchange reactions, or the formation of strongly bound inner-sphere complexes through ligand exchange reactions. Hayes and Leckie (1987) used an inner- and outer-sphere surface complexation triple-layer model (TLM) for adsorption of Pb and Cd onto goethite at several different ionic strengths. They concluded that it is possible to distinguish between inner- and outer-sphere surface complexes by studying the effects of ionic strength on Pb adsorption data.

Many researchers have used various solutions to displace or remove Pb bound to soils. The effect of competing cations varied with the types of both the adsorbent and the cation. At lower metal loadings, the amount of metal replaced by cation exchange increased (Slavek and Pickering, 1987, 1988). However, this may have been due to the fact that a lower sorption pH was used to effect lower metal loadings. On montmorillonite, 0.001 mol·L^{-1} $NaNO_3$ had little effect in displacing sorbed Pb ions. However, an increase to 0.1 mol·L^{-1} $NaNO_3$ or using 0.05 mol·L^{-1} $CaCl_2$ resulted in about 75% recovery (Farrah and Pickering, 1977a). Lead forms both soluble and insoluble complexes with many natural organic compounds. Fulvic and humic acids in soils can form strong complexes with Pb ions. Lead–fulvic acid complexes will dominate over Pb carbonate if the fulvic acid concentration is greater than 10^{-4} mol·L^{-1} (Hodson *et al.*, 1984). Organic ligands or chelators such as EDTA and NTA removed a majority of the Pb sorbed on contaminated soils (Elliot and Brown, 1989).

The overall purpose of this research was to investigate conditions that control Pb mobility in soils, primarily sorption–desorption and complexation, and to determine the effects of environmental conditions such as pH, ionic strength, and kinetics on these processes. The study will increase our understanding of Pb mobility in soils and groundwaters through both physicochemical techniques and computer modeling.

II. MATERIALS AND METHODS

A. Iron-Oxide-Coated Sand Media

Graded Ottawa sand (99.5% SiO_2) with spherical grains of 20 to 30 mesh (0.60 to 0.85 mm) was used in the preparation of iron-oxide-coated sand (IOCS). The sand was acid washed, rinsed with deionized water, and dried at 110°C. IOCS was prepared according to the method of Edwards and Benjamin (1989). Briefly, 20 g of reagent-grade ferric nitrate ($Fe(NO_3)_3 \cdot 9H_2O$) was dissolved in 50 ml deionized water and poured over 200 g of the sand in a 9 × 9-in. Pyrex baking dish and stirred thoroughly using a glass stir rod. The baking dish was covered and placed in an oven at 110°C for 16 hr. A small gap was maintained between the lid and the dish to allow evaporation during the heating process. Afterward, the IOCS was cooled, and the grains were sieved and backwashed with deionized water until the pH of the backwash water was near 7. Using this process, approximately 54% of the added iron remained attached to the sand after washing. The final Fe surface concentration was 7.4 mg·g^{-1} sand for the coatings. The BET surface area (68 m^2·g^{-1} iron oxide, $Fe_2O_3·H_2O$) was determined by N_2 gas adsorption.

B. IOCS Titration

A potentiometric titration of the IOCS suspensions in the presence of varying concentrations of background electrolyte solution was used to determine the pH of the point-of-zero-charge (pH_{pzc}). The IOCS suspensions were prepared in a jacketed 1000-ml reactor flask with 20 g IOCS and 500 ml of 0.05, 0.1, and 1.0 mol·L^{-1} $NaNO_3$ as background electrolyte solution. Suspensions were equilibrated by mixing in a N_2 atmosphere. The temperature was kept constant (25°C) with a recirculating water bath. Various volumes of 0.1 mol·L^{-1} NaOH or 0.1 mol·L^{-1} HCl were added to adjust the pH.

C. Lead Adsorption Kinetics

Batch experiments were used to study sorption kinetics and to determine equilibrium parameters for Pb adsorption. A solution of 0.1 mmol·L^{-1} Pb in 0.05 mol·L^{-1} $NaNO_3$ background electrolyte was used to determine Pb adsorption kinetics. The solution pH was buffered with piperazine-*N,N'*-bis(2-ethansulfonic acid) (PIPES) at a pH value of 6.0. PIPES was chosen because it does not adsorb to IOCS and will not affect adsorption of metals (Szecsody *et al.,* 1994). Adsorption experiments were conducted in 50-ml, screw-top plastic centrifuge tubes. Fifty milliliters of Pb solution was reacted with 0.5 g of IOICS. The solution was purged with N_2 to eliminate carbon dioxide content. Suspensions were kept well mixed by continuous shaking in a 25°C constant temperature shaker bath. After shaking for the desired reaction time, the tubes were centrifuged at an RCF of 1200 × g for 10 min and the supernatant solution concentrations were determined. The sorption kinetics were determined for 60 hr reaction time. The supernatant Pb content was analyzed using flame atomic absorption spectroscopy (AAS). Total Pb removal from solution was determined by the difference from the initial and final solution concentrations. The pH of each sample was also determined.

D. Adsorption Isotherms

Adsorption isotherms were determined in completely mixed batch reactors in a temperature-controlled chamber. The IOCS suspensions were prepared with 1 g IOCS and 50 ml of 0.05 mol·L^{-1} $NaNO_3$ background electrolyte solution. Lead was added to yield concentrations ranging between $2.5 \cdot 10^{-2}$ and $5 \cdot 10^{-1}$ $mmol \cdot L^{-1}$ (equilibrated at a constant 20°C). Acid or base (0.1 $mol \cdot L^{-1}$ HCl or NaOH) was added to adjust the solution to the desired pH. After 24 hr equilibration time, suspensions were centrifuged and Pb concentration and pH were determined in the centrate as described in the adsorption kinetic experiments. The effects of varying pH values and supporting electrolyte concentrations were also determined on Pb equilibrium isotherms.

The effects of organic ligands, EDTA and NTA, on Pb adsorption were examined in equilibrium experiments. Either varying concentrations of EDTA or NTA were preequilibrated with Pb and then added to the IOCS suspensions (Pb–ligand-first addition sequence) or Pb was preequilibrated with IOCS before addition of EDTA or NTA (metal-first addition sequence). EDTA concentrations were analyzed by HPLC (Bergers and Groot, 1994). NTA concentrations were analyzed using a colorimetric method (EPA, 1974).

III. RESULTS AND DISCUSSION

A. IOCS Titration

The point-of-zero-charge, pH_{pzc}, defined by the common intersection point of the three titration curves, was located at pH 8.2 (Fig. 1). The determined pH_{pzc} of 8.2 corresponded approximately to the pH_{pzc} of the pure solid, $Fe(OH)_3$ (amorphous, pH_{pzc} 8.5 (Stumm and Morgan, 1981)). The prepared IOCS has a net posi-

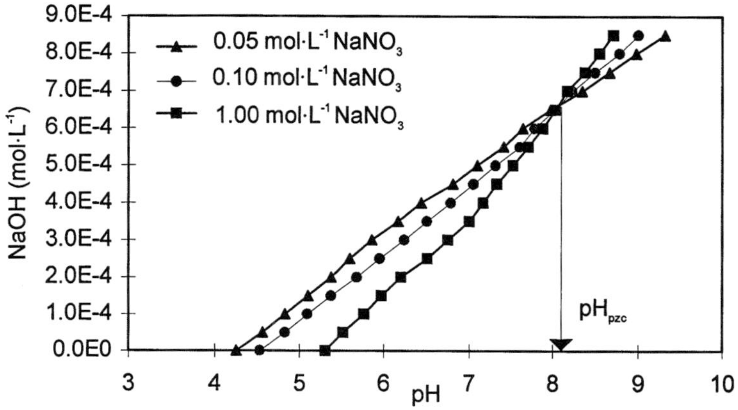

Figure 1 Iron-oxide-coated sand pH of the point-of-zero-charge (pH_{pzc}) determination using acid–base titration at various ionic strengths and 40 $g \cdot L^{-1}$ of IOCS.

tive charge below the pH_{pzc} of 8.2, and was therefore electrostatically attracted to the anionic EDTA and NTA form in its adsorption. Acid–base titration data at different sodium nitrate concentrations were used to obtain the double-layer model (DLM) and TLM parameters (Tables I and II). The ability of these models to fit the titration data over a wide range of model parameter values was examined.

B. Lead Sorption

Results indicated that equilibrium was reached in less than 24 hr in completely mixed batch systems (Fig. 2). In this experiment, 50 ml Pb solution was reacted with 0.5 g IOCS. The kinetics of metals sorption are usually found to be rapid—1 to 3 hr (Schultz *et al.,* 1987), to 24 hr (Farrah and Pickering, 1977b; Farrah *et al.,* 1980).

Equilibrium adsorption of 0.1 $mmol \cdot L^{-1}$ Pb as a function of pH (pH adsorption edge) increases dramatically with pH, from being almost undetectable at a pH of 4 to nearly 100% at a pH of 6 (Fig. 3). This corresponds closely with previous studies (Slavek and Pickering, 1986; Davis and Leckie, 1978b; Ainsworth and Pilan, 1994). Initial concentrations of 0.025, 0.05, 0.2, 0.3, and 0.5 $mmol \cdot L^{-1}$ Pb were all investigated with similar results (Fig. 3).

Adsorption edge phenomena can be explained as competition from H^+ ions for surface complexation sites on iron-oxide surfaces. At lower pH values, H^+ ions are adsorbed to the surface so that the net charge is positive. This inhibits adsorption of the positively charged Pb ions. As pH increases, the OH^- concentration increases on the surface and Pb adsorption increases dramatically. At higher pH values (>6) and Pb concentrations, precipitates may form and cannot be distin-

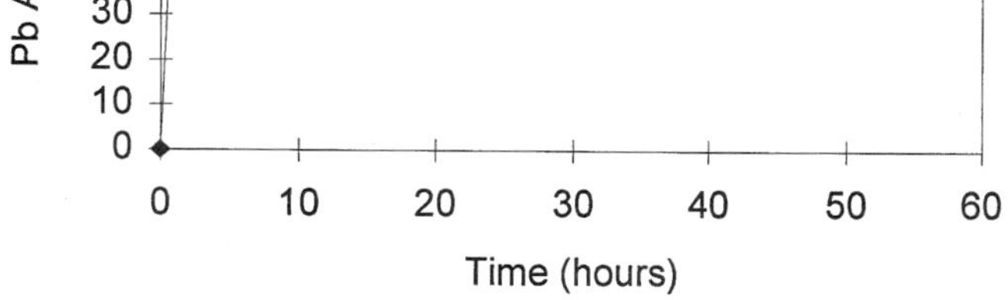

Figure 2 Lead adsorption kinetics for 60 hr in 0.05 $mol \cdot L^{-1}$ $NaNO_3$ solution with a buffered pH of 6.0, 0.1 $mmol \cdot L^{-1}$ Pb, and 10 $g \cdot L^{-1}$ IOCS.

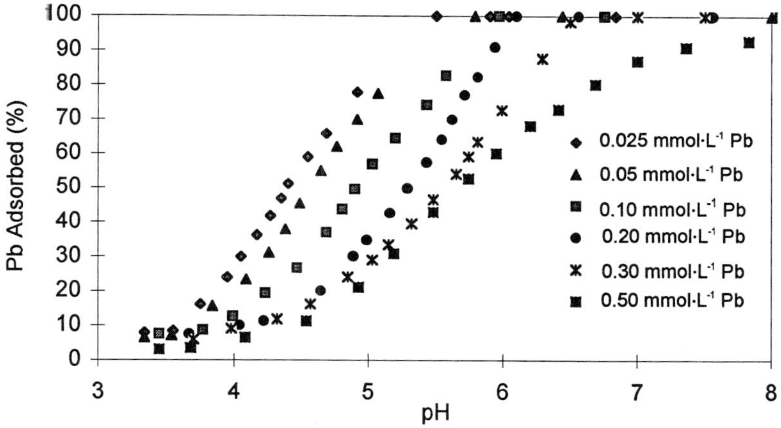

Figure 3 pH dependence of Pb^{2+} adsorption on iron-oxide-coated sand in 0.05 mol·L^{-1} $NaNO_3$ solution using varied Pb concentrations and 20 g·L^{-1} IOCS.

guished from adsorption. To avoid this problem, lead concentrations were kept lower than theoretical speciation concentrations would predict for precipitation in the pH range of 3 to 8.

The effects of Ca^{2+} and Na^+ electrolyte were examined in equilibrium experiments with a 40 ml Pb solution reacted with 2 g IOCS. Results indicated that adsorption was affected very little by competition from Ca^{2+} and Na^+ electrolytes (Figs. 4 and 5). The influence of ionic strength is attributable to two effects: direct

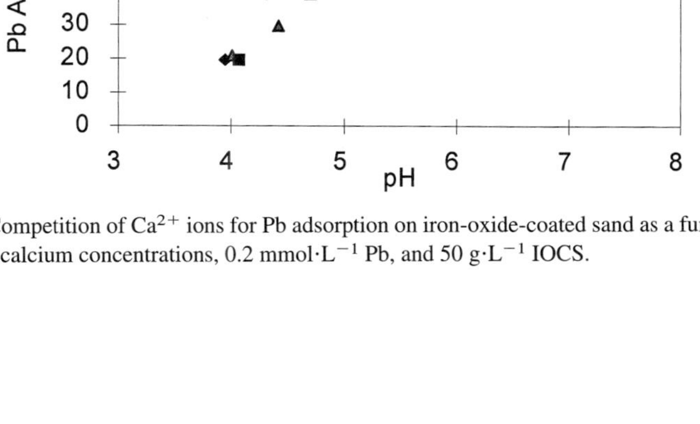

Figure 4 Competition of Ca^{2+} ions for Pb adsorption on iron-oxide-coated sand as a function of pH using varied calcium concentrations, 0.2 mmol·L^{-1} Pb, and 50 g·L^{-1} IOCS.

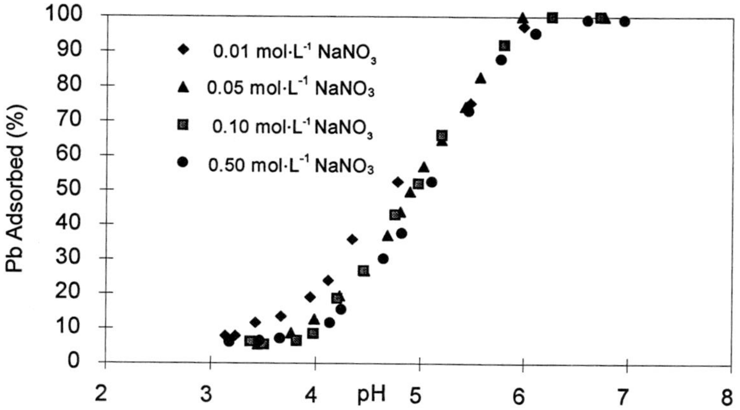

Figure 5 pH dependence of Pb adsorption on iron-oxide-coated sand using varied $NaNO_3$ concentrations, 0.1 mmol·L^{-1} Pb, and 20 g·L^{-1} IOCS.

competition of Na^+ or Ca^{2+} ions with Pb^{2+} for oxyhydroxide surface sites and a decrease in the electrostatic potential near the surface sites (Davis and Leckie, 1978b). This confirmed that Pb adsorption on IOCS corresponds to the formation of strongly bound inner-sphere complexes.

C. Effect of Organic Chelators

The adsorption of lead as a function of pH in the presence of EDTA for the Pb–ligand-first addition sequence is shown for EDTA:Pb molar ratios of 1:1, 2:1, and 5:1 in Figure 6. The Pb–$EDTA^{2-}$ complex was adsorbed by IOCS, with adsorption increasing with decreasing pH over the pH range of 3 to 10. This type of interaction has been termed a ligand-like ternary surface complex and has been observed in other studies where a metal and EDTA were preequilibrated (Bryce *et al.*, 1994; Szecsody *et al.*, 1994). At an EDTA:Pb molar ration of 1:1, about 35% of added Pb adsorbed as a Pb–$EDTA^{2-}$ complex, but at higher EDTA:Pb molar ratios of 2:1 and 5:1, Pb–$EDTA^{2-}$ adsorption decreased from 15 to near 0%, respectively (Fig. 6). This suggested that free EDTA must compete with the adsorption of the Pb–$EDTA^{2-}$ complex on IOCS.

When Pb is preequilibrated with IOCS (metal-first addition sequence), addition of 0.1 and 1.0 mmol·L^{-1} EDTA (1:1 and 10:1 molar ratios, respectively) decreased Pb adsorption on IOCS to near 0% by either exchanging EDTA for adsorbed Pb or by formation of the soluble Pb–$EDTA^{2-}$ complex (Fig. 7). Ligand-like Pb adsorption was not observed (Fig. 7). EDTA-only (0.01 mmol·L^{-1}) in 0.01, 0.05, and 0.1 mmol·L^{-1} $NaNO_3$ electrolyte solution showed typical ligand-like ad-

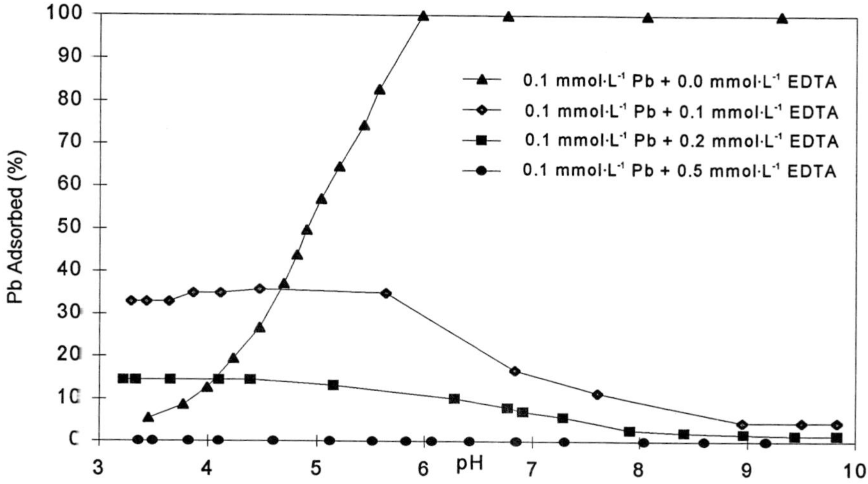

Figure 6 pH dependence of Pb adsorption on IOCS in 0.05 mol·L^{-1} $NaNO_3$ solution using varied EDTA concentrations, the preequilibrated Pb and EDTA addition sequence, and 20 g·L^{-1} IOCS.

sorption (Fig. 8). The notable influence of ionic strength on EDTA adsorption provided evidence for formation of ion-pair surface coordination (outer-sphere) complexes with surface hydroxyls of IOCS (Fig. 8). The Pb–$EDTA^{2-}$ complex was also affected by the addition of 0.5 mol·L^{-1} $NaClO_4$ electrolyte solution for adsorption on soil (Elliott and Brown, 1989).

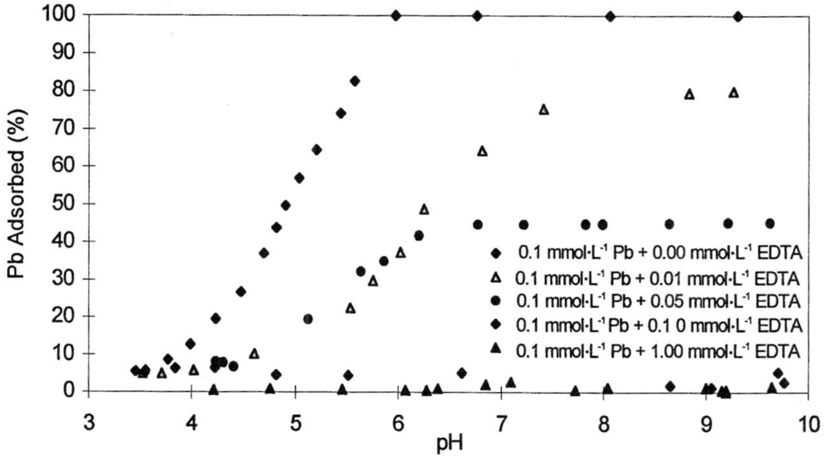

Figure 7 pH dependence of Pb adsorption on IOCS in 0.05 mol·L^{-1} $NaNO_3$ solution using varied EDTA concentrations, the metal-first addition sequence, and 20 g·L^{-1} IOCS.

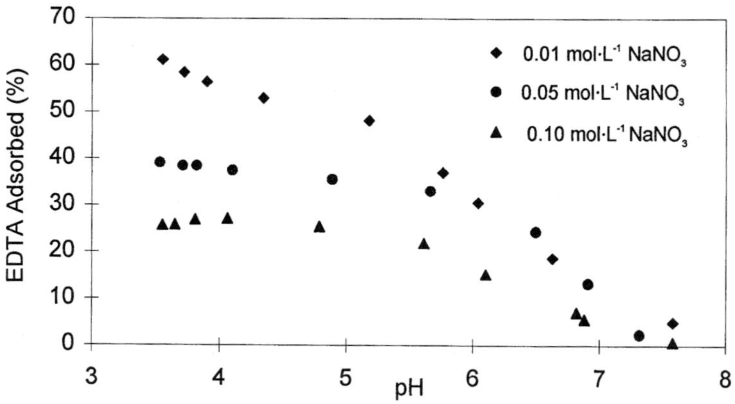

Figure 8 pH dependence of EDTA adsorption on IOCS using varied $NaNO_3$ concentrations, 0.01 $mmol \cdot L^{-1}$ EDTA, and 20 $g \cdot L^{-1}$ IOCS.

Preequilibrated Pb–NTA (Pb–ligand-first addition sequence) decreased lead adsorption on IOCS in a stoichiometric manner with near complete inhibition of adsorption at NTA:Pb molar ratios of 10:1 (Fig. 9). When Pb was preequilibrated with IOCS (metal-first addition sequence), addition of 0.05 and 0.1 $m\,mol \cdot L^{-1}$ NTA decreased Pb adsorption to 35 and 10%, respectively (Fig. 10). $Pb–NTA^-$ adsorption on IOCS was metal-like, increasing with increasing pH (Figs. 9 and 10), but adsorption of NTA-only on IOCS was ligand-like (Fig. 11). NTA adsorption on γ-

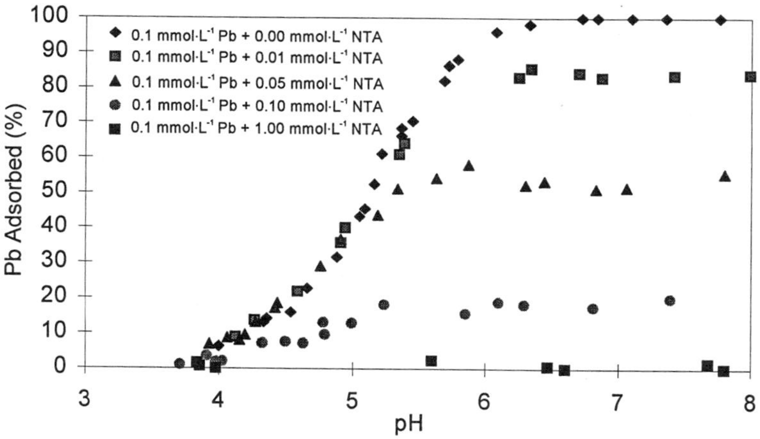

Figure 9 pH dependence of Pb adsorption on IOCS in 0.05 $mol \cdot L^{-1}$ $NaNO_3$ solution using varied NTA concentrations, 20 $g \cdot L^{-1}$ IOCS, and the preequilibrated Pb and NTA addition sequence.

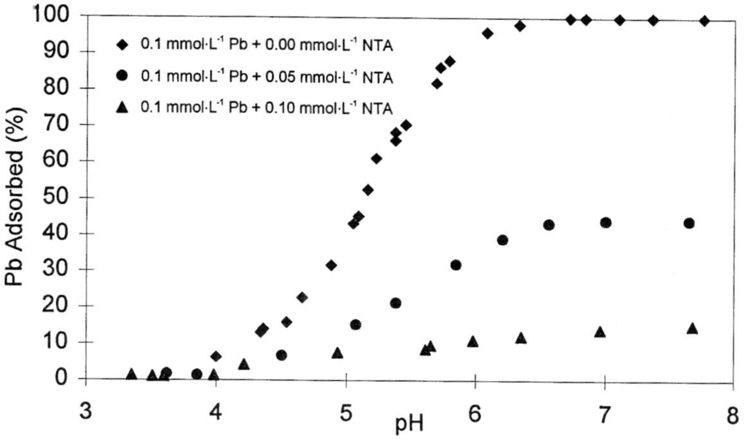

Figure 10 pH dependence of Pb adsorption on IOCS in 0.05 mol·L^{-1} $NaNO_3$ solution using varied NTA concentrations, the metal-first addition sequence, and 20 g·L^{-1} IOCS.

Al_2O_3 as a function of pH for initial NTA concentrations of 0.05 to 0.1 mmol·L^{-1} was also ligand-like over the pH range 3 to 10 (Elliott and Huang, 1979).

D. Computer Modeling

A chemical equilibrium computer program (HYDRAQL) (Papelis *et al.*, 1988), using a surface complexation triple-layer model for adsorption, was used success-

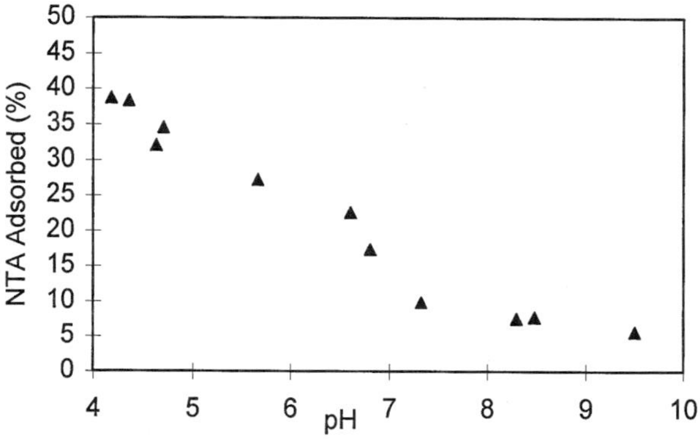

Figure 11 pH dependence of NTA adsorption on IOCS in 0.05 mol·L^{-1} $NaNO_3$ solution, 0.1 mmol·L^{-1} NTA, and 20 g·L^{-1} IOCS.

Table I

Estimates for Surface Acidity Constants for Three N_s Values

I (mol·L^{-1})	N_s [a]	log K_{a1}^{int}	log K_{a2}^{int}	V_y [b]
0.05	1.5	−6.12	−10.74	11.5
0.10	2.8	−6.34	−10.46	13.8
1.00	3.0	−6.52	−10.01	18.3

[a] N_s, site density (sites·nm^{-2}).
[b] V_y, goodness of fit parameter (values less than 20 show a good fit).

fully to simulate Pb adsorption by IOCS. Evaluation and modeling of adsorption data required an estimate of the specific surface area and the sorption site density of the IOCS, as well as a description of its acid–base surface chemistry (Dzombak and Morel, 1990). Acid–base titration data were used to obtain surface acidity constants (log K_{a1} and log K_{a2}) and the number of adsorption sites per unit surface area (N_s) from the DLM. Acid–base titrations for α-FeOOH and other metal oxides have been successfully interpreted with each of the different surface complexation models, including the DLM and TLM (Westall and Hohl, 1980). The DLM has three adjustable parameters: log K_{a1}, log K_{a2}, and N_s. Three parameters were optimized for each set of acid–base titration data. For each ionic strength data set, log K_{a1}, log K_{a2}, and N_s were optimized for varying N_s values (Table I). The fit was acceptable for N_s values less than 3, but the model did not converge when the site density increased above 5 sites·nm^{-2}. This was consistent with the finding of Sposito (1984), who obtained an estimate of 4.0 sites·nm^{-2} using the acid–base titration method which considered only Lewis acid sites of α-FeOOH. Hayes *et al.* (1991) indicated that the DLM was not sensitive to changes in N_s between 1 and 100 sites·nm^2.

Table II

Estimates for Surface Acidity Constants and Electrolyte Binding Constants for Three N_s Values

		DLM		TLM	
I (mol·L^{-1})	N_s [a]	log K_{a1}^{int}	log K_{a2}^{int}	log $K_{FeOH_2^+-NO_3^-}^{int}$	log $K_{FeO^--Na^+}^{int}$
0.05	1.5	−6.12	−10.74	6.87	−8.94
0.10	2.8	−6.34	−10.46	6.25	−8.62
1.00	3.0	−6.52	−10.01	6.17	−8.49

[a] N_s, site density (sites·nm^{-2}).

Table III

Best Estimates for Lead TLM Surface Complexation Reaction Constants Using Two Optimizable Parameters (log K and N_s)

Inner-sphere			Outer-sphere		
N_s [c]	$\log K^{int}_{FeOPb^+}$	V_y [b]	N_s [a]	$\log K^{int}_{FeO^- - Pb^{2+}}$	V_y [b]
2.84	2.54	5.74	0.102	4.21	NC[c]
3.49	2.32	3.96	0.051	4.01	424
4.00	2.19	5.04	—	—	NC[c]

[a]N_s, site density (sites·nm^{-2}).
[b]V_y, goodness of fit parameter (values less then 20 show a good fit).
[c]NC, no convergence.

The TLM has at least six adjustable parameters, log K_{a1}, log K_{a2}, log K_{NO_3}, log K_{Na}, N_s, and the inner-layer capacitance. The TLM requires electrolyte adsorption data in addition to acid–base titration data. The outer-layer capacitance is potentially variable, but it is usually set equal to 0.2 F·m^{-2} (Westall and Hohl, 1980). Using three adjustable parameters, log K_{a1}, log K_{a2}, and N_s from the DLM, electrolyte binding constants (log K_{NO_3}, log K_{Na}) for three N_s values were calculated

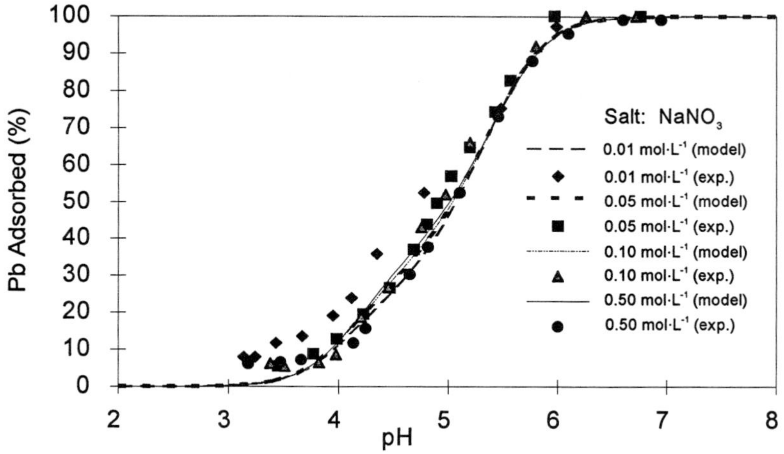

Figure 12 pH dependence of Pb adsorption on iron-oxide-coated sand at varied $NaNO_3$ concentrations. Lines correspond to the TLM calculations of Pb adsorption on IOCS as a function of pH using varied $NaNO_3$ concentrations, 0.1 mmol·L^{-1} Pb, and 20 g·L^{-1} IOCS.

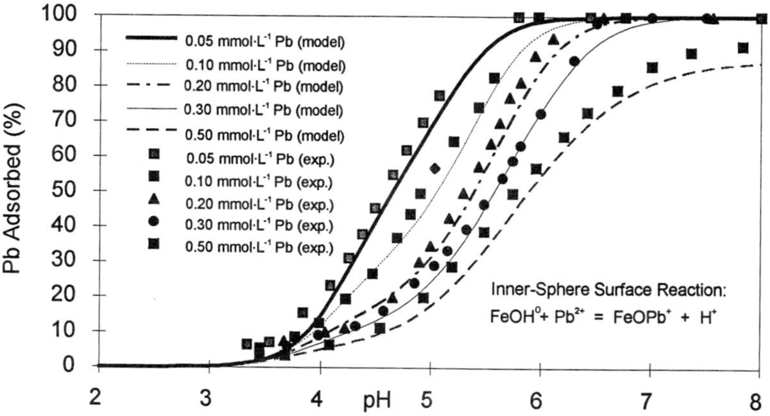

Figure 13 pH dependence of Pb adsorption on IIOCS in 0.05 mol·L^{-1} $NaNO_3$ at varied Pb concentrations. Lines correspond to the TLM calculations of Pb adsorption on IOCS as a function of pH using varied Pb concentrations and 20 g·L^{-2} IOCS.

from the TLM. The inner-layer capacitance was 1.4 F·m^{-2} based on the TLM fit to the isotherm data. Best estimates for surface acidity constants (DLM) and electrolyte binding constants (TLM) for three N_s values are given in Table II. These values were used in the TLM to estimate Pb adsorption constants, log K_{FeOPb^+} and log $K_{FeO^--Pb^{2+}}$ (Table III). The main indicator of goodness of fit is the overall variance (V_y), which is the weighted sum of squares of residuals divided by the degrees of freedom. Values of V_y between 0.1 and 20 are typical for a good fit (Westall, 1980). Using an inner-sphere Pb surface adsorption constant, the TLM was successful in describing the effect of pH and ionic strength on Pb adsorption on IOCS (Fig. 12). Additionally, the TLM accurately described the adsorption edge data at different Pb concentrations (Fig. 13).

VI. CONCLUSION

In this study, IOCS was used as a surrogate for adsorption on soils to investigate the effect of changing environmental conditions on lead adsorptive retardation and mobility. The lack of influence of competing cations on Pb adsorption on IOCS corresponded to the formation of strongly bound inner-sphere complexes. The triple-layer model, using inner-sphere Pb surface adsorption binding, was successful in describing the effect of environmental conditions such as pH and ionic strength on Pb adsorption on IOCS. Pb–EDTA^{2-} was adsorbed on the iron-oxide-coated sand by ligand-like adsorption, increasing with decreasing pH, while

Pb–NTA$^-$ adsorption on IOCS was metal-like, increasing with increasing pH. Chemically enhanced desorption of adsorbed Pb can be achieved by decreasing pH, but this may not be practical in highly buffered natural soils. Addition of complexing ligands shows promise for enhancing lead desorption, but the adsorptive behavior of Pb–ligand as a ternary complex must be carefully evaluated.

ACKNOWLEDGMENT

Funding for this study was provided by the Office of Research and Development, U.S. Environmental Protection Agency, under Agreement R-815738-01 through the Western Region Hazardous Substance Research Center. The content of this study does not necessarily represent the view of the agency.

REFERENCES

Ainsworth, C. C., and Pilan, J. L. 1994. Cobalt Cadmium and Lead Sorption to Hydrous Iron Oxide: Residence Time Effect. *Soil Sci. Am. J.* 58:1615–1623.

Anonymous. 1988. *Hazardous Waste Consultant,* January/February: 1–20.

Bergers, P. J. M., and Groot, A. C. 1994. The Analysis of EDTA in Water by HPLC. *Water Res.* 28:639–642.

Bjorklund, H., Olson, L., Seiger, A., and Hoffer, B. 1980. Chronic Lead and Brain Development: Intraocular Brain Grafts as a Method to Reveal Regional and Temporal Effects in the Central Nervous System. *Environ. Res.* 22:224–236.

Bryce, A. L., Kornicker, W. A., and Elxerman, A. W. 1994. Nickel Adsorption to Hydrous Ferric Oxide in the Presence of EDTA: Effects of Component Addition Sequence. *Environ. Sci. Technol.* 28:2353–2359.

Davis, J. A., and Leckie, J. O. 1978a. Effect of Adsorbed Complexing Ligands on Trace Metal Uptake by Hydrous Oxides. *Am. Chem. Soc.* 12:1309–1314.

Davis, J. A., and Leckie, J. O. 1978b. Surface Ionization and Complexation at the Oxide/Water Interface II. Surface Properties of Amorphous Iron Oxyhydroxide and Adsorption of Metals Ions. *J. Colloid Interface Sci.* 67:90–107.

Dzombak, D. A., and Morel, F. M. 1990. "Surface Complexation Modeling: Hydrous Ferric Oxide." Wiley, New York.

Edwards, M., and Benjamin, M. M. 1989. Adsorptive Filtration Using Coated Sand: A New Approach for Treatment of Metal-Bearing Wastes. *J. Water Pollut. Control Fed.* 61:1523–1533.

Elliott, H. A., and Brown, G. A. 1989. Comparative Evaluation of NTA and EDTA for Extractive Decontamination of Pb-Polluted Soils. *Water Air Soil Pollut.* 45:361–369.

Elliott, H. A., and Huang, C. P. 1979. The Adsorption Characteristics of Cu(II) in the Presence of Chelating Agents. *J. Colloid Interface Sci.* 70:29–45.

Environmental Protection Agency (EPA). 1974. Methods for Chemical Analysis of Water and Wastes, pp. 217–219. EPA-625-/6-74-0003, Office of Technology Transfer, Washington, DC.

Farrah, H., Hatton, D., and Pickering, W. F. 1980. The Affinity of Metal Ions for Clays Surfaces. *Chem. Geol.* 28:55–68.

Farrah, H., and Pickering, W. F. 1977a. Extraction of Heavy Metal Ions Sorbed on Clays. *Water Air Soil Pollut.* 9:491–498.

Farrah, H., and Pickering, W. F. 1977b. The Sorption of Lead and Cadmium Species by Clay Minerals. *Aust. J. Chem.* 30:1417–1422.

Hayes, K. F., and Leckie, J. O. 1987. Modeling Ionic Strength Effect on Cation Adsorption at Hydrous Oxide/Solution Interfaces. *J. Colloid Interface Sci.* 115:564–572.

Hayes, K. F., Redden, G., Ela, W., and Leckie, J. O. 1991. Surface Complexation Models: An Evaluation of Model Parameter Estimation Using FITEQL and Oxide Mineral Titration Data. *J. Colloid Interface Sci.* 142:448–469.

Hem, J. D. 1976. Geochemical Control on Lead Concentrations in Stream Water and Sediments. *Geochim. Cosmochim. Acta* 40:599–609.

Hodson, P. V., Whittle, D. M., Wong, P. T. S., Borgmann, U., Thomas, R. L., Chau, Y. K., Nriagu, J. O., and Hallett, D. J. 1984. "Lead Contamination of the Great Lakes and Its Potential Effects on Aquatic Biota," pp. 336–370. Wiley, New York.

Hohl, H., and Stumm, W. 1976. Interaction of Pb(II) with Hydrous γ-Al_2O_3. *J. Colloid Interface Sci.* 55(2):281–287.

Huang, C., and Stumm, W. 1973. Specific Adsorption of Cations on Hydrous γ-Al_2O_3. *J. Colloid Interface Sci.* 43:409–420.

King, L. D. 1988. Retention of Metals by Several Soils of the Southeastern United States. *J. Environ. Qual.* 17:239–246.

Nraigu, J. O. 1978. Lead in the Atmosphere. *In* "The Biogeochemistry of Lead in the Environment, Part A. Ecological Cycles" (J. O. Nraigu, Ed.), pp. 155–160. Elsevier/North-Holland Biomedical Press, Amsterdam.

Papelis, C., Hayes, K. F., and Leckie, J. O. 1988. HYDRAQL: A Program for the Computation of Chemical Equilibrium Composition of Aqueous Batch Systems Including Surface Complexation Modeling of Ion Adsorption at the Oxide/Solution Interface. Technical Report No. 306, Environmental Engineering and Science, Department of Civil Engineering, Stanford Univ., Stanford, CA.

Putka, G. 1992. Research on Lead Poisoning Is Questioned. *Wall Street Journal,* Mar. 6, 1992, B1.

Santillan-Medrano, J., and Jurinak, J. J. 1975. The Chemistry of Lead and Cadmium in Soil: Solid Phase Formation. *Soil Sci. Soc. Am.* 39:851–856.

Schultz, M. F., Benjamin, M. M., and Ferguson, J. F. 1987. Adsorption and Desorption of Metals on Ferrihydrite: Reversibility of the Reaction and Sorption Properties of the Regenerated Solid. *Environ. Sci. Technol.* 21:863–869.

Scokart, P. O., Meeus-Verdinne, K., and De Borger, R. 1983. Mobility of Heavy Metals in Polluted Soils Near Zinc Smelters. *Water Air Soil Pollut.* 20:451–463.

Slavek, J., and Pickering, W. F. 1986. Extraction of Metal Ions Sorbed on Hydrous Oxides of Iron(III). *Water Air Soil Pollut.* 28:151–162.

Slavek, J., and Pickering, W. F. 1987. Selective Extraction of Trace Metals Associated with Hydrous Aluminum Oxides. *Can. J. Chem.* 65:984–989.

Slavek, J., and Pickering, W. F. 1988. Metal Ion Interaction with the Hydrous Oxides of Aluminum. *Water Air Soil Pollut.* 39:201–216.

Sposito, G. 1984. "The Surface Chemistry of Soils." Oxford Univ. Press, New York.

Stumm, W., and Morgan, J. J. 1981. "Aquatic Chemistry: An Introduction Emphasizing in Natural Waters," 2nd ed. Wiley–Interscience, New York.

Szecsody, J., Zachara, J., and Bruckhart, P. 1994. Adsorption-Dissolution Reactions Affecting the Distribution and Stability of Co(II)EDTA in Iron Oxide Coated Sand. *Environ. Sci. Technol.* 28:1706–1716.

Westall, J. C. 1980. FITEQL: A Computer Program for Determination of Chemical Equilibrium Constants from Equilibrium Data, Version 2.0. Report 82-02, Oregon State Univ., Corvallis, OR.

Westall, J. C., and Hohl, H. 1980. A Comparison of Electrostatic Models for the Oxides-Solution Interface. *Adv. Colloid Interface Sci.* 12:265–294.

Chapter 7

Uranium Sorption onto Natural Sands as a Function of Sediment Characteristics and Solution pH

Jeffrey J. Rosentreter,[1] H. Swantje Quarder,[1] Robert W. Smith,[2] and Travis McLing[2]

[1]Idaho State University, Department of Chemistry, Pocatello, Idaho; [2]Idaho National Engineering Laboratory, Idaho Falls, Idaho

This chapter describes ongoing research to characterize the relationships between physical and chemical properties of selected sediments and their sorptive properties, permeability, and porosity. Specifically described are determinations of partition coefficients (K_d values) as a function of pH for uranyl sorption on 25 sediment samples collected from a sand borrow pit. In our evaluation of this data, several physical and chemical characteristics of the sediments were identified as having both significant and insignificant influences on the sediment's uranium sorptive capabilities. Significant correlations between surface area and surface area surrogates were identified. Conversely, mineralogy and grain size had essentially no influence on the sorptive capabilities of these sediments. In our evaluation of the K_d data, we were able to identify and justify a linear relationship between the log K_d values and equilibrium solution pH values. These applied equally well to all the sediments in our sample set. It eventually allowed for the development of two simple predictive equations that, based on either surface area or extractable metal concentrations, provided an effective indirect means of evaluating a sediment's sorptive behavior with respect to uranium.

Adsorption of Metals by Geomedia

I. INTRODUCTION

Mixtures of contaminants including radionuclides, toxic metals, organic solvents, and complexants have been released to the subsurface, leading to potential or actual contamination of groundwater resources. The risk to human health and the environment arising from contaminant transport in the groundwater must be clarified in order to develop cost-effective and technically defensible remediation strategies for groundwater and vadose zones and to determine sites for new waste storage and disposal facilities. Mechanistic knowledge of the interactions between aqueous adsorbing contaminants and multiple mineral surfaces (e.g., sorption) in heterogeneous media and knowledge of how rapidly water flows (residence time) through the porous media are key to understanding the fate and transport of contaminants in the unsaturated and saturated subsurface. The relationships between the flow domain and sorptive intensity of multiple mineral surfaces determine the velocity at which contaminants migrate. Although the role of physical medium heterogeneity on transport has received significant attention, similar investigation of chemical (reactive) medium heterogeneity, and the correlation of physical and chemical heterogeneity, is yet to be fully explored.

This chapter describes ongoing research to characterize the relationships between physical and chemical properties of selected sediments and their sorptive properties, permeability, and porosity. Specifically described are determinations of partition coefficients (K_d values) as a function of pH for uranyl (U^{VI}) sorption on 25 sediment samples collected from a sand borrow pit located on the southern tip of the Delmarva Peninsula in Virginia, near the village of Oyster. One criterion for the selection of the field site was that the reactivity of the sediment would be controlled by variability in the abundance of hydrous oxide coatings on the sediments. The borrow pit consists of weakly consolidated cross-bedded sand with secondary hydrous metal oxides associated with primary depositional features and is stratigraphically located in the lower portion of the Butlers Bluff Member of the Pleistocene age Nassawadox Formation. The Butlers Bluff Member averages about 18 m in thickness and consists of clean, cross-bedded, fine to coarse sand and gravel and is thought to have formed either as a barrier-spit or a shoal complex adjacent to the mouth of ancestral Chesapeake Bay (Mixon, 1985). The sediment samples were collected to span the range of variability in detrital mineralogy, amounts of secondary hydrous ion oxides (geochemistry), and grain size (surface area).

II. EXPERIMENTAL

A. Procedures for Batch Experiments

Well-established batch reaction techniques were implemented to determine solution concentrations before and after equilibration with the sediments and solid loading concentrations. Four experimental sets were completed with initial pH values of 3, 4, 5, and 6. All experimental sets included operational blanks, reagent blanks, and replicate samples.

1. Solution Preparation

All solutions were prepared using American Chemical Society (ACS) certified reagent grade chemicals. Deionized water with a resistance of 16 MΩ or higher was used as the diluent. The adjustment of pH was accomplished using trace metals grade nitric acid and reagent grade sodium hydroxide. Uranium concentrations were adjusted through the addition of an ACS certified uranium standard. The reaction solution was 0.01 $mol{\cdot}L^{-1}$ calcium perchlorate containing 250 $\mu g{\cdot}L^{-1}$ of uranium. The initial pH of the reaction solution was adjusted with additions of concentrated nitric acid and sodium hydroxide as needed. The pH adjustments provided solutions with a pH of ±0.04 pH units, which were stable within that range for a minimum of 48 hr. Operational blanks consisted of aliquots of the reaction solution that were not reacted with the geologic materials.

2. Reaction Procedure

The sorption studies were carried out using batch reaction techniques in 50-ml polyethylene centrifuge tubes. The geologic samples were subsplit to 4.0 ± 0.5 g using a riffle splitter to avoid bias. Sorption studies were carried out using 24.0 ml of reaction solution.

Mixing was achieved by rotating the sealed centrifuge tubes on a vertical rotator. No attempt was made to exclude or control CO_2 in the experiments. The samples were gently agitated at four rotations per minute for 72 hr while maintained at 25°C. The liquid phase was separated from the solid phase at the end of the equilibration period by centrifugation for 10 min at 3000 rpm in head with a center axis to mid tube radius of 10 cm. Final pH was measured and the samples were then preserved by the addition of trace metals grade nitric acid to a pH of less than 2.0.

Experimental precision in each set was determined by analysis of duplicate samples for each sediment equilibrated with reaction solution. Two types of blanks were also analyzed. The first, an operational blank, consisted of the reaction solution without sediment to account for sorption on and desorption from the container walls. The second, a reagent blank, consisted of deionized water samples used in the preparation of the reaction solutions.

3. Analytical Methods

Reaction solutions, duplicates, and blanks were analyzed for pH and uranium content just prior to and following the equilibration period. Uranium determinations were accomplished using kinetic phosphorimetry. A Chemcheck Instruments

KPA-11 Phosphorescence Analyzer provided reliable uranium detection in the sub-part-per-billion concentration range. An instrument detection limit of 0.05 $\mu g \cdot L^{-1}$ was typical in our study (with the time gate set at 240 μsec). The overall percent relative standard deviation (%RSD) for analytical replicates measured throughout the experiments ranged from 0.3 to 7.0%. Measurements of pH were performed on a Model 25 Fisher Scientific pH/Ion-meter. Using a five-point calibration curve, various pH standard solutions provided measured pH values with %RSD values of below 0.1%.

B. Geologic Materials

Characterization of the sediments included determination of bulk mineralogy, surface area, and extractable Fe^{III} and Al concentrations. Bulk mineralogy was determined by petrographic examination of acid washed and stained grain mounts. Surface area was determined by BET analysis using a Micromeritics Gemini 2360 Multipoint Rapid Surface Area Analyzer. Reducible Fe was determined using a bicarbonate buffered dithionite–citrate extraction (Mehra and Jackson, 1960). Readily extractable Al was determined using an oxalate extraction (McKeague and Day, 1966). The mineralogical, physical, and chemical characteristics of the sediment samples are listed in Table I.

III. RESULTS AND DISCUSSION

A. Determination of K_d Values and the Effect of pH

Values of K_d were calculated from the measured loss of uranium from solution using

$$K_d = (c_i - c_f) \cdot V \cdot (c_f \cdot W)^{-1}, \quad [1]$$

where c_i and c_f are the initial and final aqueous concentrations of uranium, respectively, V is the volume of the aqueous phase (24 ml), and W is the mass of the solid (~4 g). Because K_d values calculated using Eq. [1] depend only on measurements of aqueous concentrations, the values are particularly sensitive to systematic errors due to uranium losses by mechanisms other than sorption onto the sediments (e.g., sorption onto the reaction vessel). Possible systematic errors were minimized by the use of several reaction blanks with each experimental set. As used in this paper, K_d is a measure of partitioning of a solute between an aqueous and a solid phase; it does not imply linear reversible adsorption.

The effects of pH on the K_d values are shown in Figure 1. As may be seen in this figure, adsorption increases with increasing pH, consistent with behavior of

Table I

Mineralogical Characteristics and K_d Values for 25 Sediments Collected at the Oyster, Virginia, Site

Sample Number	BET surface area ($m^2 \cdot g^{-1}$)	DCB iron ($\mu mol \cdot g^{-1}$)	Oxalate aluminum ($\mu mol \cdot g^{-1}$)	Quartz (vol. %)	Potassium feldspar (vol. %)	Plagioclase feldspar (vol. %)	Lithic fragments (vol. %)	Dark minerals (vol. %)	log K_d at pH 5 ($ml \cdot g^{-1}$)	Standard deviation of log K_d
1	0.63	6.92	25.73	86.8	4.7	2.8	5.7	0.0	1.75	0.08
2	1.25	4.73	20.96	100.0	0.0	0.0	0.0	0.0	1.38	0.07
3	0.66	5.94	18.98	87.0	3.8	5.1	5.1	0.0	1.30	0.03
4	1.87	19.78	46.16	93.9	3.4	0.7	2.0	0.0	1.80	0.09
5	0.54	4.00	12.72	83.0	5.1	5.8	6.1	0.0	1.34	0.06
6	0.98	4.19	20.68	95.9	0.6	2.8	0.3	0.3	1.33	0.08
7	1.42	6.26	25.10	97.5	0.7	1.1	0.0	0.7	1.54	0.11
8	0.77	5.25	14.84	66.5	1.2	0.9	31.4	0.0	1.22	0.15
9	2.49	19.79	40.96	80.3	0.0	1.0	17.8	0.9	1.86	0.11
10	1.92	11.67	27.05	99.7	0.0	0.0	0.0	0.3	1.87	0.12
11	1.95	16.65	19.67	87.2	5.1	4.4	2.9	0.4	2.01	0.16
12	0.92	12.21	11.39	75.1	3.9	2.3	18.7	0.0	1.36	0.10
13	1.56	7.53	21.67	98.4	0.7	0.0	0.0	0.9	1.54	0.13
14	0.13	7.27	12.16	88.2	3.1	0.0	8.7	0.0	1.07	0.16
15	0.92	6.01	18.08	93.4	2.9	3.3	0.0	0.4	1.42	0.13
16	0.96	3.42	19.85	99.3	0.0	0.3	0.3	0.0	1.30	0.09
17	3.21	25.96	53.66	93.0	2.4	0.0	3.9	0.7	2.05	0.13
18	0.93	3.22	13.96	97.0	0.5	1.0	0.3	1.2	1.44	0.21
19	0.34	2.69	9.15	89.2	5.7	1.0	3.9	0.0	1.12	0.18
20	3.05	73.28	12.17	87.4	4.6	0.4	7.6	0.0	2.13	0.34
21	0.55	0.97	7.94	95.8	1.3	0.0	0.0	2.9	1.16	0.28
22	0.52	4.41	10.79	87.3	4.5	4.9	3.2	0.0	1.02	0.10
23	0.58	2.85	16.45	98.7	0.0	1.0	0.3	0.0	1.24	0.08
24	0.75	5.91	18.08	95.0	2.5	1.1	1.4	0.0	1.43	0.17
25	0.43	2.60	26.74	98.7	0.0	1.0	0.0	0.0	1.15	0.06

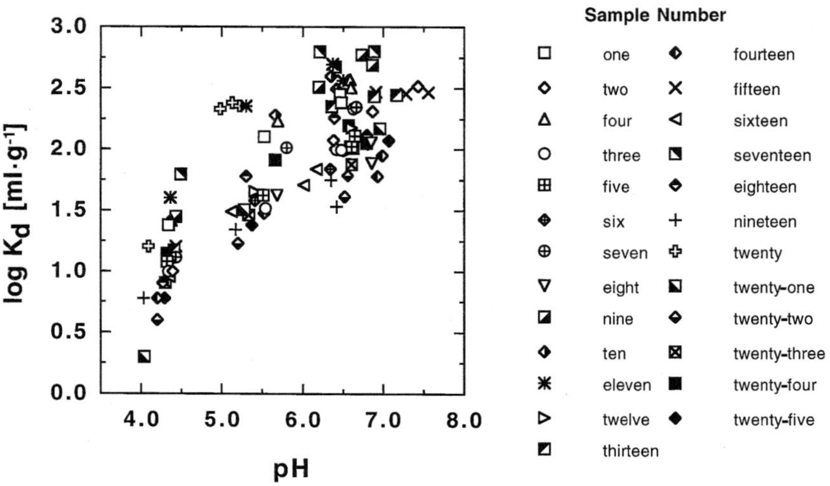

Figure 1 Measured K_d versus equilibrium pH for each of the sediment samples as recorded from the four experimental sets.

uranium in the pH range considered. This behavior is qualitatively the same as that observed by Hsi and Langmuir (1985) for uranyl adsorption onto ferric oxides, by Lieser and Thybuech (1988) for uranyl adsorption onto titanium oxide, and by Lieser *et al.* (1992) for uranyl adsorption onto hydrous silica. At higher pH values (i.e., 7), the formation of U^{VI} carbonate species results in decreasing adsorption. Because one of the goals of this study is to elucidate the role of mineralogical characteristics of the sediments in controlling K_d, it is desirable to have K_d values for each sediment at a common pH. This was accomplished using least-squared linear regression techniques. Initially, a regression of pH and log K_d values for each sample (i.e., four points) was conducted. The results of these analyses were that the slopes (i.e., d log K_d·d pH^{-1}) for all samples were similar. As a result, a regression analysis was done simultaneously on all 100 points constrained so that the slope for all samples was the same. The resulting equation is

$$\log K_{d,i} = b_i + 0.47(0.02)\cdot pH, \qquad [2]$$

where b_i is the intercept for the *ith* sample. The value in parentheses reflects the standard error associated with the fit coefficient. The similar pH dependency of log K_d for all samples is consistent with behavior that has been observed for uranyl adsorption on synthetic oxides (cf. Lieser *et al.,* 1992) and suggests that the shape of the pH versus log K_d curve is dominated by aqueous speciation and that the stoichiometry of the sorption reaction is independent of the solid.

B. Mineralogical Characteristics Affecting Sorptive Properties

A goal of this study was to define the key characterization measurements required to describe natural variation of the sorptive properties of sediments. The sediment characteristics considered included bulk mineralogy, extractable Fe^{III}, extractable Al, and BET surface area. Additional characterization to describe the relationship between physical and hydrological properties (i.e., permeability, porosity, and grain size) was also collected and is the subject of a future communication. Smith *et al.* (1996) provide a discussion of the theoretical basis of correlated physical and geochemical heterogeneities.

Values of K_d for each of the 25 samples at a pH of 5 were estimated by

$$\log K_{d,i,5} = \log K_{d,i,pH} - 0.47(0.02)\cdot[pH_i - 5]. \qquad [3]$$

The average and standard deviation for each of the samples are reported in Table I. The K_d values reported in Table I were used to evaluate the relationships between sorption and the mineralogical characteristics of the sediments.

No correlation was observed between K_d values and the bulk mineralogy of the sediments. This observation is not surprising given that a criterion for field site selection was that the reactivity be controlled by variability in the abundance of hydrous oxide coatings. A series of single-variable regression analyses were conducted using surface area and extractable Fe^{III} plus Al as independent variables. The results of these analysis are presented in Figures 2 and 3. Because surface area

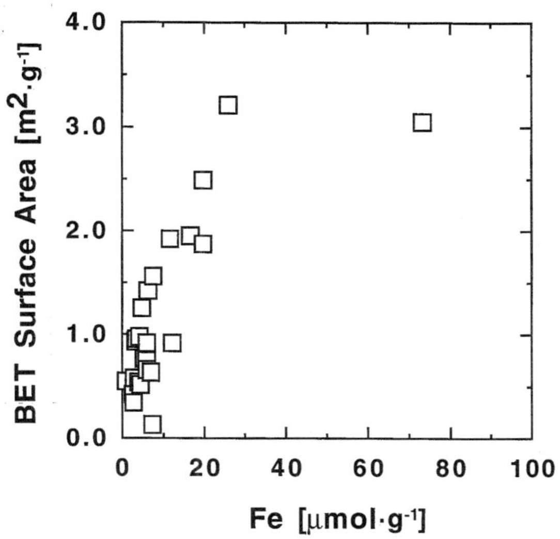

Figure 2 Measured surface area versus extractable Fe^{III} for the 25 sediment samples.

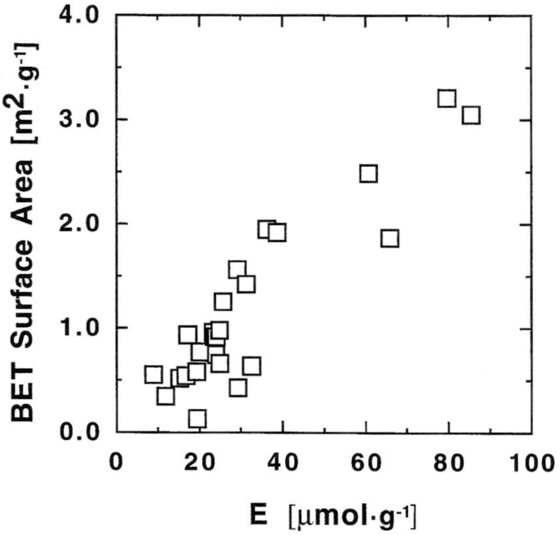

Figure 3 Measured surface area versus extractable Fe^{III} plus Al(E) for the 25 sediment samples.

and extractable metals were highly correlated (Fig. 3), regression analyses that included both surface area and the abundance of hydrous metal oxides were not included. Regression that included Al or Fe, or both Fe and Al as independent variables, resulted in poorer fits than those shown in Figures 2 and 3. In addition, an analysis that included surface area and the ratio of extractable Fe^{III} to Al was tested. The resulting coefficient for the ratio was not statistically different from 0, indicating that for the range of Fe-to-Al ratios observed for Oyster sediments, the K_d is insensitive to compositional variations in hydrous oxide coatings (i.e., the K_d for extractable Fe^{III} oxides is not significantly different than the K_d for extractable Al oxides). In order to interpret the results of these regression analyses, a simplified model of sorption at constant pH is considered.

General reactions for the sorption of uranyl ion onto sediments with variable amounts of mineralogically similar hydrous oxide coatings can be written as

$$UO_2^{2+} + nSOH \Leftrightarrow (SO)_n{-}UO_2^{(2-n)} + nH^+ \quad [4]$$

$$Q_4 = [(SO)_n{-}UO_2^{(2-n)}]\cdot[H^+]^n\cdot\{[UO_2^{2+}]\cdot[SOH]\}^{-1} \quad [4a]$$

$$SOH \Leftrightarrow SO^- + H^+ \quad [5]$$

$$Q^5 = [SO^-]\cdot[H^+]\cdot[SOH]^{-1} \quad [5a]$$

$$SOH + H^+ \Leftrightarrow SOH_2^+ \quad [6]$$

$$Q^6 = [SOH_2^+]\cdot\{[SOH]\cdot[H^+]\}^{-1} \quad [6a]$$

$$S_T = [SOH] + [SO^-] + [SOH_2^+] + n\cdot[(SO)_n\text{–}UO_2^{(2-n)}] \quad [7]$$

$$S_t = \Gamma\cdot S_a\cdot W, \quad [7a]$$

where [SOH], $[SOH_2^+]$, and $[SO^-]$ are the molar concentrations of the neutral, protonated, and deprotonated surface sites, respectively; $[(SO)_n\text{–}UO_2^{(2-n)}]$ is the molar concentration of adsorbed uranium; Q4, Q5, and Q6 are conditional equilibrium quotients (e.g., dependent on experimental conditions) for the respective reactions; S_T is the total molar concentration of reactive sites; Γ is the site density ($mol\cdot m^{-2}$); S_A is the surface area of the sorptive phase ($m^2\cdot g^{-1}$); and W is the mass of the sorptive phase per liter of aqueous phase ($g\cdot L^{-1}$). The $[UO_2^{2+}]$ term is related to the total aqueous uranium concentration, U_T, by

$$[UO_2^{2+}] = U_f\cdot U_T, \quad [8]$$

where U_f is the fraction of uranium that occurs as UO_2^{2+} and accounts for aqueous speciation. At constant pH and low surface coverage of UO_2^{2+} (i.e., $[SOH]\cdot\{1 + Q5\cdot[H^+]^{-1} + Q6\cdot[H^+]^{-1}\} >> n\cdot[(SO)_n\text{–}UO_2^{(2-n)}]$, Eq. [8] can be rearranged as

$$Q4\cdot\Gamma\cdot S_A\cdot U_f\cdot\{U_S\cdot[H^+]^n\}^{-1} = C_8\cdot U_T^{-1} = K_d\cdot 10^{-3} \quad [9]$$

$$U_S = 1 + Q5\cdot[H^+]^{-1} + Q6\cdot[H^+] \quad [10]$$

$$1000\cdot Q4\cdot\Gamma\cdot S_A\cdot U_f\cdot\{U_S\cdot[H^+]^n\}^{-1} = K_{d,a}\cdot S_A = K_d \quad [11]$$

$$\log K_d = \log K_{d,a} + \log S_A, \quad [12]$$

where U_S accounts for the surface speciation, $K_{d,a}$ is a surface-area-based distribution coefficient, and C_S is the concentration of uranium on the solid. Inspection of Eq. [11] indicates that K_d is a complex function of pH and is dependent on aqueous speciation (U_f), the surface speciation (U_S), and the stoichiometry of the adsorption reaction. In addition to that, for mineralogically similar sediments, a plot of log K_d versus log S_A (or any surrogate of S_A such as the amount of readily extractable surface coating metals) will have a slope of 1. The unit slope (calculated for regressions using the total of extractable Fe^{III} plus Al) is the result of the proportionality between surface area and the abundance of the hydrous metal oxide coatings (Fig. 3). The dependence of K_d on surface area shown in Eq. [12], and Figure 4 highlights the importance of characterizing the spatial variability of surface area, even at sites that appear mineralogically homogeneous, and indicates that surface area may be a useful surrogate for the K_d. Spatial variability of surface area is significant for describing correlations between reactive properties such as K_d and hydrologic properties such as porosity and permeability (Smith *et al.,* 1996).

A simple predictive equation describing K_d for the sediments from the Oyster site derived by linear regression of all K_d measurements is given by

$$\log K_d = -0.8(0.1) + 0.46(0.02)\cdot pH + \log S_A, \quad r = 0.938 \quad [13]$$

$$\log K_d = -2.5(0.2) + 0.47(0.02)\cdot pH + \log E, \quad r = 0.920 \quad [14]$$

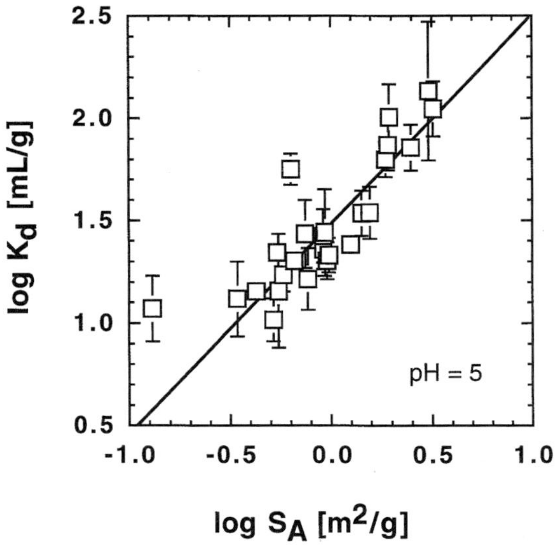

Figure 4 Measured K_d versus surface area with errors from pH adjustment to pH 5. $\log K_d = 1.0(0.1)\log S_A + 1.49(0.03)$, $r = 0.878$ (values in parentheses are the standard errors of the fit coefficients).

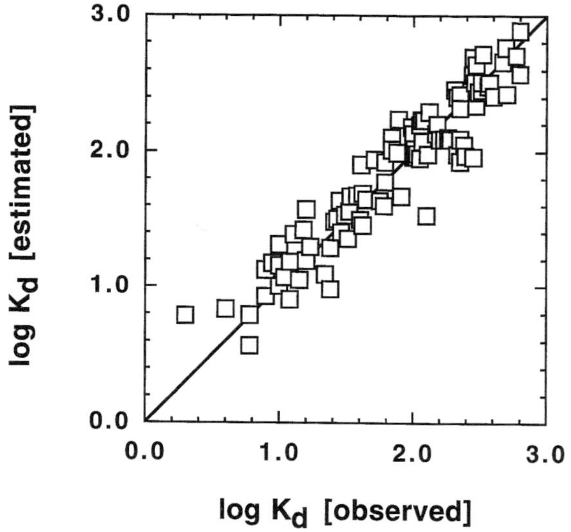

Figure 5 Multivariate predictive equation using pH and surface area. $\log K_d = 1.0(0.1)\log S_A + 0.46(0.02)\text{pH} - 0.8(0.1)$, $r = 0.941$ (values in parentheses are the standard errors of the fit coefficients).

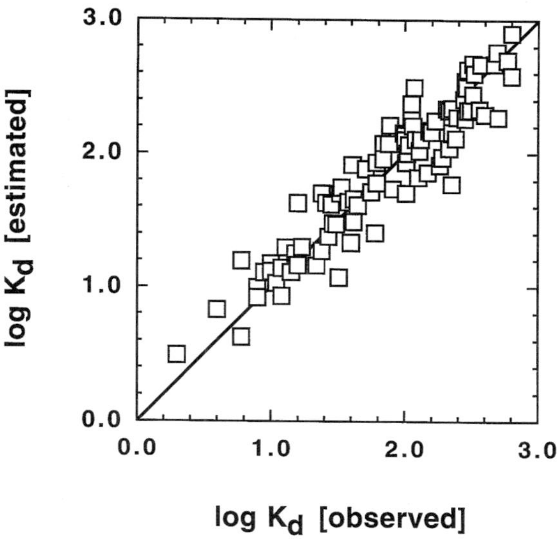

Figure 6 Multivariate predictive equation using pH and the surface area surrogate of extractable Fe plus Al. log K_d = 1.1(0.1)log E + 0.47(0.02)pH − 2.5(0.2), r = 0.938 (values in parentheses are the standard errors of the fit coefficients).

where E is the sum of the extractable Fe plus Al. A comparison between values predicted using Eqs. [13] and [14] and those measured is given in Figures 5 and 6.

IV. CONCLUSIONS

In this work, 100 partition coefficients were obtained based on 25 different sediment samples. These K_d values were obtained using four different initial reaction solution pH values. In our evaluation of this data, several physical and chemical characteristics of the sediments were identified as having both significant and insignificant influences on the sediment's uranium sorptive capabilities. Significant correlations between surface area and surface area surrogates (namely, extractable Fe^{III} and Al for these sediments) were identified. Conversely, mineralogy, which varied over only a somewhat limited range in the samples, and grain size, which varied significantly, had essentially no influence on the sorptive capabilities of these sediments. In our evaluation of the K_d data, we were able to identify and justify a linear relationship between the log K_d values and equilibrium solution pH values. These applied equally well to all the sediments in our sample set. This relationship allowed for the further evaluation of the data set as a whole. It eventually allowed for the development of two simple predictive equations that, based on either surface area or extractable metal concentrations, provided an effective indirect means of evaluating a sediment's sorptive behavior with respect to uranium. The use of such indirect indicators, which are based on rapid and reproducible laboratory procedures, may prove more useful for determining sorptive characteristics of large sample sets than traditional laboratory batch processes.

REFERENCES

Hsi, C-K., and Langmuir, D. 1985. Adsorption of Uranyl onto Ferric Oxyhydroxides: Application of the Surface Complexation Site-Binding Model. *Geochim. Cosmochim. Acta* 49:1931–1941.

Lieser, K. H., Quandt-Klenk, S., and Thybuech, B. 1992. Sorption of Uranyl Ions on Hydrous Silicon Dioxide. *Radiochim. Acta* 57:45–50.

Lieser, K. H., and Thybuech, B. 1988. Sorption of Uranyl Ions on Hydrous Titanium Dioxide. *Z. Anal. Chem.* 332:351.

McKeague, J. A., and Day, J. H. 1966. Dithionite- and Oxalate-Extractable Fe and Al as Aids in Differentiating Various Classes of Soils. *Can. J. Soil Sci.* 46:13–22.

Mehra, P., and Jackson, M. L. 1960. Iron Oxide Removal from Soils and Clays by Dithionite-Citrate System Buffer. *Clays Clay Miner.* 7:317–327.

Mixon, R. B. 1985. Stratigraphic and Geographic Framework of Uppermost Cenozoic Deposits in the Southern Delmarva Peninsula, Virginia and Maryland. Prof. Paper 1067-J. U.S. Geological Survey, Reston, VI.

Smith, R. W., Schafer, A. L., and Tompson, A. F. B. 1996. Theoretical Relationships between Reactivity and Permeability for Monomineralic Porous Media. *Mater. Res. Soc. Symp. Proc.* 412:693–700.

Chapter 8

Intraparticle Diffusion of Metal Contaminants in Amorphous Oxide Minerals

Lisa Axe[1] and Paul R. Anderson[2]
[1]Department of Civil and Environmental Engineering, New Jersey Institute of Technology, Newark, New Jersey; [2]Department of Chemical and Environmental Engineering, Illinois Institute of Technology, Chicago, Illinois

This chapter presents a review of the sorption process for metal contaminants sorbing to amorphous oxide minerals with an emphasis on intraparticle diffusion. Amorphous aluminum, iron, and manganese oxides are porous, ubiquitous sorbents that have a high affinity for inorganic contaminants. Metal contaminants sorb to the external surface of the oxide sorbent rapidly; the surface boundary condition depends on the hydraulic regime. At one extreme the surface flux may be described by film diffusion, and in well-mixed systems the surface may rapidly equilibrate with the bulk concentrations. However, in all cases, the rate of uptake is ultimately limited by intraparticle, surface diffusion. This intraparticle diffusion that has been observed in macroscopic experiments reveals a process that may take years to reach equilibrium. Surface diffusivities, which can be assessed experimentally or theoretically, are an important parameter in measuring, monitoring, and predicting metal contaminant transport in the subsurface environment.

I. INTRODUCTION

Heavy metal ion contaminants released in subsurface systems pose a threat to human health and the surrounding environment. Sources of heavy metals include landfill leachate, mining activity, urban runoff, erosion from metal-rich

Adsorption of Metals by Geomedia

soil, agricultural uses, and industrial discharge. To be able to predict the transport and fate of heavy metals in aquatic environments, sorption processes must be understood.

One major factor that determines the amount of metal sorbed to soil and sediment is the quantity and composition of the sorption substrate. Most sorption substrates include amorphous oxides, reactive particulate organic carbon, and to a smaller extent clay (Jenne and Zachara, 1987). Guy and Chakrabarti (1976) found sorbent capacities in the following order: hydrous oxides > humic substances > clay. Because amorphous oxide minerals of aluminum, iron, and manganese have large surface areas, porous structures, and an abundance of binding sites, they can control the distribution of inorganic contaminants in many aquatic environments (Jenne, 1968; Coughlin and Stone, 1995). Furthermore, although these oxides are metastable minerals, their aging to crystalline, stable minerals can be retarded by sorption of organic or inorganic species (Jenne, 1968; Schwertmann and Fisher, 1973; Axe, 1995). As a result, amorphous oxide minerals present as discrete phases and coatings on other minerals are not only persistent in aquatic environments, but are also important adsorbents for metal ion contaminants in aquatic environments (Stollenwerk and Grove, 1985; McLaren *et al.,* 1986; Jenne and Zachara, 1987; Zachara *et al.,* 1989; Harter, 1991).

The literature contains numerous examples of the importance of amorphous oxide surfaces (Balistrieri *et al.,* 1981; Lion *et al.,* 1982; Manceau and Charlet, 1992a,b; Waychunas *et al.,* 1993; Coughlin and Stone, 1995). Contaminant association with these oxide minerals occurs through a two-step process. A relatively labile fraction forms when the contaminant adsorbs to external surface sites and rapidly equilibrates with the solute in the bulk solution. The kinetics of this adsorption reaction are quite fast where, with sufficient mixing, equilibrium is reached within minutes. Under less turbulent regimes, film transfer resistance controls the external surface concentration. In the second step of the sorption process, with increasing contact time a growing fraction of the contaminant slowly diffuses through micropores of the oxide along interior surface sites. This relatively nonlabile fraction can become isolated from the bulk phase. These two fractions are described by equilibrium models and diffusion models, respectively.

Several comprehensive reviews of equilibrium models for adsorption at oxide surfaces have been written, including those by Westall (1986), Dzombak and Morel (1990), and Davis and Kent (1990). The widely applied surface complexation models, which can be interpreted as modifications of the basic Stern model, use mass balance equations and mass law expressions to describe the surface equilibrium. The significant differences between the models are in their description of the electrical double layer. All these models are based on the assumption that equilibrium conditions provide an appropriate description of the surface reactions. This assumption is not surprising because data used to develop the models come from relatively short-term experiments. Recent studies, however, demonstrate that diffusion processes can play an important role in the transport and fate of contaminants (Axe, 1995; Coughlin and Stone, 1995; Papelis *et al.*, 1995).

II. DIFFUSION PROCESSES

Mass transfer of a metallic contaminant to the sorbing phase depends on the chemical and physical properties of the solute and the medium. Also, impedances along the path the solute travels must be appropriately modeled. Intraparticle diffusion and film diffusion are the two potential, mass transfer, rate-limiting steps. By examining the hydraulic regime along with the adsorbate diffusivity, the Biot number (Bi), which is the ratio of external to internal mass transfer, can be estimated. Bird *et al.* (1960) present a correlation for convection around a sphere of diameter d; the Nusselt number, Nu, is

$$\mathrm{Nu} = k_f d/D = 2.0 + 0.6\mathrm{Re}^{0.5}\mathrm{Sc}^{0.33} \qquad [1]$$

for turbulence under forced convection, which is representative of batch studies where k_f is the mass transfer coefficient, d is the particle diameter, **D** is the diffusivity, Re = d**v**/ν is the particle Reynolds number, and Sc = ν/**D** is the Schmidt number. For a hydraulic regime in a packed bed (Cussler, 1984),

$$\mathrm{Nu} = k_f d/D = 1.17\mathrm{Re}^{-0.42}\mathrm{Sc}^{-0.67} \qquad [2]$$

for laminar flow, which is characteristic of groundwater systems. System Reynolds numbers reflect a system characteristic length as opposed to the particle length for

Table I
Characteristics of Hydraulic Regimes

		$\mathbf{D}_{bulk} = 10^{-5}\,cm^2\cdot sec^{-1}$, Sc = 10^{-3}			$\mathbf{D}_{surface} = 10^{-13}\,cm^2\cdot sec^{-1}$, Sc = 10^{11}		
Sorbent diameter (cm)	Velocity ($cm\cdot sec^{-1}$)	Re[a]	Nu	Bi = kR/**D**	Re[a]	Nu	Bi = kR/**D**
$2\cdot10^{-3}$	50[b]	10	21	10.5	10	8,809	4,404
$2\cdot10^{-3}$	150[b]	30	35	17	30	15,256	7,628
10^{-2}	50[b]	50	44	22	50	19,695	9,847
10^{-2}	150[b]	150	76	38	150	34,111	17,050
$2\cdot10^{-3}$	$3\cdot10^{-5}$[c]	$6\cdot10^{-6}$	10^{-2}	0.5	$6\cdot10^{-6}$	5	2.5
$2\cdot10^{-3}$	$3\cdot10^{-1}$[c]	$6\cdot10^{-2}$	2.24	1.12	$6\cdot10^{-2}$	976	488
10^{-2}	$3\cdot10^{-5}$[c]	$3\cdot10^{-5}$	0.03	0.015	$3\cdot10^{-5}$	12	6
10^{-2}	$3\cdot10^{-1}$[c]	$3\cdot10^{-1}$	5.7	2.85	$3\cdot10^{-1}$	2,482	1,241

[a] Reynolds number is based on particle diameter.
[b] Velocities represent batch studies (turbulent), where the system Reynolds number ranged from 10^5 to 10^7.
[c] A range of groundwater velocities: 10 to 10,000 $m\cdot yr^{-1}$.

a particle Reynolds number. The system Reynolds number is the criterion used for assessing the hydraulic regime. Table I presents a range of conditions for groundwater and batch studies. External mass transfer is not an important resistance for systems when intraparticle diffusivities are much smaller than bulk values. Under these conditions the Biot number is greater than 1 and intraparticle diffusion is the rate-limiting mechanism. However, when diffusion is equal to bulk values, external film mass transfer becomes important in groundwater systems.

A. Intraparticle Diffusion

The flux, **N,** of an ion in a supporting, electrolyte solvent is due to migration, diffusion, and advection and is defined by the equation (Prentice, 1991)

$$\mathbf{N} = -zuFS\nabla\phi - \mathbf{D}\nabla S + S\mathbf{v}, \qquad [3]$$

where z is the charge on the ion, u is the ion mobility, F is Faraday's constant, S is the ion concentration, ϕ is the electric potential, **D** is the diffusivity, and **v** is the bulk fluid velocity. The flux, potential gradient, concentration gradient, and fluid velocity are vector quantities. Generally, in groundwater systems, contaminants are present at dilute concentrations. For example, the ionic strength of the aqueous phase may be 10^{-3} while contaminants are in the range of 10^{-8} to 10^{-6} $mol{\cdot}L^{-1}$ (Morrey *et al.,* 1986; Serne *et al.,* 1990). The migration potential in one dimension can be assessed through considering a liquid junction potential for a multicomponent system. Assumptions used in the following analysis (Prentice, 1991) were (1) activity coefficients are equal to 1, and (2) concentration gradients are linear. The transference number is the fraction of current carried by an ion in the absence of a concentration gradient and is defined as

$$t_i = \frac{z_i^2 u_i S_i}{\Sigma z_j^2 u_j S_j} \qquad [4]$$

where z is the ion charge, u is the ion mobility, and S is the ion concentration. Defining the Gibbs free energy in terms of potential, E, for a 1:1 electrolyte results in $\Delta G = -FE$, where F is the Faraday constant. The movement of one Faraday is defined as

$$\Delta G = \sum \frac{t_i}{z_i} d\mu_i = RT\sum \frac{t_i}{z_i} d \ln a_i, \qquad [5]$$

where μ is the chemical potential, a is the activity, and

$$d \ln a_i = d \ln S_i = d\, S_i/S_i \qquad [6]$$

for activity coefficients equal to one, which is appropriate for dilute species. Solving for E results in

$$dE = \frac{-RT}{f}\frac{\Sigma z_i u_i dS_i}{\Sigma z_j^2 u_j S_j}, \qquad [7]$$

where the ionic mobility is related to the equivalent ionic conductance, λ, both of which effect how an ion moves through a solution:

$$u_i = \frac{\lambda_i}{|z_i|\, F^2}. \qquad [8]$$

Considering a typical contaminant concentration of 10^{-7} mol·L^{-1} over two electrolyte conditions, the liquid junction potential contribution to the estimated flux can be assessed. The ratio of diffusion-to-migration flux is presented in Table II; the upper half of the table includes calculations for an ionic strength of $3 \cdot 10^{-3}$ and the lower half reflects one of $3 \cdot 10^{-2}$. For both cases, the ratio of diffusion to migration flux is large, indicating that the contribution from the potential is insignificant.

Table II

Assessment of Migration Flux

Parameter[a]	Na^+	NO_3^-	Cd^{2+}	Sr^{2+}	NO_3^- for Cd (or Sr)
Case 1: N_{dif}/N_{migr}					
Z_i	1	−1	2	2	−1
λ_i	50.1	7.15	54	59.4	71.5
S_{i1}	$3 \cdot 10^{-3}$	$3 \cdot 10^{-3}$	10^{-7}	10^{-7}	$2 \cdot 10^{-7}$
u_i	$5.9 \cdot 10^{-9}$	$7.7 \cdot 10^{-9}$	$5.8 \cdot 10^{-9}$	$6.4 \cdot 10^{-9}$	$7.7 \cdot 10^{-9}$
dS_{i1}	0	0	10^{-7}	10^{-7}	$2 \cdot 10^{-7}$
$z_i u_i dS_i$	0.0	0.0	$-5.8 \cdot 10^{-16}$	$-6.4 \cdot 10^{-16}$	$1.5 \cdot 10^{-15}$
$z_i^2 u_i S_{i1}$	$1.6 \cdot 10^{-11}$	$1.2 \cdot 10^{-11}$	$1.2 \cdot 10^{-15}$	$1.3 \cdot 10^{-15}$	$1.5 \cdot 10^{-15}$
$\Sigma z_i u_i dS_{i1}$			$9.6 \cdot 10^{-16}$	$9.0 \cdot 10^{-16}$	
$\Sigma z_i^2 u_i S_{i1}$			$3.9 \cdot 10^{-11}$	$3.9 \cdot 10^{-11}$	
dE_1			$6.3 \cdot 10^{-4}$	$5.9 \cdot 10^{-4}$	
			21,000	22,000	
Case 2: N_{dif}/N_{migr}					
S_{i2}	$3 \cdot 10^{-2}$	$3 \cdot 10^{-2}$	10^{-7}	10^{-7}	$2 \cdot 10^{-7}$
dS_{i2}	0	0	10^{-7}	10^{-7}	$2 \cdot 10^{-7}$
$z_i u_i dS_{i2}$	0.0	0.0	$-5.8 \cdot 10^{-16}$	$-6.4 \cdot 10^{-16}$	$1.5 \cdot 10^{-15}$
$z_i^2 u_i S_{i2}$	$1.6 \cdot 10^{-10}$	$2.3 \cdot 10^{-10}$	$1.2 \cdot 10^{-15}$	$1.3 \cdot 10^{-15}$	$1.5 \cdot 10^{-15}$
$\Sigma z_i u_i dS_{i2}$			$9.6 \cdot 10^{-16}$	$9.0 \cdot 10^{-16}$	
$\Sigma z_i^2 U_i S_{i2}$			$3.9 \cdot 10^{-10}$	$3.9 \cdot 10^{-10}$	
dE_2			$6.3 \cdot 10^{-5}$	$5.9 \cdot 10^{-5}$	
			210,000	220,000	

[a]Units for parameters are λ, cm^2·Ω^{-1}·eq^{-1}; S, mol·L^{-1}; and u, cm^2·mol·J^{-1}·sec^{-1}.

Also, because fluid velocity within the particle structure is negligible, the velocity term can be neglected. As a result, Fick's law describes the species flux

$$\mathbf{N} = -\mathbf{D}\nabla S. \qquad [9]$$

Transport of an adsorbate through interior pores of an amorphous oxide is intraparticle diffusion. The International Union of Pure and Applied Chemistry (IUPAC) classification of pores delineates micropores ($d < 2$ nm), mesopores (2 nm $< d < 50$ nm), and macropores (50 nm $< d$). However, this division of pore sizes, which is based on gas–solid systems, is probably not appropriate for oxide surfaces in aquatic systems. For example, in the micropore range, the adsorbate interacts with the pore wall as a result of physical forces, such as electrostatic and van der Waals forces (Xiao and Wei, 1992). Diffusion under these conditions is known as configurational diffusion (Weisz, 1973; Froment and Bischoff, 1990; Xiao and Wei, 1992). The IUPAC classification is not applicable for larger molecules or sorbents with both charged surfaces and adsorbed water layer(s) in aqueous environments, where the micropore regime would include larger pore dimensions. For an oxide sorbent in an aqueous environment, the hydration along the pore surface combined with forces extending from the surface results in adsorption of the fluid phase. In this description, diffusion is hindered and considered configurational for pore diameters in the "mesopore" range.

Porosity and pore size distribution are two of the chemical and physical properties of the sorbent needed to understand and model the sorption process; other characteristics include surface area, site density, particle size, and shape. A variety of techniques are used to assess surface area, including potentiometric titration, N_2 gas adsorption, electron microscopy, glycol retention, and water vapor adsorption. Currently, the BET method is the standard procedure, but it probably defines an area far less than what is available when the oxide is hydrated and in an aqueous environment, where it precipitates as gel-like or water-swelled nanometer-sized particles (Dzombak and Morel, 1990) that stabilize in water as micron-sized aggregates (Axe and Anderson, 1995). While surface area is an important property of a sorbent, porosity and pore size distribution are just as important, if not more. Surface area is used under equilibrium conditions with such models as surface complexation. However, because oxides are microporous, intraparticle or surface diffusion is the rate-limiting mechanism in the sorption process and therefore requires data on pore size and porosity. From the argon and nitrogen sorption isotherms, the pore size distribution can be assessed. Mercury porosimetry is another method for assessing pore size distribution ($\geq$6 nm) as well as porosity. The shape and size of amorphous oxides can be assessed in the aqueous phase with an environmental scanning electron microscope (operates under a water vapor pressure as opposed to vacuum) and a particle size analyzer, respectively.

Intraparticle diffusion of gaseous organics in micorporous oxides, particularly zeolites, has been distinguished as configurational diffusion and examined considerably due to catalytic applications (Barrer, 1951; Gorring, 1973; Weisz, 1973;

Ruthven, 1984; Kärger and Ruthven, 1992; Xiao and Wei, 1992). Xiao and Wei (1992) present an approximate guideline for the dominance of a configurational regime when hydrocarbons diffuse in zeolites: a "ratio of the molecular diameter to the channel diameter . . . greater than 0.6 to 0.8." Gorring (1973) found that size was not necessarily the prerequisite for assessing diffusivities of n-paraffins, as n-dodecane diffused 140 times faster than n-octane molecules. He attributed the phenomenon to the molecules' orientation and to the interaction of the species with the sorbent surface.

Contaminant sorption parallels catalysis. Wu and Gschwend (1988), Olmstead and Weber (1990), Rijnaarts *et al.* (1990), and Ball and Roberts (1991) have modeled intraparticle diffusion as the rate-limiting mechanism in the sorption of organic contaminants in aqueous environments. Diffusivities in these models were orders of magnitude smaller than pore diffusion coefficients (Olmstead and Weber, 1990; Rijnaarts *et al.*, 1990; Ball and Roberts, 1991).

Similarly, transient studies involving inorganics and oxide minerals reveal this slow process where sorption is characterized by two steps (Berube *et al.*, 1967; Gadde and Laitinen, 1974; Aringhieri *et al.*, 1985; Sposito, 1986; Bruemmer *et al.*, 1988; Davis and Kent, 1990; Wehrli *et al.*, 1990; Weber *et al.*, 1991; Fuller *et al.*, 1993; Waychunas *et al.*, 1993; Papelis *et al.*, 1995; Axe and Anderson, 1995; Axe, 1995). The first step rapidly attains equilibrium between the bulk aqueous phase and the external surface of the sorbent, which includes surfaces in the macropores where bulk diffusion occurs. In the slow, second step, the contaminant adsorbed at the surface slowly diffuses along sorption sites in the micropores of the oxide particle. Examples of this observation include Berube *et al.* (1967) where the second slow step of proton sorption at the ferric oxide–aqueous solution interface was found to be proportional to the square root of time indicative of diffusion. Bruemmer *et al.* (1988) also found that sorption was linearly related to the square root of time, and concluded that intraparticle diffusion was the rate-limiting mechanism. They studied the sorption of Cd as well as Ni and Zn to goethite. They observed that cation sorption increased substantially with time and increased temperature, and that after 42 days none of the systems reached equilibrium. As a result, the slow diffusion process was responsible for isolating the adsorbate from the bulk aqueous phase.

Similarly, Gadde and Laitinen (1974) studied the effect of Pb sorbed on HFO over time. They concluded that when lead was present during the HFO aging more Pb was sorbed (30 days), which supports intraparticle diffusion as the slow sorption step, as is further demonstrated with the lack of "reversibility" of almost 50% of the lead sorbed. This lack of reversibility can be attributed to Pb being held up in the pores of the HFO. Slow desorption of metals has been observed in a number of studies, including, but not limited to, those by Di Toro *et al.* (1986), Schultz *et al.* (1987), Selim and Amacher (1988), Richard and Bourg (1991), Stahl and James (1991), and Chen *et al.* (1991).

Fuller *et al.* (1993) and Waychunas *et al.* (1993) studied arsenate sorption onto HFO at the macroscopic as well as microscopic level. Fuller *et al.* modeled the two-step sorption process where intraparticle diffusion was the rate-limiting mechanism. In their model, the adjustable parameters were porosity and diffusivity divided by the square of the particle radius. Waychunas *et al.* used X-ray absorption spectroscopy to assess the electronic structure of the sorbent and sorbate. Because As^{V} was found to have oxygen in the first shell and Fe in the second, they were able to distinguish it as adsorption and not precipitation. Further, they conducted coprecipitation studies and found As only coordinated on the sorbent surface and was not substituted for the other ions in the unit cell.

Papelis *et al.* (1995) studied sorption of Cd and selenite to three different amorphous aluminum oxides and concluded that intraparticle diffusion was the rate-limiting mechanism. Effective diffusivities ($\mathbf{D}_e = \mathbf{D}_p/\tau$, where $\mathbf{D}_p$ is the molecular diffusion coefficient and τ is the tortuosity) ranged from $1.7\cdot10^{-12}$ to $1.7\cdot10^{-11}$ $m^2\cdot sec^{-1}$. In their interpretation of the diffusivities, they credited the hindered bulk diffusion to very large tortuosities (>>10), which is beyond the applicable range identified by Satterfield (1970).

Axe and Anderson (1995) used batch experiments to investigate the sorption of strontium onto HFO. Using a material balance for the particle geometry with the initial and boundary conditions, the one unknown or fitting parameter, surface diffusivity, was found to be $4\cdot10^{-13}$ $cm^2\cdot sec^{-1}$. Axe (1995) also found similar diffusivities for Cd: at pH 7 it was 10^{-14} $cm^2\cdot sec^{-1}$, and at pH 6, $6\cdot10^{-13}$ $cm^2\cdot sec^{-1}$.

Based on the previous discussion, when metal ions diffuse in hydrated ferrihydrite micropores (Bottero *et al.,* 1993), diffusion (<<bulk) is the rate-limiting mechanism in the sorption process. Because of the oxide's charged surface with potentially layers of adsorbed water (Kärger and Ruthven, 1992), the very small intraparticle diffusivities can be ascribed as configurational.

Metals associated with amorphous oxide sorbents in subsurface environments can be divided into two fractions. A labile fraction of contaminant is adsorbed to the external surface sites, and a relatively nonlabile fraction is located at sites within the micropores of the oxide particle. The interior fraction is effectively isolated from the bulk phase. Consequently, to accurately depict leaching, it is necessary to model diffusion and reaction within the oxide particle.

B. Experimental Surface Diffusivities

The species mass balance for a porous, spherical oxide particle yields the following equation:

$$\epsilon_p \frac{\partial S}{\partial t} + \rho \frac{\partial C}{\partial t} = \frac{\rho}{\mathbf{r}^2}\frac{\partial[\mathbf{D}_s\mathbf{r}^2\,\partial C/\partial \mathbf{r}]}{\partial \mathbf{r}} + \frac{1}{\mathbf{r}^2}\frac{\partial(\mathbf{D}_e\mathbf{r}^2\,\partial s/\partial \mathbf{r})}{\partial \mathbf{r}} \quad [10]$$

In this expression, S is the contaminant concentration in the aqueous phase ($mol\cdot L^{-1}$), C is the contaminant concentration in the sorbed phase ($mol\cdot g^{-1}$HFO) and is a function of S through the adsorption process, ϵ_p is the porosity of the particle, $\mathbf{D}_s$ is the surface diffusion coefficient within the sorbent particle, ρ is the oxide bulk density, $\mathbf{D}_e$ is the pore diffusion coefficient within the sorbent, and **r** is the radial position within the sphere measured from the center. Two assumptions used in the material balance are (1) that the particles are spherical (Schwertmann and Taylor, 1989; Axe and Anderson, 1995), and (2) that the radial dimension is an independent variable. The validity of these assumptions remains to be tested.

For instantaneous adsorption and constant diffusivities (for dilute contaminant concentrations), the balance can be written as

$$\frac{\partial C}{\partial t} = \frac{[\mathbf{D}_s + (\mathbf{D}_e dS/\rho dC)]}{[1 + (\epsilon_p dS/\rho dC)]} \frac{1}{\mathbf{r}^2} \frac{\partial[\mathbf{r}^2(\partial C/\partial \mathbf{r})]}{\partial \mathbf{r}}, \tag{11}$$

where adsorption equilibrium is described by the appropriate isotherm and the initial condition is given by

$$C(\mathbf{r},0) = 0 \tag{12}$$

for reasons of symmetry, the boundary condition at the center of the particle is given by

$$\frac{\partial C}{\partial \mathbf{r}}(0, t) = 0, \tag{13}$$

and at the external oxide surface, $\mathbf{r} = R$, the concentration is either in equilibrium with the bulk aqueous phase concentration because the hydraulic regime is turbulent ($Bi \rightarrow \infty$),

$$C(R, t) = C_s, \tag{14}$$

or the flux is controlled by the film transfer resistance,

$$\frac{\partial C}{\partial \mathbf{r}}(R, t) = \frac{k_f}{\mathbf{D}_s \rho}(S - S_s), \tag{15}$$

where k_f is the external mass transfer coefficient and the external surface concentrations are represented by C_s for contaminant sorbed and S_s for contaminant in the aqueous phase. Because of the strong affinity between the adsorbate and the surface, molecular diffusion as a transport process through the oxide micropores is insignificant (Axe and Anderson, 1995). Surface diffusion dominates with the high surface loading (Axe, 1995); this topic is also discussed in such references as Froment and Bischoff (1990) and Trowbridge (1989).

For surface diffusion, the solution for Eq. [11], subject to Eqs. [12]–[14], is given by Crank (1975),

$$\frac{C}{C_s} = 1 + \frac{2R}{\pi \mathbf{r}} \sum_{n=1}^{\infty} \frac{(-1)^n}{n} \sin \frac{n\pi \mathbf{r}}{R} \exp\left(\frac{-\mathbf{D}n^2\pi^2 t}{R^2}\right), \tag{16}$$

where R is the particle radius,

$$\mathbf{D} = \frac{\mathbf{D}_s}{1 + (\epsilon_p/\rho K_i)} \qquad [17]$$

assuming a linear isotherm, the distribution coefficient, K_i, that describes the fraction of sites within the particle or internally can be measured from isotherm experiments (Axe, 1995). Surface diffusivity ($\mathbf{D}_s$) is therefore the fitting parameter in the model. Based on diffusivities of 10^{-14} to $6 \cdot 10^{-13}$ $cm^2 \cdot sec^{-1}$ for Cd diffusion in ferrihydrite at pH 7 and 6, respectively, theoretical concentration profiles

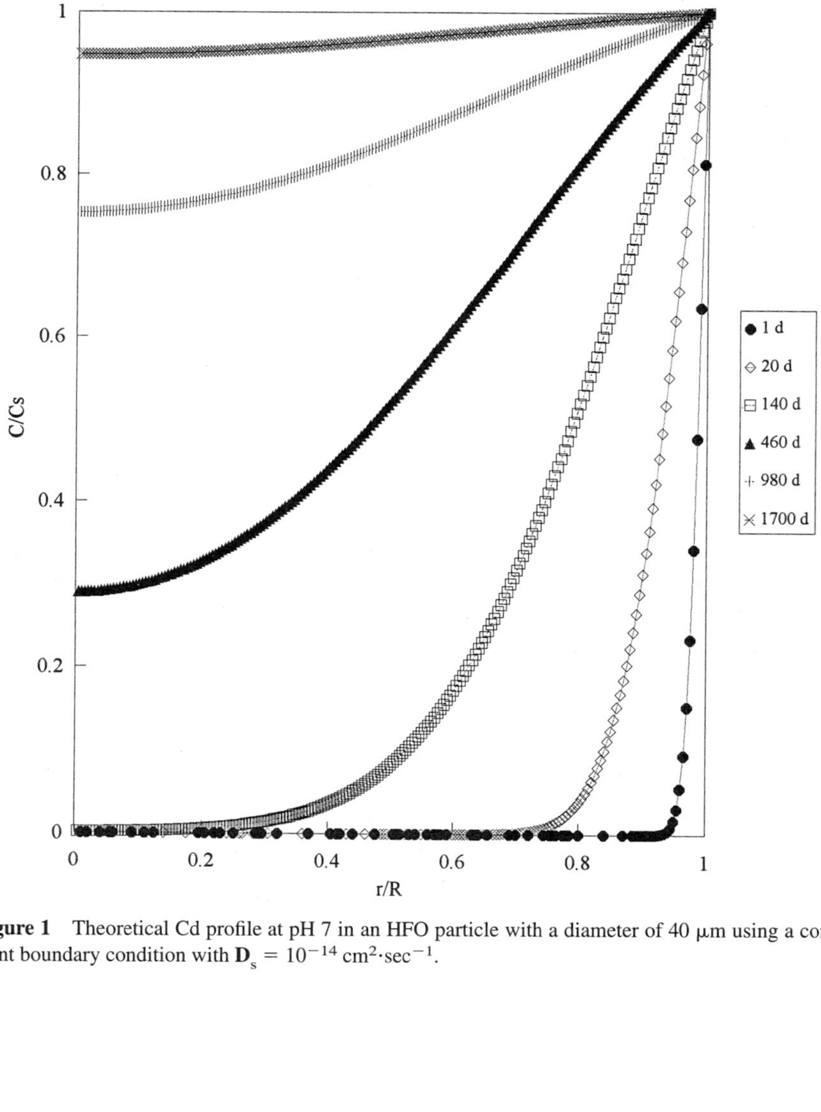

Figure 1 Theoretical Cd profile at pH 7 in an HFO particle with a diameter of 40 μm using a constant boundary condition with $\mathbf{D}_s = 10^{-14}$ $cm^2 \cdot sec^{-1}$.

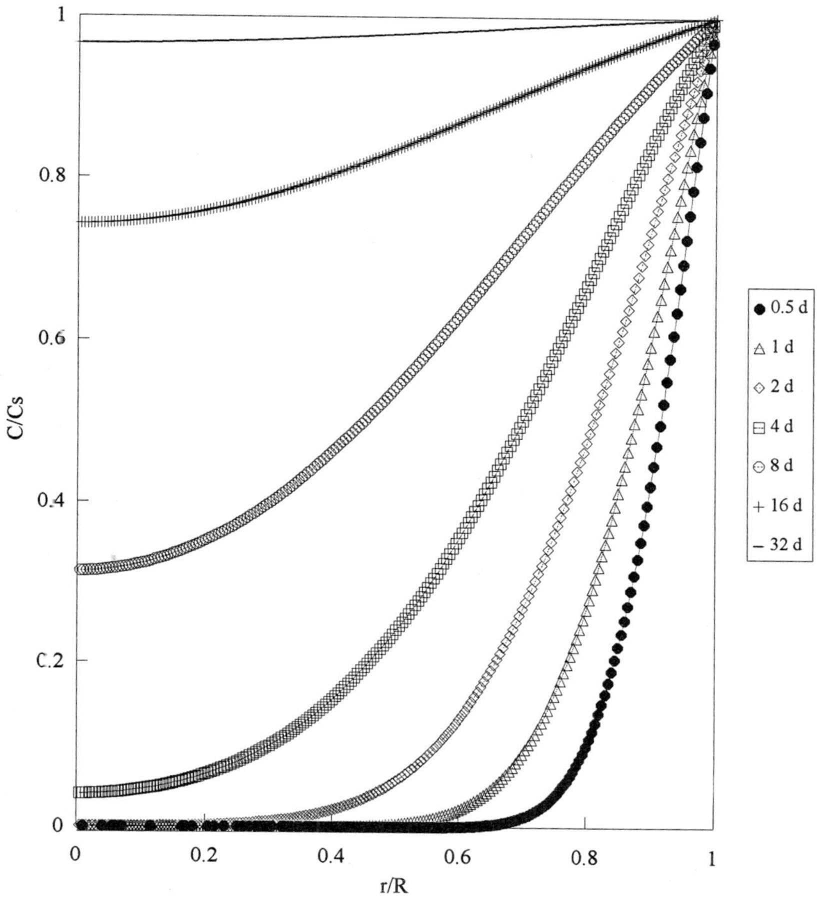

Figure 2 Theoretical Cd profile at pH 6 in an HFO particle with a diameter of 40 μm using a constant boundary condition with $\mathbf{D}_s = 6 \cdot 10^{-13}$ $cm^2 \cdot sec^{-1}$.

are shown in Figures 1 and 2 for a particle radius of 20 μm (Axe, 1995). Interestingly, the Cd diffusivities at two pH values are an order of magnitude different, and as a result for the Cd system at pH 7, equilibrium is expected to be reached in approximately 5 yr, while for Cd at pH 6, equilibrium is reached within 1 month.

The order of magnitude difference in surface diffusivity for the two different pH values is consistent with site activation theory (Kärger and Ruthven, 1992),

$$\mathbf{D}_s = \lambda^2 \nu_{vib} \exp[-V_0/R/T], \quad [18]$$

where λ is the distance between neutral sites on the surface, V_0 is the height of the energy barrier between sites, R is the gas constant, T is temperature, and ν_{vib} is the vibrational frequency of the adsorbate at the sorption site and is a function of the potential energy,

$$\nu_{vib} = \frac{1}{2\pi}[(d^2U/dx^2)/\underline{m}]^{1/2}, \qquad [19]$$

where m is the molecular mass, and U is the surface potential energy (Kärger and Ruthven, 1992). The adsorbate vibrates at a site until it has sufficient energy to hop to the next site. As the ferrihydrite pH of the zero point of charge is approximately 8, there are less neutral sites on the ferrihydrite surface at pH 6 than at pH 7. Therefore, the distance between sites is greater at pH 6, which results in a greater diffusivity.

The surface potential energy can be obtained through molecular dynamics (MD), Monte Carlo simulations, X-ray absorption spectroscopy (XAS), and simplifying descriptions of the potential energy along the surface.

Lasaga (1990) discussed molecular dynamics methods in surface studies, where "the time-averaged property of one system will be equivalent to the instantaneous ensemble-average over many systems. . . . Molecular dynamics results are based on trajectories of particles over time to simulate a macroscopic system on an atomic level (p. 59)." In the MD simulation, the number of particles remains constant. Other constraints are also used and may include constant volume, temperature, pressure, and energy. The simulation begins with an initial condition and boundary conditions imposed for the three dimensions which are a function of time. Through the equation of motion each particle's position and velocity are assessed. Interparticle potentials are evaluated through either thermodynamic relations, such as the Lennard-Jones potential used for nonpolar molecules, or theory—*ab initio,* which requires long computer times (Lasaga, 1990). According to Lasaga, "the *ab initio*-derived potential may be able to reproduce accurately interatomic forces far from the equilibrium positions where the dynamics of bond-breaking and formation actually take place (p. 61)."

Other methods for assessing the potential energy include Monte Carlo simulations and temperature studies using X-ray absorption spectroscopy. The Monte Carlo method uses statistical sampling by random numbers. The reaction and diffusion rates become probabilities and the system dynamics are developed stochastically. From XAS studies, bond distances at different temperatures furnish an assessment of theoretical force constants, which are a function of the potential energy.

Transition State Theory also provides a rigorous theory of activated diffusion (Kärger and Ruthven, 1992). For the case of a surface of homogeneous adsorption sites arranged in a square lattice, the minima of the potential energy surface are the adsorption sites, while energy maxima are the barriers between sites. When the number of adsorption sites occupied is small relative to the number of sites available, the species will not interact with itself, but will interact with the sorbent. In this case, the surface diffusivity is self-diffusion. When the adsorbate is on the adsorption site, it is in the ground state. During the time the adsorbate is jumping from one site to the next, it is in the transition state. The partition function for an atom or molecule in the ground state is the product of the separate partition functions, each representing a type of energy—electronic, translational, rotational, and vibrational. These energies are theoretically independent of one another (Laidler, 1965). The jump frequency is determined from dynamic equilibrium between the ground and the transition states.

Surface diffusion in small pores of an oxide may also be described by a simplifying assumption that a sinusoidal potential field exists. In this analysis, Kärger and Ruthven (1992) state that a "spatially varying but symmetric potential field" prevails. Under this potential field, the minimum or lowest energy locations along the surface are adsorption sites.

III. CONCLUSIONS

The rate-limiting mechanism for the transport of metal ion contaminants in the subsurface environment is the intraparticle diffusion of these contaminants in amorphous Al, Fe, and Mn oxides. In the sorption process, the particle acts as a sink for the contaminant. Likewise, when the contaminant desorbs, a constant source of contaminant exists, which is presently not included in hydrogeological transport codes and is not accounted for during site remediation. To develop an accurate description of contaminant distribution, therefore, it is necessary to incorporate the diffusion model as a subroutine into the transport code. Surface diffusivities are parameters like the equilibrium coefficients, which need to be defined for metallic contaminants present in the subsurface to understand and predict contaminant transport. From transient studies, experimental diffusivities can be determined as a fitting parameter when solving the species mass balance with the appropriate boundary conditions. Theoretically, surface diffusion coefficients can be assessed from the distance between sites and with an understanding of the potential energy on the pore surface of the oxide mineral. A convective dispersive groundwater flow model can consider the transient concentration of contaminants by using a subroutine to assess contaminant sorbed on the external surface and therefore available for leaching, as well as the contaminant diffusing in or out of the particle based on the boundary conditions.

REFERENCES

Aharoni, C., and Sparks, D. L. 1991. Kinetics of soil chemical reactions—A theoretical treatment. *In* "Rates of Soil Chemical Processes." Soil Science Society of America Special Publication no. 27. Madison, WI.

Aringhieri, R., Carrai, P., and Petruzzelli, G. 1985. Kinetics of Cu^{2+} and Cd^{2+} adsorption by an italian soil. *J. Soil Sci.* 139:197–204.

Axe, L. 1995. Sr and Cd diffusion and reaction within Fe oxides. Ph.D. thesis, Illinois Institute of Technology, Chicago, IL.

Axe, L., and Anderson, P. R. 1995. Sr diffusion and reaction within Fe oxides: Evaluation of the rate-limiting mechanism for sorption. *J. Colloid Interface Sci.* 175:157–165.

Balistrieri, L., Brewer, P. G., and Murray, J. W. 1981. Scavenging residence times of trace metals and surface chemistry of sinking particles in the deep ocean. *Deep-Sea Res.* 28A:101–121.

Ball, W. P., and Roberts, P. V. 1991. Long-term sorption of halogenated organic chemicals by aquifer material. 2. Intraparticle diffusion. *Environ. Sci. Technol.* 25:1237.

Barrer, R. M. 1951. "Diffusion in and through Solids." University Press, Cambridge.

Berube, Y. G., Onoda, G. Y., and De Bruyn, P. L. 1967. Proton adsorption at the ferric oxide/aqueous solution interface. II. Analysis of kinetic data. *Surf. Sci.* 8:448–461.

Bird, R. B., Stewart, W. E., and Lightfoot, E. N. 1960. "Transport Phenomena." John Wiley, New York.

Bottero, J. Y., Anaud, M., Villieras, F., Michot, L. J., De Donato, P., and Francois, M. 1993. Surface and textural heterogeneity of fresh hydrous ferric oxides in water and in the dry state. *J. Colloid Interface Sci.* 159:45–52.

Bruemmer, G. W., Gerth, J., and Tiller, K. G. 1988. Reaction kinetics of the adsorption and desorption of nickel, zinc and cadmium by goethite. I. Adsorption and diffusion of metals. *J. Soil Sci.* 39:37–52.

Chen, W. Y., Anderson, P. R., and Holsen, T. M. 1991. Recovery and recycle of metals from wastewater with a magnetite-based adsorption process. *Res. J. Water Pollut. Control Fed.* 63:958–964.

Coughlin, B. R., and Stone, A. T. 1995. Nonreversible adsorption of divalent metal ions (Mn, Co, Ni, Cu, Pb) onto goethite: Effects of acidification, Fe addition, and picolinic acid addition. *Environ. Sci. Technol.* 29:2445–2455.

Crank, J. 1975. "The Mathematics of Diffusion," 2nd ed. Clarendon Press, Oxford.

Cussler, E. L. 1984. "Diffusion Mass Transfer in Fluid Systems." Cambridge Univ. Press, New York.

Davis, J. A., and Kent, D. B. 1990. Surface complexation modeling in aqueous geochemistry. *In* "Mineral-Water Interface Geochemistry" (M. F. Hochella and A. F. White, Eds.), Vol. 23. Mineralogical Society of America, BookCrafters, Inc., Chelsea, MI.

Di Toro, D. M., Mahony, J. D. Kirchgraber, P. R., O'Byrne, A. L., Paquale, L. R., and Piccirilll, D. C. 1986. Effects of nonreversibility, particle concentration, and ionic strength on heavy metal sorption. *Environ. Sci. Technol.* 20:55–61.

Dzombak, D. A., and Morel, F. M. M. 1990. "Surface Complexation Modeling Hydrous Ferric Oxide." John Wiley & Sons, New York.

Froment, G. F., and Bischoff, K. B. 1990. "Chemical Reactor Analysis and Design," 2nd ed. John Wiley, New York.

Fuller, C. C., Davis, J. A., and Waychunas, G. A. 1993. Surface chemistry of ferrihydrite: Part 2. Kinetics of arsenate adsorption and coprecipitation. *Geochim. Cosmochim. Acta* 57:2271–2282.

Gadde, R. R., and Laitinen, H. A. 1974. Studies of heavy metal adsorption by hydrous iron and manganese oxides. *Anal. Chem.* 46:2022–2026.

Gorring, R. L. 1973. Diffusion of normal paraffins in zeolite T occurrence of window effect. *J. Catal.* 31:13–26.

Guy, R. D., and Chakrabarti, C. L. 1976. Studies of metal-organic interactions in model systems pertaining to natural waters. *Can. J. Chem.* 54:2600–2611.

Harter, R. D. 1991. Kinetics of sorption/desorption processes in soil. *In* "Rates of Soil Chemical Processes." Soil Science of America Special Publication no. 27, Madison, WI.

Jenne, E. A. 1968. Controls on Mn, Fe, Co, Ni, Cu, and Zn concentrations in soils and water: The significant role of hydrous Mn and Fe oxides. *In* "Trace Inorganics in Water," Advances in Chemistry. v. 73. American Chemical Society, Washington, DC.

Jenne, E. A., and Zachara, J. M. 1987. Factors influencing the sorption of metals. *In* "Fate and Effects of Sediment-Bound Chemicals," Chap. 8, Proceedings of the Sixth Pellston Workshop. Pergamon Press, New York.

Kärger, J., and Ruthven, D. M. 1992. "Diffusion in Zeolites and Other Microporous Solids." John Wiley, New York.

Laidler, K. J. 1965. "Chemical Kinetics." McGraw-Hill Book Company, New York.

Lasaga, A. C. 1990. Atomic treatment of mineral-water surface reactions. *In* "Mineral-Water Interface Geochemistry" (M. F. Hochella and A. F. White, Eds.), Vol. 23. Mineralogical Society of America, BookCrafters, Inc., Chelsea, MI.

Lion, L. W., Altmann, R. S., and Leckie, J. O. 1982. Trace-metal adsorption characteristics of estuarine particulate matter: Evaluation of contributions of Fe/Mn oxide and organic surface coatings. *Environ. Sci. Technol.* 16:660–666.

Manceau, A., and Charlet, L. 1992a. X-ray absorption spectroscopic study of the sorption of Cr(III)at the oxide-water interface. I. Molecular mechanism of Cr(III) oxidation on Mn oxides. *J. Colloid Interface. Sci.* 148:425–442.

Manceau, A., and Charlet, L. 1992b. X-ray absorption spectroscopic study of the sorption of Cr(III)at the oxide-water interface. II. Adsorption, coprecipitation, and surface precipitation on hydrous ferric oxide. *J. Colloid Interface Sci.* 148:443–458.

McLaren, R. G., Lawson, D. M., and Swift, R. S. 1986. Sorption and desorption of cobalt by soils and soil components. *J. Soil Sci.* 37:413–426.

Morrey, J. R., Kincaid, C. T., Hostetler, C. J., Yabusaki, S. B., and Vail, L. W. 1986. Geohydrochemical models for solute migration, Vol. 3, Evaluation of selected computer codes. EA-3417, Electric Power Research Institute, Palo Alto, CA.

Olmstead, K. P., and Weber, W. J. 1990. Statistical analysis of mass-transfer parameters for sorption processes and models. *Environ. Sci. Technol.* 24:1693–1700.

Papelis, C., Roberts, P. V., and Leckie, J. O. 1995. Modeling the rate of cadmium and selenite adsorption on micro- and mesoporous transition aluminas. *Environ. Sci. Technol.* 29:1099–1108.

Prentice, G. 1991. "Electrochemical Engineering Principles." Prentice Hall, Englewood Cliffs, NJ.

Richard, F. C., and Bourg, A. C. M. 1991. Aqueous geochemistry of chromium: A review. *Water Res.* 25:807–816.

Rijnaarts, H. H. M., Backmann, A., Jumelet, J. C., and Zehnder, A. J. B. 1990. Effect of desorption and intraparticle mass transfer on the aerobic biomineralization of α-hexachlorocyclohexane in a contaminated calcareous soil. *Environ. Sci. Technol.* 24:1349–1354.

Ruthven, D. M. 1984. "Principles of Adsorption and Adsorption Processes." John Wiley, New York.

Satterfield, C. N. 1970. "Mass Transfer in Heterogeneous Catalysis." Massachusetts Institute of Technology Press, Cambridge, MA.

Schwertmann, U., and Fischer, W. R. 1973. Natural "amorphous" ferric hydroxide. *Geoderma* 10:237–247.

Schwertmann, U., and Taylor, R. M. 1989. Iron oxides. *In* "Minerals in Soil Environments," 2nd ed., Chap. 8. Soil Science Society of America Book Series No. 1, Madison, WI.

Schultz, M. F., Benjamin, M. M., and Ferguson, J. F. 1987. Adsorption and desorption of metals on ferrihydrite: Reversibility of the reaction and sorption properties of the regenerated solid. *Environ. Sci. Technol.* 21:863–869.

Selim, H. M., and Amacher, M. C. 1988. A second-order kinetic approach for modeling solute retention and transport in soils. *Water Resources Res.* 24:2061–2075.

Serne, R. J., Arthur, R. C., and Krupka, K. M. 1990. Review of geochemical processes and codes for assessment of radionuclide migration potential at commercial LLW sites. NUREG/CR-558, Pacific Northwest National Laboratory, Richland, WA.

Sposito, G. 1986. Distinguishing adsorption from surface precipitation. *In* "Geochemical Processes at Mineral Surfaces," ACS Symposium Series 323, pp. 216–228. American Chemical Society, Washington, DC.

Stahl, R. S., and James, B. R. 1991. Zinc sorption by manganese-oxide-coated sand as a function of pH. *Soil Sci. Soc. Am. J.* 55:1291–1294.

Stollenwerk, K. G., and Grove, D. B. 1985. Adsorption and desorption of hexavalent chromium in an alluvial aquifer near Telluride, Colorado. *J. Environ. Qual.* 14:150–155.

Trowbridge, L. D. 1989. Isotopic selectivity of surface diffusion: An activated diffusion model. DE90 003473, U.S. DOE, Oak Ridge, TN.

Waychunas, G. A., Rea, B. A., Fuller, C. C., and Davis, J. A. 1993. Surface chemistry of ferrihydrite. Part 1. EXAFS studies of the geometry of coprecipitated and adsorbed arsenate. *Geochim. Cosmochim. Acta* 57:2251–2269.

Weber, W. J., McGinley, P. M., and Katz, L. E. 1991. Sorption phenomena in subsurface systems: Concepts, models and effects on contaminant fate and transport. *Water Res.* 25:499–528.

Wehrli, B., Ibric, S., and Stumm, W. 1990. Adsorption kinetics of vanadyl (IV) and chromium (III) to aluminum oxide: Evidence for a two-step mechanism. *Colloids Surf.* 51:77–78.

Weisz, P. B. 1973. Zeolites—New horizons in catalysis. *CHEMTECH* 3:498–505.

Westall, J. C. 1986. Reactions at the oxide-solution interface: Chemical and electrostatic models. *In* "Geochemical Processes at Mineral Surfaces" (J. A. Davis and K. F. Hayes, Eds.). American Chemical Society, Washington, DC.

Wu, S., and Gschwend, P. M. 1988. Numerical modeling of sorption kinetics of organic compounds to soil and sediment particles. *Water Resources Res.* 24:1373–1383.

Xiao, J., and Wei, J. 1992. Diffusion mechanism of hydrocarbons in zeolites. I. Theory. *Chem. Eng. Sci.* 47:1123–1141.

Zachara, J. M., Ainsworth, C. C., Cowan, C. E., and Resch, C. T. 1989. Adsorption of chromate by subsurface soil horizons. *Soil Sci. Soc. Am. J.* 53:418–428.

Chapter 9

Copper Sorption Kinetics and Sorption Hysteresis in Two Oxide-Rich Soils (Oxisols)

Effect of Phosphate Pretreatment

Luiz Roberto G. Guilherme[1,2] and Sharon J. Anderson[1,3]
[1]Contribution from Department of Crop and Soil Sciences, Michigan State University, East Lansing, Michigan; [2]Present address Departamento de Ciência de Solo, Universidade Federal de Lavras, Lavras, Brazil; [3]Present address Institute for Earth Systems Science and Policy, California State University Monterey Bay, Seaside, CA

Trace metal sorption by humic substances and metal oxides has been studied in single-sorbent and binary-sorbent systems. However, extrapolation of results for pure sorbents to natural heterogeneous systems such as soils is difficult because surface functional groups of metal oxides and soil organic matter are often intimately associated and because pure oxides likely are not good models for pedogenic oxides. In this study, the effects of phosphate pretreatment (P treatment) on the kinetics and reversibility of Cu adsorption and desorption were measured in A- and B-horizon samples of two Oxisols. Phosphate treatment caused up to a 50% increase in the initial rate of Cu adsorption, but only a 2 to 20% increase in Cu sorption capacity. The large increase in the initial sorption rate may be caused by lower activation energies for sorption on surface-phosphate groups than on surface-OH groups. Phosphate treatment also caused a marked decrease in the initial Cu desorption rate and in sorption reversibility. In P-treated soils, the fraction of Cu desorbed by $Ca(NO_3)_2$ was about half that in control soils. The effect of P treatment on sorption hysteresis was more pronounced for B horizons than for A horizons. In addition, little Cu was desorbed from P-treated soils after the initial 15-min reaction time, whereas Cu desorption increased with increasing reaction time in control soils. The decrease in sorption reversibility for P-treated soils is noteworthy because it will greatly affect the mobility of Cu in P-treated Oxisols.

I. INTRODUCTION

Adsorption–desorption reactions are among the most important controls on the mobility of Cu in the environment (James and Barrow, 1981), particularly in oxide-rich soils. Copper mobility in soils is influenced by mineralogy and organic matter content as well as by the Cu concentration and the composition of the soil solution.

Oxisols, which are highly weathered acid soils with low to medium organic matter content and variable clay content, cover about 60% of Brazilian territory. The mineralogy of the clay fraction is characterized by the predominance of Fe and Al oxides and kaolinite, all of which have low cation exchange capacities. Goethite (α-FeOOH) and hematite (α-Fe_2O_3) are the most common Fe oxides, whereas gibbsite (γ-$Al(OH)_3$) is the main Al oxide present in these soils. These oxides have pH_{PZC} values in the range 7.5 to 9.0 (Sposito, 1984), which means that they will be positively charged at the natural soil pH (4.5 to 6.0) of most Oxisols. Organic matter in Oxisol A horizons decreases the pH_{PZC} to near 4.0, so Oxisol A horizons are negatively charged at typical Oxisol pH.

A. SOURCES OF Cu IN SOILS

Annual anthropogenic inputs of Cu to soils contain a wide range of Cu concentrations and include fungicides (12,000 to 50,000 $mg \cdot kg^{-1}$), sewage sludges (50 to 17,000 $mg \cdot kg^{-1}$), phosphate fertilizers (1 to 300 $mg \cdot kg^{-1}$), lime (2 to 125 $mg \cdot kg^{-1}$), and livestock manures (2 to 172 $mg \cdot kg^{-1}$) (Alloway, 1990; Baker, 1990; Kabata-Pendias and Pendias, 1992). In addition, agricultural inputs such as lime, fertilizers, manures and sewage sludge can modify Cu sorption behavior in Oxisols by changing the composition of reactive surface sites, modifying the net surface charge, or generating mobile organic and inorganic colloids, which can enhance Cu transport through the soil profile (McCarthy and Zachara, 1989).

On a local scale, inputs from anthropogenic sources can greatly exceed natural Cu contents (Tiller and Merry, 1981). Natural soil Cu concentrations typically range from 20 to 30 $mg \cdot kg^{-1}$ (Baker, 1990), although Oxisols can contain very high Cu if the parent material has a high Cu concentration (Baker, 1990; Curi and

Franzmeier, 1987; Kabata-Pendias and Pendias, 1992). Total Cu contents over 210 $mg \cdot kg^{-1}$ have been reported for mafic-derived Oxisols in Brazil (Curi, 1983).

B. Copper Sorption by Metal Oxides and Organic Matter

In soils, the relative importance of organic matter or of a particular mineral for Cu sorption depends on the relative contribution of that sorbent to the overall surface charge density at the pH and ionic strength of interest, as well as on the affinity of each type of surface functional group for Cu. Copper concentrations in natural soil solutions typically are controlled by adsorption on surface hydroxyl groups of metal oxides and organic matter (McLaren *et al.,* 1981). Soil organic matter can provide high concentrations of sites for metal sorption (Logan, 1990; Stumm, 1992). Among divalent first-row transition metals, Cu has the greatest affinity for organic matter (Stevenson and Arkadani, 1972). The high degree of selectivity shown by organic matter for Cu is caused by the formation of inner-sphere complexes, often referred to as chemisorption or specific adsorption.

Iron and aluminum oxides also exhibit a high affinity for Cu (McBride, 1982; Forbes *et al.,* 1976; Schwertmann and Taylor, 1977). Electron spin resonance studies have shown that Al–OH groups sorb Cu mainly in inner-sphere complexes (McBride, 1982; Harsh and Doner, 1984). In addition, Cu desorption from organic matter and Fe and Al oxides is very small (McLaren *et al.,* 1983) because activation energies are large for desorption of inner-sphere complexes (McBride, 1989). The time dependence of trace-metal sorption on goethite indicates that the adsorption reaction initially involves adsorption on external surface sites, with subsequent diffusion to internal sorption sites (Bruemmer *et al.,* 1988). Such diffusional processes can contribute to desorption hysteresis and reaction irreversibility (Padmanabham, 1983a,b; Barrow, 1985). For kaolinite, ion exchange (outer-sphere complexes) may be important at low pH and low ionic strength, whereas an increase in both ionic strength and pH favor chemisorption at amphoteric surface hydroxyls (Schindler *et al.,* 1987).

C. Copper Sorption Kinetics

Most soil reactions can be described as heterogeneous solid–liquid reactions that take place by a multistep mechanism comprising transport processes as well as chemical reactions (Aharoni and Sparks, 1991). In a practical sense, the effects of transport and chemical processes are often experimentally inseparable (Sparks, 1989). The difficulty of separating the effects of transport phenomena and chemical reaction kinetics, along with the heterogeneous character of surface functional groups in soils, often restricts the use of simple kinetic models in soils (Sparks, 1989). However, attempts have been made to treat adsorption as a simple reaction in which the surface and the sorbing solute are reactants and the sorbed solute a product (Aharoni and Sparks, 1991). In one study, Cu desorption data for a Cu-contaminated soil were fit very well by a first-order kinetics equation in which sorbed Cu was considered the reactant (Jopony and Young, 1987). Aringhieri *et al.* (1985) found that Cu adsorption kinetics for an organic soil could be described by a model wherein the reaction is first-order with respect to both Cu and substrate concentration (second-order overall) and exhibits a dependence on internal diffusion. Long-term sorption reaction rates are probably mass-transfer limited on oxides (Van der Zee and Van Riemsdijk, 1991; Bruemmer *et al.*, 1988) and on highly aggregated Oxisols (Nkedi-Kizza *et al.*, 1982), although batch-shake methods may eliminate much of the mass-transfer control normally found in aggregated soils (Lima, 1995). In many batch systems, simple first-order kinetic equations may be applicable to the initial rate of adsorption or desorption. During the initial reaction, sorption sites in direct contact with bulk solution are available for adsorption or desorption; in this case, the reaction would neither be site-limited (surface-controlled) nor transport-controlled. In addition, if a reaction is far from equilibrium during the initial adsorption or desorption step, the net reaction rate equals the forward reaction rate. However, as a batch reaction approaches equilibrium, the back reaction (desorption) begins to affect the overall reaction rate.

D. Effect of Phosphate on Cu Sorption

Oxyanions such as phosphate can cause either an increase or a decrease in metal sorption by soils. When an anion and metal are added to a soil simultaneously, the effect of the anion on metal sorption depends on the net surface charge of the soil; the affinity of the soil for the metal, anion, and metal–anion complexes; and the tendency of the metal and anion to form soluble complexes. The latter depends in part upon the anion:metal charge ratio. A large excess of the anion generally suppresses metal adsorption, whereas an anion:metal ratio of 1:1 or less generally favors adsorption by ternary complex formation (McBride, 1994).

If an anion is added to the soil before a metal is added (for example, phosphate fertilization followed later by Cu fungicide application), the anion will affect metal sorption principally by converting metal–OH groups into metal–phosphate surface functional groups, altering the net surface charge, and promoting the formation of metal–phosphate–Cu ternary surface complexes. Phosphate adsorption by positively charged soils (e.g., Oxisol B horizons) causes net surface charge to become less positive (more negative) and Cu sorption to increase (Guilherme *et al.*, 1995). In soils with net negative surface charge (e.g., Oxisol A horizons), PO_4 sorption has much less effect on net charge and pH_{PZC} (Lima, 1995). Beneficial effects of PO_4 for reducing injury from Cu fungicide applications have been reported (Chaney and Giordano, 1977) and may be due to Cu sorption by metal–phosphate surface functional groups.

The objective of the research described here is to determine the effect of phosphate pretreatment on Cu adsorption–desorption kinetics and Cu sorption hysteresis in Oxisols that differ in mineralogy and organic carbon (OC) concentration. Many Oxisols are heavily fertilized with phosphate to maintain productivity, yet the effects of phosphate on the rate of trace metal sorption and on sorption hysteresis have not been studied. Copper sorption was measured at time intervals between 15 min and 18 days in batch experiments. Initial rate constants were calculated assuming a first-order reaction during the initial 15-min period.

II. MATERIALS AND METHODS

A. Soil Material

Samples of A and B horizons from two uncultivated Oxisols were collected near S. João Del Rei in the Campos das Vertentes region of Minas Gerais, Brazil (latitude 21° 20′ S; longitude 44° 30′ W). Both soils were vegetated with semideciduous tropical cerrado (tortuous trees and shrubs scattered above grass and herbaceous plants) and were underlain by mica schists of the Andrelandia group. The first soil, a Yellow-Red Latosol (very fine, allitic, isothermic Typic Hapludox), formed on nearly horizontal mica schist strata and was relatively poorly drained. The second soil, a Dark-Red Latosol (very fine, allitic, isothermic Typic Hapludox), formed on steeply inclined strata and was very well drained. Both soils have pH values near 4.5 in the A horizon and 5.5 in the B horizon. Although both soils have similar clay contents (about 700 $g \cdot kg^{-1}$) and total Fe oxide contents (165 $g \cdot kg^{-1}$), differences in drainage have caused the Yellow-Red Latosol to have greater kaolinite:gibbsite and goethite:hematite ratios than the Dark-Red Latosol (Table I). The A and B horizons of each soil differ in OC content as well as in the relative proportions of amorphous and crystalline Fe (estimated as the ratio of oxalate-extractable to dithionite-extractable Fe). Soil samples were gently crushed to break large aggregates and then were seived to obtain the <2-mm fraction, which was used in all of the experiments described below. Relevant soil properties are summarized in Table I; additional details concerning soil characterization may be found in Lima and Anderson (1997).

1. Phosphate Pretreatment

To assess the effect of phosphate pretreatment (hereafter, P treatment) on Cu sorption kinetics, samples of each soil material were reacted with 10.75 $mmol \cdot L^{-1}$ $Ca(H_2PO_4)_2 \cdot H_2O$ (21.5 $mmol \cdot L^{-1}$ P) at a soil:solution ratio of 2:3 to give a P addition rate of 1 $g \cdot kg^{-1}$ soil. The samples were shaken for 48 hr on a reciprocating

Table I

Selected Properties of A and B Horizons of Two Oxisols from Brazil

Horizon	Clay ($g \cdot kg^{-1}$ soil)	OC[a] ($g \cdot kg^{-1}$ soil)	Fe_d[b] ($g \cdot kg^{-1}$ soil)	Fe_o[b] ($g \cdot kg^{-1}$ soil)	Fe_o/Fe_d	Kt[c] ($g \cdot kg^{-1}$ clay)	Gb[c] ($g \cdot kg^{-1}$ clay)	Gt:Hm[d]	SSA[e] $m^2 \cdot g^{-1}$
Yellow-Red Latosol (KtGt)									
A	711	15.8	101	1.20	0.0108	480	375	10.2	54 ± 3
B	721	9.2	114	0.59	0.0052	480	400	8.3	54 ± 2
Dark-Red Latosol (GbHm)									
A	691	18.8	99	1.88	0.0190	350	510	5.0	64 ± 5
B	753	11.0	114	0.65	0.0057	350	510	4.1	63 ± 2

[a]OC is organic carbon measured by high-temperature combustion at 900°C.
[b]F_d and Fe_o are, respectively, Fe extracted by dithionite-citrate-bicarbonate (Mehra and Jackson, 1960) and by ammonium oxalate (pH 3.1) in the dark (Schwertmann, 1964).
[c]Kt and Gb are, respectively, kaolinite and gibbsite measured by differential thermal analysis in dithionite-treated clay samples (but expressed on a total-clay basis).
[d]Gt:Hm is the goethite:hematite ratio in the clay fraction, measured by X-ray diffraction of NaOH-treated samples (Kämpf and Schwertmann, 1982a,b).
[e]SSA is the specific surface area measured by five-point BET N_2 adsorption isotherms.

shaker (120 cycles·min^{-1}) and centrifuged to sediment colloids larger than 0.45 μm. Excess solution was removed, and the P concentration was measured colorimetrically. The samples were centrifuge-washed once with 5 mmol·L^{-1} $Ca(NO_3)_2$ (2:1 solution:soil) to remove entrained and readily desorbed P. The P concentration of the supernatant wash solutuion was measured. Adsorbed P was near 30 mmol·kg^{-1} for all samples. Soil pastes from the centrifugation step were air dried and crushed to pass a 2-mm sieve. The pH, pH_{PZSE} and DTPA (diethylenetriamine pentaacetic acid)-extractable Cu measured in subsamples of control and P-treated soils are reported in Table II. Although pH_{PZSE} values are measured by titration and are not necessarily the same as pH_{PZC} values derived from electrophoretic mobility measurements, the pH_{PZSE} values reported in Table II for control and P-treated soils are very close to pH_{PZC} values reported previously (Lima, 1995) for the clay fraction of these soils. The specific surface area of the soils, measured by N_2 adsorption (Table I) were unaffected by P-treatment (not shown).

2. pH Adjustment

Prior to measuring Cu sorption capacity or Cu sorption kinetics, suspensions of the eight soil samples (control and P-treated samples from A and B horizons of two Oxisols) were adjusted to pH 5.5 using the following procedure. Triplicate 2.5-g subsamples of each soil material were suspended in 167 ml of $Ca(NO_3)_2$ (pH 5.5; $\boldsymbol{I}$ = 15 mmol·L^{-1}) and shaken. The pH was measured and adjusted to 5.5 with either saturated $Ca(OH)_2$ or 7 mmol·L^{-1} HNO_3. All samples were shaken for 24 hr on a reciprocating shaker (120 cycles·min^{-1}), suspension pH was measured again, and acid or base was added as needed to readjust to pH 5.5. Samples were shaken again for 24 hr and pH was readjusted as necessary before the samples were shaken again for 24 hr. During the third 24-hr shaking period, suspension pH changed less than 0.05 pH units, so pH was considered to be stable after 72 hr. This 72-hr shaking time during pH adjustment effectively disaggregates all samples (Lima, 1995). Consequently, the sorption kinetics experiments described below are unaffected by the original degree of aggregation of the soils and are not confounded by initial differences in aggregation between P-treated and control samples. The pH-adjusted soil suspensions were used in the Cu sorption capacity and kinetics experiments described below and were prepared fresh immediately before each experiment.

3. Copper Sorption Capacity

Copper sorption capacity was measured by repeated reaction with 500 μmol·L^{-1} Cu. This "repeated reaction" approach is preferable to a single reaction at a high Cu concentration because Cu concentrations > 500 μmol·L^{-1} at pH 5.5 could cause Cu precipitation (Allison *et al.*, 1990). After suspension pH was adjusted as previously described, appropriate amounts of $Cu(NO_3)_2$ in $Ca(NO_3)_2$ (pH 5.5; $\boldsymbol{I}$ = 15 mmol·L^{-1}) were added to each soil suspension to give initial total Cu

Table II

Chemical Properties of Control (Ctrl) and P-Treated (P-trt) Soil Samples

	pH[a]		pH_{PZSE}[b]		DTPA-extractable Cu[c] ($mmol \cdot kg^{-1}$)		Cu sorption capacity ($mmol \cdot kg^{-1}$)	
Horizon	Ctrl	P-trt	Ctrl	P-trt	Ctrl	P-trt	Ctrl	P-trt
Yellow-Red Latosol (KtGt)								
A	4.42	4.65	4.3	3.8	0.0424 ± 0.0020	0.3691 ± 0.0150	55.0 ± 0.6	62.3 ± 2.4
B	5.42	5.83	6.2	5.3	0.0055 ± 0.0001	0.0124 ± 0.0001	49.0 ± 0.7	48.8 ± 1.4
Dark-Red Latosol (GbHm)								
A	4.48	4.70	4.3	4.0	0.0478 ± 0.0032	0.0609 ± 0.0000	53.6 ± 3.8	69.5 ± 0.3
B	5.30	5.64	6.1	5.8	0.0126 ± 0.0001	0.0238 ± 0.0006	44.4 ± 0.4	50.4 ± 0.8

[a]Suspension pH for 1:67 soil: solution in 5 $mmol \cdot L^{-1}$ $Ca(NO_3)_2$.
[b]pH value at the point of zero salt effect measured by potentiometric titration (van Raij and Peech, 1972). Uncertainty is ± 0.2 pH units.
[c]Cu extractable by DTPA (Baker and Amacher, 1982).

concentrations of 500 $\mu mol \cdot L^{-1}$ and a solution:soil ratio of 100:1. Triplicate soil suspensions were shaken with 500 $\mu mol \cdot L^{-1}$ $Cu(NO_3)_2$ in 4.5 $mmol \cdot L^{-1}$ $Ca(NO_3)_2$ (pH 5.5; $I = 15\ mmol \cdot L^{-1}$) for 72 hr and then centrifuged. The supernatant solutions were decanted and saved for Cu analysis, and the mass of entrained solution was recorded. Each soil paste was reacted again with fresh 500 $\mu mol \cdot L^{-1}$ Cu in $Ca(NO_3)_2$, and the centrifugation–decantation–Cu analysis procedure was repeated until negligible additional Cu was sorbed. Suspension pH was readjusted to 5.5 after every other cycle. Ten sorption cycles were required. The Cu sorption capacities of control and P-treated soils were reported in Table II.

B. Adsorption–Desorption Kinetics

After pH adjustment, appropriate amounts of $Cu(NO_3)_2$ in $Ca(NO_3)_2$ (pH 5.5; $I = 15\ mmol \cdot L^{-1}$) were added to each soil suspension to give initial total Cu concentrations of 50 and 150 $\mu mol \cdot L^{-1}$ and a solution:soil ratio of 100:1. Samples were shaken at 120 $cycles \cdot min^{-1}$ on a reciprocating shaker for 10 min and then were centrifuged for 10 min at 9000 rpm (RCF of 16,000 xg), which is sufficient to sediment colloids $> 0.45\ \mu m$. A 1.5-ml aliquot of supernatant solution was withdrawn from each bottle and saved for Cu analysis by flame atomic absorption spectroscopy. The soils were resuspended and shaken as previously described. The centrifuging and subsampling procedure was repeated later for total shaking times of 1 hr, 6 hr, 24 hr, 3 days, 6 days, and 18 days. After 18 days, the supernatant solution was decanted.

The soil pastes that remained in the bottles after centrifuging and decanting the adsorption solution were weighed to determine the mass of entrained solution. Copper desorption from these soil pastes was measured by adding 250 ml of 5 $mmol \cdot L^{-1}$ $Ca(NO_3)_2$ to each soil paste. Samples were shaken, centrifuged, and subsampled at the same time intervals used for adsorption kinetics. Total Cu desorbed by 5 $mmol \cdot L^{-1}$ $Ca(NO_3)_2$ at each reaction time was corrected for the amount of entrained Cu.

C. Initial Rate Constants

Initial rate constants for Cu adsorption ($k_{ads,i}$) were calculated under the assumption that the sorption process is a reversible second-order reaction (first-order with respect to both total solution-phase [Cu] and substrate concentration), as suggested by Aringhieri *et al.* (1985). The reaction rate was expressed as:

$$d[Cu]/dt = k_{ads,i}[Cu][S] - k_{des,i}[S - Cu],$$

where [Cu] equals total supernatant Cu concentration ($\mu mol \cdot L^{-1}$), [S] equals substrate concentration, taken as the Cu sorption capacity expressed per unit of volume of solution (μmol·L^{-1}), [S − Cu] equals concentration of the product of adsorption at time *t, t* = 15 min, and $k_{ads,i}$ and $k_{des,i}$ are the rate constants of the initial fast adsorption and desorption reactions, respectively. Detailed description of the integrated rate law and the determination of the rate constants can be found in Aringhieri *et al.* (1985). Although the samples were shaken for 10 min, we used 15 min as the time interval for the initial reaction time because the sorption reaction does not stop until the soil is separated from bulk solution, and 5 min (half of the centrifugation time) is sufficient to separate most soil particles from the solution.

These initial rate constants are precise in that each sample was reacted for exactly the same amount of time prior to centrifugation. However, because of differential settling during centrifugation and the impossibility of determining exactly when the "average" soil particle was no longer in contact with the bulk solution, the initial rate constants may not be appropriate for comparison with other studies that have used other methods and time scales. Nevertheless, the rate constants calculated herein are suitable for determining the effect of P treatment and soil composition on initial sorption rate constants.

III. RESULTS

A. Adsorption Kinetics

1. Initial Rate Constants

Initial Cu adsorption rate constants ($k_{ads,i}$) were up to 50% greater for P-treated soils than for control soils (Fig. 1A). Initial rate constants were greater for 50 μmol·L^{-1} Cu than 150 μmol·L^{-1} Cu for any given sample, and P treatment had the same effect on $k_{ads,i}$ at 50 μmol·L^{-1} as at 150 μmol·L^{-1}. In control samples, $k_{ads,i}$ was greater for the kaolinite- and goethite-rich soils than for the gibbsite- and hematite-rich soil. However, in the B horizon, P treatment caused a greater increase in $k_{ads,i}$ for the gibbsitic soil than for the kaolinitic soil. Consequently, for B-horizon P-treated samples, $k_{ads,i}$ for the gibbsitic soil equaled or exceeded that for the kaolinitic soil. In control soils reacted with 150 μmol·L^{-1}, initial rate constants generally were greater for the A horizons than for the B horizons, although $k_{ads,i}$ did not differ between A and B horizons of control soils reacted with 50 μmol·L^{-1} because of the large uncertainty in $k_{ads,i}$ for these samples. Phosphate treatment generally caused initial rate constants to be greater for B-horizon than for A-horizon samples.

2. Time Dependence of Cu Adsorption

The time dependence of Cu adsorption is plotted both in terms of the fraction adsorbed (i.e., Cu_{ads}/Cu_{added}) and in terms of net Cu adsorption (mmol·kg^{-1}) on the left and right axes, respectively, of Figure 2. For reaction times between 15 min

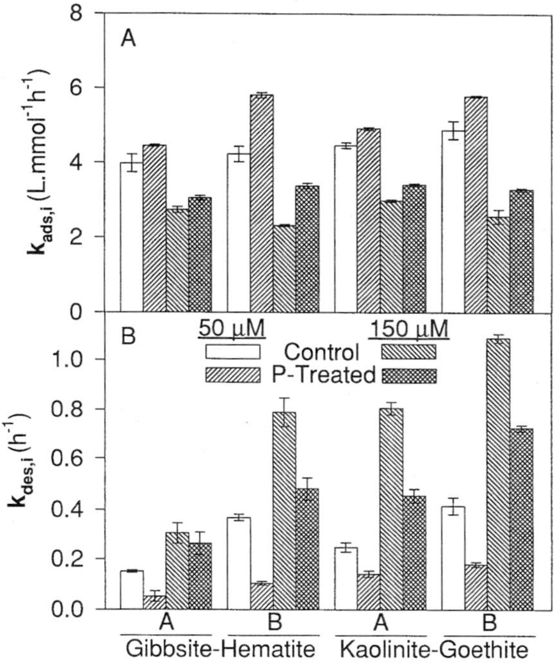

Figure 1 Initial rate constants (hr^{-1}) for (A) Cu adsorption ($k_{ads,i}$) and (B) Cu desorption ($k_{des,i}$) in control and P-treated A- and B-horizon samples of a gibbsite- and hematite-rich Oxisol (Dark-Red Latosol) and a kaolinite- and goethite-rich Oxisol (Yellow-Red Latosol).

and 18 days, the overall time dependence of net Cu sorption depended upon each of the variables included in this study: P treatment, mineralogy (kaolinite:gibbsite ratio and goethite:hematite ratio), OC content (A vs. B horizon), and initial Cu concentration. For soils reacted with 50 μmol·L^{-1} Cu (Figs. 2A and 2B), P treatment caused about a 50% increase in initial Cu adsorption, but only a 3 to 20% increase in total Cu adsorption after 18 days. For soils reacted with 150 μmol·L^{-1} Cu (Figs. 2C and 2D), P treatment caused a 20 to 25% increase in initial Cu adsorption, and a 10 to 35% increase in total (18 day) Cu adsorption (i.e., $Cu_{ads,T}$). For comparison, P treatment caused Cu sorption capacity to increase by 2.6 to 20%. The increase in Cu sorption capacity was 2.4 times greater for the gibbsitic versus the kaolinitic soil and about 3.3 times greater for A versus B horizons (Table II).

For 50 μmol·L^{-1} Cu, P-treated samples approached steady state after 24 hr, whereas control samples had not yet reached a plateau after 18 days. For 150 μmol·L^{-1} Cu, P-treated samples did not reach steady state within 18 days (Fig. 2D), whereas control samples reached steady state after 1 day (Fig. 2C). For both Cu concentrations, the kaolinitic–goethitic soil initially sorbed more Cu than did the gibbsitic–hematitic soil. However, the net sorption rate between 15 min and 6

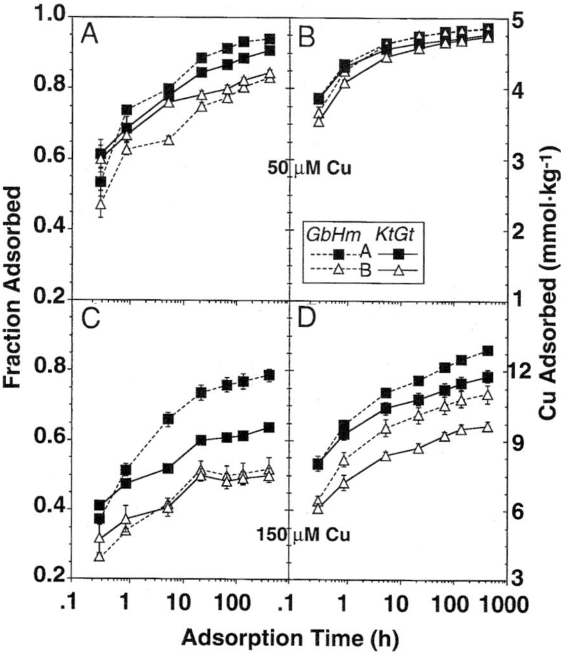

Figure 2 Time dependence of Cu adsorption in control and P-treated A- and B-horizon samples of a gibbsite- and hematite-rich Oxisol (GbHm; Dark-Red Latosol) and of a kaolinite- and goethite-rich Oxisol (KtGt; Yellow-Red Latosol). Left axis is Cu_{ads}/Cu_{added}; right axis is sorbed Cu ($mmol \cdot kg^{-1}$). (A) and (C) are control soils; (B) and (D) are P-treated soils.

hr was greater for the gibbsite-rich soil, such that net Cu sorption after 6 hr was generally greater for the gibbsitic soil (Figs. 2A–2D). The exceptions were the control B horizons reacted with either of the two Cu concentration (Figs. 2A and 2C), for which there was no difference ($P > 0.05$) between the gibbsitic and the kaolinitic soils. The time dependence of Cu sorption was similar for A and B horizons of each soil, although A-horizon samples generally sorbed more Cu than did B-horizon samples. For P-treated samples reacted with 50 $\mu mol \cdot L^{-1}$ Cu, however, Cu sorption did not differ between A and B horizons ($P < 0.05$; Fig. 2B).

B. Desorption Kinetics

1. Initial Desorption Rate Constants

Phosphate treatment, OC content, and Cu concentration had the opposite effect on $k_{des,i}$ compared with their effect on $k_{ads,i}$. Specifically, P treatment caused a 50% decrease in $k_{des,i}$ for samples reacted with 50 $\mu mol \cdot L^{-1}$ Cu and a 10 to 40% decrease in $k_{des,i}$ for 150 $\mu mol \cdot L^{-1}$ Cu (Fig. 1B). Initial desorption rate constants were greater for 150 μmol·L^{-1} Cu than for 50 μmol·L^{-1} Cu and greater for B horizons than for A horizons of both control and P-treated soils. Phosphate treatment caused a greater decrease in $k_{des,i}$ in B horizons than in A horizons. Mineralogy was the only property that had similar effects on both $k_{ads,i}$ and $k_{des,i}$. As for $k_{ads,i}$, $k_{des,i}$ was generally greater for the kaolinite–goethite-rich soil than for the gibbsite–hematite-rich soil; the effect of P treatment on $k_{des,i}$ did not differ markedly between the kaolinitc and the gibbsitic soils (Fig. 1B), whereas P treatment affected $k_{ads,i}$ more in the latter soil.

2. Time Dependence of Cu Desorption

To show the relationship between sorbed Cu and Cu desorbed, Cu desorption is plotted in terms of the Cu concentration (mmol·kg^{-1}) that remained adsorbed at each desorption sampling time (Fig. 3). The absolute concentration of Cu desorbed during the first 15 min is indicated by numbers to the left of the 15-min data points (Fig. 3). For 50 μmol·L^{-1} Cu, P treatment caused a 50% decrease in Cu

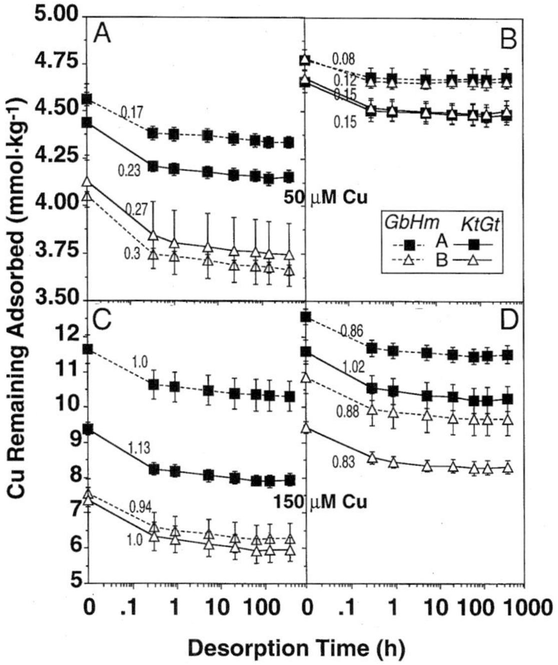

Figure 3 Time dependence of Cu remaining adsorbed (mmol·kg^{-1}) during desorption in $Ca(NO_3)_2$ for control and P-treated A- and B-horizon samples of a gibbsite- and hematite-rich Oxisol (GbHm; Dark-Red Latosol) and a kaolinite- and goethite-rich Oxisol (KtGt; Yellow-Red Latosol). Numbers near 15-min data points represent Cu desorbed during initial 15-min desorption reaction (mmol·kg^{-1}). (A) and (C) are control soils; (B) and (D) are P-treated soils.

desorption, even though the P-treated soils sorbed up to 20% more Cu than control soils (Figs. 3A and 3B). In fact, there was an inverse relationship between sorbed Cu and Cu desorbed (Figs. 3A and 3B). For 150 $\mu mol \cdot L^{-1}$ Cu (Figs. 3C and 3D), P treatment caused a 6 to 14% decrease in Cu desorption, even though P-treated soils sorbed 10 to 35% more Cu than control soils. There was not an inverse relation between sorbed Cu and Cu desorbed at 150 $\mu mol \cdot L^{-1}$, as there was at 50 $\mu mol \cdot L^{-1}$, but there were trends with mineralogy and OC concentration. The kaolinitic soil desorbed more Cu than did the gibbsitic soil, and more Cu desorbed from B than A horizons (Figs. 3C and 3D). For both control and P-treated samples at either total Cu concentration, the amount of Cu desorbed between 15 min and 18 days is very small compared with $Cu_{ads,T}$; the time dependence of Cu desorption is difficult to see in Figure 3.

The time dependence of Cu desorption can be seen more clearly when the fraction of Cu desorbed (i.e., $Cu_{des}/Cu_{ads,T}$) is plotted as a function of time, as can the effects of P treatment and soil composition on sorption hysteresis (Fig. 4). For 50 $\mu mol \cdot L^{-1}$ Cu, sorption hysteresis was greater (i.e., smaller fraction desorbed) in

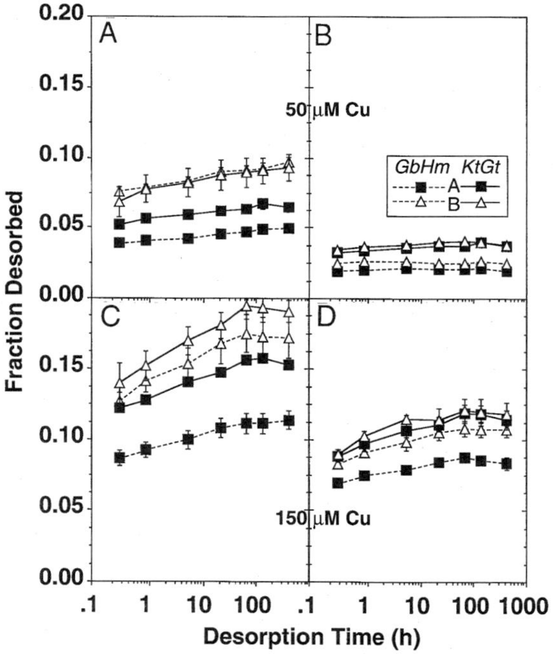

Figure 4 Time dependence of Cu fraction desorbed ($Cu_{des}/Cu_{ads,T}$) in control and P-treated A- and B-horizon samples of a gibbsite- and hematite-rich Oxisol (GbHm; Dark-Red Latosol) and a kaolinite- and goethite-rich Oxisol (KtGt; Yellow-Red Latosol). (A) and (C) are control soils; (B) and (D) are P-treated soils.

A horizons (<7% desorbed) than in B horizons (<10% desorbed; Fig. 4A). For P-treated soils at 50 $\mu mol \cdot L^{-1}$ Cu, the fraction of Cu desorbed was about half that in control soils. There was little difference between A and B horizons in P-treated soils, but hysteresis was greater for gibbsitic soils (<2.5% desorbed) than for kaolinitic soils (<4% desorbed) (Fig. 4B). An additional difference between control and P-treated soils is that net desorption increased steadily between 15 min and 18 days for control soils but not for P-treated soils, which nearly reached steady state during the initial 15-min reaction.

For 150 $\mu mol \cdot L^{-1}$ Cu, Cu desorption from control soils accounted for 10 to 20% of sorbed Cu (Fig. 4C), with greater hysteresis in A horizons and in the kaolinitic versus the gibbsitic soil. For P-treated soils, differences between A and B horizons were again less than those in control soils, and hysteresis again was greater for the gibbsitic soil (<10% desorbed) than for the kaolinitic soil (<12% desorbed) (Fig. 4D). The control soils exhibited a greater increase in net desorption between 15 min and 18 days than did P-treated soils, as previously noted for 50 $\mu mol \cdot L^{-1}$ Cu. Both control and P-treated soils exhibited greater time dependence in the fraction of Cu desorbed for 150 $\mu mol \cdot L^{-1}$ Cu than that for 50 $\mu mol \cdot L^{-1}$ Cu.

IV. DISCUSSION

In these experiments, wherein soil was reacted with P, washed to remove excess and easily desorbed P, and then dried prior to Cu sorption experiments, the solution-phase P concentration during Cu adsorption was <3 $\mu mol \cdot L^{-1}$, far smaller than the initial Cu concentrations of 50 and 150 $\mu mol \cdot L^{-1}$. Thus, P must affect Cu sorption in these experiments by providing surface-phosphate groups on which Cu can sorb and form ternary M–phosphate–Cu complexes (M = Al^{III} or Fe^{III}), not by forming soluble Cu–phosphate complexes. Phosphate treatment caused a large increase in initial adsorption rate constants of all samples. A similar increase in Cu adsorption rate constants has been reported for goethite in the presence of organic acids (Ali and Dzombak, 1996). Because surface OH groups that are coordinated to a single metal atom of Fe and Al oxides (i.e., terminal OH groups) have a much greater affinity for trace metals than do doubly coordinated (bridging) OH groups, fast metal adsorption kinetics on goethite have been attributed to sorption on singly coordinated OH groups, and slow sorption to reaction with doubly coordinated surface OH groups (Grossl and Sparks, 1995).

Phosphate treatment caused a much greater increase in $k_{ads,i}$ than in either $Cu_{ads,T}$ (Fig. 2) or the Cu sorption capacity (Table II). Thus, the effect of P treatment on $k_{ads,i}$ and on initially sorbed Cu does not depend directly on the total number of sites on which Cu can sorb (i.e., the Cu sorption capacity). Rather, the rate enhancement caused by P sorption is probably caused by either (a) an increase in the number of sites with low activation energy or (b) a decrease in the activation energy for Cu sorption at certain types of sites. Phosphate readily forms inner-sphere complexes with metal oxides by displacing protonated terminal OH groups ($M–OH_2^+$) that are singly coordinated to Fe or Al (Sposito, 1984), thereby causing pH_{PZC} to decrease (Table II). At pH 5.5, the resulting metal–phosphate surface functional groups will likely be a mixture of neutral and negatively charged sites. Because activation energies for metal sorption are probably smallest on negatively charged sites and greatest on positively charged sites, Cu sorption should be faster on metal–phosphate surface functional groups than on positively charged $M–OH_2^+$ groups.

Although P treatment is known to affect clay dispersion by changing the net surface charge, the faster sorption kinetics in P-treated soils are probably not caused by clay dispersion. First, both control and P-treated soils were disaggregated by shaking the soil suspensions 72 hr prior to the start of the sorption kinetics experiments. Previous research has shown that identical kinetics results are obtained with sonicated (dispersed) Oxisols and unsonicated Oxisols that have been shaken at least 48 hr (Lima, 1995). In addition, P treatment caused B-horizon pH_{PZSE} values (Table II) to shift closer to suspension pH, which should decrease rather than increase the amount of dispersion; P treatment had only a small effect on the pH_{PZSE} of A-horizon samples.

The greater values of $k_{ads,i}$ for A horizons in control soils reacted with 150 $\mu mol \cdot L^{-1}$ Cu may be due in part to the greater negative charge in A horizons, which contain 1.7 times more OC than do the corresponding B horizons (Table I). Similarly, greater values of $k_{ads,i}$ for the kaolinite-rich soil (35% kaolinite, 27% gibbsite) than for the gibbsite-rich soils (25% kaolinite, 37% gibbsite) may be related to the net negative charge on kaolinite at pH 5.5. The pH_{PZC} of kaolinite is 4.8 (Ferris and Jepson, 1975), whereas the pH_{PZC} of gibbsite is probably at least 8, based on data for other aluminum hydroxides. The facts that $k_{ads,i}$ values are much greater for 50 $\mu mol \cdot L^{-1}$ Cu than for 150 $\mu mol \cdot L^{-1}$ Cu, but that time dependence was greater for 150 $\mu mol \cdot L^{-1}$ Cu, may indicate that the initially sorbed Cu concentration exceeds the concentration of readily accessible, highly reactive sorption sites with low activation energy. At low Cu coverage, sites with fast sorption kinetics (greater accessibility to bulk solution or low activation energies) are available for sorption, whereas sites with progressively slower adsorption kinetics participate as surface coverage increases.

The greater effect of P treatment on $k_{ads,i}$ and on Cu sorption capacity in the gibbsitic versus the kaolinitic soil is difficult to explain, because sorbed P was nearly the same for the two soils. The greater increase in Cu sorption capacity for A versus B horizons is similarly difficult to explain because P treatment had less of an effect on pH_{PZSE} in A horizons. The magnitude of the effect of P treatment on $k_{ads,i}$ and on Cu sorption capacity may be related to the concentration of noncrystalline Fe oxides, which sorb P very strongly. Oxalate-extractable Fe (Fe_o) was 2.5

to 3 times greater in A versus B horizons and about 1.1 to 1.5 times greater in the gibbsitic versus kaolinitic soils. Although Fe_o was less than 2% of dithionite-extractable Fe, noncrystalline oxides have much greater surface area and correspondingly more surface functional groups per unit mass than do crystalline oxides. Although P treatment had a greater effect on Cu sorption capacity and kinetics in the soil samples with greater OC contents, the relationship between OC and the magnitude of P-treatment effects on Cu sorption is probably indirect; OC inhibits Fe oxide crystallization, so amorphous Fe oxide contents increase with increasing OC. Although the soils all contain much more gibbsite (on a mass basis) than goethite or amorphous Fe oxides, the external surfaces of gibbsite comprise predominantly bridging OH groups, which are much less reactive than are the terminal OH groups that predominate in goethite (McBride, 1989). Thus, Fe oxides probably play a much greater role in Cu adsorption in these Oxisols than would be expected from a simple consideration of their mass-basis concentration.

The slow Cu sorption reaction that occurs between 15 min and 18 days may be due either to sorption on sites with higher activation energies or to slow diffusion of Cu to internal sorption sites in metal oxides. The greater time dependence of Cu sorption at the greater Cu concentration may mean that the concentration of readily accessible, highly reactive sites is less than the initially sorbed Cu concentration, even in P-treated samples.

The decrease in $k_{des,i}$ caused by P treatment, especially in B-horizon samples, may be related to the greater stability (and large desorption activation energy) for M^{3+}–phosphate–Cu bonds compared with M^{3+}–O–Cu bonds.

The marked Cu sorption hysteresis observed in both control and P-treated samples (Fig. 4) is typical for adsorption–desorption of trace metals in soils (Sparks, 1985). Desorption of Cu from organic matter and Fe and Al oxides is reported to be very small (McLaren *et al.*, 1983), because activation energies are large for desorption of inner-sphere complexes (McBride, 1989). The greater hysteresis for samples reacted with 50 $\mu mol \cdot L^{-1}$ Cu (Figs. 4A and 4C) than for samples reacted with 150 $\mu mol \cdot L^{-1}$ Cu (Figs. 4B and 4D) is consistent with the hypothesis that there are a limited number of sites with very great affinity for Cu. The greater hysteresis in A horizons (Fig. 4) is consistent with the generally observed increase in sorption irreversibility as OC content increases (Sparks, 1985). The larger desorption rate constants for kaolinitic versus gibbsitic soil may be attributed to the relatively weak bonds formed between kaolinite and either Cu or phosphate. There is also a trend of increasing hysteresis with increasing Fe_o concentration. Fujii and Corey (1986) reported a decrease in isotopically exchangeable Cd and Zn with increasing Fe_o.

In summary, P treatment causes up to a 50% increase in the initial rate of Cu adsorption, but only a 2 to 20% increase in Cu sorption capacity. The large increase in the initial sorption rate is likely caused by the lower activation energies for sorption on surface-phosphate groups, which have greater negative charge than do the same sites in control samples. The most important effect of P treatment may be the roughly 50% decrease in Cu desorption, as this will affect the mobility of Cu in soils.

ACKNOWLEDGMENT

The authors acknowledge the Brazilian Government through CNPq (Brazilian Council for Scientific and Technological Development) for the Scholarship provided to the first author (Process 200863/93-2).

REFERENCES

Aharoni, C., and Sparks, D. L. 1991. Kinetics of soil chemical reactions—A theoretical treatment. *In* "Rates of Soil Chemical Processes" (D. L. Sparks and D. L. Suarez, Eds.), SSSA Special Publ. 27, pp. 1–18. SSSA, Madison, WI.

Ali, M. A., and Dzombak, D. A. 1996. Effects of simple organic acids on sorption of Cu^{2+} and Ca^{2+} on goethite. *Geochim. Cosmochim. Acta* 60:291–304.

Allison, J. D., Brown, D. S., and Novo-Gradac, K. J. 1990. MINTEQA2/PRODEFA2, a geochemical assessment model for environmental systems: Version 3.00 user's manual. EPA-600/3-91-021, USEPA, Athens, GA.

Alloway, B. J. 1990. The origin of heavy metals in soils. *In* "Heavy Metals in Soils" (B. J. Alloway, Ed.), pp. 29–39. John Wiley and Sons, New York.

Aringhieri, R., Carrai, P., and Petruzzelli, G. 1985. Kinetics of Cu^{2+} and Cd^{2+} adsorption by an Italian soil. *Soil Sci.* 139:196–204.

Baker, D. E. 1990. Copper. *In* "Heavy Metals in Soils" (B. J. Alloway, Ed.), pp. 151–176. John Wiley and Sons, New York.

Baker, D. E., and Amacher, M. C. 1982. Nickel, copper, zinc, and cadmium. *In* "Methods of Soil Analysis" (A. L. Page, Ed.), 2nd ed, Part 2, American Society of Agronomy, Madison, WI.

Barrow, N. J. 1985. Reactions of anions and cations with variable-charge soils. *Adv. Agronomy* 38:183–230.

Bruemmer, G. W., Gerth, J., and Tiller, K. G. 1988. Reaction kinetics of adsorption and desorption of nickel, zinc, cadmium by goethite. I. Adsorption and diffusion of metals. *J. Soil Sci.* 39:37–52.

Chaney, R. L., and Giordano, P. M. 1977. Microelements as related to plant deficiencies and toxicities. *In* "Soils for Management of Organic Wastes and Waste Waters" (L. F. Elliot and F. J. Stevension, Eds.), pp. 234–279. American Society of Agronomy, Madison, WI.

Curi, N. 1983. Lithosequence and toposequence of Oxisols from Goiás and Minas Gerais States, Brazil. Ph.D. thesis, Purdue Univ., W. Lafayette, IN (*Diss. Abst. Int.* 44:1674-B).

Curi, N., and Franzmeier, D. P. 1987. Effect of parent rocks on chemical and mineralogical properties of some Oxisols in Brazil. *Soil Sci. Soc. Am. J.* 51:153–158.

Ferris, A. P., and Jepson, W. P. 1975. The exchange capacities of kaolinite and the preparation of homoionic clays. *J. Colloid Interface Sci.* 51:245–259.

Forbes, E. A., Posner, A. M., and Quirk, J. P. 1976. The specific adsorption of divalent Cd, Co, Cu, Pb and Zn on goethite. *J. Soil Sci.* 27:154–166.

Fujii, R., and Corey, R. B. 1986. Estimation of isotopically exchangeable cadmium and zinc in soils. *Soil Sci. Soc. Am. J.* 50:306–308.

Grossl, P. R., and Sparks, D. L. 1995. Evaluation of contaminant ion adsorption/desorption on goethite using pressure-jump relaxation techniques. *Geoderma* 67:87–101.

Guilherme, L. R. G., Lima, J. M., and Anderson, S. J. 1995. Efeito do fósforo na adsorção de cobre em horizontes A e B de latossolos do Estado de Minas Gerais. *In* "Resumos expandidos, Congresso Bradileiro de Ciência do Solo 25th, Viçosa, 23–29 July 1995," pp. 316–318. SBCS/UFV, Viçosa, Brazil.

Harsh, J. B., and Doner, H. E. 1984. Specific adsorption of copper on an hydroxy-aluminum-montmorillonite complex. *Soil Sci. Soc. Am. J.* 48:1034–1039.

James, R. O., and Barrow, N. J. 1981. Copper reactions with inorganic components of soils including uptake by oxide and silicate minerals. *In* "Copper in Soils and Plants" (J. F. Loneragan, A. D. Robson, and R. D. Graham, Eds.), pp. 47–68. Academic Press, Sydney.

Jopony, M., and Young, D. 1987. A constant potential titration method for studying the kinetics of Cu^{2+} desorption from soil and clay minerals. *J. Soil Sci.* 38:219–228.

Kabata-Pendias, A., and Pendias, H. 1992. "Trace Elements in Soils and Plants," 2nd ed. CRC Press, Boca Raton, Florida.

Kämpf, N., and Schwertmann, U. 1982a. Quantitative determination of goethite and hematite in kaolinitic soils by x-ray diffraction. *Clay Miner.* 17:359–363.

Kämpf, N., and Schwertmann, U. 1982b. The 5-M-NaOH concentration treatment for iron oxides in soils. *Clays Clay Miner.* 30:401–408.

Lima, J. M. 1995. Relation between phosphate sorption and aggregation in Oxisols from Brazil. Ph.D. dissertation, Michigan State University, East Lansing, MI.

Lima, J. M., and Anderson, S. J. 1997. Effect of aggregation and aggregate size on extractable Fe and Al in two Brazilian Typic Hapludoxs. *Soil Sci. Soc. Am. J.* 61:965–970.

Logan, T. J. 1990. Chemical degradation of soils. *Adv. Soil Sci.* 11:187–221.

McBride, M. B. 1982. Cu^{2+} adsorption characteristics of aluminum hydroxides and oxyhydroxides. *Clays Clay Miner.* 30:21–28.

McBride, M. B. 1989. Reactions controlling heavy metal solubility in soils. *Adv. Soil Sci.* 10:1–56.

McBride, M. B. 1994. "Environmental Chemistry of Soils." Oxford Univ. Press, New York.

McCarthy, J. F., and Zachara, J. M. 1989. Subsurface transport of contaminants. *Environ. Sci. Technol.* 23:496–502.

McLaren, R. G., Swift, R. S., and Williams, J. G. 1981. The adsorption of copper by soil materials at low equilibrium solution concentrations. *J. Soil Sci.* 32:247–256.

McLaren, R. G., Williams, J. G., and Swift, R. S. 1983. Some observations on the desorption and distribution behavior of copper with soil components. *J. Soil Sci.* 34:325–331.

Mehra, O. P., and Jackson, M. L. 1960. Iron oxide removal from soils and clays by a dithionite-citrate system buffered with sodium bicarbonate. *Clays Clay Miner.* 7:317–327.

Nkedi-Kizza, P., Rao, P. S. C., Jessup, R. E., and Davidson, J. M. 1982. Ion exchange and diffusive mass transfer during miscible displacement through an aggregated Oxisol. *Soil Sci. Soc. Am. J.* 46:471–476.

Padmanabham, M. 1983a. Adsorption-desorption behavior of copper (II) at the goethite-solution interface. *Aust. J. Soil Res.* 21:309–320.

Padmanabham, M. 1983b. Comparative study of the adsorption-desorption behavior of copper (II), zinc (II), cobalt (II) and lead (II) at the goethite-solution interface. *Aust. J. Soil Res.* 21:515–525.

Schindler, P. W., Liechti, P., and Westall, J. C. 1987. Adsorption of copper, cadmium and lead from aqueous solution at the kaolinite/water interface. *Netherlands J. Agric. Sci.* 35:219–230.

Schwertmann, U. 1964. Differenzierung der eisenoxide des bondes durch extraktion mit ammoniumoxalat-losung. *Z. Pflanzenernahr.* 105:194–202.

Schwertmann, U., and Taylor, R. M. 1977. Iron oxides. *In* "Minerals in Soil Environments" (J. B. Dixon and S. B. Weed, Eds.), pp. 145–180. Soil Science Society of America, Madison, WI.

Sparks, D. L. 1985. Kinetics of ionic reactions in clay minerals and soils. *Adv. Agronomy* 38:231–266.

Sparks, D. L. 1989. "Kinetics of Soil Chemical Processes." Academic Press, San Diego.

Sposito, G. 1984. "The Surface Chemistry of Soils." Oxford Univ. Press, New York.

Sposito, G. 1994. "Chemical Equilibria and Kinetics in Soils." Oxford Univ. Press, New York.

Stevenson, F. J., and Arkadani, M. S. 1972. Organic matter reactions involving micronutrients in soils. *In* "Micronutrients in Agriculture" (J. J. Mortvedt *et al.,* Eds.), pp. 79–114. Soil Science Society of America, Madison, WI.

Stumm, W. 1992. "Chemistry of the Solid-Water Interface." John Wiley and Sons, New York.

Tiller, K. G., and Merry, R. H. 1981. Copper pollution of agricultural soils. *In* "Copper in Soils and Plants" (J. F. Loneragan, A. D. Robson, and R. D. Graham, Eds.), pp. 119–137. Academic Press, Sydney.

Van der Zee, S. E. A. T. M., and Van Riemsdijk, W. H. 1991. Model for the reaction kinetics of phosphate with oxides and soil. *In* "Interactions at the Soil Colloid-Soil Solution Interface" (G. H. Bolt *et al.,* Eds.), pp. 205–239. Kluwer Academic Publishers, Dordrecht.

van Raij, B., and Peech, M. 1972. Electrochemical properties of some Oxisols and Alfisols of the Tropics. *Soil Sci. Soc. Am. J.* 36:587–593.

Chapter 10

Influence of pH, Metal Concentration, and Soil Component Removal on Retention of Pb and Cu by an Illitic Soil

Raymond N. Yong[1] and Elaine M. MacDonald[2]
[1]Geoenvironmental Engineering Research Centre, Cardiff School of Engineering, University of Wales, Cardiff, United Kingdom; [2]Department of Civil Engineering, McGill University, Montreal, Quebec, Canada

This study presents results and analyses on the effect of pH and initial metal (Pb and Cu) concentration on retention of metals by an illitic clay soil containing some natural organic material, carbonates, and oxides. Soil suspension experiments were conducted over a wide range of metal concentrations (0.25 to 50 mmol·L^{-1}) at the natural soil pH values and at pH values below the precipitation pH of the metals. At the natural soil pH values, sorption experiments demonstrated the expected effect of the high retention capacity of the soil. At lower pH values the results obtained showed reductions in retention as high as 90% because of diminishing metal precipitation. While removal of the carbonates from the soil did not significantly change heavy metal retention, the removal of amorphous oxides from the soil resulted in increased retention capacities which were at least two to three times greater than those of the original soil. This is a result of the removal of amorphous coatings of the minerals, thus permitting adsorption on the previously masked negatively charged surfaces. In addition, the positively charged surfaces of the amorphous oxides (at pH levels below their zero point of charge) complete with the metals ions for the negatively charged sites, and have apparently severely reduced the metal retention capacity of the soil. Pb distribution among the soil components was analyzed using sequential extraction, and the carbonate component was found to contain most of the retained Pb.

I. INTRODUCTION

The presence of heavy-metal-contaminated soils and sediments has been reported in many studies (Yong *et al.,* 1992b; Allen *et al.,* 1995). Sources of such contaminants are generally anthropogenically derived. The following is a set of examples which record the many instances and sources of contamination. Ravishankar *et al.* (1994) reported Cu concentrations in Quebec sewage sludges ranging from 139 to 7510 $mg \cdot kg^{-1}$, and studies reported by Salomons and Förstner (1984) on the Ruhr river which flows through a heavily populated and industrial region in which approximately 300 metal plating plants are situated among other industries found that 55% of the heavy metals in the river came from industrial and municipal sources. Muller *et al.* (1994) reported concentrations of Pb above 10,000 $mg \cdot kg^{-1}$ in Austrian river sediments due to mining and ore smelting activities upstream. Gobiel *et al.* (1994) traced anthropogenic Pb in the St. Lawrence Estuary sediments to industrial Pb used as an additive in gasoline. Research by Shu and Liu (1994) linked Pb contamination in Taiwanese rice paddies to two plasticizer plants in the vicinity. Analyses of the Superfund Sites, where Pb and Cu have been found to be common pollutants, showed the following list of sources: landfill/chemical waste dumps, metal finishing/plating and electronics, chemical pharmaceuticals, mining/ore processing/smelting, battery recycling, oil and solvent recycling, and paint (Allen *et al.,* 1995).

Removal of heavy metals from soil porewater occurs by mechanisms such as adsorption, complexation, precipitation, and filtration. However, sediments and soils are neither permanent nor irreversible sinks for contaminants and may become sources of contamination to surface water and groundwater when changes in the local environment occur (Farrah and Pickering, 1978; Grambell *et al.,* 1991; Saeki, 1993).

The availability and mobility of heavy metals depend on their association with soil components. Some retained contaminants may still be bioavailable (Luoma, 1989; Calmano *et al.,* 1992). The two major sets of concerns in respect to assessment of soil or sediment contamination include: (1) evaluation of the "storage" capacity (for contaminants) of the soils or sediments, and the potential for "mobilization" or release of contaminants into the immediate environment, and (2) development of an efficient strategy for removal of the contaminants that would be compatible with the manner in which the contaminants are retained in soils or sediments. In both sets of interests, evaluation of the fate of the pollutants involves determination or assessment of the distribution of the contaminants, i.e., characterization of the contaminants contained in soils, and the manner in which these are "held" or distributed in the soils, i.e., "bonded" to the various soil fractions (soil minerals, amorphous materials, and soil organic matter). One recognizes that the presence of inorganic and aqueous organic ligands in the soil system will compete with the adsorption sites from the soil fractions for contaminants. In addition, changes in the soil environment can alter the fate of the contaminants in the soil, and in the case of heavy metal pollutants, it is understood that redox and pH conditions will control and affect the chemical forms and mobility of the heavy metals (Grambell *et al.,* 1991; Saeki *et al.,* 1993).

A large body of research is available on heavy metal retention mechanisms of individual soil components. However, it is not possible at present to predict adsorption in a multicomponent system (Bassett and Melchior, 1990). Honeyman and Santschi (1988) explain that retention is only predictable in simple systems such as a single oxide. Further, it is difficult to decipher between retention by precipitation and retention by adsorption (Sposito, 1989). Much effort to model the behavior of contaminants in sediments and soils has represented the multicomponent system either by some average collective property or as a collection of discrete pure solid phases, a concept referred to as adsorptive additivity (Anderson and Benjamin, 1990; Bassett and Melchior, 1990). However, many researchers have reported nonadditive results from soil adsorption studies. Honeyman found that particle interaction in binary suspensions of oxides can lead to significant deviations from the adsorptive additivity concept (Anderson and Benjamin, 1990). Anderson and Benjamin (1990) reported that in some cases the addition of a second adsorbent decreased the overall adsorption and explain that a unified model must describe both the adsorption results and the changes in the physical characteristics of the soil. They then went on to successfully model a mixed oxide system by describing particle interactions and how those interactions influence the surface chemistry of the system—an approach which has been identified as modified adsorption additivity.

Farrah and Pickering (1978) identified four possible mechanisms that may mobilize heavy metals in soils: (1) changing acidity, (2) changing the system ionic strength, (3) changing the oxidation reduction potential, and (4) the formation of complexes. These findings stress the importance of identifying and understanding the mechanism of retention in an aqueous soil environment, be it surface water sediments, soil–groundwater systems, or landfill clay liners. The mechanisms and processes involved in heavy metal retention include precipitation as a solid phase (oxide, hydroxides, carbonates), ion exchange adsorption, and complexation reactions (Harter, 1979; Yanful *et al.,* 1988; Farrah and Pickering, 1977, 1979;

Maguire *et al.,* 1981; Yong *et al.,* 1990). The research linking pH and heavy metal retention in soils is well documented and consistently shows increased retention with increased pH (Allard *et al.,* 1991; Farrah and Pickering, 1977, 1979; Frost and Griffin, 1977; Harter, 1983; Maguire *et al.,* 1981; Phadungchewit, 1990). Whereas the mechanisms of retention may not always be defined by the researchers, most recognize the importance of precipitation, showing that as pH increases heavy metal solubility decreases (Allard *et al.,* 1991; Jeffery and Uren, 1983; Harter, 1983).

The previous studies reported by Yong and Phadungchewit (1993) and Yong *et al.* (1992b) examined the distribution of heavy metal contaminants among the soil constituents using selective sequential extraction techniques. Because the association of contaminants with each soil fraction—as determined by the selective sequential extraction technique—is in essence operationally defined, the previous studies are somewhat constrained because of the extraction techniques used. In this present experimental study, heavy metal (Pb and Cu) "contamination" of an illitic silty clay soil is examined in its natural state, and also under conditions where the carbonates and oxides have been removed from the soil before "contamination." With this selective contamination technique, the contribution of soil components such as carbonates and amorphous oxides to heavy metal retention and soil properties under various initial pH conditions and metal concentrations can be examined without the constraints associated with the previous studies. Retention of the Pb and Cu by the untreated silty clay soil (no extraction of carbonates and oxides) is examined in relation to comparable retention by the soil with carbonates and oxides removed. The experiments are conducted using soil suspensions with varying initial pH conditions and initial metal concentrations. A sequential extraction procedure was used to determine the distribution of Pb among the operationally defined soil components in the untreated soil and compared to the retention obtained when soil components are extracted prior to Pb contamination.

II. MATERIALS AND METHODS

A. Soil Components and Properties

The soil used in the study was a silty clay surrogate soil obtained from a pulverized shale and marketed commercially under the trade name of Domtar Sealbond. The choice of the pulverized shale soil was predicated on the fact that it contained most of the soil fractions found in the clay soils common in Eastern Canada. Because of the need to obtain replicate samples, the pulverized shale soil also served as a very useful source that was consistent in its composition as shown by many studies which have benefited from its use (Yong *et al.,* 1990, 1992a, 1993; Yong and Phadungchewit, 1993).

Table I

Untreated Soil (Sealbond) Composition

Mineral composition (in order of decreasing abundance)	Illite, chlorite, quartz, feldspar, and calcite
Carbonates (mean % by weight with standard deviation)	5.7 ± 0.3
Fe, Al, Si, and Mn amorphous oxides, hydroxides, and oxyhydroxides (mean % by weight of oxide with standard deviation)	1.2 ± 0.03
Natural organic material (mean % by weight with standard deviation)	0.7 ± 0.09

Characterization of the surrogate soil included tests for determination of mineral composition, amorphous materials, carbonates, organic matter, pH, specific surface area (S_A), and cation exchange capacity (CEC). The X-ray diffraction studies were conducted using the techniques described by Starkley *et al.* (1984), and amorphous materials determination was implemented using oxalic acid extraction according to the method described by Kersten and Förstner, (1989). Carbonate content was determined using 1 N HCl to dissolve the carbonates, which were then titrated with NaOH (Hesse, 1971); pH was measured in a 1:10 mass ratio of soil to distilled water. The S_A was determined using ethylene glycol monoethyl ether (EGME) saturation according to the procedure described by Eltantaway and Arnold (1973), and CEC was measured using the $BaCl_2$ method described by Hendershot *et al.* (1993). The results of these characterization procedures are shown in Tables I and II.

Soil carbonates were extracted from the soil using 1 M NaOAc adjusted to pH 5 with HOAc at a 1 to 40 soil to solution ratio for a contact period of 5 hr, similar to the procedure reported by Tessier *et al.* (1979) and Kertsen and Förstner (1989). Previous studies indicate that the buffering capacity of the NaOAc–HOAc extracting solution was sufficient to dissolve all the $CaCO_3$ in carbonate-rich sediment samples. To ensure that all the carbonates had been extracted, the soils were tested for presence of carbonates after completion of the extraction procedure us-

Table II

Properties of Untreated and Treated Soils

	CEC ($1 \cdot 10^{-2} meq \cdot g^{-1}$)	SA ($m^2 \cdot g^{-1}$)	pH (1:10 soil:water)
Untreated soil	23 ± 1.3	131	8.2
Soil treated for carbonate extraction	8 ± 0.6	150	7.6
Soil treated for carbonate and oxide extractions	10 ± 0.3	116 ± 7.4	7.2

Note Values are means with standard deviations.

ing the titration method described by Hesse (1971). The effect of extraction of carbonates on measured soil properties is shown in Table II. Changes in the soil redox conditions during extraction procedures may result in the release of components that are not targeted, such as iron and manganese oxides (Rapin *et al.,* 1986). Typically this occurs when anoxic soils are exposed to atmospheric oxygen during extraction procedures. Given the oxic conditions under which the soil used in this study has been obtained, stored, and treated, this was not foreseen as a possible source of error.

In the case of the amorphous oxides, extraction was performed by the use of 0.1 $mol \cdot L^{-1}$ oxalic acid, buffered to pH 3 by ammonium oxalate, and mixed in the dark at a 1 to 40 soil to solution mass ratio for 5 hr. Kersten and Förstner (1989) described the acidified ammonium oxalate buffer solution as particularly selective toward iron oxides with an ability to attack amorphous Al compounds. To remove the extracted soil components and the extractants, the soil was washed three times with distilled water after each extraction. Liquid–solid separation was effected through centrifugation for 10 min at 10,000 rpm where the radial distance from the center of the rotor to the top of centrifuge tube was 4.5 cm and the radial distance from the center of the rotor to the bottom of centrifuge tube was 5.5 cm. The effect of extraction of carbonates and oxides on measured soil properties is shown in Table II.

B. Soil Suspension Tests

Two general types of soil suspension tests were performed as part of this study. The first involved an initial adjustment of all soil suspensions to a constant pH prior to the addition of heavy metal solutions. These tests were conducted in two steps over a 48-hr period in which the first 24 hr was designated as a pH adjustment period. During this period, the soil was mixed with distilled water and the appropriate amount of HNO_3 at a soil to solution mass ratio of 1 to 6.67 and allowed to equilibrate. The second 24-hr period was used to establish heavy metal contact, at which point the appropriate amounts of Pb or Cu nitrate solution and distilled water were added to the soil solution to bring the final soil to solution mass ratio up to 1 to 10.

The second type of soil suspension test was performed to examine the effect of final pH at a constant initial concentration of Pb. The first 24 hr was designated as a pH adjustment period; however, rather than adding the appropriate amount of acid to reach a single desired initial pH, solutions of varying HNO_3 or NaOH concentration were added to the soil to obtain final pH values ranging between 3 and 8. After completion of the 24-hr pH adjustment, an appropriate amount of Pb nitrate solution, adjusted to pH 4 with HNO_3, was added to the soil suspension to raise the final soil to solution mass ratio to 1 to 10 and the total concentration of Pb to 5 $mmol \cdot L^{-1}$.

Stock heavy metal solutions were prepared from the following nitrate forms of each metal: $Pb(NO_3)_2$ and $Cu(NO_3)_2 \cdot 2.5H_2O$. All heavy metal solutions were adjusted to pH 4 using HNO_3. To allow for escape of gases produced by carbonate dissolution in the pH-adjusted soil suspension tests on the untreated soil, jar tests were conducted using a high-speed bench stirrer set at 250 rpm to mix the soil suspensions. Soil suspension tests performed with soil treated for carbonate extraction were conducted in closed vessels mixed on a horizontal shaker. At the end of the total 48-hr period, the solution was separated through centrifugation as described in Section II.A, and the metal concentration in the supernatants was measured by atomic adsorption spectroscopy (AAS). For both types of soil suspension tests, random duplicates were conducted as indicated on the figures presenting the results.

C. Sequential Extraction Tests

The soil residue from the suspension tests performed at variable pH values but at a constant initial Pb concentration (the second type from Section II.B) was used as the material in the sequential extraction analyses. In this experiment, a sequential extraction method was used to obtain information on the distribution of Pb among the operationally defined soil components. Where possible the same extraction technique was used as that for soil components studied by extraction prior to contamination (Section II.A). Efforts were made to select extraction procedures least likely to affect the other soil components.

The soil residues were dried, and samples of approximately 0.75 g were retained for sequential extraction analysis. The samples were first washed with distilled water and centrifuged, and the wash water was discarded. The soil was then subjected to the following sequential extraction procedures:

1. Exchangeable. The soil was extracted for 1 hr at room temperature with 8 ml of magnesium chloride solution (1 $mol \cdot L^{-1}$ $MgCl_2$, pH 7) with continuous agitation (Tessier *et al.*, 1979).
2. Carbonate Bound. The residue from step 1 was leached at room temperature for a contact period of 5 hr with 30 ml of 1 $mol \cdot L^{-1}$ sodium acetate (NaOAc) adjusted to pH 5 with acetic acid (HOAc) under continuous agitation, similar to the procedure reported by Tessier *et al.* (1979).
3. Oxide Bound. The residue from step 2 was leached at room temperature using 30 ml of 0.1 $mol \cdot L^{-1}$ oxalic acid ($H_2C_2O_4$), buffered to pH 3 by ammonium oxalate ($(NH_4)_2C_2O_4$), and continuously agitated in the dark for 5 hr (Kersten and Förstner, 1989).
4. Bound by Natural Organic Matter. The residue from step 3 was leached in a boiling water bath for 30 min with 8 ml of 5% sodium hypochlorite (NaOCl) adjusted to pH 9 with hydrochloric acid (HCl) (Shuman, 1983; Kersten and Förstner, 1989).

5. Residual. The residue from step 4 was subjected to an acid digestion procedure referred to as Method 3050 by the USEPA (1986). The soil is subjected to nitric acid (HNO_3) addition and heated to 95°C, followed by the addition of hydrogen peroxide (H_2O_2) with slight warming. The final step requires the addition of hydrochloric acid (HCl) heated to the point of reflux without boiling.

The selective extractions, with the exception of the residual phase, were conducted in 40-ml centrifuge tubes to minimize the loss of materials. Between each successive extraction, supernatant separation was effected by centrifugation as described in Section II.A. The Pb content in the supernatant was analyzed by AAS. Four random duplicates were conducted.

III. RESULTS AND DISCUSSIONS

A. Soil Properties

The soil properties reported in Table II with respect to CEC, S_A, and soil pH for the untreated soil, the soil with carbonates extracted, and soil with both carbonates and oxides removed present some interesting results. It is noted that the measured CEC of the untreated soil is much higher than the CEC of the soil with carbonates extracted. It is not immediately clear if this is due to the extraction of the carbonates, and thus the loss of the CEC associated with the extracted carbonates, or to the effects of the chemical procedure used to determine the CEC on easily solubilized carbonate minerals. Since the CEC of pure carbonates is minor to insignificant, the latter explanation is preferred; however, the results obtained need further evaluation. The S_A of the untreated soil is seen to be about 10% less than the S_A of the soil with carbonates extracted. One could therefore speculate that the removal of the residual carbonate bonding (Yong and Warkentin, 1975; Mitchell, 1976; Kersten and Förstner, 1989) provided the opportunity for greater particle dispersion, thus providing for a larger S_A. Further evaluation of this speculation is needed in respect to the CEC of the two soils (untreated and carbonate removed).

The extraction of both the carbonates and the oxides resulted in a CEC which was slightly greater than the CEC of the soil with carbonates extracted (Table II). The implications of this have yet to be fully determined; however, the increase in CEC is likely due to the loss of variable charge surface associated with the natural soil amorphous oxides.

The reductive dissolution of oxides using the oxalic acid method could reduce the "structural" iron in clay minerals such as smectites. Rozenson and Heller-Kallai, as reported by Kersten and Förstner (1989), found that whereas the extractant citrate–dithionite buffer will attack iron-rich layer silicates, no such problem has been reported when oxalic acid extractant was used. Stucki *et al.* (1984) examined the reduction of structural iron in a citrate bicarbonate solution using $Na_2S_2O_4$ as a reductant and found that the structural iron was reoxidized by exposure to oxygenated water. Since the soil samples in this present study were washed three times with distilled water and air dried after each extraction, this was not seen as a possible complication. Furthermore, the CEC measurements did not increase after oxide extraction to such an extent that would suggest structural Fe had been reduced.

It is useful to note that the decrease of the S_A of the soil with extraction of both the carbonates and the oxides supports a previous experience reported by Yong *et al.* (1992a) for the same soil. In that particular instance, instead of extracting carbonates and oxides, iron and silica amorphous oxides were added to the soil such that they made up 15% of the soil composition. The addition of these oxides resulted in a 50 to 100% increase in the S_A, depending on the proportions of silica to iron oxide added. A similar increase was recorded in the CEC of the soil amorphous oxide mixture when measured at pH 7.

B. Soil Suspension Tests

The results for the metal retention soil suspension tests for the untreated soil without pH adjustment are shown in Figure 1. As can be seen, complete retention of Pb and Cu is obtained by the untreated soil without pH adjustment for original concentrations as high as 40 $mmol \cdot L^{-1}$ for Pb and 50 $mmol \cdot L^{-1}$ for Cu. This compares well with previous work performed by Yong and Phadungchewit (1993) for the same soil.

The problem of distinguishing between retention by precipitation mechanisms as opposed to retention by sorption is rendered difficult because of experimental constraints. Soil suspension studies reported previously (Farrah and Pickering, 1977, 1978, 1979; Maguire *et al.*, 1981; Yong *et al.*, 1990; Yong and Sheremata, 1991; Yong and Phadungchewit, 1993) indicate that accumulation of heavy metal contaminants increases with increasing pH, and that precipitation of the heavy metals at around neutral pH and above may result in the formation of compounds such as carbonates, hydroxides, sulfates, and chlorates species. At acidic pH values, heavy metals become mobile and may adsorb onto clay soil particles. However, adsorption becomes less effective due to competition at the exchange sites from H^+ ions. The initial pH of the untreated soil used in this study was 8.2; however, the equilibrium pH decreased with increasing initial metal concentration (Fig. 1). Distinguishing between adsorbed and precipitated Pb in the perceived "retention" of Pb also depends on the concentration of Pb introduced into the soil suspension—as shown previously by Yong (1996) for a kaolinite soil suspension in interaction with Pb (Fig. 2). As noted in the diagram, some of the precipitated Pb was "recorded" as Pb adsorbed by the soil solids at pH values above approximately 6.5.

To investigate the possibility of Cu and Pb precipitation as a function of pH, to-

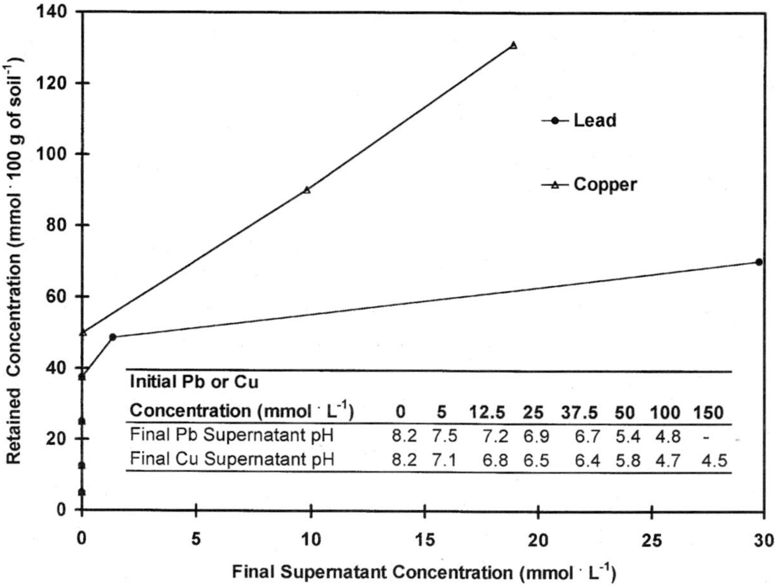

Initial Pb or Cu Concentration (mmol · L^{-1})	0	5	12.5	25	37.5	50	100	150
Final Pb Supernatant pH	8.2	7.5	7.2	6.9	6.7	5.4	4.8	-
Final Cu Supernatant pH	8.2	7.1	6.8	6.5	6.4	5.8	4.7	4.5

Figure 1 Single metal retention by untreated soil without pH adjustment.

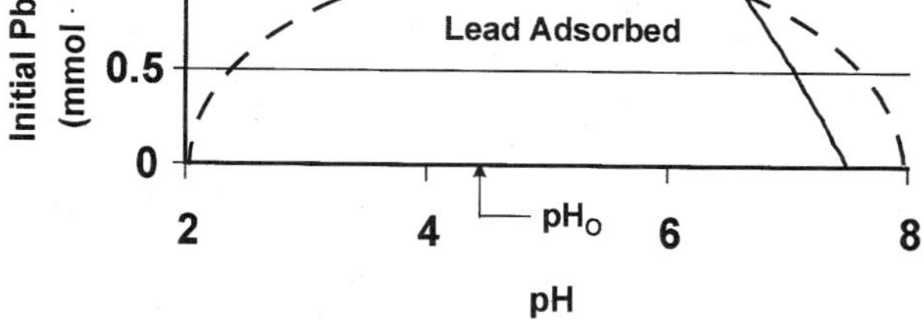

Figure 2 Pb partition in kaolinite suspension in relation to pH (Yong, 1996).

tal metal concentration, and the presence of ligands, the USEPA geochemical model MINTEQA2 was used. Metal precipitation was investigated with and without the presence of carbonates, assumed to be calcite ($CaCO_3$). To investigate the effect of calcite, CO_3^{2-} and Ca^{2+} components were set at the same concentration as measured in the soil as carbonate content (Table I). Allison *et al.* (1991) described that in theory it does not affect the final result of the model as to whether the totals for various components are specified as dissolved or as precipitated since MINTEQA2 will shift mass from the dissolved to the solid phase and vice versa as required for equilibrium. Therefore the calcite was modeled by the inclusion of the separate dissolved components of Ca^{2+} and CO_3^{2-}. The environment, with calcite or calcite-forming components, provides the opportunity for the soluble Pb concentration to be influenced and/or controlled by the precipitation of cerrusite ($PbCO_3$), as shown in Table II. The removal of the soil calcite (modeled by excluding the calcite components from the modeling process) reduces the amount of precipitation and changes the controlling species from cerrusite to $Pb(OH)_2$, as shown by Table III.

Soluble Cu precipitation in the same ion environment will be controlled by two species, malachite ($Cu_2(OH)_2CO_3$) and azurite ($Cu_3(OH)_2(CO_3)_2$), as shown in Table IV. The removal of the calcite components from the modeling exercise reduced the precipitation of Cu but not to the same extent as Pb. Tenorite (CuO) became the precipitate controlling Cu solubility when calcite was not present in the system (Table IV).

The findings of the geochemical model suggest that precipitation may have made a large contribution to the Pb and Cu retention measured when no pH adjustments were performed (Fig. 1) on the soil suspension.

Table III

Precipitation of Pb with and without the Presence of Calcite as Determined by the USEPA Geochemical Model MINTEQA2 (Version 3.11)

	Percentage of initial Pb precipitated[a]					
	as cerrusite, with calcite present, Equilibrium pH			as $Pb(OH)_2$, without calcite present, Equilibrium pH		
Initial Pb concentration ($mmol \cdot L^{-1}$)	4	5	6	4	5	6
0.5	0	92.6	99.4	0	0	98.9
5	53.2	99.2	99.9	0	0	98.5
12.5	76.4	99.6	100	0	0	97.8
25	82.0	99.7	100	0	0	95.7
50	74.7	97.4	99.7	0	0	66.8

[a]All Pb precipitation in the presence of calcite was as cerrusite ($PbCO_3$), and all Pb precipitation without calcite present was as $Pb(OH)_2$.

Table IV

Precipitation of Cu with and without the Presence of Calcite as Determined by the USEPA Geochemical Model MINTEQA2 (Version 3.11)

Initial Cu concentration (mmol·L^{-1})	Percentage of initial Cu precipitated[a]					
	as azurite and/or malachite with calcite present Equilibrium pH			as tenorite without calcite present Equilibrium pH		
	4	5	6	4	5	6
0.5	0	0	92.9 M	0	0	90.4
5	0	80.5 A	99.3 M	0	0	98.8
12.5	0	91.8 A	99.7 M	0	36.1	99.5
25	0	95.4 A	99.8 M	0	64.3	99.7
50	0	97.2 M/A	99.9 M	0	79.3	99.8

[a]All precipitation in the presence of calcite was as malachite ($Cu_2(OH)_2CO_3$) and/or azurite ($Cu_3(OH)_2(CO_3)_2$) as indicated by an M and/or A, respectively, adjacent to the percentage value. All precipitation without calcite present was as tenorite (CuO).

The adjustment of initial soil suspension pH values to levels of approximately 5 and below resulted in an overall decrease in Pb and Cu retention by as much as 90% (Figs. 3, 4, and 6) from the retention noted without initial soil pH adjustment (Fig. 1). Figures 3 and 4 show the test results for the untreated and Pb-treated soil suspensions for initial suspension pH values of 5.2 and 4.0, respectively. Results from soil suspension tests performed at a constant initial Pb concentration of 5 mmol·L^{-1} but across a range of pH values are summarized in Figure 6. As can be noted from Figures 3, 4, and 6, the effect of carbonate removal on the amount of Pb retained is relatively minor. However, when the oxides were removed, it is seen that the differences in retention of Pb are considerable, particularly for the condition when the initial soil suspension pH was 4.0 (Fig. 4) or at final pH values below 4 (Fig. 6). Similar patterns of retention behavior have been observed for Cu as the contaminant metal (Figs. 7 and 8).

The effect of removal of the carbonates from the soil can be considered to be minimal for Pb at an initial soil solution pH of 5.2. However, a slight increase in retention of Pb due to the removal of soil carbonates is apparent at an initial pH of 4.0 (Figs. 3 and 4). At an initial Pb concentration of 5 mmol·L^{-1} a slight increase in Pb retention upon the removal of soil carbonates at final pH values of approximately 4 and below was observed. However, above a final pH of 4, the soil carbonate extraction did not affect the overall quantity of Pb retained (Fig. 6). In almost all cases, final pH values, as shown in Figure 5, indicate a drop in pH from the initial soil suspension pH levels. Generally, final pH values in the

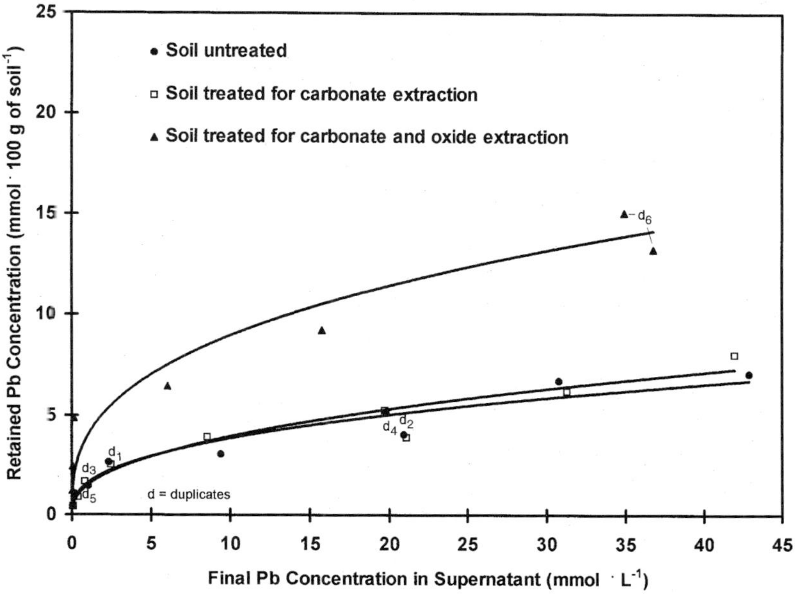

Figure 3 Pb retention at an initial soil suspension pH of approximately 5.2 for untreated soil, soil treated for carbonate extraction, and soil treated for carbonate and oxide extraction.

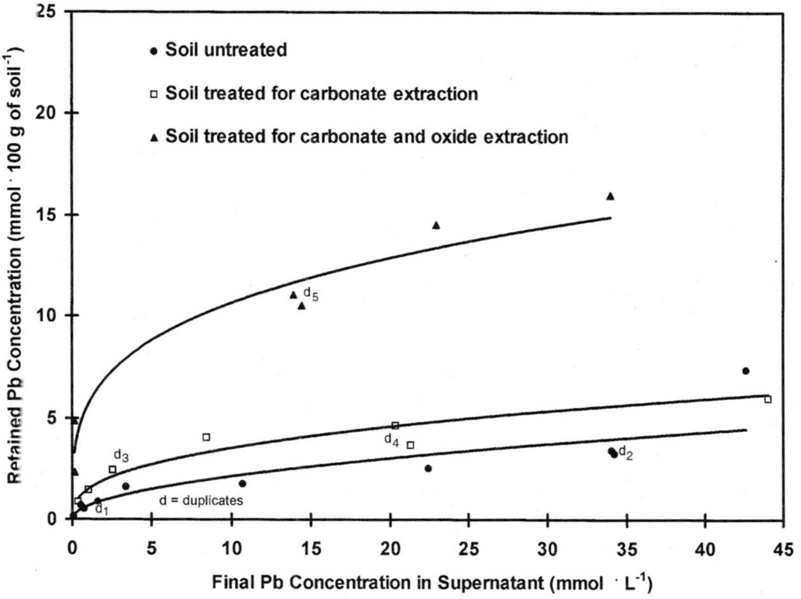

Figure 4 Pb retention at an initial soil suspension pH of approximately 4.0 for untreated soil, soil treated for carbonate extraction, and soil treated for carbonate and oxide extraction.

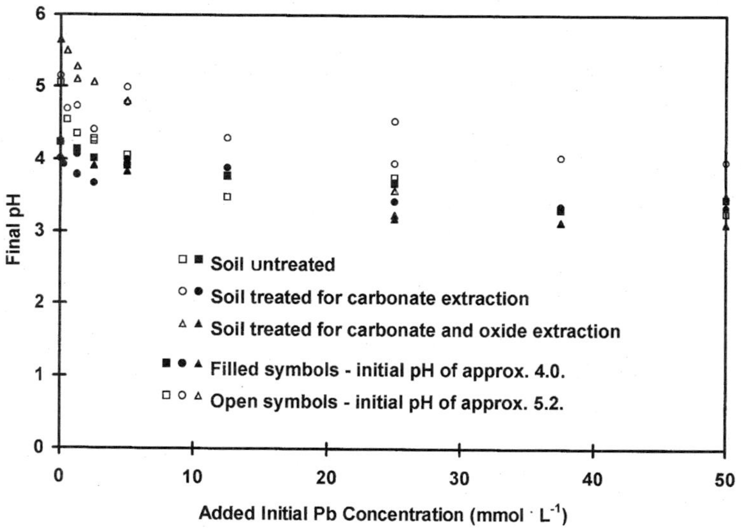

Figure 5 Final pH values of Pb soil suspension tests performed at initial pH values of approximately 4.0 and 5.2.

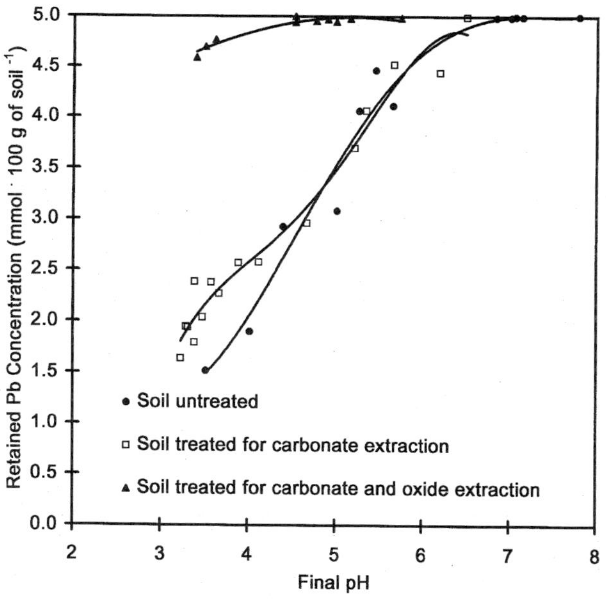

Figure 6 Pb fraction retained at an initial Pb concentration of 5 mmol·L^{-1} for untreated soil, soil treated for carbonate extraction, and soil treated for carbonate and oxide extraction.

soil suspensions which had an initial pH of 5.2 are above those with an initial soil suspension of pH of 4.0. At higher final pH values, Pb interaction with the soil carbonate results in its precipitation as cerrusite (Table III). However, as equilibrium pH values drop below 4.0, cerrusite precipitation decreases. Instead the soil carbonates are dissolved, allowing the cation component of the soil carbonate to compete with the Pb for adsorption. The removal of the soil carbonate phase and the resulting decrease in competing cations could be an explanation for the observed increase in Pb retention. A similar explanation could be considered for the minor increases in Cu retention observed at initial pH values of 4.2 given the role of soil carbonates in the precipitation of Cu as malachite and azurite. As was the case for Pb suspension tests, final pH values for Cu suspension tests indicate a drop in pH from the initial soil suspensions pH values (Fig. 9). The removal of the soil carbonates resulted in a slight decrease in the retention of Cu at an initial soil solution pH of 5.3 (Fig. 7). One could speculate that at an initial soil solution pH of 5.3 some Cu precipitation may have occurred due to complexation with anions liberated during the pH adjustment, forming Cu species of low solubility. The removal of the carbonate phase, which is conducted using a pH 5 extractant, may have removed the source of these anions, resulting in less Cu precipitation and hence less retention.

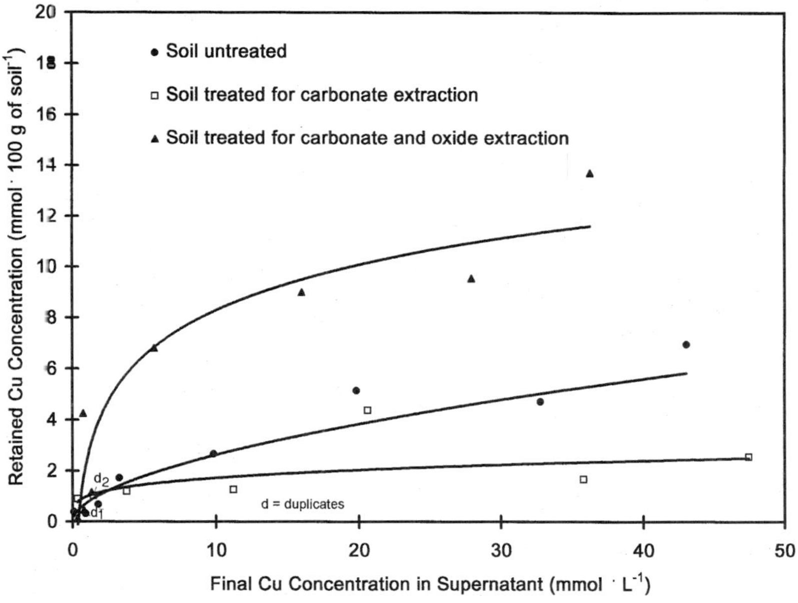

Figure 7 Cu retention at an initial soil suspension pH of approximately 5.3 for untreated soil, soil treated for carbonate extraction, and soil treated for carbonate and oxide extraction.

Changes in the mass ratios of soil components due to the removal of the soil carbonates should also be considered as a possible explanation for observed results; i.e., the extraction of carbonates from the soil will result in a slight increase in the clay minerals, oxides, and organic material present per gram of soil. Furthermore, the resulting increased exposure of S_A (Table II) may have influenced the retention when an increase has been observed.

A considerable increase in the retention of the two metals occurred with the removal of amorphous oxides (Figs. 3, 4, and 6–8). As depicted by Figure 6, complete retention of Pb on the untreated soil and the soil subjected to a carbonate extraction is obtained only when final pH values have exceeded 6. However, the soil that has undergone a carbonate and oxide extraction treatment retains over 90% of the Pb at a final pH value as low as 3.5 and reaches almost complete retention of Pb at pH values between 4 and 5. This set of results is expected inasmuch as similar observations have been made by other investigators; e.g., Elliot *et al.* (1986) found that when iron oxide was removed from the two clay soils used in their study, adsorption of heavy metal cations increased. The authors attributed the increase to two factors, the first being increased electrostatic attraction due to the reduction in the zero point of charge (ZPC) of the soil, and the second being the stripping of oxide coatings and oxides bound to structural exchange sites—thereby increasing the accessibility of the heavy metal cations to these sites.

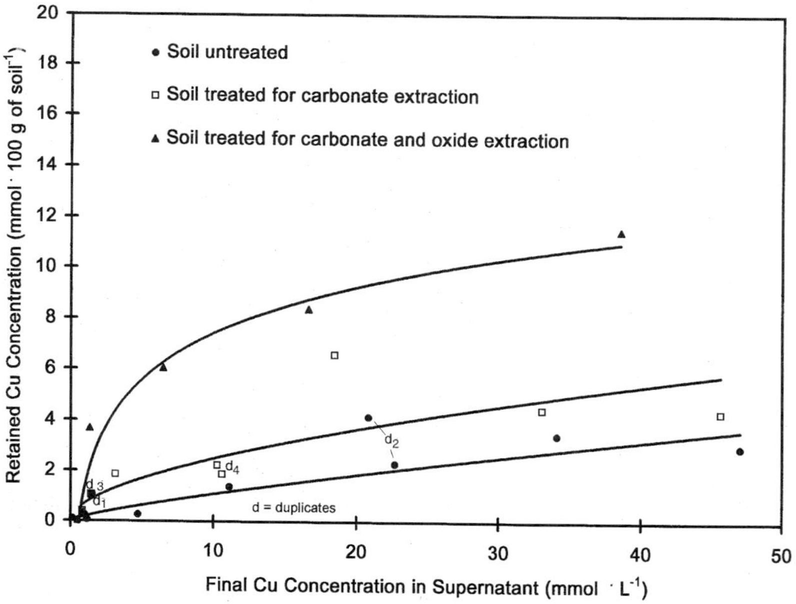

Figure 8 Cu retention at an initial soil suspension pH of approximately 4.2 for untreated soil, soil treated for carbonate extraction, and soil treated for carbonate and oxide extraction.

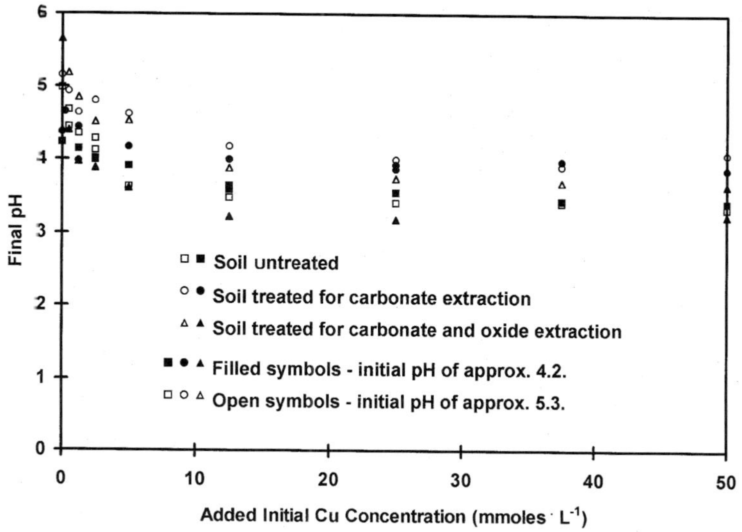

Figure 9 Final pH values of Cu soil suspension tests performed at initial pH values of approximately 4.2 and 5.3.

It should be noted that not all reports on retention of the heavy metals upon removal of amorphous oxides have shown increases in retention. For example, Cavallaro and McBride (1984a,b) have reported considerable reduction in Zn and Cu sorption following the removal of the oxide fraction from an acid subsoil, and Elliot and Liberati (1981) have also reported similar decreases in retention in their studies on Cd adsorption on soils. It is likely that the manner in which the amorphous oxides are distributed within the system can affect their capability to retain heavy metals. Yong and Ohtsubo (1987) and Ohtsubo *et al.* (1991) have shown that the ambient pH of the system is not as important as the initial pH conditions governing the contaminant–soil interaction process. Depending on the initial pH conditions and ZPC of the oxides, the oxides can either coat the particles or coexist as discrete particles at the ambient pH. Similar observations have been made by Elliot *et al.* (1986) with respect to the role of oxides in immobilizing metals.

In studies where the nature and form of mixed oxides have been examined in relation to their ability to retain metals, Meng and Letterman (1993) demonstrated that Cd and Ca adsorption on mixed oxides was affected by the nature of the oxide; e.g., adsorption by particles coated with $Al(OH)_3$ was almost identical to adsorption by $Al(OH)_3$, and that SiO_2 partially coated with $Al(OH)_3$ showed enhanced adsorption. However, Cd adsorption on mixed SiO_2 and $Fe(OH)_3$ was found to be the same as on a similar amount of pure $Fe(OH)_3$. Meng and Letterman (1993) observed the mixed oxides by transmission electron microscopy and found that unlike the $Al(OH)_3$–SiO_2 mix, $Fe(OH)_3$ formed as discrete precipitates.

As discussed under Material and Methods (Section II.B), the soil suspensions were first adjusted to an initial pH condition, and then the metal solutions were added at a pH of 4. Upon completion of the metal adsorption stage, the equilibrium pH of the system was measured, and as expected, in most cases the pH values had dropped below the initial pH of the soil suspension (Figs. 5 and 9). Equilibrium pH values ranged from close to the initial soil suspension at low initial metal concentrations to pH values of 3.2 at high initial metal concentrations. The pH drop could be a result of many reactions in the system, including but not limited to hydrogen ions released due to metal–proton exchange reactions on surface sites, hydrolysis of metals in the soil solution, precipitation of metals, and the addition of pH 4 metal solution. In Figures 10 and 11, the metal ions retained are compared to the H^+ released for soil suspension tests performed on the soil with carbonates and oxides extracted. The ratio of hydrogen ions released to Pb ions retained averaged 0.7 for both initial pH conditions but started out low (0.2 to 0.3) at initial Pb concentrations below 10 mmol·L^{-1}, reaching approximately 1.0 at concentrations of 25 mmol·L^{-1} and above. The ratio of hydrogen ions released to Cu ions retained averaged 2.5 for Cu at an initial pH of 4.2 and 1.5 at an initial pH of 5.3. In both cases, more hydrogen ions were released than Cu ions retained. Without further information it is not possible to determine the reactions

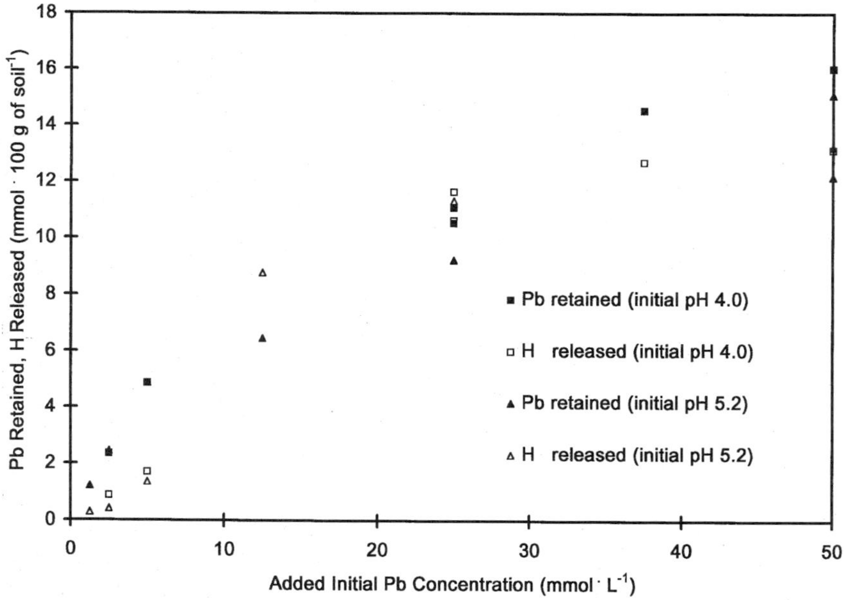

Figure 10 Pb retained and H released in soil suspension tests performed on the soil treated for carbonate and oxide extraction at initial soil suspension pH values of approximately 5.2 and 4.0.

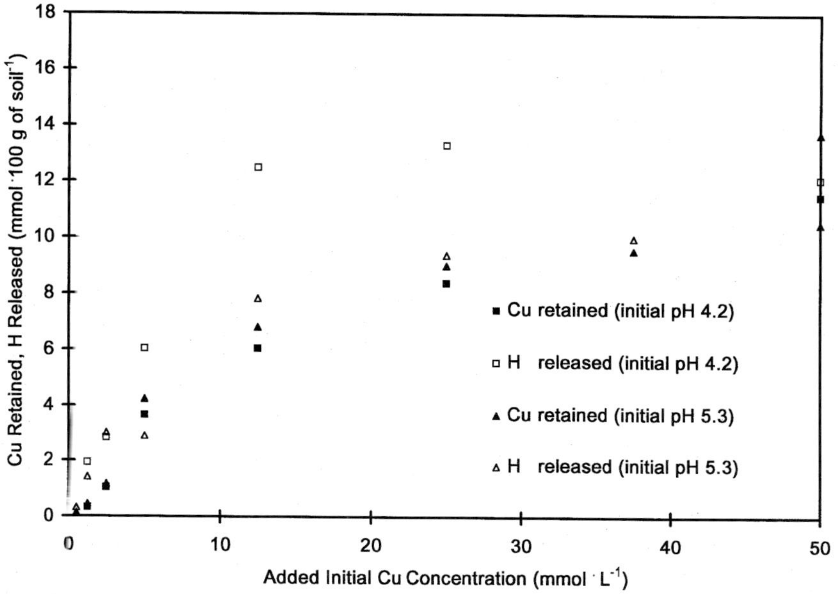

Figure 11 Cu retained and H released in soil suspension tests performed on the soil treated for carbonate and oxide extraction at initial soil suspension pH values of approximately 5.3 and 4.2.

(surface or solution) resulting in release of hydrogen ions and the corresponding change in pH, although these ratios could be used to predict surface adsorption reactions. When modeling on a microscopic scale by the use of surface complexation models, the relationship between metal adsorption and proton release is defined through the proposed reactions. However, in complex heterogeneous systems, it is often necessary to study the system on a macroscopic scale. In such systems the net proton release or consumption is due to all the chemical reactions involving proton transfer. It is not possible to determine the source of the protons other than the generic relationship between adsorption and proton activity (Honeyman and Leckie, 1986).

By comparing the Pb and Cu mineral precipitation to the equilibrium activity and final pH of the metals in supernatant of the suspension tests performed using the soil with the carbonates and oxides extracted (Fig. 12), it can be seen that the metal activities were below the saturation point of the most insoluble metal precipitates. This could be used as evidence in support of the hypothesis that adsorption and not precipitation is the dominant mechanism responsible for retaining the metals on the soil with the oxides and carbonates extracted. To calculate activity coefficients using the Davies equation, it was assumed that the extraction of the carbonates and oxides had removed the soluble soil components which may affect

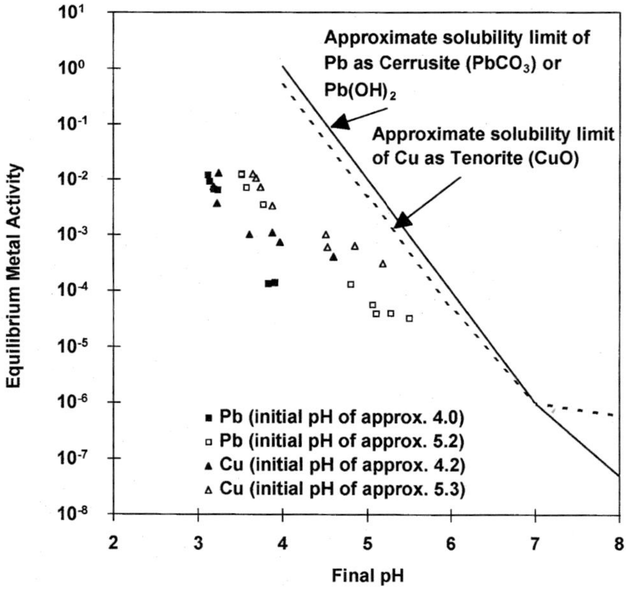

Figure 12 Final pH versus equilibrium metal activity of Pb and Cu in the supernatant of soil suspension tests performed using the soil treated for carbonate and oxide extraction with a comparison to possible mineral precipitates.

the ionic strength of the soil solution. This assumption could not be applied to the soil suspension tests performed without the carbonates and oxides extracted since the pH adjustments may dissolve some of these soil components, thus affecting the ionic strength of the soil solution. Hence, similar calculations could not be made for these soil suspension tests.

C. Sequential Extraction Tests

The distribution of Pb among soil components was examined using a sequential extraction procedure (Section II.C). Sequential extraction was performed on eight samples of the soil taken from the suspension tests performed at variable pH values but at a constant initial Pb concentration of 5 $mmol \cdot L^{-1}$. The distribution determined, as shown in Figure 13, is compared to the additional retention obtained when the carbonate and oxide components were extracted from the soil. At equilibrium pH values of 4 and less, the exchangeable and carbonate phases both played dominate roles in the retention of Pb. However, as the final pH increased, the carbonate fraction become the dominate contributor to Pb retention, with the oxide component also increasing its contribution while the exchangeable fraction became less predominant. No detectable Pb was found associated with the natural soil organic material. It is difficult to correlate these operationally defined fractions, extracted after contamination, to those components extracted from the soil prior to contamination. For example, Pb extracted with the exchangeable fraction could be associated with the clay minerals, oxides, natural organic matter, and even perhaps carbonate components (Stumm and Morgan, 1996). Furthermore, Pb may precipitate as a hydroxide and Cu as an oxide (Tables III and IV), which may not necessarily be the result of interactions with natural soil oxides, although these precipitates may be extracted with the oxide extraction technique used. As can be seen from Figure 13, both the carbonate and the oxide components play a major role in retaining the Pb. However, when these two components are extracted from the soil prior to addition of the Pb, the overall retention of Pb increases, as shown in Figure 6 and compared to the extraction results in Figure 13. This increase in retention is only apparent at final pH values below 5, since near complete retention of Pb by all soils (treated and untreated) occurs above equilibrium pH values of 5. These findings indicate that more research is required to understand the relationship between postsorption sequential extraction of soil fractions to presorption extraction of soil components.

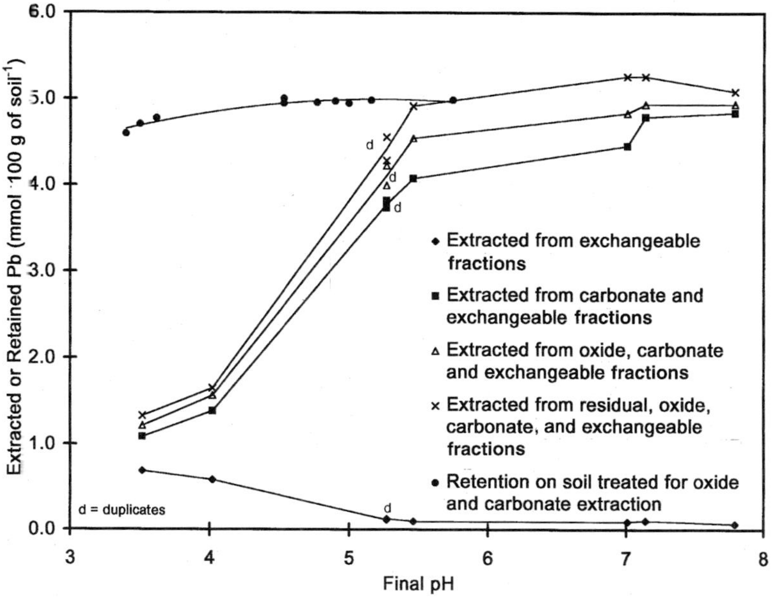

Figure 13 Sequential extraction of Pb from untreated soil added at an initial concentration of 5 mmol·L^{-1} and compared to the Pb retention on soil treated for carbonate and oxide extraction.

IV. CONCLUDING REMARKS

The soil properties reported for the untreated and treated soils presented some interesting but inconclusive results. Speculation on the decrease in the CEC with carbonate extraction evokes two possible explanations. The loss of the CEC associated with the extracted carbonates could result in a decrease in the overall CEC of the soil; however, the CEC of pure carbonates is minor to insignificant. The preferred explanation is that the decrease is due to the effects of the chemical procedure used to determine CEC on easily solubilized carbonate minerals. A similar reduction in S_A upon carbonate extraction could be due to greater particle dispersion caused by the removal of the residual carbonate bonds. The extraction of both the carbonates and the oxides resulted in a CEC slightly higher than that measured with just the soil carbonates extracted. The implications of this have yet to be fully determined. However, it is speculated that the increase in the CEC is likely due to the loss of variable charge surfaces associated with the soil amorphous oxides.

Without pH adjustment the untreated soil retains high concentrations of Pb and Cu; however, when initial soil suspension pH values were adjusted to 5 and less, metal retention decreased by as much as 90%. Analysis of Pb and Cu precipitation in the presence of calcite using the USEPA geochemical model MINTEQA2 indicates that much of this retention at unadjusted pH levels could be the result of precipitation.

The effect of removal of the carbonates from the soil can be considered to be minimal for an initial soil solution pH of approximately 5; however, a slight increase in retention of Pb and Cu was observed at initial soil solution pH values of approximately 4 and final pH values of 4 and below. The removal of the soil carbonate phase and the resulting decrease in competing cations associated with soil carbonates at low pH values could be an explanation for the observed increase in retention. The extraction of soil carbonates and oxides increased the retention of Pb by approximately threefold and doubled the Cu retention. This could be attributed to two factors, the first being increased electrostatic attraction due to the reduction in the variable charge sites on the soil surface, and second being the stripping of oxide coatings and oxides bound to structural exchange sites—thereby increasing the accessibility of the heavy metal cations to these sites. However, at this point these explanations are speculative and further research is required to determine the exact cause(s).

Upon completion of the metal adsorption stage the equilibrium pH of the system had dropped below the initial pH of the soil suspension. The pH drop could be a result of many reactions in the system (surface and solution), and for that reason, it is not conclusive evidence that adsorption has occurred. However, evidence in support of the hypothesis that adsorption and not precipitation is the dominant mechanism responsible for retaining the metals on the soil with the oxides and carbonates extracted is obtained by comparing the Pb and Cu mineral precipitation to the equilibrium activity and final pH of the metals in supernatant. For these soil suspension tests, the metal activities were far below the saturation point of the most insoluble metal precipitates.

Sequential extraction analysis determined that the Pb retained on the untreated soil from an initial concentration of 5 mmol·L^{-1} was associated predominantly with the carbonate component—with exchangeable and oxide soil components also retaining significant quantities. However, the extraction of the carbonate and oxide components prior to addition of the Pb resulted in an increase in retention. Such findings indicate that more research is required to understand the correlation, if any, between postsorption sequential extraction of soil components to presorption extraction of soil components. Continuing study with selective sequential extraction tests, similar to the ones performed in this study, will be conducted with the contaminated samples obtained using the procedures described heretofore.

ACKNOWLEDGMENTS

The authors are grateful for helpful comments and suggestions offered by three anonymous reviewers. Furthermore, a special thanks is expressed to Dr. E. A. Jenne for his editorial comments and manuscript preparation guidance. The authors wish to record their appreciation to the Natural Sciences and Engineering Research Council of Canada (NSERC) for Grant in Aid of Research A-882, which provided the financial support for studies conducted.

REFERENCES

Allard, B., Hakansson, K., Karlsson, S., and Sigas, E. 1991. A field study of diffusion controlled migration of copper, zinc and cadmium in a clay formation. *Water Air Soil Pollut.* 57/58:259–268.

Allen, H. E., Huang, C. P., Bailey, G. W., and Bowers, A. R. 1995. "Metal Speciation and Contamination of Soil," p. 7. Lewis Publishers, Boca Raton, FL.

Allison, J. D., Brown, D. S., and Novo-Gradac, K. J. 1991. MINTEQA2/PRODEFA2, a Geochemical Assessment Model for Environmental Systems: Version 3.0 User's Manual. Environmental Research Laboratory, Office of Research and Development, U.S. Environmental Protection Agency, Athens, GA.

Anderson, P. R., and Benjamin, M. M. 1990. Modeling adsorption in aluminum-iron binary oxide suspensions. *Environ. Sci. Technol.* 24:1586–1592.

Bassett, R. I., and Melchior, D. C. 1990. Chemical Modeling of Aqueous Systems: An Overview. *In* "Chemical Modeling of Aqueous Systems II" (D. L. Melchior and R. I. Bassett, Eds.), Am. Chem. Symp. Ser. No. 416, pp. 1–14. American Chemical Society, Washington, DC.

Calmano, W., Ahlf, W., and Bening, J. C. 1992. Chemical mobility and bioavailability of sediment bound heavy metals influenced by salinity. *Hydrobiologia* 235/236:605–610.

Cavallaro, N., and McBride, M. B. 1984a. Zinc and copper sorption and fixation by an acid soil clay: Effect of selective dissolution. *Soil Sci. Soc. Am. J.* 48:1050–1054.

Cavallaro, N., and McBride, M. B. 1984b. Effect of selective dissolution on charge and surface properties of an acid soil. *Clays Clay Miner.* 32:283–289.

Elliot, H. A., and Liberati, M. R. 1981. Properties Affecting Retention of Heavy Metals from Wastes to Northeastern U.S. Soils. *In* "Proceedings 13th Mid-Atlantic Waste Conference." Ann Arbor Science, Ann Arbor, MI.

Elliot, H. A., Liberati, M. R., and Huang, C. P. 1986. Effect of iron oxide removal on heavy metal sorption by acid subsoils. *Water Air Soil Pollut.* 27:379–389.

Eltantaway, J. N., and Arnold, P. W. 1973. Reappraisal of ethylene glycol monoethyl ether (EGME) method of surface area estimates of clays. *J. Soil Sci.* 24:232–238.

Farrah, H., and Pickering, W. F. 1977. The sorption of lead and cadmium species by clay minerals. *Aust. J. Chem.* 30:1417–1422.

Farrah, H., and Pickering, W. F. 1978. Extraction of heavy metal ions sorbed on clays. *Water Air Soil Pollut.* 9:491–498.

Farrah, H., and Pickering, W. F. 1979. pH effects in the adsorption of heavy metal ions by clays. *Chem. Geol.* 25:317–326.

Frost, R. R., and Griffin, R. A. 1977. Effect of pH on adsorption of copper, zinc, and cadmium from landfill leachate by clay minerals. *J. Environ. Sci. Health* 12:139–156.

Gobiel, C., Johnson, W. L., Macdonald, R. W., and Wong, C. S. 1994. Sources and burden of lead in St. Lawrence estuary sediments: Isotopic evidence. *Environ. Sci. Technol.* 29:193–201.

Grambell, R. P., Wiesepape, J. B., Parrick, W. H., and Duff, M. C. 1991. The effects of pH, redox, and salinity on metal release from a contaminated sediment. *Water Air Soil Pollut.* 57/58:359–367.

Harter, R. D. 1979. Adsorption of copper and lead by Ap and B2 horizons of several northeastern United States soils. *Soil Sci. Soc. Am. J.* 43:679–683.

Harter, R. D. 1983. Effect of soil pH on adsorption of lead, copper, zinc and nickel. *Soil Sci. Soc. Am. J.* 47:47–51.

Hendershot, W. H., Lalande, H., and Duquette, M. 1993. Ion Exchange and Exchangeable Cations. *In* "Soil Sampling and Methods Analysis," pp. 167–176. Can. Soc. Soil Sci. Press, Toronto.

Hesse, P. R. 1971. "A Textbook of Soil Chemical Analysis." William Clowes and Sons, London.

Honeyman, B. D., and Leckie, J. O. 1986. Macroscopic Partitioning Coefficients for Metal Ion Adsorption. *In* "Geochemical Processes at Mineral Surfaces" (J. A. Davis and K. F. Hayes, Eds.), Am. Chem. Symp. Ser. No. 323, pp. 162–190. American Chemical Society, Washington, DC.

Honeyman, B. D., and Santschi, P. H. 1988. Metals in aquatic systems. *Environ. Sci. Technol.* 22:862–871.

Jeffery, J. J., and Uren, N. C. 1983. Copper and zinc species in the soil solution and the effect of soil pH. *Aust. J. Soil. Res.* 21:479–488.

Kersten, M., and Förstner, U. 1989. Speciation of Trace Elements in Sediments. *In* "Trace Element Speciation, Analytical Methods, and Problems" (G. E. Batley, Ed.), pp. 245–317. CRC Press, Boca Raton, FL.

Luoma, S. N. 1989. Can we determine the biological availability of sediment-bound trace elements? *Hydrobiologia* 176/177:379–396.

Maguire, M., Slavek, J., Vimpany, I., Higginson, F. R., and Pickering, W. F. 1981. Influence of pH on copper and zinc uptake by soil clays. *Aust. J. Soil Res.* 19:217–229.

Meng, X., and Letterman, R. D. 1993. Modeling ion adsorption on aluminum hydroxide modified silica. *Environ. Sci. Technol.* 27:1924–1929.

Mitchell, J. K. 1976. "Fundamentals of Soil Behaviour." John Wiley and Sons, Toronto.

Muller, H. W., Schwaighofer, B., and Kalham, W. 1994. Heavy metal contents in river sediments. *Water Air Soil Pollut.* 72:191–203.

Ohtsubo, M., Yoshimura, A., Wada, S. I., and Yong, R. N. 1991. Particle interaction and rheology of illite-iron oxide complexes. *Clays Clay Miner.* 39:347–354.

Pankow, J. F. 1991. "Aquatic Chemistry Concepts." Lewis Publishers, Chelsea, MI.

Phadungchewit, Y. 1990. The Role of pH Buffer Capacity in Heavy Metal Retention in Clay Soils. Ph.d. Thesis, McGill Univ., Montreal, Quebec.

Rapin, F., Tessier, A., Cambell, P. G. C., and Carignan, R. 1986. Potential artifacts in the determination of metal partitioning in sediments by sequential extraction procedure. *Environ. Sci. Technol.* 20:836–840.

Ravishankar, B. R., Auclair, J. C., and Tyagi, R. D. 1994. Partitioning heavy metals in some Quebec municipal sludges. *Water Pollut. Res. J. Can.* 29:457–470.

Saeki, K., Okazaki, M., and Matsumoto, S. 1993. The chemical phase changes in heavy metals with drying and oxidation of the lake sediments. *Wat. Res.* 27:1243–1251.

Salomons, W., and Förstner, U. 1984. "Metals in the Hydrocycle." Springer-Verlag, New York.

Shu, G. Y., and Liu, J. C. 1994. Content and fractionation of heavy metals in soils of two contaminated sites in Taiwan. *Environ. Prog.* 13:89–93.

Shuman, L. M. 1983. Sodium Hypochlorite Methods for the extracting microelements associated with soil organic matter. *Soil. Sci. Soc. Am. J.* 47:656–660.

Sposito, G. 1989. "The Chemistry of Soils." Oxford Univ. Press, New York.

Starkley, H. C. Blackmon, P. D., and Hauff, P. L. 1984. The routine mineralogical analysis of clay bearing samples. *U.S. Geological Survey,* Reston, VI. 1563:2–18.

Stucki, J. W., Golden, D. C., and Roth, C. B. 1984. Effects of reduction and reoxidation of structural iron on the surface charge and dissolution of dioctahedral smectities. *Clays Clay Miner.* 32:150–156.

Stumm, W., and Morgan, J. J. 1996. "Aquatic Chemistry." John Wiley and Sons, Toronto.

Tessier, A., Campbell, P. G. C., and Bisson, M. 1979. Sequential extraction procedure for the speciation of particulate trace metals. *Anal. Chem.* 51:844–850.

United States Environmental Protection Agency (USEPA). 1986. Method 3050—Acid Digestion of Sediments, Sludges, and Soils. *In* "Test Methods for Evaluating Solid Waste: Physical Chemical Methods." SW-846, U.S. Environmental Protection Agency, Washington, DC.

United States Environmental Protection Agency (USEPA). 1991. MINTEQA2—Metal Speciation Equilibrium Model for Surface and Groundwater, Version 3.11; PRODEFA2—Problem Definition Program for MINTEQA2, Version 3.11. Center for Exposure Assessment Modeling, U.S. Environmental Protection Agency, Office of Research and Development, Environmental Research Laboratory, Athens, GA.

Yanful, E. K., Nesbitt, W. H., and Quigley, R. M. 1988. Heavy metal migration at a landfill site, Sarnia, Ontario, Canada. I. Thermodynamic assessment and chemical interpretations. *Appl. Geochem.* 3:523–533.

Yong, R. N. 1996. Waste Disposal, Regulatory Policy and Potential Health Threats. *In* "Engineering Geology of Waste Disposal" (S. P. Bentley, Ed.), Geol. Soc. Eng. Geol. Special Pub. No. 11, pp. 325–339. Geological Society of England, London.

Yong, R. N., Galvez-Coutier, R., and Phadungchewit, Y. 1993. Selective sequential extraction analysis of heavy-metal retention in soil. *Can. Geotech. J.* 30:834–847.

Yong, R. N., Mohamed, A. M. O., and Wang, B. W. 1992a. Influence of amorphous silica and iron hydroxide on interparticle action and soil surface properties. *Can. Geotech. J.* 29:803–818.

Yong, R. N., Mohamed, A. M. O., and Warkentin, B. P. 1992b. "Principles of Contaminant Transport in Soils." Elsevier, Amsterdam.

Yong, R. N., and Ohstubo, M. 1987. Interparticle action and rheology of kaolinite-amorphous iron hydroxide (ferrihydrite) complexes. *Appl. Clay Sci.* 2:63–81.

Yong, R. N., and Phadungchewit, Y. 1993. pH influence on selectivity and retention of heavy metals in some clay soils. *Can. Geotech. J.* 30:821–830.

Yong, R. N., Sheremata, T. W. 1991. Effect of chloride ions on adsorption of cadmium from a landfill leachate. *Can. Geotech. J.* 28:378–387.

Yong, R. N., and Warkentin, B. P. 1975. "Soil Properties and Behaviour." Elsevier Scientific Publishing Company, Amsterdam.

Yong, R. N., Warkentin, B. P., Phadungchewit, Y., and Galvez, R. 1990. Buffer capacity and lead retention in some clay materials. *Water Air Soil Pollut.* 53:53–67.

Chapter 11

Immobilization of Pb by Hydroxylapatite

Valérie Laperche and Samuel J. Traina
School of Natural Resources, Ohio State University, Columbus, Ohio

Previous studies have shown that lead interactions with hydroxylapatite induced hydroxypyromorphite precipitation, causing a decrease in aqueous Pb concentration. In the present study we investigated the Pb solids formed after we let Pb synthetic hydroxylapatite react at low Pb concentrations. In order to understand the mechanisms of lead immobilization, we studied the phosphates formed after reaction of aqueous lead with apatites and after coprecipitation of (Pb, Ca) phosphates. These phosphates were analyzed by different spectroscopic methods (XRD, IR, EXAFS, and SEM) to determine the structure of the precipitated Pb-phosphates and the processes of formation of reaction products. The results showed that at low Pb concentrations it was still pure pyromorphite and not (Pb, Ca)-apatite that was formed when we let aqueous lead react with apatite.

I. INTRODUCTION

Many investigations have shown the ability of apatite to immobilize dissolved Pb (Ma *et al.*, 1993) and Pb in contaminated soils (Laperche *et al.*, 1996). The mineral formed by reaction of apatite with dissolved Pb is pyromorphite, or more commonly hydroxypyromorphite (HP) or chloropyromorphite (ClP). Pyromorphite is a stable mineral at pH values >3 (log $K_{sp} = -25$ for ClP and -4 for HP); consequently it may be possible to use apatite to remediate contaminated soils. However, field-scale applications of this method require a comprehensive understand-

Adsorption of Metals by Geomedia

ing of the crystal structure and composition of the reaction products since incorporation of foreign ions into the pyromorphite structure could affect the crystallinity, morphology, lattice parameters, and consequently the geochemical stability of pyromorphites. Incorporation of foreign ions and subsequent changes in geochemical stability are well know for apatites (Bigi *et al.,* 1986, 1991). Furthermore, the formation of apatite phases from aqueous systems can be hindered by the presence of dissolved foreign ions. Le Geros *et al.* (1980) and Ma *et al.* (1994a,b) found that the formation of HP was perturbed by the presence of high concentrations of Al, Fe, Cu, Cd, and CO_3^{2-}.

Two processes for the reaction of dissolved Pb with apatite may be proposed. First, Pb can react with apatite through hydroxylapatite (HA, $Ca_{10}(PO_4)_6(OH)_2$) dissolution (1) followed by precipitation (2) of pure hydroxypyromorphite ($Pb_{10}(PO_4)_6(OH)_2$) as described by Ma *et al.* (1993):

$$Ca_{10}(PO_4)_6(OH)_2(s) + 14\,H^+\,(aq) \underset{\text{Precipitation}}{\overset{\text{Dissolution}}{\rightleftharpoons}} 10Ca^{2+}(aq)\;6H_2PO_4^-(aq) + 2H_2O \quad [1]$$

$$10Pb^{2+}(aq) + 6H_2PO_4^-(aq) + 2H_2O \leftrightharpoons Pb_{10}(PO_4)_6(OH)_2(s) + 14\,H^+(aq). \quad [2]$$

Alternatively, Pb can substitute for Ca in apatite. This chemical equation is given as

$$Ca_{10}(PO_4)_6(OH)_2(s) + xPb^{2+}(aq) \overset{\text{Adsorption or coprecipitation}}{\leftrightharpoons} (Ca_{(10-x)}Pb_x)(PO_4)_6(OH)_2(s) + xCa^{2+}(aq). \quad [3]$$

Thus, (Ca, Pb) apatites could potentially be formed by adsorption of Pb or by dissolution of HA followed by coprecipitation of mixed apatites. Finally, Pb ions could potentially adsorb onto the surfaces of apatite particles. Whereas Ma and coworkers have clearly shown that the reaction of dissolved Pb with hydroxylapatite results in hydroxypyromorphite precipitation (Ma *et al.,* 1993), the adsorption of Pb at concentrations below the detection limits of X-ray diffraction (XRD) could not be ruled out.

In a study of apatite interactions with specimen Pb solids, Pb-contaminated soil, and soil material, Laperche *et al.* (1996) observed conversion of "native" Pb forms to pyromorphites. XRD and scanning electron microscope (SEM) data were found to be consistent with equilibrium thermodynamic modeling.

The aim of the present work was to characterize the minerals formed after reaction of hydroxylapatite with $Pb(NO_3)_2$ at low concentrations of dissolved Pb at acidic pH values. A specific effort was made to distinguish between the formation of comparatively pure HP crystals versus the formation of (Pb, Ca) phosphate solid solutions or absorbed Pb. To achieve this objective, we prepared two sample series of Pb-phosphates at low Pb concentrations. The first series, designated as the "sorption samples," was prepared by reacting $Pb(NO_3)_2$ solutions with HA at pH 5. The second series, designated as the "coprecipitation samples," was made by coprecipitating (Ca, Pb) apatites with varying solid-phase Pb/Ca ratios. These materials were examined with Fourier transform infrared (FTIR), XRD, SEM, and extended X-ray absorption fine structure (EXAFS) spectroscopy in an effort to better ascertain the mechanism of aqueous-Pb immobilization by HA.

Extended X-ray absorption fine structure spectroscopy is one of the very few methods capable of selectively probing the structural environment of a dilute atom. EXAFS was used in the present study to examine the chemical state of Pb in phosphates. EXAFS spectroscopy consists of recording and analyzing the variations of the absorption coefficient above an absorption edge for a given element (Teo, 1986). Variations in the absorption coefficient originate from interferences between the outgoing electronic wave from the X-ray absorber and the incoming wave that is backscattered by the electrons of neighboring atoms (backscatterers). Full analysis of the absorption modulations provides information concerning the local environment of the adsorber atom in terms of interatomic distances and the type and number of neighboring atoms in the two or three nearest atomic shells. Thus, EXAFS spectroscopy has the potential to differentiate the various possible structural solutions previously envisaged for the location of Pb into pyromorphite. Additionally, its greater sensitivity should allow for a differentiation between incorporation of Pb into pyromorphites versus adsorption of Pb onto apatite surfaces.

II. MATERIALS AND METHODS

A. Sample Preparation

Sorption samples were prepared by reacting 2 g of synthetic hydroxylapatite (BioRad Laboratories, Bio Gel HTP; see Xu and Schwartz, 1994, for description of this material) with 50-ml solutions of $Pb(NO_3)_2$ (24 $\mu mol \cdot L^{-1}$ to 24 $mmol \cdot L^{-1}$ Pb) at pH 5 for a period of 2 hr (Table I). After the reaction period, the solids were separated from the solution phase by filtration and washed with H_2O, and the solids and filtrates were saved for subsequent analyses.

The preparation of hydroxylapatite, hydroxypyromorphite, and their solid solutions was based on the reaction (Narasaraju *et al.* 1972)

$$10M(NO_3)_2 + 6(NH_4)_2HPO_4 + 2H_2O \rightarrow M_{10}(PO_4)_6(OH)_2 + 12NH_4NO_3 + 8HNO_3,$$

where M = Ca or Pb for the end-member and (Ca + Pb) for the solid solutions, the proportion of Ca to Pb being varied as desired (Table I). Aqueous solutions of the reactants containing stoichiometric quantities required for the formation of approximately 5 g of the desired product were prepared as next described.

Solution I contained a mixture of $Ca(NO_3)_2$ and $Pb(NO_3)_2$ in the desired proportions. Solution II contained a stoichiometric amount of diammonium hydrogen phosphate corresponding to the amount of nitrate(s) in solution I. The pH was adjusted in each solution to pH 12 by addition of ethylenediamine. Solutions I and II were added dropwise simultaneously into approximately 100 ml of water main-

Table I
Samples Obtained by Reaction of $Pb(NO_3)_2$ with Hydroxylapatite and by Coprecipitation of (Ca, Pb) Phosphates

	Sorption samples (SS)			Coprecipitated samples (CS)		
	$[Pb]_{solution}$ ($mmol \cdot L^{-1}$)				Pb and Ca content (%) in initial solution	
Sample[a]	Initial	Final	%HP formed	Sample[a]	Pb	Ca
SS0.00007	0.024	0.0	0.0066	CS0.00	0	100
SS0.0007	0.270	0.0	0.068	CS0.03	5	95
SS0.0013	0.502	0.0	0.128	CS0.08	10	90
SS0.006	2.423	0.0	0.614	CS0.12	15	85
SS0.013	5.237	0.0	1.338	CS0.27	25	75
SS0.029	12.124	0.520	2.912	CS0.64	50	50
SS0.058	23.891	0.778	5.802	CS1.00	100	0

[a]Moles of Pb by unit cell divided by the sum of moles of Pb and Ca by unit cell in the final product.

tained as before at pH 12. In order to obtain apatitic compounds in the complete range, it was necessary to add $Ca(NO_3)_2$ and $Pb(NO_3)_2$ aqueous solutions (maintained at pH 12) separately to $(NH_4)_2HPO_4$ (also maintained at pH 12).

The resulting solutions were allowed to sit undisturbed overnight. The precipitates were then removed by filtration, washed with water, and dried at 60°C.

The samples are named by the abbreviation of "sorption sample" or "coprecipitation sample," followed by a number indicating the proportion of Pb in solid (mol Pb by unit cell/(mol Pb + Ca by unit cell)).

B. Chemical Analysis

All samples were digested using HF–HCl/HNO_3 in a Parr Bomb (Bernas, 1968). Flame atomic absorption spectrometry (Perkin Elmer 3030B) was used to quantify total Pb and Ca concentrations in the digests. Total phosphorus was measured colorimetrically with a Beckman DU-6 spectrophotometer.

C. Infrared Spectrometry

FTIR spectra were recorded under vacuum with a Bruker IFS 113v spectrometer equipped with a glower source and a TGS detector. One milligram of sample was carefully mixed with 150 mg of KBr (infrared grade). All infrared (IR) spectra were obtained from KBr pellets and were recorded between 4000 and 450 cm^{-1} as an average of 200 scans at a 2 cm^{-1} resolution.

D. X-Ray Diffraction

All XRD measurements were made with a Philips X-ray diffractometer using Cu Kα radiation at 35 kV and 20 mA. Data were collected using a step scanning technique with a fixed time of 4 sec per 0.01°2θ. The XRD patterns were obtained from 20 to 35°2θ. All XRD analyses were performed using back-filled, randomly oriented mounts.

E. Spectral Analysis

Curve fits of the FTIR and XRD spectra were obtained with the GRAMS/386 (Galactic Industries) Curvefit module. Second-order derivatives were used to identify approximate peak positions of superimposed peaks and the number of overlapping peaks present. Curve fitting was employed to obtain information on peak intensities and positions. Composite peaks were fit with Cauchy–Lorentzian profiles, using nonlinear least-squares refinement procedures based on a finite difference Levenberg–Marquardt algorithm. The details of these curve-fitting procedures were described by Gillette *et al.* (1982) and Maddoms (1980).

F. Scanning Electron Microscopy

The same samples used in the FTIR and XRD analyses were also examined with a scanning electron microscope (JEOL JSM-A2O). The samples were mounted on carbon stubs using double stick tape and then carbon or gold coated.

G. Extended X-Ray Absorption Fine Structure

All EXAFS experiments were conducted on a wiggler beamline IV-3 at Stanford Synchrotron Radiation Laboratory (SSRL), at ambient temperature. Data were collected using a Si(111) monochromator crystal. The monochromator was detuned by 50% at the highest scanning energy to minimize the transmission of higher order harmonics. The Pb–L(III)-edge spectra of the samples were collected in the fluorescence mode using a Lytle-type detector. A germanium filter was positioned between sample and detector. The synchrotron electron beam energy was 3.0 GeV and the beam current ranged from 20 to 90 mA. The energy calibration was monitored with a Pb metal foil. The threshold energy was set as 13,035 eV, which was chosen at the first inflection point of the Pb–L(III) edge. The data were collected for the energy range 12,800 to 14,050 eV. Powdered samples were mounted between two thin kapton tape sheets which were supported by plastic slide holders. Concentrated samples were diluted with boron nitride, while low-concentration samples were run without dilution. The number of scans (individual spectra collected) varied from sample to sample depending on the absorber concentration. There were usually 6 to 10 scans.

III. RESULTS AND DISCUSSION

A. Assignments of the Infrared Absorption Bands

The infrared spectra of aqueous, simple solid phosphates and apatites are well established and absorption bands have been assigned (Chapman and Thirlwell, 1964; Stutman *et al.,* 1965; Baddiel and Berry, 1966; Levitt and Condrate, 1970). The free orthophosphate ion PO_4^{3-} has tetrahedral symmetry (T_d) and exhibits four normal infrared absorption modes (Fig. 1).

In the absence of environmental perturbations, the triply antisymmetric P–O stretching mode (ν_3) is situated in the 1100 to 1000 cm^{-1} region, the symmetric P–O stretching mode (ν_1) at 940 cm^{-1}, the triply degenerate antisymmetric P–O bending mode (ν_4) in the 570 to 500 cm^{-1} region, and the symmetric O–P–O bending mode (ν_2) between 420 and 350 cm^{-1} (Baddiel and Berry, 1966). Modes ν_1 and ν_2 are active in the Raman only, and modes ν_3 and ν_4 are active in the Raman and the infrared. The complexation of the oxyanions (PO_4^{3-}) with cations (case of crystalline phosphate) affects the force fields of oxyanion tetrahedra and lowers their symmetry from T_d to either C_{3v} (corner-sharing), C_{2v} (edge-sharing, didentate binuclear), or C_1 (corner-sharing, edge-sharing, didentate binuclear, multidentate) (Nakamoto, 1986). In the case of hydroxylapatite, the site symmetry of the phosphate ion is reduced from T_d to C_1. Modes ν_1 and ν_2 become infrared active because of the loss of symmetry. Furthermore, ν_2, ν_3, and ν_4 lose their degenerate character and eight bands should be observed instead of three. In total, nine internal modes of the phosphate ion become active in the infrared.

Baddiel and Berry (1966) assigned the absorption bands as follows: the band at 962 cm^{-1} is attributed to mode ν_1; mode ν_3 of the free phosphate has split into three unequally spaced compounds, two of which are quite distinct (1094 and 1037 cm^{-1}) and the third appearing as a shoulder at 1065 cm^{-1}; and the absorption bands at 603 and 564 cm^{-1} and the shoulder at 575 cm^{-1} are assigned to the former mode ν_4. Fowler (1968) and Klee (1970) assigned the absorption bands at 471 cm^{-1} to mode ν_2. Two other absorption bands observed on the spectra have been assigned to the OH stretch at 3571 cm^{-1} and the OH libration at 635 cm^{-1} (Cant *et al.,* 1971). As in the case of the hydroxylapatite, the symmetry of the phosphate ion in hydroxypyromorphite is reduced, all the degeneracies are removed, and all nine internal modes of the ion become active in the infrared (Adler, 1968; Levitt and Condrate, 1970). Following the assignment of Levitt and Condrate (1970) in

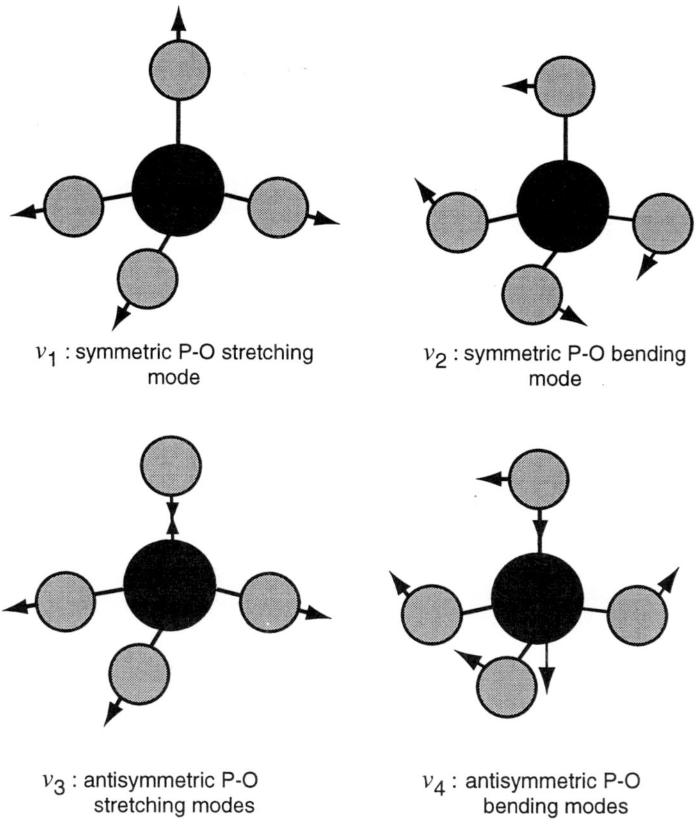

Figure 1 Vibrational modes of the PO_4^{3-} ion.

the case of the chloropyromorphite, we assigned the absorption bands of hydroxy-pyromorphite as follows: the band at 922 cm^{-1} is attributed to mode ν_1; mode ν_3 has split into three unequally spaced compounds, two of which are quite distinct (1042 and 975 cm^{-1}) and the third appearing as a very small shoulder at 993 cm^{-1}; the absorption bands at 576, 551, and 538 cm^{-1} are assigned to mode ν_4, and the band at 425 cm^{-1} is attributed to mode ν_2. The absorption band at 3560 cm^{-1} is assigned to the OH stretch, but the libration mode of OH was not observed and probably occurs below 400 cm^{-1}.

The infrared absorption spectra of calcium–lead apatites reveal a shift of the OH stretch and the internal modes of the PO_4^{3-} ions toward lower frequencies as the lead content increases, in agreement with the data of Andres-Verges *et al.* (1983) and Bigi *et al.* (1991) (Table II).

Table II

Infrared Frequencies (in cm^{-1}) and Assignments for Hydroxylapatite (HA) and Calcium–Lead Apatites Obtained by Coprecipitation (CS) from 4000 to 400 cm^{-1}

Samples	Chemical analysis	ν_{OH}	$\nu_1(PO_4)$	$\nu_2(PO_4)$	$\nu_3(PO_4)$	$\nu_4(PO_4)$
HA (BioRad)	$Ca_{10}(PO_4)_6(OH)_2$	3570, 634	962	471	1094, 1067, 1034	604, 576, 564
CS0.00	$Ca_{10}(PO_4)_6(OH)_2$	3570, 635	962	471	1093, 1067, 1038	604, 576, 564
CS0.03	$(Ca_{9.67}Pb_{0.33})(PO_4)_6(OH)_2$	3567, 632	961	468	1093, 1064, 1033	603, 575, 563
CS0.08	$(Ca_{9.24}Pb_{0.76})(PO_4)_6(OH)_2$	3565, 632	959	468	1093, 1063, 1032	602, 574, 562
CS0.12	$(Ca_{8.78}Pb_{1.22})(PO_4)_6(OH)_2$	3565	959	465	1091, 1063, 1031	601, 574, 562
CS0.27	$(Ca_{7.28}Pb_{2.72})(PO_4)_6(OH)_2$	3563	953	455	1090, 1062, 1032	599, 574, 560
CS0.64	$(Ca_{3.60}Pb_{6.40})(PO_4)_6(OH)_2$	3562	922	446	1092, 1039, 978	594, 576, 550
CS1.00	$Pb_{10}(PO_4)_6(OH)_2$	3560	922	425	1042, 993, 975	576, 551, 538

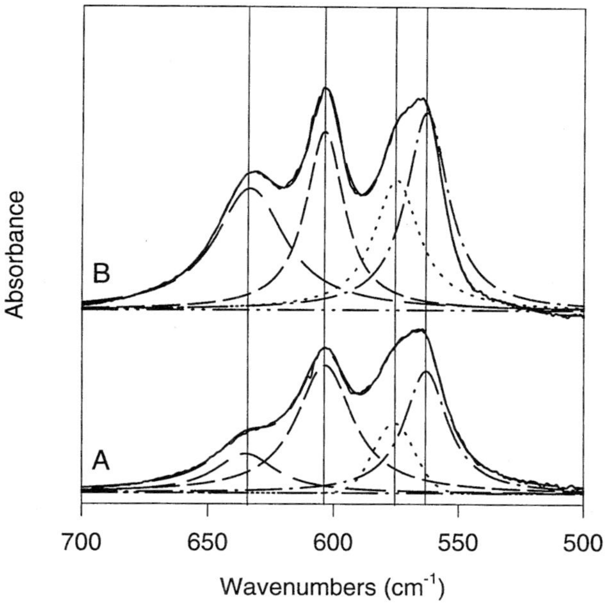

Figure 2 FTIR spectra of pure hydroxylapatite (HA) synthesized at pH 12 (A) and HA from BioRad (B).

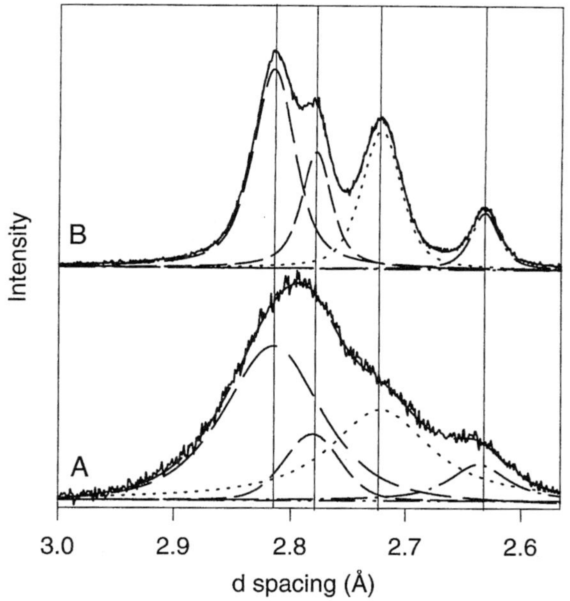

Figure 3 X-ray diffraction patterns of pure hydroxylapatite (HA) synthesized at pH 12 (A) and HA from BioRad (B).

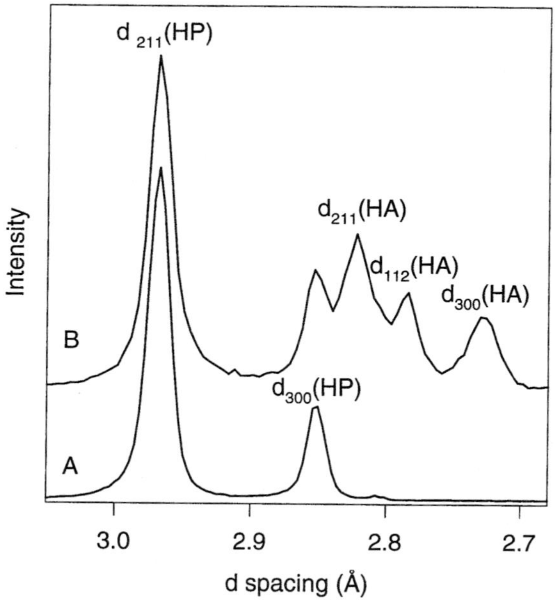

Figure 4 X-ray diffraction patterns of hydroxypyromorphite synthesized at pH 12 (A) and HA reacted with $Pb(NO_3)_2$ at pH 5 for 2 hr (B).

The preparation of the synthetic (Ca, Pb) apatites at pH 12 was chosen for the relative simplicity of the method and also to prepare homogeneous samples in large quantities. The fact that this method is fast and at ambient temperature did not favor the synthesis of large crystals. A comparison of the IR (Fig. 2) and XRD (Fig. 3) spectra and SEM pictures (Fig. 5) between HA synthesized at pH 12 (Figs. 2A and 3A) and HA from the BioRad Company (Figs. 2B and 3B), and also between HP synthesized at pH 12 (Fig. 4A) and HP precipitated at pH 5 (Fig. 4B), showed a difference only in the width of the IR bands and XRD peaks and not in the IR position. The broadness of the peaks is likely due to the small crystal size of the phophates synthesized at pH 12 (Figs. 5 and 6).

B. Immobilization of Pb from Solution at pH 5

The XRD patterns of the reaction products of aqueous Pb with HA at pH 5 are presented in Figure 7. At pH 5, the removal of aqueous Pb from solution was

Figure 5 SEM micrograph of hydroxylapatite (HA) synthesized at pH 12 (A) and HA from BioRad (B).

Figure 5

Figure 6

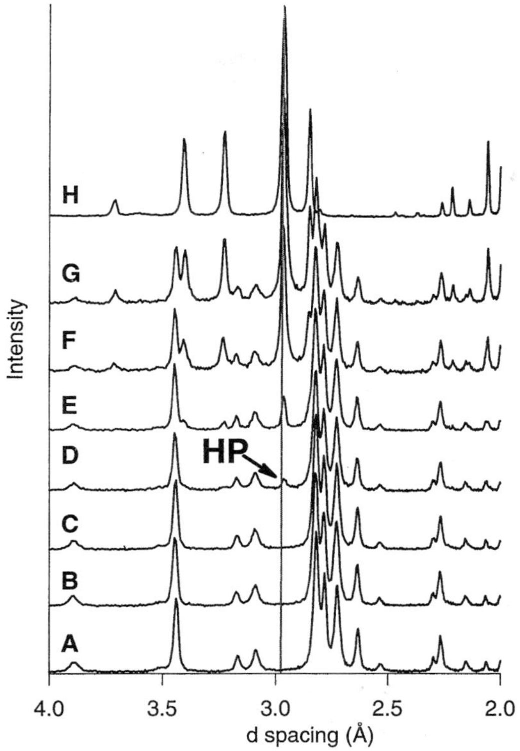

Figure 7 X-ray diffraction patterns of pure hydroxylapatite (HA) and HA reacted with $Pb(NO_3)_2$ at pH 5 for 2 hr: pure HA (A); sample SS0.00.7 (B); sample SS0.0013 (C); sample SS0.006 (D); sample SS0.013 (E); sample SS0.029 (F); sample SS0.058 (G); and pure hydroxypyromorphite (H).

>99.99% in less than 2 hr. For the solutions with an amount of Pb < 2.4 mmol·L^{-1}, the XRD patterns of the solids formed are similar to those obtained from pure HA. This is not surprising since the total amount of HP which could have formed in these samples would have amounted to <0.6 wt % of the total solid mass, and the nominal detection limit in powder XRD is approximately 2 to 3 wt %. A new peak at 2.96 Å was detected on sample SS0.006 when the initial concentration of dissolved Pb was 2.4 mmol·L^{-1} (Fig. 7D). Only when the initial concentration of dissolved Pb exceeded 12 mmol·L^{-1} were all of the HP peaks detectable in the XRD patterns. Hydroxylapatite was very easy to detect by XRD

Figure 6 SEM micrograph of hydroxypyromorphite synthesized at pH 12 (A) and HA reacted with $Pb(NO_3)_2$ at pH 5 for 2 hr (B).

when it was present at solid-phase concentrations ≥ 3 wt %. Sharp, narrow XRD peaks for HP indicated a high degree of crystallinity for these precipitates.

The FTIR spectra of the reaction products of aqueous Pb and HA are presented in Figure 8. All of the absorption bands of SS0.0013 and SS0.006 (B and C in Fig. 8) are identified as those of HA (spectrum A). On the spectra of SS0.029 and SS0.058, we observed two shoulders around 980 and 540 cm^{-1} (D and E) not present in the HA spectrum. The fitted curve of the IR spectra between 700 and 500 cm^{-1} (range of bending modes, ν_4) of HA, SS0.029, and HP (Fig. 9) showed that the shoulder around 540 cm^{-1} is due to the presence of HP absorption bands at 551 and 538 cm^{-1}. The intensity of the absorption band at 576 cm^{-1} increased due the presence of a HP absorption band. No new absorption bands were seen that would suggest the formation of mixed (Ca, Pb) apatites as solid solutions.

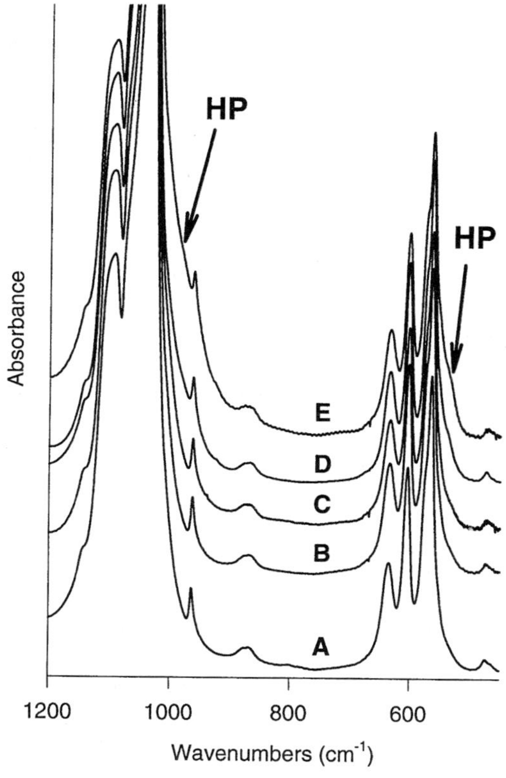

Figure 8 FTIR spectra of pure hydroxylapatite (HA) and HA reacted with $Pb(NO_3)_2$ at pH 5 for 2 hr: pure HA (A); sample SS0.0013 (B); sample SS0.006 (C); sample SS0.029 (D); and sampler SS0.058 (E).

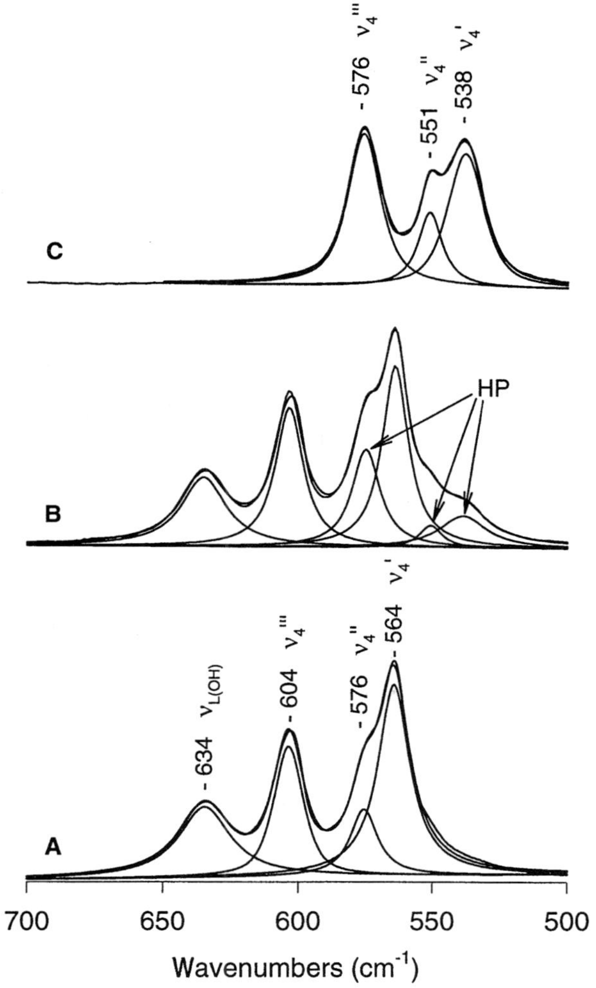

Figure 9 Curve-fitted FTIR spectra of pure HA (A), SS0.029 (B), and pure HP (C). Data presented for the spectral range of 700 to 500 cm^{-1}.

The SEM micrographs (Figs. 5B and 6B) show that HA crystals (flakes) are covered by needle-shaped HP crystals formed from reacted HA with aqueous Pb. Typically dimensions of HP needles are about 2.2 by 0.2 μm, but the size of the largest HP needles can reach 4.6 by 0.45 μm. No other solid-phase forms of Pb are visible in these micrographs.

C. Immobilization of Pb by Coprecipitation

The XRD patterns from the coprecipitation samples are presented in Figure 10. The diffraction lines of the apatite phase were highly broadened by the presence of lead in the structure. All spectra were fit (data not shown) to determine the exact position of the peaks, in particular, d_{211}, and these results were compared with the data of Bigi *et al.* (1989). Bigi *et al.* (1989) reported that Pb can substitute for

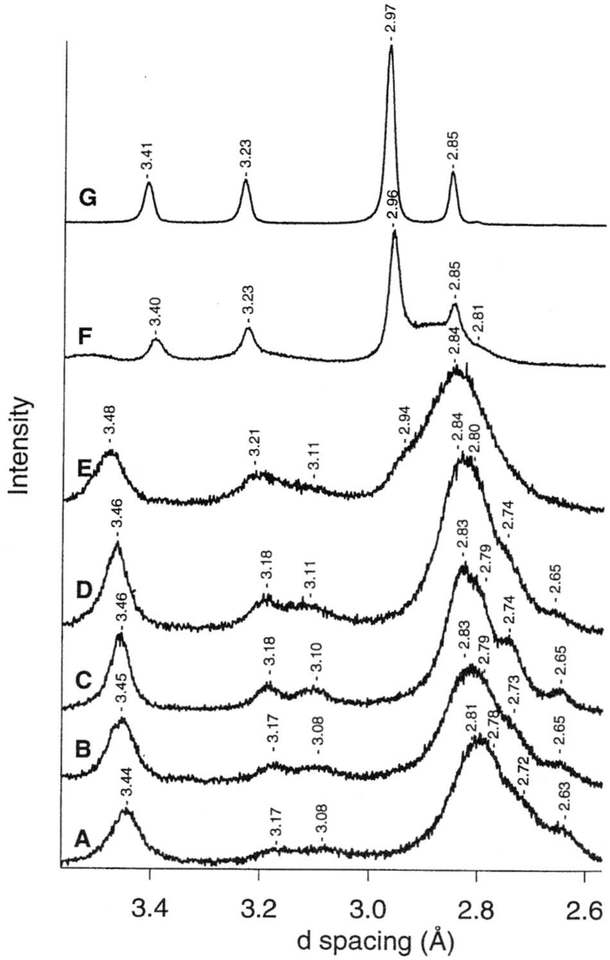

Figure 10 X-ray diffraction patterns of pure hydroxylapatite (A) and the coprecipitated samples CS0.03 (B), CS0.08 (C), CS0.12 (D), CS0.27 (E), CS0.64 (F), and CS1.00 (G).

Ca in hydroxylapatite, forming a continuous series of solid solutions in the whole range of compositions obtained. The peaks shift linearly with lead content. The similarity between the data from the present study and that reported by Bigi *et al.* (1989) suggests that our coprecipitation samples do indeed represent Pb–Ca solid solutions. As the Pb content increased (CS0.27 and CS0.64), some significant differences became apparent. Sample CS0.27 showed a peak at 2.94 Å (shoulder) corresponding to a pyromorphite phase and a peak at 2.84 Å corresponding to an apatite phase. Sample CS0.64 showed a peak at 2.96 Å corresponding to a pyromorphite phase and a peak at 2.81 Å (shoulder) corresponding to an apatite phase. Clearly both apatite ad pyromorphite were present in the samples. It could not be determined from the present XRD data and the total chemical analysis of the solids if these phases were pure or if trace quantities of Ca were present in the pyromorphite and trace quantities of Pb in the apatite.

The infrared absorption spectra of the coprecipitated samples (Fig. 11; only the ν_4 bending modes are presented) revealed a shift of the internal vibration modes of PO_4^{3-} groups toward lower frequencies as the Pb content increased, in agreement with the data of Andres-Verges *et al.* (1983). The intensity of the absorption band of Ca–OH at 635 cm^{-1} decreased as the Pb content increased. All these results indicated that the coprecipitated samples were similar to those prepared by other investigators and did indeed represent a Ca–Pb solid solution.

The SEM micrographs (Figs. 5A and 6A) show that the size of the apatite crystals (flakes) decreased as the Pb content increased. The mean flake size is 10 by 5 μm for HA (Fig. 5A) and 0.5 by 0.3 μm for CS1.00 (Fig. 6A).

D. Determination of the Process of Pb Immobilization at Low Pb Concentration

The FTIR spectra, the SEM micrographs, and the XRD data all showed that only HP was formed from the reaction of aqueous Pb with HA. The absence of other solid phases supports the hypothesis of Ma *et al.* (1993) that Pb immobilization was due to HA dissolution and HP precipitation. But these methods are not sensitive enough for low concentrations of Pb in solid phase, since >1 wt % of a solid is required for detection by XRD and >3 wt % is needed for FTIR.

To determine if the mechanism of immobilization of Pb was the same at low total and dissolved concentrations of Pb, we examined the two series of samples with EXAFS spectroscopy.

The hexagonal unit cell of HA and HP contains 10 cations (Ca or Pb) arranged on two nonequivalent sites (Fig. 12): 4 on site (1), and 6 on site (2). Cations in site (1) are surrounded by three O(1), three O(2), and three O(3) oxygen atoms; cations in site (2) are surrounded by one O(1), one O(2), and four O(3) oxygen atoms, and one (in the case of Ca) or two (in the case of Pb) OH. In the (Ca, Pb) solid solu-

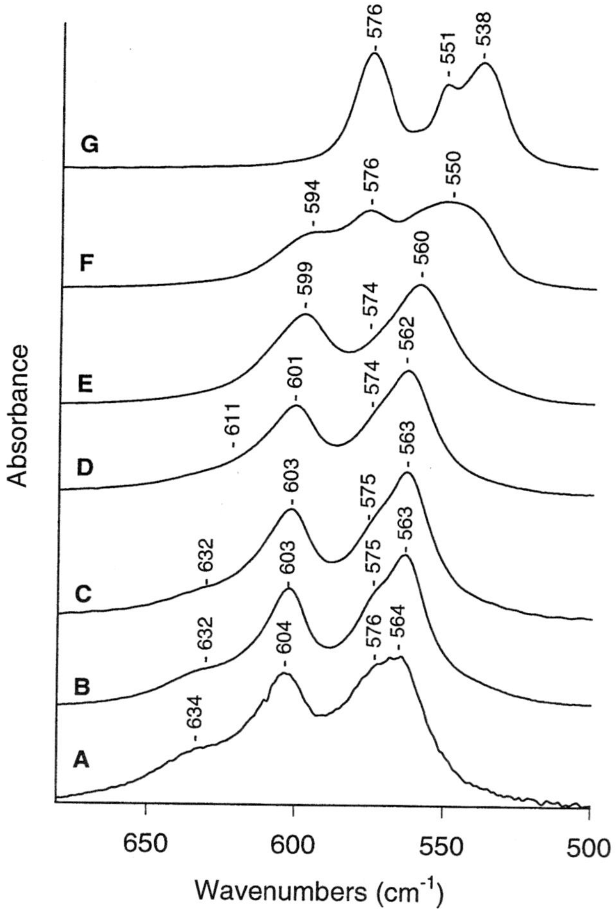

Figure 11 FTIR spectra of pure hydroxylapatite (A) and the coprecipitated CS0.03 (B), CS0.08 (C), CS0.12 (D), CS0.27 (E), CS0.64 (F), and CS1.00 (G).

tions, the site-occupancy factors of Pb atoms indicate a clear preference of Pb for site (2) of the apatite structure (Engel *et al.,* 1975; Andres-Verges *et al.,* 1983). When Pb substitutes for Ca in the apatite structure, it fills site (2), until the (Pb/(Ca+Pb)) ratio is higher than 0.45, after which it also begins to fill site (1) (Bigi *et al.,* 1989).

The content in samples SS0.00007, SS0.0007, and SS0.0013 was too low to allow detection of HP with either XRD or IR; thus these samples were examined with EXAFS in an attempt to ascertain the local coordination environment of Pb in the solid phase. Complete analysis of the EXAFS spectrum should facilitate distinction of Pb absorbed onto the surface of apatite particles, Pb surrounded by Ca atoms as second-shell neighbors (as in a solid solution where the fractional Pb content < 0 45), or Pb surrounded by Pb atoms as second-shell neighbors (as in a pure HP). The data obtained for samples SS0.00007 and SS0.0007 were too noisy (low signal/noise), and the analysis was done only on sample SS0.0013. For samples SS0.0013 and CS0.08, the solid-phase Pb content was <0.45, meaning that if a solid solution did form (instead of precipitation of a HP phase), the Pb atoms should have been exclusively in site (2).

The baseline-corrected, Fourier transforms of the EXAFS spectra from CS1.00, SS0.0013, and CS0.08 are presented in Figure 13, without phase corrections. Multiple peaks are present in the SS0.0013 spectra at k values > 3 Å, inconsistent with the presence of Pb on the surface of apatite as isolated, adsorbed ions. It is apparent that the radial structure functions (RSF) of CS1.00 (the pure pyromorphite) and SS0.0013 (spectra A and B) are similar and present three intense and narrow peaks at coincident positions. This is strong evidence that a pyromorphite-like phase has formed in sorption sample SS0.0013 that is detectable by EXAFS. It should be not-

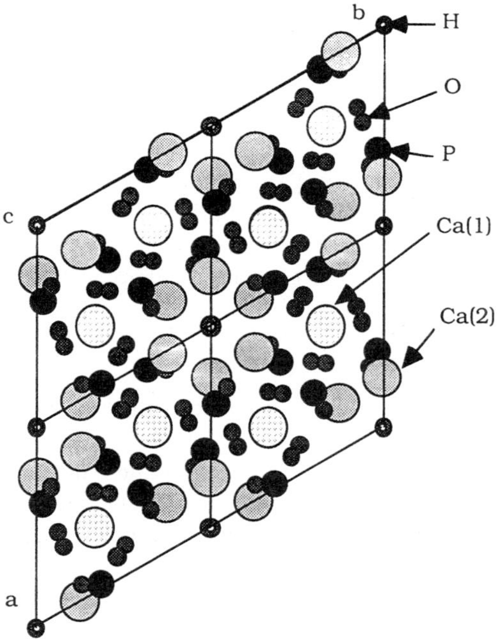

Figure 12 Projection of the hydroxylapatite (plan 001) structure along the c axis (Hughes *et al.*, 1989).

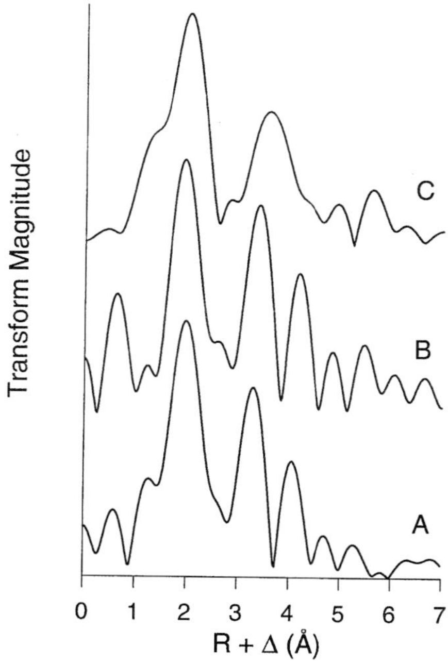

Figure 13 EXAFS radial structure functions (uncorrected) of CS1.00 (A), SS0.0013 (B), and CS0.08 (C).

ed that this phase is detectable by EXAFS below the detection limits of conventional powder XRD and FTIR. Our preliminary interpretation of the radial structure from CS1.00 assigned the first peak at 2 Å (uncorrected) to the Pb–O shell, the second peak at 3.6 Å to Pb(1)–Pb(1), and the third peak at 4 Å to Pb(1)–Pb(2) and Pb(2)–Pb(2). These positions, which are similar to those in SS0.0013, are indicative of pyromorphite, and inconsistent with the substitution of Pb into the structure of HA.

A very different RSF was obtained from the coprecipitation sample CS0.08. At this concentration of solid-phase Pb, all of the Pb would occupy site (2), as described by Bigi *et al.* (1989), if it were substituted into the apatite structure. The RSF from this sample only shows one broad peak in the second shell which is centered at 4 Å and would be consistent with a Pb(2)–Ca(2) and Pb(2)–Pb(2) distance if Pb were indeed substituted into the hydroxylapatite structure.

The present interpretation of the EXAFS spectra is clearly preliminary and must be reevaluated after careful examination of spectra from a number of model compounds along with theoretical EXAFS calculations and data fitting. Nevertheless, it is apparent from Figure 13 that the RSF of sample SS0.0013 is very similar to that obtained from pure pyromorphite, and quite different from that obtained from CS0.08. Apparently, attenuation of aqueous Pb by apatite results from the precipitation of pyromorphite for all ranges of Pb concentration examined in this study.

IV. CONCLUSION

Results of chemical, XRD, IR, SEM, and EXAFS analyses strongly support the mechanism of dissolution of HA and precipitation of HP at acidic pH subsequent to the reaction of apatites with aqueous Pb. Thus, it appears that reaction of dissolved Pb with apatites generally results in the dissolution of apatite and the subsequent precipitation of Pb phosphates rather than in the adsorption or absorption of Pb by apatite particles.

REFERENCES

Adler, H. H. 1968. Infrared spectra of phosphate minerals: Splitting and frequency shifts associated with substitution of PO_4^{3-} for AsO_4^{3-} in mimetite ($Pb_5(AsO_4)_3Cl_2$). *Am. Miner.* 53:1740–1744.

Andres-Verges, M., Higes-Rolando, F. J., Valenzuela-Calahorro, C., and Gonzales-Diaz, P. F. 1983. On the structure of calcium-lead phosphate apatite. *Spectrochim. Acta Part A* 39(12):1077–1082.

Baddiel, C. B., and Berry, E. E. 1966. Spectra structure correlations in hydroxy and fluorapatite. *Spectrochim. Acta* 22:1407–1416.

Bernas, B. 1968. A new method for decomposition and comprehensive analysis of silicates by atomic absorption spectrometry. *Anal. Chem.* 40:1682–1686.

Bigi, A., Foresti, E., Ripamonti, A., and Noveri, N. 1986. Fluoride and carbonate incorporation into hydroxyapatite under condition of cyclic pH variation. *J. Inorg. Biochem.* 27:31–39.

Bigi, A., Gandolfi, M., Gazzano, M., Ripamonti, A., Roveri, N., and Thomas, A. 1991. Structural modifications of hydroxyapatite induced by lead substitution for calcium. *J. Chem. Soc. Dalton Trans.*, 2883–2886.

Bigi, A., Ripamonti, A., Gazzano, M., Roveri, N., and Thomas, S. A. 1989. Structure refinement of lead substituted calcium hydroxyapatite by X-ray powder fitting. *Acta Crystallogr. B* 45(128):247–251.

Cant, N. W., Bett, J. A. S., Wilson, G. R., and Hall, W. K. 1971. The vibrational spectrum of hydroxyl groups in hydroxyapatites. *Spectrochim. Acta Part A* 27:425–439.

Chapman, A. C., and Thirlwell, L. E. 1964. Spectra of phosphorus compounds. I. The infra-red spectra of orthophosphates. *Spectrochim. Acta* 20:937–947.

Engel, G., Krieg, F., and Reif, G. 1975. Mischkristallbildung und kationenordnung im system bleihydroxylapatit-calciumhydroxylapatit. *J. Solid-State Chem.* 15:117–126.

Fowler, B. O. 1968. *In* Brown, W. E., and Young, R. A. (Eds.), "Proceedings of International Symposium on Structural Properties of Hydroxylapaptite and Related Compounds." Gaithersburg, MD, Chap. 7, pp. 8–9.

Gillette, P. C., Lando, J. B., and Koenig, J. L. 1982. Band shape analysis of Fourier transform infrared spectra. *Appl. Spectrosc.* 36:401–404.

Hughes, J. M., Cameron, M., and Crowley, K. D. 1989. Structural variations in natural F, OH and Cl apatites. *Am. Miner.* 74:870–876.

Klee, W. E. 1970. The vibrational spectra of the phosphate ions in fluorapatite. *Z. Kristallogr.* 131:95–102.

Laperche, V., Traina, S. J., Gaddam, P., and Logan, T. J. 1996. Chemical and mineralogical characterizations of Pb in a contaminated soil: Reactions with synthetic apatite. *Environ. Sci. Technol.* 30:3321–3326.

Le Geros, R. Z., Taheri, M. M., Quirolgico, G. B., and Le Geros, J. P. 1980. Formation and stability of apatites: Effects of some cationic substituents. "Proc. 2nd Int. Phosphorus, Boston," pp. 89–103. IMPHOS, Paris.

Levitt, S. R., and Condrate, R. A. 1970. The vibrational spectra of lead apatites. *Am. Miner.* 55:1562–1575.

Ma, Q. Y., Traina, S. J., Logan, T. J., and Ryan, J. A. 1993. In situ lead immobilization by apatites. *Environ. Sci. Technol.* 27:1803–1810.

Ma, Q. Y., Traina, S. J., Logan, T. J., and Ryan, J. A. 1994b. Effects of aqueous Al, Cd, Cu, Fe(II), Ni and Zn on Pb immobilization by hydroxyapatite. *Environ. Sci. Technol.* 28:1219–1228.

Ma, Q. Y., Logan, T. J., Traina, S. J., and Ryan, J. A. 1994a. Effects of NO_3^-, Cl^-, SO_4^{2-} and CO_3^{2-} on Pb immobilization by hydroxyapatite. *Environ. Sci. Technol.* 28:408–418.

Maddoms, W. F. 1980. The scope and limitations of curve fitting. *Appl. Spectrosc.* 34:245–267.

Nakamoto, K. 1986. "Infrared and Raman Spectra of Inorganic and Coordination Compounds," p. 483. John Wiley, New York.

Narasaraju, T. S. B., Singh, R. P., and Rao, V. L. N. 1972. A new method of preparation of solid solutions of calcium and lead hydroxyapatite. *J. Inorg. Nucl. Chem.* 34:2072–2074.

Stutman, J. M., Termine, J. D., and Posner, A. S. 1965. Vibrational spectra and structure of the phosphate ion in some calcium phosphates. *Trans. N. Y. Acad. Sci.* 27:669–675.

Teo, B. K. 1986. "EXAFS: Basic Principles and Data Analysis." Springer-Verlag, Berlin.

Xu, Y., and Schwartz, F. W. 1994. Lead immobilization by hydroxyapatite in aqueous solutions. *J. Contam. Hydrol.* 15:187–206.

Chapter 12

Effect of the Solid:Liquid Ratio on the Sorption of Sr^{2+} and Cs^{+} on Bentonite

Dennis W. Oscarson and Harold B. Hume
AECL, Whitehall Laboratories, Pinawa, Manitoba, Canada

Several factors can affect the magnitude of the distribution coefficient, K_d, for contaminants with earthen materials, but one of the most puzzling is the commonly observed decrease in K_d with an increase in the solid:liquid ratio or particle concentration, C_p—the so-called particle concentration effect. The main explanations advanced for the phenomenon are the presence of nonsettling particles in solution and the effects of particle–particle interactions. Here we examine the effect of C_p on K_d for Sr^{2+} and Cs^{+} with unconfined and compacted Avonlea bentonite. The solution phase was either a synthetic groundwater (a Na–Ca–Cl-dominated solution with an effective ionic strength of 0.22 $mol \cdot L^{-1}$) (Cs experiments) or 0.1 $mol \cdot L^{-1}$ NaCl (Sr experiments). The particle concentration ranged from 5.2 to 41.6 $g \cdot L^{-1}$ for Sr and from 2.4 to 18.6 $g \cdot L^{-1}$ for Cs. For both sorbates, K_d decreased with increasing C_p. Moreover, the effect was more pronounced for the compacted clay system. The difference observed between the unconfined and the compacted clay systems is likely due to the presence of more occluded pores or very small pores in compacted clay that the sorbates cannot enter; thus they cannot access as many of the sorption sites on the clay surface. The results indicate that the particle concentration effect is largely due to particle–particle interactions, or the blocking of some sorption sites, and not to the presence of nonsettling particles.

Adsorption of Metals by Geomedia

I. INTRODUCTION

The distribution or partition coefficient, K_d, measured at equilibrium, is a useful concept that expresses the relative affinity for a sorbate in solution to sorb to a particular solid. The parameter has been used extensively in models to predict the behavior of contaminants in the environment. The batch method is often used to determine this coefficient in the laboratory and the results extrapolated to field situations.

The K_d is defined as

$$K_d = \frac{\text{(mass sorbed/mass of solid)}}{\text{(mass in solution/volume of solution)}} \qquad [1]$$

or, alternatively,

$$K_d = \frac{\text{(mass sorbed)}}{\text{(mass in solution)}C_p}, \qquad [2]$$

where C_p is the particle concentration in mass per unit volume.

The magnitude of K_d is a function of the properties of the solid phase, such as mineralogical composition, particle size, and organic carbon content, and of the solution phase, like pH, ionic strength or salinity, and concentration of complexing ligands. The influence of these factors can generally be rationalized within the frameworks of solution and surface chemistry. However, another, more puzzling, factor that can affect K_d values is the solid:liquid ratio or C_p. In theory, at trace sorbate concentrations, K_d should not depend on C_p. Commonly, though, measured K_d values decrease with increasing C_p (Honeyman and Santschi, 1988; Meier *et al.,* 1987; Di Toro *et al.,* 1986; Gschwend and Wu, 1985; Voice *et al.,* 1983; O'Connor and Connolly, 1980)—the so-called particle concentration effect.

Two schools of thought have evolved over the interpretation of the particle concentration effect: (1) it is essentially an experimental artifact caused by colloids that are operationally defined as part of the dissolved fraction but are nonetheless sorbing solutes (an implicit assumption is that the concentration of the sorbing colloids in solution is proportional to C_p), and (2) the phenomenon is the consequence of physical-chemical processes such as particle–particle interactions that affect the equilibrium distribution of solutes.

Clearly, if K_d values determined by the batch method are to be used in models to predict the fate of contaminants in natural environments, the particle concentration effect must be clarified. Here we examine the effect of particle concentration on K_d for Sr and Cs with Avonlea bentonite in both an unconfined and a compacted state.

We are interested in bentonite and the K_d parameter because compacted bentonite has been proposed for use as a barrier material in the disposal of nuclear fuel waste in many countries, and the K_d is a common component of mass transport models used to predict the migration of trace levels of radionuclides in these compacted bentonitic barriers (Johnson *et al.,* 1994).

II. MATERIALS AND METHODS

A. Clay

Avonlea bentonite from the Bearpaw Formation of Upper Cretaceous age in southern Saskatchewan, Canada, was used. It contains about 80 wt % dioctahedral smectite, 10 wt % illite, 5 wt % quartz, and minor amounts of gypsum, feldspar, and carbonate (Oscarson and Dixon, 1989). The clay has a cation exchange capacity of 60 $cmol \cdot kg^{-1}$ and a specific surface area of $480 \cdot 10^3$ $m^2 \cdot kg^{-1}$, and Na^+ is the main exchangeable cation.

The Avonlea bentonite contains about 0.2 $g \cdot kg^{-1}$ of Sr (Oscarson and Dixon, 1989). If soluble, this could lead to an increase in Sr concentration in the experimental systems to a level where precipitation as well as sorption reactions occur. [The likely solid Sr phase to form in these systems is strontianite, $SrCO_3$ (Benson and Teague, 1980).] Because of this native Sr, for the Sr experiments the clay was washed several times with a 1 $mol \cdot L^{-1}$ solution of NaCl. Afterward the clay was dialyzed against deionized water for several days to remove excess salt, and then freeze-dried. Subsequent analysis showed that this treatment removed about half of the Sr in the clay. After this treatment, the clay's exchange complex is, of course, also saturated with Na^+.

B. Solutions

A synthetic groundwater (SGW) solution was used in the Cs/bentonite experiments; it has an effective ionic strength of 0.22 $mol \cdot L^{-1}$, and the following composition (in $mmol \cdot L^{-1}$): Na, 83; K, 0.36; Mg, 2.5; Ca, 53; Cl, 170; and SO_4, 12. It was chosen to represent groundwater found deep in granitic rock in the Canadian Shield where a nuclear fuel waste disposal vault could be sited (Johnson *et al.,* 1994).

The SGW solution contains some Sr, derived from impurities in some of the salts used in its preparation; this source of Sr could also affect the K_d measurement as already discussed. Consequently, a 0.1 $mol \cdot L^{-1}$ solution of high-purity NaCl was substituted for the SGW solution in the Sr/Na–bentonite experiments.

The salinity of the solution phase in both the Sr and the Cs experiments was such that the ionic strength and pH—9.0 for the Sr and 7.8 for the Cs experiments—were effectively constant throughout the reaction period. Thus, during the course of the experiments, the sorption reactions were not influenced by changes in the solution phase.

C. Sorption Experiments

1. Unconfined Clay

Bentonite was suspended in 0.1 L of a 0.1 mol·L^{-1} NaCl solution containing ^{85}Sr, or 0.5 L of the SGW solution containing ^{137}Cs. The initial ^{85}Sr concentration was about 0.1 nmol·L^{-1}, and that of ^{137}Cs, 1 nmol·L^{-1}; the total initial Sr concentration was, however, about 2.1 μmol·L^{-1} because of the soluble native Sr present in the clay and the NaCl solution. The clay or particle concentration ranged from 5.2 to 41.6 g·L^{-1} for Sr and from 2.4 to 18.6 g·L^{-1} for Cs. All experiments were conducted in triplicate in polyethylene containers. The suspensions were gently shaken periodically. Every few days, a 1-ml aliquot of solution was carefully withdrawn without disturbing the sedimented clay and analyzed for ^{85}Sr or ^{137}Cs by liquid scintillation counting. Sorption equilibrium was assumed when the sorbate solution activity was constant with time; this occurred in less than a week for both sorbates.

2. Compacted Clay

For comparison with unconfined clays, sorption experiments were conducted with ^{137}Cs and compacted bentonite using specially designed cells (Fig. 1). A hydraulic press was used to statically compact the clay in the stainless-steel rings of the sorption cells to target dry densities, ρ, ranging from 0.5 to 1.5 Mg·m^{-3}; in the sorption experiments these densities gave particle concentrations that cover the same range as those used in the Cs/unconfined clay system. The rings containing the compacted clays were placed in the sorption cells and the cells immersed in 0.5 L of SGW solution in polycarbonate containers. When the cells are immersed, Cs diffuses through the porous disks of the cell and into the pores of the clay, where some of it is sorbed; hence its concentration in the contacting solution decreases. The activity of ^{137}Cs in solution was monitored periodically by liquid scintillation counting. The experiment was continued until the activity of ^{137}Cs in solution became constant with time, indicating that sorption equilibrium had been established. The reaction period depended on the clay density—the greater the density, the longer the reaction period—but was generally about 600 days. The uniform distribution of ^{137}Cs within the clay plugs after the reaction periods also indicated that equilibrium was achieved in these systems (Oscarson *et al.,* 1994).

For both the compacted and the unconfined clay systems, control experiments, conducted identically but without clay, showed that no detectable amounts of Sr or Cs sorbed on the walls of the polyethylene or polycarbonate containers, nor on

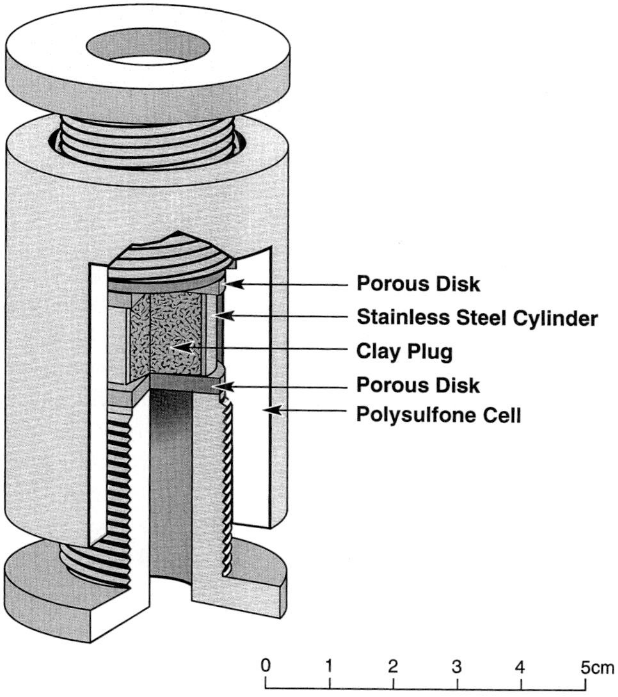

Figure 1 Diagram of the sorption cell.

any of the components of the sorption cell. All experiments were conducted at 22 ± 2°C.

D. Distribution Coefficients

Distribution coefficients were calculated from

$$K_d = \frac{A_i V_i - A_e V_e}{A_e V_e C_p}, \qquad [3]$$

where A_i is the net (total activity minus background activity) initial activity and A_c is the net equilibrium activity of ^{85}Sr or ^{137}Cs in solution [the activities were corrected for the radioactive decay of ^{85}Sr ($t_{1/2}$ = 64 days) and ^{137}Cs ($t_{1/2}$ = 30 yr) that occurred during the experiments], V_i is the initial volume of solution, V_e is the solution volume when sorption equilibrium was established, and C_p is the particle concentration at sorption equilibrium.

E. Pore Parameters of Compacted Bentonite

Pore parameters of compacted bentonite ($\rho = 1.3\ Mg \cdot m^{-3}$) were measured to provide supporting evidence that colloidal material does not move in these compacted clays, and hence there should be little or no colloidal material in the solution phase in the experiments with compacted clay. Therefore, if the particle concentration effect was observed in these systems, it could not be attributed to suspended colloids.

The pore parameters were obtained as follows. A clay plug was sandwiched between porous disks and constrained in a specially designed stainless-steel holder. The plug was saturated by immersing it in the SGW solution for 4 weeks; it was then slowly dried—to minimize disruption of the pore structure—in chambers having progressively drier atmospheres. The relative humidities (over H_2SO_4 solutions) in the chambers were 81, 58, and 3.2%. The plug was kept at a given humidity for at least two weeks. The clay was then examined by Hg intrusion porosimetry to 414 MPa. At the maximum intrusion pressure of 414 MPa, Hg can theoretically enter pores with a diameter as small as 3.6 nm. Before the measurement, the sample was outgassed at 6.5 Pa at room temperature.

III. RESULTS AND DISCUSSION

The K_d values for Sr and Cs obtained at various C_p are given in Table I. The greater values for Cs than for Sr are attributed, at least in part, to minor amounts

Table I

Distribution Coefficients, K_d, for Sr and Cs with Unconfined Avonlea Bentonite

Sr		Cs	
C_p ($kg \cdot L^{-1} \cdot 10^3$)	K_d ($L \cdot kg^{-1}$)	C_p ($kg \cdot L^{-1} \cdot 10^3$)	K_d ($L \cdot kg^{-1}$)
5.2	83.7 ± 9.60[a]	2.4	1840 ± 56
8.1	73.7 ± 1.08	3.5	1700 ± 14
10.4	71.1 ± 2.12	11.2	1390 ± 17
11.1	70.3 ± 0.64	16.7	1290 ± 31
15.8	65.4 ± 2.52	18.6	1200 ± 15
19.6	64.8 ± 1.50		
20.8	63.9 ± 1.40		
21.9	63.1 ± 1.40		
41.6	54.9 ± 1.93		

[a]Arithmetic mean ± one standard deviation of three replicates.

of illitic material present in Avonlea bentonite—about 10 wt % (Oscarson and Dixon, 1989): illite has a unique affinity for certain alkali metals, including Cs^{+} (Sawhney, 1972). Also evident in Table I, consistent with the observations reported by many others, is a decrease in K_d with increasing C_p—the particle concentration effect. Figure 2 shows log K_d–log C_p plots of the data. O'Connor and Connolly (1980) noted that the particle concentration effect is generally more pronounced for sorbates with higher K_d values. This is not the case for our data, despite the fact that K_d is more than an order of magnitude greater for Cs; the slope of the log–log plot for both Sr and Cs is −0.19.

As noted, the particle concentration effect has been largely attributed to either nonsettling particles or to particle–particle interactions. The compacted clay experiment was designed, in part, to test these hypotheses.

A log K_d–log C_p plot for Cs with unconfined and compacted bentonite is shown in Figure 3. About one-third less Cs is sorbed on compacted versus un-

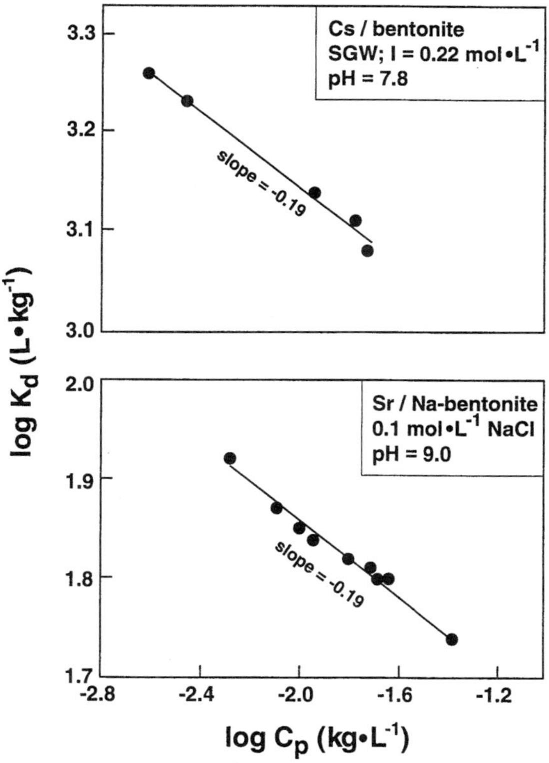

Figure 2 K_d versus C_p for Sr and Cs with unconfined bentonite. I, effective ionic strength.

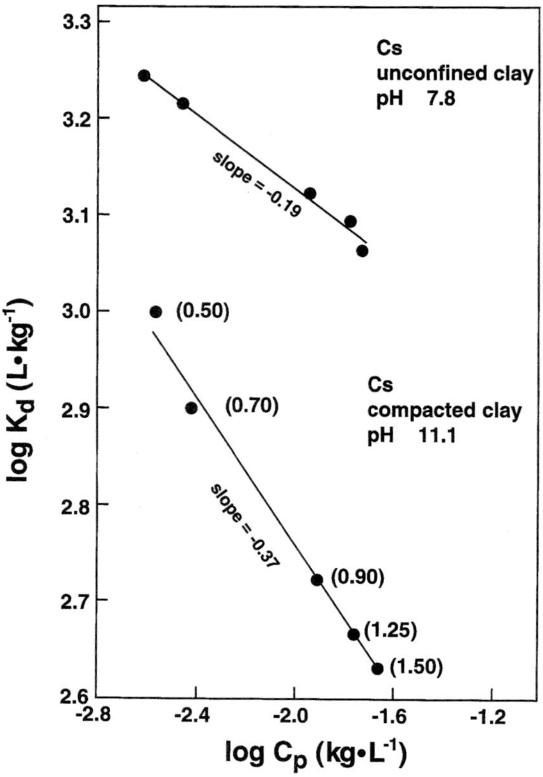

Figure 3 K_d versus C_p for Cs with unconfined and compacted bentonite. The values in parentheses for the compacted clay are the clay dry densities in $Mg \cdot m^{-3}$.

confined bentonite, and the slope of the plot for Cs with compacted clay is significantly greater than that obtained with unconfined clay. Hence, the particle concentration effect is more pronounced for the compacted clay system. These data are confounded to some extent by the fact that the solution pH in the compacted clay system (pH = 11.1) is greater than that of the unconfined clay (pH = 7.8). The K_d for Cs with bentonite, however, increases with increasing pH (Fig. 4). Therefore, the higher pH for the compacted clay system cannot explain the lower K_d measured in this system (Fig. 3). [We have determined that the relatively high pH in the compacted clay system is due to the porous Ni disk used in the sorption cell. Because of the effect of the porous disk on pH, we have not yet determined the sorption of Sr on compacted clay: at this high pH, the concentration of Sr in solution would exceed the solubility limit of $SrCO_3(s)$ (Benson and Teague, 1980). Before testing other sorbates with compacted clay, other porous materials that do not significantly alter the solution chemistry need to be identified for use in the sorption cell.]

It is likely that Cs cannot access as many of the sorption sites on the compacted clay as it can on the unconfined clay. This exclusion may be due, at least in part, to the presence of more extremely small pores, occluded pores, or "ink bottle" pores in compacted clay that the Cs ions cannot enter—and the higher the density, the greater the number of inaccessible pores.

The particle concentration effect observed for the compacted clay system (Fig. 3) is not due to nonsettling colloids. Based on the following arguments, there should be few, if any, colloidal particles in the solution phase because such colloids cannot move from the compacted clay into the solution.

The migration of colloidal particles in compacted bentonite is extremely slow, if they can move at all. To our knowledge, no reliable measurements of diffusion coefficients for colloidal material in compacted bentonite have been obtained, although attempts have been made. In a study of colloid and radionuclide transport in compacted bentonite ($\rho = 0.9\ Mg \cdot m^{-3}$), for example, Nowak (1984) reported that colloidal gold (mean particle diameter = 16.4 nm) was completely excluded from compacted bentonite. Moreover, Eriksen and Jacobsson (1982) found no diffusion of large organic molecules (molecular weight ranged from 350 to 30,000 $g \cdot mol^{-1}$) in compacted bentonite, and concluded that diffusion coefficients for these solutes must be $<10^{-15}\ m^2 \cdot sec^{-1}$. (Because of size similarities, diffusion measurements of large organic molecules in compacted bentonite are relevant to colloid transport.) With a diffusion coefficient of $<10^{-15}\ m^2 \cdot sec^{-1}$, a colloidal particle could only migrate <0.2 mm in compacted bentonite within the time of the Cs sorption experiments with compacted clay, about 600 days; this distance is

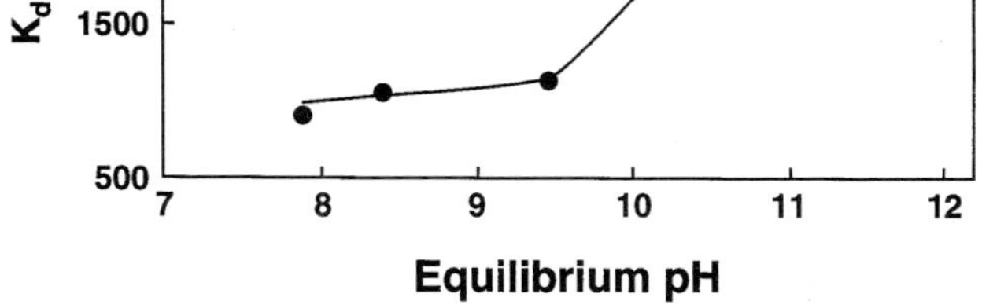

Figure 4 K_d versus equilibrium pH for Cs with unconfined bentonite.

<2% of the total plug length. Therefore, only colloidal material at the interface of the clay and porous disk could possibly migrate into the solution phase. Technical data obtained from the supplier of the Ni disks used in the sorption cells indicate that their average pore diameter is 25 μm. Thus colloidal transport through even the porous disks would be very slow.

This immobility of colloids in compacted bentonite is supported by the measurement of the pore size distribution of compacted bentonite (Fig. 5). The diameter of nearly all pores is close to 0.01 μm, and there are few, if any, larger pores. The associated pore parameters obtained from Hg intrusion porosimetry are total Hg intrusion volume, 141 $cm^3 \cdot kg^{-1}$; median pore diameter, 14.1 nm; and average pore diameter, 11.6 nm. These values are close to those obtained for compacted Na- and Ca-bentonite, and the pore size distribution shown in Figure 5 is essentially the same as that of Na-bentonite compacted to a similar density (Choi and Oscarson, 1996).

In the compacted clay system, therefore, there are likely few, if any, colloidal particles in the solution phase. Even if some are present, they are almost certainly at a much lower concentration than in the unconfined clay systems, yet the particle concentration effect is more pronounced for the compacted clay system. We conclude, therefore, that the particle concentration effect observed in Fig. 3 is largely due to particle interactions, through the blocking of some sorption sites, rather than nonsettling colloids in the solution phase.

Di Toro *et al.* (1986) also concluded that particle interactions were the main reason for the particle concentration effect they observed in a study of the sorption of heavy metals on clay and quartz. Their experiments also virtually eliminated the possibility of nonsettling colloids as being responsible for the effect. On the other hand, Gschwend and Wu (1985) and Voice *et al.* (1983) present convincing evidence that nonsettling (or nonfilterable) colloids were the cause of the particle con-

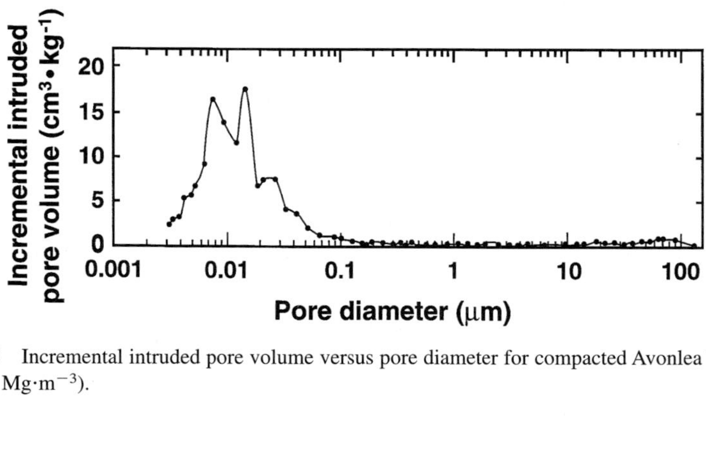

Figure 5 Incremental intruded pore volume versus pore diameter for compacted Avonlea bentonite ($\rho = 1.3\ Mg \cdot m^{-3}$).

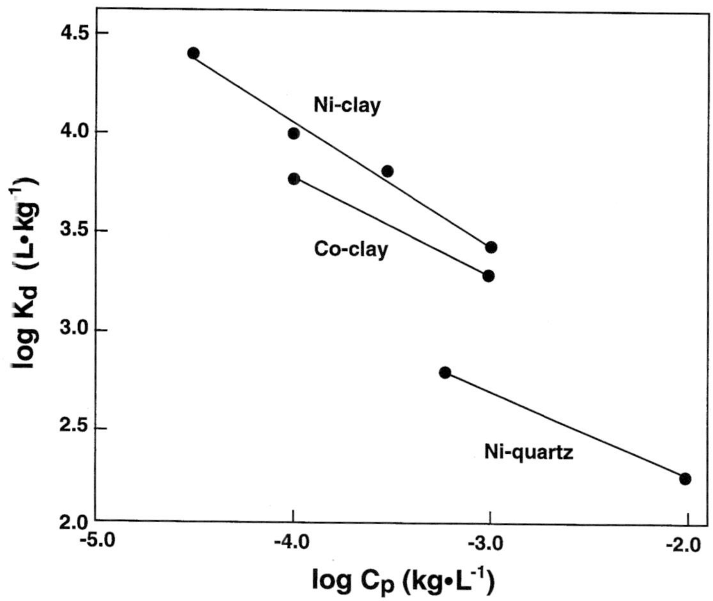

Figure 6 K_d versus C_p for Ni and Co with montmorillonite and quartz at constant equilibrium metal concentration (40 $\mu g \cdot L^{-1}$) and pH 9 (data adapted from Di Toro *et al.*, 1986).

centration effect in studies of the sorption of hydrophobic organic compounds on sediments.

McKinley and Jenne (1991) reported that K_d for Cd with Fe oxide was independent of the particle concentration. They concluded that the particle concentration effect is caused in some cases by inappropriate experimental design or errors of data interpretation. They state, "If all three variables (meaning C_p, and the mass sorbed and the mass in solution at equilibrium) are computationally unconstrained, the value of K_d is, likewise, not constrained to a meaningful value. If C_p is to be treated as a variable, then K_d has a valid definition only at constant $[M^{2+}]$ (p. 2086)." (In their notation, $[M^{2+}]$ is the equilibrium sorbate concentration.) We cannot test this assertion from our data as sorption isotherms at different C_p are not available. Di Toro *et al.* (1986), however, present sorption isotherms for Ni and Co with montmorillonite and quartz at different C_p (Fig. 3 in their paper). From these data, we calculated K_d values at a constant equilibrium sorbate concentration for the C_p values used by these authors; the log K_d–log C_p plots are shown in Figure 6. A similar plot can be constructed from the data on Kepone sorption on sediments reported by O'Connor and Connolly (1980). Clearly, even at a constant equilibrium sorbate concentration, the particle concentration effect is evident in these systems. At our present level of understanding, it is not clear why the particle concentration effect should be observed in some systems, but apparently not in others.

IV. SUMMARY AND CONCLUSIONS

A decrease in K_d with an increase in C_p—the particle concentration effect—is commonly observed for a variety of sorbate–sorbent systems, and our data for Sr and Cs sorption on both unconfined and compacted bentonite also exhibit this phenomenon. Although extensively studied, the reason for the particle concentration effect remains obscure. Main explanations involve nonsettling colloids and particle–particle interactions. Our data for Sr and Cs sorption on both unconfined and compacted bentonite suggest that particle–particle interactions play a predominant role.

Despite the particle concentration effect, K_d can be a useful parameter in models used to predict contaminant behavior at trace levels ($<\mu M$) in natural environments. However, K_d values should be measured under conditions as close as possible to those in the environment of interest; these conditions include particle concentration, solution composition, and sorbate concentration.

ACKNOWLEDGMENTS

The Canadian Nuclear Fuel Waste Management Program is funded jointly by AECL and Ontario Hydro under the auspices of the CANDU Owners Group.

REFERENCES

Benson, L. V., and Teague, L. S. 1980. A tabulation of thermodynamic data for chemical reactions involving 58 elements common to radioactive waste package systems, LBL-11448. Lawrence Berkeley Laboratory, Livermore, CA.

Choi, J.-W., and Oscarson, D. W. 1996. Diffusive transport through compacted Na- and Ca-bentonite. *J. Cont. Hydrol.* 22:189–202.

Di Toro, D. M., Mahony, J. D., Kirchgraber, P. R., O'Byrne, A. L., Pasquale, L. R., and Piccirilli, D. C. 1986. Effects of nonreversibility, particle concentration, and ionic strength on heavy metal sorption. *Environ. Sci. Technol.* 20:55–61.

Eriksen, T. E., and Jacobsson, A. 1982. Diffusion of hydrogen, hydrogen sulfide, and large molecular weight anions in bentonite. Swedish Nuclear Fuel Supply Company. SKBF-KBS-TR-82-17 Stockholm, Sweden.

Gschwend, P. M., and Wu, S. 1985. On the constancy of sediment-water partition coefficients of hydrophobic organic pollutants. *Environ. Sci. Technol.* 19:90–96.

Honeyman, B. D., and Santschi, P. H. 1988. Metals in aquatic systems. *Environ. Sci. Technol.* 22:862–871.

Johnson, L. H., LeNeveu, D. M., Shoesmith, D. W., Oscarson, D. W., Gray, M. N., Lemire, R. J., and Garisto, N. C. 1994. The disposal of Canada's nuclear fuel waste: The vault model for postclosure assessment. Atomic Energy of Canada, Ltd., AECL-10714, COG-93-4. Pinawa, Manitoba.

McKinley, J. P., and Jenne, E. A. 1991. Experimental investigation and review of the "solids concentration" effect in adsorption studies. *Environ. Sci. Technol.* 25:2082–2087.

Meier, H., Zimmerhackl, E., Zeitler, G., Menge, P., and Hecker, W. 1987. Influence of liquid/solid ratios in radionuclide migration studies. *J. Radioanal. Nucl. Chem.* 109:139–151.

Nowak, E. J. 1984. Diffusion of colloids and other waste species in brine-saturated backfill materials. *In* "Symposium Proceedings on the Scientific Basis for Nuclear Waste Management" (G. L. McVay, Ed.), Vol. 26, pp. 59–68. Materials Research Society, Pittsburgh, Pennsylvania.

O'Connor, D. J., and Connolly, J. P. 1980. The effect of concentration of adsorbing solids on the partition coefficient. *Water Res.* 14:1517–1523.

Oscarson, D. W., and Dixon, D. A. 1989. Elemental, mineralogical, and pore-solution composition of selected Canadian clays. Atomic Energy of Canada, Ltd., AECL-9891. Pinawa, Manitoba.

Oscarson, D. W., Hume, H. B., and King, F. 1994. Sorption of cesium on compacted bentonite. *Clays Clay Miner.* 42:731–736.

Sawhney, B. L. 1972. Selective sorption and fixation of cations by clay minerals: A review. *Clays Clay Miner.* 20:93–100.

Voice, T. C., Rice, C. P., and Weber, W. J. 1983. Effect of solids concentration on the sorptive partitioning of hydrophobic pollutants in aquatic systems. *Environ. Sci. Technol.* 17:513–518.

Chapter 13

Adsorption of U^{VI} and Citric Acid on Goethite, Gibbsite, and Kaolinite

Comparing Results for Binary and Ternary Systems

George D. Redden, Jinhe Li, and James Leckie
Department of Civil and Environmental Engineering, Stanford University, Stanford, California

Liquid radioactive wastes containing heavy metals and transuranic elements are often associated with metal-complexing organic cocontaminants. Predicting radionuclide transport in contaminated groundwater will require understanding how these ligands, and naturally occurring compounds, affect the partitioning of metals between the solution phase and the solid surfaces.

A study has been conducted on the adsorption of uranyl ion (UO_2^{2+}) on goethite, gibbsite, and kaolinite in the presence of varying concentrations of citric acid. Adsorption of either solute separately, as a function of pH, is generally typical of the interaction between cations or anions and protolyzable metal oxides. Differences in adsorption behavior reflect differences in the properties of surface functional groups of the solid phases.

At equimolar concentrations of UO_2^{2+} and citrate, the sorption of each solute was similar to behavior in the single sorbate systems. However, at higher citrate concentrations, UO_2^{2+} sorption on goethite dramatically increases, with nearly complete UO_2^{2+} sorption occurring even at relatively low pH values. In the case of kaolinite, citric acid significantly decreases UO_2^{2+} sorption while little effect was observed for

Adsorption of Metals by Geomedia

gibbsite. These results for citrate are similar to the results from experiments performed previously involving UO_2^{2+}, hematite, and EDTA as the complexing ligand. A stable bridging structure involving UO_2^{2+} and citrate is implied for the iron phase whereas solution complexes appear to be favored in the kaolinite system. These results have significant implications for modeling radionuclide transport where complexing cocontaminants or cosolutes may increase or decrease sorption of UO_2^{2+} depending on the solid phases present and the concentrations of the solutes.

I. BACKGROUND

Adsorption of metal cations or anions on mineral oxides has been extensively studied over the years in an attempt to develop chemical models that will simulate solute partitioning. Combined with models of groundwater dynamics, it should be possible to improve the predictions of contaminant transport in groundwater. Work has gradually progressed in two directions: (1) increasingly detailed studies of the nature of solute–solution interfaces and the molecular-scale mechanisms by which solutes bind to surfaces, and (2) the empirical simulation of more complex systems that are better representations of real field conditions. An example of an intermediate level of complexity is in the attempts to understand the influence of synthetic and natural organic ligands, especially multifunctional chelating compounds, on the adsorption of heavy metals on different mineral surfaces. The following summarizes the experimental phase of an investigation of the influence of citrate on U^{VI} sorption to goethite (α-FeOOH), gibbsite (γ-Al(OH)$_3$), and kaolinite.

Organic metal-complexing compounds of synthetic origin (e.g., EDTA, NTA) are often present in hazardous complex waste mixtures (Swisher *et al.,* 1974). An important example are the wastes that have been generated in conjunction with nuclear weapons and nuclear fuel production programs. Review of the known inventory of chemical wastes at DOE facilities reveals wastes that are mixtures of radionuclides, heavy metals, fuels, oils, organic solvents, and a variety of organic metal-complexing agents (Riley and Zachara, 1992). These wastes are a serious environmental risk due to the large inventory, the complexity of the mixtures, the chemical reactions taking place, and the long half-lives of the radioactive components. In many cases, these wastes have already contaminated soils and groundwater at storage and disposal sites (Killey *et al.,* 1984; Riley and Zachara, 1992). Naturally occurring organic metal-complexing compounds are also common (Morel and Hering, 1993), and the multiple interactions between radionuclides, organic ligands, and mineral surfaces will influence the mobility of metal contaminants. This will have a direct impact on the accuracy of predictive transport models needed to evaluate long-term risk of human exposure, and on the effectiveness of remediation strategies.

Several studies have illustrated how soluble organic ligands have increased mobility of metal ions and radionuclides in groundwater. Killey *et al.* (1984) found Co-60 to be rapidly leached from a waste disposal infiltration pit at the Chalk River Nuclear Laboratories. The largest fraction of mobilized cobalt apparently was present as a hydrophilic organic complex. Since tartaric acid and oxalic acid had been added to the waste pit prior to an observed increase in beta activity in the groundwater, the implication is that formation of organic cobalt complexes was responsible for a dramatic increase in the rate of leaching from the pit. Elliott and Denneny (1982) observed similar behavior for cadmium in soil columns in the presence of a suite of organic ligands that included oxalate, EDTA, and NTA. Reduction of cadmium adsorption (increased mobility) was directly related to the number of coordinating functional groups available from the organic ligands, with EDTA having the largest effect. In this study an important point made was that Al^{III} and Fe^{III} (i.e., products of mineral dissolution) also formed stable complexes with the organic ligands and were capable of displacing cadmium from the organic ligands.

Although these studies and others (Means *et al.,* 1978; Girvin *et al.,* 1993; Fletcher and Beckett, 1987; Tam and McColl, 1991) have demonstrated the ability of metal–organic complexes to solubilize and mobilize metal ions, organic ligands can also enhance sorption of metals on mineral surfaces (Bowers and Huang, 1986; Schindler and Stumm, 1987; Girvin *et al.,* 1993). A study by Zachara *et al.* (1995) demonstrated the ability of EDTA to enhance the adsorption of Co-60 to goethite in the pH range where cobalt, in the absence of EDTA, alone does not appreciably adsorb. Above this pH, Co adsorption was decreased in the presence of EDTA. While the decrease in Co adsorption could be explained and modeled using a small set of stoichiometric reactions including the Co–EDTA solution complex, an explanation for enhanced sorption (in the lower pH range) was more difficult to simulate. Competition by Fe^{III} and Al^{III} from dissolution of the solids, and the Ca^{2+} added as the electrolyte, was found to be significant. The kinetics of goethite dissolution and ligand exchange was correlated with the extent and time dependence of cobalt adsorption. For a range of soil samples, the role of the mineral phase properties was evident. The stability of a Co–EDTA surface complex decreased in the general order iron oxide > aluminum oxide, clay > organic coatings. Iron oxides and hydroxides are common as discrete minerals and surficial coatings and are therefore considered to play a significant role in ionic solute transport (Schwertmann, 1988; Schindler and Stumm, 1987).

The purpose of the present study was to determine the influence of citric acid on the adsorption of UO_2^{2+} on representative mineral surfaces. Citric acid occurs naturally in soils and groundwater as a result of microbial activity (Cambier and Sposito, 1991), is used as a cleaning agent in industrial applications, and is found as a component of radioactive waste mixtures (Riley and Zachara, 1992). Goethite is a common constituent of groundwater systems and is often present as coatings on mineral surfaces (Zachara *et al.*, 1995). As a result, the role of goethite in solute partitioning can greatly exceed its relative mass abundance in aquifer material. Kaolinite, a simple 1:1 clay mineral with relatively low permanent charge, is an end-member mineral with properties common to other clay minerals. Isomorphic substitution of Al for Si in the siloxane layer results in a finite negative structural charge (Schindler and Stumm, 1987). Gibbsite is a stable aluminum hydroxide mineral phase resulting from weathering reactions at low temperatures, and has structural similarities to the Al sites of kaolinite. In addition, dissolution of kaolinite under acidic conditions (pH 3 to 6) may be followed by adsorption and precipitation of Al^{III} resulting in a gibbsite phase on the kaolinite substrate (Wieland and Stumm, 1992).

Systems involving these sets of components have not yet been studied, but the results from analogous systems are beginning to yield a degree of consistency. Numerical values for adsorption model parameters will eventually provide input for predictive transport models and aid in the development or optimization of remediation strategies such as "soil washing." Only the experimental results are presented here.

II. EXPERIMENTAL SETUP

A. Materials and Methods

1. Goethite

Goethite was prepared using a slightly modified version of the method described by Atkinson *et al.* (1967). Reagent-grade ferric nitrate ($Fe(NO_3)_3 \cdot 9H_2O$) was dissolved in low-CO_2 water in a polypropylene container under an argon atmosphere. Millipore Milli-Q system water was used throughout the study. Low-carbonate 5 $mol \cdot L^{-1}$ NaOH was used to adjust the pH to around 12. The solution was heated at 60°C for 24 hr. The precipitate was placed in trace-metal-free dialysis tubing (Spectra por 7) and dialyzed against Milli-Q water. The water was changed once or twice a day until the measured conductivity matched that of fresh Milli-Q water (1 to 2 μS). A stock slurry of goethite was stored under an argon atmosphere at 4°C. A working slurry of goethite (29.0 $g \cdot L^{-1}$) was prepared by diluting the stock slurry. An XRD spectrum confirmed the synthesized material as goethite. The specific surface area was measured by the BET nitrogen adsorption method on the dried powdered goethite.

2. Gibbsite

Gibbsite was obtained from ALCOA Chemicals (Pittsburgh, PA). Chemical analysis revealed a high purity of the bulk phase with minor impurities of Fe (less than 0.006% was Fe_2O_3), Si (less than 0.07% as SiO_2), and Na (less than 0.42% as Na_2O) (Wieland *et al.*, 1993). No specific pretreatment was carried out prior to adsorption studies. The surface area was determined to be 11.2 $m^2 \cdot g^{-1}$ using the BET procedure (Wieland *et al.*, 1993). The bulk crystalling structure was confirmed by XRD analysis.

3. Kaolinite

Kaolinite (KGa-1) was provided by the Clay Minerals Society (Columbia, MO). For pretreatment, 60 g of kaolinite was suspended in 4 L of Milli-Q water and stirred for 1 hr. The pH was adjusted to 9.5 using 1.0 $mol \cdot L^{-1}$ carbonate-free NaOH. The suspension was ultrasonicated at 300 watts for 15 min to detach adhered fine particles, stirred, and allowed to stand for 8 min to obtain a suspension with <20 μm particle size (procedure developed by J. M. Zachara, personal communication). The suspended portion was separated and centrifuged at 110 RCF for 5 min using an IEC DPR-6000 rotor (Needham, MA). The supernatant was retained, and the pellet was resuspended and recentrifuged. All the supernatants were collected and centrifuged again at 4750 RCF for 12 min. The pellets were washed at pH 3 using a mixed solution of 1 $mol \cdot L^{-1}$ NaCl + 0.001 $mol \cdot L^{-1}$ HCl. This was followed by continuous washing with a 0.1 $mol \cdot L^{-1}$ NaCl solution until the pH reached 5.8 (a minimum of five washes). The cleaned kaolinite was kept in a bottle with argon headspace at 4°C. The total surface area of the pretreated kaolinite determined from the BET method was 15.5 $m^2 \cdot g^{-1}$. Scanning electron microscopy showed a homogeneous distribution of kaolinite crystals.

4. UO_2^{2+} Stock Solution

The radioisotope ^{232}U stock sample (0.025 μCi activity, Isotope Products Laboratories, Burbank, CA) was purified by separating the ^{228}Th daughter product from the stock sample by an anion exchange technique, as described by Berman *et al.* (1960) using AG1-X8 100 to 200 mesh anion exchange resin, BioRad, Richmond, CA. The purity of the separated ^{232}U fraction was assessed by analysis of its alpha emission spectrum. A UO_2^{2+} working stock solution was prepared by mixing aliquots of 1000 ppm (4.2 $mmol \cdot L^{-1}$) Specpure uranyl nitrate standard solution (Johnson Matthey Catalog Co., Ward Hill, MA), purified ^{232}U solution, 0.1 $mol \cdot L^{-1}$ HCl, and Milli-Q water to give a final UO_2^{2+} concentration of $1.0 \cdot 10^{-4}$ $mol \cdot L^{-1}$ with a specific alpha activity of approximately 34,000 $dpm \cdot ml^{-1}$.

5. Citric Acid

Citrate stock solutions were prepared from reagent-grade sodium citrate (J. T. Baker Chemical Co., Phillipsburg, NJ). The solutions were kept at 4°C in the dark.

Carbon-14-labeled citrate was obtained from Amersham (Arlington Hts., IL) at a specific activity of 50 $\mu Ci \cdot ml^{-1}$ and 114 $mCi \cdot mol^{-1}$ as [1,5-^{14}C]citric acid. The working stock solution, with a specific beta activity of $2.22 \cdot 10^6$ $dpm \cdot ml^{-1}$, was made by 1:50 dilution of the above stock with Milli-Q water and ethanol (2%, as sterilizer). Depending on the desired citrate concentration, an appropriate amount of citrate stock solution was combined with 100 μl of ^{14}C-labeled citrate solution prior to the adsorption experiment.

AR-grade NaCl (Mallinckrodt Chemical, Inc., Paris, KY) was used as the solution electrolyte.

B. Adsorption Experiments

Adsorption of UO_2^{2+} and citrate was determined in batch-type experiments. Polycarbonate reaction vessels and centrifuge tubes (Nalge Company, Rochester, NY) were selected for use in adsorption experiments since adsorption of UO_2^{2+} or citrate onto the container walls was minimal over the working pH range. When estimating blank corrections for these type of batch sorption experiments, it is necessary to consider the container walls to be competing sorbing surfaces. Making simple corrections based on the amount of sorbate retained by container walls in the absence of the solid suspension will be correct only when the effective binding constant for the container material, within the studied pH range, significantly exceeds that of the suspension. Otherwise, a simple blank correction would be an overestimate and could potentially contribute to significant errors. The polycarbonate containers used in this study were evaluated as a competing sorbent. Since the capacity of the container walls to sorb uranyl or citrate was low compared to that of the suspensions (<6% of the sorbate the lowest sorbate concentrations), and the effective binding strength was found to be lower than that of the suspensions, a blank correction was found to have no effect on the sorption results.

All labware was washed in 1 $mol \cdot L^{-1}$ NaOH and 2 $mol \cdot L^{-1}$ HCl or 10% HNO_3, rinsed with Milli-Q water, and dried prior to use. An Orion EA940 pH meter (Orion Research Inc., Boston, MA) with a Ross combination electrode (Orion Research Inc.) was used to measure pH. The electrode was calibrated using National Institute of Standards and Technology reference buffers.

An aliquot of the mineral working suspension was combined with argon purged (i.e., low CO_2) Milli-Q water and a specified amount of NaCl to give a desired mass concentration and ionic strength. The suspension was equilibrated for 2 hr with stirring at approximately 22°C and then transferred to a 500-ml polycarbonate jar with a cap designed to accommodate an argon inlet, a pH electrode, and sampling port. To remove possible carbonate species from the solution, the suspension was adjusted to pH 3.5 to 4.0 and purged with argon for 2 hr.

The initial pH of the suspensions was adjusted to about 3 for the UO_2^{2+} adsorption experiments, and about 10 to 11 for the citrate or UO_2^{2+}-citrate experiments. Since all experiments were performed with an initial UO_2^{2+} concentration of 10^{-5} mol·L^{-1}, UO_2^{2+} was undersaturated with respect to β-$UO_2(OH)_2$ (schoepite) over the entire pH range. Appropriate amounts of UO_2^{2+} and citrate stock solutions were added and the pH was incrementally adjusted in steps by addition of small amounts of 0.1 or 1.0 mol·L^{-1} HCl or CO_2-free NaOH. At each pH step, 8 ml of the suspension was withdrawn and transferred to a polycarbonate centrifuge tube. The tubes were flushed with argon, capped, and equilibrated overnight with mild shaking. Preliminary experiments were conducted at a pH of 4 to 5 for uranyl and a pH of 8 to 10 for citrate to determine the time required to establish stable solution concentrations after sorption. The pH values were chosen such that the final sorbate concentration would be approximately 50% of the initial concentration for each solid. Less than 2 hr were required to establish stable UO_2^{2+} concentrations, and about 4 hr were required to establish stable citrate concentrations. Therefore, an equilibration period of about 24 hr was considered adequate.

After equilibration, the final pH of each tube was measured while stirring with an argon purge. The samples were then centrifuged for 45 min at 1700 RCF. A mixture of 3 ml of the supernatant and 15 ml of Ultima Gold XR liquid scintillation solution (Packard, Meriden, CT) was prepared in a Packard 20-ml super polyethylene scintillation vial for radioactivity counting with a Packard Tri-Carb 2500 TR/AB liquid scintillation analyzer. Quench corrections for the ^{14}C spectra were made using the Packard system software utility that compares the characteristic shape parameters of the full sample spectrum to those of an unquenched standard spectrum. Tests were also performed to determine whether sample pH or ionic strength affected the spectra for ^{232}U or ^{14}C. The response of the Ultima Gold XR scintillation solution was not significantly affected by pH or ionic strength. (However, the performance of some other solutions tested could be significantly affected. See Ong *et al.* (1995) for results with ^{232}U.) The adsorbed fractions of UO_2^{2+} or citrate were determined by comparison with reference samples adjusted to pH ~1.5, where all UO_2^{2+} is desorbed, or pH 12, where all citrate is desorbed (Kohler *et al.,* 1993). The α spectra for ^{232}U and β spectra for ^{14}C were found not to interfere.

For selected samples, the mass balance was checked by centrifuging the sample, removing the supernatant, and resuspending the solid residual at pH 1.5 or 12 for UO_2^{2+} or citrate, respectively, in order to extract the remaining adsorbed species. Results of the mass balance tests indicated that both UO_2^{2+} and citrate were being quantitatively recovered.

Total dissolved Al and Fe were analyzed spectroscopically with a Varian SpectrAA-640 flameless atomic absorption spectrometer (Palo Alto, CA). The samples were collected during the course of a potentiometric titration or an adsorption experiment. The samples were centrifuged, and the supernatants were filtered with a

0.1-μm polycarbonate membrane (Nuclepore Corp., Pleasanton, CA) prior to analyses. Filtration was found to be necessary to remove fine suspended particles. The total dissolved iron and aluminum measured in the low and high pH ranges was approximately equal to the values predicted based on equilibrium calculations. However, the detection limits of the analytical technique did not allow for precise evaluation of the metal concentrations. Dissolved metal in the middle pH ranges (approximately pH 5 to 10) was below the dection limits.

For discussion purposes, values for the pH_{pzc} of the three solids are listed in Table I. The pH_{pzc} is defined as the pH at which net surface charge due to proton exchange is zero (pzc stands for "point of zero charge"). This is assumed to be equivalent to the pH were moderate changes in ionic strength have no effect on the pH of a suspension of the solid. The pH_{pzc} for goethite, 8.9, was determined by potentiometric titrations under low-CO_2 conditions (Robertson, 1996). The pH_{pzc} for the Alcoa gibbsite was determined by potentiometric titrations using the same technique as for goethite but also accounted for proton consumption due to dissolution of the solid and was found to be about 9.4. Hiemstra *et al.* (1987) reported a slightly higher pH_{pzc} of 10.0 for gibbsite. The structure of kaolinite makes assessment of a pH_{pzc} more difficult since several distinct sets of functional groups are present: a basal siloxane layer, a basal gibbsite-like layer, and edges composed of aluminol and silanol groups. Since solute adsorption is site specific, it is reasonable that a mineral such as kaolinite should be viewed as an assemblage of different surfaces. Carroll-Webb and Walther (1988) found that the proton-promoted dissolution rate of kaolinite could be explained better with a two-site model than with a one-site model for surface proton reactions. Wieland and Stumm (1992) expanded the description of the kaolinite surface chemistry to include ion exchange

Table I
Properties of Solids

			Kaolinite[e]	
	Goethite	Gibbsite	Edge	Al layer
Surface area ($m^2 \cdot g^{-1}$)	68[a]	11.2	2	6.5
pH_{pzc}[b]	8.4[a]	9.4	8.0	6.0
Estimated site density (nm^{-2})	2.3 to 7[c]	4[d]	6[d]	4[d]
Estimated site concentrations (mol^{-3})	0.19 to 0.57	0.37	0.02	0.06

[a]Robertson (1996).
[b]pH_{pzc}, pH at which net surface charge due to proton exchange is zero.
[c]Estimates range from 2.3 (Davis and Kent, 1990) to 7 sites·nm^{-2} (Hayes, 1987).
[d]Wieland *et al.* (1993).
[e]Surface areas are based on 15.5 m^2/g^{-1} and an estimated surface area distribution of 13–20% edge and 80–87% for Al and Si basal planes (Sposito, 1984).

reactions at fixed charge sites in the siloxane layer, and separate acid–base reactions for the basal gibbsite-like plane and the edge aluminol groups. Treated separately, the pH_{pzc} of the gibbsite layer and of the edge sites was estimated to be 5.9 and 7.5, respectively, for a kaolinite sample obtained from English Clays Lovering Pochin & Co. These values are based on an analysis of the proton chemistry and interpreted in terms of the Constant Capacitance surface complexation model. Our reevaluation of the data in terms of the Triple-Layer Model (e.g., Davis and Leckie, 1978; Davis and Kent, 1990) gives pH_{pzc} values of 6.0 and 8.0. Although, in principle, solute sorption can occur at all site types (including ion exchange), it has been proposed that metal cations preferentially adsorb to the aluminol edge sites, or a combination of edge sites and fixed charge sites depending on pH, ionic strength, electrolyte type, and sorbate concentration (Schindler *et al.,* 1987; Zachara *et al.,* 1988, 1992; Goldberg and Glaubig, 1986; Cowan *et al.,* 1992). For comparison, Sverjensky (1994) calculated pH_{pzc} values based on ideal crystal structure and solvation theory. The values for goethite and gibbsite were 9.4 and 9.84, respectively. An apparently averaged value for kaolinite was 4.66, which would presumably reflect the contribution from the silanol (quartz-like) functionality. The brief summary of pH_{pzc} values for the three solids in Table I is provided to provide a sense of the level of consistency, and also the range of values that will depend upon the method of evaluation and likely upon the variability in individual samples. (The actual range of reported values is much greater and would include earlier data where experimental techniques had been less well developed.)

III. RESULTS AND DISCUSSION

A. Equilibrium Solution Speciation Projections

Using the set of thermodynamic constants in Table II, species distributions were estimated for selected UO_2^{2+}–citrate–solid phase systems using the equilibrium speciation program HYDRAQL (Papelis *et al.,* 1988) (see Figs. 1 through 6).

For simplicity, only results for the most significant or relevant species are displayed although all relevant reactions from Table II were included in the calculations. For several sets of calculations involving both UO_2^{2+} and citrate, the total citrate concentration was two orders of magnitude greater than that of UO_2^{2+}. This is the more extreme of the experimental conditions tested, but helps to emphasize possible complications associated with dissolution of the solid phases. Silicate speciation has not been included in the calculations. These model projections must be treated with caution since it is unlikely that equilibrium with respect to the solid phases was achieved. They are, however, useful for identifying the probable initial distribution of UO_2^{2+} and citrate species and for predicting long-term condi-

TABLE II

Thermodynamic Constants Used in Speciation Calculations

Reaction	log K (I=0, 25°C)
$H_2O = OH^- + H^+$	−13.997
$Cit^{3-} + H^+ = HCit^{2-}$	6.396
$Cit^{3-} + 2H^+ = H_2Cit^-$	11.157
$Cit^{3-} + 3H^+ = H_3Cit$	14.285
$UO_2^{2+} + H_2O = UO_2OH^+ + H^+$	−5.2
$UO_2^{2+} + 2H_2O = UO_2(OH)_2 + 2H^+$	−10.3
$UO_2^{2+} + 3H_2O = UO_2(OH)_3^- + 3H^+$	−19.2
$2UO_2^{2+} + H_2O = (UO_2)_2OH^{3+} + H^+$	−2.7
$2UO_2^{2+} + 2H_2O = (UO_2)_2(OH)_2^{2+} + 2H^+$	−5.62
$3UO_2^{2+} + 4H_2O = (UO_2)_3(OH)_4^{2+} + 4H^+$	−11.9
$3UO_2^{2+} + 5H_2O = (UO_2)_3(OH)_5^+ + 5H^+$	−15.55
$UO_2^{2+} + Cit^{3-} = UO_2(Cit)^-$	8.69
$2UO_2^{2+} + 2Cit^{3-} = (UO_2)_2(Cit)_2^{2-}$	20.8
$Al^{3+} + H_2O = AlOH^{2+} + H^+$	−4.99
$Al^{3+} + 2H_2O = Al(OH)_2^+ + 2H^+$	−10.13
$Al^{3+} + 3H_2O = Al(OH)_3 + 3H^+$	−16.76
$Al^{3+} + 4H_2O = Al(OH)_4^- + 4H^+$	−23.3
$Al^{3+} + Cit^{3-} = Al(Cit)$	9.91
$Al^{3+} + 2Cit^{3-} = Al(Cit)_2^{3-}$	14.1
$Al^{3+} + H^+ Cit^{3-} = Al(HCit)^+$	12.894
$Fe^{3+} + H_2O = FeOH^{2+} + H^+$	−2.187
$Fe^{3+} + 2H_2O = Fe(OH)_2^+ + 2H^+$	−5.67
$Fe^{3+} + 3H_2O = Fe(OH)_3 + 3H^+$	−13.6
$Fe^{3+} + 4H_2O = Fe\ (OH)_4^- + 4H^+$	−21.6
$2Fe^{3+} + 2H_2O = Fe_2(OH)_2^{4+} + 2H^+$	−2.854
$3Fe^{3+} + 4H_2O = Fe_3(OH)_4^{5+} + 4H^+$	6.288
$Fe^{3+} + Cit^{3-} = Fe(Cit)$	13.1
$Fe^{3+} + H^+ + Cit^{3-} = Fe(HCit)^+$	8.6
$Fe^{3+} + H_2O + Cit^{3-} = FeOH(Cit)^{2-} + H^+$	10.3
$2Fe^{3+} + 2H_2O + 2Cit^{3-} = Fe_2(OH)_2(Cit)_2^{2-} + 2H^+$	5.09
$Fe^{3+} + 3H_2O = \alpha\text{-}FeOOH(s) + 3H^+$ (geothite)	−0.5
$Al^{3+} + 3H_2O = Al(OH)_3(s) + 3H^+$ (gibbsite)	−8.11
$UO_2^{2+} + 2H_2O = \beta\text{-}UO_2(OH)_2(s) + 2H^+$	−4.8

Note. Higher order polymers have been omitted as insignificant at the experimental concentrations used. UO_2^{2+} hydrolysis constants are from Grenthe *et al.* (1992). Remaining constants are from Smith and Martell (1993).

tions where Al^{III} and Fe^{III} from dissolution of the solids may play significant roles. (It should also be kept in mind that species that are prevalent in solution do not necessarily remain intact when adsorbed unless adsorption occurs as a result of weaker electrostatic interactions.)

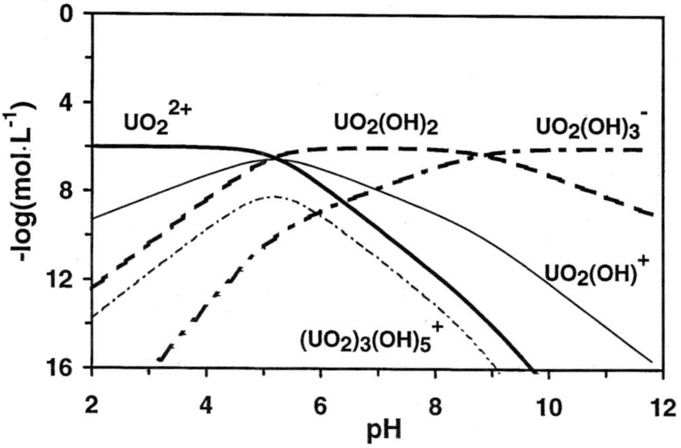

Figure 1 UO_2^{2+} hydrolysis speciation. Total $UO_2^{2+} = 10^{-6}$ mol·L^{-1}. Citrate = 0 mol·L^{-1}. NaCl = 0.1 mol L^{-1}.

B. Adsorption Results

1. Binary Systems

Adsorption results for UO_2^{2+} in the binary systems are shown in Figure 7. The UO_2^{2+} sorption behavior for all three solids is typical of cation adsorption on metal oxide solids with pH-dependent surface charge. Since the transition from 0 to

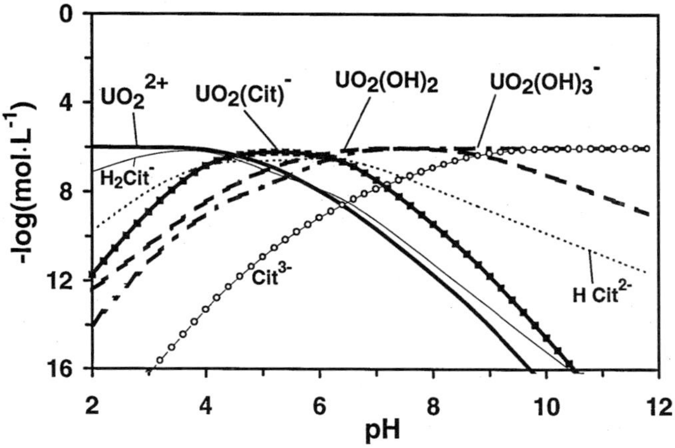

Figure 2 UO_2^{2+} speciation in the presence of citric acid. Total UO_2^{2+} = citrate = 10^{-6} mol·L^{-1}. NaCl = 0.1 mol·L^{-1}.

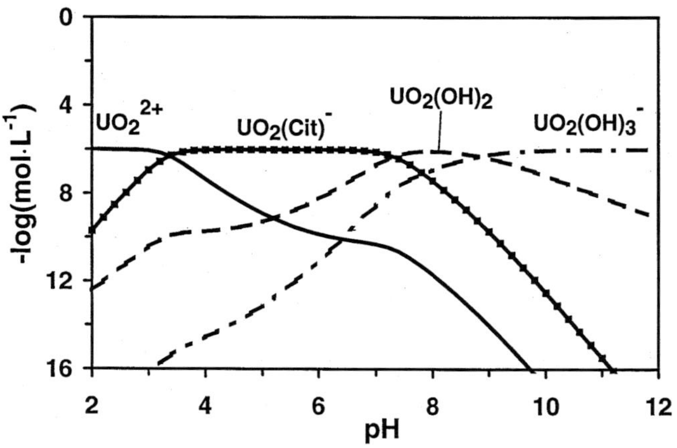

Figure 3 UO_2^{2+} speciation in the presence of citric acid. Total $UO_2^{2+} = 10^{-6}$ mol·L^{-1}. Citrate = 10^{-4} mol·L^{-1}. NaCl = 0.1 mol·L^{-1}. (Citric acid species not shown.)

nearly 100% adsorption occurs in pH ranges where the surfaces have net positive charge (see pH_{pzc} values in Table I), and the dominant uranyl solution species is the UO_2^{2+} ion, an inner-sphere surface complex is implied.

Earlier studies in our laboratory (unpublished data, Kohler *et al.,* 1994) for similar systems where the ionic strength was varied support this conclusion since the electrolyte concentration was found to have no significant effect on the sorption

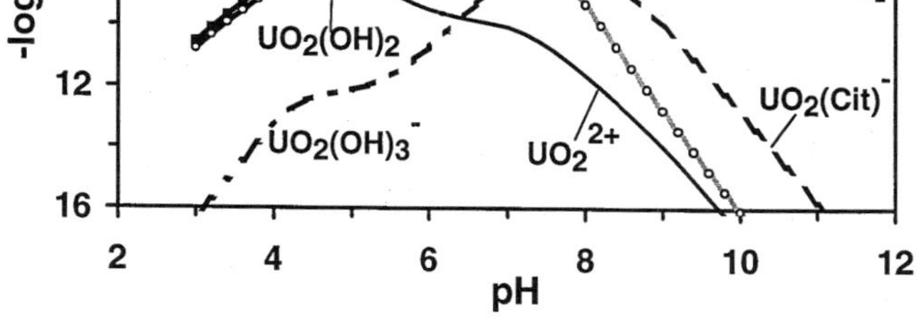

Figure 4 UO_2^{2+} and citrate speciation in the presence of kaolinite. Total $UO_2^{2+} = 10^{-6}$ mol·L^{-1}. Citrate = 10^{-4} mol·L^{-1}. NaCl = 0.1 mol·L^{-1}.

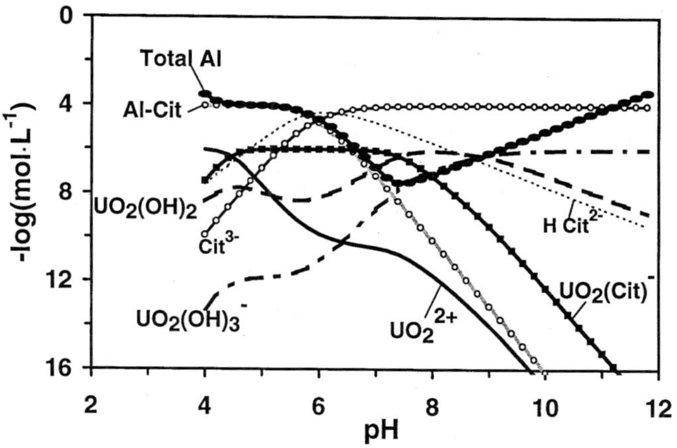

Figure 5 UO_2^{2+} and citrate speciation in the presence of gibbsite. Total $UO_2^{2+} = 10^{-6}$ mol·L^{-1}. Citrate = 10^{-4} mol·L^{-1}. NaCl = 0.1 mol·L^{-1}.

behavior. An ionic strength effect is considered to be indicative of weaker outer-sphere complexes that are bound primarily by electrostatic interactions (Hayes and Leckie, 1987).

Sorption of citric acid for a range of citrate concentrations is shown in Figures 8, 9, and 10 for the kaolinite, gibbsite, and goethite systems. The adsorption of citrate is typical of anion adsorption on a surface with pH-dependent surface charge.

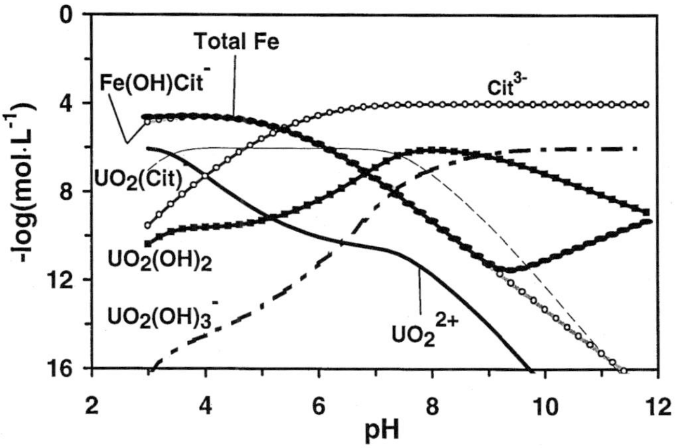

Figure 6 UO_2^{2+} and citrate speciation in the presence of goethite. Total U = 10^{-6} mol·L^{-1}. Citrate = 10^{-4} mol·L^{-1}. NaCl = 0.1 mol·L^{-1}.

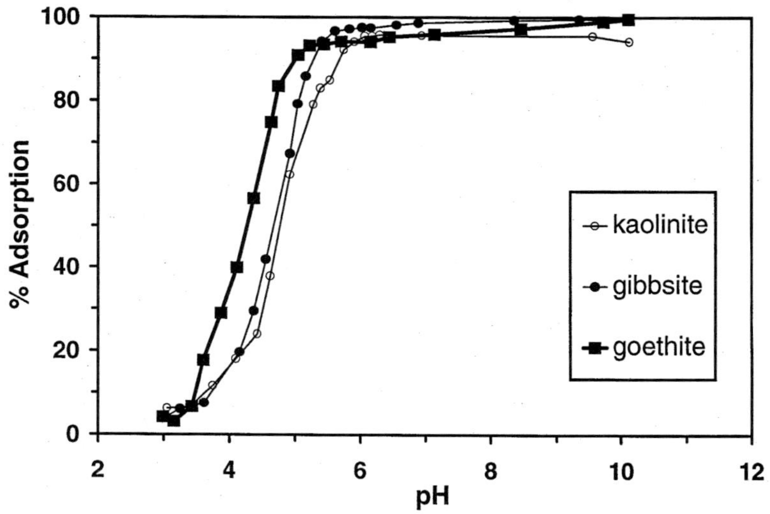

Figure 7 UO_2^{2+} sorption on gibbsite, kaolinite, and goethite. U = 10^{-6} mol·L^{-1}. Kaolinite = 1.2 g·L^{-1}. Gibbsite = 5 g·L^{-1}. Goethite = 1 g·L^{-1}. NaCl = 0.1 mol·L^{-1}.

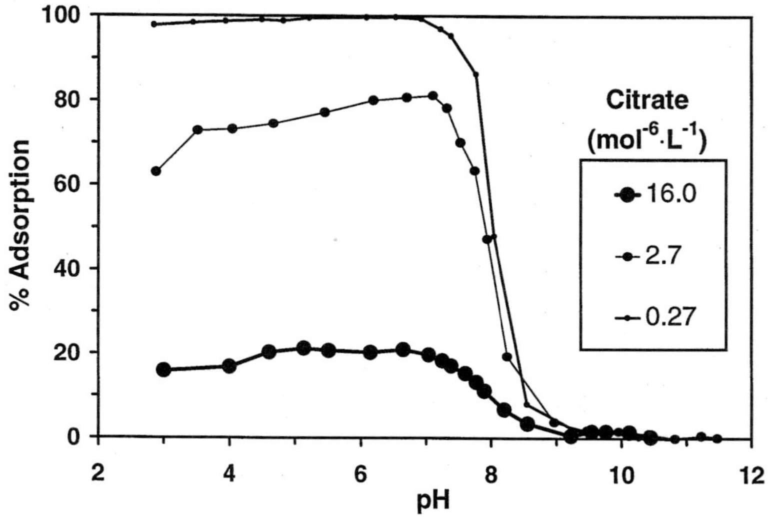

Figure 8 Citrate adsorption on kaolinite. Kaolinite = 1.2 g·L^{-1}. NaCl = 0.1 mol·L^{-1}.

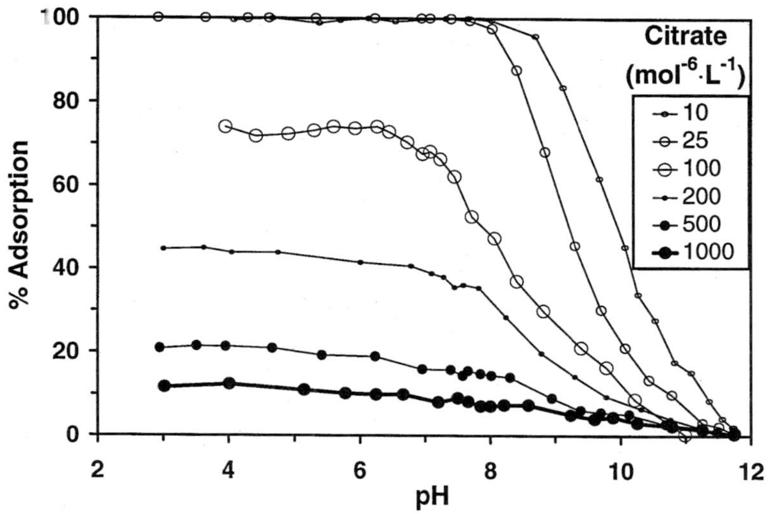

Figure 9 Citrate adsorption on gibbsite. Gibbsite = 5 g·L^{-1}. NaCl = 0.1 mol·L^{-1}.

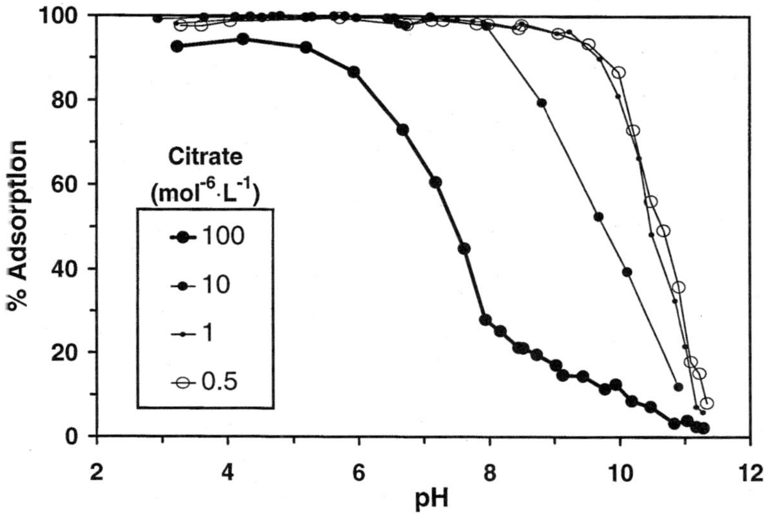

Figure 10 Citrate adsorption on goethite. Goethite = 1 g·L^{-1}. NaCl = 0.1 mol·L^{-1}.

Citrate also appears to form a strong surface complex since sorption occurs in the pH range where the net surface charge is negative and citrate exists as the triply charged anion. This is consistent with the observation that ionic strength (0.01 to 0.10 mol·L^{-1} NaCl) had little effect on adsorption of citrate for all solids tested. That the maximum citrate adsorption is less than 100% for the higher concentrations of citrate suggests that available surface sites are limiting in these systems.

2. Ternary Systems

The effect of citrate on the sorption of UO_2^{2+} on kaolinite is shown in Figure 11. As the total concentration of citrate increases, UO_2^{2+} partitioning is shifted to the solution phase. The pH region where UO_2^{2+} adsorption is reduced corresponds to the range where the 1:1 UO_2^{2+}–citrate solution complex is a significant uranyl species (see Figs. 2 and 3) if the solution citrate concentration is sufficiently high.

It can be seen that the concentrations of citrate remaining in solution are sufficient to produce the UO_2^{2+}–citrate complex in the pH region where reduction of UO_2^{2+} adsorption occurs. In this case, the effect of citrate is similar to that in the reports mentioned earlier that demonstrate the ability of chelating ligands to stabilize metal cations in solution (increasing mobility). The corresponding effect of UO_2^{2+} on citric acid adsorption (Fig. 12) is more difficult to interpret. Above pH 8, UO_2^{2+} appears to enhance adsorption of citrate for the cases where the citrate

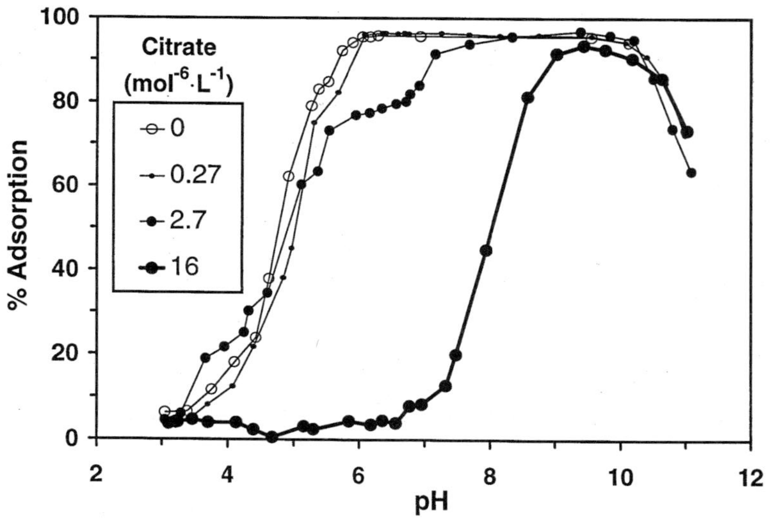

Figure 11 UO_2^{2+} adsorption on kaolinite—effect of citric acid. U = 10−^{6} mol·L^{-1}. Kaolinite = 1.2 g·L^{-1}. NaCl = 0.1 mol·L^{-1}.

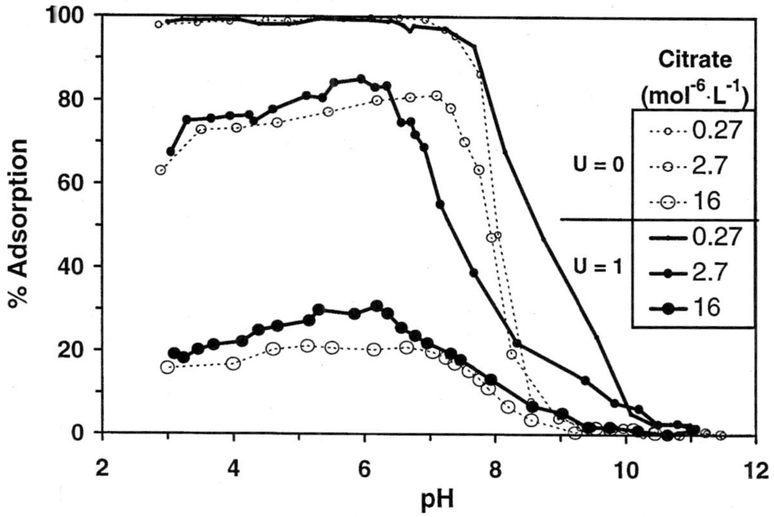

Figure 12 Citrate adsorption on kaolinite in the presence of UO_2^{2+}. Total UO_2^{2+} = 10^{-6} mol·L^{-1}. Kaolinite = 1.2 g·L^{-1}. NaCl = 0.1 mol·L^{-1}.

concentrations are comparable to those of UO_2^{2+}. Since UO_2^{2+} is nearly 100% sorbed in this pH range, it is possible that an association between UO_2^{2+} and citrate exists on the kaolinite surface. The same apparent effect was seen at the lower pH range as well; however, this is more difficult to explain since the results shown in Figure 11 indicate that UO_2^{2+} should be stabilized in solution as a result of an association with citrate, and therefore a decrease in sorption of both UO_2^{2+} and citrate would be expected. Hypotheses for this data at the moment are being deferred pending further work. Although quench corrections for the ^{14}C spectra were performed, and the scintillation solution was tested for variable spectral response to sample pH and ionic strength, errors due to counting efficiencies can still be considered a possibility. If dissolution of kaolinite had been more extensive in the binary citrate–kaolinite system, the formation of an Al–citrate solution complex could reduce the sorption of citrate (see Fig. 4) without significantly affecting the UO_2^{2+}–citrate species distribution.

The effect of citrate on UO_2^{2+} sorption for the gibbsite system is qualitatively similar to the behavior observed for kaolinite (Fig. 13). The sorption of UO_2^{2+} is slightly decreased in the presence of increasing citrate concentration. Although the effect does not appear to be as pronounced for the UO_2^{2+}:citrate ratios tested. The greatest effect in the kaolinite system occurs at the highest citrate concentration where the capacity of the kaolinite surface for citrate appears to be exceeded. Similar data for gibbsite were not obtained. As was found in the kaolinite system, the

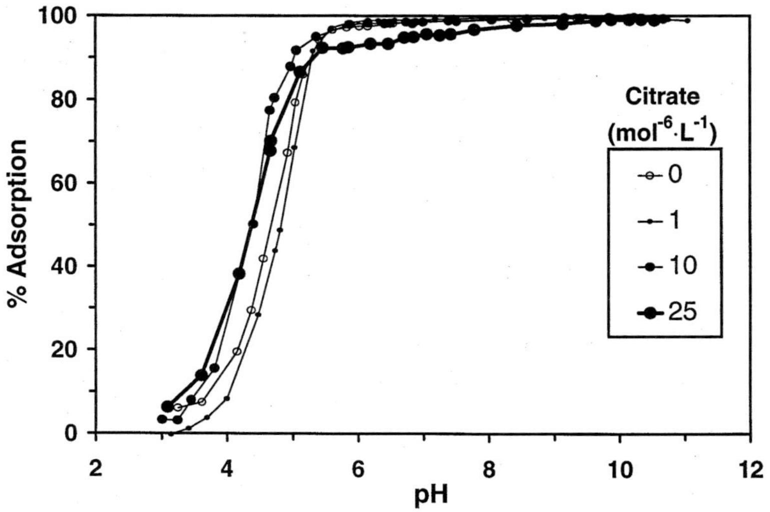

Figure 13 UO_2^{2+} adsorption on gibbsite—effect of citric acid. Total $UO_2^{2+} = 10^{-6}$ mol·L^{-1}. Gibbsite = 5 gl·L^{-1}. NaCl = 0.1 mol·L^{-1}.

presence of UO_2^{2+} on the gibbsite surface appears to increase the adsorption of citrate above pH 8 where the transition from zero to maximum citrate adsorption occurs (Fig. 14).

The goethite system (Figs. 15 and 16) exhibited distinctly different behavior than either the kaolinite or the gibbsite system. The most obvious difference is that UO_2^{2+} adsorption in the goethite system increases as the citrate concentration increases. At the highest citrate concentration, UO_2^{2+} is predominantly sorbed over the entire pH range studied. As for the kaolinite system, the effect of citrate on UO_2^{2+} in the goethite system is not pronounced until the apparent capacity of the goethite surface for citrate has been exceeded. The citrate:UO_2^{2+} ratio may be less of a factor than the citrate:surface ratio.

While it is possible that UO_2^{2+} had some effect on the sorption of citrate in the kaolinite and gibbsite systems, there appears to be little effect of UO_2^{2+} on the sorption of citrate in the goethite system. At equilibrium, goethite is considerably less soluble than either gibbsite or kaolinite in terms of total metal species released to solution. Although the rates of dissolution have not yet been examined and the total solution Fe or Al were not accurately known, it would be reasonable to assume that dissolved Fe was a minor constituent given the relatively short times allowed for equilibration. If this was the case, interference from Fe would be less of a factor.

A few general conclusions concerning the effect of citrate on UO_2^{2+} sorption

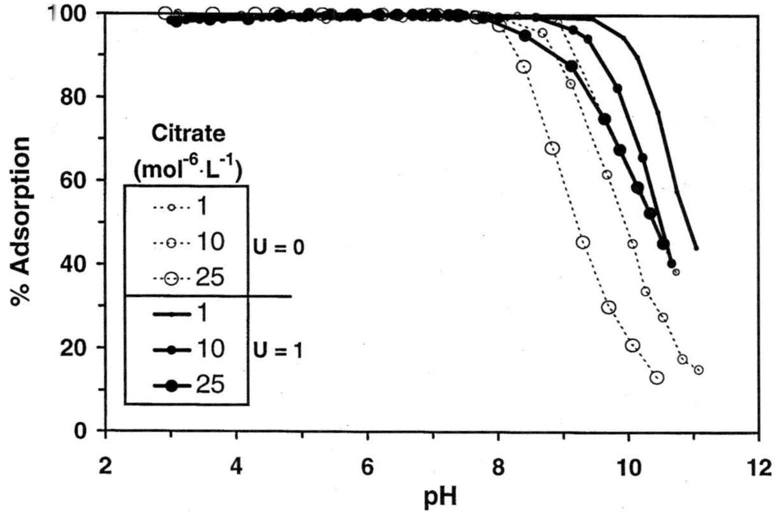

Figure 14 Citrate adsorption on gibbsite in the presence of UO_2^{2+}. Total UO_2^{2+} = 10^{-6} mol·L^{-1} when present. Gibbsite = 5 g·L^{-1}. NaCl = 0.1 mol·L^{-1}.

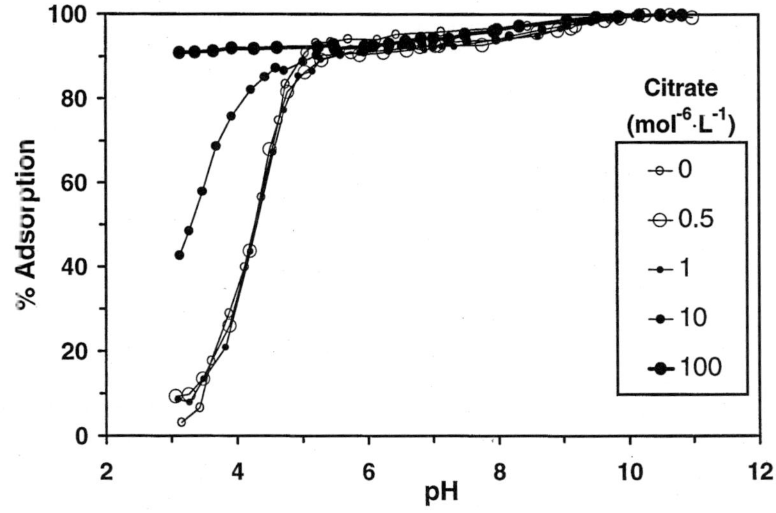

Figure 15 UO_2^{2+} adsorption on goethite—effect of citric acid. Total UO_2^{2+} = 10^{-6} mol·L^{-1}. Goethite = 1 g·L^{-1}. NaCl = 0.1 mol·L^{-1}.

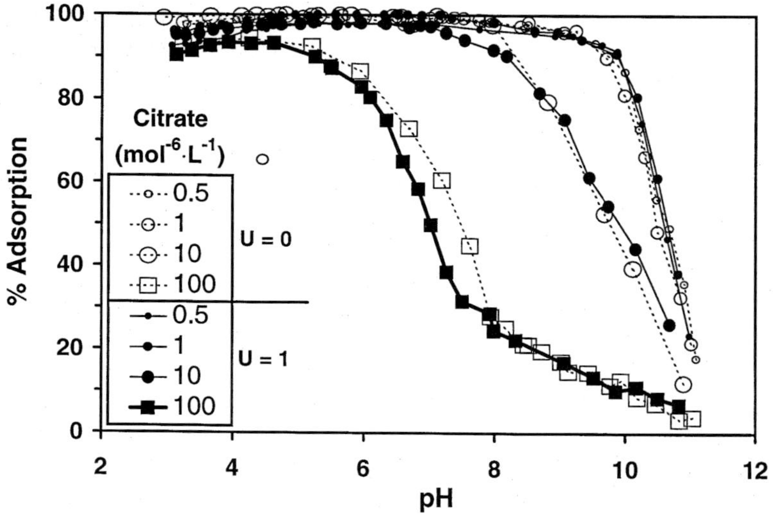

Figure 16 Citrate adsorption on goethite in the presence of UO_2^{2+}. Total UO_2^{2+} = 10^{-6} mol·L^{-1}. Goethite = 1 g·L^{-1}. NaCl = 0.1 mol·L^{-1}.

can be drawn from these experimental observations. It would appear that citrate and UO_2^{2+} do not appreciably interact at the surface given a sufficient excess of available surface sites. The surface complexes for each sorbate in these systems are more stable than the solution complexes between UO_2^{2+} and citric acid. However, as the surface capacity for citrate is exceeded, UO_2^{2+} partitioning is shifted to the solution phase for kaolinite (and possibly for gibbsite), and to the bound state for goethite.

These observations lead to some hypotheses concerning the interactions between a multidentate organic ligand and a metal oxycation. There are several ways to approach modeling for the association between a multifunctional anion and a metal oxide surface. It was noted earlier that sorption of citrate on mineral surfaces was typical for the sorption of an anion on surfaces that have pH-dependent surface charge. As the net surface charge becomes more positive, the sorption of an anionic species should increase. This has been modeled satisfactorily for many simple anionic species such as aresenate, selenate, and selenite (e.g., Davis and Leckie, 1980). The reverse would be true for a cation. Although semiempirical, the current complexation models utilize simple stoichiometric functions involving single protolyzable surface sites, the anions, and protons (or hydroxide). A more complex, multifunctional species such as citrate, however, presents some interesting problems.

Citrate sorption to oxide surfaces is unlikely to follow such simple stoichiometries, especially at high surface coverages. Each functional group can potentially bind to separate surface sites. This scenario was proposed by Cornell and Schindler (1980) in a spectroscopic study of citrate adsorption on goethite. Given the size and relatively flexibility of the citrate molecule, the orientation of citrate on a nonhomogeneous surface (i.e., with multiple crystal faces, defects, step features, etc.) is not likely to be unique. Numerous orientations should be possible, each of which could have relatively small differences in net free energy. In addition, the availability of unoccupied adjacent sites suitable for the formation of a multidentate surface complex will diminish as the surface concentration increases. Nevertheless, bi- and monodentate complexes should still be possible.

If attachment of citrate can be seen as a stepwise sequence of bond formations between the carboxyl groups and metal–oxygen functional groups from the oxide surface, the individual bond strength for the "first" attached functional group will be stronger than those of the next two. (It has been suggested by Cornell and Schindler (1980) that the hydroxyl group can also participate in bonding.) A tridentate structure would still be thermodynamically preferred due to the net enthalpy and high positive entropy advantage, through replacement of solvation water but the proportion of incomplete coordination configurations would increase as sorption coverage increased. If the surface is not uniform, and suitable arrangements of three or more surface sites occur in isolated clusters as opposed to a completely uniform distribution of sites (e.g., a checkerboard), the transition from tridentate structures to lower dentate structures would occur at surface coverages that would be lower than some estimations of the available number of surface sites.

With respect to modeling citrate adsorption according to a surface complexation convention, the preceding arguments would mean that 100% sorption of citric acid could be composed of a distribution of tri-, bi-, and monodentate complexes. Regions where the amount of citrate appears to exceed the capacity of the surface for tridentate binding could have increasing proportions of the bi- and monodentate complexes. Presumably, the uncomplexed carboxyl groups would be available for binding UO_2^{2+}. This bridging structure (described as "Type B" by Schindler and Stumm, 1987) is proposed as one explanation for the observations in the goethite system where UO_2^{2+} is sorbed at more than 90% over the entire pH range studied when in the presence of citrate that is in apparent excess of available surface capacity.

That the same phenomenon is not observed for the kaolinite and gibbsite systems can be explained by the solution chelate complex with UO_2^{2+} being more stable than the bridging arrangement in the case of the Al-containing solids. The log of the 1:1 citrate–metal formation constants for UO_2^{2+}, Al^{3+}, and Fe^{3+} are 8.69, 9.91, and 13.1, respectively. While the Al^{3+} complex is moderately stronger than the complex with UO_2^{2+}, the complex with Fe^{3+} is significantly greater than either one (see Table II). This should be reflected in the strength of citrate complexes with the mineral surfaces. Inspection of sorption profiles for citrate on the three solids (Figs. 8, 9, and 10) shows that the distance from the pH_{pzc} to the sorption edges for citrate at the lowest citrate concentrations is greatest for goethite and approximately equal for gibbsite and kaolinite. This condition generally translates into larger formation constants when viewed in the context of surface complexation models. The requirement is that the monodentate complex between goethite and citrate (that results from insufficient sites for the tridentate complex) is thermodynamically preferred over the UO_2^{2+}–citrate solution complex at least in the time frame of the experiments.

An important consideration following from the above argument is that, given sufficient time, dissolution products from the solids (Fe^{3+}, Al^{3+}) could eventually change the solute partitioning. Exchange reactions for multidentate complexes are known to be kinetically slow (Morel and Hering, 1993). The species distribution in Figure 6 shows an equilibrium concentration of total Fe that is higher than that for goethite as a result of ligand-promoted dissolution. (In these calculations additional considerations such as surface complexation with the dissolution products were not included.) In the case of goethite, the Fe^{3+} complex with citrate is considerably stronger than that with UO_2^{2+}. The long-term consequence would be displacement of UO_2^{2+} by Fe^{3+} and the subsequent release of UO_2^{2+} to solution. In this case, the profiles shown in Figure 15 for the higher citrate concentrations might tend toward the profiles for lower citrate concentrations as formation of Fe^{3+}–citrate complexes effectively sequester the available citrate. Given the stronger complex formed between the citrate and Al^{3+} as opposed to UO_2^{2+}, the same situation might be expected for the gibbsite and kaolinite systems.

IV. CONCLUSIONS

Citric acid was found to significantly affect adsorption of UO_2^{2+} on mineral oxide surfaces. For gibbsite and kaolinite it appears that the formation of the stable soluble UO_2^{2+}–citrate complex reduces UO_2^{2+} adsorption, while for goethite citrate enhances adsorption of UO_2^{2+}. In all cases, the effect of citrate is not observed until the citrate:UO_2^{2+} ratio is much greater than one, or until the citrate:surface ratio is high enough to potentially exceed the capacity of the oxide surface to adsorb citrate. The effect of citrate on adsorption of UO_2^{2+} to goethite would be consistent with the possibility that citrate is incompletely coordinated to the oxide surface at high citrate loadings, and that a bridging oxide–citrate–UO_2^{2+} structure is formed. There is also a possibility that at high pH values, adsorbed UO_2^{2+} may enhance the adsorption of citrate via an oxide–UO_2^{2+}–citrate structure. Elucidation of the actual mechanisms will require additional spectroscopic information and confirmation by complexation model simulations that account for competition by products of oxide dissolution.

The results of surface complexation modeling for these systems is to be presented elsewhere; however, preliminary analyses using the “Triple Layer Model” for surface complexation has resulted in good fits of the experimental data for the binary systems involving UO_2^{2+}. This is not unexpected since the shape of the UO_2^{2+} sorption profiles is very similar to those for other metal cations on metal oxide surfaces. However, the binary citrate systems and the ternary systems will require more analysis for several reasons. First, the previous arguments dictate that additional experimental information will be required to determine whether the systems have reached equilibrium (or meta-stable) states with respect to solid dissolution and precipitation, and with respect to metal exchange with citrate. In addition, spectroscopic information will be essential for placing constraints on the set of reactions used to define the system. However, the most important hurdle, at this point, is that current models are not amenable to the particular problem associated with multidentate complexes.

ACKNOWLEDGMENTS

Funding for this project was provided by The U.S. Department of Energy Subsurface Science Program and Battelle Pacific Northwest Laboratories. John Zachara at Battelle Pacific Northwest Laboratories provided helpful advice and guidance for the project.

REFERENCES

Atkinson, R. J., Posner, A. J., and Quirk, J. P. 1967. Adsorption of potential-determining ions at the ferric oxide-aqueous electrolyte interface. *J. Phys. Chem.* 71:550–558.

Berman, S. S., McKinney, L. E., and Bednas, M. E. 1960. The separation of carrier-free ^{234}Thorium (UX_1) from uranium by anion-exchange. *Talanta* 4:153–157.

Bowers, A. R., and Huang, C. P. 1986. Adsorption characteristics of metal-EDTA complexes on hydrous oxides. *J. Colloid Interface Sci.* 110:575–590.

Cambier, P., and Sposito, G. 1991. Adsorption of citric acid by synthetic pseudoboehmite. *Clays Clay Miner.* 39:369–374.

Carroll-Webb, S., and Walther, J. V. 1988. A surface complex reaction model for the pH-dependence of corundum, and kaolinite dissolution rates. *Geochim. Cosmochim. Acta* 52:2609–2623.

Cornell, R. M., and Schindler, P. W. 1980. Infrared study of the adsorption of hydroxycarboxylic acids on alpha-FeOOH and amorphous Fe(III)hydroxide. *Colloid Polym. Sci.* 258:1171–1175.

Cowan, C. E., Zachara, J. M., Smith, S. C., and Resch, T. C. 1992. Individual sorbent contributions to cadmium sorption on ultisols of mixed mineralogy. *Soil Sci. Soc. Am. J.* 56:1084–1094.

Davis, J. A., James, R. O., and Leckie, J. O. 1977. Surface ionization and complexation at the oxide/water interface. 1. Computation of electrical double layer effects. *J. Colloid Interface Sci.* 63:480–499.

Davis, J. A., and Leckie, J. O. 1978. Surface ionization and complexation at the oxide/water interface. II. Surface properties of amorphous iron oxyhydroxide and adsorption of metal ions. *J. Colloid Interface Sci.* 67:90–107.

Davis, J. A., and Kent, D. B. 1990. Surface complexation modeling in aqueous geochemistry. In: "Mineral–Water Interface Geochemistry, Reviews in Mineralogy" (M. F. Hochella and A. F. White, Eds.) pp. 177–248. Mineralogical Society of America, Washington, DC.

Davis, J. A., and Leckie, J. O. 1980. Surface ionization and complexation at the oxide/water interface. 3. Adsorption of anions. *J. Colloid Interface Sci.* 74:32–43.
Elliott, H. A., and Denneny, C. M. 1982. Soil adsorption of cadmium from solutions containing organic ligands. *J. Environ. Qual.* 11:658–663.
Fletcher, P., and Beckett, P. H. T. 1987. The chemistry of heavy metals in digested sewage sludge. II. Heavy metal complexation with soluble organic matter. *Water Res.* 21:1163–1172.
Girvin, D. C., Gassman, P. L., and Bolton, J. H. 1993. Adsorption of aqueous cobalt ethylenediaminetetraacetate by d-A12O3. *Soil Sci. Soc. Am. J.* 57:47–57.
Goldberg, S., and Glaubig, R. A. 1986. Boron adsorption and silicon release by the clay minerals kaolinite, montmorillonite, and illite. *Soil Sci. Soc. Am. J.* 50:1442–1448.
Gregg, S. J., and Sing, K. S. W. 1982. "Adsorption, Surface Area and Porosity." Academic Press, Orlando.
Grenthe, I., *et al.* 1992. "Chemical Thermodynamics of Uranium," Vol. 1. Elsevier Science, Amsterdam.
Hayes, K. F. 1987. Equilibrium, spectroscopic, and kinetic studies of ion adsorption at the oxide/aqueous interface. Ph.D. Thesis, Stanford University, Stanford, CA.
Hayes, K. F., and Leckie, J. O. 1987. Modeling ionic strength effects on cation adsorption at hydrous oxide/solution interfaces. *J. Colloid Interface Sci.* 115:564–572.
Hiemstra, T., van Riemsdijk, W. H., and Bruggenwert, M. G. 1987. Proton adsorption mechanism at the gibbsite and aluminium oxide solid/solution interface. *Netherlands J. Agric. Sci.* 35:281–293.
Killey, R. W. D., McHugh, J. O., Champ, D. R., Cooper, E. L., and Young, J. L. 1984. Subsurface cobalt-60 migration from a low-level waste disposal site. *Environ. Sci. Technol.* 18:148–157.
Kohler, M., Honeyman, B. D., and Leckie, J. O. 1994. Uranyl interactions in the goethite/solution interphase region: Formation of binary and ternary surface complexes. Submitted for publication.
Kohler, M., Honeyman, B. D., van Geen, A., and Leckie, J. O. 1993. Neptunium(V) sorption on hematite (α-Fe_2O_3) in aqueous suspension: Effects of carbonate and EDTA. Submitted for publication.
Means, J. L., Crerar, D. A., and Duguid, J. O. 1978. Migration of radionuclide wastes: Radionuclide mobilization by complexing agents. *Science* 200:1477–1486.
Morel, F. M. M., and Hering, J. G. 1993. "Principles and Applications of Aquatic Chemistry." John Wiley, New York.
Ong, C. G., Prasad, A., and Leckie, J. O. 1995. Effect of pH, NaCl and cocktail selection on ^{232}U liquid scintillation spectra. *Anal. Chem.* 67:3893–3896.
Papelis, C., Hayes, K. F., and Leckie, J. O. 1988. HYDRAQL: A program for the computation of chemical equilibrium composition of aqueous batch systems including surface-complexation modeling of ion adsorption at the oxide/solution interface. Technical Report 306, Environmental Engineering and Science, Department of Civil Engineering, Stanford, CA.
Riley, R. G., and Zachara, J. M. 1992. Chemical contaminants on DOE lands and selection of contaminant mixtures for subsurface science research. U.S. Department of Energy, Washington DC.
Robertson, A. P. 1996. Goethite/humic acid interactions and their effects on copper(II) binding. Ph.D. Thesis, Stanford University, Stanford, CA.
Schindler, P. W., Liechti, P., and Westall, J. C. 1987. Adsorption of copper, cadmium and lead from aqueous solution to the kaolinite/water interface. *Netherlands J. Agric. Sci.* 35:219–230.
Schindler, P. W., and Stumm, W. 1987. The surface chemistry of oxides, hydroxides and oxide minerals. *In* "Aquatic Surface Chemistry" (W. Stumm, Ed.), pp. 83–125. John Wiley, New York.
Schwertmann, U. 1988. Occurrence and formation of iron oxides in various pedoenvironments. *In* "Iron in Soils and Clay Minerals" (J. W. Stucki, B. A. Goodman, and U. Schwertmann, Ed.), NATO ASI Series C., Vol 217, pp. 267–308. NATO, Netherlands.
Smith, R. M., and Martell, A. E. 1993. NIST Critical Stability Constants of Metal Complexes Database, NIST Standard Reference Database 46. U.S. Department of Commerce, Gaithersburg, MD.

Sposito, G. 1984. "The Surface Chemistry of Soils." Oxford University Press, New York.
Sverjensky, D. A. 1994. Zero-point-of-charge prediction from crystal chemistry and solvation theory. *Geochim. Cosmochim. Acta* 58:3123–3129.
Swisher, R. D., Taulli, T. A., and Malec, E. J. 1974. *In* "Trace Metals and Metal-Organo Interactions in Natural Waters" (P. C. Singer, Ed.), pp. 237–264. Ann Arbor Science, Ann Arbor.
Tam, S.-C., and McColl, J. G. 1991. Aluminum-binding ability of soluble organics in Douglas fir litter and soil. *Soil Sci. Soc. Am. J.* 55:1421–1427.
Wieland, E., Kohler, M., and Leckie, J. O. 1993. Adsorption of neptunium (Np(V)) on mineral surfaces: Modeling of surface complexation. Submitted for publication.
Wieland, E., and Stumm, W. 1992. Dissolution kinetics of kaolinite in acidic aqueous solutions at 25°C. *Geochim. Cosmochim. Acta* 56:3339–1755.
Zachara, J. M., Cowan, C. E., Schmidt, R. L., and Ainsworth, C. C. 1988. Chromate adsorption by kaolinite. *Clays Clay Miner.* 36:317–326.
Zachara, J. M., Gassman, P. L., Smith, S. C., and Taylor, D. 1995. Oxidation and adsorption of Co(II)EDTA^{2-} complexes in subsurface materials with iron and manganese oxide grain coatings. *Geochim. Cosmochim. Acta* 59:4449–4463.
Zachara, J. M., Smith, S. C., Resch, T. C., and Cowan, C. E. 1992. Cadmium adsorption to soil separates containing layer silicates and iron and aluminium oxides. *Soil Sci. Soc. Am. J.* 56:1076–1084.

Chapter 14

Surface and Solution Speciation of Ag^{I} in a Heterogeneous Ferrihydrite-Solution System with Thiosulfate

Colin G. Ong and James O. Leckie
Environmental Engineering & Science, Department of Civil and Environmental Engineering, Stanford University, Stanford, California

The presence of thiosulfate modifies the adsorptive characteristics of Ag^{I} on ferrihydrite through the formation of a ternary surface complex which adsorbs in a ligand-like manner, directly overmapping the adsorption edge profile of thiosulfate itself. Experimental data are modeled using the Triple-Layer Surface Complexation Model implemented in the HYDRAQL code. Modeling included the evaluation of several alternative reaction stoichiometries for both inner-sphere and outer-sphere surface complexes, taken singly and two at a time. Modeling results suggest that $AgS_2O_3^-$ is bound as an inner-sphere ternary complex at low pH, overlapping the monodentate Ag^{I} outer-sphere surface complex which increases in a metal-like manner with increasing pH. A mechanism is proposed for the formation of the ternary inner-sphere surface complex based on a condensation reaction leading to the transition from an outer-sphere to an inner-sphere ternary surface species. The proposed mechanism is initiated by the approach of the $AgS_2O_3^-$ species toward the protolyzed surface moiety and the formation of hydrogen bonds, which is enhanced by charge stabilization of the S_2O_3 molecule by Ag. Coordination of an O atom and subse-

Adsorption of Metals by Geomedia

quent bond formation facilitates H_2O as a leaving group completing the dehydration reaction.

I. ENVIRONMENTAL CHEMISTRY AND THE FATE OF SILVER

Silver is a metal of varied environmental toxicity risk. Although reported cases of human toxicosis by Ag are rare, acute toxicity to animals and fish has been found to cover a wide range of Ag concentrations, although without carcinogenic effects. In laboratory tests, silver speciation was found to greatly influence the metal toxicity, where silver nitrate was found to be the most toxic soluble form of silver to freshwater plants, while silver thiosulfate and silver chloride were the least toxic (Petering and McClain, 1991). Generally, however, in the environment, and especially in those places with high chloride concentrations (e.g., estuarine systems), inorganic silver chloride complexes would dominate over organic complexes and other inorganic complexes (Miller and Bruland, 1995). Thiosulfate is produced as a significant reaction intermediate from the oxidation of FeS(s) under estuarine intertidal redox conditions (Nelson, 1978) with slow decomposition rates (Xu and Schoonen, 1995), and potentially may contribute to silver immobilization through ternary surface complexation.

The environmental biogeochemical cycle of silver is not understood in detail. This metal is introduced into the environment principally by waste discharge from industrial and military photoprocessing operations, and has been found to be a contaminant in the San Francisco Bay estuarine system. In this region, natural and industrial discharged silver was found to be efficiently removed by suspended particulate matter (Smith and Flegal, 1993), which led to a large pool of silver in the benthic environment.

Secondary biological waste treatment results in effective removal and recovery of the metal as solid Ag_2S in the sludge (Cooley *et al.,* 1988), and only small quantities pass through and are released from treatment facilities in a particulate, insoluble form (Bard *et al.,* 1976). Despite regulation and decreased discharge in recent years, remobilization of the silver pool threatens the San Francisco Bay's environmental health. Concern has been expressed over the contamination of the benthic sediments (Luoma and Phillips, 1988), with respect to detrimentally affecting benthic organisms such as bivalves, which are capable of bioaccumulating silver. These benthic organisms consequently have been shown to provide a better indication of the extent of environmental silver enrichment than do the sediments (Luoma *et al.,* 1990).

II. SILVER AND SILVER–LIGAND COMPLEX SORPTION TO MINERALS

Silver(I) ion sorption onto amorphous iron hydroxide, $Fe(OH)_3$(a) (also known as ferrihydrite, $FeOOH{\cdot}H_2O$), was reported to exhibit a metal-characteristic sorption edge beginning at about pH 4 and reaching a maximum at about pH 8 (Davis and Leckie, 1978a). When thiosulfate ion is added, desorption, characteristic of anion-regulated sorption, of Ag^I ion from $Fe(OH)_3$(a) is observed between pH 4 and 7. Above pH 7, Ag^I ion reverses its trend and the sorbed quantity increases. Further, when Ag^I and $S_2O_3^{2-}$ are present in equimolar concentrations, a local sorption maximum, with approximately 80% Ag^I_T sorbed, is reached at about pH 5, and Ag^I increasingly desorbs from $Fe(OH)_3$(a) with decreasing pH. The data points making up the sorption edge below pH 4.5 were omitted in previous modeling efforts (Davis and Leckie, 1979).

Similar experimental sorption work has shown that increased thiosulfate shifts the sorption edge of Cd toward higher pH for a variety of minerals, including amorphous iron hydroxide (Benjamin and Leckie, 1982). Qualitatively, however, the characteristic metal sorption edge was maintained, and ligand-mediated sorption at low pH was not observed. Thus, the AgS_2O_3 ternary surface complex provides a unique opportunity to evaluate a ternary complex for an anion-like adsorption.

Recently, a new chemical mechanism for surface complexation of Ag^I onto $Fe(OH)_3$(a) in the presence of thiosulfate ion was interpreted from modeling results using the computer program HYDRAQL (Papelis *et al.*, 1988) with the triple-layer model (TLM). The capacitance values, acidity constants, and bulk electrolyte binding constants used in this effort were previously reported (Davis and Leckie, 1978b). The complete experimental data sets were obtained from Davis (1977). This effort extends previous discussions and offers reinterpretation of data by simulating the experimental sorption data inclusive of the data points between pH 4 and 4.5. The approach was initiated by the simulation of the simpler binary complexation of Ag^I or $S_2O_3^{2-}$, and from that basis, the ternary surface complexation system was constructed.

III. APPLICATION OF THE TRIPLE-LAYER SURFACE COMPLEXATION MODEL

A. The Triple-Layer Surface Complexation Model

The TLM differentiates between weakly and strongly binding ions, and is posed to allow for the spatial distinction between inner-sphere and outer-sphere surface complex formation. Stern (1924) recognized this need and introduced the first three-plane model for the mercury electrode–aqueous interface. The Stern model equivalent for the oxide–aqueous interface is shown in Figure 1. As shown, there are two adsorption planes: a surface plane for inner-sphere or coordination complexes (o-plane) and a second plane for more weakly (ionically) bound outer-sphere surface complexes (β-plane), followed by a boundary defining the diffuse layer (d-plane). The three planes are separated by "Helmholtz" capacitances defining the electrostatic relationships of the inner phase region of the solid–solution interface (Fig. 1A). The first surface complexation model (SCM) to incorporate a three-plane structure for the oxide–solution interface was that of Yates (1974) as applied by Davis and Leckie (1978b), and modified by Hayes and Leckie (1987). The molecular hypotheses and constraint equations of the TLM are summarized in Table I. The parameters include surface protolysis constants (K_a), electrolyte surface binding constants (K_A, K_C), capacitances (C), and total site concentration (S_T), as well as the fitting parameters of metal- and ligand-binding constants (K_{Me}, K_L).

The most important differences between the TLM and the simpler one- and two-plane SCMs are the additional adsorption plane and the type of surface complexes allowed in the model structure. In the TLM SCM version of Davis and Leckie (1978b) and Hayes and Leckie (1987), counter ion binding is incorporated directly into the model structure. The counter ions form ion-pair outer-sphere surface complexes as shown in Figure 1B. By accounting specifically for counter ion binding via the formation of ion-pair surface complexes, the experimental observation of an increase in surface charge as a function of ionic strength is recognized as a

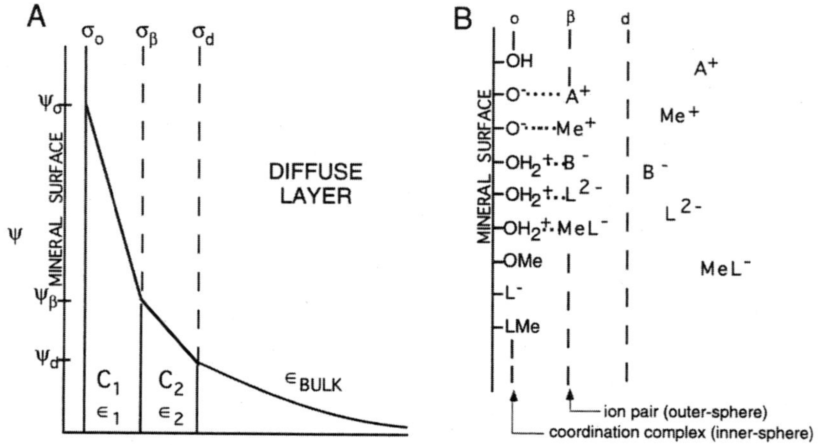

Figure 1 Schematic representations of (A) the electrostatic relationships of the inner-phase region of the solid–solution interface, and (B) the possible configurations of surface complexes in the three-layer model for a Me^{+}–L^{2-} system in 1:1 electrolyte AB.

Table I

Triple-Layer SCM Constraints and Molecular Hypotheses for a Me^+–L^{2-} System

System
- Hydrous oxide; univalent adsorbing metal ion; divalent adsorbing ligand; 1:1 symmetrical electrolyte

Species
- Aqueous: Me^+, H^+, OH^-, H_2O, L^{2-}, A^+, B^-
- Surface: SOH, SOH_2^+, SO^-, SOMe, SL^-, SMeL (coordination complexes)
- β-plane: SOH_2^+–B^-, SO^-–A^+, SO^-–Me^+, SOH_2^+–L^{2-}, SOH_2^+–MeL^- (ion pairs)

Model parameters
- S_T, $^{int}K_{a1}$, $^{int}K_{a2}$, $^{int}K_{Me}$, $^{int}K_{L,1}$, $^{int}K_{L,2}$, $^{int}K_A$, $^{int}K_C$, C_1, C_2

Constraints
- Surface mass balance
- Surface charge balance
- Electroneutrality condition: $\sigma_o + \sigma_\beta + \sigma_d = 0$
- Equilibrium equations for surface species

Molecular hypotheses
- Amphoteric surface sites
- Unidentate surface complexes
- Three planes in interfacial region
  - —Surface plane for adsorption
  - —β-plane
  - —Diffuse layer plane
- Charge (σ)/potential (ψ) relationship
  - —$\sigma_o = -0.117^{3/2}\sinh(ZF\psi_d/2RT)$
  - —$\psi_0 - \psi_\beta = \sigma_o/C_1$; $\psi_\beta - \psi_d = (\sigma_o + \sigma_\beta)/C_2$
- Ion association model: ionic aqueous and surface species, single ion activity coefficients
  - —Davies equation for aqueous-phase species
  - —Adsorbing species activity correction

simple consequence of an increase in counter ion binding with increasing counter ion concentration.

B. Model Input Data and Parameters

Preparation and characterization of the $Fe(OH)_3$(a) mineral phase was described by Davis and Leckie (1978a). All the experimental data values used to model surface complexation were obtained from Davis (1977). Reaction stoichiometries and parameters employed in the surface complexation model are listed in Table II. Properties of the solid, including capacitance values, acidity constants, and bulk electrolyte binding constants, have been previously reported by Davis and Leckie (1978b). Solution species hydrolysis, acidity, and complex formation constants were obtained from Smith and Martell (1976). The Ag–S_2O_3 species formation constants were available for six complexes at 25°C, 4.0 mol·L^{-1}

Table II

Model Parameters

Surface area	600 $m^2 \cdot g^{-1}$
Inner-layer capacitance	140 $\mu F \cdot cm^{-2}$
Outer-layer capacitance	20 $\mu F \cdot cm^{-2}$
Site-density	3.3 $sites \cdot nm^{-2}$
	pK
$>SOH + H^+ = >SOH_2{}^+$	+5.1
$>SOH = >SO^- + H^+$	−10.7
$>SOH + Na^+ = >SO^-–Na^+ + H^+$	−9.0
$>SOH + NO_3^- + H^+ = >SOH_2^+–NO_3^-$	+6.9
$>SOH + S_2O_3^{2-} + H^+ = >SS_2O_3^- + H_2O$	+7.0
$>SOH + S_2O_3^{2-} + H^+ = >SOH_2^+–S_2O_3^{2-}$	+9.7
$>SOH + Ag^+ = >SOAg^0 + H^+$	−3.5
$>SOH + Ag^+ = >SO^-–Ag^+ + H^+$	−3.9
$>SOH + S_2O_3^{2-} + Ag^+ + H^+ = >SS_2O_3Ag^0 + H_2O$	+16.8

ionic strength (also listed are constants for three complexes at 20°C, zero ionic strength), but no conversion was performed to obtain constants at zero ionic strength. It was assumed that differences in the values would be minor, as is similarly found in the extrapolated values published by Morel and Hering (1993). HYDRAQL treated all the constants as those for 25°C, zero ionic strength, and calculated the appropriate 0.1 $mol \cdot L^{-1}$ ionic strength value as shown in Table III.

The solution species listed in Table III reflect the chemistry of Ag and S_2O_3. Silver(I) exists as the free ion, Ag^+, and its hydrolysis products, $AgOH^0$ and $Ag(OH)_2^-$. Thiosulfate exists in solution as the diprotic acid, $H_2S_2O_3^0$, and its conjugate bases, $HS_2O_3^-$ and $S_2O_3^{2-}$. The complexes considered include the 1:1 species ($AgS_2O_3^-$), the mononuclear-polyligand species ($Ag(S_2O_3)_2^{3-}$ and $Ag(S_2O_3)_3^{5-}$), and the polynuclear-polyligand species ($Ag_2(S_2O_3)_4^{6-}$, $Ag_3(S_2O_3)_5^{7-}$, and $Ag_6(S_2O_3)_8^{10-}$). Speciation calculations with HYDRAQL showed that Ag is dominantly in the Ag^+ form, while thiosulfate is principally in the fully deprotonated form. The concentrations of both these solution species were invariant across the pH range studied. Formation of $AgS_2O_3^-$ and $Ag(S_2O_3)_2^{3-}$ complexes is appreciable across the pH range of interest, while other complexes, notably the polynuclear species, were negligible at the concentrations used in the study.

C. Model Simulations of Experimental Data

A subjective approach was taken to obtain model fits. Model fits were initially optimized by matching the model and experimental sorption edges (i.e., surface

Table III

Calculated Acidity, Hydrolysis, and Complex Formation Equilibria for Silver(I) and Thiosulfate

Equilibrium	pK (0.1 mol·L^{-1} ionic strength)
$Ag^+ + OH^- \Leftrightarrow AgOH^0$	11.90[a]
$Ag^+ + 2OH^- \Leftrightarrow Ag(OH)_2^-$	23.79[a]
$H^+ + S_2O_3^{2-} \Leftrightarrow HS_2O_3^-$	1.17[a]
$2H^+ + S_2O_3^- \Leftrightarrow H_2S_2O_3^0$	0.56[a]
$Ag^+ + S_2O_3^{2-} \Leftrightarrow AgS_2O_3^-$	6.93[b]
$Ag^+ + 2S_2O_3^- \Leftrightarrow Ag(S_2O_3)_2^{3-}$	12.72[b]
$Ag^+ + 3S_2O_3^- \Leftrightarrow Ag(S_2O_3)_3^{5-}$	14.78[b]
$2Ag^+ + 4S_2O_3^- \Leftrightarrow Ag_2(S_2O_3)_4^{6-}$	28.23[b]
$3Ag^+ + 5S_2O_3^- \Leftrightarrow Ag_3(S_2O_3)_5^{7-}$	42.58[b]
$6Ag^+ + 8S_2O_3^- \Leftrightarrow Ag_6(S_2O_3)_8^{10-}$	85.23[b]

Note. Values used in calculation are from Smith and Martell (1976).
[a]Reference data for 25°C, 0 mol·L^{-1} ionic strength.
[b]Reference data for 25°C, 4.0 mol·L^{-1} ionic strength assumed equivalent to 0 mol·L^{-1} ionic strength value.

concentration of a sorbing species as a function of pH) at the 50% uptake point. Parameters for the modeled sorption edge were then adjusted slightly to increase goodness of fit between modeled and experimental data. In the binary surface complex cases, both inner-sphere and outer-sphere configurations were used in HYDRAQL. Combinations of inner-sphere and outer-sphere surface complexes, including binary and ternary, were tested when modeling the Ag with the thiosulfate system. In determining a suitable set of model reactions, the number of reaction stoichiometric expressions needed to achieve an acceptable goodness of fit was kept to a minimum.

1. Ag Sorption

In the original effort, Davis (1977) employed the formation of $>SO^-{-}Ag^+$ and $>SO^-{-}AgOH^0$ surface complex species (where the dash, –, indicates outer-sphere type complex) to fit the experimental data of Ag sorption onto $Fe(OH)_3$(a). Using the TLM, a reasonable fit was obtained using the former complex alone (Fig. 2). Furthermore, Davis (1977) only presented a fit for the pH range 4 to 9, thereby omitting additional data at higher pH values. The TLM model result does not pass through the data above pH 9 either but, like the experimental data, did not reach 100% Ag uptake. Employing an inner-sphere $>SOAg^0$ surface complex species, a good fit was obtained in the pH 6.5 to 8 range; the high-pH range was overestimated, while the low-pH range was underestimated. Silver ion is presumed

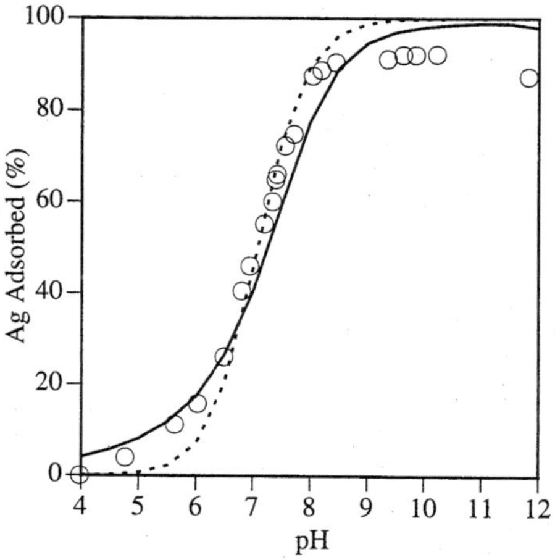

Figure 2 TLM model pH–sorption edge using inner-sphere (dashed line) and outer-sphere (solid line) reaction stoichiometries for the sorption of Ag onto $Fe(OH)_3$(a).

to be tightly coordinated with aquoligands, which supports the use of an outer-sphere complex in this case.

2. S_2O_3 Sorption

The tetrahedral structure of the thiosulfate molecule is similar to that of the sulfate molecule with a substituted oxo-group (Benjaminm and Leckie, 1982; Shriver *et al.,* 1994). Thiosulfate can act as a monodentate η^1–S ligand, a monohapto bidentate bridging ligand (μ,η^1–S), or a dihapto chelating η^2–S,O ligand (Wells, 1975). The central atomic location is occupied by a S molecule bound by double bonds to another S atom and an O atom, and by single bonds to two other O atoms. The singly bound O atoms each possess a single negative charge. Since bond-forming electrons are dynamic, the double-bond bridge between the central S atom and the O atom is shared among all O atoms. It is also possible that electrons forming the S=S bond may be, with some probability, shared with S–O bonds.

Surface complex species describing S_2O_3 sorption were not obtained by Davis (1977). Due to the bulkiness of the thiosulfate anion, an outer-sphere surface complex species of the type $>SOH_2^+–S_2O_3^{2-}$ would be expected. A good fit is obtained with this configuration using the TLM (Fig. 3). Use of an inner-sphere complex produces a sorption–pH gradient which is too steep. The outer-sphere behavior of thiosulfate is also consistent when compared to selenate ion, a species of similar chemical structure, whose sorption edge has been shown to be affected by ionic strength when sorbed to goethite (Hayes *et al.,* 1987), indicative of outer-sphere complex formation.

3. Ag with S_2O_3 Sorption

Davis (1977) proposed that this system would produce sorption of the Ag^+ ion and an $AgS_2O_3^-$ complex. In his presentation of the modeling fit, data below pH 4.55 were omitted in the $[S_2O_3^{2-}]_T = 4 \cdot 10^{-7}$ mol·L^{-1} data set. These data points reflected a local sorption maximum of about 80% uptake at around pH 4.85.

Sorption modeling using the constants for 20°C, zero ionic strength resulted in a ternary surface complex formation constant (pK = 18.0) about 1.2 log units greater than 25°C, 4 mol·L^{-1} ionic strength solution species formation constants. Although both sets of solution species formation constants resulted in similarly shaped predicted sorption edges, the latter set of formation constants provided a markedly better optimum fit, especially around the aforementioned local sorption maximum at low pH. The 25°C, 4 mol·L^{-1} ionic strength set of constants also provided a smaller residual in the range above pH 8. Thus, a discussion of the complexation chemistry follows which is based on the predicted sorption edge using the 25°C, 4.0 mol·L^{-1} ionic strength Ag–S_2O_3 solution species complex formation constants.

A combination of the $>SO^-$–Ag^+ and $>SS_2O_3^-$ complex species in the TLM model resulted in the best simulation of the uptake reversal (Fig. 4). Use of the outer-sphere configuration for S_2O_3 sorption, as found appropriate previously in the binary complex simulation, did not do well in the ternary system. The simultaneous use of both inner- and outer-sphere S_2O_3 surface complex configurations showed that the inner-sphere complex clearly dominated. However, the effect of an inner-sphere complex in the binary system was not observed, although the sur-

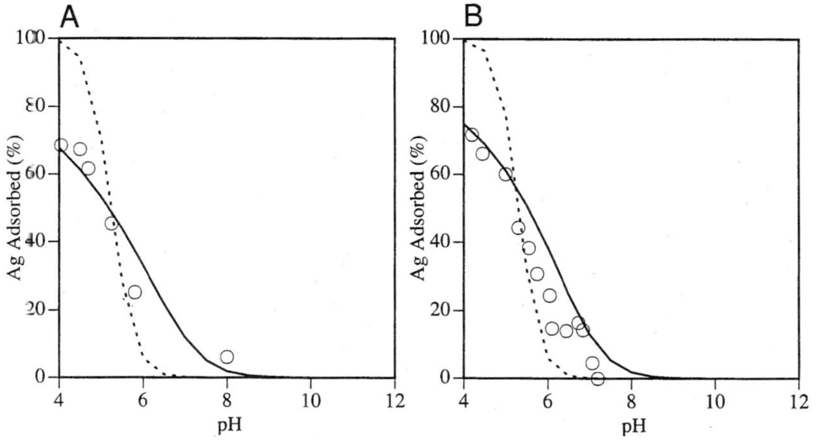

Figure 3 TLM model pH-sorption edge using inner-sphere (dashed line) and outer-sphere (solid line) reaction stoichiometries for the sorption of (A) $4 \cdot 10^{-6}$ and (B) $4 \cdot 10^{-7}$ mol·L^{-1} S_2O_3 onto $Fe(OH)_3$(a).

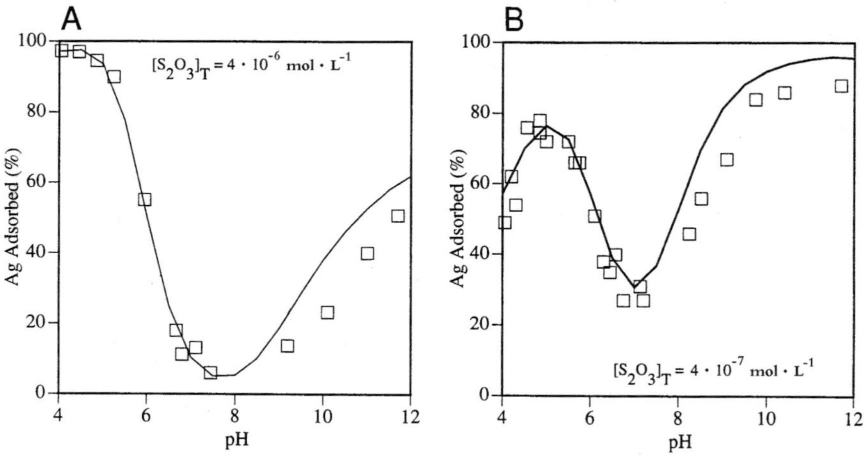

Figure 4 TLM model pH-sorption edge using $>SO^{-}–Ag^{+}$ and $>SS_2O_3^{-}$ reaction stoichiometries along with an inner-sphere ternary surface complex formation stoichiometry for the sorption of $4 \cdot 10^{-7}$ mol·L^{-1} Ag with (A) $4 \cdot 10^{-6}$ and (B) $4 \cdot 10^{-7}$ mol·L^{-1} S_2O_3 onto $Fe(OH)_3$(a).

face complex may have been present at low concentrations ($\leq 1\%$), and therefore did not make a noticeable contribution to the sorption trend. Formation of the inner-sphere S_2O_3 complex also leads to the possible formation of a $>SS_2O_3Ag$ ternary surface complex which is consistent with the best fit.

The typical approach to modeling such sorption data is based on reproducing macroscopic observed trends using the fewest stoichiometries. However, this example with S_2O_3 surface complexation clearly demonstrates that modeling results and their interpretation are not inherently absolute, and such results should only be used insofar as to suggest possible scenarios which are consistent with a data set. Therefore, confirmation of such observations with direct methods is highly desirable, and may be provided by spectroscopic analysis.

D. Possible Reaction Pathways

The inner-sphere S_2O_3 surface complex appears to form in the presence of Ag (Fig. 5). A mechanism for this to occur can be postulated. First, the chemistry reflected by the reaction stoichiometries in the best fit will be described. In the initial uptake from pH 4 to 4.55, the $>SS_2O_3Ag$ surface complex is the principle configuration leading to Ag sorption. The Ag desorption observed in the range pH 4.55 to 7 closely follows the thiosulfate pH sorption edge, and reflects a decrease in ligand-mediated Ag sorption. Another reversal at about pH 7 reflects the increasing dissociation of $S_2O_3^{2-}$ from the surface because of amphoteric surface changes, and the subsequent formation of the $>SO^-–Ag^+$ complex. As in the binary Ag sorption case, Ag uptake is overestimated. Use of the inner-sphere Ag complex yields worse predicted uptake.

The inner-sphere $S_2O_3^{2-}$ complex might be initiated by the outer-sphere association of a $AgS_2O_3^-$ complex with the protonated surface group $>SOH_2^+$. It is possible that one of the O atoms on the thiosulfate anion coordinates through hydrogen bonding with the two protons on the surface. Either of the S atoms (refer to the central S atom as S_c, and to the other as S_e) could then coordinate with the surface O atom (hereafter referred to as O_s). Presumably, this interaction weakens the hydrogen bonds on O_s, and may result in the release of a H_2O molecule (similar to condensation in some organic polymer formation reactions). Condensation of a water molecule is favored by the mix of single- and double-bonded O atoms, and is possibly enhanced by charge stabilization by association of one single-bonded O with an Ag^+ ion. Under low-pH conditions, the condensation step may be enhanced by equilibrium formation of the hydronium ion, H_3O^+. Charge stabilization may be most important in terms of decreasing the distance of closest approach of the thiosulfate molecule to the surface moiety. The reduction may come about

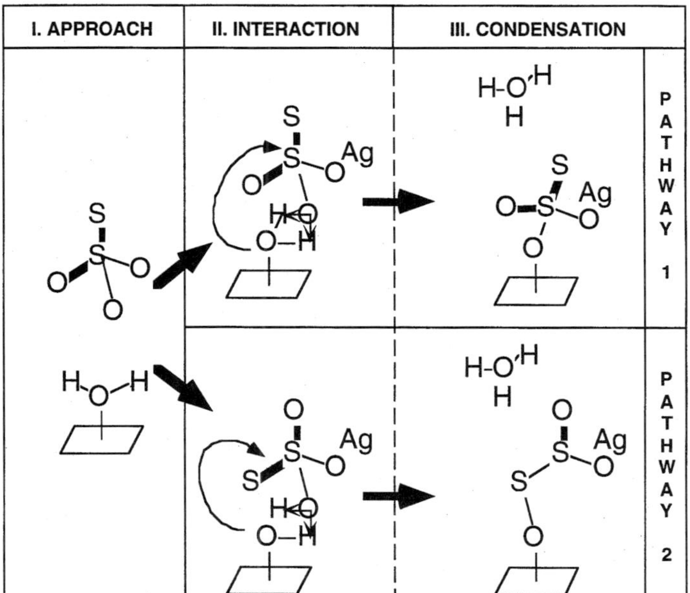

Figure 5 Postulated pathways for the formation of an inner-sphere ternary surface complex from the sorption of Ag and S_2O_3 onto $Fe(OH)_3$(a).

from directionally concentrating negative charge on the O atoms closest to the Ag^+ ion. This effectively would reduce bond delocalization between S^c and the three O atoms, permitting interaction of either S_c or S_e with O_s. It is also likely that hydrogen bonding between surface protons and a thiosulfate O atom precedes interaction of O_s with S_c or S_e, as it would probably be necessary for O_s to be destabilized beforehand.

Although Na^+ is present as part of the supporting electrolyte, inner-sphere $S_2O_3^{2-}$ complexation is not observed in the absence of Ag because formation of a $NaS_2O_3^-$ complex is minor relative to formation of an $AgS_2O_3^-$ complex (Fig. 6). The former reaches maximum distribution above pH 10, and is negligible in the species distribution below pH 5. Therefore, Na^+ simply is not bound to $S_2O_3^{2-}$ when the surface hydroxyl group is protonated and ready for $S_2O_3^{2-}$ binding.

The preceding binding and condensation scenario is proposed but not confirmed, but it does present the question of which S is involved in coordinating with O_s. A plausible mechanism involving either S atom can be postulated. In pathway 1 (Fig. 5), the surface site is involved in nucleophilic attack on the central S atom, similar to the first step in aldol condensation (Fig. 7).

Pathway 1 would also be favored if the silver atom is covalently bound to and occupies S_e, permitting only O_s bonding with S_c. However, no direct evidence of the covalent bond between Ag and S_e in the $AgS_2O_3^-$ complex has been demonstrated in aqueous solution, although this possibility cannot be overlooked because

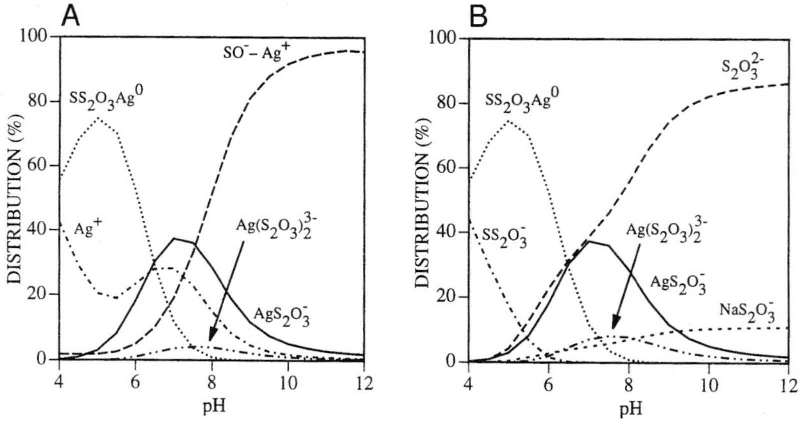

Figure 6 Equilibrium distribution of (A) Ag and (B) S_2O_3 species in a HYDRAQL simulation of the adsorption of $4{\cdot}10^{-7}$ mol·L^{-1} Ag and $4{\cdot}10^{-7}$ mol·L^{-1} on 10^{-3} mol·L^{-1} $Fe(OH)_3$(a) in 0.1 mol·L^{-1} $NaNO_3$. THe following species were also considered in the model calculation, but distribution percentage values were negligible: $AgOH^0$, $Ag(OH)_2^-$, $Ag(S_2O_3)_3^{5-}$, $Ag_2(S_2O_3)_4^{6-}$, $Ag_3(S_2O_3)_5^{7-}$, $Ag_6(S_2O_3)_8^{10-}$, $H_2S_2O_3^-$, and $H_2S_2O_3^0$.

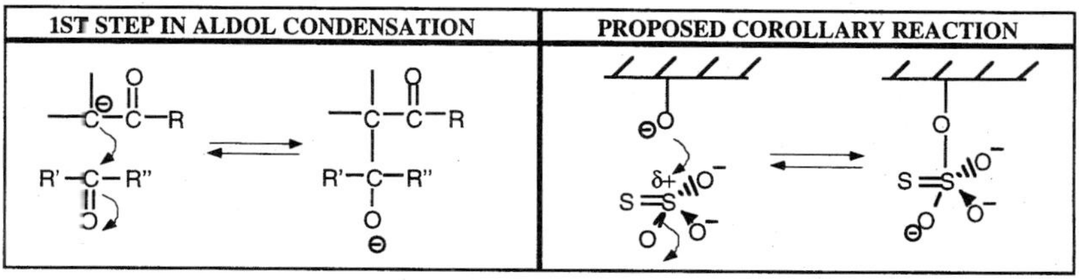

Figure 7 Schematic of the first step in the aldol condensation reaction, and a proposed corollary reaction for the binding of a thiosulfate molecule to an oxide surface with a fully deprotonated site.

of such bonding between these two atoms in other instances. Firstly, bonding of silver atoms to the external S atom of the thiosulfate ligand has been described for the molecular structure of the $NaAgS_2O_3H_2O$ crystal (Cavalca *et al.*, 1970). Further, the organic reaction, sufonatothio-de-halogenation, also known as Bunte salt formation (March, 1985),

$$RX + S_2O_3^{2-} \underset{1}{\rightarrow} R\text{–}S\text{–}SO_3^- + X^-,$$

might serve as a potential parallel in the formation of an $Ag\text{–}S\text{–}SO_3^-$ solution complex (R = primary or secondary alkyl group, X = halide).

Should Ag^+ form an ion pair with the thiosulfate molecule through a negatively charged O atom, pathway 2 may occur where the electron-rich surface O bonds to the external S atom, saturating the S=S double bond. The external S atom is relatively unshielded compared to the central S atom, and may be a preferred bridging point in the formation of a surface complex. As schematically illustrated in Figure 7, however, the central S atom would undergo a change in valence state. A redox step is possible, and may be exemplified by the decomposition of thiosulfate to sulfite, sulfate, and tetrathionate, which has been extensively studied, and even shown to be catalyzed by pyrite (Xu and Schoonen, 1995).

IV. DETAILING PATHWAYS USING SPECTROSCOPY

Spectroscopic analysis might be utilized to elucidate the structure of this surface complex. The S_e could serve as a bridging atom, reducing the double bond to S_c to a single bond. Also, S_e may have relatively less shielded access to O_s. On the other hand, S_c could be the participating atom since it would simply be exchanging one S–O bond for another.

The spectroscopic evidence that would be needed would include the presence of S–S bond stretch (difference between a single and double bond) and S–O decoupling; in one configuration, S_c is bonded to three O atoms and S_e to none, and in the other, S_c is bonded to two O atoms and S_e to one. The pattern of bond bending spectral information may be different due to the iso structure of one configuration and the tert structure of the other. The maximum bonded chain length may also be evident in spectral data (the S_c bonded configuration has a chain length of three atoms, while the S_e bonded configuration has a four-atom chain length).

Attenuated total reflectance Fourier transform infrared spectroscopy (ATR-FTIR) was recently applied to the confirmation of inner-sphere complexation of $B(OH)_3^0$ and $B(OH)_4^-$ to various minerals, including amorphous iron hydroxide (Su and Suarez, 1995). By systematically subtracting reference spectra from the spectra of more complex samples, ATR-FTIR information about the surface coordination of salicylic acid on aluminum and iron(III) oxides was obtained (Biber and Stumm, 1994).

In addition, *in situ* extended X-ray absorption fine structure (EXAFS) data can be applied to the distinction of inner- and outer-sphere complexes of binary and ternary surface complexes, as was shown in a study on the sorption of selenium oxyanions on goethite (Hayes *et al.,* 1987).

If the inner-sphere $>SS_2O_3Ag$ and/or $>SS_2O_3^-$ complexes are indeed confirmed by spectroscopic or other means, then it would be demonstrated that ternary surface complex formation would not necessarily be simply derived from the nature of the complexes formed in the binary systems. That is, a solution complex could initially be bound to the surface as an ion pair at the β-plane, but could eventually be bound at the o-plane through a chemical rearrangement at the mineral–solution interface.

REFERENCES

Bard, C. C., Murphy, J. J., Stone, D. L., and Terhaar, C. J. 1976. Silver in photoprocessing effluents. *J. Water Pollut. Control Fed.* 48:389–394.

Benjamin, M. M., and Leckie, J. O. 1982. Effects of complexation by Cl, SO_4, and S_2O_3 on adsorption behavior of Cd on oxide surfaces. *Environ. Sci. Technol.* 16:162–170.

Biber, M. V., and Stumm, W. 1994. An in situ ATR-FTIR study: The surface coordination of salicylic acid on aluminum and iron (III) oxides. *Environ. Sci. Technol.* 28:763–768.

Cavalca, L., Mangia, A., Palmieri, C., and Pelizzi, G. 1970. The crystal and molecular structure of $NaAgS_2O_3H_2O$. *Inorg. Chim. Acta* 4:299–304.

Cooley, A. C., Dagon, P. W., Jenkins, P. W., and Robillard, K. A. 1988. Silver and the environment. *J. Imaging Technol.* 14:183–189.

Davis, J. A. 1977. Adsorption of trace metals and complexing ligands at the oxide/water interface. Ph.D. Thesis. Stanford University, Stanford, CA.

Davis, J. A., and Leckie, J. O. 1978a. Effect of adsorbed complexing ligands on trace metal uptake by hydrous oxides. *Environ. Sci. Technol.* 12:1309–1315.

Davis, J. A., and Leckie, J. O. 1978b. Surface ionization and complexation at the oxide/water interface. II. Surface properties of amorphous iron oxyhydroxide and adsorption of metal ions. *J. Colloid Interface Sci.* 67:90–107.

Davis, J. A., and Leckie, J. O. 1979. Speciation of adsorbed ions at the oxide/water interface. *In* "Chemical Modeling in Aqueous Systems" (E. A. Jenne, Ed.), ACS Symp. Ser. 93, pp. 299–317. American Chemical Society, Washington, DC.

Hayes, K. F., and Leckie, J. O. 1987. Modeling ionic strength effects on cation adsorption at hydrous oxide/solution interfaces. *J. Colloid Interface Sci.* 115:564–572.

Hayes, K. F., Roe, A. L., Brown, G. E., Hodgson, K. O., Leckie, J. O., and Parks, G. A. 1987. In situ X-ray absorption study of surface complexes: Selenium oxyanions on α-FeOOH. *Science* 238:783–786.

Luoma, S. N., Dagovitz, R., and Axtmann, E. 1990. Temporally intensive study of trace metals in sediments and bivalves from a large river-estuarine system: Suisun Bay/delta in San Francisco Bay. *Sci. Total Environ.* 97/98:685–712.

Luoma, S. N., and Phillips, D. J. H. 1988. Distribution, variability, and impacts of trace elements in San Francisco Bay. *Marine Pollut. Bull.* 19:413–425.

March, J. 1985. "Advanced Organic Chemistry," 3rd ed. Wiley, New York.

Miller, L. A., and Bruland, K. W. 1995. Organic speciation of silver in marine waters. *Environ. Sci. Technol.* 29:2616–2621.

Morel, F. M. M., and Hering, J. G. 1993. "Principles and Applications of Aquatic Chemistry." Wiley, New York.

Nelson, M. B. 1978. Kinetics and mechanisms of the oxidation of ferrous sulfide. Ph.D. Thesis, Stanford University, Stanford, CA.

Papelis, C., Hayes, K. F., and Leckie, J. O. 1988. HYDRAQL: A Program for the Computation of Chemical Equilibrium Composition of Aqueous Batch Systems Including Surface-Complexation Modeling of Ion Adsorption at the Oxide-Solution Interface. Technical Report No. 306, Department of Civil Engineering, Stanford University, Stanford, CA.

Petering, H. G., and McClain, C. J. 1991. Silver. *In* "Metals and Their Compounds in the Environment" (E. Merian, Ed.), pp. 1191–1202. VCH Publishers, New York.

Shriver, D. F., Atkins, P., and Langforf, C. H. 1994. "Inorganic Chemistry," 2nd ed. W. H. Freeman, New York.

Smith, G. J., and Flegal, A. R. 1993. Silver in San Francisco Bay estuarine waters. *Estuaries* 16:547–558.

Smith, R. M., and Martell, A. E. 1976. "Critical Stability Constants," Vol. 4, "Inorganic Complexes." Plenum, New York.

Stern, O. 1924. Zur theory der electrolytischer doppelschicht. *Z. Elektrochem.* 30:508–516.

Su, C., and Suarez, D. L. 1995. Coordination of adsorbed boron: A FTIR spectroscopic study. *Environ. Sci. Technol.* 29:302–311.

Wells, A. F. 1975. "Structural Inorganic Chemistry," 4th ed. Oxford Univ. Press, Oxford.

Xu, Y., and Schoonen, M. A. A. 1995. The stability of thiosulfate in the presence of pyrite in low-temperature aqueous solutions. *Geochim. Cosmochim. Acta* 59:4605–4622.

Yates, D. E. 1974. Site-binding model of the electrical double-layer at the oxide-water interface. *J. Chem. Soc. Faraday Trans. 1* 70:1807–1818.

Chapter 15

Extended X-Ray Absorption Fine Structure (EXAFS) Analysis of Aqueous Sr^{II} Ion Sorption at Clay–Water Interfaces

Chia-Chen Chen,[1] Charalambos Papelis,[2] and Kim F. Hayes[1]

[1]Environmental and Water Resources Engineering, Department of Civil and Environmental Engineering, The University of Michigan, Ann Arbor, Michigan; [2]Desert Research Institute, Water Resources Center, University and Community College System of Nevada, Las Vegas, Nevada

The fate and transport of strontium (Sr) in the environment have attracted considerable attention during the last two decades because of the potential for radioactive Sr migration away from waste disposal and nuclear test sites into the accessible environment. In the past, estimates of Sr retardation were based on laboratory sorption studies and field experiments. Because ion-exchange processes are typically assumed to account entirely for Sr retardation, clay minerals with high surface area and high cation-exchange capacity are likely to control the migration of Sr. The objective of this study was to test the hypothesis that Sr forms only mononuclear, outer-sphere surface complexes on clay minerals, regardless of clay type, sorption site type, or solution conditions. This hypothesis was tested by collecting and analyzing X-ray absorption spectroscopy data of Sr sorbed on clay minerals with different physiochemical characteristics, namely, kaolinite, illite, hectorite, and montmorillonite, under a variety of geochemical conditions. All clays were purified to recover the clay size fraction and to re-

Adsorption of Metals by Geomedia

move carbonates, iron and manganese oxides, and organic carbon. The experimental conditions included sample collection at pH values of 4.5 to 10, ionic strengths of 0.1, 0.01, and 0.001 mol·L^{-1}, and $NaNO_3$ and $CaCl_2$ as background electrolytes. Extended X-ray absorption fine structure analysis was consistent with formation of Sr outer-sphere, mononuclear complexes regardless of clay mineral and solution composition. For all sorption samples, the structural parameters (Sr–O bond distances and coordination numbers) were similar, namely, a Sr–O distance of 2.57 ± 0.02 Å and a coordination number of 6 ± 20%. The similarity of these results to the geometry of the aqueous Sr ion and the absence of any evidence for second-shell neighbors suggest the formation of Sr outer-sphere, mononuclear complexes. These results have significant implications for the migration of Sr in the environment and can be used to constrain the choice of possible surface reactions used in surface complexation modeling.

I. INTRODUCTION

Strontium-90 is one of the most frequently released radionuclides in the environment. With a half-life of 28.8 yr, it is the most long-lived strontium (Sr) radioisotope (Lederer and Shirley, 1978) and the main reason for the increased interest in Sr fate and transport in the environment during the last two decades. Strontium-90 can be found in low-level radioactive waste, but substantially higher activities are present in military and civilian high-level nuclear waste, where it is produced either by neutron activation or by fission of transuranic isotopes. Health-hazard-related concerns are not based on the toxicity of Sr, but rather on the biogeochemical similarity of this element to calcium, a fact which can potentially result in accumulation of this radioisotope in bones (Laws, 1993).

Burial of radioactive waste in geological repositories is presently considered the most likely method for disposal of radioactive waste. In addition, underground nuclear tests conducted in the United States between 1953 and 1992 resulted in additional production of radioisotopes, including Sr. Regardless of origin, migration of radioactive Sr from nuclear waste disposal sites or nuclear test sites could potentially contaminate adjacent aquifers. In order to assess the potential of Sr migration from waste disposal and nuclear test sites into the accessible environment, investigations of Sr interactions with soils and aquifer materials are required.

Extensive studies of Sr sorption on several types of mineral phases have been conducted, including studies with oxides and hydroxides (Huang and Stumm, 1973; Kinniburgh *et al.*, 1976), pure clay minerals (Rafferty *et al.*, 1981; Konishi *et al.*, 1988; Czurda and Wagner, 1991; Adeleye *et al.*, 1994; Grutter *et al.*, 1994) and natural soils or clay mixtures (Yanagi *et al.,* 1989; Ohnuki, 1994; Ahmad, 1995; Yasuda *et al.,* 1995). In addition, estimation of Sr retardation was attempted in several field studies (Pickens *et al.,* 1981; Reynolds *et al.,* 1982; Jackson and Inch, 1983; Gillham *et al.,* 1990). In most Sr sorption studies with natural soils or clay minerals it is assumed that ion-exchange reactions are solely responsible for Sr retardation (e.g., Ohnuki, 1994). It is recognized, however, that confirmation of a particular sorption mechanism cannot be obtained without detailed information on the molecular-level structure of sorption complexes, which can only be obtained from appropriate spectroscopic experiments (Sposito, 1986). This detailed, molecular-level information for Sr complexes is to date very limited.

X-ray absorption spectroscopy (XAS) has been used successfully in the past to determine the local molecular structure of complexes sorbed at mineral–water interfaces (Hayes *et al.,* 1987; Bassett and Brown, 1990; Brown, 1990; Fendorf *et al.,* 1994; O'Day *et al.,* 1994a,b; Papelis and Hayes, 1996). In this study, we used XAS and, specifically, extended X-ray absorption fine structure (EXAFS) analysis to enhance our understanding of Sr sorption on clay mineral surfaces in general and, specifically, to test the often made assumption that Sr forms only outer-sphere complexes on clay mineral surfaces.

Metal ions are believed to sorb on clay minerals either at permanent-charge (ion-exchange) sites or at surface-hydroxyl (amphoteric) sites (edge sites where the crystal structure is interrupted) (Inskeep and Baham, 1983; Schindler *et al.,* 1987; Fletcher and Sposito, 1989; Cowan *et al.,* 1992; Hayes *et al.,* 1995; Papelis and Hayes, 1996). Strontium sorption in subsurface environments may be controlled by reactions with either of these site types. In this study, Sr sorption on four different clay minerals with different physiochemical characteristics was studied to assess the effects of mineral structure and geochemical solution conditions on the type of Sr sorption complexes formed. Clay minerals were used as sorbents because they posses permanent charge sites which are generally thought to be primarily responsible for Sr retardation in subsurface environments. High-specific-surface-area clay minerals, such as those selected for this study, are representative of the most important types typically present in soils, and usually account for a substantial fraction of the cation-exchange capacity (CEC) of soils. The characterization of Sr sorption on surface hydroxyl sites, which assumes greater importance for Sr retention when high-CEC clay minerals are absent, is the subject of a future paper (Hayes *et al.,* unpublished data).

The following clay minerals were used: kaolinite, illite, hectorite, and montmorillonite. These clay minerals were specifically selected to represent clays of different types, with varying cation-exchange capacity, surface charge, and structure. For example, kaolinite is a 1:1 clay mineral with essentially no layer charge, a small specific surface area, and a very low CEC, whereas the other three are 2:1 clay minerals and have substantially higher layer charges, specific surface areas, and cation-exchange capacities. Hectorite is trioctahedral, whereas the other three are dioctahedral. Kaolinite has essentially no layer charge, hectorite and montmorillonite have an intermediate layer charge, and illite has the highest layer charge. Finally, hectorite and montmorillonite are swelling clays (smectites), whereas kaolinite and illite are nonswelling (Bohn *et al.,* 1985; Dixon and Weed, 1989; Sposito, 1989; Velde, 1992).

The combination of the selected clay mineral sorbents, along with a range of geochemical solution conditions, was designed to represent a broad range of conditions likely to be found in natural soil environments and to promote Sr sorption on different sorption site types. By collecting and analyzing XAS spectra under those conditions we were able to test the hypothesis that Sr forms only mononuclear, outer-sphere surface complexes on clay mineral surfaces, regardless of clay type, sorption site type, or solution conditions. The results from this study are significant, not only because they provide direct spectroscopic evidence for the type of Sr sorption complexes formed on clay minerals, but also because they can be used to impose constraints on the representation of Sr sorption complexes in surface complexation models (Hayes and Katz, 1996).

II. EXPERIMENTAL

A. Materials and Methods

The adsorbents, Georgia kaolinite (KGa-1), Montana illite (IMt-1), California hectorite (SHCa-1), and Arizona montmorillonite (SAz-1), were obtained from the Source Clays Repository of the Clay Minerals Society at the University of Missouri (Columbia, MO). The clays were purified according to the procedure described in Kunze and Dixon (1986). The purification steps included sedimentation, to recover the clay size fraction (<2 μm), as well as steps designed to remove carbonates, iron and manganese oxides, and organic carbon.

A summary of the purification steps follows. Clay separation was accomplished by repeated cycles of washing with sodium carbonate followed by centrifugation and decanting of the supernatant. Carbonate removal was accomplished by reaction with sodium acetate and pH adjustment with glacial acetic acid. Iron and manganese oxides were removed by reaction with a dithionate–citrate–bicarbonate reagent. Finally, organic matter was removed by a combination of washing with 30% hydrogen peroxide, heating the solution to 80°C, and centrifuging and decanting the supernatant.

The purified clay was then converted to the Na form by reacting with 4 N sodium chloride. The excess salt was removed by washing several times with Milli-Q water. The resulting clay slurry was stored in Milli-Q water in a 0.01 mol·L^{-1} sodium nitrate solution. The Ca-montmorillonite was prepared using the same procedure by reacting with 2 mol·L^{-1} calcium chloride and was stored in a 0.005

mol·L^{-1} calcium chloride solution. The reported surface areas for kaolinite, illite, hectorite, and montmorillonite are 16, 100, 480, and 820 m^2·g^{-1}, respectively (Fripiat and Van Olphen, 1979; Schwarzenbach *et al.*, 1993).

Strontium(II) aqueous solutions were prepared from crystalline strontium nitrate obtained from Fisher Scientific (Fair Lawn, NJ). Strontium standard solutions used for atomic absorption analysis were also obtained from Fisher Scientific. Sodium nitrate or calcium chloride was used to adjust the ionic strength, and nitric acid (from Fisher Scientific) and sodium hydroxide (J.T. Baker Inc., Phillipsburg, NY) were used to adjust the pH of the solution. All reagents used in this study were of ACS reagent-grade quality or better and all solutions were prepared using CO_2-free Milli-Q water.

Samples were prepared in individual 250-ml polypropylene centrifuge bottles. The samples were equilibrated on a shaker table at 25°C for at least 24 hr. Following equilibration, the final pH of the suspension was measured and the samples were centrifuged at a relative centrifugal force of 13,180 for 30 min to achieve solid–liquid separation. The supernatant solution was removed and samples were loaded as wet pastes into aluminum sample holders and sealed with Kapton tape. A 1-ml aliquot was removed from the supernatant and was used to measure the concentration of Sr with a Perkin-Elmer 1100B Atomic Absorption Spectrophotometer in graphite furnace mode. Fractional uptake was calculated by comparing the Sr concentration in the supernatant to the concentration of a control sample (a sample to which all reagents, except for the clay mineral adsorbent, were added).

B. XAS Data Analysis

X-ray absorption spectra were collected at the Stanford Synchrotron Radiation Laboratory (SSRL) using wiggler beam lines 4-3 and 7-3 with Si(220) monochromators, a beam energy of 3.0 GeV, and a beam current ranging from 40 to 100 mA. Spectra were collected in fluorescence mode for all samples using a 13-element Ge array detector. Beam energy was calibrated by assigning the first inflection point on the absorption edge of the Bi foil to an energy of 16,388 eV. Multiple scans (4 to 12) were collected, depending on Sr concentration, to obtain a signal-to-noise ratio sufficiently high for EXAFS analysis. To further improve the signal-to-noise ratio, by minimizing thermal disorder, data were collected at low temperature (approximately 20 K) using a liquid-helium cryostat. Comparison between data collected at ambient temperature and data collected at liquid-helium temperature suggests that the Sr coordination environment is not significantly altered at liquid-helium temperatures.

Acceptable spectra were averaged and the background absorption was subtracted by a straight line through the pre-edge region and a fifth-order, three-segment spline fit in the region above the edge. The background-subtracted spectra were normalized using a Victoreen polynomial and tabulated McMaster atomic absorption fall-off coefficients (McMaster *et al.,* 1969). The spectra were then converted from energy to frequency space using the photoelectron wave factor, k, which is defined by

$$k = \sqrt{\frac{2m_e}{h^2}(E-E_o)}, \qquad [1]$$

where m_e is the mass of the electron, h is Planck's constant, E is the kinetic energy of the electron, and E_0 is the energy of the photoelectron at $k = 0$. The EXAFS spectra were subsequently multiplied (weighted) by k^3 to compensate for the decrease in the amplitude of EXAFS oscillations with increasing k (Teo, 1986; Brown *et al.,* 1988).

The normalized, background-subtracted, k^3-weighted EXAFS spectra for reference compounds and the sorption samples were filtered over similar k ranges and Fourier-transformed with an unsmoothed window to obtain the radial structure functions (RSF), a pair correlation between the central absorbing atom and its nearest neighbors. Individual major peaks in the Fourier transform were back-transformed to produce filtered EXAFS spectra in order to isolate individual contributions to the EXAFS (Brown *et al.,* 1988; Sayers and Bunker, 1988). Structural information for the sorption complexes was then determined by fitting the filtered unknown spectra with nonlinear least-squares methods using phase and amplitude parameters obtained from analysis of the reference compounds.

Least-squares fitting of each shell results in the determination of two parameters, namely, the coordination number (CN) and the bonding distance (R). The Debye–Waller factor, σ^2, and E_0, defined as the energy at which $k = 0$, were also allowed to vary during the optimization of CN and R. As long as the differences between σ^2 and E_0 and the values for the reference compound (reported as $\Delta\sigma^2$ and ΔE_0, respectively) are relatively small (e.g., ± 0.001 and ± 2, respectively), the fit is considered valid. The resulting CN and R, based on this fitting procedure, are expected to be accurate to $\pm 20\%$ and ± 0.02 Å, respectively, for the first shell (Brown, 1990).

In this study, SrO was used as a reference compound because it has a simple, well-defined structure with a single first-neighbor Sr–O distance and no interference from second-neighbor shells. Empirically determined phase and amplitude parameters, obtained from appropriate reference compounds, are often preferable to theoretically derived parameters because uncertainties in methods for background removal, normalization, and estimation of the electron mean free path are automatically accounted for when reference compounds and unknown samples are analyzed in exactly the same fashion (Koningsberger and Prins, 1988). A reference compound is considered ideal when both of the following conditions are satisfied: the reference compound contains the same central-atom-backscatterer pair as the unknown, and the distance and coordination environment are also about the same

(Sayers and Bunker, 1988). The similarities between the first-shell Sr coordination environment of SrO and that of the unknown samples (see the following section and Table II for details) make SrO a particularly appropriate reference compound for analysis of these samples.

III. RESULTS AND DISCUSSION

Strontium sorption experiments were conducted at different solution conditions using purified kaolinite, illite, hectorite, and montmorillonite. Spectroscopic data were collected at different pH values, using background electrolytes of different compositions and concentrations, and with a 10^{-4} mol·L^{-1} total Sr^{II} concentration. A summary of some characteristic experimental conditions for the XAS samples is given in Table I. The background-subtracted, normalized, k^3-weighted EXAFS spectra for two reference compounds, SrO and $SrCO_3$, and six sorption samples are shown in Figure 1. Figure 2 shows the radial structure functions resulting from the Fourier transformation of the corresponding EXAFS spectra. It should be noted that several additional spectra under different pH and background electrolyte conditions were collected but are not shown in Figures 1 and 2 because of space considerations and because they were essentially indistinguishable from the spectra shown.

The results of XAS data analysis, including coordination numbers, bond distances (R (Å)), Debye–Waller factors ($\Delta\sigma^2$ (Å^2)), and ΔE_0 (eV), are listed in Table II. The CN and bond distance determine the coordination environment in a sorption sample or a reference compound; $\Delta\sigma^2$ is a measure of the difference in ther-

Table I

Summary of XAS Experimental Conditions

Clay mineral	Background electrolyte cation concentration (mol·L^{-1})	pH	Uptake (%)	Surface coverage, Γ (10^{-3} mol·L^{-1}·m^{-2})
Kaolinite	0.01 Na	9.96	20	1.25
Illite	0.01 Na	4.55	42	0.42
Illite	0.01 Na	9.01	43	0.43
Hectorite	0.01 Na	9.85	40	0.09
Hectorite	0.001 Na	9.97	80	0.19
Montmorillonite	0.01 Na	9.75	77	0.19
Montmorillonite	0.01 Na	7.60	77	0.19
Montmorillonite	0.01 Ca	8.95	80	0.21
Montmorillonite	0.1 Ca	9.05	40	0.09

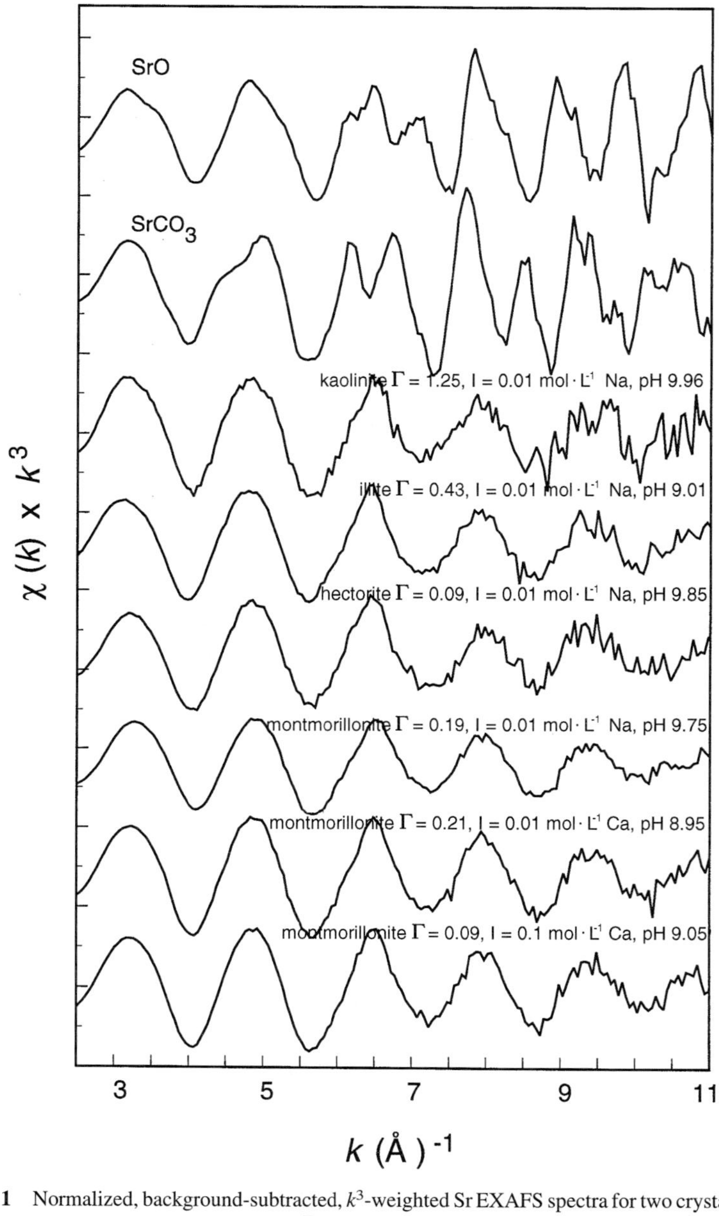

Figure 1 Normalized, background-subtracted, k^3-weighted Sr EXAFS spectra for two crystalline reference compounds, SrO and $SrCO_3$, and Sr^{II} sorption samples on kaolinite, illite, hectorite, and montmorillonite at a pH of approximately 9, under different surface coverages and background electrolyte conditions. Only the reference compound samples show "beat patterns" suggesting the presence of more than one coordination shell surrounding the central Sr atom. In all sorption samples, including samples not shown here, only one frequency is evident, suggesting the presence of a single oxygen shell surrounding Sr and the formation of outer-sphere, mononuclear complexes.

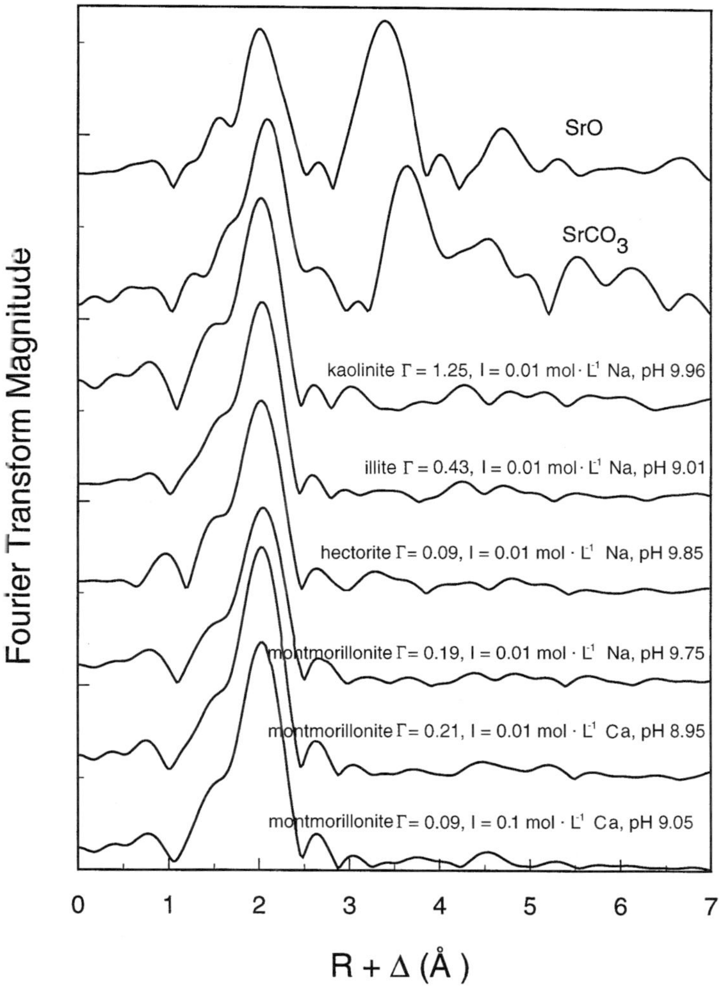

Figure 2 Radial structure functions produced by Fourier-transforming the EXAFS spectra of the crystalline reference compounds and the sorption samples shown in Figure 1. Only one peak is visible in the spectra of the sorption samples, at approximately 2 Å—uncorrected for phase shift—which corresponds to a shell of oxygens surrounding the central Sr atom. The Sr–O distance is approximately constant in all sorption samples.

mal and static disorder effects between the reference compound and the unknown sample being analyzed; and ΔE_0 accounts for any differences in the coordination environment (and therefore binding energy) between the reference compound and the unknown sample being analyzed.

Inspection of the EXAFS spectra for all sorption samples in Figure 1 clearly indicates the presence of a single frequency. This frequency is essentially the same in all sorption samples, suggesting the presence of the same single shell of neighbors surrounding the central absorbing atom at approximately the same distance. The spectra of the reference compounds, however, clearly show "beat patterns," the result of multiple frequencies, suggesting the presence of more than one shell of backscattering atoms, as expected, based on the structure of the crystalline compounds. These results are in agreement with the Fourier transforms of these spectra, shown in Figure 2. A single peak, corresponding to a single shell of neighbors (backscattering atoms) is evident for all sorption samples at approximately 2 Å (uncorrected for phase shift), whereas several peaks, corresponding to multiple shells surrounding Sr, are evident for both reference compounds.

Least-squares fit analysis of this single backscattering shell, using phase and amplitude parameters derived from the SrO reference compound, is consistent with a Sr–O shell, as expected. For all Sr sorption samples on different clay minerals under similar solution conditions ($\boldsymbol{I} = 0.01$ mol·L^{-1} $NaNO_3$ and pH values of 8.95 to 9.96), the Sr–O bond distances are 2.57 to 2.58 ± 0.02 Å, and the coordination numbers vary between approximately 5.7 and 7.2 (Table II). The bond distances remain essentially the same within the accuracy of the method (±0.02

Table II
Results of XAS Analysis

Clay mineral	Background electrolyte concentration (mol·L^{-1})	pH	CN	*R* (Å)	$\Delta\sigma^2$ (Å^2)	ΔE_o (eV)
Kaolinte	0.01 Na	9.96	7.192	2.57	0.00856	−0.8
Illite	0.01 Na	4.55	6.015	2.58	0.00736	−2.06
Illite	0.01 Na	9.01	6.335	2.58	0.00685	−1.36
Hectorite	0.01 Na	9.85	7.012	2.57	0.00869	−0.984
Hectorite	0.001 Na	9.97	7.195	2.56	0.00827	0.489
Montmorillonite	0.01 Na	9.75	5.655	2.58	0.00773	1.73
Montmorillonite	0.01 Na	7.6	5.836	2.58	0.00621	1.508
Montmorillonite	0.01 Ca	8.95	5.727	2.57	0.00548	0.864
Montmorillonite	0.1 Ca	9.05	5.838	2.57	0.00600	−0.323
$SrCO_3$[a]			6.7384	2.62	0.00698	2.4785
$SrCO_3$[b]			9	2.636		
SrO[b]			6	2.575		

[a]Structural parameters obtained using phase and amplitude parameters obtained from SrO reference compound spectra.
[b]The structure refinement parameters were obtained for crystalline $SrCO_3$ from de Villiers (1971), as reported in Pingitore *et al.* (1992), and for SrO from Ewald and Hermann (1931).

Å), regardless of pH (Table II), while the coordination numbers show a slightly higher fluctuation as a function of sorbent; i.e., for montmorillonite coordination numbers tend to be lower, approximately 5.7; for illite they tend to be slightly higher, approximately 6.1; and for hectorite and kaolinite they tend to be the highest, approximately 7.2

These coordination numbers, however, are still within the possible margin of error (±20%) associated with CN determinations from EXAFS analysis. In the case of Sr, the uncertainty in CN determinations may be even higher than the average ±20%. As can be seen from Table II, based on phase and amplitude parameters derived from SrO, the estimated average number of O atoms coordinated to Sr in the $SrCO_3$ model compound is 6.74, while the true CN is 9 (de Villiers, 1971, as reported in Pingitore *et al.*, 1992), an error of 25%. This larger error associated with the $SrCO_3$ CN using phase and amplitude parameters derived from SrO was also reported by Pingitore *et al.* (1992). D'angelo *et al.* (1996) reported that due to the evidence for Sr double-excitation K edges, the conventional analysis methods may result in systematic errors on the structural parameters, and in particular in underestimation of the CN.

Thus, the variation in coordination numbers among Sr sorption samples on different adsorbents may not be significant. Furthermore, these structural parameters are similar to those of Sr in aqueous solutions (D'angelo *et al.*, 1996). We may conclude that Sr maintains its aqueous solution geometry upon sorption at clay–water interfaces no matter what the solution conditions or adsorbents are, which is consistent with formation of mononuclear outer-sphere complexes. The results also show that different background electrolyte conditions (montmorillonite samples with different background electrolytes and different ionic strengths, or hectorite samples with different ionic strengths) have no effect on the coordination environment of Sr sorption complexes. In conclusion, the bond distances appear to be constant for all samples, within the accuracy of the method, and the coordination numbers seem to vary slightly as a function of adsorbent, although these differences may not be statistically significant.

No obvious beat patterns can be observed in any of the sorption samples in Figure 1, regardless of experimental conditions. Although the amplitude (and quality) of the EXAFS spectra is severely reduced beyond k values of approximately 9 $Å^{-1}$, because of the high disorder in the coordination environment of aqueous Sr, it is believed that the presence of second-neighbor Sr would be obvious in the first half of the spectrum (at k values smaller than 8 $Å^{-1}$; compare the EXAFS spectra of the reference compounds). The XAS study of Sr coprecipitated with calcite conducted by Pingitore *et al.* (1992) did indeed show distinctive beat patterns (Sr–Sr interactions) at low-k ranges. Sr ($Z = 38$) is a strong enough backscatterer to allow detection of second-neighbor Sr (Teo and Lee, 1979). This is evident from the spectra of the Sr reference compounds in Figures 1 and 2.

It is also believed that the formation of inner-sphere complexes or polynuclear species would result in decreased disorder in the Sr aqueous coordination environment, thereby allowing the detection of additional shells beyond the first. For example, the XAS studies of Co^{II} adsorption on kaolinite and γ-Al_2O_3 showed second-shell peaks for samples with low surface coverage when Co formed inner-sphere complexes on solid surfaces (Chisholm-Brause *et al.,* 1990; O'Day *et al.,* 1994b). In a previous study of Co sorption on montmorillonite (Papelis and Hayes, 1996), under similar conditions, a Co–Co second shell was observed, even at relatively low surface coverages, suggesting that the formation of polynuclear complexes or surface precipitates of divalent cations should also be observable. Thus, by analogy with past work on Co, a second peak would be expected for Sr sorption samples, even at low surface coverages, if inner-sphere or polynuclear complexes were formed. In this study, however, no second peak was observed, regardless of surface coverage.

It is conceivable for Sr, however, that additional neighbors (second or further shells) are not observed because of structural disorder caused by a wide distribution of Sr–surface metal ion distances. This scenario was, in part, tested by collecting the spectrum of a freshly precipitated, presumably disordered, Sr phase. Based on known solubilities of Sr phases, the precipitated phase must have been a disordered carbonate solid, but it was not characterized, nor is it shown. For this sample, however, additional backscattering peaks beyond the first shell (presumably due to Sr backscattering) were apparent. This suggests that structural disorder alone is probably not responsible for the absence of backscattering peaks from second or more distant shells. Hence, the XAS data are most consistent with the formation of outer-sphere, mononuclear complexes, regardless of experimental conditions and adsorbates.

Despite the similar EXAFS spectra for all solution conditions and clay mineral samples, certain conclusions can be drawn based on the systems investigated and experiments performed. For example, the similar structural environment of Sr^{II} sorbed on montmorillonite and illite suggests that regardless of the extent and location of interlayer charge, Sr^{II} retains its primary hydration sheath. We speculated that a relatively large divalent cation like Sr (ionic radius of 0.118 nm compared to Co^{II} with an ionic radius of 0.075 nm) might have a tendency to dehydrate upon sorption to an interlayer fixed-charge site in a swelling clay like montmorillonite or to a relatively highly charged site like that present on illite. This was not the case, however, regardless of pH, ionic strength, or whether the major exchanging cation was Na^{I} or Ca^{II} (Table II and data not shown). In contrast, we correctly anticipated that Sr^{II} would retain its primary hydration sheath and only form outer-sphere complexes over a range of pH and ionic strength values with the surface hydroxyl sites on kaolinite and on the edge sites of hectorite and montmorillonite at high pH and ionic strength (Table II and data not shown). By absence of any second-shell features, the EXAFS analysis confirms, within the limitations previously pointed out, that Sr^{II} forms only outer-sphere complexes at clay–water interfaces, regardless of solution pH and ionic strength for a variety of clay types.

IV. CONCLUSIONS

This study included Sr XAS data collection and analysis using a variety of clay mineral adsorbents, namely, kaolinite, illite, hectorite, and montmorillonite. The geochemical experimental conditions included sample collection at pH values of 4.5 to 10, ionic strengths of 0.1, 0.01, and 0.001 $mol{\cdot}L^{-1}$, and sodium nitrate and calcium chloride background electrolytes. For all sorption samples, the structural parameters (Sr–O bond distances and coordination numbers) were similar, namely, a Sr–O distance of 2.57 ± 0.02 Å and a coordination number of 6 ± 20%. The similarity of these results to the geometry of the aqueous Sr ion and the absence of any evidence for second-shell neighbors suggest the formation of Sr outer-sphere, mononuclear complexes on clay surfaces regardless of clay mineral adsorbent and solution composition.

This study provides direct evidence to support the generally accepted hypothesis of Sr outer-sphere complex formation on mineral surfaces. These results have significant implications for the migration potential of Sr in the environment. Formation of outer-sphere complexes on clay mineral surfaces exclusively, regardless of clay type and geochemical conditions, suggests that major cations in solution would be able to compete effectively with Sr for sorption sites, resulting in increased Sr mobility. In addition, these results can be used to constrain the choice of possible surface reactions used in surface complexation modeling. Based on this study, Sr should be modeled as forming outer-sphere, mononuclear complexes with ion-exchange sites.

ACKNOWLEDGMENTS

Financial support was provided by the National Science Foundation through grant BCS-8958407 and from a gift from the Procter and Gamble Company. Graham George (SSRL) is gratefully acknowledged for providing the EXAFS data reduction program. We thank Eric J. Wight and Lynn E. Katz for assistance with XAS data collection. The facilities and staff of the SSRL and the Biotechnology Program were essential to the success of this project. SSRL is operated by the U.S. Department of Energy. The SSRL Biotechnology Program is supported by NIH and DOE.

REFERENCES

Adeleye, S. A., Clay, P. G., and Oladipo, M. O. A. 1994. Sorption of cesium, strontium and europium ions on clay-minerals. *J. Mater. Sci.* 29:954–958.

Ahmad, S. 1995. Competitive adsorption of Sr-90 on soil sediments, pure clay phases and feldspar minerals. *Appl. Radiat. Isotopes* 46:287–292.

Bassett, W. A., and Brown, G. E., Jr. 1990. Synchrotron radiation: Applications in the earth sciences. *Annu. Rev. Earth Planet. Sci.* 18:387–447.

Bohn, H. L., McNeal, B. L., and O'Connor, G. A. 1985. "Soil Chemistry." Wiley, New York.

Brown, G. E., Jr. 1990. Spectroscopic studies of chemisorption reaction mechanisms at oxide-water interfaces. *In* "Mineral-Water Interface Geochemistry" (M. F. Hochella, Jr., and A. F. White, Eds.), pp. 309–363. Mineralogical Society of America, Washington, DC.

Brown, G. E., Jr., Calas, G., Waychunas, G. A., and Petiau, J. 1988. X-ray absorption spectroscopy and its applications in mineralogy and geochemistry. *In* "Mineral-Water Interface Geochemistry" (F. C. Hawthorne, Ed.), pp. 431–512. Mineralogical Society of America, Washington, DC.

Chisholm-Brause, C. J., O'Day, P. A., Brown, G. E., and Parks, G. A. 1990. Evidence for multinuclear metal-ion complexes at solid/water interfaces from x-ray absorption spectroscopy. *Nature* 348:528–531.

Cowan, C. E., Zachara, J. M., Smith, S. C., and Resch, C. T. 1992. Individual sorbent contributions to cadmium sorption on ultisols of mixed mineralogy. *Soil Sci. Soc. Am. J.* 56:1084–1094.

Czurda, K. A., and Wagner, J.-F. 1991. Cation transport and retardation processes in view of the toxic waste deposition problem in clay rocks and clay liner encapsulation. *Eng. Geol.* 30:103–113.

D'angelo, P., Nolting, H. F., and Pavel, N. V. 1996. Evidence for multielectron resonances at the Sr K-edge. *Phys. Rev. A* 53:798–805.

De Villiers, J. P. R. 1971. Crystal structures of aragonite, strontianite, and witherite. *Am. Mineral.* 56:758–767.

Dixon, J. B., and Weed, S. B. 1989. "Minerals in Soil Environments," Soil Science Society of America Book Series. Soil Science Society of America, Madison, WI.

Ewald, P. P., and Hermann, C. 1931. "Strukturbericht, 1913–1928." Akademische Verlagsgesellschaft, Leipzig.

Fendorf, S. E., Sparks, D. L., Lamble, G. M., and Kelley, M. J. 1994. Applications of x-ray absorption fine structure spectroscopy to soils. *Soil. Sci. Soc. Am. J.* 58:1583–1595.

Fletcher, P., and Sposito, G. 1989. The chemical modeling of clay/electrolyte interactions for montmorillonite. *Clay Miner.* 24:375–391.

Fripiat, J. J., and Van Olphen, H. 1979. "Data Handbook for Clay Materials and Other Non-metallic Minerals." Pergamon, New York.

Gillham, R. W., Robin, M. J. L., and Ptacek, C. J. 1990. A device for in situ determination of geochemical transport parameters. 1. Retardation. *Ground Water* 28:666–672.

Grutter, A., Vongunten, H. R., Rossler, E., and Keil, R. 1994. Sorption of strontium on unconsolidated glaciofluvial deposits and clay-minerals—Mutual interference of cesium, strontium and barium. *Radiochim. Acta* 64:247–252.

Hayes, K. F., Chen, C.-C., and McAvoy, D. C. 1995. Quaternary ammonium surfactant effects on sorption of trace metals onto quartz and aluminosilicates. *Soil Sci. Soc. Am. J.* 59:380–387.

Hayes, K. F., and Katz, L. E. 1996. Application of x-ray absorption spectroscopy for surface complexation modeling of metal ion sorption. *In* "The Physics and Chemistry of Mineral Surfaces" (P. V. Brady, Ed.), pp. 145–220. CRC Press, Boca Raton, FL.

Hayes, K. F., Roe, A. L., Brown, G. E., Jr., Hodgson, K. O., Leckie, J. O., and Parks, G. A. 1987. In situ x-ray absorption study of surface complexes: Selenium oxyanions on α-FeOOH. *Science* 238:783.

Huang, C. P., and Stumm, W. 1973. Specific adsorption of cations on hydrous γ-Al_2O_3. *J. Colloid Interface Sci.* 43:409–420.

Inskeep, W. P., and Baham, J. 1983. Adsorption of Cd(II) and Cu(II) by Na-montmorillonite at low surface coverage. *Soil Sci. Soc. Am. J.* 47:660–665.

Jackson, R. E., and Inch, K. J. 1983. Partitioning of strontium-90 among aqueous and mineral species in a contaminated aquifer. *Environ. Sci. Technol.* 17:231–237.

Kinniburgh, D. G., Jackson, M. L., and Syers, J. K. 1976. Adsorption of alkaline earth, transition, and heavy metal cations by hydrous oxide gels of iron and aluminum. *Soil Sci. Soc. Am. J.* 40:796–799.

Koningsberger, D. C., and Prins, R. 1988. "X-Ray Absorption: Principles, Applications, Techniques of EXAFS, SEXAFS, and XANES." Wiley, New York.

Konishi, M., Yamamoto, K., Yanagi, T., and Okajima, Y. 1988. Sorption behavior of cesium, strontium and americium ions on clay materials. *J. Nucl. Sci. Technol.* 25:929–933.

Kunze, G. W., and Dixon, J. B. 1986. Pretreatment for mineralogical analysis. *In* "Methods of Soil Analysis, Part 1. Physical and Mineralogical Methods" (A. Klute, Ed.), pp. 91–100. American Society of Agronomy/Soil Science Society of America, Madison, WI.

Laws, E. A. 1993. "Aquatic Pollution: An Introductory Text." Wiley, New York.

Lederer, C. M., and Shirley, V. S. 1978. "Table of Isotopes." Wiley, New York.

McMaster, W. H., Del Grande, N. K., Mallett, J. H., and Hubbell, J. H. 1969. Compilation of x-ray cross-sections III. UCRL-50174, U.S. Atomic Energy Commission.

O'Day, P. A., Brown, G. E., Jr., and Parks, G. A. 1994a. X-ray absorption spectroscopy of cobalt(II) multinuclear surface complexes and surface precipitates on kaolinite. *J. Colloid Interface Sci.* 165:269–289.

O'Day, P. A., Parks, G. A., and Brown, G. E., Jr. 1994b. Molecular structure and binding sites of cobalt(II) surface complexes on kaolinite from x-ray absorption spectroscopy. *Clays Clay Miner.* 42:337–355.

Ohnuki, T. 1994. Sorption characteristics of strontium on sandy soils and their components. *Radiochim. Acta* 64:237–245.

Papelis, C., and Hayes, K. F. 1996. Distinguishing between interlayer and external sorption sites of clay minerals using x-ray absorption spectroscopy. *Colloids Surf. A* 107:89–96.

Pickens, J. F., Jackson, R. E., and Inch, K. J. 1981. Measurement of distribution coefficients using a radial injection dual-tracer test. *Water Resources Res.* 17:529–544.

Pingitore, N. E., Lytle, F. W., Davis, B. M., Eastman, M. P., Eller, P. G., and Larson, E. M. 1992. Model of incorporation of Sr^{2+} in calcite: Determination by x-ray absorption spectroscopy. *Geochim. Cosmochim. Acta* 56:1531–1538.

Rafferty, P., Shiao, S. Y., Binz, C. M., and Meyer, R. E. 1981. Adsorption of Sr(II) on clay minerals: Effects of salt concentration, loading and pH. *J. Inorg. Nucl. Chem.* 43:797–806.

Reynolds, W. D., Gillham, R. W., and Cherry, J. A. 1982. Evaluation of distribution coefficients for the prediction of strontium and cesium migration in a uniform sand. *Can. Geotechnol. J.* 19:92–103.

Sayers, D. E., and Bunker, B. A. 1988. Data analysis. *In* "X-Ray Absorption: Principles, Applications, Techniques of EXAFS, SEXAFS and XANES" (D. C. Koningsberger and R. Prins, Eds.), pp. 211–253. Wiley, New York.

Schindler, P. W., Liechti, P., and Westall, J. C. 1987. Adsorption of copper, cadmium and lead from aqueous solution to the kaolinite/water interface. *Netherlands J. Agric. Sci.* 35:219–230.

Schwarzenbach, R. P., Gschwend, P. M., and Imboden, D. M. 1993. "Environmental Organic Chemistry." Wiley, New York.

Sposito, G. 1986. Distinguishing adsorption from surface precipitation. *In* "Geochemical Processes at Mineral Surfaces" (J. A. Davis and K. F. Hayes, Eds.), pp. 217–228. American Chemical Society, Washington, DC.

Sposito, G. 1989. "The Chemistry of Soils." Oxford Univ. Press, New York.

Teo, B. K. 1986. "EXAFS: Basic Principles and Data Analysis," Vol. 9, "Inorganic Chemistry Concepts." Springer-Verlag, New York.

Teo, B. K., and Lee, P. A. 1979. Ab initio calculations of amplitude and phase functions for extended x-ray absorption fine structure spectroscopy. *J. Am. Chem. Soc.* 101:2815–2832.

Velde, B. 1992. "Introduction to Clay Minerals." Chapman & Hall, London.
Yanagi, T., Watanabe, M., and Yamamoto, K. 1989. Sorption behavior of cesium and strontium ions on mixtures of clay sorbents. *J. Nucl. Sci. Technol.* 26:861–864.
Yasuda, H., Uchida, S., Muramatsu, Y., and Yoshida, S. 1995. Sorption of manganese, cobalt, zinc, strontium, and cesium onto agricultural soils—Statistical-analysis on effects of soil properties. *Water Air Soil Pollut.* 83:85–96.

Chapter 16

Structure and Composition of Uranium^{VI} Sorption Complexes at the Kaolinite–Water Interface

H. A. Thompson, G. A. Parks, and G. E. Brown, Jr.
Department of Geological and Environmental Sciences, Stanford University, Stanford, California

We have investigated the nature of U^{VI} sorption by kaolinite by applying X-ray absorption spectroscopy (XAS) to batch sorption samples ($2 \cdot 10^{-5}$ $mol \cdot L^{-1}$ total U; 0.5 $g \cdot L^{-1}$ kaolinite; 1.5 $mmol \cdot L^{-1}$ $NaNO_3$; pH 6.0 to 7.5; presence and absence of ambient CO_2; 24-hr equilibration; 1.2 to 1.7 $\mu mol \cdot m^{-2}$). Under these conditions, XAS shows that the U sorption complex always contains a uranyl moiety (UO_2^{2+}) surrounded by five equatorial O atoms. Si (and/or Al) is present as a second neighbor under all of the conditions studied. At the lowest pH conditions ($7.0 > pH \geq 6.0$) in the presence of air, mononuclear U species dominate. At higher pH values ($7.5 > pH \geq 7.0$) in air, small, multinuclear U species dominate. We believe that all of the U species formed in the presence of air are sorption complexes bound in an inner-sphere fashion to kaolinite. The number of U atoms constituting each sorption complex is similar to that of the dominant species in aqueous solution under similar pH and ionic strength conditions. In the absence of CO_2 at slightly higher pH values ($8.0 \geq pH \geq 7.1$), dominant U species contain at least three U atoms. This is indicative of the formation of multinuclear sorption complexes or surface precipitates. In either case, the kaolinite surface appears to enhance clustering of U atoms in the absence of CO_2 relative to homogeneous solution. We consider the formation of a homogeneous precipitate, either a pure U phase or a mixed U–Si/Al phase, unlikely in these systems.

Adsorption of Metals by Geomedia

I. INTRODUCTION

Uranium contamination of surface pond waters is one of the environmental legacies of uranium processing in the United States. Sorption of uranium on a high-surface-area solid followed by flocculation and removal of the solid has been proposed as one method for removing radionuclides from pond waters. Clay minerals are candidates for the sorbent owing to their high surface areas (Van Olphen and Fripiat, 1979), negative surface charge and hence propensity for sorption of cations, abundance, and low cost. Several studies have demonstrated U sorption by kaolinite in aqueous media (Muto *et al.,* 1965; Giblin, 1980; Sekine *et al.,* 1991). The resulting U–kaolin complexes have not been characterized, however. We therefore know little about the structure of the complexes as a function of changing solution conditions.

In this study we characterize U sorption complexes on kaolinite as influenced by solution pH and carbonate content. Published uranium sorption studies suggest that pH and dissolved carbonate concentration are the dominant solution variables in controlling uptake by clay minerals (Muto *et al.,* 1965; Giblin, 1980; Sekine *et al.,* 1991), thus we vary these parameters near values that maximize U uptake. We use X-ray absorption spectroscopy (XAS) as a direct molecular probe to characterize sorbed uranium species, specifically to confirm that U is associated with the sorbent and to establish the mode of sorption, e.g., inner- vs. outer-sphere, mono- vs. multinuclear, or surface precipitation.

II. BACKGROUND

A. Past Sorption Studies

Researchers have consistently assumed (Tripathi, 1984; Hsi and Langmuir, 1985) and demonstrated that the uranyl moiety is preserved upon sorption of U^{VI} by a variety of solid metal oxides (Maya, 1982; Ho and Miller, 1986; Combes, 1988; Dent *et al.,* 1992; Manceau *et al.,* 1992; Chisholm-Brause *et al.,* 1994; Waite *et al.,* 1994). The number and position of more distant atoms in U^{VI} sorption complexes and the mode of sorption have been subjects of conflicting interpretation. In systems containing relatively few ligands, aqueous carbonate concentration and solution pH have been shown to affect the extent of U^{VI} uptake by solids. It is suspected that these parameters affect uptake by influencing aqueous U^{VI} speciation. If this is the case, then aqueous carbonate concentration and pH can reasonably be expected to affect the structures of the resulting sorption complexes and possibly the mode of sorption.

Uranium^{VI} uptake by kaolinite increases with pH up to pH $\approx$ 7.5 above which it decreases in the presence of air (Muto *et al.,* 1965; Giblin, 1980; Sekine *et al.,*

1991) (Fig. 1). This behavior has consistently been attributed to the presence of carbonate in solution because no decrease in uptake is observed in the absence of CO_2. The often unstated assumption that accompanies this explanation is that U that desorbs is bound as an inner-sphere sorption complex; carbonate destroys or precludes formation of this inner-sphere complex by strongly complexing U, effectively excluding the solid and other potential ligands from the inner coordination shell of U. If U were bound as an outer-sphere complex, carbonate ligands could complex U without displacing solid ligands, and therefore would not likely cause its return to the solution phase through carbonate–surface ligand exchange, although it might result in desorption if bonding is largely electrostatic. If carbonate is responsible for U desorption by ligand exchange, then one would expect to find U solution species at pH values > 7.5 that contain carbonate group(s) in the inner coordination shell. This is consistent with calculations of dominant solution speciation (Grenthe *et al.*, 1992) and has been spectroscopically observed (Basile *et al.*, 1978; Madic *et al.*, 1983; Morse *et al.*, 1984; Allen *et al.*, 1995; Clark *et al.*, 1995).

The observation that carbonate "removes" U from the solid does not necessarily preclude carbonate from an inner-sphere U sorption complex, however. Uranium^{VI} has four to six equatorial ligand positions. In cases in which U^{VI} has been shown to adsorb to the solid in an inner-sphere fashion, two sorbent O atoms are bonded directly to U (Combes, 1988; Manceau *et al.*, 1992; Waite *et al.*, 1994). If

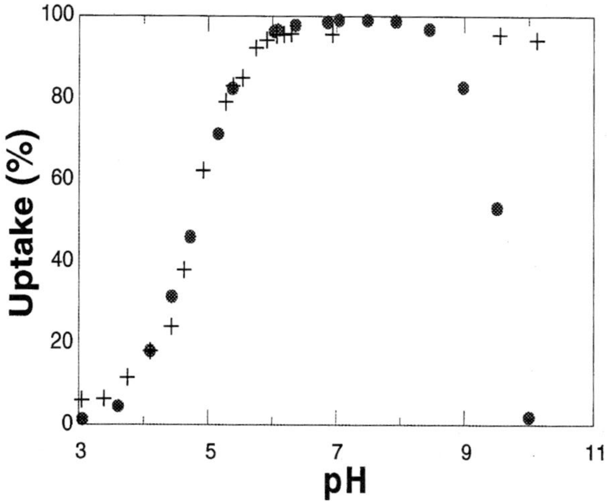

Figure 1 Uptake of U^{VI} by kaolinite as a function of pH in the presence (circles; Sekine *et al.*, 1991) and absence (crosses; Redden *et al.*, 1998) of carbonate. Note the drop in uptake above pH 8 in the presence of carbonate.

this is consistent for U sorbed on a variety of solids, two to four equatorial positions within the inner-sphere U complex remain available for bonding to carbonate or other ligands.

Inner-sphere U sorption on solids has been observed directly for iron oxide sorbents, but not for aluminosilicate sorbents. Several researchers (Combes, 1988; Manceau *et al.*, 1992; Waite *et al.*, 1994) have used XAS to demonstrate that U^{VI} is bound to hydrous iron oxides in a bidentate, inner-sphere fashion at pH values from 5 to 6. In spectra of the sorption samples, they were able to identify contributions from Fe neighbors located at distances that would only be possible if U were bound as an inner-sphere complex. Neither Dent *et al.* (1992) nor Chisholm-Brause *et al.* (1994) saw spectral contributions from Si or Al using the same technique to study U sorption by montmorillonite (both) and silica (Dent), however. These XAS results may indicate that U sorbs to montmorillonite and silica in an outer-sphere fashion, suggesting that the mode of U sorption by solids depends strongly on the composition and structure of the solid. Alternatively, the absence of Si and Al from the XAS spectra could be attributable to their weak scattering cross sections or to high levels of structural disorder in the sorption complex, in which case U could be bound as an inner-sphere complex (discussed further in Section II.B.2).

Solution pH modulates the extent of sorption in a variety of ways. Solid surface charge and the distribution of U among a variety of aqueous species are both pH-sensitive; thus pH affects the relative activities of the reactants involved in sorption, and consequently affects the composition of the sorption product. Because carbonate is the conjugate of a weak acid, solution pH affects the relative activities of carbonate species and therefore the distribution of U among hydroxo and carbonato complexes (if carbonate is present), thereby potentially affecting the composition of the sorption complex. Of course, this assumes that the composition of U sorption complexes is influenced by U solution speciation, a logical starting point, but one that has not been demonstrated for U sorption. It is also reasonable to suspect that complexes in the vicinity of the solid–water interface might have different thermodynamic formation constants than those in bulk solution. Specifically, formation of multinuclear sorption complexes might be expected in circumstances in which they do not form in bulk solution (James and Healy, 1972; Dillard and Koppelman, 1982). This has been observed in a variety of systems, both in solids dried after they have been equilibrated with aqueous solutions of the sorbing cation [Co^{II} on goethite (Schenck *et al.*, 1983), Cu^{II} on gibbsite (McBride *et al.*, 1984), Mn^{II} on boehmite (Bleam and McBride, 1985), and U^{VI} on zeolites (Bartlett and Cooney, 1989)] and *in situ* [Co^{II} on γ-Al_2O_3 (Chisholm-Brause *et al.*, 1990) and Co^{II} on kaolinite (O'Day *et al.*, 1994)]. Of the studies employing *in situ* methods to characterize U sorption complexes (Combes, 1988; Chisholm-Brause *et al.*, 1994; Dent *et al.*, 1992; Manceau *et al.*, 1992; Morris *et al.*, 1994), none have observed sorbed multinuclear species.

B. X-Ray Absorption Spectroscopy

1. Technique Capabilities for Sorption Studies

X-ray absorption spectroscopy has proven to be particularly valuable in studies of sorption at the mineral–water interface for several reasons: (1) its absorber specificity allows study of a selected element in the presence of many elements; (2) backscattering atoms that differ in atomic number (Z) by more than 3 from that of the absorber atom can be distinguished and roughly identified, and their number can be approximated; (3) interatomic distances can be determined very accurately ($\pm$0.02 to 0.04 Å for most backscattering atoms); (4) it is a nonvacuum technique (for absorber $Z > 20$) usable at ambient temperatures and is therefore applicable to solutions and suspensions as well as solids; and (5) it is sensitive to relatively low concentrations of the absorbing atom (Brown *et al.,* 1988).

In the best situations, XAS can verify that sorption has occurred and identify the mode of sorption and structure of the sorption complex (see Brown and Parks, 1989, for a review). To verify sorption, one or more atoms unique to the sorbent must be identified in the coordination environment around the sorbed atom. The number of sorbent atoms should be small enough that absorption or diffusion into the solid can be discounted. Combining interatomic distance information resulting from XAS with a knowledge of the structure of the solid sorbent, one can identify the mode of sorption (inner- vs. outer-sphere, mono- vs. multidentate) and in some cases what type of solid site is occupied by the sorbed atom. Number, distance, and identity of neighboring atoms provide information regarding the structure of the sorption complex. Polymerization or precipitation is indicated by detection of one or more sorbate ions in the coordination environment of the adion.

2. Uranium XAS Limitations

Limitations of XAS in studies of sorbed species are not insignificant. Because XAS samples all absorber atoms in the bulk sample, the resulting information represents a sum of all atomic environments in which the absorbing atom is present. If the absorbing atom is present only in sorption complexes at the solid–water interface, however, then XAS measurements are quite sensitive to the interfacial region. Numerical interpretation of XAS data relies on theoretical or experimental model compounds, thus the accuracy of resulting numbers is necessarily a function of whether or not a single structure predominates in the sample and how well the model compound represents that unknown structure. Since XAS is insensitive to low Z backscattering atoms under ambient conditions, one cannot use it to discern the arrangement of water molecules at the solid–water interface, except as they can be traced through their oxygen atoms in the near vicinity of an absorber atom.

To establish the limitations of XAS applied specifically to U sorption, we conducted a separate study (Thompson *et al.,* 1997) in which we analyzed XAS data for a number of U^{VI}-containing solids of known composition and structure. These "model compounds" contain U in the range of environments we might expect for the U–water–kaolinite system, including isolated U monomers and U in a multinuclear species or precipitate, each of which might include Si or Al (due to sorption or coprecipitation), carbonate, or nitrate. The results are briefly summarized in Table I, in which we identify the range of interatomic distances and coordination numbers for which certain atoms in the U environment can be detected by XAS. In addition to the coordination number and distance conditions specified in Table I, we would generally expect XAS to be capable of detecting the same atoms at shorter distances and/or larger coordination numbers.

Particular to the study of uranium sorption at the kaolinite–water interface, there are several implications of the results in Table I. Foremost, Si/Al atoms at distances that might be indicative of inner-sphere U sorption may be difficult to detect. This may explain their lack of detection in studies of U sorption on silica and montmorillonite, discussed earlier (Section II.A). If XAS succeeds in detecting them at all, it cannot distinguish between Al and Si backscatterers because of their similar scattering amplitudes and frequencies, thus likely precluding determination of the type of solid site occupied by the sorbed U atom. Because N and C are difficult to detect, it would be difficult to say whether nitrate or carbonate groups occupy equatorial positions. Nearest-neighbor U atoms contribute strongly to XAS spectra, suggesting that multinuclear species, when present, should be readily detectable. Conversely, an XAS spectrum that lacks a second-

Table I

EXAFS Detection Limits in the Uranyl Environment

Element	Number[a]	Distance (Å)[b]	Comment[c]
O_{ax}	2	1.8	Definite
O_{eq}	3	2.3	Definite
O_{eq}	2	2.5	Definite
N	2	2.9	Difficult
C	2	2.9	Difficult
Si	1	3.2	Possible
P	4	3.6	Possible
U	2	4.0	Definite
U	2	4.3	Definite
U	2	4.9	Probable
U	4	5.2	Probable

[a]Number of atoms of the given element at the reported distance.

[b]Distance between U and the neighboring atom.

[c]Certainty of detection, in descending order, is definite, probable, possible, and difficult.

neighbor U contribution is indicative of the predominance of mononuclear species, even if minor multiple species coexist. Because of the difficulty in detecting more distant U atoms, however, we probably could not discern between a multinuclear complex and a precipitate that have similar local coordination environments.

III. EXPERIMENTAL

A. Materials

We used KGa-2, a poorly crystallized kaolinite standard from Georgia, as received, as the sorbent. The XRD powder spectrum of KGa-2 confirmed its identity as kaolinite. Field emission scanning electron microscope (FE-SEM) images of KGa-2 revealed predominantly subhedral clay particles from <1 to 10 μm in diameter.

We dissolved U_3O_8 (NBS 950b) in dilute HNO_3 to prepare the U stock solution for sorption sample UK3. NIST uranium standard solution (10,000 $mg \cdot L^{-1}$ U in 5% HNO_3) was used in preparation of all other sorption samples. All other reagents, including sodium nitrate, nitric acid, and sodium hydroxide, were reagent grade. All water used in the preparation of sorption samples was singly deionized.

B. Methods

1. XAS Sorption Sample Preparation

Relevant quantities and measurement results are presented in Table II. With the exception of UK3, which was prepared with Ar-purged water in an Ar-filled glovebox to exclude CO_2, all samples were exposed to laboratory air, and hence to CO_2. During the course of their preparation, sorption samples contacted containers made of glass, Teflon, "heavy plastic" [probably polyethylene (HDPE)], LDPE, polypropylene, and Mylar.

In each sample vessel, sodium nitrate was added to ~5 L of U^{VI} stock solution to bring ionic strength to the values specified in Table II. Sodium hydroxide was added to each vessel to increase solution pH to "initial" pH values recorded in Table II. Final solution stocks contained 5 L of $2 \cdot 10^{-5}$ $mol \cdot L^{-1}$ total U^{VI}.

Kaolinite (2.5 g KGa-2) was added to each sample with mixing. Stirring of the resulting suspension continued for 24 hr, during which pH was adjusted with HNO_3 or NaOH to maintain the initial pH value (with the exception of UK3, in which no pH adjustment occurred). "Final" pH was measured after 24 hr for all

Table II

Sorption Sample Preparation Conditions

Sample ID	Atmosphere	Ionic strength (mmol·L^{-1})	Initial pH[a]	Post KGa-2 pH[b]	Final pH[c]	Γ^{d} (μmol·m^2)
UK3	CO_2-free	1.5	8.2	—[e]	7.5[f]	1.62 ± 0.03
UK12	Air	4.0	7.5	7.9	7.1	1.72 ± 0.03
UK10	Air	4.0	7.0	7.2	7.0	1.64 ± 0.03
UK8	Air	4.0	6.6	6.6	6.7	1.30 ± 0.03
UK6	Air	4.0	6.0	6.1	6.2	1.11 ± 0.03

[a]pH measured prior to addition of KGa-2.
[b]pH measured after adjustment following KGa-2 addition.
[c]pH measured at end of 24-hr equilibration, prior to centrifugation.
[d]Sorption density.
[e]Not measured.
[f]pH measured at end of 24-hr equilibration, after centrifugation and filtration.

samples but UK3. Sample supernatants were decanted and centrifuged at 5000 rpm for 1 hr (UK3) or 12,000 rpm for 10 min (all other samples). Resulting centrifugates were filtered through a 0.45-μm Millipore filter. Sorption density (Γ) for UK3 was calculated from neutron activation analysis (NAA) measurements of the U solution stock (pre-KGa-2 addition) and the centrifuged, filtered supernatant. All other sorption densities were calculated from measurements of XAS U L_{III}-edge jump for each sorption sample, calibrated using samples of known U concentration in a kaolinite (KGa-2) matrix (Fig. 2).

To make XAS samples, the solid centrifuge residue was loaded without drying into a 3-mm-thick polyethylene holder with Mylar windows. Several milliliters of the filtered supernatant solution from preparation of UK12 were separately loaded into a Teflon holder with Mylar windows for XAS data collection.

2. XAS Data Collection

Uranium L_{III}-edge XAS spectra were collected over the energy range ~17 to 18 keV at the Stanford Synchrotron Radiation Laboratory (SSRL) operating at 3 GeV and 40 to 90 mA on wiggler beam line IV-2. The X-ray beam was unfocused on Si(220) monochromator crystals, cut #2 ($\phi = 90°$). Sorption sample and supernatant solution spectra were collected in fluorescence mode with the sample oriented 45° to the incident beam. Fluorescence detection was accomplished with a 13-element Ge-array detector. Harmonic rejection was effected by 40 to 80% detuning of the incident beam. Three to sixteen scans were collected for each sample.

3. XAS Data Analysis

We provide here a description of the extended X-ray absorption fine structure (EXAFS) data analysis procedure, focusing on details specific to this study. Numerous review articles provide more complete accounts (Cramer and Hodgson, 1979; Sayers and Bunker, 1988). Unless otherwise stated, we conducted all of the data analysis using programs contained in EXAFSPAK, written by G. George of SSRL (George and Pickering, 1993).

Raw data files for each scan were calibrated individually by setting the position of the first inflection point of the calibration standard absorption edge equal to 17,166 eV, the nominal L_{III}-edge energy for elemental uranium (Vaughan, 1986). Although all of the uranium in this study is expected to be in the +VI oxidation state, not elemental, consistency rather than absolute energy calibration was the objective. The position of the first inflection varied by less than 2.5 eV among scans for a single sample. We calculated a weighted average of all calibrated scans for each sample to produce a single averaged spectrum for that sample. Weighting was proportional to the square of signal-to-noise, where the signal is defined as the magnitude of the edge jump and the noise level is determined by application of a high-pass filter to the data.

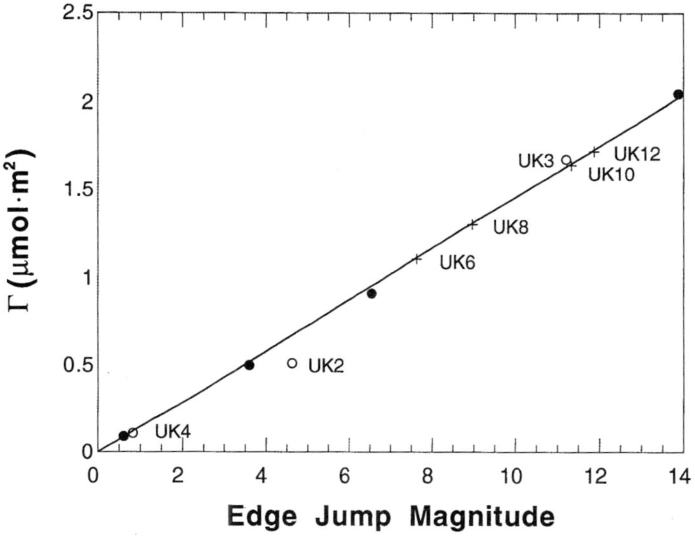

Figure 2 Calibration curve for determining surface coverage (Γ) from XAS edge-jump magnitude. A linear least-squares fit of data from calibration standards (solid circles) and sorption samples whose supernatants were analyzed by neutron activation analysis (open circles) provided an equation for calculating Γ for sorption samples with previously unknown Γ (crosses) from their edge-jump magnitudes.

We used first-order polynomials to approximate and subtract the pre-edge background. Splines consisting of 3 or 4 regions of fourth-order polynomials, the number of regions depending primarily upon the data range, were fit to and subtracted from the extended X-ray absorption fine structure (EXAFS) regions of the spectra (>17,200 eV). Resulting spectra were normalized using the absorption cross section for uranium measured at 17,200 eV and extrapolated through the EXAFS regions using the Victoreen equation. The resulting EXAFS spectra were then transformed over the k range (k ($Å^{-1}$) $\propto$ $E(eV)^{0.5}$) where data quality was high, but never starting below k = 3 $Å^{-1}$, to produce Fourier transform (FT) spectra.

Based on qualitative similarities among EXAFS spectra for uranophane [$Ca(UO_2)_2(SiO_3OH)_2 \cdot 5H_2O$ (s)], a uranyl silicate mineral, and the sorption samples, phase shift and amplitude parameters from FEFF6 calculations based on the uranophane structure were used to fit the sorption data (see Thompson *et al.,* 1997 for details of FEFF6 calculations). For shells of backscattering atoms that could be isolated in the uranophane experimental EXAFS spectrum [i.e., axial O (O_{ax}) and U], empirical phase shift and amplitude parameters were extracted using a Gaussian window. These were alternately substituted for FEFF6-generated parameters to fit similar shells in sorption sample EXAFS data, in order to corroborate the coordination number (N), interatomic distance (R), and disorder parameter (σ^2) values resulting from use of the FEFF6 parameters.

Unfiltered sorption data were fit without further refinement (e.g., deglitching) using a multishell approach. In accordance with the findings of Thompson *et al.* (1997), the value of S_o^2 (an amplitude reduction factor in FEFF6) was fixed to 1.0 in all fits. The same value of ΔE_0 was assigned to each coordination shell. This value was allowed to adjust while remaining equal for all shells. The axial shell was fit by fixing $N_{O_{ax}} = 2$ and floating R and σ^2. Because the equatorial O (O_{eq}) feature in our sorption sample Fourier transforms appears to split into two subfeatures, spectra for each sample were alternately fit assuming one or two equatorial oxygen shells. When assuming a single shell, $N_{O_{eq}}$ was fixed at 5, and R and σ^2 were allowed to vary from initial values of 2.4 Å and 0.01 $Å^2$, respectively. When assuming two equatorial subshells, N and σ^2 values could not be varied simultaneously without underdefining the system, thus two fitting strategies were employed. In the first, both σ^2 values were fixed to 0.002 $Å^2$ (taken from the uranophane fit), and both N and R values were varied. Using the second strategy, the sum of the two N values was fixed at 5 (average of expected values) and both R and σ^2 values were allowed to vary.

Fourier transform features ultimately attributed to Si/Al neighbors were isolated using a Gaussian window and fit in the absence of other spectral components.

The amplitude envelope of the windowed EXAFS function was examined for the characteristic shape of Si/Al, which peaks around k = 8 Å^{-1} when multiplied by k^3. Values of N and σ^2 for Si/Al resulting from the filtered fit were allowed to adjust in a multishell fit of the unfiltered spectra; R remained fixed at the value determined during the fit of filtered data. Unlike Si/Al, U contributions could be fit on unfiltered spectra because of their larger amplitude. For each U shell, N_U, R, and σ^2 were allowed to adjust in a multishell fit while parameters for other shells were held fixed.

IV. RESULTS

A. Qualitative XAS Analysis

Within the signal-to-noise limitations of the data, EXAFS spectra for all of the sorption samples are qualitatively similar to each other (Fig. 3). Multiple frequencies constitute the EXAFS spectra, and there is correspondingly more than one feature in each FT. Visual comparison of these Fourier transforms with those of uranyl model compounds (Thompson *et al.,* 1997) suggests that the sorption sample FT peaks at 1.3 and 1.9 Å are attributable to axial and first equatorial shell

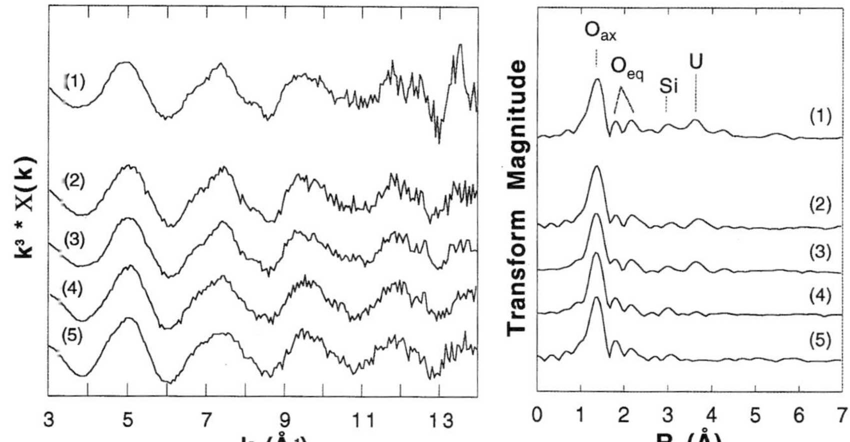

Figure 3 EXAFS spectra (left) and Fourier transforms of those spectra (right) for samples prepared in the absence (1) and presence (2–5) of CO_2, with pH increasing from bottom to top. Spectra correspond to samples (1) UK3, (2) UK12, (3) UK10, (4) UK8, and (5) UK6. In the Fourier transforms, features are denoted with the atoms to which they correspond.

ligands, respectively.[1] Further inspection of sorption Fourier transforms reveals that the equatorial contribution appears to break into two distinct shells (1.9 and 2.2 Å), as was found to occur for uranophane.

Features indicative of more distant backscatterers are present in all of the Fourier transforms. A peak that is clearly significant above background is present at 3.0 Å. Based on distance and known sample constituents, the most likely backscattering atoms responsible are Si or Al, as Si is located 3.1 to 3.8 Å from U in uranyl silicate minerals, and it (or Al) might be expected to be found at a similar distance if U is sorbed as an inner-sphere complex by kaolinite or coprecipitated with Si or Al. The peak is too distant for O_{eq}, nitrate-N, or carbonate-C, and too close for a nearest-neighbor U atom. A more distant FT peak is discernible above background at 3.8 Å in UK3, UK10, and UK12 spectra. This distance is typical of U neighbors. Furthermore, the prominence of this peak at a relatively large distance suggests that the backscattering atom(s) must have a significant backscattering cross section. Of the atoms present in significant concentrations in these samples, this is uniquely true for U.

The spectrum from a representative supernatant solution failed to reveal a U X-ray absorption edge, indicating solution-phase U concentrations too low to contribute to EXAFS spectra. This confirms that sorption sample spectra do not reflect species in the small amount of solution phase contained in the samples.

B. Quantitative XAS Analysis

EXAFS spectra fit results are reported in Table III. The uranyl U–O_{ax} distance was found to be within 0.01 Å of 1.80 Å for all of the sorption samples. Equatorial oxygens in all sorption samples were best fit using two subshells. The two O_{eq} shell-fitting strategies, i.e., fixing both O_{eq} σ^2 values to 0.002 Å^2 and fixing the sum of O_{eq} N values to five oxygens, resulted in fits of equal quality. The balance between N and σ^2 for each shell necessarily varied depending on the fitting strategy employed; however, R values did not vary beyond the error limits of the method (±0.02 Å). Because the R values did not vary significantly, we conclude that both resulting sets of N, R, and σ^2 parameters are equally valid. Among samples, parameters can only be compared within a single strategy, however. Because our knowledge basis for fixing N values (numerous uranyl compounds) is significantly greater than that for fixing σ^2 values (one model compound, uranophane), and the first fitting strategy consistently underestimates by two to three the expected average O_{eq} coordination number of five, we focus on the parameters resulting from the second strategy (last three columns in Table III).

[1]In this and the following sections, FT peak positions are reported uncorrected for phase shift, which typically moves the peak to higher R than observed by 0.2 to 0.4 Å. Distances that result from fitting the spectra are corrected for phase shift, however.

Table III

Results of Sorption Sample EXAFS Analysis

		Fixed O_{eq} σ^2 values[a]			Fixed O_{eq} N values[a]		
SAMPLE	LIGAND	N[b]	R (Å)[c]	σ^2(Å^2)[d]	N	R (Å)	σ2 (Å^2)
UK3	O_{ax}	2	1.80	0.0029	2	1.80	0.0027
	$O_{eq(1)}$	1.3	2.29	0.002	2.1	2.28	0.0039
	$O_{eq(2)}$	1.4	2.48	0.002	2.9	2.49	0.0048
	U	2	3.87	0.0045	2	3.87	0.0045
UK12	O_{ax}	2	1.80	0.0032	2	1.80	0.0032
	$O_{eq(1)}$	0.9	2.28	0.002	2.0	2.26	0.0066
	$O_{eq(2)}$	1.0	2.48	0.002	3.0	2.48	0.0092
	U	1	3.87	0.0043	1	3.87	0.0041
UK10	O_{ax}	2	1.80	0.0034	2	1.80	0.0034
	$O_{eq(1)}$	1.0	2.29	0.002	2.1	2.27	0.0059
	$O_{eq(2)}$	1.2	2.48	0.002	2.9	2.49	0.0077
	U	1	3.88	0.0051	1	3.88	0.0043
UK8	O_{ax}	2	1.80	0.0029	2	1.80	0.0029
	$O_{eq(1)}$	1.1	2.30	0.002	2.1	2.29	0.0068
	$O_{eq(2)}$	1.3	2.47	0.002	2.9	2.48	0.0077
UK6	O_{ax}	2	1.79	0.0032	2	1.79	0.0032
	$O_{eq(1)}$	1.1	2.30	0.002	2.0	2.28	0.0068
	$O_{eq(2)}$	1.5	2.47	0.002	3.0	2.47	0.0076

[a]See text for explanation.
[b]Number of neighboring atoms.
[c]Distance between U and the neighboring atom.
[d]Debye–Waller factor, a disorder term.

Fixing the sum of equatorial oxygens to five, the distance from U to the closer equatorial subshell ($R_{U\text{-}O_{eq(1)}}$) lies between 2.26 and 2.29 Å, and the longer distance ($R_{U\text{-}O_{eq(2)}}$) lies between 2.47 and 2.49 Å. Rounding to whole numbers of atoms, equatorial oxygens consistently divide into two at the shorter distance and three at the longer distance.

Unfortunately, the source of the 3.0 Å FT peak does not produce discernible features in the corresponding EXAFS spectra. Thus, the source of this peak must be determined from filtered spectra, but it can be evaluated on the basis of its improved modeling of unfiltered FT features. The amplitude function of the filtered spectrum (for the 3.0 Å peak) is characteristic of atoms with atomic numbers near Si/Al (Teo *et al.,* 1977). The filtered spectra are well fit by 0.3 to 1.0 Si/Al atoms located approximately 3.3 Å from U ($\sigma^2 = 0.001$ to 0.005 Å^2) in all samples. These same N, R, and σ^2 values can be used to fit the 3.0 Å peak in the unfiltered spectra, but it is desirable to allow N and σ^2 to adjust because they can be under- or overestimated when filtering is done in a region not isolated from other backscattering atoms, as is the case for the region under discussion. Unfortunately, allowing these values to vary simultaneously tends to result in values that are clearly outside of reasonable ranges. This is not surprising given that the Si/Al contribution to the spectra at this distance is relatively small, and therefore has a minor effect on the least-squares fit. For these reasons we consider the $N_{Si/Al}$ and $\sigma^2_{Si/Al}$ values to be very approximate, and they are not reported in Table III. Distance values should not be significantly affected by the filtering process, unless more than one Si/Al atom contributes to the spectrum from slightly different distances from U or large termination ripples from O_{eq} or U overlap with the Si/Al feature. Both of these possibilities are quite realistic; therefore we have not reported $R_{Si/Al}$ in Table III. We are nonetheless confident that the 3.0 Å FT features correspond to Si/Al neighbors. The only alternative explanation, given constraints on the structure placed by U^{VI} coordination chemistry, is that the 3.0 Å peak is a multiple-scattering (MS) feature. This is highly unlikely given the asymmetric nature of sorption sites at the solid–water interface. Furthermore, we found in our model compound study that MS contributions to uranyl EXAFS beyond $k = 3$ Å^{-1}, while occasionally visible, do not generally cause features of this magnitude (Thompson *et al.,* 1997).

The 3.8 Å peak in Fourier transforms of UK3, UK10, and UK12 spectra is well fit using phase shift and amplitude parameters for U, resulting in $N_U = 1$ (fixed) at 3.87 to 3.88 Å in UK10 and UK12 and $N_U = 2$ at the same distance in UK3. Disorder parameters for the U shells are similar for all three fits (0.0041 to 0.0045 Å^2) and very close to those found for two U neighbors at 4.3 and 3.9 Å in rutherfordine [UO_2CO_3 (s)] and uranophane [$Ca(UO_2)_2(SiO_3OH)_2 \cdot 5H_2O$ (s)], respectively. In fact, the number of U atoms was set to two for UK3 because the disorder parameter was unusually small (0.0013 Å^2) for such a distant neighbor when N_U was set equal to 1.0. The similarity in both N_U and σ^2_U values between uranophane and UK3 suggests that the values obtained for UK3 are reasonable. For the same reason, the discrepancy between N values, but not σ^2 values, for uranophane and UK10 and UK12 may suggest that both N and σ^2 values for UK10 and UK12 should be larger or smaller than the values in Table III. Without additional information, it is difficult to ascertain more accurate values for either.

Sorption sample UK3 has additional structure on the high R side of the 3.8 Å peak and around 5.5 Å in its FT. These features are most likely attributable to additional U neighbors because, of the elements present in the sample, only U is expected to scatter strongly enough to be seen at these distances. Due to significant overlap with the 3.8 Å U–U peak and the relatively small size of the 5.5 Å feature, this additional structure has been difficult to fit.

Disorder parameter (σ^2) values for all of the shells are generally consistent with those found for similar backscattering shells in the model compound study. Sorbate O_{ax} σ^2 values are at the high end of the range found for model compounds, suggesting slightly weaker U–O_{ax} bonds (higher vibrational disorder) and/or a greater difference between the two U–O_{ax} bond lengths (higher static disorder) for the uranyl moiety at the solid–water interface. (This provides an additional argument against MS as the source of spectral features.) Among other explanations, the former could be caused by stronger equatorial bonds or participation of the axial oxygens in a second bond (e.g., a hydrogen bond), and the latter by unequal environments for the two O_{ax} atoms.

Sorbate O_{eq} subshell σ^2 values are similar to each other within a reasonably narrow range. Compared to uranophane, sorbate σ^2 values are larger, but of the same order of magnitude. Consistently larger σ^2 values in the sorbates are indicative of more O_{eq} shell disorder (static and/or vibrational) than in uranophane.

V. DISCUSSION

Over the range of solution conditions explored in this study, the sorbate is characterized by a uranyl moiety ($R_{U\text{-}O_{ax}} \approx 1.80$ Å) surrounded by five equatorial O atoms for which the U–O bond lengths separate into two groups of three long (≈2.48 Å) and two short (≈2.28 Å) bonds. Silicon (and/or Al) is present in the U coordination environment under all of the conditions studied, although the number and location of Si/Al neighbors cannot be well determined. Samples in contact with air at $6.0 \leq \text{pH} \leq 7.0$ contain predominantly monomeric uranyl species. For samples prepared in contact with air at solution pH values $7.0 \leq \text{pH} \leq 7.9$, the dominant U species is multinuclear. In the absence of CO_2 at pH 7.5 to 8.2, the sorbate contains at least three U atoms.

The effect of pH on U sorbates exposed to air is consistent with that observed for U solution complexes under conditions of similar initial U concentration. The N_U increases with pH over the range 6.0 to 7.5 in the sorbate and in aqueous solution (Fig. 4). Furthermore, comparison of sorbate N_U with those of dominant solution species expected at each pH reveals a remarkable similarity, assuming that precipitation did not occur in our samples (Fig. 4). Under final solution conditions, however, mononuclear U species dominate in the aqueous system without a sorbent. This recalls inferences (James and Healy, 1972) and observations (Dillard and Koppelman, 1982; Chisholm-Brause *et al.*, 1990; O'Day *et al.*, 1994) that Co^{II} sorption behavior parallels its solution behavior, with an apparent enhancement of Co^{II} polymerization by the solid.

In the absence of carbonate, the similarities between sorbates and solution complexes are less apparent. In UK3, $N_U \geq 3$, whereas the dominant solution species under similar conditions contains one U atom (Fig. 5). It would appear that in the absence of carbonate, kaolinite enhances U polymerization in the short term. The multinuclear sorbates could be precursors to the U precipitates that dominate the aqueous system at equilibrium.

Although we have addressed the short-range environment of U in our sorption

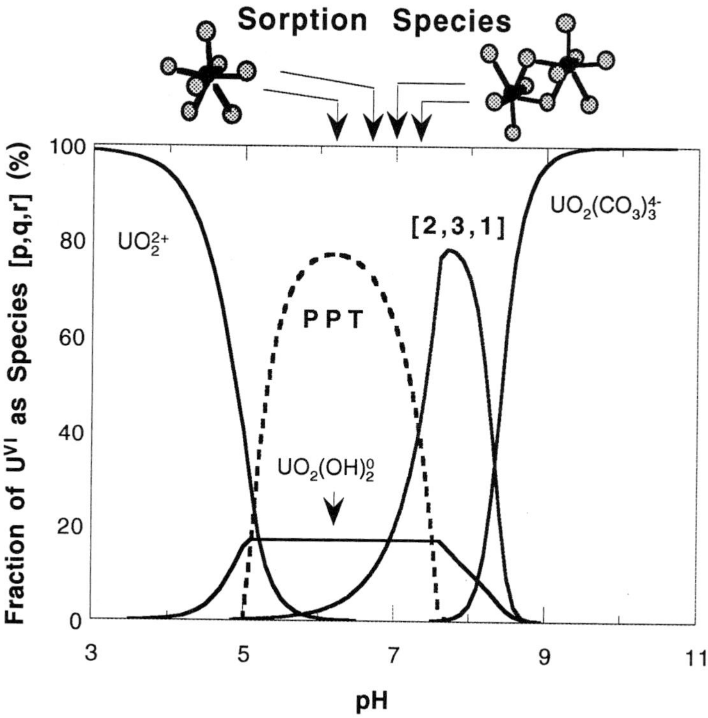

Figure 4 Comparison of structure and composition of sorption products with dominant solution speciation for air-exposed samples. Equilibrium solution speciation was calculated for initial conditions of experiments, allowing kaolinite to dissolve, based on the thermodynamic data base of Grenthe *et al.* (1992). [p,q,r] denotes aqueous uranyl complexes of the type $(UO_2)_p(OH)_q(CO_3)_r$. PPT denotes $UO_3 \cdot 2H_2O$ or $\beta\text{-}UO_2(OH)_2$, both of which are supersaturated. Dark atoms in sorption species denote U; therefore $p = 1$ at $pH < 7.0$ and $p = 2$ at $pH \geq 7.0$.

samples, we have yet to discuss how U is associated with kaolinite. Given the composition of the XAS samples (primarily solid with a small amount of supernatant), U could be found in one or more of three different phases: dissolved in aqueous solution, precipitated from solution, and sorbed by kaolinite (including surface precipitation). We have ruled out the first possibility by finding no X-ray absorption edge in XAS spectra of the supernatants.

Precipitation of a U solid phase must be seriously considered, because solutions from which U was sorbed were supersaturated with respect to uranyl trioxide dihydrate ($UO_3 \cdot 2H_2O$) and a uranyl hydroxide [$\beta\text{-}UO_2(OH)_2$] (Grenthe *et al.*, 1992), and UK3 was also supersaturated with respect to soddyite, a uranyl silicate, assuming that an equilibrium amount of kaolinite dissolved (Nguyen *et al.*, 1992). Furthermore, we have observed $N_U > 0$ in some of our samples, which could be indicative of precipitation. Under final solution conditions, however, all samples except UK6 were undersaturated with respect to all of the solids in Table IV; UK6 remained supersaturated with respect to uranyl dihydrate and hydroxide.

With the exception of UK3, precipitation of a U phase probably did not occur in our samples for several reasons. For UK6, whose solution was supersaturated with respect to uranyl solids at all times during the experiment, $N_U = 0$. Of the samples for which $N_U > 0$ (UK10 and UK12), both contain Si/Al second neighbors, which is inconsistent with the structures of the dihydrate and the hydroxide phases (Christ and Clark, 1960; Roof *et al.*, 1964). Furthermore, N_U in the sorbates (1) is much lower than that in the uranyl solids (4 or 6). This raises the question of a mixed U and Si/Al solid phase, but of those that have been reported and

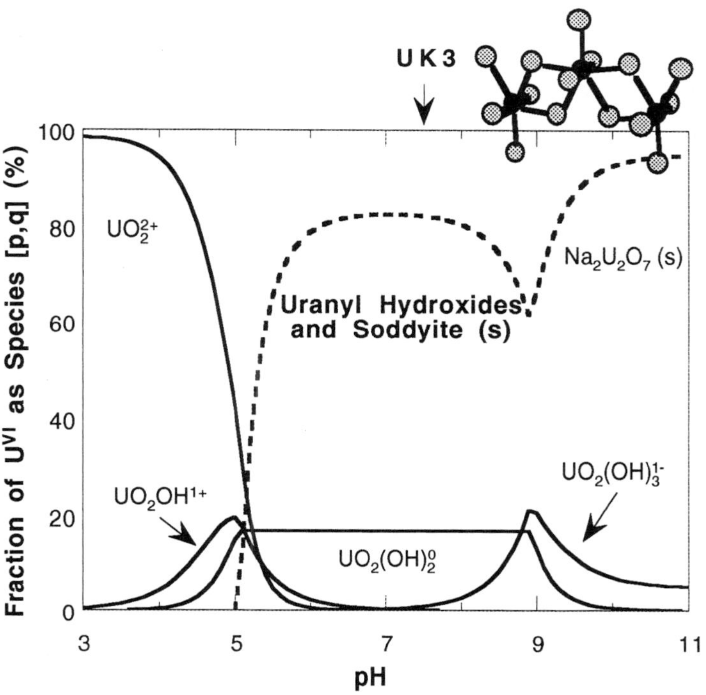

Figure 5 Comparison of structure and composition of sorption product with dominant solution speciation in the absence of CO_2. Equilibrium solution speciation was calculated for initial conditions of experiments, allowing kaolinite to dissolve, based on the thermodynamic data base of Grenthe *et al.* (1992). [p,q] denotes aqueous uranyl complexes of the type $(UO_2)_p(OH)_q$. PPT denotes $UO_3{\cdot}2H_2O$, β-$UO_2(OH)_2$ or soddyite $[(UO_2)_2SiO_4{\cdot}2H_2O]$, each of which is supersaturated. Dark atoms in sorption species denote U; therefore p = 3 (actually p $\geq$ 3).

Table IV

Solution and Solid Equilibrium Constants Used in Calculating Speciation Diagrams

log K[a]	Solution complex[b]	log K[a]	Solid
−5.20	$[1,1,0]^{+}$	−9.20	α-$UO_3 \cdot 0.9\ H_2O$
−10.30	[1,2,0]	−4.95	β-$UO_2(OH)_2$
−19.20	$[1,3,0]^{-}$	−4.83	$UO_3 \cdot 2H_2O$
−33.00	$[1,4,0]^{2-}$	−11.98	$UO_2(NO_3)_2$
−2.70	$[2,1,0]^{3+}$	−8.51	$UO_2(NO_3)_2 \cdot H_2O$
−5.62	$[2,2,0]^{2+}$	−4.92	$UO_2(NO_3)_2 \cdot 2H_2O$
−11.90	$[3,4,0]^{2+}$	−3.67	$UO_2(NO_3)_2 \cdot 3H_2O$
−15.55	$[3,5,0]^{+}$	−2.25	$UO_2(NO_3)_2 \cdot 6H_2O$
−31.00	$[3,7,0]^{-}$	−30.18	α-Na_2UO_4
−21.90	$[4,7,0]^{+}$	−22.70	$Na_2U_2O_7$
9.68	[1,0,1]	14.47	UO_2CO_3
16.94	$[1,0,2]^{2-}$	26.94	$Na_4UO_2(CO_3)_3$
21.60	$[1,0,3]^{4-}$	17.57[c]	$(UO_2)_2SiO_4 \cdot 2H_2O$
−0.77	$[2,3,1]^{-}$	17.24[c]	$NaH_3OUO_2SiO_4 \cdot H_2O$
0.74	$(UO_2)_3O(OH)_2(HCO_3)^{+}$	136.35[c]	$Na_2(UO_2)_2(Si_2O_5)_3 \cdot 7H_2O$
36.96	$[11,12,6]^{2-}$		
54.00	$[3,0,6]^{6-}$	38.41[d]	Kaolinite-$Al_2Si_2O_5(OH)_4$
0.30	$UO_2NO_3^{+}$	−8.11[e]	Gibbsite-$Al(OH)_3$

[a]Unless otherwise noted, based on the data base of Grenthe *et al.* (1992). Log K values are the equilibrium constants for the formation reactions (cf. Papelis *et al.*, 1988).
[b]Solution complexes are expressed as [p,q,r] for $(UO_2)_p(OH)_q(CO_3)_r^{2p-q-2r}$.
[c]Based on Nguyen *et al.* (1992).
[d]Based on Helgeson *et al.* (1978) and May *et al.* (1986).
[e]Based on May *et al.* (1979).

for which formation energies have been determined (soddyite, sodium boltwoodite, and sodium weeksite), none should form under initial or final experimental conditions (Nguyen *et al.*, 1992). This is due, in part, to the limited solubility of kaolinite; even at equilibrium, which we almost certainly have not achieved, the amount of Si in solution is approximately one order of magnitude lower than dissolved U concentrations (Helgeson *et al.*, 1978; May *et al.*, 1986). We can furthermore preclude the possible formation in UK10 and UK12 of an amorphous solid phase, which would be more likely to form than a crystalline phase during the relatively short duration of the sorption experiments, because its formation would require greater oversaturation than crystalline phases of similar composition (Stumm and Morgan, 1981; Morse and Casey, 1988). Although unlikely given the short duration of our experiments (Carroll-Webb and Walther, 1988), enough kaolinite could have dissolved to allow gibbsite precipitation (at the kaolinite surface or in solution), thus providing a second adsorbent, and causing an increased aqueous Si concentration. If this occurred, the constituents of a U–Si coprecipitate might become available (d'Espinose de la Caillerie *et al.*, 1995; Scheidegger *et al.*, 1996; Towle *et al.*, 1996).

Sorption by kaolinite is therefore the most likely fate of U in all of our air-exposed sorption samples. Silicon and/or Al has been fit as a backscatterer in the coordination environment of U for all of these, thus confirming inner-sphere complexation of U with either a kaolinite surface site or a coadsorbed Si or Al species. Although we are unable to constrain $N_{Si/Al}$, the small size of the Si/Al contribution to the spectra precludes the possibility that U is absorbed into the kaolinite structure.

It appears quite likely that we may be sampling a precipitate (surface or homogeneous), or precursor multinuclear species, in the absence of carbonate. Spectra for UK3 definitely indicate the presence of close U neighbors (≈3.9 Å) and suggest the possibility of more distant U neighbors, despite a lower surface coverage than the two air-exposed samples that contain U second neighbors. This suggests we are observing larger U species in UK3 owing to the absence of carbonate rather than an increase in surface coverage. We can rule out formation of $UO_3 \cdot 2H_2O$ (s) and β-$UO_2(OH)_2$ (s) based on the presence of Si/Al at 3.3 Å and the small number of U neighbors in UK3. Without a structure determination for soddyite, however, we cannot establish whether soddyite has formed in UK3. It is also possible that a previously undocumented precipitate, coprecipitate with Si and/or Al, or precipitate precursor has formed in solution or at the solid interface. Because final solution conditions were undersaturated with respect to known solid phases, a precipitate or multinuclear species may have formed rapidly at the higher initial U concentration that would ultimately dissolve to produce sorption species more similar to dominant solution species.

VI. CONCLUSIONS

In this study, EXAFS analysis of samples in which aqueous U^{VI} has been equilibrated with kaolinite has provided unique structural and compositional information on U species sorbed at the kaolinite–water interface. In the presence and absence of carbonate over a range of pH values, the atomic environment of U^{VI} is similar out to approximately 3.5 Å. In the presence of carbonate, U is sorbed to kaolinite as an inner-sphere complex, and N_U increases with pH, similar to the trend observed in aqueous solution under similar conditions. In the absence of carbonate, U is present as either an inner-sphere multinuclear sorption complex or surface precipitate or a U–Si/Al coprecipitate of unknown identity. In either case, the presence of kaolinite appears to promote short-term U cluster formation in the absence of carbonate. This promotion is not observed (or is much weaker) in the presence of carbonate, probably because the carbonate group dominates other uranyl species, but not solid surface O atoms, in competing for complexation of uranyl under these conditions. Thus although we cannot directly observe equatorial carbonate ligands using XAS, we infer their presence in U sorption complexes prepared in the presence of air. Further changes with longer equilibration times appear likely.

With respect to the use of kaolinite as a sorbent for removing uranium from ponded waters, these results are promising. The formation of multinuclear sorption species, even in the presence of carbonate, increases kaolinite's capacity for uranium over what might be possible were only mononuclear species to form. Furthermore, inner-sphere sorption results in fairly strong retention of sorbed species; thus we would not expect the uranium to readily desorb, unless solution conditions were significantly altered.

ACKNOWLEDGMENTS

We acknowledge Ines Triay and co-workers at Los Alamos National Laboratory (LANL), Chemical Science and Technology Division, for preparing the sorption samples. Financial support for this study came from a U.S. Department of Energy (DOE) contract through LANL (9-X42-6947E-1) and a separate DOE grant (DE-FG03-93ER14347). We thank the staff of the Stanford Synchrotron Radiation Laboratory for their help. SSRL is supported by DOE and NIH. F. Farges facilitated the XAS U L_{III}-edge jump measurements. The manuscript benefited significantly from comments by an anonymous reviewer.

REFERENCES

Allen, P. G., Bucher, J. J., Clark, D. L., Edelstein, N. M., Ekberg, S. A., Gohdes, J. W., Hudson, E. A., Kaltsoyannis, N., Lukens, W. W., Neu, M. P., Palmer, P. D., Reich, T., Shuh, D. K., Tait, C. D., and Zwick, B. D. 1995. Multinuclear NMR, Raman, EXAFS, and X-ray diffraction studies of uranyl carbonate complexes in near-neutral aqueous solution. X-ray structure of $[C(NH_2)_3]_6[(UO_2)_3(CO_3)_6]\ 6.5H_2O$. *Inorg. Chem.* 34:4797–4807.

Bartlett, J. R., and Cooney, R. P. 1989. Raman spectra of zeolites exchanged with uranyl (VI) cations. I. Zeolite Y. *Spectrochim. Acta Part A* 45:541–547.

Basile, L. J., Ferraro, J. R., Mitchell, M. L., and Sullivan, J. C. 1978. The Raman scattering of actinide (VI) ions in carbonate media. *Appl. Spectrosc.* 32:535–537.

Bleam, W. F., and McBride, M. B. 1985. Cluster formation versus isolated-site adsorption. A study of Mn(II) and Mg(II) adsorption on boehmite and goethite. *J. Colloid Interface Sci.* 103:124–132.

Brown, G. E., Jr., and Parks, G. A. 1989. Synchrotron-based x-ray absorption studies of cation environments in earth materials. *Rev. Geophys.* 27:519–533.

Brown, G. E., Jr., Calas, G., Waychunas, G. A., and Petiau, J. 1988. X-ray absorption spectroscopy and its applications in mineralogy and geochemistry. *In* "Spectroscopic Methods in Mineralogy and Geology" (F. C. Hawthorne, Ed.), pp. 431–512. Mineralogical Society of America, Washington, DC.

Carroll-Webb, S. A., and Walther, J. V. 1988. A surface complex reaction model for the pH-dependence of corundum and kaolinite dissolution rates. *Geochim. Cosmochim. Acta* 52:2609–2623.

Chisholm-Brause, C., Conradson, S. D., Buscher, C. T., Eller, P. G., and Morris, D. E. 1994. Speciation of uranyl sorbed at multiple binding sites on montmorillonite. *Geochim. Cosmochim. Acta* 58:3625–3631.

Chisholm-Brause, C. J., O'Day, P. A., Brown, G. E., Jr., and Parks, G. A. 1990. Evidence for multinuclear metal-ion complexes at solid/water interfaces from X-ray absorption spectroscopy. *Nature* 348:528–530.

Christ, C. L., and Clark, J. R. 1960. Crystal chemical studies of some uranyl oxide hydrates. *Am. Mineral.* 45:1026–1061.

Clark, D. L., Hobart, D. E., and Neu, M. P. 1995. Actinide carbonate complexes and implications for actinide environmental chemistry. *Chem. Rev.* 95:25–48.

Combes, J.-M. 1988. Evolution de la structure locale des polymeres et gels ferriques lors de la cristallisation des oxydes de fer. Application au piegeage de l'uranium. Docteur, Sciences de la Terre, l'Universite Pierre et Marie Curie, Paris.

Cramer, S. P., and Hodgson, K. O. 1979. X-ray absorption spectroscopy: A new structural method and its applications to bioinorganic chemistry. *Prog. Inorg. Chem.* 25:1–39.

Dent, A. J., Ramsay, J. D. F., and Swanton, S. W. 1992. An EXAFS study of uranyl ion in solution and sorbed onto silica and montmorillonite clay colloids. *J. Colloid Interface Sci.* 150:45–60.

d'Espinose de la Caillerie, J.-B., Kermarec, M., and Clause, O. 1995. Impregnation of γ-alumina with Ni(II) or Co(II) ions at neutral pH: Hydrotalcite-type coprecipitate formation and characterization. *J. Am. Chem. Soc.* 117:11471–11481.

Dillard, J. G., and Koppelman, M. H. 1982. X-ray photoelectron spectroscopic (XPS) surface characterization of cobalt on the surface of kaolinite. *J. Colloid Interface Sci.* 87:46–55.

George, G. N., and Pickering, I. J. 1993. EXAFSPAK: A suite of computer programs for analysis of X-ray absorption spectra. SSRL Report, Stanford. CA.

Giblin, A. M. 1980. The role of clay adsorption in genesis of uranium ores. *In* "International Uranium Symposium on the Pine Creek Geosyncline, Vienna," pp. 521–529. International Atomic Energy Agency, Vienna.

Grenthe, I., Fuger, J., Konings, R. J. M., Lemire, R. J., Muller, A. B., Nguyen-Trung, C., and Wanner, H. 1992. "Chemical Thermodynamics of Uranium." North-Holland, Amsterdam.

Helgeson, H. A., Delaney, J. M., Nesbitt, H. W., and Bird, D. K. 1978. Summary and critique of the thermodynamic properties of rock-forming minerals. *Am. J. Sci.* 278A:1–229.

Ho, C. H., and Miller, N. H. 1986. Adsorption of uranyl species from bicarbonate solution onto hematite particles. *J. Colloid Interface Sci.* 110:165–171.

Hsi, C. D., and Langmuir, D. 1985. Adsorption of uranyl onto ferric oxyhydroxides: Application of the surface complexation site-binding model. *Geochim. Cosmochim. Acta* 49:1931–1941.

James, R. O., and Healy, T. W. 1972. Adsorption of hydrolyzable metal ions at the oxide-water interface. I. Co(II) adsorption on SiO_2 and TiO_2 as model systems. *J. Colloid Interface Sci.* 40:42–52.

Madic, C., Hobart, D. E., and Begun, G. M. 1983. Raman spectrometric studies of actinide(V) and -(VI) complexes in aqueous sodium carbonate solution and of solid sodium actinide(V) carbonate compounds. *Inorg. Chem.* 22:1494–1503.

Manceau, A., Charlet, L., Boisset, M. C., Didier, B., and Spadini, L. 1992. Sorption and speciation of heavy metals on hydrous Fe and Mn oxides. From microscopic to macroscopic. *Appl. Clay Sci.* 7:201–223.

May, H. M., Helmke, P. A., and Jackson, M. L. 1979. Gibbsite solubility and thermodynamic properties of hydroxy-aluminum ions in aqueous solution at 25°C. *Geochim. Cosmochim. Acta* 43:861–868.

May, H. M., Kinniburgh, D. G., Helmke, P. A., and Jackson, M. L. 1986. Aqueous dissolution, solubilities and thermodynamic stabilities of common aluminosilicate clay minerals: Kaolinite and smectites. *Geochim. Cosmochim. Acta* 50:1667–1677.

Maya, L. 1982. Sorbed uranium (VI) species on hydrous titania, zirconia, and silica gel. *Radiochim. Acta* 31:147–151.

McBride, M. B., Fraser, A. R., and McHardy, W. J. 1984. Cu^{2+} interaction with microcrystalline gibbsite. Evidence for oriented chemisorbed copper ions. *Clays Clay Miner.* 32:12–18.

Morris, D. E., Chisholm-Brause, C. J., Barr, M. E., Conradson, S. D., and Eller, P. G. 1994. Optical spectroscopic studies of the sorption of UO_2^{2+} species on a reference smectite. *Geochim. Cosmochim. Acta* 58:3613–3623.

Morse, J. W., and Casey, W. H. 1988. Ostwald processes and mineral paragenesis in sediments. *Am. J. Sci.* 288:537–560.

Morse, J. W., Shanbhag, P. M., Saito, A., and Choppin, G. R. 1984. Interaction of uranyl ions in carbonate media. *Chem. Geol.* 42:85–99.

Muto, T., Kirono, S., and Kurata, H. 1965. Some aspects of fixation of uranium from natural waters. *Min. Geol. (Tokyo)* 15:287–298.

Nguyen, S. N., Silva, R. J., Weed, H. C., and Andrews, J. E., Jr. 1992. Standard Gibbs free energies of formation at the temperature 303.15 K of four uranyl silicates: Soddyite, uranophane, sodium boltwoodite, and sodium weeksite. *J. Chem. Thermodyn.* 24:359–376.

O'Day, P. A., Brown, G. E., Jr., and Parks, G. A. 1994. X-ray absorption spectroscopy of cobalt(II) multinuclear surface complexes and surface precipitates on kaolinite. *J. Colloid Interface Sci.* 165:269–289.

Papelis, C., Hayes, K. F., and Leckie, J. O. 1988. HYDRAQL: A program for the computation of chemical equilibrium composition of aqueous batch systems including surface-complexation modeling of ion adsorption at the oxide/solution interface. Technical Report 306, Stanford University, Stanford, CA.

Redden, G. D., Li, J., and Leckie, J. O. 1998. Adsorption of U^{VI} and citric acid on goethite, gibbsite, and kaolinite: Comparing results for binary and ternary systems. *In* "Adsorption of Metals by Geomedia" (E. A. Jenne, Ed.). Academic Press, San Diego.

Roof, R. B., Cromer, D. T., and Larson, A. C. 1964. The crystal structure of uranyl dihydroxide, $UO_2(OH)_2$. *Acta Crystallogr.* 17:701–705.

Sayers, D. E., and Bunker, B. A. 1988. Data analysis. *In* "X-Ray Absorption: Principles, Applications, Techniques of EXAFS, SEXAFS and XANES" (D. C. Koningsberger and R. Prins, Eds.), pp. 211–253. Wiley, New York.

Scheidegger, A. M., Lamble, G. M., and Sparks, D. L. 1996. Investigation of Ni sorption on pyrophyllite: An XAFS study. *Environ. Sci. Technol.* 30:548–554.

Schenck, C. V., Dillard, J. G., and Murray, J. W. 1983. Surface analysis and the adsorption of Co(II) on goethite. *J. Colloid Interface Sci.* 95:398–409.

Sekine, K., Payne, T. E., Waite, T. D., and Davis, J. A. 1991. International Alligator Rivers analogue project (18): Experimental study of uranium adsorption on kaolinite-pH dependence in air-equilibrated system. JAERI Memo 03-036, Japan Atomic Energy Research Institute.

Stumm, W., and Morgan, J. J. 1981. "Aquatic Chemistry," 2nd ed. Wiley, New York.

Teo, B.-K., Lee, P. A., and Simons, A. L. 1977. EXAFS: Approximation, parameterization, and chemical transferability of amplitude functions. *J. Am. Chem. Soc.* 99:3854–3856.

Thompson, H. A., Brown, G. E., Jr., and Parks, G. A. 1997. XAFS spectroscopic study of uranyl coordination in solids and aqueous solution. *Am. Mineral.* 82:483–496.

Towle, S. N., Bargar, J. R., Brown, G. E., Jr., and Parks, G. A. 1996. Surface precipitation of Co(II)(aq) on Al_2O_3. *J. Colloid Interface Sci.* 187:62–82.

Tripathi, V. S. 1984. Uranium(VI) transport modeling: Geochemical data and submodels. Ph.D. Thesis, Stanford University, Stanford, CA.

Van Olphen, H., and Fripiat, J. J. (Eds.). 1979. "Data Handbook for Clay Materials and Other Nonmetallic Minerals." Pergamon Press, New York.

Vaughan, D. (Ed.). 1986. "X-Ray Data Booklet," 2nd ed. Lawrence Berkeley Laboratory, Berkeley.

Waite, T. D., Davis, J. A., Payne, T. E., Waychunas, G. A., and Xu, N. 1994. Uranium(VI) adsorption to ferrihydrite: Application of a surface complexation model. *Geochim. Cosmochim. Acta* 58: 5465–5478.

Chapter 17

Surface Charge and Metal Sorption to Kaolinite

Patrick V. Brady, Randall T. Cygan, and Kathryn L. Nagy
Geochemistry Department, Sandia National Laboratories, Albuquerque, New Mexico

Kaolinite surface charge can be described by proton donor and acceptor reactions occurring on edge Al sites, and by proton donor reactions occurring on edge Si sites. Edge Al sites are hypothesized to control pH-dependent sorption of metals, though there also appears to be minor pH-independent sorption of hard acids (Cs^+ and Sr^{2+}) on basal planes. Scanning force microscopy images indicate a higher percentage of edges, relative to basal planes, than previously observed. Independent molecular modeling results point to both the elevated reactivity of Al edge sites, relative to Si, and the weak sorption on basal planes.

I. INTRODUCTION

Clay mineral surface reactivity is critically important to a large number of geochemical processes. Clay surfaces are often able to regulate the compositions of soil waters by sorbing metals, protons and/or hydroxyls, and organic molecules. Clays also constitute potentially important transport vehicles for sorbed contaminants in groundwaters. Because of the fundamental importance of clay reactivity in the subsurface, it is important that the molecular origins of clay surface reactivity be understood. The task is complicated by the wide range of compositions seen in nature. Nevertheless, if the controls on clay surface reactivity can be un-

Adsorption of Metals by Geomedia

raveled at the atomic level for a series of well-characterized clays, reactivity might be reliably predicted for the others. The sorptive capacity of clays depends largely on surface-charging reactions occurring at frayed edge sites, and on ion exchange reactions occurring along basal surface and interlayer planes possessing a permanent charge. The first set of reactions is pH-dependent and arises from the hydration of broken bonds, which then can accept or donate protons and/or hydroxyls. The second set of reactions occurs to balance a permanent charge arising from heterovalent cation substitution in the clay structure. Note that the net sorptive capacity of a given clay mineral for a sorbent will depend upon both (1) the intrinsic affinity of the metal for sorption and exchange, and (2) the concentration of edge and basal plane sites. Neither effect can be predicted from first principles and, therefore, must be measured.

Here we examine first the molecular controls on proton donor–acceptor reactions at kaolinite, $Al_2Si_2O_5(OH)_4$, surfaces. Kaolinite is a useful model mineral to examine because: (1) it contains octahedral Al and tetrahedral Si, the two primary functional group oxides exposed at most clay surfaces, (2) it contains no easily leached cations, (3) its (hydr)oxide components exist in a variety of distinctly different structural environments (e.g., basal planes and edges), and (4) there is minimal substitution of variable-valence cations that would promote a pH-independent exchange capacity. We used potentiometric titrations to measure kaolinite proton acceptor–donor capacity at 25, 50, and 70°C and to formulate a surface complexation model for the kaolinite–solution interface. The derived site acidities were then interpreted at the molecular level using empirical atomistic methods (Cygan *et al.*, 1998). Relative surface areas were then estimated using scanning force microscopy (SFM) images to establish the distribution of edge and basal plane sites.

The second part of this chapter will describe preliminary measurements of metal cation adsorption to kaolinite. Metal cation sorption to edge sites is generally observed to be pH-dependent, reflecting the pH dependence of edge surface charging. Metal cation association with basal planes is less pH-dependent and may occur due to heterovalent substitutions in the lattice. By measuring sorption of a variety of metals having different chemical reactivity, we were able to: (1) examine the chemical reactivity of basal planes and (2) isolate the chemical controls on the solution side of the sorption process.

II. EXPERIMENTAL METHODS

pH- and temperature-dependent surface charge was measured on a well-crystallized kaolinite, KGa-1 (Clay Minerals Society source clay from Washington County, Georgia, U.S.A.), having a surface area of 10 $m^2 \cdot g^{-1}$ (van Olphen and Fripiat, 1979), in a background electrolyte of 0.1 $mol \cdot L^{-1}$ NaCl. Metal (Sr^{2+},

Cs^+, and Cd^{2+}) sorption was measured using the same background electrolyte, except for the case of Cd^{2+}. Because Cd^{2+} forms strong ion pairs with Cl^-, 0.1 mol·L^{-1} $NaClO_4$ solutions were used instead. No extensive washing or aging of the kaolinite was performed in order to avoid potentially complicating effects of aluminum and or silica coatings observed to form in electrokinetic studies of clay minerals (see Scales *et al.,* 1990).

Four grams of kaolinite was titrated in 50 ml of electrolyte using a Mettler DL 12 titrator. Solutions were continuously stirred and N_2 gas was continuously pumped directly into the solution for 15 min before and during each titration to purge the solution of CO_2. Temperature was held to ±2°C by manually adjusting a hot plate and water jacket in which the reaction vessel was immersed. Generally, 1 to 5 min elapsed between titrant additions, allowing pH readings to stabilize. Adsorption equilibrium was presumed to exist when the pH changed by less than ~0.01 units per minute. After each titration the solutions were filtered (0.45 μm) and analyzed with a directly coupled plasma (DCP) spectrometer for Al and Si. For kaolinite the 0.45-μm filter technique gives the same results as a 0.05-μm filter (Nagy *et al.,* 1991). Dissolved levels of Al and Si were too small (10^{-4} mol·L^{-1}) to significantly impact the charge balance calculations below.

pH-dependent excess surface charge was calculated at each temperature using

$$\sigma = (F/A_s)([C_a - C_b] - [H^+] + [OH^-]), \qquad [1]$$

where F is Faraday's constant; σ is the proton surface charge density (C·m^{-2}); A_s is mineral surface area (m^2·L^{-1}); and C_a and C_b are the amount of acid or base titrant added (mol·L^{-1}), respectively.

For the metal sorption measurements the same fluid/mineral ratio was used and a spike of measured metal-bearing solution was added. Periodically small volume extractions were made and filtered and analyzed by DCP (or ICP-MS in the case of Cs). Sorption was calculated by difference. Scanning force microscopy images were obtained in air using a Park Scientific Instrument Autoprobe LS in contact mode using PSI Ultralever tips.

III. KAOLINITE SURFACE CHARGE

Measured kaolinite surface charge is heterogeneous in that it is the net result of a number of reactions occurring on a variety of crystallographically distinct sites. The kaolinite unit cell is, simplistically, a gibbsite-like sheet of dioctahedrally coordinated Al bound to a sheet of silica tetrahedra through shared oxygens. Kaolinite cells stack in booklets that are loosely attached by way of hydrogen bonding between hydroxyl sites on the gibbsite basal plane and the highly electronegative oxygens of the siloxane sheet (Fig. 1).

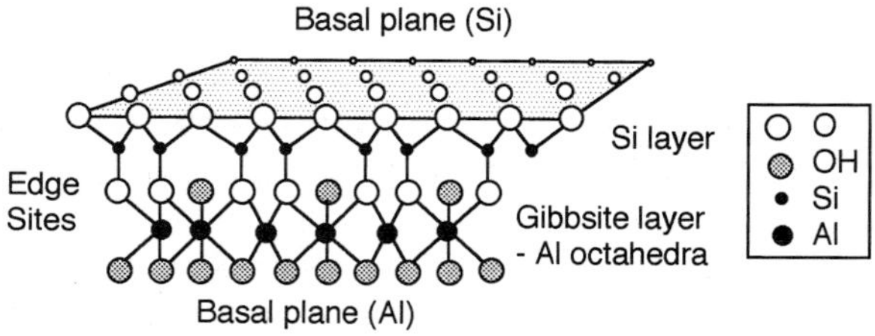

Figure 1 Schematic of kaolinite crystal structure.

Kaolinite surface charge has been ascribed to a variety of crystallographically distinct sites. They include the following:

1. *Edges.* Al and Si centers at kaolinite edges are terminated by hydroxyls (Schofield and Samson, 1954) and, in the case of the aluminol, accept or donate protons. Since alumina surfaces are positively charged at pH < 8.7 (e.g., Huang and Stumm, 1973), whereas silica groups are anionic to pH ~ 2 (e.g., Parks, 1965), positive edge charge probably arises from gibbsite-like edge sites (Ferris and Jepson, 1975; Bolland *et al.,* 1976). Negative surface charge is thought to originate at high pH (pH > 9) primarily from proton donor reactions on Si sites (Ferris and Jepson, 1975; Bolland *et al.,* 1976; Riese, 1982). Adsorption of water onto aluminol sites also gives rise to Lewis acid sites (Sposito, 1984, p. 18).
2. *The Si basal plane.* Substitution of Al or Fe^{3+} for Si in the tetrahedral sheet would cause a permanent (non-pH-dependent) negative charge on the Si basal plane (Schofield and Sampson, 1954; Cashen, 1959).
3. *Basal plane hydroxyl groups.* Hydroxyl groups on the octahedral plane are coordinated to two underlying aluminum atoms, and are thought to be appreciably less reactive than edge aluminols and silanols (Sposito, 1984, p. 18). Hydroxyl groups may exist locally on the siloxane sheet as well (e.g., Wieland and Stumm, 1992). Zhou and Gunter (1992) compared the measured kaolinite surface charge with that of the constituent oxides SiO_2 and Al_2O_3 to argue for ionization of basal planes.

A variety of surface complexation models have been used to explain pH-dependent charge and the electrokinetic behavior of the kaolinite surface, including semiempirical power exchange models, constant capacitance models, and triple-layer models (see Riese, 1982; Schindler *et al.,* 1987; Carroll-Webb and Walther, 1988; Zachara and Smith, 1988; Motta and Miranda, 1989; Xie and Walther, 1992; Wieland and Stumm, 1992). As the cation exchange capacity of KGa-1 kaolinite is minimal at pH 4, $\sim 10^{-3}$ meq·g^{-1} (Riese, 1982), and minor compared to the measured proton surface charge density, we initially neglected any permanent, pH-independent, negative charge on the Si basal plane. In effect we modeled kaolinite surface acidity solely by modeling the amphoteric properties of aluminol and silanol sites on the edges and on the basal planes (Brady *et al.,* 1996). The small basal plane charge of the KGa-1 total surface charge does play a role in metal sorption, as we shall discuss below.

We included basal planes in the model since Zhou and Gunter (1992) have argued that at best only a tenth of the pH-dependent surface charge can be fit onto the kaolinite edges, which requires the basal planes to have an appreciable pH-dependent negative surface charge as well. If the acidity of the edges is different from that of the basal planes then the distribution between the two must be known to characterize surface charge.

The program FITEQL (Westall, 1982) was used to regress proton surface charge density measurements (calculated from Eq. [1]) and to provide optimal best-fit values of surface site acidity constants as well as total site density, S_T. The latter were calculated using FITEQL's default analytical errors (relative errors of 0.01 units for pH and 0.02303 for $[H^+]$). FITEQL iteratively performs a nonlinear least-squares optimization calculation using an assumed chemical model of the solid–solution interface. A constant capacitance model (see Stumm and Morgan, 1996) was used to link surface charge, σ, and surface potential, ψ. The latter model was used for its simplicity in providing a direct comparison of site acidities and differential metal binding behavior. The capacitance was set to 4.0 $F \cdot m^{-2}$ to minimize the variance in the fit to the 25°C data.

The acidities of the edges and bases could not be differentiated. This was done by fitting the data to a four-site model ($Al_{edge} + Al_{basal} + Si_{edge} + Si_{basal}$), each with its own acidity constant. This gave no appreciable improvement in the fit to the titration data. We therefore modeled surface charge development through reactions that were assumed initially to occur equivalently on edges and basal planes. Because of the difference in the behavior of their respective (hydr)oxides, we explicitly differentiated between aluminol and silanol groups.

For aluminol groups,

$$\begin{aligned} &>\text{Al–O–H}_2^+ \leftrightarrow >\text{Al–O–H} + \text{H}^+, \\ K_{Al,1} &= [>\text{Al–O–H}]\exp(F\psi/RT)a_{H^+}/[>\text{Al–O–H}_2^+] \end{aligned} \quad [2]$$

$$\begin{aligned} &>\text{Al–O–H} \leftrightarrow >\text{Al–O}^- + \text{H}^+, \\ K_{Al,2} &= [>\text{Al–O}^-]\exp(-F\psi/RT)a_{H^+}/[>\text{Al–O–H}]. \end{aligned} \quad [3]$$

For silanol groups,

$$\begin{aligned} &>\text{Si–O–H} \leftrightarrow >\text{Si–O}^- + \text{H}^+ \\ K_{Si,2} &= [>\text{Si–O}^-]\exp(-F\psi/RT)a_{H^+}/[>\text{Si–O–H}]. \end{aligned} \quad [4]$$

Bracketed terms are concentrations ($mol \cdot m^{-2}$). R and T are, respectively, the gas constant and temperature in Kelvin. Note that in choosing the previous reactions, we assume that silanol groups do not protonate at the pH values of the experiments. The very low pH_{zpc} of SiO_2 (pH 2 to 3) supports this approach.

Independent evidence for crystallographic controls on edge acidity come from molecular electrostatic potential (MEP) calculations (see Cygan *et al.,* 1998). The latter calculation uses the van der Waals radii of atoms to generate a solid surface about the kaolinite molecule, and can therefore be used to predict the optimal sites on a mineral surface for reactions driven by long-range electrostatic forces. MEP calculations point to a stronger Lewis base behavior of Al sites relative to edge silanol sites (Cygan *et al.,* 1998).

In Figure 2, pH-dependent kaolinite surface charge at 25, 50, and 70°C is plotted, along with best fits to the data from FITEQL. The regressed surface equilibrium constants and site densities used to generate the lines in Fig. 2 are listed as a function of temperature in Table I. Note that best fits require roughly 50% more Si sites than Al sites at all temperatures. There is no way to independently confirm this deviation from bulk stoichiometry as the actual ratio of the two sites exposed to solution is likely to be a complex function of particle–particle interactions and nonstoichiometric dissolution prior to the titrations. The difference cannot be accounted for by nonstoichiometric leaching during the titrations since dissolved Al levels are too low ($<10^{-4}$ mol·L^{-1}) (ac-

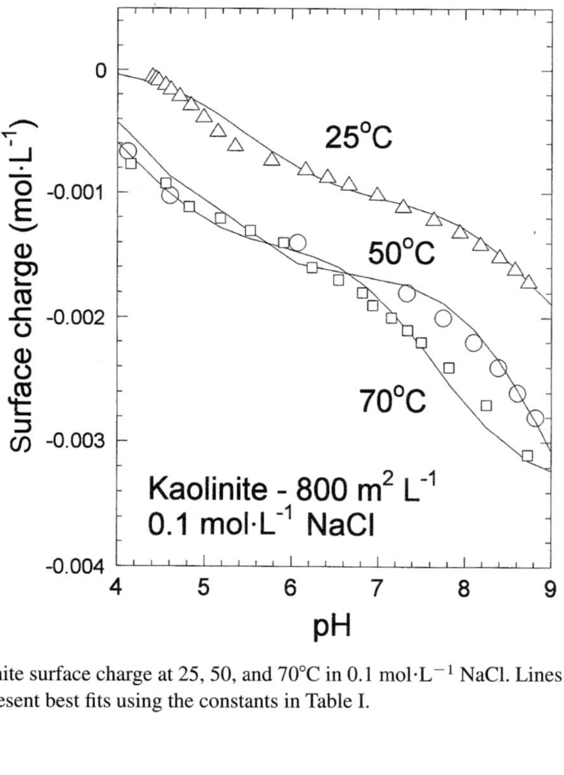

Figure 2 Kaolinite surface charge at 25, 50, and 70°C in 0.1 mol·L^{-1} NaCl. Lines are best fits from FITEQL and represent best fits using the constants in Table I.

Table I
Model Fit Parameters

T (°C)	$pK_{Al,1}$	$pK_{Al,2}$	$pK_{Si,2}$	$[>Al–O–H]_T$[a]	$[>Si–O–H]_T$[a]
25	−2.33	5.28	8.23	1.3	2.3
50	−2.54	4.20	7.98	2.0	7.8
70	−2.43	3.80	6.75	1.8	2.8

[a] $\mu mol \cdot m^{-2}$.

cording to Nagy *et al.* (1991), XPS analyses indicate minimal nonstoichiometric dissolution of kaolinite).

Total site densities calculated from Table I and the experimental surface area (800 $m^2 \cdot g^{-1}$) range from 2.2 to 5.9 sites·nm^{-2} and are so large as to point to participation of the basal planes in surface charging. Edge site densities are estimated to be about 8 sites·nm^{-2} (see Sposito, 1984, p. 40), yet the latter are generally assumed to make up only 7 to 14% of the kaolinite surface (Zhou and Gunter, 1992), and the basal planes are routinely assumed to contribute little to pH-dependent surface charge (e.g., Riese, 1982). In other words, the total site density averaged over the whole surface should be ~0.8 sites·nm^{-2}, or around an order of magnitude less than the postulated edge density. Because the measured values are somewhat higher, either the relatively hydrophobic basal planes accept and donate protons, or the total number of edge sites is greater than heretofore assumed. If the 50°C value is ignored, our results require between 29 and 37% edge surface area by the latter approach (the 50°C value is anomalously high—almost three times the size of the 25 and 70°C values—and believed to be an artifact of the fewer number of 50°C data points). The necessary edge surface area is calculated assuming 8 sites·nm^{-2} on the edges. We use SFM images of the KGa-1 surface to argue for a higher edge surface area. Scanning force microscopy imaging may have the advantage over previous measurement techniques of being better able to resolve steps and edges on basal planes. In Figure 3 is shown a typical SFM image of the KGa-1 sample. The high edge surface area (13 to 47%) relative to basal planes supports the hypothesis that edge surface areas have generally been underestimated. It also allows us to explain our surface-charging results without calling on proton adsorption–desorption onto the basal planes.

In Figure 4 are shown the thermodynamic constants (K) from Table I as a function of reciprocal temperature. Enthalpies for the various surface proton donor reactions have been calculated with the van't Hoff equation and are shown in Figure 4. The enthalpy for reaction [2] is nearly identical to that measured for the deprotonation of $>Al–O–H_2^+$ sites measured on γ-Al_2O_3 by Brady (1994), lending some support to the identification of Al sites as the source of positive charge at low pH.

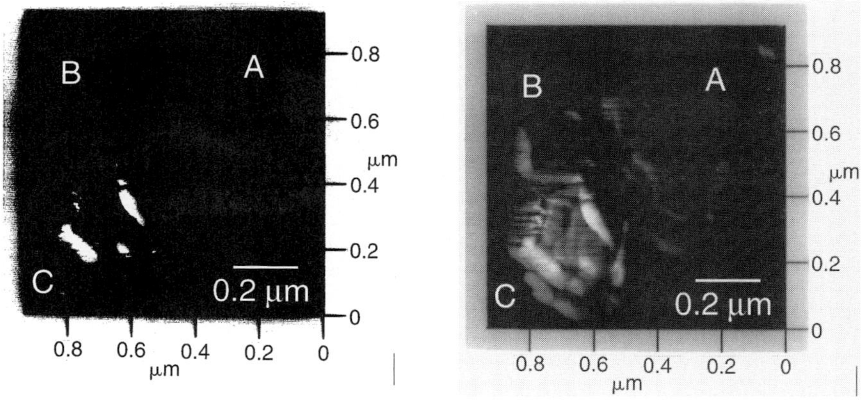

Figure 3 Scanning force microscopy images of kaolinite basal planes and edges. The percent edge areas are (A) 13, (B) 33, and (C) 47%.

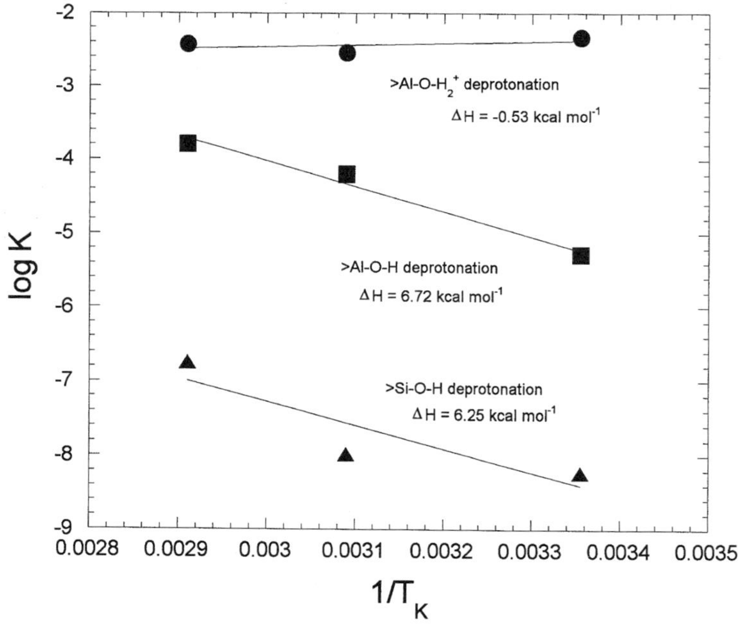

Figure 4 Temperature dependence of surface proton donor reactions. Enthalpies were calculated by the van't Hoff equation. Solid lines are Arrhenius fits.

IV. METAL SORPTION

Figure 5 shows measured adsorption of Cd, Sr, and Cs to kaolinite at 25°C (>Me denotes adsorbed metal). The equilibrium constants and site densities calculated previously were used as input to FITEQL, and the binding constants for unidentate sorption to Al sites (after Zachara and Smith, 1988) were then adjusted to determine optimal values. Poorer fits resulted for each run when unidentate sorption to Si sites was modeled. Unidentate adsorption was presumed in the fitting procedure (as opposed to bidentate adsorption) since unidentate adsorption of metals is generally observed on hydrous oxides (e.g., Dzombak and Morel, 1990). A capacitance of 4 $F \cdot m^{-2}$ was used to fit metal sorption results. Speciation calculations indicate that the bare cations were the predominant species in each titra-

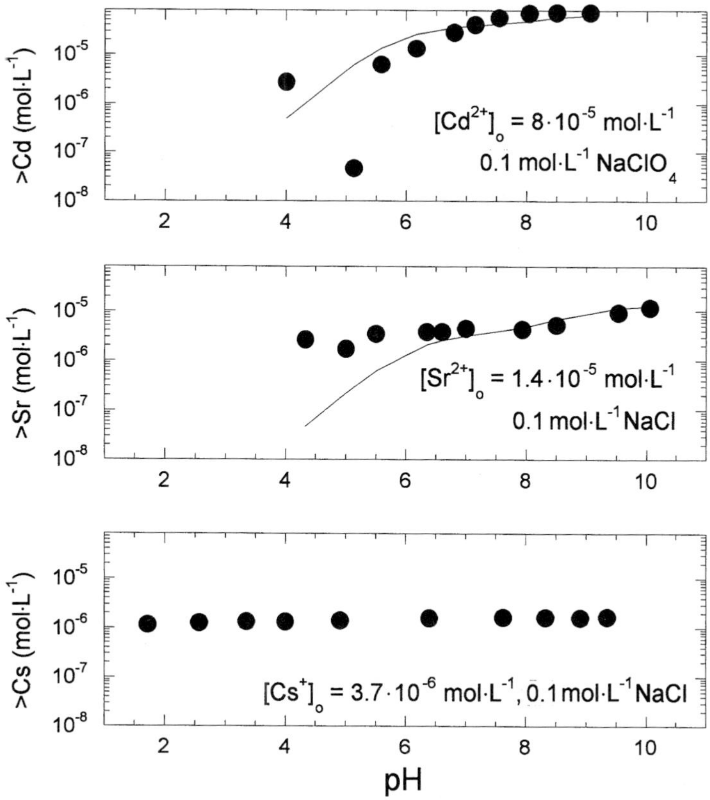

Figure 5 Adsorption of Sr and Cs onto kaolinite from 0.1 M NaCl solutions at 25°C. Cadmium sorption was measured in 0.1 M $NaClO_4$. Solid lines are best fits.

tion. The only exception was the Cd titration. Cd carbonate complexes were favored to form at high pH (pH > 8) at the very end of the titration.

Note first of all that Cd sorption is stronger than Sr sorption; at 25°C, $K_{Cd}^{DL} = 10^{-3.19}$ vs. $K_{Sr} = 10^{-3.84}$. Cesium sorption is far less pH-dependent than that of the divalent metals. Total site densities are on the order of 3 $\mu mol \cdot m^{-2}$, maximum sorption is generally an order of magnitude or less (Sr, 0.015 $\mu mol \cdot m^{-2}$; Cd, 0.09 $\mu mol \cdot m^{-2}$; Cs, 0.0025 $\mu mol \cdot m^{-2}$). Kaolinite edges are thought to be relatively soft Lewis bases (e.g., Puls and Bohn, 1988). According to the hard–soft acid base principle, soft Lewis acids tend to complex with soft Lewis bases, and likewise, hard Lewis acids prefer to associate with hard Lewis bases. "Hard" molecules are those having high electronegativity, low polarizability, and relatively small ionic size. "Soft" molecules are the opposite (see Huheey, 1972, p. 226). Cadmium is a soft acid; Sr and Cs are hard acids. The stronger binding of Cd to hydroxylated edge sites on kaolinite might therefore be explained as the interaction of a soft Lewis base with a soft Lewis acid (e.g., Puls and Bohn, 1988).

A second important feature of the sorption curves shown in Figure 5 is that, at least for Sr and Cs, there appears to be a significant, pH-independent sorbed component ($1.5 \cdot 10^{-5}$ $meq \cdot g^{-1}$). This is particularly apparent for Cs. Note in particular the measurable adsorption of both cations near the pH_{zpc} (pH ~4). The simplest explanation for these results is that cation binding occurs because of a permanent charge arising from minor heterovalent substitution in the lattice. The amount of metal adsorbed is well within the estimated basal plane charge estimated by Riese (1982) of 10^{-3} $meq \cdot g^{-1}$. Cadmium may sorb less strongly to the basal plane than Sr and Cs. The fact that the Cd experiments were done with higher starting Cd levels because of analytical limitations makes it difficult to compare the relative affinity of the different cations for the basal plane. Zinc, which is harder than Cd, was observed in preliminary experiments to sorb on the basal plane. Riese (1982) observed similar adsorption of Ra and Ca on KGa-1 as well. The trends therefore suggest that the harder cations are more likely to sorb to the basal planes than soft cations with easily deformable electron clouds. This may indicate basal plane sites which are relatively hard. Ongoing work is centered on this question.

V. CONCLUSIONS

Our results point to the following conclusions about kaolinite surface chemistry.

1. Al sites are the primary proton acceptor sites.
2. Al sites on a kaolinite edge are appreciably more acidic than are Al sites exposed on Al-(hydr)oxides. Si sites on kaolinite edges appear to be no more acidic than their counterparts on pure silica.

3. Large edge surface areas make a sizable contribution to net proton donor/acceptor capacity.
4. Hard acids (Cs^+ and Sr^{2+}) sorb less strongly to edge sites when modeled as sorbing in a unidentate fashion, but possibly more strongly to basal planes than soft Lewis bases such as Cd^{2+}.

ACKNOWLEDGMENTS

We greatly appreciate the financial support of the U.S. Nuclear Regulatory Commission and the US-DOE/OBES-Geosciences under contract DE-ACO4-94AL85000, and the very helpful suggestions of two anonymous reviewers.

REFERENCES

Bolland, M. D. A., Posner, A. M., and Quirk, J. P. 1976. Surface charge on kaolinites in aqueous suspension. *Aust. J. Soil. Sci.* 14:197–216.

Brady, P. V. 1994. Alumina surface chemistry at 25, 40, and 60°C. *Geochim. Cosmochim. Acta* 58:1213–1217.

Brady, P. V., Cygan, R. T., and Nagy, K. L. 1996. Molecular controls on kaolinite surface charge. *J. Colloid Interface Sci.* 183:356–364.

Carroll-Webb, S. A., and Walther, J. V. 1988. A surface complex reaction model for the pH-dependence of corundum and kaolinite dissolution. *Geochim. Cosmochim. Acta* 52:2609–2623.

Cashen, G. H. 1959. Electric charges on kaolinites in aqueous suspension. *Trans. Faraday Soc.* 55:477–486.

Cygan, R. T., Nagy, K. L., and Brady, P. V. 1998. Molecular models of cesium sorption on kaolinite. *In* "Adsorption of Metals by Geomedia" (E. A. Jenne, Ed.). Academic Press, San Diego.

Dzombak, D. A., and Morel, F. M. M. 1990. "Surface Complexation Modeling—Hydrous Ferric Oxide." Wiley, New York.

Ferris, A. P., and Jepson, W. B. 1975. The exchange capacities of kaolinite and the preparation of homionic clays. *J. Colloid Interface Sci.* 51:245–259.

Huang, C. P., and Stumm, W. 1973. Specific adsorption of cations on hydrous Al_2O_3. *J. Colloid Interface Sci.* 43:409–420.

Huheey, J. E. 1972. "Inorganic Chemistry: Principle of Structure and Reactivity." Harper & Row, New York.

Motta, M. M., and Miranda, C. F. 1989. Molybdate adsorption on kaolinite, montmorillonite, and illite: Constant capacitance modeling. *Soil Sci. Soc. Am. J.* 53:380–385.

Nagy, K. L., Blum, A. E., and Lasaga, A. C. 1991. Dissolution and precipitation of kinetics of kaolinite at 80°C and pH 3. *Am. J. Sci.* 291:649–686.

Parks, G. A. 1965. The isoelectric points of solid oxides, solid hydroxides, and aqueous hydroxo complex systems. *Chem. Rev.* 65:177–197.

Puls, R. W., and Bohn, H. L. 1988. Adsorption of cadmium, nickel, and zinc by kaolinite and montmorillonite suspensions. *Soil. Sci. Soc. Am. J.* 52:1289–1292.

Riese, A. C. 1982. Adsorption of Radium and Thorium onto Quartz and Kaolinite: A Comparison of Solution/Surface Equilibria Models. Ph.D. Thesis, Colorado School of Mines, Golden, CO.

Scales, P. J., Griesser, F., and Healy, T. W. 1990. Electrokinetics of the muscovite mica-aqueous solution interface. *Langmuir* 6:582–589.

Schindler, P. W., Liechti, P., and Westall, J. C. 1987. Adsorption of copper, cadmium and lead from aqueous solution to the kaolinite/water interface. *Netherlands J. Agric. Sci.* 35:219–230.

Schofield, R. K., and Samson, H. R. 1954. Flocculation of kaolinite due to the attraction of oppositely charged crystal surfaces. *Discuss. Faraday Soc.* 18:135–143.

Sposito, G. 1984. "The Surface Chemistry of Soils." Oxford University Press, New York.

Stumm, W., and Morgan, J. J. 1996. "Aquatic Chemistry." Wiley, New York.

Van Olphen, H., and Fripiat, J. 1979. "Data Handbook for Clay Materials and Other Non-metallic Minerals." Pergamon, Oxford.

Westall, J. C. 1982. FITEQL, a computer program for determination of chemical equilibrium constants from experimental data. Report 82-01, Chemistry Dept., Oregon State University, Corvallis, OR.

Wieland, E., and Stumm, W. 1992. Dissolution kinetics of kaolinite in acid aqueous solutions at 25°C. *Geochim. Cosmochim. Acta* 55:3339–3356.

Xie, Z., and Walther, J. V. 1992. Incongruent dissolution and the surface area of kaolinite. *Geochim. Cosmochim. Acta* 55:3357–3364.

Zachara, J. M., and Smith, S. C. 1988. Edge complexation reactions of cadmium on speciment and soil-derived smectite. *Soil Sci. Soc. Am. J.* 58:762–769.

Zhou, Z., and Gunter, W. D. 1992. The nature of the surface charge of kaolinite. *Clays Clay Miner.* 40:365–368.

Chapter 18

Molecular Models of Cesium Sorption on Kaolinite

Randall T. Cygan, Kathryn L. Nagy, and Patrick V. Brady
Geochemistry Department, Sandia National Laboratories, Albuquerque, New Mexico

Molecular models of interactions between kaolinite mineral surfaces and water molecules, free ions, and ionic complexes are modeled by evaluating all possible intermolecular and intramolecular forces. Computer simulations incorporate force fields, based on quantum mechanical and spectroscopic data, that account for bonding forces such as bond stretching, angle bending, and torsion. Nonbonded terms, including short-range repulsive, van der Waals attractive, and Coulombic energies, are evaluated to account for interactions between atoms at the mineral surface with those of the solution species. Molecular dynamics simulations indicate that Cs is preferentially sorbed at edge sites on the (010) surface of kaolinite as an inner-sphere complex. In contrast, the calculations suggest that Cs is sorbed on the (001) aluminol basal surface of kaolinite as an outer-sphere complex. Electrostatic calculations based on the charge distribution of kaolinite also strongly support the significant role of kaolinite edge sites for cation sorption. These model results agree with recent Cs isotherm, surface charge, and AFM data.

I. INTRODUCTION

The fate of radionuclide and chemical wastes in the environment is intimately linked to the ability of subsurface materials to attenuate and immobilize the con-

Adsorption of Metals by Geomedia

taminants by chemical sorption and precipitation processes. Our present ability to evaluate these processes is based on a variety of experimental and analytical methods, including surface charge titrations, sorption isotherm experiments, atomic force microscopy (AFM), and X-ray absorption spectroscopies (EXAFS and XANES). However, due to the complexities in structure and composition of minerals such as clays, and the inherent uncertainties of the experimental methods, it is critical to apply atomistic models for a better understanding and interpretation of these chemical processes, particularly, mechanisms of metal sorption. In this chapter we provide a theoretical basis for determining the fundamental mechanisms of how clay surfaces and clay interlayers control and bind metals. We emphasize the sorption behavior of Cs due to the inclusion of ^{135}Cs and ^{137}Cs in numerous inventories of radioactive wastes in need of remediation. Kaolinite provides a relatively simple substrate for modeling surface sorption due to its lack of interlayer metal sites. Such atomistic simulations provide a basis for predicting efficient methods for waste treatment and improved immobilization of contaminants.

The sorption of Cs on kaolinite surfaces has been experimentally examined by Westrich *et al.* (1995). Sorption isotherms suggest a pH-dependent affinity of Cs for the Al octahedral edge sites of the kaolinite crystal, with possibly weak sorption on the Si tetrahedral edge sites only at high pH values. Binding of Cs to the basal Al octahedral surface appears to be independent of pH. These results are supported by acid titrations of kaolinite powders and the subsequent modeling of surface charge as a function of pH using the model FITEQL (Brady *et al.*, 1996, 1998). These results point to control of the net surface charge by Al–O surface species, rather than Si–O species, up to a pH of 8 for kaolinite. Brady *et al.* (1996) also obtained AFM images of individual kaolinite grains (KGa-1 standard clay from the Clay Minerals Society). The images show grain diameters of approximately 0.2 μm and evidence for significant edge surface areas, often up to 47% of the total area. These AFM images provide additional support for the strong influence of kaolinite edge sites on Cs sorption.

II. KAOLINITE STRUCTURE AND SURFACE HYDROLYSIS

To examine specific molecular controls on Cs sorption and to evaluate the relative acidity of crystalline sites at the kaolinite–water interface, we use an atomistic model of the bulk crystalline structure of kaolinite that includes the modification of exposed surfaces due to hydration. Figure 1 provides a schematic of the kaolinite structure based on the Bish (1993) refinement of neutron powder diffraction data and a reanalysis of the original Young and Hewatt (1988) data. Kaolinite

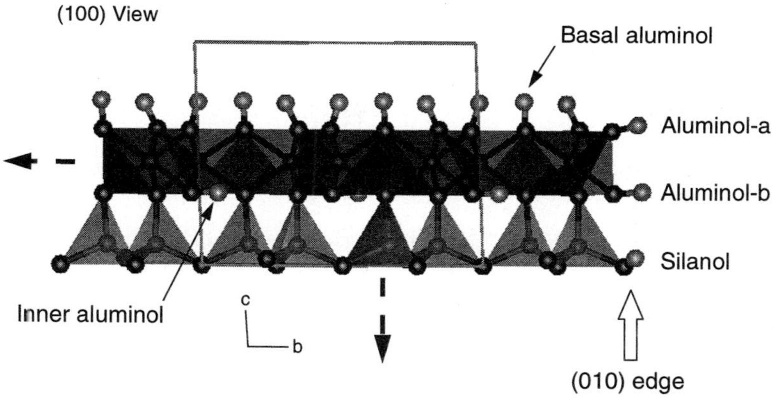

Figure 1 Crystal structure of kaolinite (Bish, 1993) as modified for hydrolysis of the (010) surface at pH 4. The unit cell for bulk structure is indicated by the gray outline. Unique structural and surface hydroxyls are indicated.

($Al_2Si_2O_5(OH)_4$), one of the simplest clay minerals, is characterized by Si in tetrahedral (T) sheets and Al in octahedral (O) sheets. The Al octahedral sheet includes hydroxyl groups that form hydrogen bonds with a Si tetrahedral sheet in the next TO layer repeated in the (001) direction. Termination of the structure at the (010) crystalline edge allows for hydrolysis of the dangling O bonds. Single protonation of these oxygens provides a surface that is representative of kaolinite in equilibrium with a solution of approximately pH 4 (Brady *et al.,* 1996). Five different types of hydroxyl groups exist in this hydrolyzed form of the kaolinite structure: silanol, basal aluminol, aluminol-a (neighboring the basal aluminols), aluminol-b (neighboring the silanols), and an inner hydroxyl (aluminol) associated with the vacant octahedral site. Similar hydrolysis surfaces exist for the (100) and (110) morphological forms that are common to kaolinite.

III. THEORETICAL BASIS FOR COMPUTER SIMULATIONS

A. Empirical Molecular Models

Molecular modeling techniques require a complete mathematical description of all interactions and energies among the constituent atoms of the chemical system. Although quantum mechanics attempts to solve the Schrödinger equation to obtain the exact or approximate potential energy surface of the molecule, these approaches are limited to fairly small numbers of atoms and electrons due to the immense computational effort required. Empirical techniques, however, provide a more efficient method wherein the potential energy surface is fit to a particular set of analytical functions. These energy functions, referred to collectively as force fields, are of a specific mathematical form and require parameters that are adjusted in an empirical fit of the potential energy surface.

B. Interatomic Potentials

The computer simulations that we incorporate in this molecular modeling study are all based on the calculation of the total energy of the chemical system using the empirical approach. The computer "experiment" calculates the energy and spatial energy derivatives of the system based on the atomic coordinates of the atoms, taking into account the partial charges and bonding characteristics. Energy contributions from bond stretching, bond angle bending, and bond torsion are evaluated based on the particular force field being used. Additionally, nonbonded interactions such as Coulombic (electrostatic) interactions, short-range repulsions, and van der Waals attraction are incorporated. The nonbonded energy contributions play a particularly significant role in modeling metal sorption. Examples of the analytical expressions in a force field include a Morse function for the bond stretch term and a Lennard–Jones potential function for the short-range and van der Waals nonbonded term. The empirical parameters for a force field are derived from experimental measurements, primarily through various spectroscopic methods, or from molecular orbital calculations. Typically, the empirical data are obtained for a small set of molecules and applied in the computer simulation to a larger set of related molecules or structures.

The nonbonded interaction terms are the critical energy contributions that will control how a molecule, ion, or ion complex will approach a mineral surface. The Coulombic energy will dominate the overall nonbonded interactions, but ultimately with very small interatomic distances the short-range terms will be significant. The analytical expression (Eq. [1]) for the nonbonded energies includes the Coulombic and Lennard–Jones potential terms that we use in our energy calculations. The potential energy E_{ij} for the interaction between atom i and atom j at a distance r_{ij} is given by

$$E_{ij} = \frac{z_i z_j e^2}{r_{ij}} + \epsilon_{ij}\left[\left(\frac{r^*_{ij}}{r_{ij}}\right)^{12} - 2\left(\frac{r^*_{ij}}{r_{ij}}\right)^{6}\right], \qquad [1]$$

where z_i is the charge on the atom and e is the electronic charge. The Lennard–Jones parameters ϵ_{ij} and r^*_{ij} represent, respectively, the potential well depth (in $kcal \cdot mol^{-1}$) and the interatomic distance (in Å) at the energy minimum. The first term represents the Coulombic energy that can be either positive or negative depending on the charges of the interacting atoms or ions, and is the dominant nonbonded energy term. The last two terms of Eq. [1] account for, respectively, the positive energy contribution as two atoms repel each other at short range and the negative van der Waals attraction energy. Figure 2 provides the energies of the Lennard–Jones potential as a function of interatomic distance between the atoms. The minima in the energy curves represent the equilibrium bond distances in the absence of Coulombic forces. The similarity of the Cs–O and O_{wat}–O_{wat} curves is related to the comparable values for their ionic radii. Forcefield parameters for the energy calculations are based on the consistent valence force field (CVFF) of Dauber-Osguthorpe *et al.* (1988). Cesium nonbonded interaction terms are taken from Smith and Dang (1994) and D. E. Smith (personal communication). Table I provides a summary of the short-range interaction parameters and the partial charges. The partial charges for the water H and O are derived from the observed dipole moment for molecular water; the partial charges for the kaolinite atoms are discussed in Section IV.A.

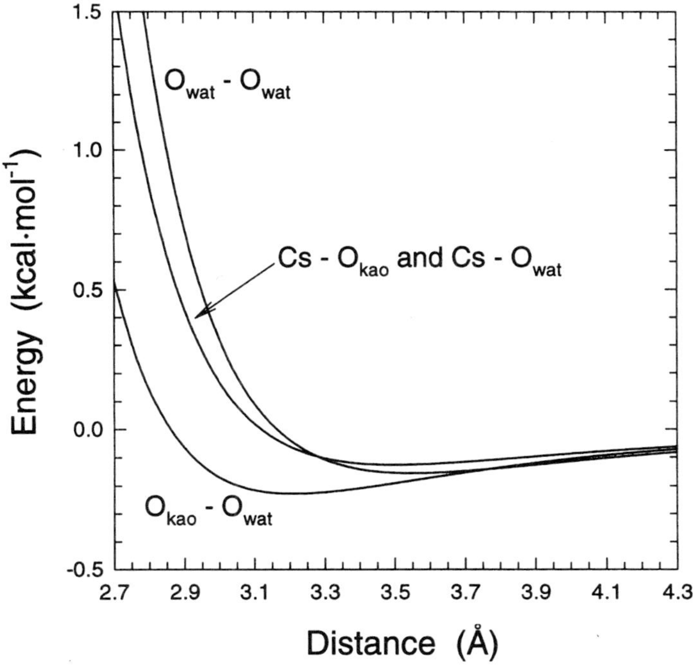

Figure 2 Lennard–Jones potential energy as a function of interatomic distance for Cs and O interactions. No Coulombic interactions are included. The different O components include O_{kao} for those associated with the kaolinite and O_{wat} for O in the free water molecules.

Table I

Short-Range Lennard–Jones Potential and Partial Charge Parameters

Atomic pair	r^*_{ij} (Å)	E_{ij} (kcal·mol^{-1})	z_i	z_j
Cs–O_{wat}	3.50	0.126	1.00	−0.82
Cs–O_{kao}	3.50	0.126	1.00	−1.20
O_{wat}–O_{wat}	3.55	0.155	−0.82	−0.82
O_{kao}–O_{wat}	3.21	0.228	−1.20	−0.82
H_{wat}–O_{wat}	—	—	0.41	−0.82
H_{kao}–O_{wat}	—	—	0.56	−0.82

C. Energy Minimization

We incorporate several different molecular modeling tools in our analysis of kaolinite and the sorption of Cs. An energy minimization approach provides the optimized geometry of a molecule based on obtaining a configuration of atoms, with or without constraints, that is associated with a minimum on the potential energy surface. Several minimization algorithms, such as steepest descents, conjugate gradients, and Newton–Raphson methods, are employed to ensure that a true global minimum and not a local minimum is achieved. Energy minimization assumes that the molecular configuration is at absolute zero temperature with no vibrational or entropic component, and therefore emphasizes the energy and static properties of the system.

D. Molecular Dynamics

A molecular dynamics (MD) approach solves Newton's equation of motion for a chemical system based on the force related to the spatial derivative of the total energy E,

$$-\frac{dE}{dr} = m\frac{d^2r}{dt^2}, \qquad [2]$$

where r is the spatial distance, m is the atomic mass, and t is time. A Maxwell–Boltzmann distribution of thermal energy is imparted on the system for a particular temperature to obtain initial atomic velocities, and then Newton's equation is evaluated for each time step, usually 1 fsec. The MD technique, in solving the classical equations of motion for a system of atoms interacting according to a potential energy force field, provides a means of examining the time evolution of a molecular system and the various conformational and momentum changes that occur.

Although computationally more intensive than energy minimization methods, the MD approach allows for a direct simulation of temperature effects related to the partitioning of kinetic energy among vibrational, rotational, and translational modes of the system. We use the so-called Verlet leapfrog algorithm (Hockney, 1970) for integrating the equations of motion for the MD simulations in this study. The calculations are performed within the constraints of a constant volume, constant temperature (NVT) thermodynamic ensemble.

Only recently have MD methods been used to investigate the interactions of solution complexes with mineral surfaces (e.g., Lupkowski and Pabalan, 1994; Sposito and Skipper, 1996). A third atomistic modeling approach is the Monte Carlo sampling of a large number of molecular geometries and configurations. The most stable energy configurations for a chemical system are chosen based on meeting a critical energy criterion. Monte Carlo methods have been used successfully to investigate interlayer waters and the swelling properties of complex clays (Skipper *et al.,* 1991, 1995a,b; Delville, 1991, 1992). We do not address Monte Carlo techniques any further in this chapter. However, Allen and Tildesley (1987) provide specifics of Monte Carlo approaches, as well as a comprehensive review of atomistic computational models.

IV. RESULTS AND DISCUSSION

A. Molecular Electrostatic Potential Surface

An additional method that helps to identify the specific crystallographic control of site acidity at the kaolinite–water interface is the calculation of the charge distribution of the atomic partial charges on the mineral surface. This approach relies on the net summation of charges at a given distance to provide a molecular electrostatic potential (MEP) surface. The MEP calculation generates a solid surface about the kaolinite molecule using the van der Waals radii of the atoms; these radii can be considered as the physical extent of the electron cloud for an unbonded atom. The surface is triangulated and the electrostatic potential V is evaluated at each vertex $\mathbf{R}$ of the polygonal surface,

$$V = \sum_{i=1}^{i=N} \frac{z_i}{|\mathbf{R} - \mathbf{r}_i|}, \qquad [3]$$

where N is the total number of atoms (ions), z_i is the charge of the *ith* atom, and $\mathbf{r}_i$ (a vector) is the location of the *ith* atom. The MEP surface provides an indication of the electrostatic potential at the van der Waals surface assuming that the atomic charge distribution is represented as a point charge centered at the nucleus (i.e., a spherical distribution of charge). The MEP surface can therefore be used to predict the most favorable site on a mineral surface for a reaction that is initially driven by relatively long-range electrostatic forces, for example, the sorption of ions and ionic complexes from solution.

A molecular representation of kaolinite including the basal and edge surfaces as presented in Figure 1 was constructed to perform the MEP calculations. The molecule is composed of approximately 10 unit cells (450 atoms) to provide a large double-layer crystallite structure. Atomic charges are assigned using those obtained by the Mulliken analysis and periodic *ab initio* molecular orbital calculations of kaolinite by Hess and Saunders (1992). Silicon has a charge of 2.5; Al, 2.1; O, -1.2; and H, 0.56. The atomic charge for edge protons is assumed to be identical to that of the structural (inner and basal) hydrogens.

Prior to the MEP calculation, the H positions were energy-optimized while constraining the heavy atom positions to those observed by Bish (1993). This was accomplished by evaluating the nonbonded interactions and Al–Si–O–H bond energies using an energy minimization approach. Forcefield parameters for the energy calculations are based on the CVFF (Section III.B) and the partial charges assigned for the MEP surface determination. All calculations were performed on a Silicon Graphics UNIX workstation using the Discover and MEP codes provided through the Catalysis and Sorption Project software consortium and Molecular Simulations Incorporated (San Diego).

The MEP calculation suggests a significant difference in the distribution of charge between the edge and basal surfaces of the kaolinite (Fig. 3). The (010) edge sites exhibit an extreme electrostatic potential that is approximately -0.9 charge$\cdot$Å^{-1} whereas the (001) surface is characterized by an extreme value of -0.3 charge$\cdot$Å^{-1}. This sizable difference in electrostatic potential between the two kaolinite surfaces is related to the internal structure of kaolinite and the charge (ion) distribution between the octahedral and tetrahedral sheets. The MEP surface for the (010) surface of kaolinite exhibits the most negative potential at the edge of the Al octahedral sheet between two of the aluminol-a sites. This result suggests a stronger Lewis base behavior for these Al sites relative to the (001) basal aluminol groups. The exposed edge of the Si tetrahedral sheet exhibits a negative net charge but smaller absolute value (-0.6 charge$\cdot$Å^{-1}) than that observed for the Al sites. The differences in electrostatic potential exhibited in the MEP surface underscore the relative affinity of the different kaolinite surfaces and sites for subsequent reaction with an aqueous cation or positively charged complex. The spatial distribution of charge on the kaolinite surface will guide the initial approach of these species as driven by the long-range electrostatic forces.

Asymmetries in the electrostatic potential distribution observed in the MEP surface are related to the crystallographic differences among the cleaved surfaces and to edge effects as limited by size of the kaolinite crystallite that is practical for the computational effort. Additionally, the surface H positions determined by the en-

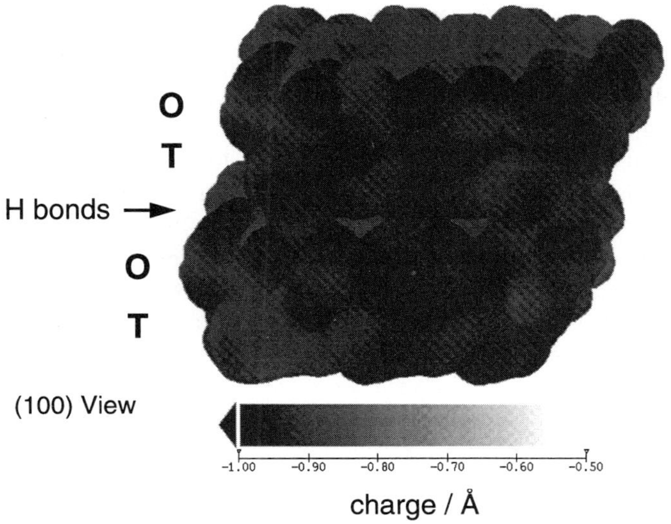

Figure 3 MEP surface for the H-optimized double-layer structure of kaolinite indicating electrostatic preference for cation sorption on edge sites relative to the basal (001) surface. Octahedral (O) and tetrahedral (T) sheets and the intermediate basal plane are indicated. Modified from Brady *et al.* (1996). A color version of this figure that more clearly demonstrates the electrostatic potential differences among the sites can be obtained from the first author.

ergy minimization calculations are controlled by hydrogen bonding with oxygens between the tetrahedral and octahedral sheets and between kaolinite layers.

B. Simulation of the Aqueous Cesium Ion Complex

The solvation of a bare Cs ion was investigated to obtain an equilibrium configuration of Cs and water molecules and to test the suitability of the Smith and Dang (1994) forcefield parameters. The computer simulation consisted of a 15 × 15 × 15-Å box containing 100 water molecules and a single Cs ion. Energy minimization calculations were performed using the parameters in Table I to obtain the stable configuration of water about Cs. Periodic boundary conditions were imposed to preserve the bulk water concentration by allowing water molecules to enter and leave the simulation volume. An Ewald summation method (Heyes, 1981) was therefore used to ensure the proper convergence of the Coulombic energy (varies as 1/r) under these periodic conditions. The radial distribution function of water oxygens about the Cs in the energy optimized configuration indicates a coordination of seven water molecules in the first coordination sphere at a mean distance of 2.86 Å from the Cs ion. X-ray and neutron diffraction studies (Ohtaki and Radnai, 1993) suggest a coordination of either six or eight, with Cs–O distances of, respectively, 3.12 and 2.95 Å, based on two separate studies. The second hydration sphere from the computer simulation results is composed of 15 water molecules positioned about 4.5 Å from the Cs.

MD simulations of the Cs–water system were also performed starting with the energy minimized configuration and the same periodic boundary conditions. For a simulation at 300 K involving the 10,000 time steps of 1 fsec each, energy equilibration occurred within 3 psec. The equilibrated structure is characterized by a water coordination about the Cs ion that has increased from seven to eight with a final mean Cs–O distance of 3.03 Å. This equilibrium distance and coordination number are in fairly good agreement with the X-ray diffraction data noted in Ohtaki and Radnai (1993) (see above). This increase of the Cs–O distance with water coordination number is expected as more solvent molecules are packed about the isolated Cs ion. The agreement of the aqueous simulation results with the experimental data provides some validation in the use of the selected forcefield parameters.

C. Simulation of Cesium on Kaolinite Surface

1. Sorption Sites on the Kaolinite (010) Surface

Energy minimization and MD calculations were performed to simulate the interaction of the solvated Cs ion with the kaolinite surfaces. A three-layer crystallite model of the kaolinite structure composed of 15 unit cells (570 atoms) was generated to represent the cleaved (010) substrate. Single protonation of the exposed oxygens, as discussed earlier, was used to represent the hydrolysis state of the kaolinite surface at a pH of 4. All heavy atoms of the kaolinite were constrained to the observed structure of Bish (1993), allowing the inner aluminol hydrogens and surface hydroxyl hydrogens to freely translate. Although beyond the scope of the present simulations, surface relaxation of kaolinite may occur and lead to local changes in the surface structure beyond the translation of hydrogens (Nagy *et al.*, 1995). As with the MEP calculation, partial charges of the kaolinite atoms were assigned based on the molecular orbital analysis of Hess and Saunders (1992). The optimum configuration for a Cs ion and 26 water molecules, obtained from the solvation calculations and the first two hydration spheres of Cs (Section IV.B), initially was used for examining the interaction with the kaolinite surface. Several different initial positions of the Cs–water assembly relative to the kaolinite (010) surface were selected in order to avoid energy minima that would not represent the true stable configuration.

The results of the energy minimization simulations indicate that the Cs ion prefers to sorb directly to the Al octahedra on the (010) kaolinite edge, being coordinated to four aluminols to form an inner-sphere complex. The mean Cs–O distances are 2.47 ± 0.06 Å (2) and 4.11 ± 0.32 Å (2), with the Cs being bound more closely to the two diagonally situated aluminols. The relatively small Cs–O distance (2.47 Å), compared to that expected for just the simple Lennard–Jones interactions (approximately 3.3 Å; see Fig. 2), indicates the strong influence of electrostatics in compressing the sorption distance between the free Cs ion and the kaolinite oxygens. At least five water molecules coordinate about the Cs ion; the remaining water molecules coordinate to silanol and other aluminol groups on the kaolinite edge with alignment of their dipoles (hydrogens pointing toward the kaolinite and oxygens away) in response to the local negative charge on the surface. Although hydrogen bonds are not explicitly included in the forcefield functionality, these water interactions with the charged kaolinite surface are the basis of hydrogen bonding.

Molecular dynamics simulations were performed at 300 K using a 1-fsec time step. All water molecules, Cs, and kaolinite H were allowed to freely translate. Figure 4 presents a snapshot of the equilibrated configuration for one of the MD simulations performed for a total time of 20 psec (over 20,000 time steps). The time evolution of the total energy for the simulation is presented in Figure 5. The initial increase in the total energy is related to the reequilibration of the 0 K energy minimized configuration that was used to start the MD simulation. As with the en-

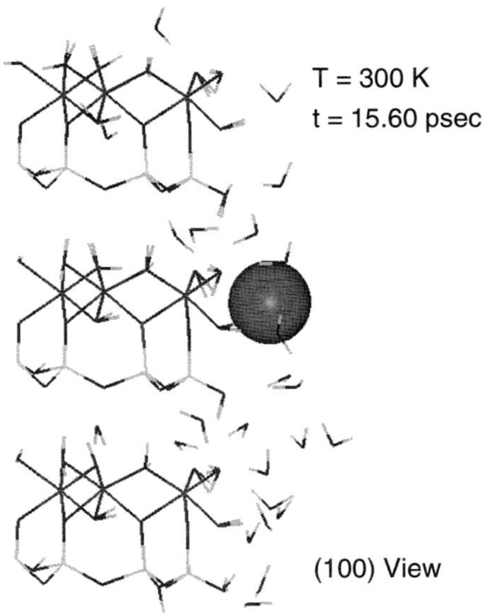

Figure 4 Snapshot of the equilibrated configuration for the MD simulation of Cs sorption on the (010) surface of a kaolinite crystallite. Cesium is bound as an inner-sphere complex. The kaolinite surface is singly protonated although it appears otherwise in the projection view.

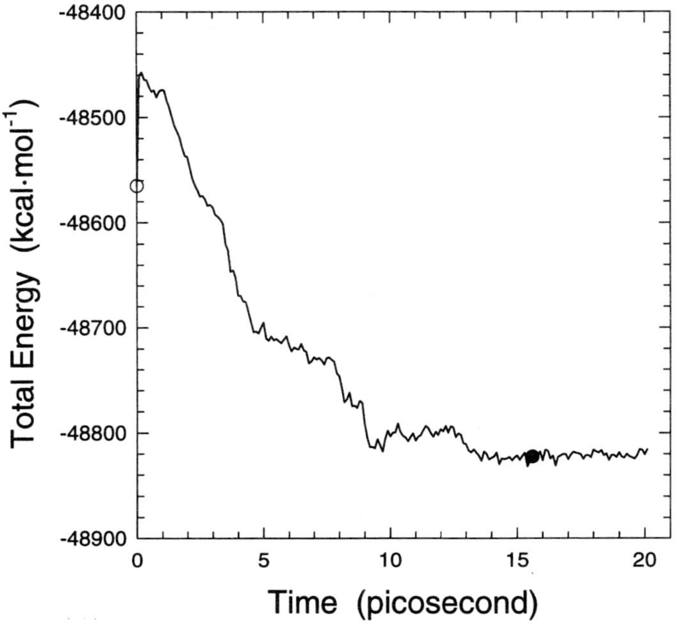

Figure 5 Total energy as a function of time for the MD simulation of Cs sorption on the (010) surface of kaolinite. The black circle denotes the energy corresponding to the molecular configuration presented in Figure 4. The open circle indicates the potential energy of the 0 K energy minimized configuration used to initiate the MD simulation.

ergy minimization calculation, the MD results suggest that Cs is bound to the four aluminol groups on the edge. Water molecules redistribute themselves across all the surfaces in order to reduce the total energy during the MD simulation. By incorporating thermal motions for the principal components, we are able to directly evaluate the variability of Cs coordination on the (010) edge site. Figure 6 demonstrates the variation in the Cs–O distances for the four binding aluminol oxygens. A critical evaluation of the energy minimization results and MD simulations requires a detailed surface analysis of Cs interactions on kaolinite using X-ray absorption spectroscopy. Such an experimental study would provide additional validation of the atomistic models used here beyond the comparison made with aqueous complex structures as discussed in Section IV.B.

2. Sorption Sites on the Kaolinite (001) Aluminol Surface

Additional energy minimization and MD calculations were performed to examine the specific interactions and binding site of Cs sorption on the (001) aluminol surface of kaolinite. This basal surface is characterized by the structural hydroxyls associated with the Al octahedral sheet. A two-layer crystallite model of the kaolinite structure, similar to the one used for the MEP calculation, was created. It was composed of 10 unit cells (450 atoms) and represents the cleaved (001) substrate. Identical partial charges, bonding conditions, and calculation schemes as incorporated with the (010) surface calculations were used to describe the Cs–water interactions with the (010) surface. The energy minimized configuration for the (001) surface is characterized by Cs positioned considerably removed from the basal aluminols. The optimal Cs–O distance is 4.54 Å, with several water molecules coordinated to the Cs but situated between the metal and the basal aluminol. The stability of an outer-sphere complex is in marked contrast to the more tightly bound Cs that occurs as an inner-sphere complex on the (010) edge surface of kaolinite.

Figure 7 provides a snapshot of one of the early configurations (within 1 psec) resulting from the MD simulation of Cs on the kaolinite (001) surface. A similar distance of approximately 4.5 Å was observed between Cs and the basal aluminol O. However, after 3 psec of simulation time, once the waters had fully equilibrat-

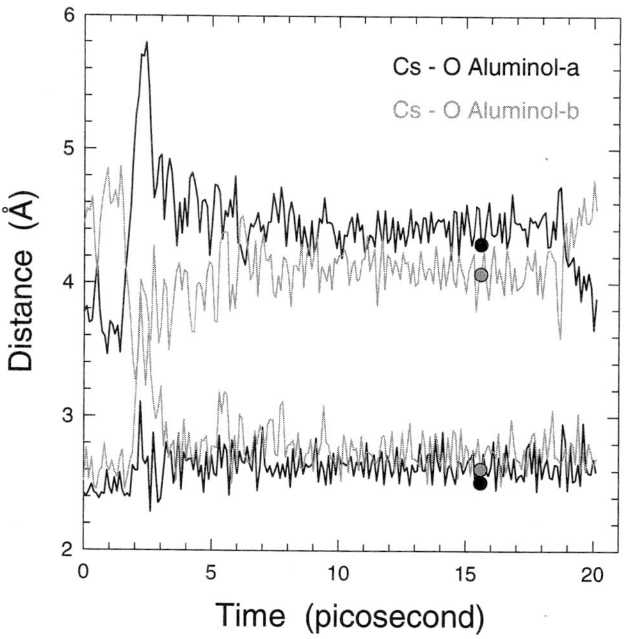

Figure 6 Interatomic distances between Cs and each of the four coordinating oxygens for the MD simulation of Cs sorption on the (010) surface of kaolinite. The black and gray circles denote the distances corresponding to the molecular configuration presented in Figure 4.

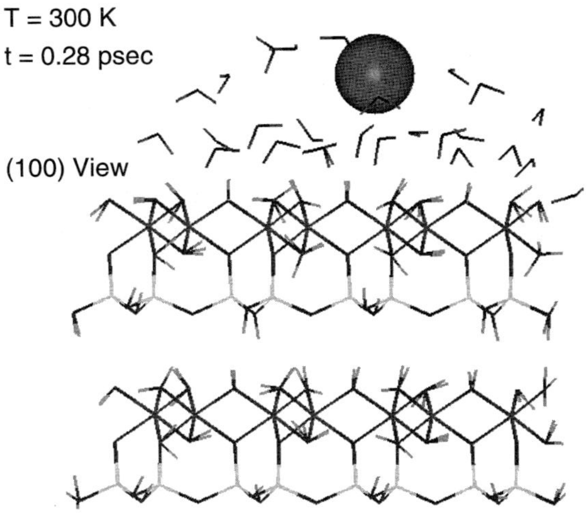

Figure 7 Snapshot of the equilibrated configuration for the MD simulation of Cs sorption on the (001) surface of a kaolinite crystallite. The kaolinite surface is singly protonated although it appears otherwise in the projection view. Cesium is loosely sorbed as an outer-sphere complex.

ed with the (001) surface, the Cs ion migrated toward the edge of the (001) surface and repositioned itself directly on the (010) surface, coordinated tightly to four aluminol groups. These results were confirmed by additional MD simulations using several different initial configurations of the Cs–water complex on the (001) surface. Note that these simulations depict a vastly different surface environment than what would be expected for Cs sorbed as an interlayer complex in a permanently charged clay structure such as montmorillonite.

D. Limitations and Concerns Associated with Atomistic Simulations

It should be noted that the MD simulations of the interaction of Cs with the kaolinite surfaces were limited to a finite number of water molecules and a finite-sized kaolinite molecule. In contrast to a fully periodic boundary MD analysis, these calculations are unable to fully simulate the solution behavior several layers of water molecules into the bulk solution. We have chosen instead in this study to emphasize the details of the bonding at the kaolinite surface. One would expect slight changes in the $Cs–O_{kao}$ bond distances if the electrostatics of the water molecules in the bulk solution were incorporated into the model. However, these long-range electrostatics are relatively small in comparison to the electrostatics and nonbonded forces associated locally with the kaolinite surface, particularly those of the (010) surface.

In addition, the calculations evaluate the Cs ion as a simple point charge associated with the interaction energies defined by a Lennard–Jones function. No electronic polarization is incorporated in the simulations. Typically, one would expect to see a decrease in total energy due to polarization of the fairly large and polarizable Cs electron cloud as it interacts with the surrounding electrostatic field. The polarization contribution is fairly minor for a symmetrical electrostatic distribution (e.g., the solvation of the free ion), but it may be significant for a Cs ion interacting with the asymmetrical field generated by the kaolinite surface. A refinement of the present calculations would be the incorporation of a point polarizable Cs ion (and O) as characterized by an electronic polarizability (see Smith and Dang, 1994).

In this study, we have only emphasized simulations of the hydroxylated surfaces of kaolinite (i.e., those that are affected by solution pH). Minor substitution of Al^{3+} or Fe^{3+} for Si^{4+} may create a small permanent, and non-pH-dependent, negative charge on the Si basal surface (Schofield and Samson, 1954). Brady *et al.* (1998) suggest that the (001) basal surface of kaolinite associated with the Si tetrahedral sheet controls the pH-independent part of the Cs sorption observed for kaolinite powders. We are currently using energy minimization and MD simulations to examine this aspect of metal ion sorption on kaolinite.

V. CONCLUSIONS

The results from our molecular models provide a strong fundamental atomistic basis for the interpretation of experimental surface titrations, Cs adsorption isotherm measurements, and atomic force microscope observations. Kaolinite crystal and surface structures play significant roles in the control of surface processes, including surface charge distribution, the relative importance of metal sorption sites, and sorption mechanisms. Specifically, we observed different surface charge distributions among the basal aluminol sites and the edge aluminol and silanol sites. The MEP surface calculation clearly shows a strong Lewis base behavior for the aluminol edge sites relative to the edge silanols and basal aluminols. Energy minimization and MD simulations suggest that Cs sorption also occurs preferentially at the (010) aluminol sites with the formation of a stable inner-sphere complex. Cesium is coordinated to four hydroxyl groups associated with two shared Al octahedra in the gibbsite sheet. In contrast, Cs is much more weakly bound on the kaolinite (001) surface (in the absence of heterovalent substitutions). The energy minimization simulations suggest that sorbed Cs exists as an outer-sphere complex; however, the MD simulation predicts the transport of the Cs to an energetically more favored site at the edge associated with the (010) surface.

More refined calculations, including those with periodic boundary conditions, a greater number of water molecules, and additional solute ions, would help to fully evaluate the mechanisms of Cs sorption. Ultimately, with continued validation, the computer simulations could be expanded to examine more sophisticated mineral surfaces and waste components.

ACKNOWLEDGMENTS

This contribution was significantly improved by the comments and suggestions provided by four anonymous reviewers. Also, we greatly appreciate the input of Henry Westrich on the application of molecular modeling toward understanding mineral sorption processes. Steve Levine of Molecular Simulations Incorporated provided invaluable guidance on the use of the software in the early stages of the study. This work was supported by the U.S. Nuclear Regulatory Commission and the U.S. Department of Energy under contract DE-AC04-94L85000.

REFERENCES

Allen, M. P., and Tildesley, D. J. 1987. "Computer Simulation of Liquids." Oxford University Press, New York.

Bish, D. L. 1993. Rietveld refinement of the kaolinite structure at 1.5 K. *Clays Clay Miner.* 41:738–744.

Brady, P. V., Cygan, R. T., and Nagy, K. L. 1996. Molecular controls on kaolinite surface charge. *J. Colloid Interface Sci.* 183:356–364.

Brady, P. V., Cygan, R. T., and Nagy, K. L. 1998. Surface charge and metal sorption to kaolinite. *In* "Adsorption of Metals by Geomedia" (E. A. Jenne, Ed.). Academic Press, San Diego.

Dauber-Osguthorpe, P., Roberts, V. A., Osguthorpe, D. J., Wolff, J., Genest, M., and Hagler, A. T. 1988. Structure and energetics of ligand binding to proteins: *E. coli* dihydrofolate reductase-trimethoprim, a drug-receptor system. *Prot. Str. Func. Genet.* 4:31–47.

Delville, A. 1991. Modeling the clay-water interface. *Langmuir* 7:547–555.

Delville, A. 1992. Structure of liquids at a solid interface: An application to the swelling of clay by water. *Langmuir* 8:1796–1805.

Hess, A. C., and Saunders, V. R. 1992. Periodic ab initio Hartree-Fock calculations of the low-symmetry mineral kaolinite. *J. Phys. Chem.* 96:4367–4374.

Heyes, D. M. 1981. Electrostatic potentials and fields in infinite point charge lattices. *J. Chem. Phys.* 74:1924–1929.

Hockney, R. W. 1970. The potential calculation and some applications. *Methods Comput. Phys.* 9:136–211.

Lupkowski, M., and Pabalan, R. T. 1994. Molecular dynamics simulation of uranyl sorption on mineral surfaces. *Eos Trans. Am. Geophys. Union* 75:138.

Nagy, K. L., Cygan, R. T., and Brady, P. V. 1995. Kaolinite morphology: AFM observations, model predictions, and adsorption site density. *Clay Min. Soc. Program Abstracts* 32:95.

Ohtaki, H., and Radnai, T. 1993. Structure and dynamics of hydrated ions. *Chem. Rev.* 93:1157–1204.

Schofield, R. K., and Samson, H. R. 1954. Flocculation of kaolinite due to the attraction of oppositely charged crystal surfaces. *Discuss. Faraday Soc.* 18:135–143.

Skipper, N. T., Chang, F. C., and Sposito, G. 1995a. Monte Carlo simulation of interlayer molecular structure in swelling clay minerals. 1. Methodology. *Clays Clay Miner.* 43:285–293.

Skipper, N. T., Refson, K., and McConnell, J. D. C. 1991. Computer simulation of interlayer water in 2:1 clays. *J. Chem. Phys.* 94:7434–7445.

Skipper, N. T., Sposito, G., and Chang, F. C. 1995b. Monte Carlo simulation of interlayer molecular structure in swelling clay minerals. 2. Monolayer hydrates. *Clays Clay Miner.* 43:294–303.

Smith, D. E., and Dang, L. X. 1994. Computer simulations of cesium-water clusters: Do ion-water clusters form gas-phase clathrates? *J. Chem. Phys.* 101:7873–7881.

Sposito, G., and Skipper, N. 1996. Computer simulation of interfacial molecular structure on layer aluminosilicates. *Mater. Res. Soc.* 317.

Westrich, H. R., Cygan, R. T., Brady, P. V., Nagy, K. L., Anderson, H. L., Kim, Y., and Kirkpatrick, R. J. 1995. The sorption behavior of Cs and Cd onto oxide and clay surfaces. *In* "Proceedings of the Waste Management Conference, WM'95," 24-4. Folio Infobase.

Young, R. A., and Hewatt, A. W. 1988. Verification of the triclinic crystal structure of kaolinite. *Clays Clay Miner.* 36:225–232.

Chapter 19

Sorption of Molybdenum on Oxides, Clay Minerals, and Soils

Mechanisms and Models

Sabine Goldberg,[1] Chunming Su,[1,2] and Harold S. Forster[1]
[1]USDA-ARS, U.S. Salinity Laboratory, Riverside, California; [2]Presently at USEPA, National Risk Management Research Laboratory, Ada, OK

The constant capacitance model was well able to describe molybdate sorption on aluminum and iron oxides, clay minerals, and soils as a function of pH. The triple-layer model contained in version 3.1 of the program FITEQL was used to simultaneously optimize Mo surface complexation constants to Mo sorption data on goethite, gibbsite, δ-Al_2O_3, kaolinite, montmorillonite, and two arid-zone soils as a function of pH (3 to 10.5) and ionic strength (0.01, 0.1, and 1.0 mol·L^{-1} NaCl). This is a new capability for the FITEQL program, and the triple-layer model was well able to describe the ionic strength effects on all materials. Because of model sensitivity to the surface site density parameter, we used a surface site density value of 2.31 sites·nm^{-2}, which had been recommended by Davis and Kent (1990. *Rev. Mineral.* 23:117–260) for natural materials. Triple-layer modeling of Mo sorption was successful on all materials using this site density value. Use of a consistent site density value will facilitate the development of a self-consistent thermodynamic database, especially for heterogeneous natural sorbents such as clay minerals and soils.

I. INTRODUCTION

Molybdenum is a trace element required for both plant and animal nutrition. Molybdenum deficiencies are reported throughout the world for many agronomic crops, especially legumes (Murphy and Walsh, 1972). Molybdenum in its anionic form is readily taken up by forage plants and can accumulate to levels detrimental to grazing ruminant animals (Reisenauer *et al.,* 1962). To evaluate plant availability of Mo in soils, knowledge of its adsorption chemistry is required.

Molybdate adsorption has been investigated for a variety of soil minerals and soils. These adsorbent surfaces include aluminum oxides (Jones, 1957; Reisenauer *et al.,* 1962; Ferreiro *et al.,* 1985; Vordonis *et al.,* 1990; Spanos *et al.,* 1990a,b; Bibak and Borggaard, 1994; Spanos and Lycourghiotis, 1995; Goldberg *et al.,* 1996), iron oxides (Jones, 1957; Reisenauer *et al.,* 1962; Reyes and Jurinak, 1967; Kyriacou, 1967; McKenzie, 1983; Ferreiro *et al.,* 1985; Zhang and Sparks, 1989; Bibak and Borggaard, 1994; Goldberg *et al.,* 1996), clay minerals (Jones, 1957; Phelan and Mattigod, 1984; Mikkonen and Tummavuori, 1993a; Motta and Miranda, 1989; Goldberg *et al.,* 1996), and soils (Jones, 1957; Reisenauer *et al.,* 1962; Barrow, 1970; Theng, 1971; Gonzalez *et al.,* 1974; Jarrell and Dawson, 1978; Karimian and Cox, 1978; Roy *et al.,* 1986, 1989; Xie and MacKenzie, 1991; Xie *et al.,* 1993; Mikkonen and Tummavuori, 1993b; Goldberg *et al.,* 1996).

Molybdate adsorption on all of the above adsorbents increased with increasing solution pH from pH values of 2 to 4, exhibited a peak near pH 4, and decreased with increasing pH above pH 4. Aluminum and iron oxides represent important molybdate adsorbing surfaces in soils. Molybdate adsorption on soil clays dominant in kaolinite and illite was decreased by removal of amorphous aluminum and iron oxides (Theng, 1971). Molybdate adsorption in soils was highly correlated with extractable aluminum (Barrow, 1970) and iron (Gonzalez *et al.,* 1974; Jarrell and Dawson, 1978; Karimian and Cox, 1978), and drastically reduced after removal of amorphous iron oxides (Jones, 1957).

Ligand exchange is suggested to be the mechanism of molybdate adsorption on aluminum and iron oxide minerals (Jones, 1957; Ferreiro *et al.,* 1985). By this mechanism ions become adsorbed specifically as inner-sphere surface complexes. Inner-sphere surface complexes, by definition, contain no water molecules between the adsorbing ion and the surface functional group (Sposito, 1984). The point of zero charge (PZC) is defined as the pH value where there is no net particle charge (Sposito, 1984). Specific adsorption of anions onto variable-charge minerals, such as oxides, shifts the PZC to a more acid pH value. Molybdate adsorption lowered the PZC of goethite (McKenzie, 1983), amorphous iron oxide, δ-Al_2O_3, gibbsite, amorphous aluminum oxide, and kaolinites (Goldberg *et al.,* 1996), indicating specific adsorption on these minerals.

The dependence of ion adsorption on the effect of ionic strength has been used to distinguish between inner- and outer-sphere surface complexes (Hayes and Leckie, 1987; Hayes *et al.,* 1988). Outer-sphere surface complexes, by definition, contain at least one water molecule between the adsorbing ion and the surface functional group (Sposito, 1984). Hayes *et al.* (1988) suggested that selenite, which showed little ionic strength dependence in its adsorption behavior, was specifically adsorbed on goethite in an inner-sphere surface complex, while selenate, which showed great ionic strength dependence, was adsorbed nonspecifically in an outer-sphere surface complex. Similarly, Zhang and Sparks (1989) found little ionic strength dependence of molybdate adsorption on goethite and interpreted this result as supporting evidence for inner-sphere surface complex formation.

The constant capacitance model (Stumm *et al.,* 1980) and the triple-layer model (Davis *et al.,* 1978) are chemical surface complexation models of the oxide–solution interface that use a ligand exchange mechanism to describe specific anion adsorption. These models explicitly define inner-sphere surface complexes and chemical reactions and consider the charge on both the adsorbate anion and the adsorbent surface. The constant capacitance model has been used successfully to describe molybdate adsorption on various aluminum and iron oxides via ligand exchange with surface hydroxyl groups and on various clay minerals via ligand exchange with aluminol groups as a function of solution pH (Goldberg *et al.,* 1996). The constant capacitance model has successfully described Mo adsorption on clay minerals as a function of equilibrium Mo concentration (Motta and Miranda, 1989). The constant capacitance model was unable to describe Mo adsorption on three soils as a function of solution pH (Goldberg *et al.,* 1996). The triple-layer model has been used successfully to describe molybdate adsorption on goethite as a function of solution pH (Zhang and Sparks, 1989).

The present study was initiated to reevaluate the ability of the constant capacitance model to describe molybdate adsorption behavior previously determined by Goldberg *et al.* (1996) on a variety of soils and soil minerals—Al oxide, Fe oxide, and clay minerals—as a function of solution pH. Molybdate adsorption on various adsorbents was evaluated as a function of solution pH and ionic strength in an effort to deduce adsorption mechanisms. The ability of the triple-layer model to describe all ionic strength data simultaneously with one set of molybdate surface complexation constants will also be investigated.

II. MATERIALS AND METHODS

Molybdenum adsorption behavior as a function of solution pH and ionic strength was studied on various adsorbents. δ-Al_2O_3, under the trade name Aluminium Oxid C, was obtained from Degussa (Teterboro, NJ). Goethite, α-FeOOH, was synthesized as described by McLaughlin *et al.* (1981). Gibbsite, γ-$Al(OH)_3$, was synthesized according to the procedure of Kyle *et al.* (1975). No trace contaminants were observed in the oxides using X-ray diffraction analysis. Samples of KGa-1 kaolinite and SWy-1 montmorillonite were obtained from the Clay Minerals Society's Source Clays Repository (Univ. of Missouri, Columbia) and used without pretreatment. Surface samples of the Pachappa (coarse-loamy, mixed, thermic Mollic Haploxeralf) and Porterville (fine, montmorillonitic, thermic Typic Chromoxerert) soil series consisted of the <2-mm fraction. Organic and inorganic carbon analyses were carried out using the method of Nelson and Sommers (1982). Free aluminum and iron oxides were extracted as described by Coffin (1963).

Trace impurities in the oxides and clay minerals were determined using X-ray diffraction powder mounts (see Table I). To obtain dominant clay mineralogy of the soils, X-ray diffraction peak areas obtained using oriented mounts were converted directly to clay mineral contents as described by Klages and Hopper (1982). Specific surface areas of the clay minerals and oxides were determined with a single-point BET N_2 adsorption isotherm obtained using a Quantasorb Jr. surface area analyzer (Quantachrome Corp., Syosset, NY). Specific surface areas of the soil samples were obtained using ethylene glycol monoethylether (EGME) adsorption as described by Cihacek and Bremner (1979). Points of zero charge and electrophoretic mobilities were determined for all oxides and kaolinite by microelectrophoresis as described by Goldberg *et al.* (1996). Table I presents point of zero charge and specific surface area data for the oxides and clay minerals. Table II presents chemical, mineralogical, and specific surface area data for the soils.

Molybdate adsorption experiments were carried out in batch systems to determine adsorption envelopes (amount of Mo adsorbed as a function of solution pH per fixed total Mo concentration). Samples of adsorbent were added to 50-ml polypropylene centrifuge tubes or 250-ml centrifuge bottles and equilibrated with aliquots (see Table III for solids concentration) of a 0.01, 0.1, or 1.0 mol·L^{-1} NaCl solution by shaking for 20 hr on a reciprocating shaker at 23 ± 1°C. This solution

Table I

Characterization of Oxides and Clay Minerals

Solid	Surface area ($m^2 \cdot g^{-1}$)	Point of zero charge
Goethite	63.7	8.82
δ-Al_2O_3	102.9	9.30
Gibbsite	56.5	9.41
KGa-1 kaolinite	9.14	2.88
SWy-1 montmorillonite	18.6	

Table II

Characterization of Soils

Soil	Inorganic carbon (%)	Organic carbon (%)	Free aluminum (%)	Free iron (%)	Surface area ($m^2 \cdot g^{-1}$)	Dominant minerals
Pachappa	0.010	0.49	0.067	0.76	36.3	Illite, kaolinite, montmorillonite
Porterville	0.023	0.84	0.090	1.07	172.2	Kaolinite, illite, montmorillonite

Table III

Solids Concentrations and Intrinsic Surface Complexation Constants Obtained with the Constant Capacitance Model

Solid	Solids concentration ($g \cdot L^{-1}$)	log $K_+(int)$	log $K_-(int)$	log $K^1_{Mo}(int)$	log $K^2_{Mo}(int)$
Iron oxides					
Hematite	5.0	7.31	−8.80	8.71	3.44
Goethite	1.25	7.31	−8.80	8.54	1.13
Poorly crystallized goethite	0.64	7.31	−8.80	9.51	3.10
Amorphous iron oxide	0.64	7.31	−8.80	9.72	2.75
Average (iron oxides)				8.99 ± 0.64	2.33 ± 1.05
Aluminum oxides					
δ-Al_2O_3	4.0	7.38	−9.09	9.61	—[a]
Gibbsite	1.25	7.38	−9.09	8.93	1.60
Amorphous aluminum oxide	0.35	7.38	−9.09	9.37	2.89
Average (aluminum oxides)				9.30 ± 0.34	2.25 ± 0.91
Average (oxides)				9.15 ± 0.49	2.29 ± 0.87
Clays					
KGa-1 kaolinite	200	4.95	−9.09	7.68	—
KGa-2 kaolinite	100	5.99	−9.09	8.69	—
SWy-1 montmorillonite	30	7.38	−9.09	8.43	0.62
SAz-1 montmorillonite	40	3.83	−1.26	8.36	—
STx-1 montmorillonite	40	2.81	−9.09	5.41	—
IMt-illite	50	7.38	−9.09	8.79	—
Average (clays)				7.89 ± 1.28	
Soils					
Hesperia soil	200	7.35	−8.95	8.28	0.63
Pachappa soil	200	7.35	−8.95	9.36	1.20
Porterville soil	200	7.35	−8.95	9.00	2.04
Average (soils)				8.88 ± 0.55	1.29 ± 0.71

[a]No convergence.

contained 0.292 mol·m^{-3} of Mo from $Na_2MoO_4 \cdot 2H_2O$ (Mallinckrodt, Inc., St. Louis, MO) and had been adjusted to the desired pH values using 1.0 mol·L^{-1} HCl or 1.0 mol·L^{-1} NaOH additions that changed the total volume by ≤2%. Experiments were carried out at 0.292 mol·m^{-3} of Mo to avoid the formation of Mo polymers in solution (Carpéni, 1947). The samples were centrifuged at a relative centrifugal force of 7800*g* for 20 min. The decantates were analyzed for pH, filtered through a 0.45-μm Whatman filter, and analyzed for Mo concentration with inductively coupled plasma (ICP) emission spectrometry.

In situ attenuated total reflectance Fourier transform infrared (ATR-FTIR) spectroscopy and diffuse reflectance infrared Fourier transform (DRIFT) spectroscopy were used to study sorption of Mo on a synthetic amorphous iron oxide, $Fe(OH)_3$(a), described by Su and Suarez (1995). Amorphous iron oxide was chosen because it is known to sorb large amounts of Mo, and it does not exhibit strong infrared (IR) absorbance in the region where Mo shows characteristic IR bands. Amorphous iron oxide has a BET N_2 surface area of 250 m^2·g^{-1}. The suspensions of $Fe(OH)_3$(a) were prepared by adding 2.0 g of solids to each 20 ml of 1.0 mol·L^{-1} NaCl (pH 6), 0.05 mol·L^{-1} Na_2MoO_4 + 1.0 mol·L^{-1} NaCl (pH 6), or 1.0 mol·L^{-1} Na_2MoO_4 + 1.0 mol·L^{-1} NaCl (pH 6). Suspension pH was maintained at 6 by frequent additions of 1.0 mol·L^{-1} NaOH or 1.0 mol·L^{-1} HCl. The suspensions were shaken for 24 hr at ambient temperature and centrifuged. Fifteen milliliters of the supernatant were removed and a 3.5-ml subsample was used as the reference in the ATR-FTIR study. The solid was resuspended in the remaining 5 ml of supernatant and used as the sample. A solid concentration of 400 g·L^{-1} was achieved in the sampling ATR reservoir. Concentrations of Mo in the supernatant were determined by ICP.

Infrared spectra of the aqueous solutions of 1.0 mol·L^{-1} NaCl (pH 6) and 0.05 and 0.10 mol·L^{-1} Na_2MoO_4 + 1.0 mol·L^{-1} NaCl (pH 6), and the suspensions of $Fe(OH)_3$(a) reacted with Mo, were recorded in the 4000 to 7000 cm^{-1} range using a Bio-Rad Digilab FTS-7 spectrometer (Bio-Rad Digilab Div., Cambridge, MA). The ATR accessory consisted of a horizontal reservoir and a ZnSe crystal rod with a 45° angle of incidence. Single-beam IR spectra were obtained from 2000 scans using a resolution of 4 cm^{-1}. All final spectra were the results of subtracting the spectrum of the supernatant or of 1.0 mol·L^{-1} NaCl (pH 6) from the spectrum of the $Fe(OH)_3$(a) suspensions or of 0.05 and 0.10 mol·L^{-1} Na_2MoO_4 + 1.0 mol·L^{-1} NaCl (pH 6), respectively. A subtraction factor of unity was always used.

A subsample of 1.0 ml of the solid suspension (400 g·L^{-1}) was washed twice with 30 ml of deionized water and air dried before DRIFT spectroscopic analysis. DRIFT spectra of samples diluted with KBr (5-mg sample in 95 mg KBr) were recorded from 4000 to 200 cm^{-1} at 4 cm^{-1} resolution over 500 scans. A subsample of the washed solids was examined by X-ray diffraction and subsequently dissolved in 0.5 mol·L^{-1} HNO_3 and analyzed for Mo concentration by ICP.

A. Constant Capacitance Modeling

The constant capacitance model (Stumm *et al.*, 1980) was used to describe molybdate adsorption behavior on the adsorbents as a function of solution pH in a background electrolyte of 0.1 $mol \cdot L^{-1}$ NaCl. The computer program FITEQL, version 3.1 (Herbelin and Westall, 1994), had been used to fit intrinsic molybdate surface complexation constants to the experimental adsorption data as presented in Goldberg *et al.* (1996). In the present study the data from Goldberg *et al.* (1996) are reanalyzed using Mo (ads) as a "dummy" component. Using this procedure, we obtained an improved fit and were able to use the preprocessor graphing routine to check the goodness-of-fit of the model to the data. Additional explanation on the use of the adsorbed ion as a "dummy" component is provided by Herbelin and Westall (1994).

Molybdenum occurs as MoO_4^{2-} over most of the pH range. The acid–base reactions undergone by molybdic acid are

$$H_2MoO_{4(aq)} \rightleftarrows H^+_{(aq)} + HMoO^-_{4(aq)} \qquad [1]$$

$$H_2MoO_{4(aq)} \rightleftarrows 2H^+_{(aq)} + MoO^{2-}_{4(aq)}, \qquad [2]$$

with pK values of 4.00 and 8.24, respectively (Lindsay, 1979).

In the constant capacitance model the protonation and dissociation reactions for the surface functional group, SOH (where SOH represents a reactive surface hydroxyl bound to a metal ion, S (Al or Fe), in the oxide mineral or an aluminol at the clay mineral edge), are defined as

$$SOH_{(s)} + H^+_{(aq)} \rightleftarrows SOH^+_{2(s)} \qquad [3]$$

$$SOH_{(s)} \rightleftarrows SO^-_{(s)} + H^+_{(aq)}. \qquad [4]$$

The constant capacitance model contains the assumption that all surface complexes are inner-sphere. Therefore the surface complexation reactions for molybdate adsorption are defined as

$$SOH_{(s)} + H_2MoO_{4(aq)} \rightleftarrows SHMoO_{4(s)} + H_2O \qquad [5]$$

$$SOH_{(s)} + H_2MoO_{4(aq)} \rightleftarrows SMoO^-_{4(s)} + H^+_{(aq)} + H_2O. \qquad [6]$$

The intrinsic equilibrium constants for protonation and dissociation reactions of the surface functional group are

$$K_+(\text{int}) = \frac{[SOH_2^+]}{[SOH][H^+]} \exp(F\psi_o/RT) \qquad [7]$$

$$K_-(\text{int}) = \frac{[SO^-][H^+]}{[SOH]} \exp(-F\psi_o/RT), \qquad [8]$$

where F is the Faraday constant ($C \cdot mol_c^{-1}$), ψ_o is the surface potential (V), o refers to the surface plane of adsorption, R is the molar gas constant ($J \cdot mol^{-1} \cdot K^{-1}$), T is the absolute temperature (K), and square brackets represent concentrations ($mol \cdot L^{-1}$). The intrinsic equilibrium constants for the molybdate surface complexation reactions are

$$K^{1is}_{Mo}(int) = \frac{[SHMoO_4]}{[SOH][H_2MoO_4]} \quad [9]$$

$$K^{2is}_{Mo}(int) = \frac{[SMoO_4^-][H^+]}{[SOH][H_2MoO_4]} \exp(-F\psi_o/RT), \quad [10]$$

where the subscript "is" refers to inner-sphere surface complexation.

The mass balance expression for the surface functional group is

$$[SOH]_T = [SOH] + [SOH_2^+] + [SO^-] + [SHMoO_4] + [SMoO_4^-] \quad [11]$$

where $[SOH]_T$ is related to the surface site density, N_s, by

$$[SOH]_T = \frac{S_A C_p 10^{18}}{N_A} N_s, \quad [12]$$

where S_A is the surface area ($m^2 \cdot g^{-1}$), C_p is the solids concentration ($g \cdot L^{-1}$), N_A is Avogadro's number, and N_s has units of sites$\cdot nm^{-2}$.

The charge balance expression is

$$\sigma_o = [SOH_2^+] - [SO^-] - [SMoO_4^-], \quad [13]$$

where σ_o represents the surface charge ($mol_c \cdot L^{-1}$). The relationship between surface charge and surface potential is

$$\sigma_o = \frac{\mathbf{C} S_A C_p}{F} \psi_o \quad [14]$$

where $\mathbf{C}$ is the capacitance ($F \cdot m^{-2}$).

In our application of the constant capacitance model, the surface site density was treated as molybdate reactive site density and set to the maximum Mo adsorption obtained in our experiments. Numerical values of the intrinsic protonation constant, $K_+(int)$, and the intrinsic dissociation constant, $K_-(int)$, were obtained from the literature compilation of experimental values for aluminum and iron oxides of Goldberg and Sposito (1984a). The intrinsic protonation and dissociation constants were initially fixed at $\log K_+(int) = 7.31$ and $\log K_-(int) = -8.80$ for goethite; $\log K_+(int) = 7.38$ and $\log K_-(int) = -9.09$ for gibbsite and the clays (Goldberg and Sposito, 1984a); and $\log K_+(int) = 7.35$ and $\log K_-(int) = -8.95$ for the soils (Goldberg and Sposito, 1984b). For the kaolinites and two of the montmorillonites it was subsequently necessary to optimize $\log K_+(int)$ or $\log K_+(int)$ and $\log K_-(int)$ as well as the molybdate surface complexation constants using the FITEQL program. The capacitance density was fixed at $\mathbf{C} = 1.06$ $F \cdot m^{-2}$, considered optimum for γ-Al_2O_3 by Westall and Hohl (1980). It is preferable to minimize the number of adjustable parameters by obtaining values of log K_+(int) and log K_-(int) experimentally from titration data when available.

The goodness-of-fit criterion is the overall variance, V, in Y (Herbelin and Westall, 1994),

$$V_Y = \frac{SOS}{DF}, \quad [15]$$

where SOS is the weighted sum of squares of the residuals and DF is the degrees of freedom.

B. Triple-Layer Modeling

The triple-layer model allows ion adsorption as either inner-sphere or outer-sphere surface complexes. In addition to the protonation–dissociation reactions, Eqs. [3] and [4], the triple-layer model considers outer-sphere surface complexation reactions for the background electrolyte:

$$SOH_{(s)} + Na^+_{(aq)} \rightleftarrows SO^-{-}Na^+_{(s)} + H^+_{(aq)} \quad [16]$$

$$SOH_{(s)} + H^+_{(aq)} + Cl^-_{(aq)} \rightleftarrows SOH_2^+{-}Cl^-_{(s)}. \quad [17]$$

In the triple-layer model, inner-sphere surface complexation reactions and intrinsic equilibrium constant expressions for molybdate are given by Eqs. [5], [6], [9], and [10], as for the constant capacitance model. The outer-sphere surface complexation reactions for molybdate adsorption are

$$SOH_{(s)} + H_2MoO_{4(aq)} \rightleftarrows SOH_2^+{-}HMoO^-_{4(s)} \quad [18]$$

$$SOH_{(s)} + H_2MoO_{4(aq)} \rightleftarrows SOH_2^+{-}MoO^{2-}_{4(s)} + H^+_{(aq)}. \quad [19]$$

The intrinsic equilibrium constants for outer-sphere surface complexation are

$$K_{Na+}(int) = \frac{[SO^- {-}Na^+][H^+]}{[SOH][Na^+]} \exp[F(\psi_\beta - \psi_o)/RT] \quad [20]$$

$$K_{Cl-}(int) = \frac{[SOH_2^+{-}Cl^-]}{[SOH][H^+][Cl^-]} \exp[F(\psi_o - \psi_\beta)/RT] \quad [21]$$

$$K^{1os}_{Mo}(int) = \frac{[SOH_2^+{-}HMoO_4^-]}{[SOH][H_2MoO_4]} \exp[F(\psi_o - \psi_\beta)/RT] \quad [22]$$

$$K^{2os}_{Mo}(int) = \frac{[SOH_2^+{-}MoO_4^{2-}][H^+]}{[SOH][H_2MoO_4]} \exp[F(\psi_o - 2\psi_\beta)/RT], \quad [23]$$

where β refers to the plane of outer-sphere adsorption and the subscript "os" refers to outer-sphere surface complexation.

The mass balance for the surface functional group is

$$[\mathrm{SOH}]_T = [\mathrm{SOH}] + [\mathrm{SOH_2^+}] + [\mathrm{SO^-}] + [\mathrm{SHMoO_4}] + [\mathrm{SMoO_4^-}] + [\mathrm{SOH_2^+}\text{–}\mathrm{HMoO_4^-}] + [\mathrm{SOH_2^+}\text{–}\mathrm{MoO_4^{2-}}] + [\mathrm{SO^-}\text{–}\mathrm{Na^+}] + [\mathrm{SOH_2^+}\text{–}\mathrm{Cl^-}]. \quad [24]$$

The charge balance expressions are

$$\sigma_o + \sigma_\beta + \sigma_d = 0 \quad [25]$$

$$\sigma_o = [\mathrm{SOH_2^+}] + [\mathrm{SOH_2^+}\text{–}\mathrm{HMoO_4^-}] + [\mathrm{SOH_2^+}\text{–}\mathrm{MoO_4^{2-}}] + [\mathrm{SOH_2^+}\text{–}\mathrm{Cl^-}] - [\mathrm{SO^-}] - [\mathrm{SOMoO_4^-}] - [\mathrm{SO^-}\text{–}\mathrm{Na^+}] \quad [26]$$

$$\sigma_\beta = [\mathrm{SO^-}\text{–}\mathrm{Na^+}] - [\mathrm{SOH_2^+}\text{–}\mathrm{HMoO_4^-}] - 2[\mathrm{SOH_2^+}\text{–}\mathrm{MoO_4^{2-}}] - [\mathrm{SOH_2^+}\text{–}\mathrm{Cl^-}]. \quad [27]$$

The relationships between the surface charges and the surface potentials are

$$\sigma_o = \frac{\boldsymbol{C}_1 S_A C_p}{F} (\psi_o - \psi_\beta) \quad [28]$$

$$\sigma_d = \frac{\boldsymbol{C}_2 S_A C_p}{F} (\psi_d - \psi_\beta) \quad [29]$$

$$\sigma_d = \frac{S_A C_p}{F} (8\epsilon_o DRT\boldsymbol{I})^{1/2} \sinh(F\psi_d/2RT), \quad [30]$$

where $\boldsymbol{C}_1$ and $\boldsymbol{C}_2$ are capacitances, d refers to the plane of the diffuse ion swarm, ϵ_o is the permittivity of vacuum, D is the dielectric constant of water, and $\boldsymbol{I}$ is the ionic strength.

The surface site density was set at a value of 2.31 sites·nm^{-2}. This value had been recommended by Davis and Kent (1990) for natural materials. Numerical values for the intrinsic protonation and dissociation constants and the surface complexation constants were obtained from the literature. For goethite these constants were log K_+(int) = 4.3, log K_-(int) = −9.8, log K_{Na^+}(int) = −9.3, and log K_{Cl^-}(int) = 5.4, as obtained by Zhang and Sparks (1990). For aluminum oxides, clays, and soils these constants were log K_+(int) = 5.0, log K_-(int) = −11.2, log K_{Na^+}(int) = −8.6, and log K_{Cl^-}(int) = 7.5, as obtained by Sprycha (1989a,b) on γ-Al_2O_3. Molybdate surface complexation constants were fit simultaneously to the adsorption data at three different ionic strengths using either inner-sphere or outer-sphere adsorption mechanisms. For δ-Al_2O_3 and clays, and gibbsite with an outer-sphere adsorption mechanism, it was subsequently necessary to optimize log K_{Na^+}(int) and log K_{Cl^-}(int) as well as the molybdate surface complexation constants using the FITEQL program. The capacitances were fixed at $\boldsymbol{C}_1$ = 1.2 F·m^{-2} and $\boldsymbol{C}_2$ = 0.2 F·m^{-2}, values considered optimum for goethite by Zhang and Sparks (1990). It is preferable to minimize the number of adjustable parameters by obtaining values of log K_+(int), log K_-(int), log K_{Na^+}(int), and log K_{Cl^-}(int) experimentally from titration data when available.

III. RESULTS AND DISCUSSION

The fit of the constant capacitance model to the data of Goldberg *et al.* (1996) using Mo(ads) as a "dummy" component is indicated in Figures 1 to 5. Molybdate adsorption on all materials exhibited a maximum at low pH (3 to 5). With increasing solution pH, adsorption decreased rapidly, with little adsorption occurring above pH values of 7 to 8. The constant capacitance model was well able to describe molybdate adsorption on all iron and aluminum oxides studied, with some deviations occurring at low and high pH values (Figs. 1 and 2). Use of Mo(ads) as a "dummy" component improved the model fit for all oxides except amorphous iron and aluminum oxide (compare Figs. 1 and 2 to Figs. 4 and 5 of Goldberg *et al.* (1996)). Molybdate adsorption on the oxides was described with the model when only the molybdate surface complexation constants, K^1_{Mo}(int) and K^2_{Mo}(int), were optimized.

The constant capacitance model was well able to describe molybdate adsorp-

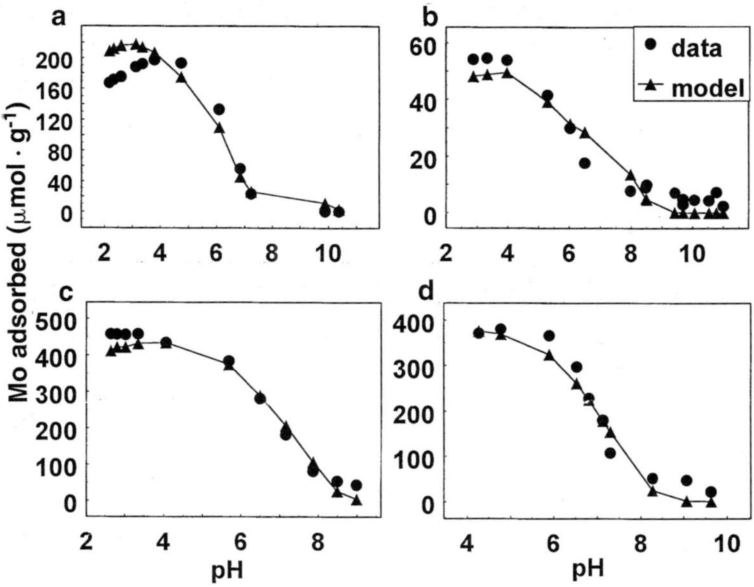

Figure 1 Molybdate adsorption on iron oxides as a function of solution pH: (a) goethite, $V_Y = 106$; (b) hematite, $V_Y = 312$; (c) poorly crystallized goethite, $V_Y = 85.9$; (d) amorphous iron oxide, $V_Y = 112$. Circles represent experimental data. Model results are represented by triangles and solid lines.

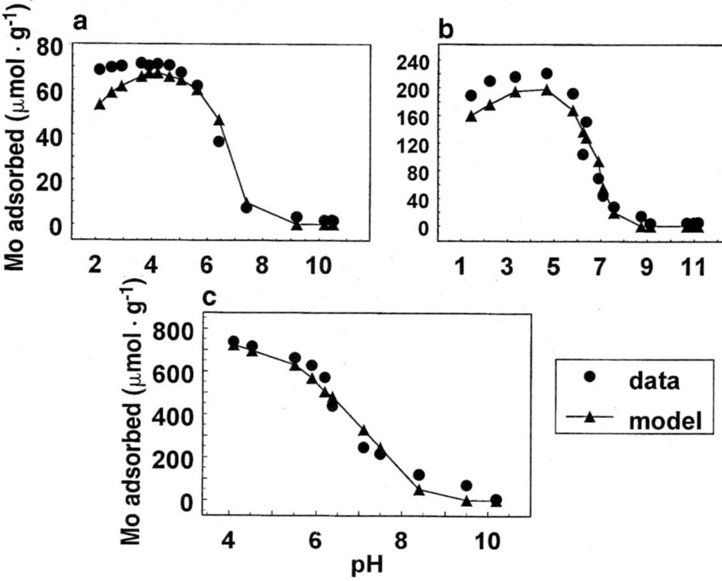

Figure 2 Molybdate adsorption on aluminum oxides as a function of solution pH: (a) δ-Al_2O_3, V_Y = 67.3; (b) gibbsite, V_Y = 92.5; (c) amorphous aluminum oxide, V_Y = 111. Circles represent experimental data. Constant capacitance model results are represented by triangles and solid lines.

tion on the kaolinites with no improvement observed by use of the "dummy" component (compare Fig. 3 to Figs. 6a and 6b of Goldberg *et al.* (1996)). The ability of the model to describe molybdate adsorption on 2:1 clay minerals is indicated in Figure 4. Use of the "dummy" component improved the fit for SAz-1 and STx-1 montmorillonite, and degraded it for SWy-1 montmorillonite and IMt-1 illite (compare Fig. 4 to Figs. 6c, 6d, 6e, and 6f of Goldberg *et al.* (1996)). However, this latter comparison is deceptive because unlike in the previous study (Goldberg *et al.*, 1996), log K_+(int) and log K_-(int) were not optimized for SWy-1 montmorillonite and IMt-1 illite. log K_+(int) was optimized along with the molybdate surface complexation constants in describing adsorption on the kaolinites and the STx-1 montmorillonite. For the SAz-1 montmorillonite, log K_-(int) was optimized as well.

The constant capacitance model was able to describe molybdate adsorption on the three soils studied with some deviations occurring especially at low pH values (Fig. 5). The model was unable to describe molybdate adsorption on the Hesperia soil below pH 4 where large deviations from the data were observed. In their application of the model, Goldberg *et al.* (1996) were unable to describe any of the soil data since convergence of the FITEQL computer program (Herbelin and Westall, 1994) either could not be obtained or provided a very bad fit.

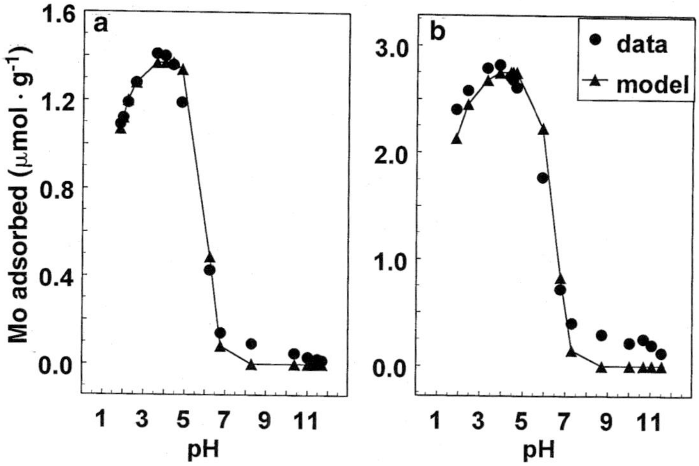

Figure 3 Molybdate adsorption on kaolinites as a function of solution pH: (a) KGa-1 kaolinite, $V_Y = 38.7$; (b) KGa-2 kaolinite, $V_Y = 155$. Circles represent experimental data. Constant capacitance model results are represented by triangles and solid lines.

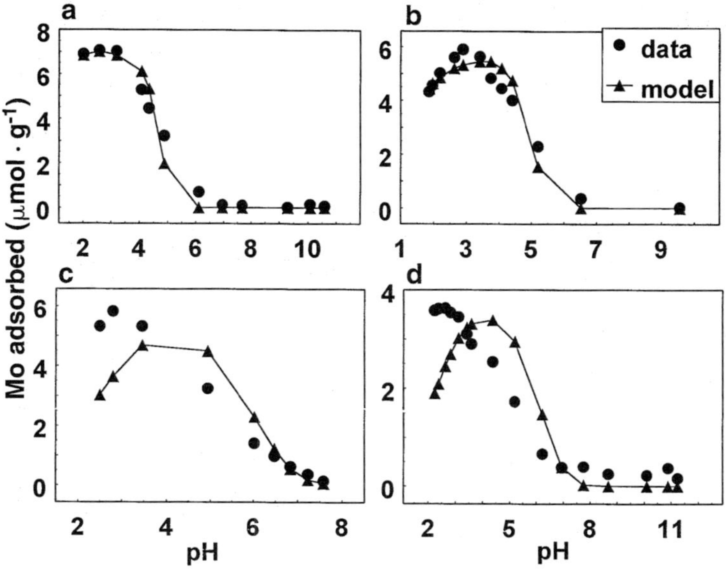

Figure 4 Molybdate adsorption on 2:1 clay minerals as a function of solution pH: (a) SAz-1 montmorillonite, $V_Y = 96.1$; (b) STx-1 montmorillonite, $V_Y = 64.4$; (c) SWy-1 montmorillonite, $V_Y = 271$; (d) IMt-1 illite, $V_Y = 327$. Circles represent experimental data. Constant capacitance model results are represented by triangles and solid lines.

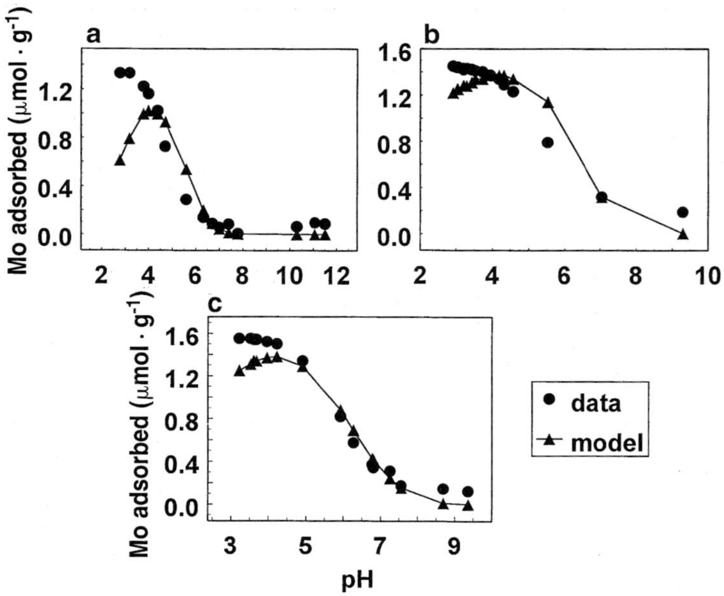

Figure 5 Molybdate adsorption on soils as a function of solution pH: (a) Hesperia, $V_Y = 324$; (b) Pachappa, $V_Y = 147$; (c) Porterville, $V_Y = 125$. Circles represent experimental data. Constant capacitance model results are represented by triangles and solid lines.

Table III provides values of the molybdate surface complexation constants obtained using the constant capacitance model in the FITEQL program for all materials. To fit the molybdate adsorption data we never optimized more than two adjustable parameters for any adsorbent. This number of adjustable parameters compares very favorably with the empirical Langmuir and Freundlich adsorption isotherm approach. The good fit of the model to molybdate adsorption on oxides, kaolinites, two of the clay minerals, and two of the soils suggests that inner-sphere surface complexation is the appropriate adsorption mechanism for these materials. The numerical values of the molybdate surface complexation constants for these materials were similar in magnitude (see Table III). Averages for the molybdate surface complexation constants obtained with the constant capacitance model for aluminum oxides, iron oxides, clay minerals, and soils were not statistically different at the 95% level of confidence, suggesting a common adsorption mechanism.

The effect of ionic strength on molybdate adsorption on a variety of adsorbents is indicated in Figures 6 to 12. Solution ionic strength was varied by two orders of magnitude, from 0.01 to 1.0 mol·L^{-1} NaCl. On all materials, adsorption of molybdate was consistently lowest for the highest ionic strength. The adsorbents exhibited diverse behavior in their ionic strength dependence. Goethite and montmorillonite showed relatively little ionic strength dependence, suggesting the formation of inner-sphere surface complexes. An inner-sphere adsorption mechanism for goethite had already been indicated by the electrophoretic mobility measurements of Goldberg *et al.* (1996). The soils showed little ionic strength dependence except at both low and high pH values. Gibbsite, δ-Al_2O_3, and kaolinite exhibited obvious ionic strength dependence. However, this behavior is in contradiction to the inner-sphere adsorption mechanism suggested for these materials by the electrophoretic mobility results of Goldberg *et al.* (1996). This discrepancy highlights the limitation in relying on macroscopic chemical information to deduce ion adsorption mechanisms.

The triple-layer model was used to describe the adsorption of molybdate on goethite, gibbsite, δ-Al_2O_3, KGa-1 kaolinite, SWy-1 montmorillonite, and two soils as a function of solution pH and ionic strength. All ionic strength and pH data were optimized simultaneously. Figures 6 to 12 indicate the ability of the triple-layer model to describe molybdate adsorption using both an inner-sphere and an outer-sphere adsorption mechanism. The surface site density and the capacitances were set at identical values for all adsorbents. Log K_+(int) and log K_-(int) values were obtained from the literature and set at identical values for all materials having AlOH as the reactive functional group (i.e., gibbsite, δ-Al_2O_3, kaolinite, montmorillonite, and soils). Different values were used for the goethite having FeOH as the reactive functional group.

The ability of the triple-layer model to describe molybdate adsorption on

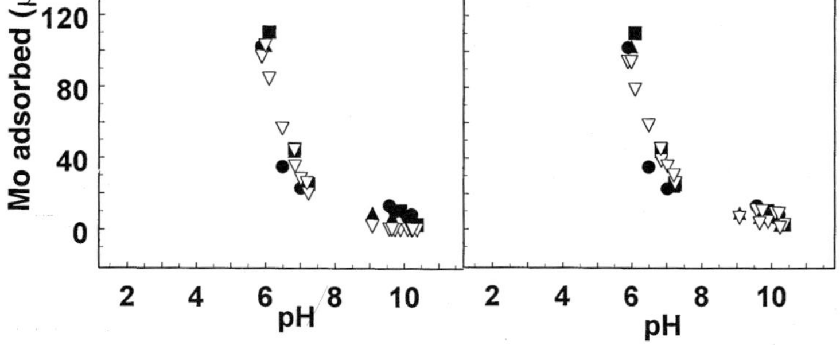

Figure 6 Molybdate adsorption on goethite as a function of solution pH and ionic strength. Filled symbols represent experimental data. Triple-layer model results are represented by open triangles. V_Y = 39.7 for inner-sphere adsorption and V_Y = 29.1 for outer-sphere adsorption.

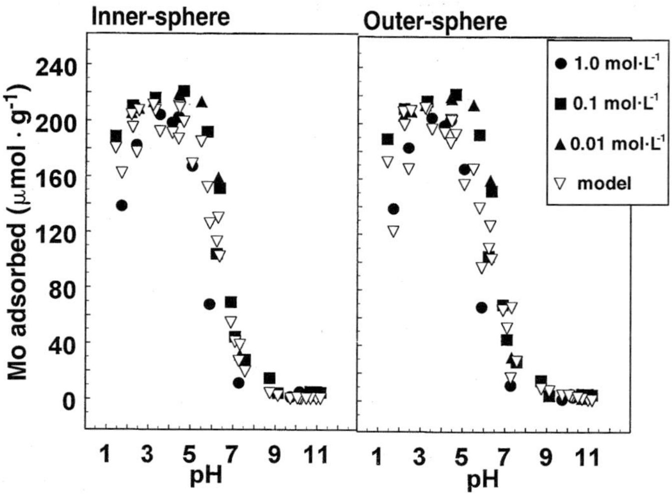

Figure 7 Molybdate adsorption on gibbsite as a function of solution pH and ionic strength. Filled symbols represent experimental data. Triple-layer model results are represented by open triangles. $V_Y = 83.9$ for inner-sphere adsorption and $V_Y = 81.8$ for outer-sphere adsorption.

goethite and gibbsite is presented in Figures 6 and 7, respectively. For both oxides the fit of the model was improved using the inner-sphere adsorption mechanism. For goethite this result is in agreement with the electrophoretic mobility and ionic strength results. For gibbsite the ionic strength effect data suggest an outer-sphere adsorption mechanism, in contrast to the modeling and electrophoretic mobility results. For δ-Al_2O_3 the quality of the triple-layer model fit is comparable for inner-sphere and outer-sphere adsorption mechanisms (Fig. 8).

Figures 9 and 10 present the ability of the triple-layer model to describe molybdate adsorption on kaolinite and montmorillonite, respectively. For kaolinite the quality of the model fit is improved using an inner-sphere adsorption mechanism (Fig. 9). This result is in agreement with the electrophoretic mobility results but contradicts the ionic strength dependence data. For montmorillonite an acceptable model fit could be obtained only using an outer-sphere adsorption mechanism (Fig. 10). This finding is in contradiction with the inner-sphere adsorption mechanism implied by the small ionic strength dependence of the adsorption data. For the soils (Figs. 11 and 12), the quality of the model fit is slightly better for the inner-sphere adsorption mechanism, in agreement with the small ionic strength dependence in the intermediate pH range.

Table IV provides values of the molybdate inner- and outer-sphere surface complexation constants obtained using the triple-layer model in the FITEQL program for all materials. For goethite with both mechanisms, and for gibbsite using an in-

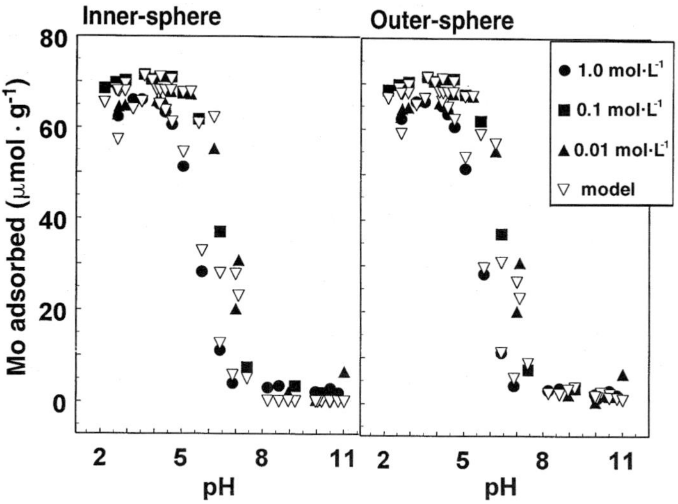

Figure 8 Molybdate adsorption on δ-Al_2O_3 as a function of solution pH and ionic strength. Filled symbols represent experimental data. Triple-layer model results are represented by open triangles. V_Y = 44.3 for inner-sphere adsorption and V_Y = 20.7 for outer-sphere adsorption.

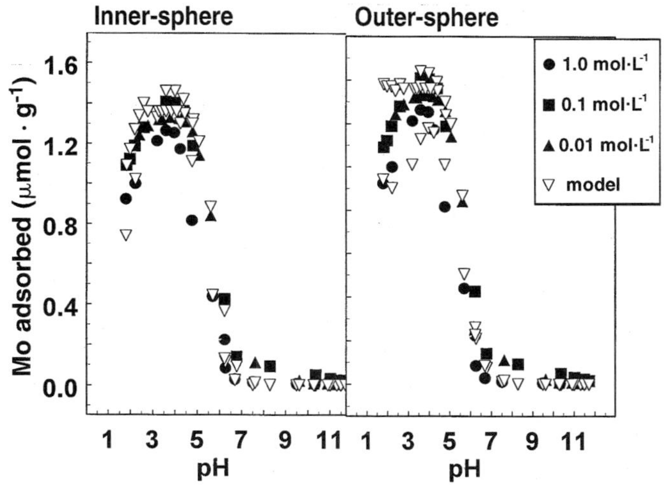

Figure 9 Molybdate adsorption on KGa-1 kaolinite as a function of solution pH and ionic strength. Filled symbols represent experimental data. Triple-layer model results are represented by open triangles. V_Y = 35.6 for inner-sphere adsorption and V_Y = 47.0 for outer-sphere adsorption.

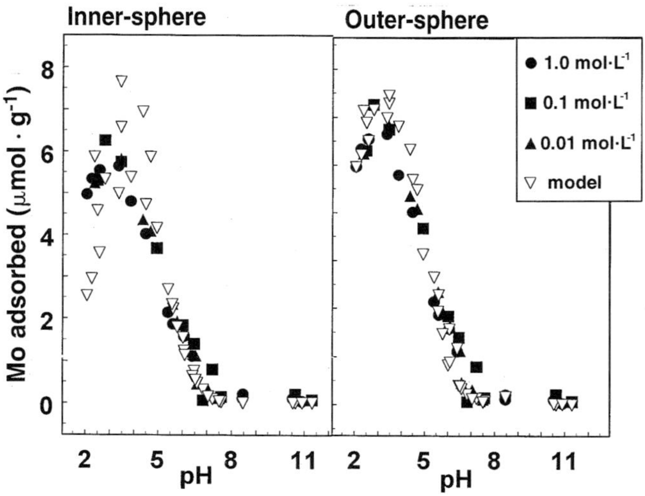

Figure 10 Molybdate adsorption on SWy-1 montmorillonite as a function of solution pH and ionic strength. Filled symbols represent experimental data. Triple-layer model results are represented by open triangles. $V_Y = 146$ for inner-sphere adsorption and $V_Y = 76.0$ for outer-sphere adsorption.

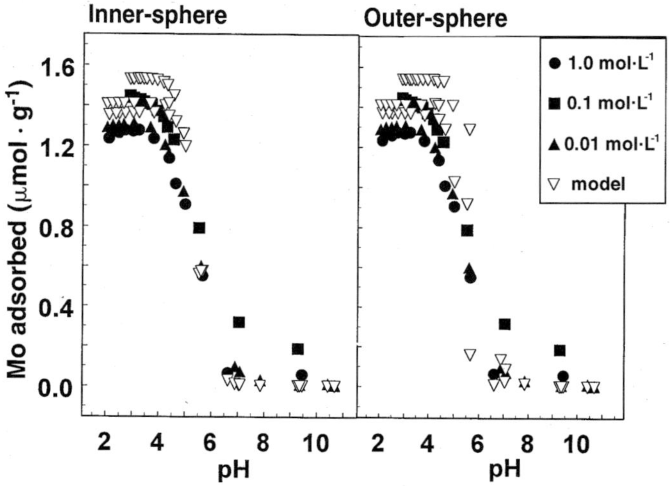

Figure 11 Molybdate adsorption on Pachappa soil as a function of solution pH and ionic strength. Filled symbols represent experimental data. Triple-layer model results are represented by open triangles. $V_Y = 93.0$ for inner-sphere adsorption and $V_Y = 131$ for outer-sphere adsorption.

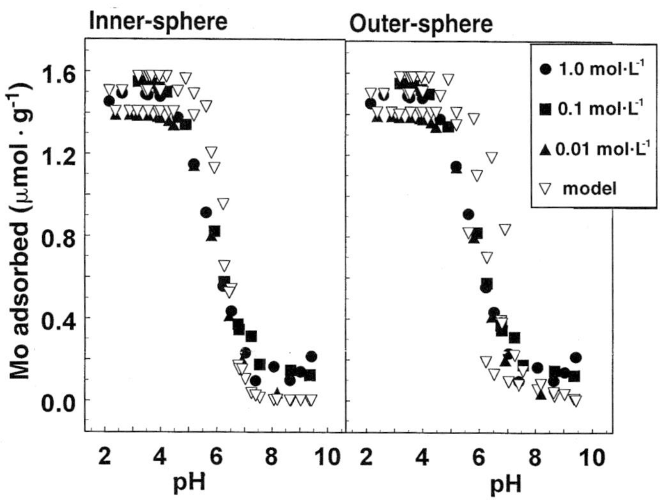

Figure 12 Molybdate adsorption on Porterville soil as a function of solution pH and ionic strength. Filled symbols represent experimental data. Triple-layer model results are represented by open triangles. $V_Y = 176$ for inner-sphere adsorption and $V_Y = 207$ for outer-sphere adsorption.

ner-sphere adsorption mechanism, only two adjustable parameters were optimized, the molybdate surface complexation constants. This number of adjustable parameters, again, compares very favorably with the empirical Langmuir and Freundlich adsorption isotherm approach. For all other materials we optimized the surface complexation constants for Na^+ and Cl^- along with the molybdate surface complexation constants, increasing the number of adjustable parameters to four. The surface site density suggested by Davis and Kent (1990) for natural materials was found appropriate for modeling molybdate adsorption on a variety of oxides, clay minerals, and soils. The triple-layer model was able to simultaneously fit molybdate adsorption at several ionic strengths with one set of surface complexation constants.

Triple-layer modeling suggests an inner-sphere adsorption mechanism for goethite, gibbsite, kaolinite, and the soils, and an outer-sphere adsorption mechanism for montmorillonite. Lack of agreement between various indirect methodologies of inferring adsorption mechanisms—zero point of charge shifts, ionic strength effects, and surface complexation modeling—underscores the necessity for direct spectroscopic elucidation of adsorption mechanisms. It is best to obtain spectroscopic evidence for the presence of particular surface complexes prior to postulating them in surface complexation models.

The advantage of the constant capacitance model over the triple-layer model is its simplicity and small number of adjustable parameters. The advantage of the triple-layer model is its ability to describe ion adsorption as a function of solution

Table IV

Solids, Concentrations, and Intrinsic Surface Complexation Constants Obtained with the Triple-Layer Model

	Inner-sphere mechanism				Outer-sphere mechanism			
Solid	log K^1_{Mo}(int)	log K^2_{Mo}(int)	log K_{Na^+}(int)	log K_{Cl^-}(int)	log K^1_{Mo}(int)	log K^2_{Mo}(int)	log K_{Na^+}(int)	log K_{Cl^-}(int)
Goethite	7.28	3.82	−9.3	5.4	9.66	4.92	−9.3	5.4
δ-Al_2O_3	8.93	3.92	−9.3	5.4	11.76	5.38	−11.23	9.64
Gibbsite	9.26	0.52	−8.52	13.00	8.97	2.09	−6.87	8.11
KGa-1 kaolinite	5.01	−0.87	−7.48	10.08	7.23	−0.17	−37.20	4.58
SWy-1 montmorillonite	—[a]	−0.92	−6.47	9.66	—	−0.47	−6.47	1.77
Pachappa soil	—	−2.09	−9.3	5.4	12.08	−0.61	−9.3	5.4
Porterville soil	—	−1.49	−9.3	5.4	12.24	−0.13	−9.3	5.4

[a]No convergence.

ionic strength and to consider both inner-sphere and outer-sphere surface complexes. The user must weigh improved chemical reality against increased complexity.

In the IR study high concentrations of Mo had to be used due to limitations in instrument sensitivity. Unfortunately, at these high concentrations polymeric Mo species are present in solution and their adsorption on the oxide surface cannot be ruled out. The kinetics and equilibria for processes occurring as basic solutions of MoO_4^{2-} are acidified are very complex. The polymolybdate anions consist primarily of octahedral MoO_6 groups, so that the conversion of MoO_4^{2-} into polyanions requires an increase in coordination number. Polynuclear Mo^{VI} species contain seven and eight Mo atoms in solution, such as $Mo_7O_{24}^{6-}$ and $Mo_8O_{26}^{4-}$ (Cotton and Wilkinson, 1980). Three IR bands at 933, 885, and 835 cm^{-1} were observed for the aqueous Mo anionic species at 0.05 and 0.1 $mol{\cdot}L^{-1}$ (Fig. 13). Assignments of IR bands for polymolybdate anions are not available in the literature.

Molybdate sorbed at the interface of $Fe(OH)_3$(a) and water exhibited two IR bands at 925 and 880 cm^{-1}, with greater band intensities for the higher initial 0.1 $mol{\cdot}L^{-1}$ Na_2MoO_4 concentration than for 0.05 $mol{\cdot}L^{-1}$ Na_2MoO_4 (Fig. 14). This is consistent with the higher sorption of Mo (975 $mmol{\cdot}kg^{-1}$) at 0.1 $mol{\cdot}L^{-1}$

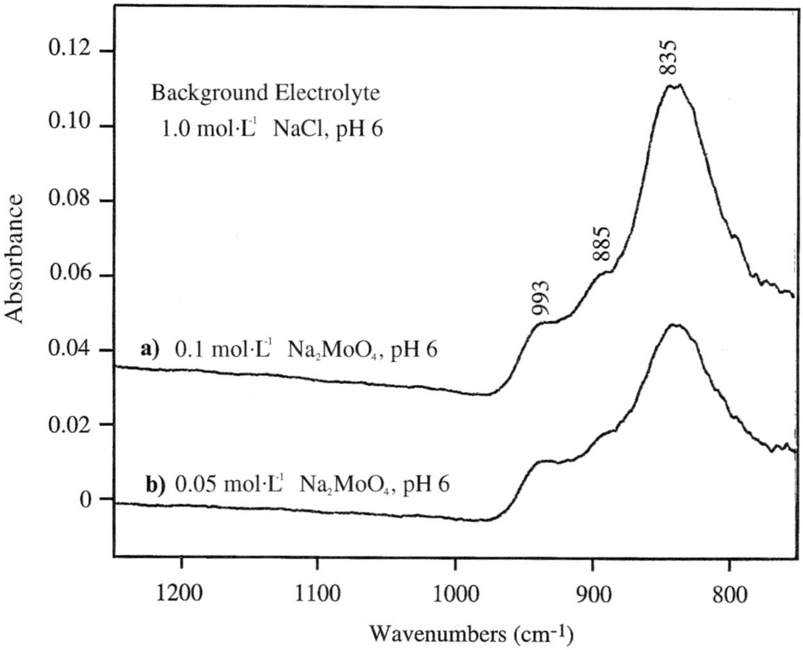

Figure 13 ATR-FTIR difference spectra of Na_2MoO_4 (pH 6) solutions: (a) 0.1 $mol{\cdot}L^{-1}$; (b) 0.05 $mol{\cdot}L^{-1}$. The reference spectrum was 1.0 $mol{\cdot}L^{-1}$ NaCl (pH 6).

Na_2MoO_4 compared to that at 0.05 mol·L^{-1} Na_2MoO_4 (499 mmol·kg^{-1}), corresponding to 99.1 and 99.9% of sorption of added Mo, respectively. The molar ratio of released OH to sorbed Mo was 1.36 at 0.1 mol·L^{-1} Na_2MoO_4 and 1.81 at 0.05 mol·L^{-1} Na_2MoO_4, suggesting a mixture of monodentate and bidentate species of complexed Mo at the mineral surface. It is evident that ligand exchange (inner-sphere complexation) is a mechanism for Mo sorption on $Fe(OH)_3$(a). X-ray diffractograms of $Fe(OH)_3$(a) after Mo sorption showed no crystalline solid phase. This microscopic result is in agreement with the macroscopic results obtained for the iron oxide, goethite, using zero point of charge shifts, ionic strength effects, and triple-layer modeling.

Two washings with deionized water desorbed 14 and 28% of the initially sorbed Mo from $Fe(OH)_3$(a) for the 0.05 and 0.1 mol·L^{-1} Na_2MoO_4 treatments, respectively. DRIFT difference spectra show that at least three IR bands can be identified in the range 1000 to 700 cm^{-1} for sorbed Mo (Figs. 15a and 15b). In com-

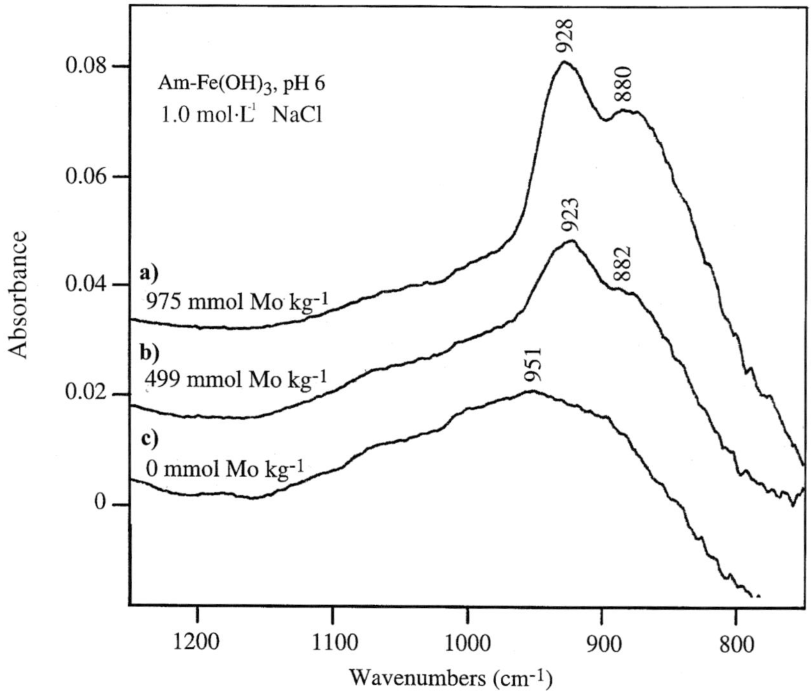

Figure 14 ATR-FTIR difference spectra of $Fe(OH)_3$(a) (the spectrum of the supernatant was subtracted from the spectrum of the solid suspension) as affected by Mo sorption. Solid concentration was 400 g·L^{-1}. The amount of sorbed Mo was (a) 975, (b) 499, and (c) 0 mmol·kg^{-1} of Mo for initial Mo concentrations of 0.1, 0.05, and 0 mol·L^{-1} Na_2MoO_4, respectively, in 1.0 mol·L^{-1} NaCl (pH 6).

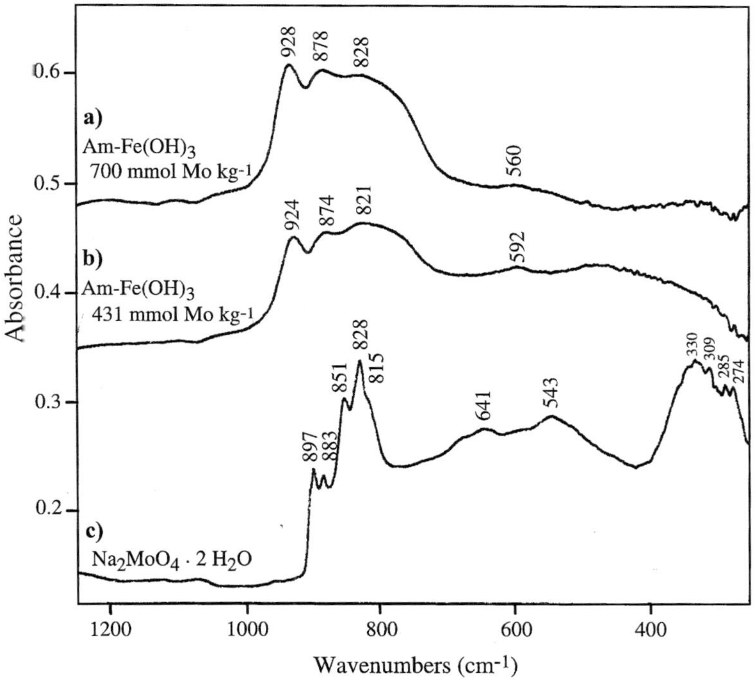

Figure 15 DRIFT difference spectra of Mo sorbed on $Fe(OH)_3$(a) after two washings with deionized water and air-drying: (a) 700 mmol·kg^{-1} of Mo for the initial 0.1 mol·L^{-1} Na_2MoO_4 treatment, and (b) 431 mmol·kg^{-1} of Mo for the initial 0.05 mol·L^{-1} Na_2MoO_4 treatment. (c) DRIFT spectrum of reagent-grade $Na_2MoO_4 \cdot 2H_2O$. All samples were diluted with KBr (5 mg in 95 mg KBr).

parison, more bands were observed for the reagent-grade $Na_2MoO_4 \cdot 2H_2O$ (Fig. 15c). A model monodentate complex $[Co(NH_3)_5MoO_4]Cl$ shows three IR bands at 910, 877, and 833 cm^{-1}, whereas a bidentate chelate $[Co(NH_3)_4MoO_4]NO_3$ exhibits four bands at 920, 868, 845, and 795 cm^{-1} (Ross, 1972). An estimation of the relative distribution of mono- and bidentate complexes is difficult due to overlapping bands of the complexes. Nevertheless, it is highly possible that both types of complexes exist on the surface of $Fe(OH)_3$(a), as supported by both IR spectra and the molar ratio of OH released to Mo sorbed.

IV. SUMMARY

Molybdate adsorption on all materials exhibited a maximum at low pH (3 to 5). With increasing solution pH, adsorption decreased rapidly, with little adsorption occurring above pH values of 7 to 8. Molybdate adsorption was lowest for the highest solution ionic strength. Ionic strength dependence of molybdate adsorption was slight on goethite, montmorillonite, and soils, suggesting an inner-sphere adsorption mechanism. Ionic strength dependence of molybdate adsorption was obvious on gibbsite, Aluminium Oxid C, and kaolinite, suggesting an outer-sphere adsorption mechanism.

The constant capacitance model was able to describe molybdate adsorption on the oxides, clay minerals, and soils as a function of solution pH. Averages for the molybdate surface complexation constants obtained with the constant capacitance model for oxides, clay minerals, and soils were not statistically different at the 95% level of confidence.

The triple-layer model was able to describe molybdate adsorption on goethite, gibbsite, Aluminium Oxid C, kaolinite, montmorillonite, and two soils as a function of solution pH and ionic strength using a universal site density value of 2.31 sites·nm^2 (recommended for natural materials). Good fits of the model to the data were obtained using both inner-sphere and outer-sphere adsorption mechanisms for all materials except montmorillonite, where an acceptable fit was obtained only with an outer-sphere mechanism.

Results from the FTIR spectroscopy indicate that ligand exchange is a mechanism for Mo adsorption on amorphous iron hydroxide. Hydroxyl release suggests a mixture of monodentate and bidentate Mo surface complexes.

REFERENCES

Barrow, N. J. 1970. Comparison of the adsorption of molybdate, sulfate and phosphate by soils. *Soil Sci.* 109:282–288.

Bibak, A., and Borggaard, O. K. 1994. Molybdenum adsorption by aluminum and iron oxides and humic acid. *Soil Sci.* 158:323–327.

Carpéni, G. 1947. Sur la constitution des solutions aqueuses d'acide molybdique et de molybdates alcalins. IV.—Conclusions générales. *Bull. Soc. Chim.* 14:501–503.

Cihacek, L. J., and Bremner, J. M. 1979. A simplified ethylene glycol monoethyl ether procedure for assessing soil surface area. *Soil Sci. Soc. Am. J.* 43:821–822.

Coffin, D. E. 1963. A method for the determination of free iron oxide in soils and clays. *Can. J. Soil Sci.* 43:7–17.

Cotton, F. A., and Wilkinson, G. 1980. "Advanced Inorganic Chemistry," 4th ed., pp. 852–856. Wiley, New York.

Davis, J. A., James, R. O., and Leckie, J. O. 1978. Surface ionization and complexation at the oxide/water interface. I. Computation of electrical double layer properties in simple electrolytes. *J. Colloid Interface Sci.* 63:480–499.

Davis, J. A., and Kent, D. B. 1990. Surface complexation modeling in aqueous geochemistry. *Rev. Mineral.* 23:117–260.

Ferreiro, E. A., Helmy, A. K., and de Bussetti, S. G. 1985. Molybdate sorption by oxides of aluminium and iron. *Z. Pflanzenernaehr. Bodenk.* 148:559–566.

Goldberg, S., Forster, H. S., and Godfrey, C. L. 1996. Molybdenum adsorption on oxides, clay minerals, and soils. *Soil Sci. Soc. Am. J.* 60:425–432.

Goldberg, S., and Sposito, G. 1984a. A chemical model of phosphate adsorption by soils. I. Reference oxide minerals. *Soil Sci. Soc. Am. J.* 48:772–778.

Goldberg, S., and Sposito, G. 1984b. A chemical model of phosphate adsorption by soils. II. Noncalcareous soils. *Soil Sci. Soc. Am. J.* 48:779–783.

Gonzalez, B. R., Appelt, H. Schalscha, E. B., and Bingham, F. T. 1974. Molybdate adsorption characteristics of volcanic-ash-derived soils in Chile. *Soil Sci. Soc. Am. Proc.* 38:903–906.

Hayes, K. F., and Leckie, J. O. 1987. Modeling ionic strength effects on cation adsorption at hydrous oxide/solution interfaces. *J. Colloid Interface Sci.* 115:564–572.

Hayes, K. F., Papelis, C., and Leckie, J. O. 1988. Modeling ionic strength effects on anion adsorption at hydrous oxide/solution interfaces. *J. Colloid Interface Sci.* 125:717–726.

Herbelin, A. L., and Westall, J. C. 1994. FITEQL: A computer program for determination of chemical equilibrium constants from experimental data. Rep. 94-01, Version 3.1, Dept. of Chemistry, Oregon State University, Corvallis, OR.

Jarrell, W. M., and Dawson, M. D. 1978. Sorption and availability of molybdenum in soils of western Oregon. *Soil Sci. Soc. Am. J.* 42:412–415.

Jones, L. H. P. 1957. The solubility of molybdenum in simplified systems and aqueous soil suspensions. *J. Soil Sci.* 8:313–327.

Karimian, N., and Cox, F. R. 1978. Adsorption and extractability of molybdenum in relation to some chemical properties of soil. *Soil Sci. Am. J.* 42:757–761.

Klages, M. G., and Hopper, R. W. 1982. Clay minerals in northern plains coal overburden as measured by x-ray diffraction. *Soil Sci. Soc. Am. J.* 45:415–419.

Kyle, J. H., Posner, A. M., and Quirk, J. P. 1975. Kinetics of isotopic exchange of phosphate adsorbed on gibbsite. *J. Soil Sci.* 26:32–43.

Kyriacou, D. 1967. The pH-dependence of adsorption of metallic oxyanions by ferric oxide powder. *Surf. Sci.* 8:370–372.

Lindsay, W. L. 1979. "Chemical Equilibria in Soils." Wiley, New York.

McKenzie, R. M. 1983. The adsorption of molybdenum on oxide surfaces. *Aust. J. Soil Res.* 21:505–513.

McLaughlin, J. R., Ryden, J. C., and Syers, J. K. 1981. Sorption of inorganic phosphate by iron and aluminium-containing components. *J. Soil Sci.* 32:365–377.

Mikkonen, A., and Tummavuori, J. 1993a. Retention of vanadium(V), molybdenum(VI) and tungsten(VI) by kaolin. *Acta Agric. Scand. Sect. B Soil Plant Sci.* 43:11–15.

Mikkonen, A., and Tummavuori, J. 1993b. Retention of molybdenum(VI) by three Finnish mineral soils. *Acta Agric. Scand. Sect. B Soil Plant Sci.* 43:206–212.

Motta, M. M., and Miranda, C. F. 1989. Molybdate adsorption on kaolinite, montmorillonite, and illite: Constant capacitance modeling. *Soil Sci. Soc. Am. J.* 53:380–385.

Murphy, L. S., and Walsh, L. M. 1972. Correction of micronutrient deficiencies with fertilizers. *In* "Micronutrients in Agriculture" (J. H. Mortvedt, P. M. Giordano, and W. L. Lindsay, Eds.). Soil Science Society of America, Madison, WI.

Nelson, D. W., and Sommers, L. E. 1982. Total carbon, organic carbon, and organic matter. *In* "Methods of Soil Analysis" (A. L. Page *et al.,* Ed.), 2nd ed., Part 2, Chap. 9, pp. 539–579. American Society of Agronomy, Madison, WI.

Phelan, P. J., and Mattigod, S. V. 1984. Adsorption of molybdate anion (MoO_4^{2-}) by sodium-saturated kaolinite. *Clays Clay Miner.* 32:45–48.

Reisenauer, H. M., Tabikh, A. A., and Stout, P. R. 1962. Molybdenum reactions with soils and the hydrous oxides of iron, aluminum and titanium. *Soil Sci. Soc. Am. J.* 26:23–27.

Reyes, E. D., and Jurinak, J. J. 1967. A mechanism of molybdate adsorption on αFe_2O_3. *Soil Sci. Soc. Am. J.* 31:637–641.

Ross, S. D. 1972. "Inorganic Infrared and Raman Spectra," p. 217. McGraw–Hill, London.

Roy, W. R., Hassett, J. J., and Griffin, R. A. 1986. Competitive interactions of phosphate and molybdate on arsenate adsorption. *Soil Sci.* 142:203–210.

Roy, W. R., Hassett, J. J., and Griffin, R. A. 1989. Quasi-thermodynamic basis of competitive-adsorption coefficients for anionic mixtures in soils. *J. Soil Sci.* 40:9–15.
Spanos, N., and Lycourghiotis, A. 1995. Codeposition of $Mo^{(VI)}$ species and Ni^{2+} ions on the γ-alumina surface: Mechanistic model. *J. Colloid Interface Sci.* 171:306–318.
Spanos, N., Vordonis, L., Kordulis, Ch., and Lycourghiotis, A. 1990a. Molybdenum-oxo species deposited on alumina by adsorption. I. Mechanism of the adsorption. *J. Catal.* 124:301–314.
Spanos, N., Vordonis, L., Kordulis, Ch., Koutsoukos, P. G., and Lycourghiotis, A. 1990b. Molybdenum-oxo species deposited on alumina by adsorption. II. Regulation of the surface Mo^{VI} concentration by control of the protonated surface hydroxyls. *J. Catal.* 124:315–323.
Sposito, G. 1984. "The Surface Chemistry of Soils." Oxford University Press, New York.
Sprycha, R. 1989a. Electrical double layer at alumina/electrolyte interface. I. Surface charge and zeta potential. *J. Colloid Interface Sci.* 127:1–11.
Sprycha, R. 1989b. Electrical double layer at alumina/electrolyte interface. II. Adsorption of supporting electrolytes. *J. Colloid Interface Sci.* 127:12–25.
Stumm, W., Kummert, R., and Sigg, L. 1980. A ligand exchange model for the adsorption of inorganic and organic ligands at hydrous oxide interfaces. *Croat. Chem. Acta* 53:291–312.
Su, C., and Suarez, D. L. 1995. Coordination of adsorbed boron: A FTIR spectroscopic study. *Environ. Sci. Technol.* 29:302–311.
Theng, B. K. G. 1971. Adsorption of molybdate by some crystalline and amorphous soil clays. *New Zealand J. Sci.* 14:1040–1056.
Vordonis, L., Koutsoukos, P. G., and Lycourghiotis, A. 1990. Adsorption of molybdates on doped γ-aluminas in alkaline solutions. *Colloids Surf.* 50:353–361.
Westall, J., and Hohl, H. 1980. A comparison of electrostatic models for the oxide/solution interface. *Adv. Colloid Interface Sci.* 12:265–294.
Xie, R. J., and MacKenzie, A. F. 1991. Molybdate sorption-desorption in soils treated with phosphate. *Geoderma* 48:321–333.
Xie, R. J., MacKenzie, A. F., and Lou, Z. J. 1993. Causal modeling pH and phosphate effects on molybdate sorption in three temperate soils. *Soil Sci.* 155:385–397.
Zhang, P. C., and Sparks, D. L. 1989. Kinetics and mechanisms of molybdate adsorption/desorption at the goethite/water interface using pressure-jump relaxation. *Soil Sci. Soc. Am. J.* 53:1028–1034.
Zhang, P., and Sparks, D. L. 1990. Kinetics of selenate and selenite adsorption/desorption at the goethite/water interface. *Environ. Sci. Technol.* 24:1848–1856.

Chapter 20

Nonequilibrium and Nonlinear Sorption during Transport of Cadmium, Nickel, and Strontium through Subsurface Soils

Wei-Zi Wang,[1] Mark L. Brusseau,[1,2] and Janick F. Artiola[1]
[1]Department of Soil, Water and Environmental Sciences, University of Arizona, Tucson, Arizona; [2]Department of Hydrology and Water Resources, University of Arizona, Tucson, Arizona

The sorption and transport of Cd, Ni, and Sr in two subsurface soils (Hayhook series and Borden) were investigated using batch and miscible displacement experiments. The sorption isotherms of Cd, Ni, and Sr were strongly nonlinear and well described by the Freundlich equation ($S = K_F C^{n_F}$). Electrostatic adsorption was found to be the major contributor to the sorption of these elements. The breakthrough curves of Cd, Ni, and Sr exhibited nonideal behavior (asymmetry and tailing), and were analyzed using a model that included various combinations of instantaneous, rate-limited, linear, and nonlinear sorption. The model incorporating both nonlinear and rate-limited sorption provided the best simulation of the data. Flow-interruption experiments were conducted to help identify the controlling process. The results indicated that nonlinear sorption was the major cause of nonideal transport, and that rate-limited sorption was a relatively minor contributor.

Adsorption of Metals by Geomedia

I. INTRODUCTION

The fate of heavy metals emanating from hazardous waste sites, landfills, and other such sites has become a major environmental issue. A survey of 395 remedial action sites revealed that heavy metals were the single most prevalent class of contaminants (EPA, 1984). Heavy metals sorbed to soil can be gradually released into groundwater through the action of abiotic or biotic processes, thereby posing a threat to drinking water supplies. For this reason, the transport of heavy metals is of great interest.

Rate-limited and nonlinear reaction processes, as well as the presence of material heterogeneities, can cause transport of contaminants in soils and aquifers to often be nonideal (e.g., Brusseau, 1994). For example, asymmetrical and nonsigmoidal breakthrough curves have been reported for inorganic compounds (Pickens *et al.,* 1981; Parker and Jardine, 1986; Jardine *et al.,* 1985, 1993; Schulin *et al.,* 1986; Selim *et al.,* 1989; Selim, 1992; Hinz and Selim, 1994; Kookana *et al.,* 1994), as well as for many organic compounds (e.g., Brusseau and Rao, 1989). It is possible, in many situations, that more than one factor may contribute to nonideal transport of heavy metals (Grove and Stollenwerk, 1985). Such factors may include nonlinear sorption, rate-limited sorption–desorption, and structured/locally heterogeneous porous media. It appears that little information has been reported to evaluate the combined effects of rate-limited and nonlinear sorption on transport of heavy metals. The objectives of this chapter are to evaluate these effects and to identify which factor controls nonideal behavior in soil systems designed to exclude the effect of structured porous media.

II. MATERIALS AND METHODS

A. Materials

Two soils were used in this study. One is a sandy subsoil collected from the Canadian Air Force base in Borden, Ontario. The second (Hayhook series, AZ), which was collected from a 25- to 50-cm depth in the C horizon, is classified as a coarse-loamy, mixed, nonacid thermic typic torriorthent. The soils were dried at room temperature in a low-humidity environment and then passed through a 2-mm sieve. The <2-mm fraction was used for the experiments. Selected properties of the soils are listed in Table I.

B. Batch Experiments

Sorption isotherms for Cd, Ni, and Sr were measured at 25°C using batch methods. All experiments were conducted with 0.01 mol·L^{-1} KNO_3 as an electrolyte

Table I
Properties of Soil

Soil	O.C. (%)	Sand (%)	Silt (%)	Clay (%)	pH	CEC ($meq \cdot 100\ g^{-1}$)	Fe_2O_3 (%)
Hayhook[a]	0.11	86	4	10	7.47	6.30	0.52
Borden	0.03	98	1	1	8.32	1.15	ND[b]

[a]Major clay mineral composition of Hayhook soil was mica, smectite, and a small portion of kaolinite.
[b]Not determined.

solution. Duplicate 2.5-g samples of soil were placed in preweighed polypropylene tubes, mixed with 20 ml of 1.0 $mol \cdot L^{-1}$ KNO_3, and shaken overnight. The suspensions were centrifuged at 10,000 rpm (12,100*g* relative centrifugal force), and the supernatants discarded. The remaining soil was washed twice with 0.01 $mol \cdot L^{-1}$ KNO_3. Finally, the tubes containing soil and residual solution were weighed to calculate residual solution. Aliquots (20 ml) containing different initial concentrations of metals (0.15, 0.25, 0.40, 0.50, 0.75, 1.0, 1.5, and 2.0 $mmol \cdot L^{-1}$ of Cd, Ni, or Sr), which were prepared with 0.01 $mol \cdot L^{-1}$ KNO_3, were then added to the tubes. Preliminary experiments showed that sorption of the metals was completed within 2 hr. Hence, the solutions were shaken for 2 hr and the resulting suspensions were centrifuged at 10,000 rpm for 10 min. The supernatants were decanted and collected for determination of metals concentrations and pH. The amounts of metal adsorbed to the soil were estimated from the differences between the added amounts and the remaining amounts of metals in the final supernatant solutions. The standard solutions of Cd, Ni, and Sr used in these experiments were prepared by using AR reagent-grade $Cd(NO_3)_2$, $Ni(NO_3)_2$, and $Sr(NO_3)_2$, respectively. Concentrations of Cd, Ni, and Sr were determined by use of flame atomic absorption spectroscopy (Perkin-Elmer 503).

C. Miscible Displacement Experiments

Miscible displacement experiments were conducted to investigate transport behavior of Cd, Ni, and Sr in the soils. Plexiglass columns (2.77 cm i.d., and 10.0 cm length) were homogeneously packed in small increments to a known bulk density. Prior to the start of displacement experiments, the columns were wetted from the bottom with 0.01 $mol \cdot L^{-1}$ KNO_3 at a slow rate using a precision piston pump. This was done to remove the entrapped air and to ensure soil–water saturation. During the saturation period, the fluid flow rate was increased incrementally to 1.0 $ml \cdot min^{-1}$. The columns were conditioned by pumping at least 200 pore volumes of electrolyte solution through them.

To examine the hydrodynamic properties of the columns, KNO_3 solutions containing nonsorbing and nonreactive tracers (tritiated water or pentafluorobenzoate) were introduced into the columns. For tritiated water, the effluent was collected by a fraction collector and the activity of tritiated water in the effluent fraction was measured by radioassay using liquid scintillation counting techniques. For PFBA, the effluent concentration was monitored at a wavelength of 250 nm with a flow-through UV detector connected to a strip chart recorder (linear; Model 1200).

Solutions (1.0 $mmol \cdot L^{-1}$) of Cd, Ni, and Sr were prepared using 0.01 $mol \cdot L^{-1}$ KNO_3 solution as the background electrolyte. These solutions were introduced individually into the columns. The column experiments were run until the effluent concentrations reached 100% of the influent metal levels, and then the input solutions were switched to background electrolyte solution (0.01 $mol \cdot L^{-1}$ KNO_3). The column effluents were collected with a fraction collector (ISCO Retriever 500) for determination of metals and pH. The flow-interruption method was used to investigate possible nonequilibrium transport (Brusseau, *et al.,* 1989b). The influence of potential competitive adsorption effects on metal transport was also studied by using solutions containing 1.0 $mol \cdot L^{-1}$ Cd plus 1.0 $mmol \cdot L^{-1}$ Ni.

D. Modeling Transport of Metals

One-dimensional, equilibrium and nonequilibrium, and linear and nonlinear sorption models were used to simulate the transport behavior of Cd, Ni, and Sr.

1. Equilibrium Model

The advective–dispersive transport equation for reactive solute based on assumptions of homogeneous porous media, steady-state flow, and local equilibrium can be described as

$$R_f \frac{\partial C}{\partial t} = D_e \frac{\partial^2 C_{aq}}{\partial z^2} - v \frac{\partial C_{aq}}{\partial z}, \qquad [1]$$

where C_{aq} is aqueous concentration ($M \cdot L^{-3}$), v is average pore-water velocity ($L \cdot T^{-1}$), t is time (T), D_e is the diffusion/dispersion coefficient ($L^2 \cdot T^{-1}$), z is the distance from injection point (L), and R_f is the retardation factor.

For linear sorption, the retardation factor, R_f, is defined as

$$R_f = 1 + K_d \frac{\rho}{\Theta}, \qquad [2]$$

where K_d is the distribution coefficient, Θ is the volumetric water content ($L^3 \cdot L^{-3}$), and ρ is the dry bulk density of the soil system ($M \cdot L^{-3}$). For nonlinear sorption, R_f is concentration-dependent and is given by the relationship

$$R_f = 1 + n_F K_F C^{n_F - 1} \frac{\rho}{\theta}, \quad [3]$$

where K_F and n_F are the Freundlich coefficients.

2. Nonequilibrium Model (Two Domains)

A bicontinuum model based on first-order mass transfer was also used to analyze the transport of metals through the soil columns. With the first-order model, sorption is conceptualized to occur in two domains,

$$C_{aq} \rightleftarrows S_1 \rightleftarrows S_2 \quad [4]$$

$$S_1 = F_{S1} K_F C_{aq}^{n_F} \quad [5]$$

$$\frac{dS_2}{dt} = k_1 S_1 - k_2 S_2, \quad [6]$$

where S_1 is the sorbed-phase concentration ($M \cdot M^{-1}$) in the "instantaneous" domain, S_2 is the sorbed-phase concentration ($M \cdot M^{-1}$) in the rate-limited domain, F_{S1} is the fraction of sorbent for which sorption is instantaneous, t is time, and k_1 and k_2 are forward and reverse first-order rate coefficients (T^{-1}), respectively.

The following equations describe the transport of solute undergoing nonlinear, rate-limited sorption during one-dimensional, steady water flow in a homogeneous porous medium (Brusseau *et al.*, 1989b):

$$\frac{\partial C^*}{\partial T} + (\beta R_f - 1) n_F C^{*n_F - 1} \frac{\partial C^*}{\partial T} + (1 - \beta) R_f \frac{\partial S^*}{\partial T} = \frac{1}{P} \frac{\partial^2 C^*}{\partial X^2} - \frac{\partial C^*}{\partial X} \quad [7]$$

$$(1 - \beta) R_f \frac{\partial S^*}{\partial T} = \omega (C^* - S^*). \quad [8]$$

The variables for the model are obtained as

$$\omega = k_2 (1 - \beta) R_f L/v$$

$$T = vt/L$$

$$X = z/L$$

$$P = vL/D_e$$

$$C^* = C_{aq}/C_0$$

$$\beta = [1 + F_{S1}(\rho/\theta) K_F C_0^{n_F - 1}]/R_f$$

$$R_f = 1 + \frac{\rho}{\theta} K_F C_0^{n_F - 1}$$

$$S^* = S_2/(1 - F_{S1})K_f C_0^{n_F},$$

where C_0 is the input concentration ($M \cdot L^{-3}$), P is the Peclet number, β is the fraction of instantaneous retardation, and ω is the Damkohler number, which is the ratio of hydrodynamic residence time to the characteristic time for sorption.

For both nonlinear, equilibrium and nonlinear, nonequilibrium models, the Peclet number P was obtained from tritium or PFBA tracer breakthrough curves in each experiment. The retardation factor R_f was calculated from K_F and n values measured in the batch experiments. For the nonlinear, nonequilibrium model, the fraction of instantaneous retardation β and the Damkohler number ω were optimized by the use of a nonlinear, least-squares optimization program (Jessup *et al.,* 1989). The parameters used for simulation of breakthrough curves are given in Tables II and III.

III. RESULTS AND DISCUSSION

A. Sorption Isotherms and Kinetics

Sorption isotherms for Cd in Hayhook soil are shown in Figure 1 for four times. There are no significant differences among the isotherms for 2-, 4-, 8-, and 24-hr reaction times. This suggests that sorption of Cd was rapid and essentially completed with 2 hr. Similar results were obtained for Ni and Sr (data not shown).

Table II

Physical Properties of Soil Columns Used in the Miscible Displacement Experiments

Column no.	Tracer	ρ ($g \cdot cm^{-3}$)	θ ($cm^3 \cdot cm^{-3}$)	P (Peclet)	R_f	v ($cm \cdot hr^{-1}$)
			Hayhook soil			
201	H_2^3O	1.77	0.334	28.8	0.96	31.6
202	PFBA	1.78	0.340	36.5	1.06	30.2
203	H_2^3O	1.83	0.329	26.4	1.05	31.8
204	H_2^3O	1.71	0.355	30.2	1.06	28.4
205	H_2^3O	1.84	0.343	27.4	1.13	29.3
206	H_2^3O	1.76	0.331	20.2	1.09	22.2
			Borden soil			
301	PFBA	1.75	0.341	98.4	0.99	36.5

Note. ρ is the bulk density of the soil column. θ is the water content of the soil column. P is the Peclet number of the soil column. R_f is the retardation factor. v is the velocity of water through the soil column.

Table III

Input Parameters for Simulation Using the Nonlinear, Equilibrium/Nonequilibrium Models

Metal	P	$L^{n_F}\cdot kg^{-1}$	n_F	R_f	β	ω
			Hayhook soil			
Cd	25	246 ± 15	0.334 ± 0.016	57.5	0.17 ± 0.03	6.16 ± 0.93
Ni	25	159 ± 5.8	0.373 ± 0.012	69.0	0.36 ± 0.06	4.63 ± 1.9
Sr	25	15.3 ± 13	0.419 ± 0.024	61.2	0.03 ± 0.1	8.12 ± 1.9
			Borden soil			
Cd	98	113 ± 3.9	0.159 ± 0.01	10.3	0.793 ± 0.01	0.30 ± 0.07

Note. K_F is the sorption coefficient of metals in the Freundlich equation. n_F is the exponent constant in the Freundlich equation. P is the Peclet number of the soil column. R_f is the retardation factor of metals. β is the fraction of instantaneous retardation. $\dot{\omega}$ is the Damkohler number, which is the ratio of hydrodynamic residence time to the characteristic time for sorption of metals.

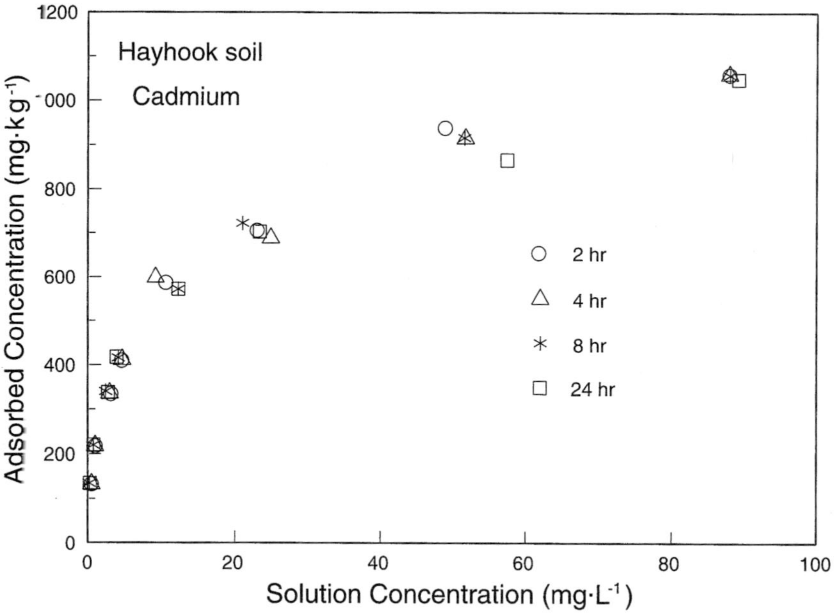

Figure 1 The sorption isotherms of Cd at 2-, 4-, 8-, and 24-hour reaction times in the Hayhook soil.

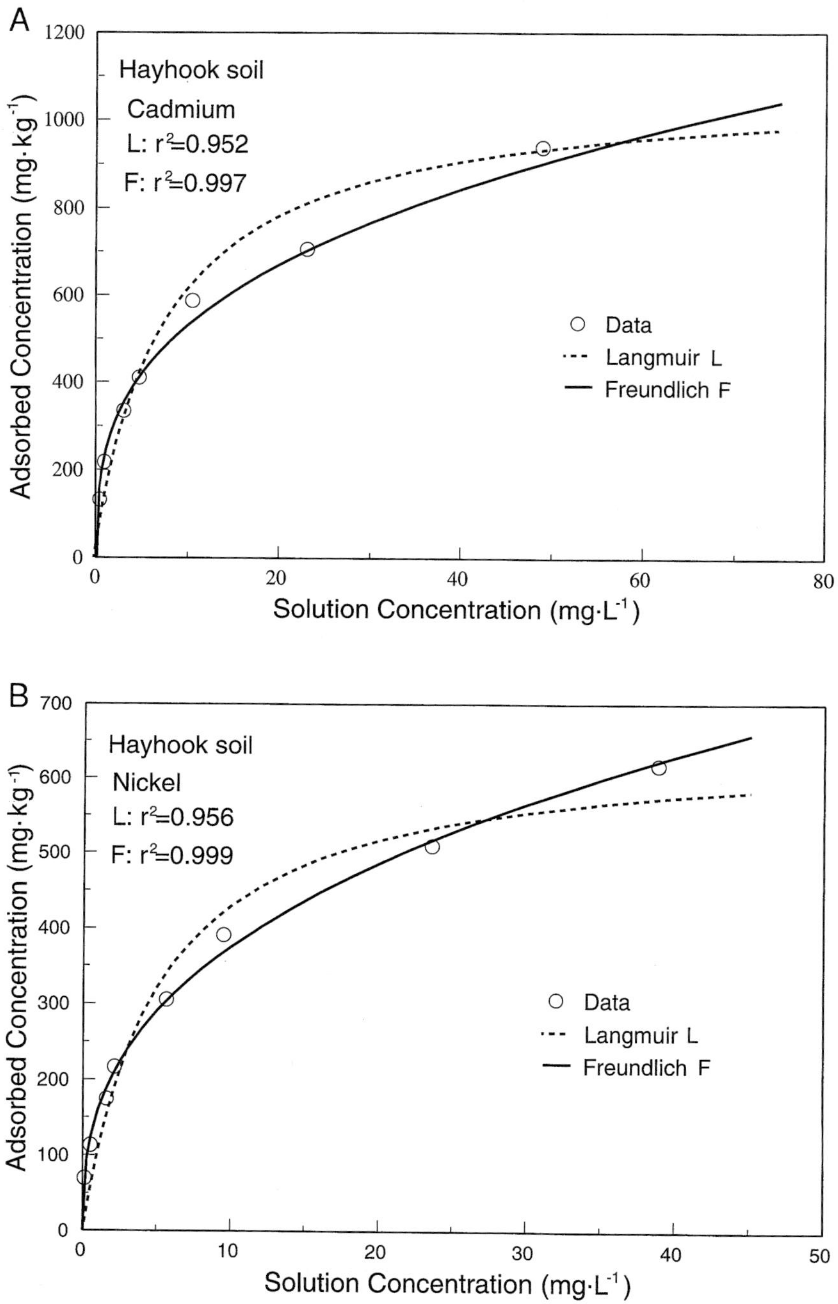

Figure 2 Sorption isotherms of Cd (A), Ni (B), and Sr (C) in the Hayhook soil and Cd (D) in the Borden soil, fitted with both Langmuir (dashed line) and Freundlich (solid line) equations; the pH values at equilibrium were 6.10 ± 0.16 for Cd, 6.00 ± 0.50 for Ni, and 6.15 ± 0.18 for Sr.

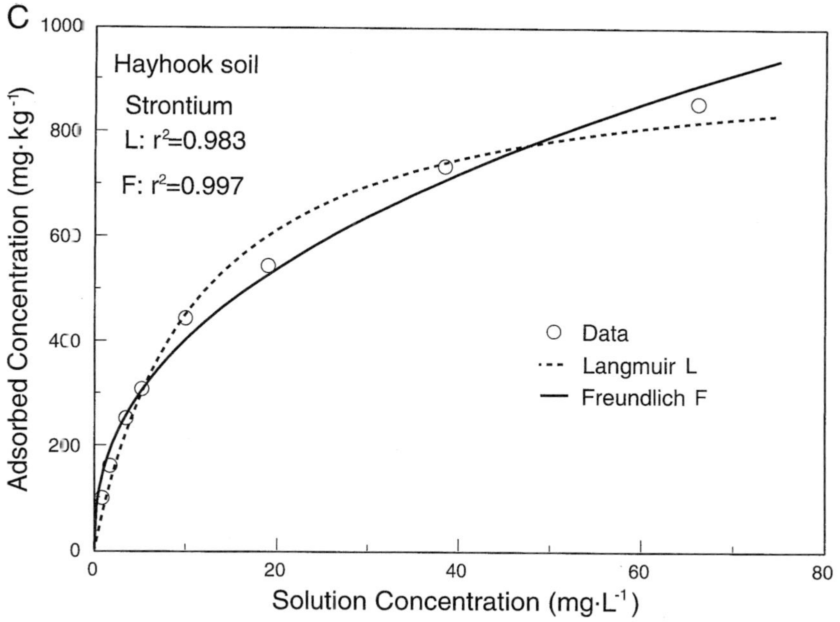

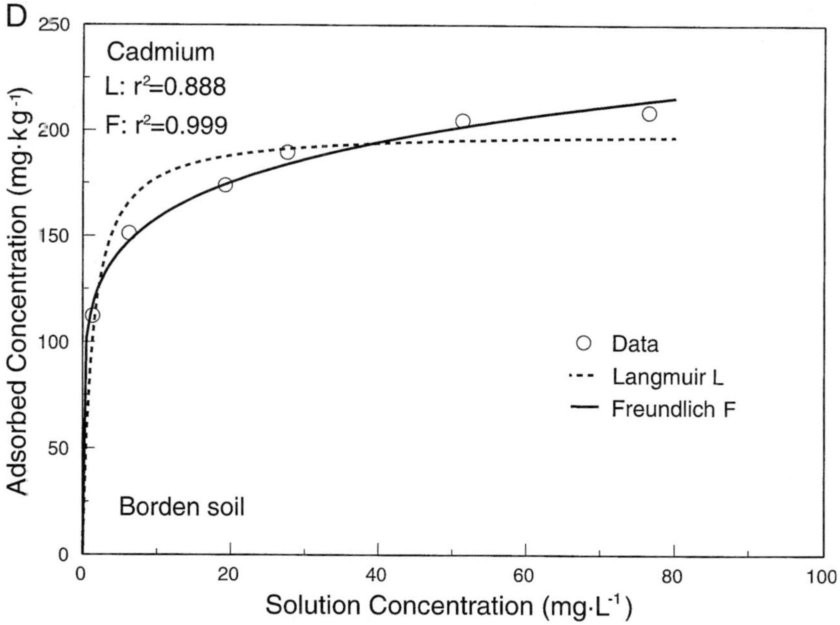

Figure 2 Continued

The sorption isotherms for Cd, Ni, and Sr in soils were strongly nonlinear (Fig. 2). The Freundlich equation provided a better fit of the isotherms than did the Langmuir equation. Straight lines were obtained with log(S)–log(C_{aq}) plots (figures not shown). As expected, there is greater mass sorption of Cd in the soil than of Sr and Ni due to the larger molecular weight of Cd. The degree of nonlinearity of sorption was Cd>Ni>Sr (Table III). The degree of nonlinearity of sorption is dependent on the initial concentration of metals used in solution. For low concentrations (<1 mg·L^{-1}), sorption has often been treated as linear (Kookana *et al.,* 1994; Jardine *et al.,* 1985, 1993). However, the concentration of metals in solution may be relatively high in heavily contaminated soils, and sorption of metals may then show significant nonlinearity.

B. Properties of Soil Columns

The breakthrough curves obtained from displacement of tritiated water and PFBA through the Hayhook and Borden soil columns, respectively, are presented in Figures 3 and 4. The breakthrough curves were symmetrical and sigmoidal in

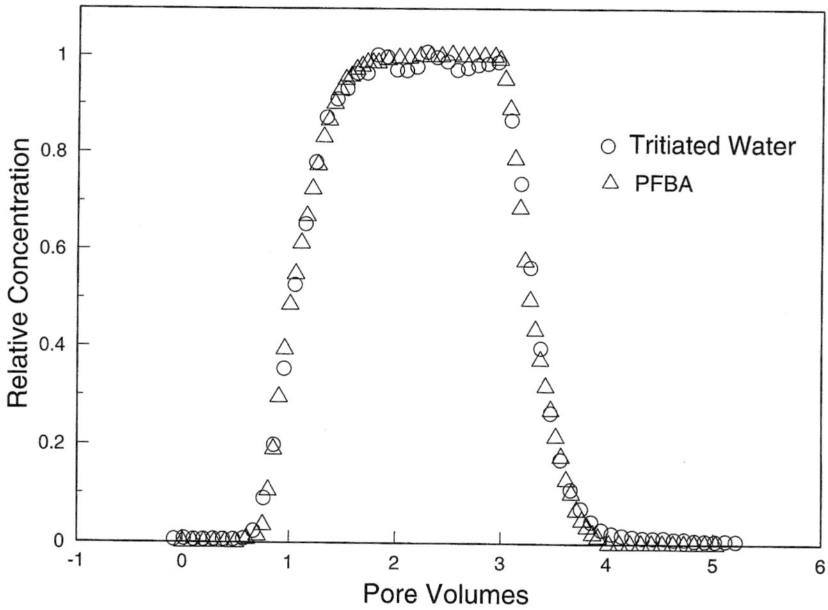

Figure 3 Breakthrough curves for transport of tritiated water and PFBA through the Hayhook soil.

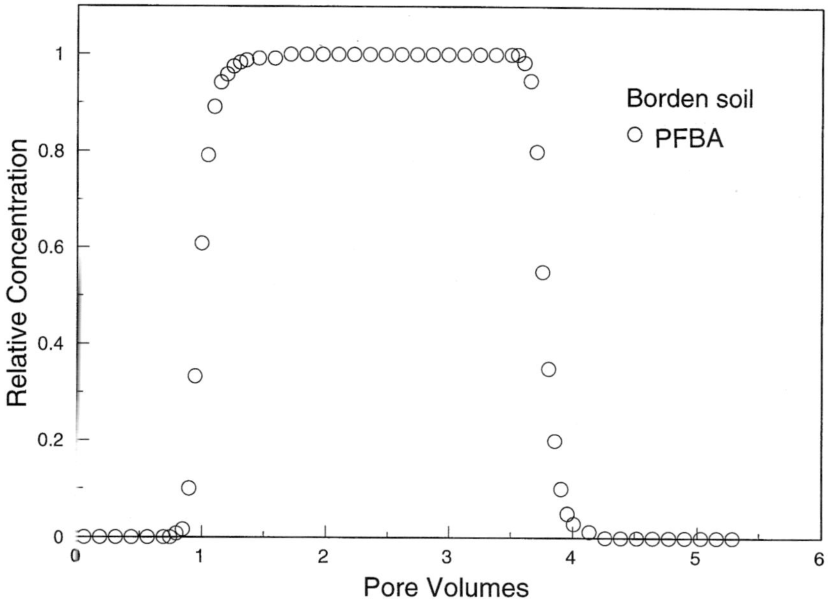

Figure 4 Breakthrough curve for transport of PFBA through the Borden soil.

shape, and the local equilibrium advection–dispersion model was successfully used to simulate the curves. The values of the retardation factor for tritiated water and PFBA were very close to 1.0. The breakthrough curves for tritiated water and PFBA were essentially identical (Fig. 3). All of these results suggest that the packed columns were homogeneous and that physical nonequilibrium processes were absent. The properties of the soil columns are listed in Table II.

C. Transport Behavior of Cd, Ni, and Sr

Measured breakthrough curves for single-metal experiments with Cd, Ni, and Sr experiments in two soils are given in Figures 5–8. The breakthrough curves were significantly retarded relative to tritiated water and PFBA. The retardation of Cd in Hayhook soil is larger than that in Borden soil, which coincides with results from batch experiments.

The Hayhook soil has significantly higher clay content than the Borden soil. The clay fraction is the dominant sorbent for metals due to the low organic content ($<0.3\%$) of the soils (Bohn *et al.*, 1985; Sposito, 1989). The results of mass-balance analysis of the breakthrough curves showed that approximately 70% of

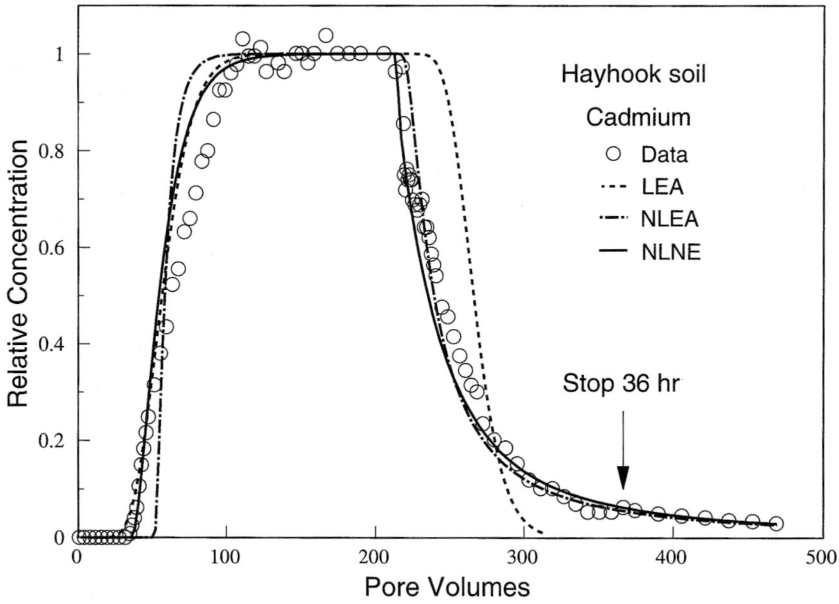

Figure 5 Breakthrough curves for transport of Cd through the Hayhook soil. The curve is predicted by using the linear sorption, local equilibrium model (LEA), the nonlinear, local equilibrium model (NLEA), and nonlinear sorption, nonequilibrium (NLNE) model (2-parameter fit).

Cd and Ni entering the columns was recovered by the end of the experiments. Note that the concentration of metals in the effuent did not reach zero at the end of experiments. This was attributed to two factors, one being the use of a monovalent electrolyte solution and the other involving chemisorption of metals. Other experiments have shown that about 90% of adsorbed Cd and Ni can be recovered by using divalent Ca instead of monovalent K, which indicates cation exchange is the major retention mechanism for Cd, Ni, and Sr in these soils (Wang *et al.,* 1997). However, the remaining 10% appears to be involved in some type of specific sorption process (Wang *et al.,* 1997).

The breakthrough curves for Cd, Ni, and Sr were asymmetrical and showed significant tailing (Figs. 5–8). The breakthrough curves were simulated with the local equilibrium, advection–dispersion model using independently measured parameters from the batch experiments under the assumption of linear sorption. Obviously, the linear, equilibrium advection–dispersion model failed to describe the observed breakthrough curves of Cd, Ni, and Sr. Application of the nonlinear, equilibrium, advection–dispersion model to the observed data significantly improved the description of metal transport. The extensive tailing was adequately predicted with the nonlinear, equilibrium model, which suggests that the tailing associated with the measured breakthrough curves may be caused by nonlinear sorption. The observed breakthrough curves of Cd, Ni, and Sr were also simulated using the nonlinear, nonequilibrium (two-domain) model. The simulation of breakthrough curves obtained by using the nonlinear, two-domain, nonequilibrium model was slightly better than that obtained by using the nonlinear, local equilibrium model.

To examine the relative influence of nonlinear and rate-limited sorption, the effect of flow interruption was investigated in these soils. A small increase in effluent concentration (<7%) was observed for the Hayhook soil (Figs. 5–7), and no change in effluent concentration was observed for the Borden soil (Fig. 8), when the flow was stopped for 36 hr during elution. This suggests that rate-limited sorption had minimal impact on transport for both soils.

Results of both flow-interruption experiments and the model simulations suggest that rate-limited sorption may have influenced the transport of these metals in Hayhook soil. However, the nonideal transport of the metals (extensive tailing)

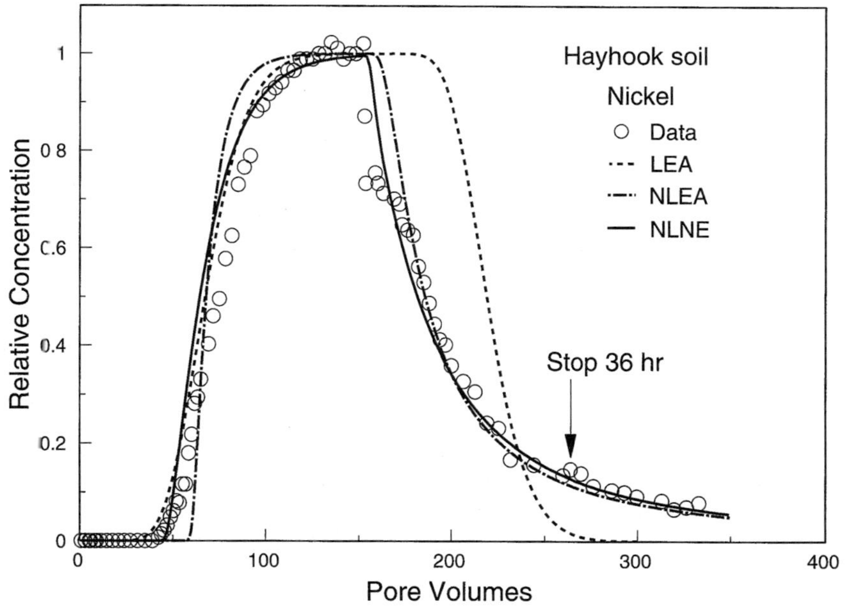

Figure 6 Breakthrough curves for transport of Ni through the Hayhook soil. The curve is predicted by using the linear sorption, local equilibrium model (LEA), the nonlinear, local equilibrium model (NLEA), and nonlinear sorption, nonequilibrium (NLNE) model (2-parameter fit).

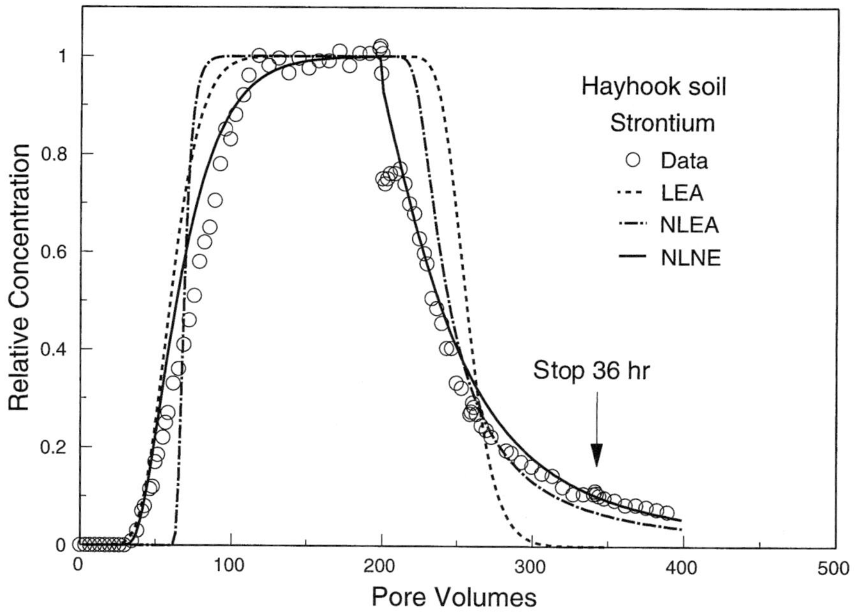

Figure 7 Breakthrough curves for transport of Sr through the Hayhook soil. The curve is predicted by using the linear sorption, local equilibrium model (LEA), the nonlinear, local equilibrium model (NLEA), and nonlinear sorption, nonequilibrium (NLNE) model (2-parameter fit).

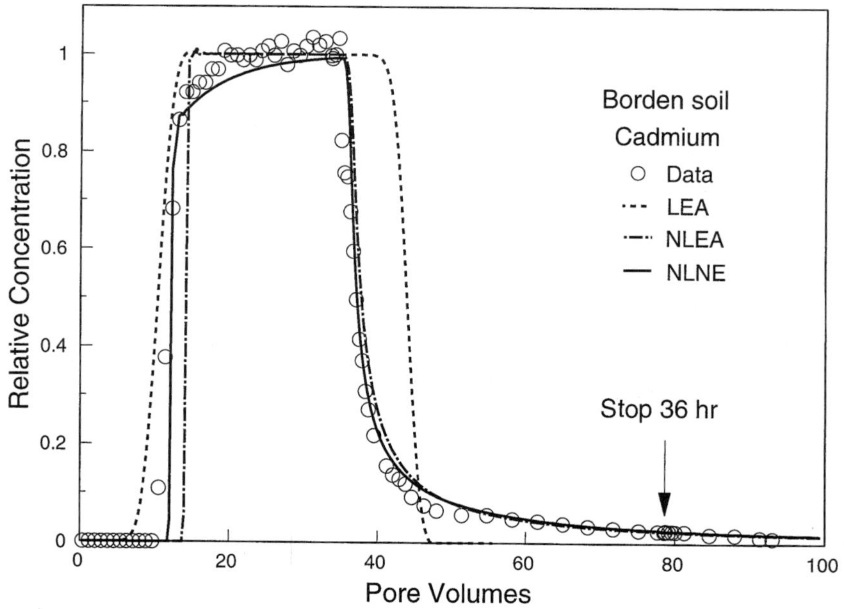

Figure 8 Breakthrough (BTC) curves for transport of Cd through the Borden soil. The BTC of Cd is predicted by using the linear sorption, local equilibrium model (LEA), the nonlinear, local equilibrium model (NLEA), and nonlinear sorption, nonequilibrium (NLNE) model (2-parameter fit).

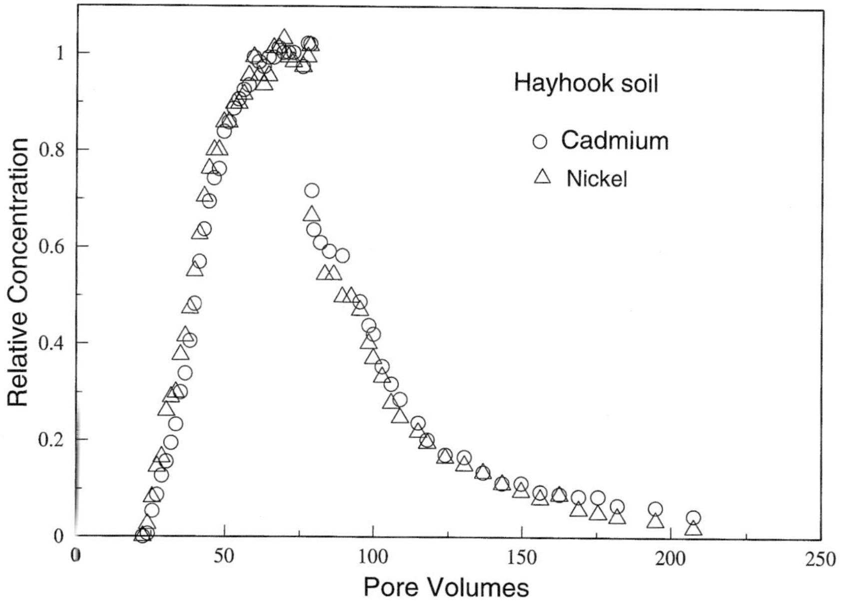

Figure 9 Breakthrough curves for transport of binary Cd–Ni solution through the Hayhook soil.

was due primarily to nonlinear sorption, and the impact of rate-limited sorption was relatively minor.

Many experiments have shown that adsorption of inorganic cations (e.g., Ca, K, NH_4, Mg, and Al) is related predominantly to the 2:1 mineralogy of clay fractions (e.g., Keay and Wild, 1961; Hsu and Bates, 1964). According to Jardine *et al.* (1988, 1993), chemical diffusion of metals into the interlayers of the 2:1 clay might result in the rate-limited sorption. Rate-limited sorption in Borden soil was not found due to the small clay fraction present.

The transport of a binary Cd–Ni solution through the Hayhook soil column was also investigated in this study (Fig. 9). The retardation of both Cd and Ni was reduced in comparison with the single-solute experiments (Figs. 5 and 6). This results from competition of Cd and Ni for adsorption sites. The breakthrough curves of Cd and Ni overlap due to their similar sorption properties.

IV. CONCLUSIONS

The sorption of Cd, Ni, and Sr by two soils was nonlinear and was well described by the Freundlich equation. Transport of Cd, Ni, and Sr through columns packed with the soils was retarded primarily by ion exchange. The retardation of Cd was significantly higher in the Hayhook soil than in the Borden soil due to a higher clay content in the Hayhook soil.

The shape of breakthrough curves from column experiments provided evidence that transport of Cd, Ni, and Sr was nonideal. The simulations obtained by using models incorporating both nonlinear and rate-limited sorption indicated that nonlinear sorption was the major cause of nonideal transport, with rate-limited sorption a minor component. This conclusion was supported by the results obtained from flow-interruption experiments. The parameters from batch and tracer experiments were used successfully to predict Cd, Ni, and Sr transport in the soils. The transport of metal was significantly affected by the presence of another metal with similar sorptive properties.

ACKNOWLEDGMENTS

This research was supported in part by grants from the U.S. EPA Exploratory Research Program and the NIEHS Superfund Research Program.

REFERENCES

Bohn, H. L., McNeal, B. L., and O'Connor, G. A. 1985. "Soil Chemistry." Wiley, New York.

Brusseau, M. L. 1994. Transport of reactive contaminants in heterogeneous porous media. *Rev. Geophys.* 32:285–313.

Brusseau, M. L., Jessup, R. E., and Rao, P. S. C. 1989a. Modeling the transport of solutes influence by multiprocess nonequilibrium. *Water Res.* 25:1971–1988.

Brusseau, M. L., and Rao, P. S. C. 1989. Sorption nonideality during contaminant transport in porous media. *CRC Crit. Rev. Environ. Control.* 19:33–99.

Brusseau, M. L., Rao, P. S. C., Jessup, R. E., and Davidson, J. M. 1989b. Flow interruption: A method for investigating sorption nonequilibrium. *J. Contam. Hydrol.* 4:223–240.

Crittenden, J. C., Hutzler, N. J. J., Geyer, D. G., Oravitz, J. L., and Friedman, G. 1986. Transport of organic compounds with saturated groundwater flow: Model development and parameter sensitivity. *Water Resources Res.* 22:271–284.

Environmental Protection Agency. 1984. Summary Report: Remedial Response at Hazardous Waste Sites. EPA-540/2-84-002a March 1984. United States Environmental Protection Agency, Office of Emergency and Remedial Response. Washington, DC.

Grove, D. B., and Stollenwerk, K. D. 1985. Modeling the rate-controlled sorption of hexavalent chromium. *Water Resources Res.* 21:1703–1709.

Hinz, C., and Selim, H. M. 1994. Transport of zinc and cadmium in soil: Experimental evidence and modeling approaches. *Soil Sci. Soc. Am. J.* 58:1316–1327.

Hsu, P. H., and Bates, T. F. 1964. Fixation of hydroxy-aluminum polymers by vermuclite. *Soil Sci. Soc. Am. Proc.* 28:763–766.

Jardine, P. M., Jacobs, G. K., and Wilson, G. V. 1993. Unsaturated transport processes in undisturbed heterogeneous porous media. I. Inorganic contaminants. *Soil Sci. Soc. Am. J.* 57:945–953.

Jardine, P. M., Parker, J. C., and Zelazny, L. W. 1985. Kinetics and mechanisms of aluminum adsorption on kaolinite using a two-site nonequilibrium transport model. *Soil Sci. Soc. Am. J.* 49:867–872.

Jardine, P. M., Wilson, G. V., and Luxmoore, R. J. 1988. Modeling the transport of inorganic ions through undisturbed soil columns from two contrasting watersheds. *Soil Sci. Soc. Am. J.* 52:1252–1259.

Jessup, R. E., Brusseau, M. L., and Rao, P. S. C. 1989. Modeling solute transport. Florida Agricultural Experiment Station Report, University of Florida, Gainesville, FL.

Keay, J., and Wild, A. 1961. The kinetics of cation exchanges in vermiculite. *Soil Sci.* 92:54–60.

Kookana, R. S., Naidu, R., and Tiller, K. G. 1994. Sorption non-equilibrium during cadmium transport through soils. *Aust. J. Soil Res.* 32:635–651.

Parker, J. C., and Jardine, P. M. 1986. Effects of heterogeneous adsorption behavior on ion transport. *Water Resources Res.* 22:1334–1340.

Pickens, J. F., Jackson, R. E., Inch, K. J., and Merritt, W. F. 1981. Measurements of distribution coefficients using a radial injection dual-tracer test. *Water Resources Res.* 17:529–544.

Schulin, R., Fluhler, H., Mansell, R. S., and Selim, H. M. 1986. Miscible displacement of ions in aggregated soils. *Geoderma* 38:311–322.

Selim, H. M. 1992. Modeling the transport and retention of inorganics in soils. *Adv. Agronomy* 47:331–389.

Selim, H. M., Amacher, M. C., and Iskandar, I. K. 1989. Modeling the transport of chromium (VI) in soil columns. *Soil Sci. Soc. Am. J.* 53:996–1004.

Sposito, G. 1989. "The Chemistry of Soils." Oxford University Press, London.

Wang, W., Bresseau, M. L., and Artiola, J. F. 1997. The use of calcium to facilitate desorption and removal of cadmium and nickel in subsurface soils. *J. Contam. Hydrol.* 25:325–336.

Chapter 21

Fluorescence Quenching and Aluminum Adsorption to Organic Substances

D. Scott Smith and James R. Kramer
Department of Geology, McMaster University, Hamilton, Ontario, Canada

Site-specific Al speciation with Suwannee River fulvic acid is determined using total luminescence spectroscopy. There are two fluorescent sites in Suwannee River fulvic acid that interact with Al. One site with a fluorescence excitation wavelength of 250 nm and an emission at 440 nm has a conditional stability constant, log K′, of 5.6 and a concentration of 0.43 $\mu mol \cdot mg^{-1}$. Another site with an excitation at 330 nm and an emission at 440 nm has a log K′ of 5.1 and a concentration of 0.018 $\mu mol \cdot mg^{-1}$. The equations used to fit fluorescence data are modified to apply to fluorescence enhancement and reactions with protonated ligand. The mathematical approach is tested with the model ligand salicylic acid, and it is found that for one fluorescent site fitting is independent of whether quenching or enhancement data are used.

I. INTRODUCTION

The toxicity of aqueous Al is speciation dependent (Dobbs *et al.*, 1989; Parker *et al.*, 1989). In particular, the presence of dissolved organic matter (DOM) can reduce Al toxicity (Hue *et al.*, 1986; Kramer *et al.*, 1986; Gjessing *et al.*, 1989). Dissolved organic matter is a complicated mixture (Christman and Gjessing, 1983; Frimmel *et al.*, 1988; Senesi and Miano, 1994) that is very important in Al binding. In order to properly address the complex heterogeneous nature of metal binding sites within DOM, a sensitive and selective methodology should be used. There are many methods available to speciate Al in the presence of DOM; excellent reviews are given by Bloom and Erich (1989) and Clarke (1994). In general, these methods assume a one to one complex between Al and DOM and cannot easily resolve complexation at discrete sites. Fluorescence measurements have been shown to be able to resolve site-specific Al binding (Patterson *et al.,* 1992). Total luminescence spectroscopy (TLS) is a type of fluorescence measurement that is especially useful in resolving site-specific Al speciation (Yang *et al.,* 1994; Luster *et al.,* 1996). This study uses TLS to investigate Al binding to Suwannee River fulvic acid.

Aluminum binding to DOM has been investigated by various methods (Stevenson and Vance, 1989; Clarke, 1994), including fluorescence spectroscopy (Shotyk and Sposito, 1990; Tam and Sposito, 1993). In general, Al binds to carboxyl and hydroxyl groups, and the binding strength is stronger for bi- and tridentate ligands (Orvig, 1993). Model compounds can be used to represent DOM in studying Al binding (Öhman and Sjöberg, 1988). In fact, Al speciation for several model compounds, such as citrate (Motekaitis and Martell, 1984; Öhman, 1988), catechol (Öhman and Sjöberg, 1983; Motekaitis and Martell, 1984), and salicylic acid (Öhman, 1991), has been determined potentiometrically, but little work, other than L-tyrosine complexation by Cu^{2+} (Ryan and Weber, 1982), has been done with model compounds and speciation as defined by a fluorescence methodology. In this study, the interaction of Al with the model ligand salicylic acid is investigated to further verify the fluorescence methodology, and in particular, the application of the method to fluorescence enhancement measurements in an Al–DOM system.

Dissolved organic matter can be fluorescent (Goldberg and Weiner, 1993). This fluorescence results from fluorescent groups within DOM. These fluorophores may occur as free molecules or as molecular building blocks of macromolecules. Probable fluorophores in DOM include salicylate, catecholate, flavones, chromone derivatives, and quinolines (Senesi, 1990). Fluorescence measurements can be used to characterize DOM (Coble *et al.,* 1990; Belin *et al.,* 1993) because different fluorophores may have different fluorescent properties that can be resolved experimentally; such properties include lifetime (Power *et al.,* 1986; Cook and Langford, 1995), polarization (Lapen and Seitz, 1982; Goldberg and Negomir, 1989; Lakshman *et al.,* 1996), and excitation and emission wavelengths. Traditionally, fluorescence is observed as excitation, emission, or synchronous spectra (Guilbault, 1990), but the dimensionality of fluorescence measurements, and thus its selectivity, can be increased by using TLS (Goldberg and Weiner, 1993). Total luminescence spectroscopy is performed by measuring the entire excitation versus emission fluorescence surface. Fluorescence excitation versus emission surfaces for DOM can be used to fingerprint samples from different locations (Goldberg and Weiner, 1993; Provenzano and Sposito, 1994) and different fractions within samples (Belin *et al.*, 1993; Campanella *et al.*, 1994). In addition, possible classes of fluorophores can be suggested from TLS studies (Coble *et al.*, 1990; Tam and Sposito, 1993; Provenzano and Sposito, 1994).

The fluorescence of DOM may change upon metal binding to it (Cabaniss, 1992). If the change in fluorescence is assumed to be proportional to the amount of metal bound, then stability constants (Stern and Volmer, 1919) and total ligand concentrations can be determined (Ryan and Weber, 1982). Numerous studies have investigated metal–DOM binding using excitation and/or emission spectra (Ghosh and Schnitzer, 1981; Tam and Spositio, 1993) and synchronous fluorescence spectra (Cabaniss, 1992). Luster *et al.* (1996) emphasize the fact that results obtained from such studies are not necessarily independent of wavelength choice. It is necessary to find a point on the excitation versus emission surface where changes in fluorescence are caused by complexation at only one site. The use of TLS improves confidence in wavelength choices because the effect of the metal across the entire fluorescence surface can be observed and regions of change assigned to each fluorophore (Luster *et al.*, 1994). The result is site-specific speciation (Luster *et al.*, 1994, 1996).

Metal binding sites have been resolved in DOM using TLS; spectra have been reported with and without added metal at fixed pH (Provenzano and Sposito, 1994; Yang, *et al.*, 1994). Recently, this has been taken one step further and TLS was performed during a titration of juniper leaf litter extract with Al (Luster *et al.*, 1994, 1996) and Cu (Luster *et al.*, 1996) at fixed pH. The approach was validated for Al using an equilibrium ion exchange method. The determination of conditional stability constants at different sites was performed using the approach of Ryan and Weber (1982). This involves fitting the fluorescence data from a titration with metal at fixed pH to three parameters in a nonlinear equation. The approach has been criticized in that the three parameters may not be independent (Cabaniss and Shuman, 1988; Ryan and Ventry, 1990), but several of the studies already mentioned have used the equation and obtained physically reasonable results from Al–DOM fluorescence measurements. The method has also been validated using independent speciation measurements (Shotyk and Sposito, 1988; Luster *et al.*, 1996).

In an effort to further validate the Ryan–Weber equation, this study examines fitting results for both a model ligand (salicylic acid) and a reference DOM (Suwannee River fulvic acid) using a one-parameter equation and a two-parameter equation, in addition to the usual three-parameter Ryan–Weber equation. The one-parameter equation is a modification of the Stern–Volmer equation (Stern and Volmer, 1919) that can be applied to fluorescence enhancement measurements; it was originally derived only to apply to fluorescence quenching measurements. The two-parameter equation is a modification of the Ryan–Weber equation using a simplifying assumption to remove one parameter. In addition, the Ryan–Weber equation was originally derived assuming that the free ligand was the fluorescent species, but a modified equation is proposed here, assuming that the fluorescent species is the protonated ligand.

The Ryan–Weber equation is a general equation applicable to positive or negative fluorescence quenching (enhancement), but as Luster *et al.* (1996) point out, fitting results for metal–DOM systems are not independent of wavelength. Salicylic acid data are fit across the entire emission spectra to assess whether the wavelength dependence is an artifact of the mathematical approach. For example, with a single fluorophore, does quenching data yield different results from enhancement data, or is the observed dependence in natural DOM systems due to multiple fluorophores affecting the observed fluorescence at a given point on the fluorescence surface?

II. EQUATIONS TO INTERPRET FLUORESCENCE MEASUREMENTS

There are two effects that metal (M) binding may have on a fluorescent ligand (***L***). The metal may partially or completely suppress the ligand's fluorescence, or the new species (ML) may be fluorescent itself. In general, if there is an effect, paramagnetic species, such as Cu^{2+}, quench fluorescence (Guilbault, 1990) and diamagnetic species, such as Al^{3+}, form new fluorescent species (Djurdjevic *et al.*, 1995). For Al, the new species are generally a more efficient fluorophore, as demonstrated by the large number of fluorescent analytical reagents used for Al determination (Snell, 1978).

When fluorescence quenching is observed, the metal ion inhibits the radiative transition that causes fluorescence emission. Thus, the entire spectrum will be observed to decrease in intensity (Provenzano and Sposito, 1994). If the bound metal forms a species with different fluorescent properties from the free ligand, then the observed fluorescence will be a combination of the spectral contributions from both the ligand and the metal–ligand species. Thus, in the case of diamagnetic ions, suppression or enhancement can be observed depending on which part of the fluorescence emission spectra is monitored (Cabaniss, 1992; Tam and Sposito, 1993).

The Stern–Volmer (Eq. [1]) and Ryan–Weber (Eq. [2]) equations have been proposed to interpret fluorescence suppression results during a titration of fluorescent ligand with quencher at fixed pH. Equation [1] is a simple linear equation and will yield a conditional stability constant (K′). Equation [2] will also yield a value for K′; in addition, values for the total ligand concentration ($\boldsymbol{L}_T$) and the fluorescence intensity when all the ligand is bound to the metal ($\boldsymbol{F}_{ML}$) are also determined. The fluorescence signal is defined as the fluorescence emission intensity (in arbitrary units) at fixed excitation and fixed emission wavelengths.

The Stern–Volmer equation,

$$\frac{F_i}{F} = 1 + K'M_T, \quad [1]$$

is a linear equation. The conditional stability constant is the slope when the initial fluorescence (F_i) over the fluorescence at any point during the titration (F), is plotted against the total metal concentration (M_T).

Equation [2] is a nonlinear equation in which three parameters must be fit. The parameters are K′, L_T, and (F_{ML}). The equation can be written

$$F = \left(\frac{F_{ML} - F_i}{2K'L_T}\right)\left(K'L_T + K'M_T + 1 - \sqrt{(K'L_T + K'M_T + 1)^2 - 4(K')^2 L_T M_T}\right) + F_i. \quad [2]$$

Equations [1] and [2] both assume that the metal and ligand form a one to one complex according to the reaction

$$M' + L' = ML' \qquad \text{(reaction (I))},$$

with

$$K' = \frac{[ML']}{[L'][M']}, \quad [3]$$

where [ML′] is the total concentration of all ligand-bound metal species, [L′] is the total concentration of all free ligand species, and [M′] is the total concentration of all metal species not bound to the ligand. The conditional constant is only applicable for the experimental conditions at which it is measured. Two important variables affecting the conditional constant are pH and ionic strength. The value of the constant should be independent of total ligand amounts, as Shotyk and Sposito (1990) showed for Al with chestnut leaf litter extract, and total metal concentration unless polynuclear species are forming.

Equations [1] and [2] both assume that fluorescence response increases linearly with fluorophore concentration ([Fluor]):

$$F = k[\text{Fluor}]. \quad [4]$$

This is a reasonable assumption in dilute solutions (Guilbault, 1990). The constant, k, is specific for each fluorescent species.

The equation derivations differ in that Eq. [1] makes two simplifying assumptions that Eq. [2] does not. These assumptions are:

1. ML species do not fluoresce
2. the total concentration of metal not bound to the ligand ([M′]) is approximately equal to the total metal (M_T)

Equation [2] assumes residual fluorescence from the bound species, and does not assume that M_T is equal to [M′].

In order for the parameters measured to be physically meaningful, both equations assume that it is a ground-state association between the metal and the fluorophore that changes the fluorescence. In addition, both equations assume that the initial total metal is negligible.

A. Modified Stern–Volmer Equation

An equation similar to Eq. [1] can be derived for fluorescence enhancement; the assumptions are the same except as further noted.

For reaction (1) the mass balance for ligand can be written

$$L_T = [L'] + [ML']. \qquad [5]$$

Dividing both sides of the expression by [ML′] and substituting [L′] from Eq. [3] yields

$$\frac{L_T}{[ML']} = \frac{1}{[M']K'} + 1. \qquad [6]$$

Further, assume that [M′] is approximately equal to M_T, as in Eq. [1]; consequently [L′] should not change at fixed pH; therefore, the fluorescence change $(F - F_i)$ is proportional only to [ML′]. With the additional assumption that the fluorescence proportionality constants, as shown in Eq. [4], are equal for free and bound ligand species, the equation can finally be written as

$$\frac{F_i}{F - F_i} = \frac{1}{M_T K'} + 1. \qquad [7]$$

A plot of F_i over $(F - F_i)$ versus the reciprocal of the total metal concentration will yield a linear plot with slope 1/K′. The equation is not defined when M_T equals zero, because of division by zero.

B. Two-Parameter Ryan–Weber Equation

The derivation of the modified Ryan–Weber equation is given next. The objective is to derive a two-parameter equation intermediate between the simplicity of Eq. [1] and the complexity of Eq. [2]. The total fluorescence at any point during the titration is a sum of fluorescence contributions from both free and bound ligand:

$$F = k_L[L'] + k_{ML}[ML']. \qquad [8]$$

The proportionality constants, k_L and k_{ML}, can be rewritten in terms of the fluorescence response when all the ligand is free ($\boldsymbol{F}_i$) and when all the ligand is bound ($\boldsymbol{F}_{ML}$):

$$\boldsymbol{F} = \left(\frac{\boldsymbol{F}_i}{\boldsymbol{L}_T}\right)[L'] + \left(\frac{\boldsymbol{F}_{ML}}{\boldsymbol{L}_T}\right)[ML']. \qquad [9]$$

Using the mass balance expression for the ligand (Eq. [5]) and rearranging yields the following relation:

$$\frac{[ML]}{\boldsymbol{L}_T} = \frac{\boldsymbol{F}_i - \boldsymbol{F}}{\boldsymbol{F}_i - \boldsymbol{F}_{ML}}. \qquad [10]$$

If the simplifying assumption, M_T equals [M′], from Eq. [1] is used then the stability constant can be written

$$K' = \frac{[ML']}{M_T(\boldsymbol{L}_T - [ML'])}. \qquad [11]$$

This expression is then rearranged to solve for the ratio [ML′] over $\boldsymbol{L}_T$ and this ratio is set equal to Eq. [10]. The result after rearrangement is a two-parameter (K′ and $\boldsymbol{F}_{ML}$) equation:

$$\boldsymbol{F} = \boldsymbol{F}_i + \left(\frac{K'M_T(\boldsymbol{F}_{ML} - \boldsymbol{F}_i)}{1 + K'M_T}\right). \qquad [12]$$

C. Aluminum Reaction with Protonated Ligand

In the previous equations the reaction between Al and organic ligand is of the form described by reaction (I). In the derivations presented so far it is assumed that the free ligand is the fluorescent species. This is not necessarily a good assumption; in the pH range of these experiments it is likely that the protonated ligand is the fluorescent species. If the reaction is assumed to be of the form

$$M' + HL' = ML' + H' \qquad \text{(reaction (II))},$$

and K′ is redefined accordingly, then the three proposed equations can be rewritten. Equation [1] becomes

$$\frac{\boldsymbol{F}_i}{\boldsymbol{F} - \boldsymbol{F}_i} = \frac{[H]}{M_T K'} + 1, \qquad [13]$$

Eq. [2] becomes

$$\boldsymbol{F} = \left(\frac{\boldsymbol{F}_{ML} - \boldsymbol{F}_i}{2K'\boldsymbol{L}_T}\right)(K'\boldsymbol{L}_T + K'M_T + [H] - \sqrt{(K'\boldsymbol{L}_T + K'M_T + [H])^2 - 4(K')^2\boldsymbol{L}_T M_T}) + \boldsymbol{F}_i, \qquad [14]$$

and Eq. [12] becomes

$$F = F_i + \left(\frac{K'M_T(F_{ML} - F_i)}{[H] + K'M_T}\right). \quad [15]$$

D. Summary of Derived Equations

There are five equations derived in this study that are modifications of Eq. [1] and Eq. [2]. Equation [1] has been modified to apply to fluorescence enhancement (Eq. [8]) and to the reaction of metal with protonated ligand (Eq. [13]). The three-parameter equation (Eq. [2]) has been modified to apply to reactions with protonated ligand (Eq. [14]) and a two-parameter equation has been derived (Eq. [12]). The two-parameter equation has also been modified for a reaction with HL (Eq. [15]). Note that Eq. [2] and all modifications apply equally well to quenching or enhancement measurements.

III. EXPERIMENTAL METHOD

Organic ligand solutions were titrated with Al at fixed pH and ionic strength and fluorescence was monitored. In the salicylic acid titrations the Al concentration was in the range 0.7 $\mu mol \cdot L^{-1}$ to 0.4 $mmol \cdot L^{-1}$ for a total of 11 solutions. The Suwannee River fulvic acid titration was in the range 0.2 $\mu mol \cdot L^{-1}$ to 0.06 $mmol \cdot L^{-1}$ for a total of 5 solutions. The reproducibility error in salicylic acid fluorescence measurements was $<1\%$. The total salicylate concentration was 26.4 $\mu mol \cdot L^{-1}$ and the total fulvic acid concentration was 8.4 $mg \cdot L^{-1}$. The organic ligands used were salicylic acid (Fisher Scientific, Fair Lawn, NJ) after recrystallization, and Suwannee River fulvic acid (International Humic Society Standard). Aluminum stock solution (1.18 $mmol \cdot L^{-1}$) was made from Al-nitrate (BDH Chemicals, Toronto, Canada) with pH adjusted to 4.0 with sodium hydroxide. The pH after Al addition was adjusted with dilute NaOH or HCl as necessary to achieve a final pH of 4.0. The ionic strength was held fixed at 0.01 $mol \cdot L^{-1}$ using potassium nitrate.

Solutions were equilibrated for 2 days before fluorescence measurement. This was determined to be long enough for equilibrium because at acidic pH, Al reaction times with salicylic acid (Plankey *et al.,* 1986) and fulvic acid (Plankey and Patterson, 1987; Plankey *et al.,* 1995) are on the order of minutes, and although reactions to form aluminum polynuclear species may occur on the order of days, these species were not expected to occur according to equilibrium calculations using Öhman's constants (1991). Fluorescence excitation versus emission surfaces were measured using a Perkin-Elmer LS-5 spectrophotometer with fixed excitation and emission slits at 5 nm. The size of the surface was limited by the controlling software to an excitation range of 240 to 340 nm and an emission range of 350 to 500 nm. Emission spectra were recorded for every 10 nm of excitation. For salicylic acid, the excitation light source was a General Electric 110-855 bulb and a 7-54 Corning glass filter combined with a $NiSO_4$ solution (Murov, 1973) to achieve a constant excitation wavelength of 300 nm with a bandwidth of about 10 nm. Emission light was collected by a fiber optic probe down the sample axis and detected by an InstaSpecIV charge coupled detector (Oriel Corporation, Stratford, CT).

Nonlinear regression analysis was performed using the simplex function minimization algorithm in Matlab (The Mathworks Inc., South Natick, MA). The sum of the absolute value of the difference between data points and the fit curve was the minimized function. The minimization was set to stop once the difference reached a total relative error of 10^{-4}. Error estimates in the fit parameters were obtained by taking the average and standard deviation of the parameters over the applicable range of the emission spectra.

IV. RESULTS

A. Salicylic Acid Titrations

Fluorescence emission spectra from the salicylic acid titration are shown in Figure 1. The curves in region B have predominantly salicylic acid contributions and the curves in region A have a stronger Al–salicylate contribution. Equilibrium calculations, using Öhman's constants (1991), show that at pH 4 the only bound Al species is AlL^+ and, at maximum Al addition, 90% of the total ligand is complexed. Therefore, the observed fluorescence spectra are a combination of AlL^+ and HL^- fluorescence spectra only, where region B is mostly HL^- and region A is mostly AlL^+. The HL^- spectra have a maximum emission at 410 nm and the AlL^+ spectra have an emission maximum around 380 nm. The curves between region A and region B are a combination of HL^- and AlL^+ spectra. Fluorescent enhancement is observed in the region around 380 nm and fluorescence quenching in the region around 420 nm.

The Stern–Volmer plot for fluorescence quenching and enhancement is shown in Figure 2. The plot for quenching is not linear but the plot for enhancement is linear. The stability constants determined differ by two orders of magnitude: log K' is 3.2 for quenching versus 5.2 for enhancement. It is interesting to note that a linear plot is observed for enhancement, but the Eq. [7] assumption that the Al is predominantly free is not valid in this case except at very high total Al; when the metal and ligand are one to one, almost all the Al is complexed. Another consequence of the inappropriate assumption is that the intercept is not 1 in the enhancement plot, as it should be according to Eq. [7].

Quenching and enhancement measured at 420 and 380 nm, respectively, are

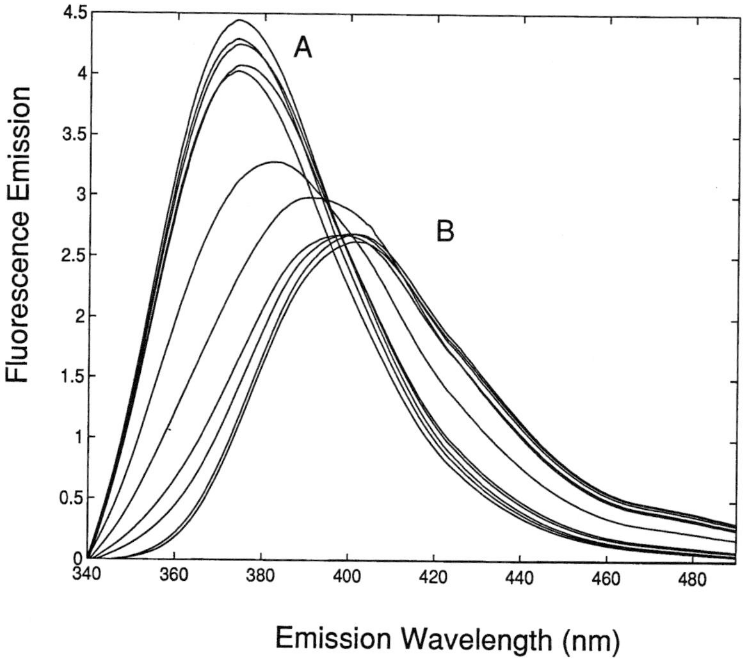

Figure 1 Fluorescence emission spectra, in arbitrary units, for salicylic acid titration with Al at pH 4.0. Region A corresponds to high amounts of total Al, and region B corresponds to low and zero amounts of total Al.

shown in Figure 3, along with fit curves from Eq. [2] and Eq. [14]. It can be seen that both equations fit the data well except in the enhancement case when the observed fluorescence starts to decrease with increasing Al and the fit curve does not. The observed decrease cannot be accounted for using equilibrium calculations; it is possibly the result of collisional deactivation of the fluorescent species in the more concentrated solutions, so-called dynamic quenching (Guilbault, 1990).

The results of parameter fitting of the salicylic acid fluorescence data are shown in Table I. The quenching results are calculated at 420 nm and the enhancement results at 380 nm. For comparison, the literature value of the conditional stability constant calculated from thermodynamic stability constants reported by Öhman (1991) and the actual values for $\boldsymbol{L}_{\mathrm{T}}$ and $\boldsymbol{F}_{\mathrm{ML}}$ are also given.

The actual conditional stability constant for reactions (I) and (II) is underestimated by all of the equations except Eq. [2] and Eq. [14] which, within experimental error, recover the actual values. Equations [12] and [15] underestimate K′ by almost an order of magnitude, but Eq. [7] and Eq. [13] yield values only 0.3 of a log unit under the true value. For fluorescence enhancement, the calculated $\boldsymbol{F}_{\mathrm{ML}}$ is very close to the observed value, but for quenching the calculated value is twice

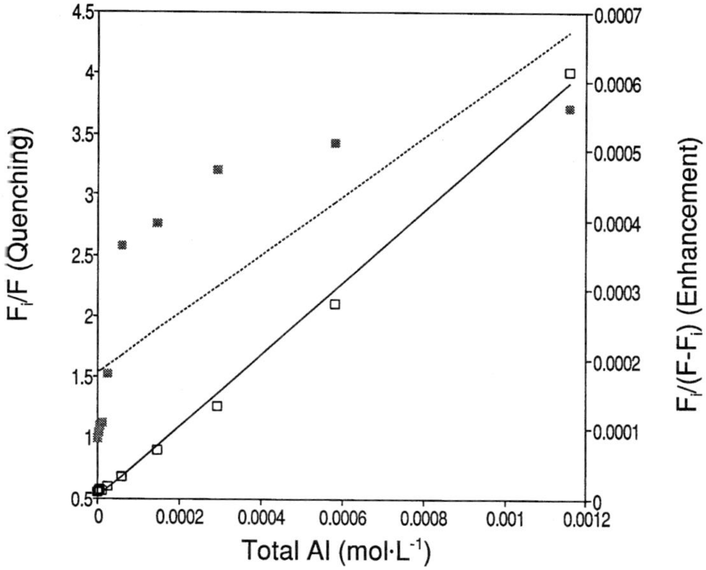

Figure 2 Stern–Volmer plot for fluorescence quenching and enhancement of salicylic acid with Al at pH 4.0. The open squares and the right axis correspond to enhancement data (at 380 nm) and Eq. [7]. The filled squares and the left axis correspond to quenching data (at 420 nm) and Eq. [1]. The enhancement data have been transformed to plot versus Al_T instead of $1/Al_T$.

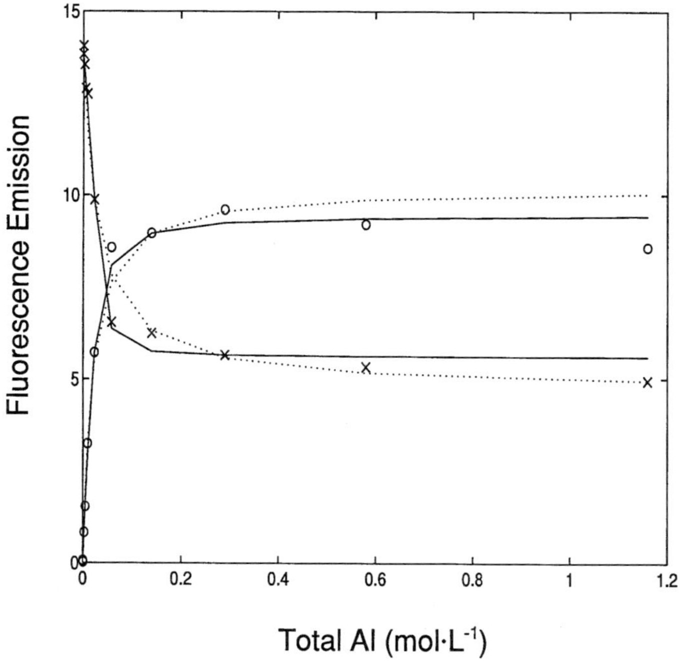

Figure 3 Fluorescence data at 380 nm (○) and 420 nm (×) in the salicylic acid titration with Al at pH 4.0. The solid lines correspond to fitting with Eq. [2] and the dotted lines to fitting with Eq. [14].

Table I
Summary of Parameter Fitting Results from Salicylic Acid Titration

Equation[a]	Reaction (I) log K′	Reaction (II) log K′	F_{ML} (arbitrary units)	L_T ($\mu mol \cdot L^{-1}$)
2,14,E	5.4 ± 0.1	2.0 ± 0.1	4.4	22 ± 1
2,14,Q	5.4 ± 0.1	1.4 ± 0.1	0.5	40 ± 1
12,15,E	4.7 ± 0.1	0.5 ± 0.1	4.6	na
12,15,Q	4.6 ± 0.1	0.25 ± 0.1	0.5	na
7,13,E	5.2 ± 0.1	1.1 ± 0.1	na	na
1,Q	3.2 ± 0.9	−0.8 ± 0.9	na	na
Actual	5.5	1.5	4.5 (E), 0.25 (Q)	21.6

[a]The equation numbers refer to equations in the text, and the E or Q refers to enhancement or quenching measurements, respectively.

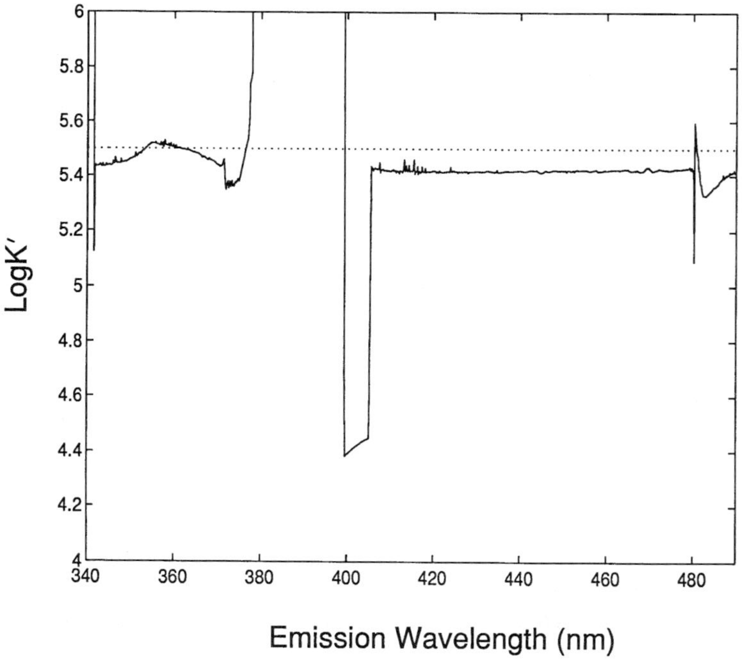

Figure 4 Log K′ fitting results across the entire emission spectrum for salicylic acid titration with Al at pH 4.0. The dotted line corresponds to the actual stability constant and the solid line is the result of fitting Eq. [2] to the titration data.

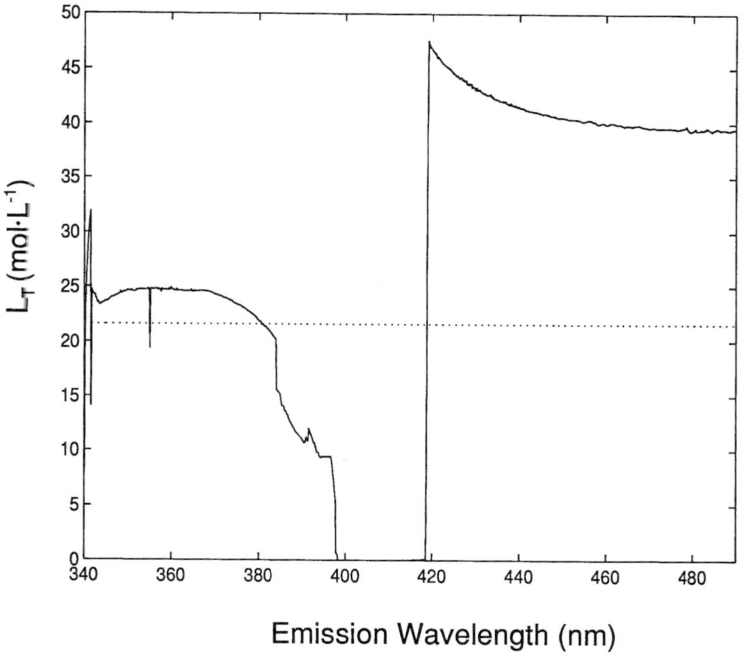

Figure 5 L_T fitting results across the entire emission spectrum for salicylic acid titration with Al at pH 4.0. The dotted line corresponds to the actual ligand concentration and the solid line is the result of fitting Eq. [2] to the titration data.

the observed value. It should be noted that the observed fluorescence is for a solution with 90% of the salicylic acid complexed, but the calculated value assumes that it is all bound. The total ligand concentration determined from fitting enhancement measurements to Eqs. [2] and [14] yields a value, within experimental error, identical to the true value. The same equations overestimate $\boldsymbol{L}_T$ by two times when quenching observations are used.

To access the effect of wavelength choice, parameter fitting can be performed across the entire fluorescence spectrum. For every set of emission measurements the constants can be determined. The results of fitting the three parameters in Eq. [2] across the entire spectrum, along with the actual values, are shown for K′ in Figure 4, for $\boldsymbol{L}_T$ in Figure 5, and for $\boldsymbol{F}_{ML}$ in Figure 6. The agreement between the actual values and the fit parameters is very good except in the region, around 400 nm, where the AlL^+ spectra and the HL^- spectra coincide. To the left of 400 nm, enhancement is observed, and to the right, quenching is observed. There is no systematic difference in fitting the quenching or enhancement data except that estimates of $\boldsymbol{L}_T$ from quenching yield values almost twice as large as those for enhancement.

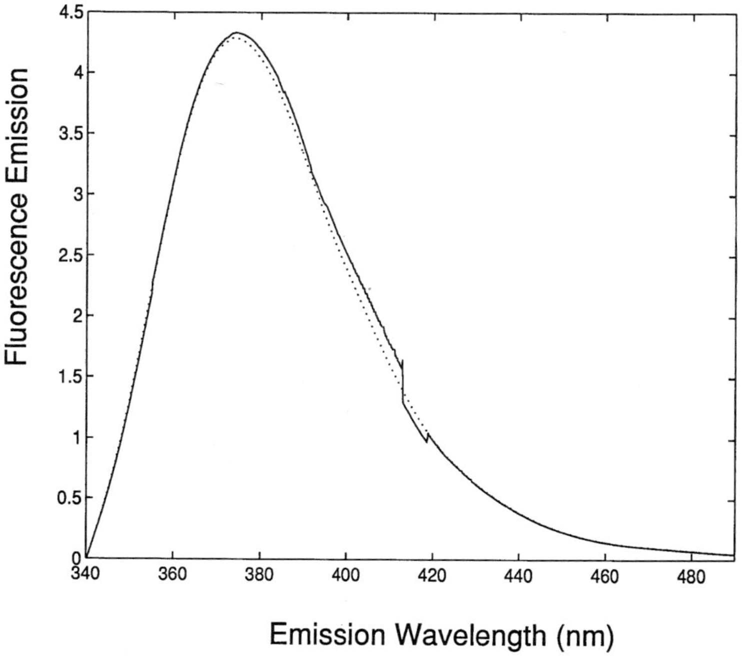

Figure 6 F_{ML} fitting results across the entire emission spectrum for salicylic acid titration with Al at pH 4.0. The dotted line corresponds to the maximum observed fluorescence during the titration and the solid line is the result of fitting Eq. [2] to the titration data.

B. Suwannee River Fulvic Acid Titration

The excitation versus emission surface for Suwannee River fulvic acid is shown in Figure 7 as a surface plot and as a contour plot. It can be seen that there are two peaks, both around an emission wavelength of 440 nm and with an excitation wavelength of either 255 or 350 nm. The resolution of this plot is only every 10 nm for excitation so the exact location of the peaks on this surface cannot be determined. The surface was determined with higher resolution by Goldberg and Weiner (1994). They concluded that the two peaks are indeed two different fluorophores. The large peak at the 500-nm emission wavelength is an artifact from the instrument that is not reproducible and therefore cannot be background subtracted away. The small peaks around the 320-nm excitation wavelength are the Raman peaks for water.

Fluorescence increased with the addition of Al. Figure 8 shows the difference surface resulting from subtracting the fulvic acid spectrum from the spectrum with 0.059 mmol·L^{-1} total Al. There are two peaks, resulting from Al binding, in the

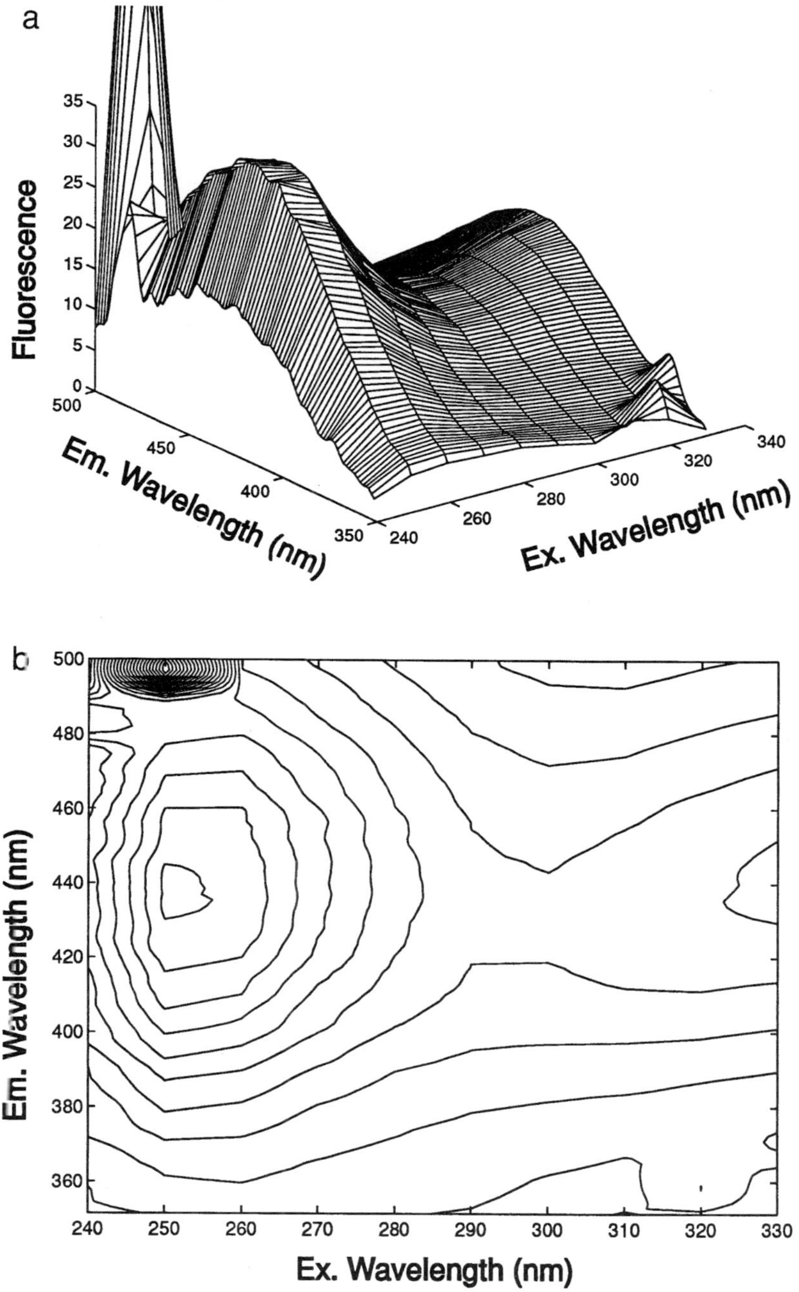

Figure 7 Excitation versus emission surface for Suwannee River fulvic acid at pH 4.0; (a) is a perspective drawing and (b) is a contour map.

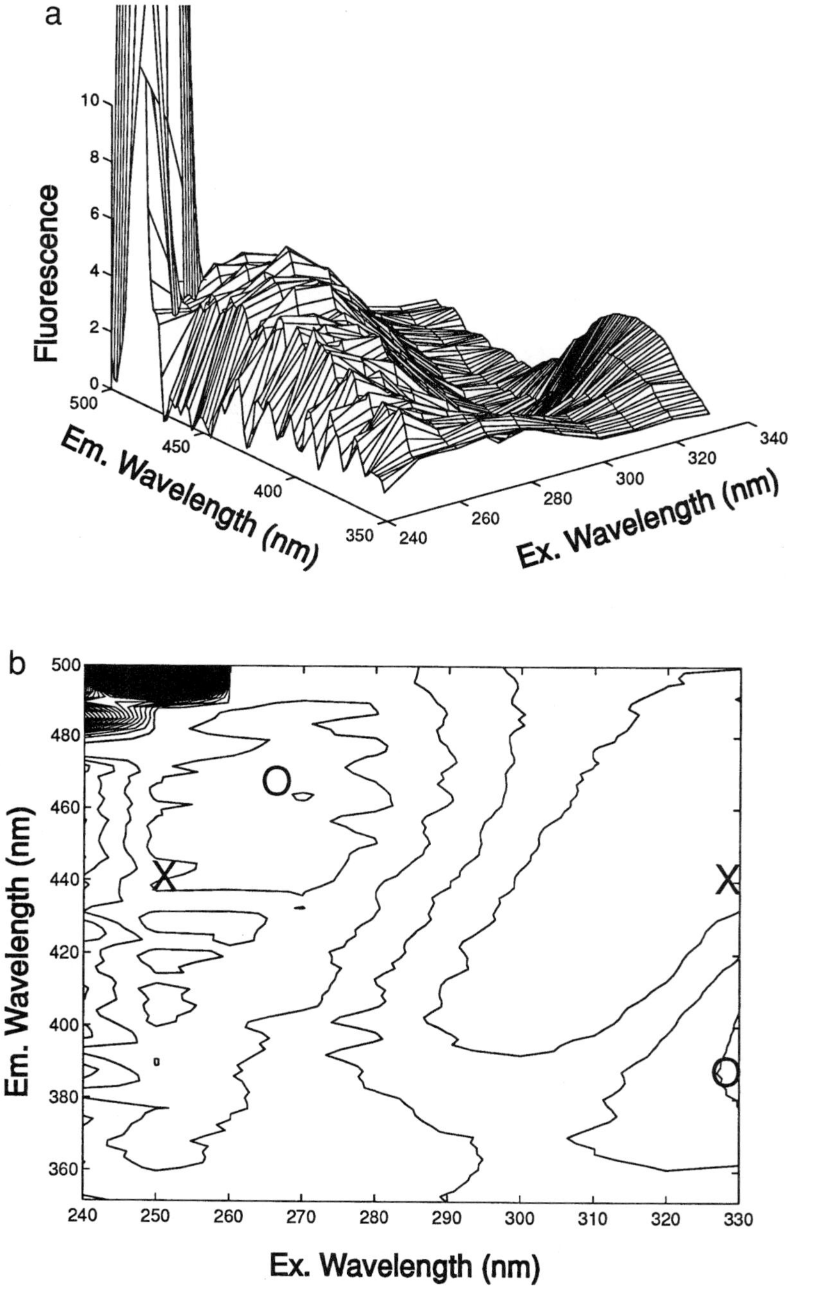

Figure 8 Excitation versus emission difference surface for Suwannee River fulvic acid with Al (0.06 mmol·L^{-1}) minus Suwannee River fulvic acid without Al at pH 4.0. Plot (a) is a perspective drawing and plot (b) is a contour map. In (b) the peaks in the fulvic acid spectrum (×) and in the difference surface (○) are indicated.

difference spectrum, one at the 270-nm excitation and 460-nm emission wavelengths, and one at the 330-nm excitation and 380-nm emission wavelengths.

C. Fulvic Acid Excitation at 270 nm

The parameter fitting was performed over the entire set of emission spectra for excitation at 270 nm. The parameters are reported for the regions of maximum fluorescence increase in Table II; because no significant quenching was observed, only enhancement data were interpreted.

The logarithm of the conditional stability constants for reaction with deprotonated ligand, reaction (I), compare well for both Eq. [2] and Eq. [12], 5.7 versus 5.3, respectively, but Eq. [7] yields a value over an order of magnitude smaller. The agreement for the alternate reaction (reaction (II)) is just over an order of magnitude different, 2.6 versus 1.3 for Eq. [14] and Eq. [15], respectively, and is lower (0.4) for Eq. [8]. The fluorescence emission peak and intensity are essentially identical for both equations. The ligand concentration of 3.4 $\mu mol \cdot L^{-1}$ corresponds to 0.43 $\mu mol \cdot mg^{-1}$ of fulvic acid.

D. Fulvic Acid Excitation at 330 nm

The parameter fitting was performed over the entire set of emission spectra for excitation at 330 nm. The parameters are reported in Table III, for the region of maximum fluorescence increase; because no significant quenching was observed, only enhancement data were interpreted.

The parameter fitting for the 330-nm excitation wavelength yielded the same values for all parameters within experimental uncertainty. The ligand concentration of 0.15 $\mu mol \cdot L^{-1}$ corresponds to 0.019 $\mu mol \cdot mg^{-1}$ of fulvic acid.

Table II

Summary of Results for Excitation at 270 nm of Suwannee River Fulvic Acid

Equation[a]	Reaction (I) log K′	Reaction (II) log K′	F_{ML} (arbitrary units)	L_T ($\mu mol \cdot L^{-1}$)
2,14,E	5.7 ± 0.1	2.6 ± 0.1	23	3.4 ± 0.7
12,15,E	5.3 ± 0.1	1.3 ± 0.1	25	na
7,13,E	4.4 ± 0.9	0.4 ± 0.9	na	na

[a]The equation numbers refer to equations in the text, and E refers to enhancement measurements.

Table III
Summary of Results for Excitation at 330 nm of Suwannee River Fulvic Acid

Equation[a]	Reaction (I) log K′	Reaction (II) log K′	F_{ML} (arbitrary units)	L_T ($\mu mol \cdot L^{-1}$)
2,14,E	5.3 ± 0.3	1.3 ± 0.3	17	0.15 ± 0.05
12,15,E	5.3 ± 0.3	1.3 ± 0.3	17	na
7,13,E	4.9 ± 0.9	0.9 ± 0.9	na	na

[a]The equation numbers refer to equations in the text, and E refers to enhancement measurements.

V. CONCLUSIONS

The two fluorophores in Suwannee River fulvic acid reported by Goldberg and Weiner (1994) both react with Al, and their concentrations and stability constants can be estimated from TLS performed during a titration with Al at fixed pH. Information about these fluorophores, and salicylic acid for comparison, is reported in Table IV.

The conditional stability constant of both groups is very close to the value for salicylic acid. Therefore, it is likely that the binding sites are like the bidentate, 1-carboxyl-2-hydroxyl group in salicylate. The fluorophore labeled as site 2 is actually similar to salicylic acid in its fluorescent properties as well. The shift in wavelength on adding Al is the same for both salicylic acid and Suwannee River fulvic acid, but the fulvic acid is excited and emits light of 30 nm longer wavelength then free salicylate.

The values determined for log K′ in this study, in the range 5.1 to 5.6, compare within a log unit with conditional stability constants obtained in other studies performed around pH 4.0. Pott *et al.* (1985), using a cation exchange method and

Table IV
Summary of Two Al Reactive Sites in Suwannee River Fulvic Acid

Ligand	No Al (ex/em)[a]	With Al (ex/em)[a]	log K′ for reaction (I)	L_T ($\mu mol \cdot mg^{-1}$)
Site 1	250/440	270/450	5.6	0.43
Site 2	330/440	330/380	5.1	0.018
Salicylic acid	300/410	300/380	5.5	na

[a]The region of the excitation versus emission surface is indicated by the wavelength couple in nanometers.

Aldrich humic acid, obtained a log K′ of around 6.8, and studies using bog water and a potentiometric method (Lövgren *et al.*, 1987) in the pH range 3.0 to 4.2 yielded a log K′ between 4.1 and 4.7. The TLS study of Luster *et al.* (1996) resolved two sites in a juniper leaf litter extract at pH 5 with log K′ values of 5.79 and 8.05. The lower strength site is similar in strength to salicylic acid but the other site is two orders of magnitude stronger than any sites observed in this study. The calculated L_T for Suwannee River fulvic acid can be accounted for within the total carboxyl group content of 6 μmol·mg^{-1} (Leenheer *et al.*, 1995); this corresponds to 7.5% of the total carboxyl content being sites available for Al.

The modified equations derived in this study reinforce the Ryan–Weber equation's validity by obtaining similar values. The modified equations present no advantages over the usual equation, except modification for reaction (II) yields stability constants for reaction with protonated ligand. The Stern–Volmer equation modified for fluorescence enhancement yields K′ values surprisingly consistent, within less than an order of magnitude, with those obtained using the Ryan–Weber equation. The usual Stern–Volmer equation applied to fluorescence quenching is not useful in Al–DOM systems because the data are not linear, and therefore do not yield reasonable results.

Fitting the three parameters in the Ryan–Weber equation across the entire emission spectrum for salicylic acid demonstrated that for one fluorophore, the fitting results do not depend on whether quenching or enhancement data are used as long as the fluorescences of the metal–ligand and ligand species do not coincide. It is important in multiple-fluorophore systems that the data for fitting are taken where the fluorescence is only influenced by changes in one site.

ACKNOWLEDGMENTS

The authors acknowledge an Ontario Graduate Scholarship to D. S. Smith and funding from the National Science and Engineering Research Council of Canada.

REFERENCES

Belin, C., Quellec, C., Lamotte, M., Ewald, M., and Simon, P. H. 1993. Characterization by fluorescence of the dissolved organic matter in natural water. Application to fractions obtained by tangential ultrafiltration and XAD resin isolation. *Environ. Technol.* 14:1131–1144.

Bloom, P. R., and Erich, M. S. 1989. The quantitation of aqueous aluminum. *In* "The Environmental Chemistry of Aluminum" (G. Sposito, Ed.), p. 1–28. CRC Press, Boca Raton, FL.

Cabaniss, S. E. 1992. Synchronous fluorescence spectra of metal-fulvic acid complexes. *Environ. Sci. Technol.* 26:1133–1139.

Cabaniss, S. E., and Shuman, M. S. 1988. Fluorescence quenching measurements of copper-fulvic acid binding. *Anal. Chem.* 60:2418–2421.

Campanella, L., Petronio, B. M., and Braguglia, C. 1994. Study of humic fractions from water of an antarctic lake. *Int. J. Environ. Anal. Chem.* 60:49–60.

Christman, R. F., and Gjessing, E. T. (Eds.). 1983. "Aquatic and Terrestrial Humic Materials." Ann Arbor Science, Ann Arbor, MI.

Clarke, N. 1994. Speciation of aluminum and iron in natural fresh waters. Ph.D. Thesis, Dept. of Chem., Div. of Anal. Chem., The Royal Institute of Technology, Stockholm, Sweden.

Coble, P. G., Green, S. A., Blough, N. V., and Gagosian, R. B. 1990. Characterization of dissolved organic matter in the Black Sea by fluorescence spectroscopy. *Nature* 348:432–435.

Cook, R. L., and Langford, C. H. 1995. Metal ion quenching of fulvic acid fluorescence intensities and lifetimes: Nonlinearities and a possible three-component model. *Anal. Chem.* 67:174–180.

Djurdjevic, P. T., Jelikic-Stankov, M., and Stankov, D. 1995. Fluorescence reaction and complexation equilibria between norfloxacin and aluminum(III) in chloride medium. *Anal. Chim. Acta* 300:253–259.

Dobbs, A. J., French, P., Gunn, A. M., Hunt, D. T. E., and Winnard, D. A. 1989. Aluminum speciation and toxicity in upland waters. *In* "Environmental Chemistry and Toxicology of Aluminum" (T. E. Lewis, Ed.), pp. 209–228. Lewis, Chelsea, MI.

Frimmel, F. H., Christman, R. F., and Bracewell, J. M. (Eds.). 1988. "Humic Substances and Their Role in the Environment." Wiley, New York.

Ghosh, K., and Schnitzer, M. 1981. Fluorescence excitation spectra of humic substances. *Soil Sci. Soc. Am. J.* 45:25–29.

Gjessing, E. T., Riise, G., Petersen, R. C., and Andruchow, E. 1989. Bioavailability of aluminum in the presence of humic substances at low and moderate pH. *Sci. Total Environ.* 81/82:683–690.

Goldberg, M. C., and Negomir, P. M. 1989. Characterization of aquatic humic acid fractions by fluorescence depolarization spectroscopy. *In* "Luminescence Applications in Biological, Chemical, Environmental and Hydrological Sciences," pp. 180–205, "Am. Chem. Symp. Ser." (M. C. Goldberg, Ed.), No. 383. Am. Chem. Soc., Washington, DC.

Goldberg, M. C., and Weiner, E. R. 1993. Fluorescence spectroscopy in environmental and hydrological sciences. *In* "Fluorescence spectroscopy: New Methods and Application" (O. S. Wolfbeis, Ed.), pp. 213–241. Springer-Verlag, New York.

Goldberg, M. C., and Weiner, E. R. 1994. Fluorescence measurements of the volume, shape, and fluorophore composition of fulvic acid from the Suwannee River. *In* "Humic Substances in the Suwannee River, Georgia: Interactions, Properties and Proposed Structure" (R. C. Averett, J. A. Leenheer, D. M. McKnight, and K. A. Thorn, Eds.), Water Supply Paper 2373, U.S. Geological Survey, Reston, VI.

Guilbault, G. G. 1990. "Practical Fluorescence," 2nd ed. Dekker, New York.

Hue, H. V., Craddock, G. R., and Adams, F. 1986. Effect of organic acids on aluminum toxicity in subsoils. *Soil Sci. Soc. Am. J.* 50:28–34.

Kramer, J. R., Hummel, J., and Gleed, J. 1986. Speciation of aluminum and its toxicity to fish. *In* "Proceedings of the International Conference on Chemistry and the Environment" (J. N. Lester, R. Perry, and R. M. Sterritt, Eds.), pp. 636–641. Selper Ltd., London.

Lakshman, S., Mills, R., Fang, F., Patterson, H., and Cronan, C. 1996. Use of fluorescence polarization to probe the structure and aluminum complexation of three molecular weight fractions of a soil fulvic acid. *Anal. Chim. Acta* 321:113–119.

Lapen, A. J., and Seitz, W. R. 1982. Fluorescence polarization studies of the conformation of soil fulvic acid. *Anal. Chim. Acta* 134:31–38.

Leenheer, J. A., Wershaw, R. L., and Reddy, M. M. 1995. Strong-acid, carboxyl-group structures in fulvic acid from the Suwannee River, Georgia. 1. Minor structures. *Environ. Sci. Technol.* 29:393–398.

Lövgren, L., Hedlund, T., Öhman, L. O., and Sjöberg, S. 1987. Equilibrium approaches to natural water systems. 6. Acid-based properties of a concentrated bog-water and its complexation reactions with aluminum(III). Water Res. 21:1401–1407.

Luster, J., Lloyd, T., and Sposito, G. 1994. Aluminum(III) complexation by an aqueous leaf litter extract: Quantitative characterization by molecular fluorescence spectrometry. *In* "Humic Substances in the Global Environment and Implications on Human Health" (N. Senesi and T. M. Miano, Eds.), pp. 335–342. Elsevier, Amsterdam.

Luster, J., Lloyd, T., and Sposito, G. 1996. Multi-wavelength molecular fluorescence spectrometry for quantitative characterization of copper(II) and aluminum(III) complexation by dissolved organic matter. *Environ. Sci. Technol.* 30:1565–1574.

Motekaitis, R. J., and Martell, A. E. 1984. Complexes of aluminum(III) with hydroxy carboxylic acids. *Inorg. Chem.* 23:18–23.

Murov, S. 1973. "Handbook of Photochemistry." Dekker, New York.

Öhman, L.-O. 1988. Equilibrium and structural studies of silicon(IV) and aluminum(III) in aqueous solution. 17. Stable and metastable complexes in the system H^+-Al^{3+}-citric acid. *Inorg. Chem.* 27:2565–2568.

Öhman, L.-O. 1991. Equilibrium and structural studies of silicon(IV) and aluminum(III) in aqueous solution. 27. Al^{3+} complexation to monocarboxylic acids. *Acta Chem. Scand.* 45:258–264.

Öhman, L.-O., and Sjöberg, S. 1983. Equilibrium and structural studies of silicon(IV) and aluminum(III) in aqueous solution. 10. A potentiometric study of aluminum(III) pyrocatecholates and aluminum hydroxo pyrocatecholates in 0.6M Na(Cl). *Polyhedron* 2:1329–1335.

Öhman, L.-O., and Sjöberg, S. 1988. Thermodynamic calculations with special reference to the aqueous aluminum system. *In* "Metal Speciation: Theory, Analysis and Application" (J. R. Kramer and H. F. Allen, Eds.), pp. 1–33. Lewis, Chelsea, MI.

Orvig, C. 1993. The aqueous coordination chemistry of aluminum. *In* "Coordination Chemistry of Aluminum" (G. H. Robinson, Ed.), pp. 85–121. VCH Publishers, New York.

Parker, D. R., Zelazny, L. W., and Kinraide, T. B. 1989. Chemical speciation and plant toxicology of aqueous aluminum. *In* "Environmental Chemistry and Toxicology of Aluminum" (T. E. Lewis, Ed.), pp. 209–228. Lewis, Chelsea, MI

Patterson, H. H., Cronan, C. S., Lakshman, S., Plankey, B. J., and Taylor, T. A. 1992. Comparison of soil fulvic acids using synchronous scan fluorescence spectroscopy, FTIR, titration and metal complexation kinetics. *Sci. Total Environ.* 113:179–196.

Plankey, B. J., and Patterson, H. H. 1987. Kinetics of aluminum-fulvic acid complexation in acidic waters. *Environ. Sci. Technol.* 21:595–601.

Plankey, B. J., Patterson, H. H., and Cronan, C. S. 1986. Kinetics of aluminum-salicylic acid complexation. *Environ. Sci. Technol.* 20:160–165.

Plankey, B. J., Patterson, H. H., and Cronan, C. S. 1995. Kinetic analysis of aluminum complex formation with different soil fulvic acids. *Anal. Chim. Acta* 300:227–236.

Pott, D., Alberts, J. J., and Elzerman, A. W. 1985. The influence of pH on the binding capacity and conditional stability constants of aluminum and naturally occurring organic matter. *Chem. Geol.* 48:293–304.

Power, J. F., LeSage, R., Sharma, D. K., and Langford, C. H. 1986. Fluorescence lifetimes of the well characterized humic substance, Armadale fulvic acid. *Environ. Technol. Lett.* 7:425–430.

Provenzano, M. R., and Sposito, G. 1994. Application of two-dimensional fluorescence spectroscopy to the study of pine litter in different ecosystems. *In* "Humic Substances in the Global Environment and Implications on Human Health" (N. Senesi and T. M. Miano, Eds.), pp. 335–342. Elsevier, Amsterdam.

Ryan, D. K., and Ventry, L. S. 1990. Exchange of comments on fluorescence quenching measurements of copper fulvic acid binding. *Anal. Chem.* 62:1523–1526.

Ryan, D. K., and Weber, J. H. 1982. Fluorescence quenching titration for determination of complexing capacities and stability constants of fulvic acid. *Anal. Chem.* 54:986–990.

Senesi, N. 1990. Molecular and quantitative aspects of the chemistry of fulvic acid and its interactions with metal ions and organic chemicals. Part II. The fluorescence spectroscopy approach. *Anal. Chim. Acta* 232:77–106.

Senesi, N., and Miano, T. M. (Eds.). 1994. "Humic Substances in the Global Environment and Implications for Human Health." Elsevier, Amsterdam.

Shotyk, W., and Sposito, G. 1988. Fluorescence quenching and aluminum complexation by a chestnut leaf litter extract. *Soil Sci. Soc. Am. J.* 52:1293–1297.

Shotyk, W., and Sposito, G. 1990. Ligand concentration effects on aluminum complexation by a chestnut leaf litter extract. *Soil Sci. Soc. Am. J.* 54:933–935.

Snell, F. D. 1978. "Photometric and Fluorometric Methods of Analysis of Metals, Part 1." Wiley, New York.

Stern, O., and Volmer, M. 1919. The extinction period of fluorescence. *Phys. Z.* 20:183–188.

Stevenson, F. J., and Vance, G. F. 1989. Naturally occurring aqueous aluminum-organic complexes. *In* "The Environmental Chemistry of Aluminum" (G. Sposito, Ed.), pp. 117–146. CRC Press, Boca Raton, FL.

Tam, S.-C., and Sposito, G. 1993. Fluorescence spectroscopy of aqueous pine litter extracts: Effects of humification and aluminum complexation. *J. Soil Sci.* 44:513–524.

Yang, A., Sposito, G., and Lloyd, T. 1994 Total luminescence spectroscopy of aqueous pine litter (O horizon) extracts: Organic ligands and their Al or Cu complexes. *Geoderma* 62:327–344.

Chapter 22

Modeling of Competitive Ion Binding to Heterogeneous Materials with Affinity Distributions

Michal Borkovec,[1] Ursula Rusch,[1] and John C. Westall[2]
[1]Institute of Terrestrial Ecology, Swiss Federal Institute of Technology, Switzerland;
[2]Department of Chemistry, Clarkson University, Potsdam, New York

Affinity distribution functions offer a promising approach for describing the binding of simple ions to heterogeneous environmental materials. However, applications of this approach have generally been limited to noncompetitive binding in systems in which the concentration of only one component is varied. In this chapter, we discuss the extension of this approach to competitive binding in systems in which concentrations of two components are varied simultaneously. We show that experimental binding data can usually be rationalized by simple affinity distributions that are functions of a single variable. These distributions are based on different types of exchange sites and noncompetitive Langmuir sites; competitive Langmuir sites, which would require affinity distributions that would be functions of two variables, are not necessary. The present framework has, in spite of its empirical character, substantial predictive capabilities and appears to be very useful for comparison of binding properties of different materials. The flexibility of this approach is illustrated through analysis of competitive binding data for protons and various metals on substrates such as humic acids and oxide surfaces.

Adsorption of Metals by Geomedia

I. INTRODUCTION

Models of ion binding processes are essential for the interpretation of a wide variety of environmental phenomena such as transport of ions in natural porous media or scavenging of metal ions in surface waters by sedimenting particles (Stumm and Morgan, 1996). Such models of ion binding are commonly based on laboratory partitioning experiments, in which one measures the amounts of all components of interest bound to the separable solid matrix as a function of their free concentrations (activities). Due to extensive efforts in the past, there has been much progress in the quantitative understanding of ion binding processes in better defined systems such as proteins, polyelectrolytes and oxides (Hiemstra *et al.*, 1989; Bashford and Karplus, 1990; Smits *et al.*, 1993; Sverjensky, 1993). However, the extension of these findings to actual environmental materials such as humic substances, soil particles, and aquifers materials remains very difficult. These difficulties are related to the heterogeneity of these materials and to the lack of detailed structural information on the molecular level.

In this chapter we present a promising modeling approach that is based on the representation of the heterogeneous material as a collection of independent binding sites. The application of modern numerical techniques (constrained, regularized least-squares) makes the construction of such models from experimental data a simple and quasi-automatic process. While such an empirical modeling approach does not provide much insight into the molecular-level mechanism of the binding process, its simplicity and flexibility make it an appealing alternative to existing modeling approaches. If such an empirical model is calibrated with appropriate experimental data, in most applications (e.g., transport modeling) its performance will be equivalent to a binding model with a more mechanistic basis.

The experimental data sets that we consider can be classified according to their dimensionality, that is, the number of components whose free concentrations are varied in the course of the experiment. For example, a classical acid–base titration curve at fixed salt concentration would yield a one-dimensional data set, while a collection of these titration curves at various salt levels would yield a two-dimensional data set. Such one- and two-dimensional data sets are most common. Additional examples of such two-dimensional data sets include metal cation binding as a function of pH and classical ion exchange data for clays with a pair of major cations. Data sets in which concentrations of three or more components are varied are less common. The dimensionality of a data set has a great influence on the structure of the binding models that are needed to rationalize the data set. After a brief review of the classical case in which a single concentration is varied, we shall discuss the extension of the empirical binding models to cases in which two concentrations are varied simultaneously. The work presented here builds on our earlier work in this area; for more details on model definition and development, see Černík *et al.* (1995a) and Černík *et al.* (1996).

II. VARIABLE CONCENTRATION FOR A SINGLE COMPONENT

The basic idea behind the present approach dates far back. Langmuir (1918) proposed modeling gas adsorption isotherms in terms of a collection of independent binding sites. Essentially the same approach was applied more recently to model acid–base titration curves at fixed salt concentrations or other ion binding data with a single variable concentration (Leuenberger and Schindler, 1986; Brassard *et al.*, 1990; Nederlof *et al.*, 1992; Stanley and Guiochon, 1993; Černík *et al.*, 1995a); computer programs are available to solve the inversion problem (Černík *et al.*, 1995b). Since the developments involving a single variable concentration are central to our later discussion of the case of two variable concentrations, let us briefly review the basic principles.

A. Independent Binding Sites

Our approach will be illustrated with an example. Consider the acid–base titration data of purified peat humic acid in 0.1 mol·L^{-1} KNO_3 as reported by Milne *et al.* (1995). The experimental data points are shown in Figure 1A. Such a titration curve can be interpreted as the result of binding of H^+ to independent binding sites. The binding process at one of these sites can be described with the phenomenological reaction (Černík *et al.*, 1995a)

$$H + X_i \rightleftharpoons HX_i, \quad [1]$$

where X_i represents hypothetical binding sites to which the proton binds with an affinity constant K_i. Development of the model then amounts to determining the set of sites i, and the associated affinity constants and total site concentrations, that describes the experimental data. Electrostatic effects are not expressed explicitly in these models, and we have dropped the ionic charges for simplicity. Electrostatics do enter implicitly, of course, through the affinity constants that are derived. Clearly the constraints of electroneutrality must apply to the systems that we consider, but we make no effort to specify the magnitude of the charge associated with the heterogeneous material itself nor the location and charge of various co- and counterions.

The mass-action and mass-balance laws, which follow from Eq. [1], lead to a Langmuir adsorption isotherm (Černík *et al.*, 1995a). Therefore, we shall refer to these sites as Langmuir sites. The experimental titration data shown in Figure 1A are now interpreted as a combination of such sites. The mathematical optimization problem is set up with a discrete distribution of affinity constants K_i for the reaction in Eq. [1] (i.e., log K_i = 2.0, 2.5, 3.0, . . . , 12.0), and solved by determining the corresponding total site concentrations, subject to the constraint that the total

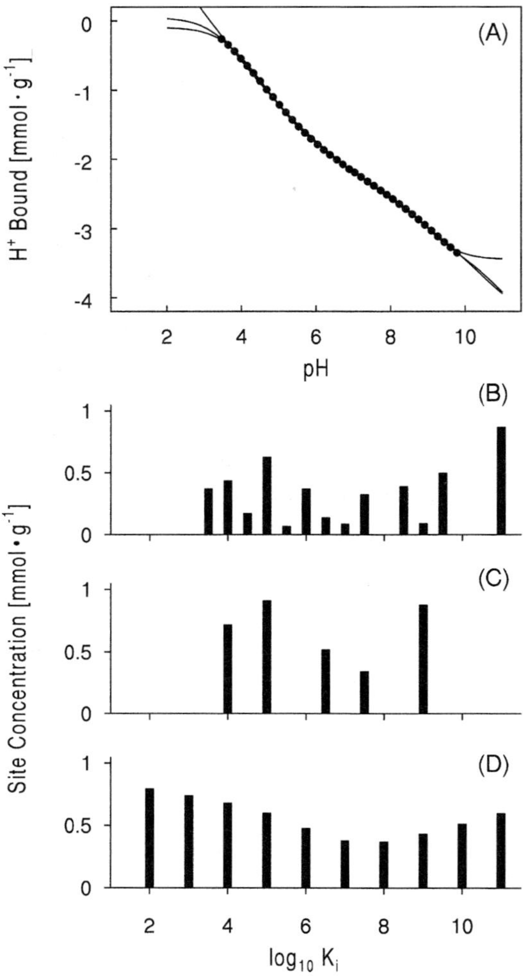

Figure 1 Acid–base titration of peat humic acid in 0.1 mol·L^{-1} KNO_3 and a humic acid concentration of ~4 g·L^{-1}. (A) Experimental data points are taken from Milne *et al.* (1995). Lines represent best fits with the multisite binding model with different regularization methods. Corresponding affinity distributions are given below: (B) no regularization, (C) regularization for a small number of sites, and (D) regularization for smoothness. While the affinity distributions have rather different appearances, the fits of the experimental data shown in (A) are similar for all three distributions. For (B) and (C) the initial grid included two sites per decade, and for (D), one site per decade (Černík *et al.*, 1996).

site concentrations are either positive or zero. Such optimization techniques are now readily available and can be easily implemented (Press *et al.*, 1992; Černík *et al.*, 1995b). The resulting proton binding isotherm also involves an additive constant, which is related to the initial degree of protonation; this constant is treated as an additional adjustable parameter. The result of such an optimization calculation is shown in Figure 1B. We can see that the algorithm has selected 13 sites to represent the data, and the total concentrations of the remaining ones are set to zero. As evident from Figure 1A, the fit of the data with this model is excellent. If one plots the residuals, no trends are apparent.

However, the resulting affinity distribution shown in Figure 1B looks rather "noisy" and thus may not be considered to be a completely satisfactory result. Moreover, a more serious disadvantage of this inversion scheme is not immediately obvious—if one would repeat the analysis with data from a second experimental run, which would deviate only slightly from the data shown here, the resulting affinity distribution could turn out to be rather different from the one shown here. In other words, this numerical procedure is unstable (Press *et al.*, 1992).

B. Regularization

The problem just discussed arises because the precise form of the affinity distribution is extremely sensitive to small changes in the experimental data. This problem can be partly circumvented by specifying how the resulting distribution is supposed to appear. This objective is achieved with an additional mathematical technique, so-called regularization (Press *et al.*, 1992). Such regularization methods allow the affinity distribution to be tuned toward the desired properties, but in a way that the good fit of the data is effectively maintained. Any regularization will stabilize the numerical inversion procedure and an analysis of a second experimental run will give results very similar to the analysis of the first.

Let us skip the technical details, which are discussed by Černík *et al.* (1995a), and describe two regularization methods that appear most promising in the present context. The first method is regularization for *small number of sites,* in which we decrease the number of sites as long as the fit of the experimental data remains satisfactory. Applied to the titration curve shown in Figure 1A, the resulting affinity distribution is shown graphically in Figure 1C. As is obvious from Figure 1A, the titration curve can be represented with just 5 sites essentially as accurately as by the previous model with 13 sites. Since this model contains a small number of sites it can be easily summarized in a tabular form (see Table I). This representation is interesting if one would like to parameterize a given binding isotherm with a small number of adjustable parameters but in a fashion compatible with common solution speciation schemes (Allison *et al.*, 1991). Such a model is easily used in a transport calculation, for example.

Table I

Multisite Binding Models Regularized for Small Number of Sites for Data Sets Discussed in the Text

Figure	Material	Reactions	$\log_{10}K^a$	$s(\mathrm{mmol\cdot g^{-1}})^b$
1	Humic acid[c]	$H + X_i \rightleftharpoons HX_i$	4.000	0.714
			5.000	0.908
			6.500	0.517
			7.500	0.337
			9.000	0.876
2	Humic acid[d]	$H + X_i \rightleftharpoons HX_i$	4.584	0.359
			5.965	0.145
			7.955	0.114
			9.534	0.220
		$H + KY_i \rightleftharpoons K + HY_i$	3.055	0.844
			4.675	0.634
			6.483	0.490
			8.233	0.685
4	Rutile[e]	$H + X_i \rightleftharpoons HX_i$	5.089	0.043
			6.572	0.041
			7.999	0.071
		$H + CdY_i \rightleftharpoons Cd + HY_i$	2.953	0.021
			4.488	0.109
			7.700	14.76
		$2H + CdZ_i \rightleftharpoons Cd + H_2Z_i$	7.154	0.038
			8.986	0.057
			10.99	0.055
			14.70	13.63

[a]In the mass-action laws, activities of solution species are given in $\mathrm{mol\cdot L^{-1}}$, while activities of the bound species are taken as mole fractions.
[b]Total site concentrations in $\mathrm{mmol\cdot g^{-1}}$.
[c]Additive constant of -3.445 $\mathrm{mmol\cdot g^{-1}}$ was added to the proton isotherm.
[d]Additive constant of -3.636 $\mathrm{mmol\cdot g^{-1}}$ was added to the proton isotherm.
[e]Additive constant of -42.477 $\mathrm{mmol\cdot g^{-1}}$ was added to the proton isotherm.

Another interesting possibility is regularization for *smoothness*. Such smooth distributions might appeal to our intuitive picture of the properties of the material. The result is shown in Figure 1D. Since we have used a rather wide grid of the logarithms of the affinity constants, the number of parameters involved in the model is still reasonable but, of course, larger than in the case of regularization for small number of sites. Note that the present distribution looks somewhat different from those derived in previously proposed inversion schemes (Nederlof *et al.*, 1992; Stanley and Guiochon, 1993; Černík *et al.*, 1995a). In the present case, we do not require that the distribution tend to zero outside the experimental window. The advantage of this modification of the regularization for smoothness is that such affinity distributions are optimally stable even upon changes of the size of the experimental data window. This aspect is very important if one would like to compare such affinity distributions originating from experimental data of different sources. We note that both regularization schemes (i.e., for small number of sites and for smoothness) effectively maintain the good quality of the fit.

III. VARIABLE CONCENTRATIONS FOR TWO COMPONENTS

The previous example pertained to a one-dimensional data set. However, multicomponent effects are often important, and numerous experimental binding data involve two or more concentrations that are varied simultaneously. Therefore, it is essential to extend these methods to a larger number of components. Here, we shall discuss the first necessary step in this direction: the case of two components with variable concentrations (Černík *et al.*, 1996).

A. Multisite Binding Models

Let us again consider the acid–base titration of the humic acid by Milne *et al.* (1995) However, we now consider six titration curves at different KNO_3 concentrations between 0.001 and 0.3 mol·L^{-1} simultaneously. The entire experimental data set is shown in Figure 2A; the titration curve at 0.1 mol·L^{-1} was considered separately in the previous section.

The extension of the approach to two variable concentrations is based on the idea that Eq. [1] is still used to model binding of H^+ and that the effect of salt on H^+ binding can be viewed as competition between H^+ and K^+. The simplest model of a competitive reaction is an ion exchange reaction. Thus, let us represent the competitive processes in the case of the salt-dependent acid–base titration curves as exchange between H^+ and K^+. Recall that an ion exchange reaction involves two ions but only a single equilibrium constant. This observation simplifies the situation for the case of two variable concentrations substantially; the resulting affinity distributions will remain a function of a single affinity constant.

These ideas can indeed be used for the quantitative interpretation of the experimental titration data shown in Figure 2A. In the simplest model we consider two types of sites. The first type of ion binding sites obey the reaction

$$H + X_i \rightleftharpoons HX_i \quad [2]$$

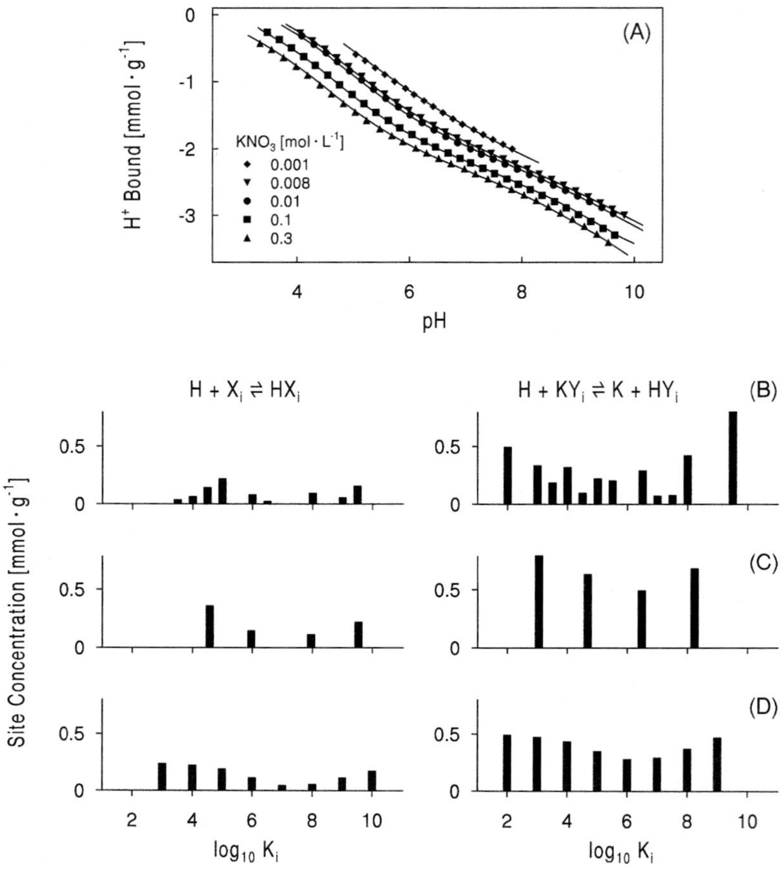

Figure 2 Acid–base titrations of peat humic acid at various concentrations of KNO_3 and a humic acid concentration of ~4 $g \cdot L^{-1}$. (A) Experimental data points are taken from Milne *et al.* (1995). Solid lines represent the best fit with the multisite binding model. Pairs of affinity distributions for the corresponding types of sites are given below: (B) no regularization, (C) regularization for a small number of sites, and (D) regularization for smoothness. The fit of the data in (A) is shown for the affinity distribution in (B); the corresponding fits based on distributions in (C) and (D) are virtually identical. For (B) a fixed grid with two sites per decade was used; for (C) a floating grid with a minimum interval of one decade; and for (D) a fixed grid with one site per decade (Černík *et al.,* 1996).

where the Langmuir site X_i binds no other ions, as already discussed. The second type of binding sites obey a simple one-to-one exchange reaction

$$H + KY_i \rightleftharpoons K + HY_i \qquad [3]$$

where Y_i denotes a one-to-one exchange site. As before, we have dropped the ionic charges for simplicity. We now fit the experimental data in a similar way as in the previous section; activity corrections were made according to the method of Pitzer (1991). The result of the data inversion without any regularization is shown in Figure 2B. Again the distributions look "noisy," but the result is very encouraging. As shown in Figure 2A, an excellent fit can be obtained with this model. This two-dimensional data set can indeed be represented with a combination of Langmuir sites and exchange sites. The major advantage of this representation is that the competitive nature of the ion binding process can be represented in terms of two affinity distributions, each of which is a function of a single affinity constant only.

For the same reasons previously discussed, we introduce regularization. The affinity distributions regularized for small number of sites are shown in Figure 2C. The model is also summarized in Table I. The model contains four Langmuir sites and four exchange sites. Alternatively, we can regularize for smoothness, which yields the distributions shown in Figure 2D. Fits of the experimental data by the regularized models (not shown) are essentially equivalent to the fit for the nonregularized model shown in Figure 2A.

This example has shown that the extension of the affinity distribution description to cases with two variable concentrations is indeed possible and provides an interesting modeling tool for competitive ion binding equilibria. The appealing aspect of this modeling framework is its flexibility; many different experimental data sets can be rationalized in this fashion. The additional examples to be discussed also illustrate this aspect.

B. Developing a "Fingerprint" of Binding Properties

A simple yet important application of any ion binding model is to parameterize the binding properties of a given material in a coherent fashion and provide a comprehensive means for comparing the binding capacities and intensities between various materials. The flexibility and stability of the multisite models discussed here make them rather promising for this task. Particularly the affinity distributions regularized for smoothness provide a unique fingerprint of the ion binding properties of the materials considered.

In the case where only a single concentration is being varied, one may argue that the experimental record of the binding isotherm already represents a sufficient representation of the material binding properties and that a transformation of the data to the affinity distribution is not really necessary. One may indeed just compare the titration curves instead of the affinity distributions. However, in the case of two variable concentrations the situation is already rather different. The experimental binding data can be given in entirely different representations; their direct comparison can turn out to be very difficult, if not impossible. In this case, the affinity distributions provide a very useful means for comparison.

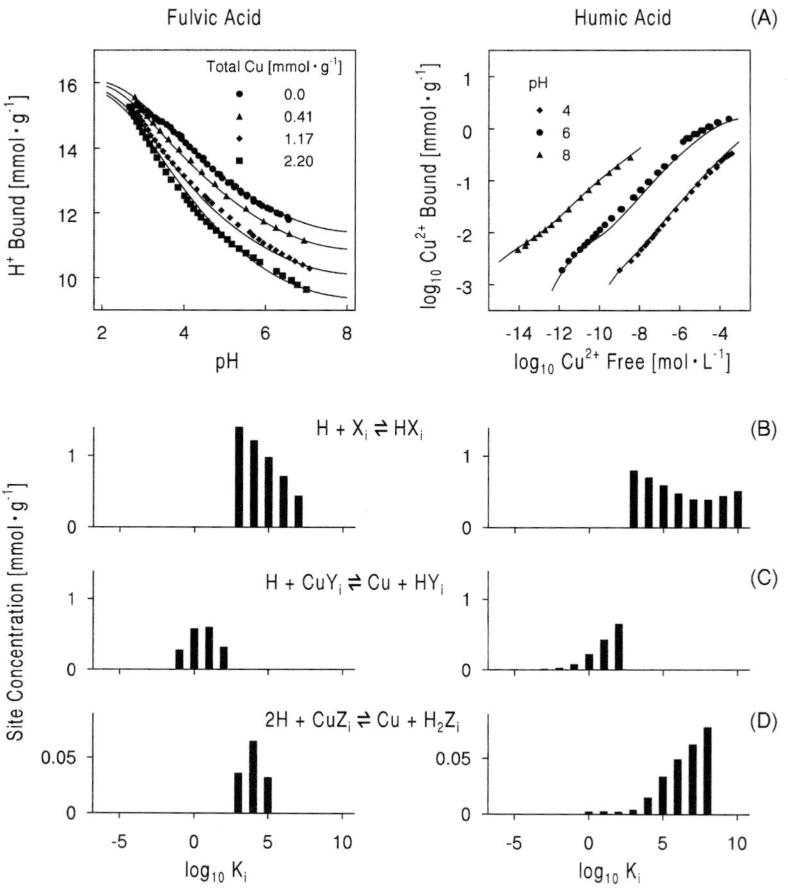

Figure 3 Comparison of binding properties for Cu^{2+} and H^+ in 0.1 mol·L^{-1} KNO_3 between a fulvic acid (left column) and a humic acid (right column). (A) Experimental data (points) together with the fit with the multisite binding model (solid lines). For the fulvic acid, the bound amount of H^+ is given as a function of pH at various total Cu^{2+} concentrations and a fulvic acid concentration of ~1.5 g·L^{-1} (Ephraim *et al.*, 1986; Ephraim and Marinsky, 1986). For the humic acid, the amount of Cu^{2+} bound is given as a function of free Cu^{2+} concentration in solution for different pH values and a humic acid concentration of 5.3 g·L^{-1} (Milne *et al.*, 1995; Benedetti *et al.*, 1995). The corresponding affinity distributions for three types of sites are given below: (B) Langmuir sites for protons, (C) one-to-one exchange sites, and (D) one-to-two exchange sites. Regularization for smoothness was performed on a fixed grid with one site per decade (Černík *et al.*, 1996).

Let us illustrate this aspect with an example of competitive binding of H^+ and Cu^{2+}. The aim is to compare the binding properties of these ions between a fulvic acid and a humic acid. We use experimental binding data in 0.1 mol·L^{-1} KNO_3 and consider the two experimental data sets shown in Figure 3A. For the fulvic acid, the amount of H^+ bound is given as a function of pH for different total Cu^{2+} concentrations (Ephraim *et al.,* 1986; Ephraim and Marinsky, 1986). For the humic acid, we know Cu^{2+} bound as a function of free Cu^{2+} at various fixed pH values (Benedetti *et al.,* 1995). The data set for the humic acid is augmented by the titration curve shown in Figure 1A. While both experimental data sets contain essentially all the necessary information about binding of H^+ and Cu^{2+}, the data are given in entirely different forms. Thus, direct comparison of these data sets is basically impossible.

The comparison of these two data sets becomes straightforward with the multisite model. This model can rationalize both data sets. The only difference to the previous discussion is that three types of sites are needed in this case. The first type obeys the reactions

$$H + K_i \rightleftharpoons HX_i, \qquad [4]$$

where X_i are specific Langmuir sites. The second type of sites obeys

$$H + CuY_i \rightleftharpoons Cu + HY_i, \qquad [5]$$

where Y_i are one-to-one exchange sites. Finally, the third type of sites obeys

$$2H + CuZ_i \rightleftharpoons Cu + H_2Z_i, \qquad [6]$$

where Z_i are one-to-two exchange sites. In the mass-action laws, which are derived on the basis of the reactions given, we always use mole fractions to express the activities of the bound species. Note again that the main reason for introducing the reactions in Eqs. [4]–[6] is to incorporate the necessary stoichiometry in the model; as already discussed, charges are not represented explicitly. The experimental data in Figure 3 are analyzed in terms of such a multisite binding model regularized for smooth distributions. The results are conditional to the appropriate ionic strength. As evident from Figure 3, the model is able to fit both experimental data sets in a satisfactory fashion. In spite of the different character of the experimental data, the resulting affinity distributions shown in Figure 3 turn out to be similar. The main differences are in the concentrations of the Langmuir sites and in the shape of distributions of the one-to-two exchange sites.

Such comparisons would certainly be most interesting for a wider range of similar materials as they provide a simple way of classifying the binding properties of actual, environmental materials. With a larger number of such fingerprints for similar materials, such comparisons would also supply an empirical way to estimate binding properties of materials not investigated in the laboratory.

C. Predictive Capabilities

Due to the empirical character of the multisite binding models, any predictive capabilities of such models may seem surprising. These models, however, do have substantial predictive capabilities within a given data set. To test these predictive capabilities it is often possible to calibrate such a model only with a subset of an appropriate experimental data set, and predict the rest of the data set without any parameter adjustment. The following example serves to illustrate this aspect.

Let us consider the adsorption data of Cd^{2+} and H^+ on rutile in 0.2 mol·L^{-1} in KNO_3 and at 20°C by Fokkink (1987). The entire data set shown in Figure 4 is rather complete and contains the following data: in Figure 4A the amount of H^+ bound as a function of pH (acid–base titration curve), in Figure 4B the amount of

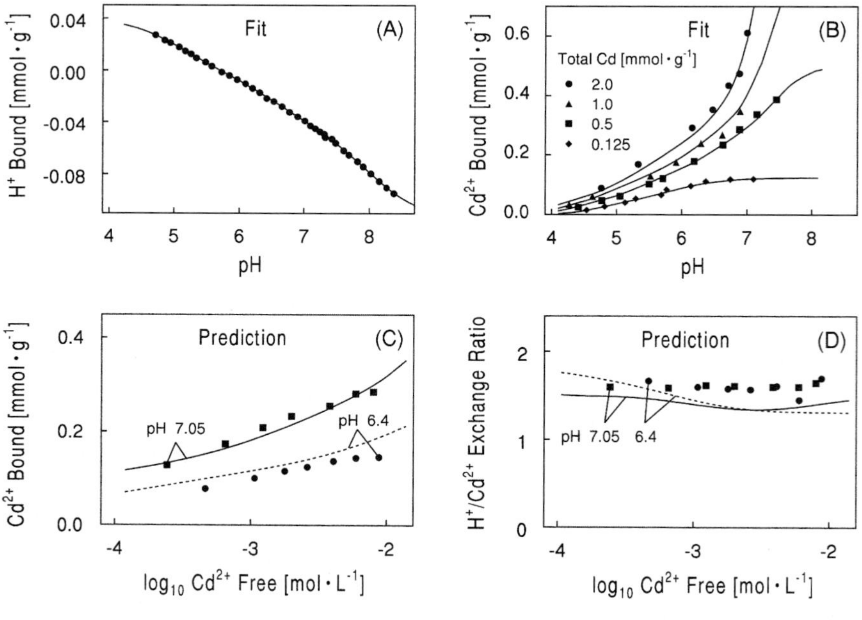

Figure 4 Predictive properties of the multisite binding model illustrated with competitive binding of Cd^{2+} and H^+ to rutile in 0.2 mol·L^{-1} KNO_3 and at 20°C by Fokkink (1987): (A) acid–base titration curve without Cd^{2+}; (B) amount of Cd^{2+} bound as a function of pH at various total concentrations of Cd^{2+}; (C) amount of Cd^{2+} bound as a function of the free concentration of Cd^{2+} for two different pH values (isotherms); and (D) differential exchange ratios between H^+ and Cd^{2+} as a function of the free concentration of Cd^{2+} for two different pH values. The rutile concentration for the data shown in (B) is 40 g·L^{-1}; all others are at 20 g·L^{-1}. The model (solid lines) is constructed on the basis of the experimental data shown in (A) and (B) and is used without further adjustment to predict the data shown in (C) and (D). Regularization from small number of sites was performed on a floating grid with a minimum interval of one decade (Černík *et al.*, 1996).

Cd^{2+} adsorbed as a function of pH at various total concentrations of Cd^{2+} (adsorption edges), in Figure 4C the amount of Cd^{2+} adsorbed as a function of free Cd^{2+} concentration in solution for two different pH values (binding isotherms), and in Figure 4D the exchange ratio between H^+ and Cd^{2+} as a function of free Cd^{2+} concentration in solution, again, for two different pH values. We now construct the multisite binding model on the basis of a subset of these data, namely, on the basis of the data shown in Figures 4A and 4B. To obtain a satisfactory fit of these data, we must include three types of sites. This model is entirely analogous to the model of the competitive binding of Cu^{2+} and H^+ by the fulvic acid and humic acid already discussed. We introduce a distribution of different sites, each of which obeys one of the three following reactions:

$$H + X_i \rightleftharpoons HX_i \qquad [7]$$

$$H + CdY_i \rightleftharpoons Cd + HY_i \qquad [8]$$

$$2H + CdZ_i \rightleftharpoons Cd + H_2Z_i. \qquad [9]$$

This model is able to represent the experimental data in a satisfactory fashion. In this case, we have applied the regularization for small number of sites. The resulting model involves 10 sites and is summarized in Table I. Again the results are conditional to the appropriate ionic strength.

We now use this model to predict the remaining part of the experimental data. First, we predict both binding isotherms. The calculated isotherms are shown in Figure 4C and agree with the experimental data in an acceptable fashion. This satisfactory result may not seem surprising; the isotherms shown in Figure 4C are nothing but cuts through the adsorption edges shown in Figure 4B. Thus, any model which provides a smooth interpolation of the adsorption edges should automatically be able to do a good job for the isotherms. Secondly, we use the same model to predict the exchange ratio data in Figure 4D. The prediction of the model is again rather satisfactory. This result may be more surprising, as the model is apparently able to pick up the competition between H^+ and Cd^{2+} without having "seen" any experimental data about the influence of Cd^{2+} on the amount of protons adsorbed.

This exercise demonstrates that once the model is properly calibrated, the model captures all features of the competitive binding process quantitatively. Quite surprisingly, classical surface-complexation models, which have been applied widely for several decades, have only rarely been tested for similar predictive capabilities.

IV. DISCUSSION

We have summarized some features of empirical multisite binding models in situations where concentrations of one or two components are being varied. Various experimental data sets can be rationalized with a superposition of independent noncompetitive Langmuir sites and exchange sites; competitive Langmuir sites are not really necessary. Such a binding model has the basic advantage that all data sets, in which concentrations of two component are varied simultaneously, can be represented in terms of affinity distributions, which are functions of a single affinity constant. If competitive Langmuir sites would be considered, distributions that would depend on two affinity constants would be needed. Note that such models based on independent binding sites are consistent with common solution speciation schemes.

The corresponding affinity distributions can be calculated by constrained and regularized least-squared methods in a simple, quasi-automatic fashion. Two regularization schemes of the resulting affinity distributions are most promising: regularization for small number of sites or regularization for smoothness. Regularization for small number of sites leads to a very simple binding model which contains a small number of adjustable parameters.

Both regularization schemes have their advantages and disadvantages. The most obvious advantages of the regularization for small number of sites is the rather small number of adjustable parameters and the ease with which the model can be summarized (see Table I). The disadvantages are numerous. The regularization for small number of sites is not optimally stable, the resulting isotherms may tend to oscillate, and the model may not be well suited for fingerprinting. Comparison of the resulting distributions is rather difficult. On the other hand, the regularization for smoothness provides an almost optimally stable inversion method, the resulting isotherms are sufficiently smooth, and the resulting affinity distributions are well suited for comparison. The major disadvantage is the larger number of sites that is introduced by the model, an aspect which makes it more difficult to incorporate such a model into popular speciation programs. But the latter difficulty could easily be overcome by minor modifications of the routines. A similar modification has been implemented in MINTEQ2A to incorporate continuous distributions (Allison *et al.,* 1991).

We have also seen that such empirical models can have substantial predictive capabilities in the case where concentrations of two (or more) components are varied. It is often possible to calibrate the model with a limited experimental data set, and the model is then able to predict the ion partitioning in situations where no experimental data were used to calibrate the model. The predictive capability of these empirical models is closely related to a fundamental thermodynamic consistency relation (Kemball *et al.,* 1948; Franses *et al.,* 1995). For equilibrium data, this consistency relation provides a mathematical relation between individual binding isotherms. In other words, complete experimental records of all binding isotherms as a function of all solution concentrations contains redundant information. For example, if we measure Cd^{2+} bound as a function of Cd^{2+} and H^+ in solution and H^+ in the absence of Cd^{2+}, this information is fully sufficient to predict the amount of

H^+ bound with Cd^{2+} present in solution, and thus the exchange ratio. In spite of its empirical character, the multisite binding model is thermodynamically consistent and is therefore able to make predictions within experimental data sets. However, the thermodynamic consistency is no speciality of the multisite binding model; any binding model, which was consistently derived from mass-action and mass-balance relationships, will satisfy the thermodynamic consistency relations as well.

One should also stress that the sites introduced in the multisite binding model are fictitious sites, and should not be interpreted to correspond to any real binding sites in the materials. They must be viewed as a collection of sites whose behavior is virtually identical to the sites of the real material. In this respect, the multisite binding model is no different from any other ion binding model. To obtain molecular information about binding mechanisms from ion binding data alone is practically impossible; such data must be augmented with spectroscopic and structural information about the material of interest. With the heterogeneous materials discussed in this paper, such information would be extremely complex and prohibitively difficult to obtain.

ACKNOWLEDGMENTS

The work was supported by the Swiss National Science Foundation and the Swiss part of EU-Environment Project EV5V-CT94-0536. J.C.W. acknowledges support from the U.S. Department of Energy (DOE) under Contract DE-AC06-76RLO 1830 as part of the DOE's Subsurface Science Program and the Co-contaminant Chemistry Subprogram.

REFERENCES

Allison, J. D., Brown, D. S., and Novo-Gradac, K. 1991. MINTEQ2A/PRODEF2A, a geochemical assessment model for environmental systems, version 3.0. EPA/600/3-91/021, U.S. Environmental Protection Agency, Athens, GA.

Bashford, D., and Karplus, M. 1990. pK_a's of ionizable groups in proteins: Atomic detail from a continuum electrostatic model. *Biochemistry* 29:10219–10225.

Benedetti, M. F., Milne, C. J., Kinniburgh, D. G., Van Riemsdijk, W. H., and Koopal, L. K. 1995. Metal ion binding to humic substances: Application of the non-ideal competitive adsorption model. *Environ. Sci. Technol.* 29:446–457.

Brassard, P., Kramer, J. R., and Collins, P. V. 1990. Binding site analysis using linear programming. *Environ. Sci. Technol.* 24:195–201.

Černík, M., Borkovec, M., and Westall, J. C. 1995a. Regularized least-squares methods for the calculation of discrete and continuous affinity distributions for heterogeneous sorbents. *Environ. Sci. Technol.* 29:413–425.

Černík, M., Borkovec, M., and Westall, J. C. 1996. Affinity Distribution Description of Competitive Ion Binding to Heterogeneous Materials. *Langmuir,* 12:6127–6137.

Černík, M., Rusch, U., and Borkovec, M. 1995b. QUASI, version 1.0. Institute of Terrestrial Ecology, Swiss Federal Institute of Technology, Schlieren, Switzerland.

Ephraim, J., Alegret, S., Mathuthu, A., Bicking, M., Malcolm, R. L., and Marinski, J. A. 1986. A united physicochemical description of the protonation and metal ion complexation equilibria of natural organic acids (humic and fulvic acids). 2. Influence of polyelectrolyte properties and functional group heterogeneity on the protonation equilibria of fulvic acid. *Environ. Sci. Technol.* 20:354–366.

Ephraim, J., and Marinsky, J. A. 1986. A united physicochemical description of the protonation and metal ion complexation equilibria of natural organic acids (humic and fulvic acids). 3. Influence of polyelectrolyte properties and functional heterogeneity on the copper ion binding equilibria in an armadale horizons Bh fulvic acid sample. *Environ. Sci. Technol.* 20:367–376.

Fokkink, L. G. J. 1987. Ph.D. Thesis, Agricultural University of Wageningen, Wageningen, The Netherlands.

Franses, E. I., Siddiqui, F. A., Ahn, D. J., Chang, C.-H., and Wang, N.-H. L. 1995. Thermodynamically consistent equilibrium adsorption isotherms for mixtures of different-size molecules. *Langmuir* 11:3177–3183.

Hiemstra, T., Van Riemsdijk, W. H., and Bolt, G. H. 1989. Multisite proton adsorption modeling at the solid/solution interface of (hydr)oxides; A new approach. *J. Colloid Interface Sci.* 133:91–104.

Kemball, C., Rideal, E. K., and Guggenheim, E. A. 1948. Thermodynamics of monolayers. *Trans. Faraday Soc.* 44:948–954.

Langmuir, I. 1918. The adsorption of gases on plane surfaces of glass, mica and platinum. *J. Am. Chem. Soc.* 40:1361–1403.

Leuenberger, B., and Schindler, P. W. 1986. Application of integral pK spectrometry to the titration curve of fulvic acid. *Anal. Chem.* 58:1471–1474.

Milne, C. J., Kinniburgh, D. G., De Wit, J. C. M., Van Riemsdijk, W. H., and Koopal, L. K. 1995. Analysis of proton binding by a peat humic acid using a simple electrostatic model. *Geochim. Cosmochim. Acta* 59:1101–1112.

Nederlof, M. M., Van Riemsdijk, W. H., and Koopal, L. K. 1992. Comparison of semianalytical methods to analyze complexation with heterogeneous ligands. *Environ. Sci. Technol.* 26:763–771.

Pitzer, K. S. 1991. Ion interaction approach: Theory and data correlation. *In* "Activity Coefficients in Electrolyte Solutions" (K. S. Pitzer, Ed.), pp. 75–153. CRC Press, Ann Arbor, MI.

Press, W. H., Teukolsky, S. A., Vetterling, W. T., and Flannery, B. P. 1992. "Numerical Recipes in FORTRAN: The Art of Scientific Computing," 2nd ed. Cambridge University Press, Cambridge.

Smits, R. G., Koper, G. J. M., and Mandel, M. 1993. The influence of nearest- and next-nearest-neighbor interactions on the potentiometric titration of linear poly(ethylenimine). *J. Phys. Chem.* 97:5745–5751.

Stanley, B. J., and Guiochon, G. 1993. Numerical estimation of adsorption energy distributions from adsorption isotherm data with the expectation-maximization method. *J. Phys. Chem.* 97: 8098–8104.

Stumm, W., and Morgan, J. J. 1996. "Aquatic Chemistry." Wiley, New York.

Sverjensky, D. A. 1993. Physical surface-complexation models for sorption at the mineral-water interface. *Nature* 364:776–780.

Chapter 23

Ion Binding to Humic Substances

Measurements, Models, and Mechanisms

D. G. Kinniburgh,[1,*] W. H. van Riemsdijk,[1] L. K. Koopal,[2] and M. F. Benedetti[3]

[1]Department of Soil Science and Plant Nutrition, Wageningen Agricultural University, The Netherlands; [2]Laboratory for Physical and Colloid Science, Wageningen Agricultural University, The Netherlands; [3]Laboratoire de Géochimie et Métallogénie, Université P. M. Curie, Paris, France

The development and application of models for ion binding by humic substances are discussed. The scope of these models ranges from those describing proton binding and the development of charge in rather simple systems to more general models which can describe metal ion binding in multimetal environments. The common underlying theme is that for such models to be successful over a broad range of solution conditions, the binding must consider the interaction of ions with surface sites having a broad range of affinities. This site heterogeneity can be successfully and succintly described by analytical models which assume a continuous distribution of affinities. These distributions can be either monomodal or bimodal in character, depending on the ion chosen and the range of conditions considered. A number of variations on this theme are discussed. Electrostatic interactions which account for the nonspecific binding of counterions and for the magnitude of specific binding through their effect on the electrostatic Boltzmann factor can be included either by considering the humic to be an impenetrable particle of simple spherical or cylindrical geometry or by assuming that

*On leave from the British Geological Survey, Wallingford, UK.

Adsorption of Metals by Geomedia

the humic can be treated as an electrically neutral phase in which all of the counterions are confined (the Donnan model). Additional complications which can be included are the nonideality of binding, the binding of metal ions to bidentate sites, and the binding of hydrolyzed metal ion species. These concepts are illustrated with data for ion binding to a purified peat humic acid. Additional multimetal ion (competition) data are required to test the various options more thoroughly.

I. INTRODUCTION

Humic substances can play an important role in determining the mobility and bioavailability of both organic and inorganic trace components in the environment. The pH and free metal ion activity are the most important factors affecting the binding of metal ions to humic substances. The relevant pH range in environmental systems is very broad. For example, fly-ash leachates may be in the pH range 10 to 12, and hypertrophic surface waters can have a pH of 10 or more during the assimilation of carbon dioxide by algae. Low-pH waters are also quite common. Acidic sandy forest soils may have a pH in the range of 3 to 4, and acidic mine drainage may have an even lower pH, sometimes even a negative pH.

The relevant free metal ion activity in the aqueous phase can also vary greatly. The activity of Cu^{2+} in a freshwater lake with a pH of 8 has been shown to have a pCu of around 15, whereas the total dissolved Cu in the same lake had a pCu_T of 8 (Xue and Sigg, 1993). Similar differences are found in seawater. The total dissolved Cu in an acidic forest soil may have a pCu_T of 6 and an estimated pCu of 11 (Benedetti *et al.,* 1996b). The concentrations of metals in mine drainage and waste streams can be much greater still. Therefore, in order to be able to interpret the mobility, bioavailability, and toxicity of metal ions in the environment, it is important to have models available that can describe metal ion binding over a very wide range of conditions. Humic and fulvic acids are probably the dominant humic substances involved in metal ion binding.

One of the complicating factors is that humic substances are structurally heterogeneous (Hayes *et al.,* 1989). This structural diversity is reflected by binding heterogeneity and it is dealing with this heterogeneity that poses one of the principal challenges in modeling ion binding by humics over a wide range of solution conditions.

A further challenge in modeling these interactions is to demonstrate that parameter estimates derived from laboratory data using purified humic substances can be applied to real-world natural systems. In other words, in our opinion, the challenge is not merely to fit a (limited) set of laboratory data to a certain model, but to develop a model that makes it possible to go "from the lab to the field." Several research groups have been active in developing such ion binding models

(Bartschat *et al.*, 1992; Tipping and Hurley, 1992; Benedetti *et al.*, 1995; Tipping *et al.*, 1995; Westall *et al.*, 1995).

In our own research we have also used various model concepts. The aim of this chapter is to review these and to outline their possibilities and limitations.

II. BASIC CHARGING BEHAVIOR OF HUMIC SUBSTANCES

A. Proton Binding

Both humic and fulvic acids show a gradual increase in negative charge with increasing pH due to the dissociation of protons. It is now well established that the carboxylic-type groups are mainly responsible for this behavior over the lower pH range, say below pH 7. The phenolic-type groups that are also known to be present are expected to contribute more to the charging behavior at higher pH values. The negative charge that develops leads to the development of a double layer which shows its effect in the salt dependence of the charging curves. Univalent counterions such as Na^+ and K^+ have a low affinity for the reactive groups of humic substances but can be present in the double layer as counterions. The total number of weak acid groups for fulvic and humic acids varies between 3.5 and 8.5 $eq \cdot kg^{-1}$ humic substance, with the higher densities found for the fulvic acids (Tipping and Hurley, 1992; de Wit *et al.*, 1993a). At near-neutral pH values, approximately half of these groups are dissociated. Therefore the net negative charge and associated cation binding is on the order of 3 $eq \cdot kg^{-1}$, or three times greater than that of a smectite clay on a weight for weight basis.

Modern physical chemical models for ion binding to humic substances consist of two parts. One part describes the specific binding of protons to reactive sites on the particle, macromolecule, or gel, and the other part describes the properties of the double layer. However, these two parts interact: the type of double-layer model used affects the way that the binding of the protons (and metal ions if present) is considered through the electrical potential that is implicit in the model. Unfortunately the double-layer potential cannot be measured experimentally, which means that direct tests of the various double-layer models are difficult. Each of the double-layer models that is chosen implies one or more assumptions. Next we discuss several possible approaches and outline the assumptions involved.

B. Random versus Patchwise Heterogeneity

A first assumption is whether all of each particle or molecule is assumed to be the same or whether there are different "patches" within a particle (van Riemsdijk *et al.*, 1991). The most common and simplest assumption is to treat all molecules or particles as the same and to assume that one can use an averaged "smeared-out" potential for all particles. We will call this assumption the "random" assumption. The interesting aspect of this simplifying assumption is that it gives the possibility of separating the effect of the variable potential on the basic charging behavior from the effect of the chemical heterogeneity. With chemical heterogeneity we mean that there may be binding sites that differ in their intrinsic chemical affinity for the binding of protons and other ions. The random assumption means that the electrostatic part of the electrochemical interaction energy is the same for all sites on all particles for a given pH and ionic strength. The alternative is to assume "patchwise" heterogeneity where each patch type can have its own associated potential at a certain pH and ionic strength. This approach requires more information. For instance, modeling a mixture of fulvic and humic acids using a different electrostatic model (or model parameters) for each type is an example of patchwise heterogeneity.

C. Location of Charge: Particle Surface or Bulk Volume

A second assumption is necessary to calculate the charge density from the charge that can be measured from an acid–base titration. Titration measurements give information about the charge per kilogram of humic substance or per kilogram of humic carbon. One can assume either that the humic substance is a particle, in which case all of the charge is assumed to reside on the particle "surface," or that the humic substance is a "gel," in which case the particle charge is contained within a three-dimensional volume. The "surface" approach requires information on the specific surface area ($m^2 \cdot g^{-1}$) in order to transform the measurable charge per gram into the required charge density (charge per unit area). Unfortunately the direct measurement of the specific surface area of humic substances as they occur in aqueous solution is not possible. However, the specific surface area can easily be calculated if one makes assumptions about the shape of the molecules and their average dimensions and if one assumes a particular particle density (de Wit *et al.,* 1993a).

For the application of the gel concept one needs to know the specific gel volume ($L \cdot kg^{-1}$) in order to transform the measured charge to the required units (charge per unit volume of aqueous phase inside the gel). This specific gel volume can in principle be measured by, for instance, viscometry or size exclusion chromatography. If it has not been measured, the gel volume can, and must, be treated as an adjustable parameter.

Both of these assumptions are reasonable although they are clearly quite different and require considerable simplifications of reality. So to some extent the favored approach will depend on how well the different approaches work and on practical considerations such as computational efficiency. Both assumptions lead to a Boltzmann factor and an implicit electrostatic potential.

With the random heterogeneity assumption, the electrostatic potential enables bulk solution concentrations to be converted to local concentrations close to the binding sites. Since the net charge and electrostatic potential are normally negative, the local concentration of cations tends to be greater than the bulk solution concentration, often by up to one order of magnitude (Bartschat *et al.*, 1992; de Wit *et al.*, 1993a). This electrostatic potential is implicit in the Boltzmann term, which accounts for the electrostatic part of the electrochemical affinity and which can be calculated from the assumed double-layer model. The double-layer model relates the charge density to the electrostatic potential for a given ionic strength and salt composition.

D. The "Master Curve"

Once a double-layer model has been selected, one can test its applicability by applying what has been called the "master curve approach" to the acid–base data measured at a series of ionic strengths of an indifferent electrolyte (de Wit *et al.*, 1990, 1993a). In this approach, the effect of the electrostatic energy on the binding is removed by plotting the protons bound (or residual charge) as a function of $p[H_s]$, the negative log proton concentration at the location of the charged sites in the humic phase, rather than as a function of the negative log proton concentration in the bulk solution, p[H]. The relationship between the two is given by

$$p[H_s] = p[H] + \log e\,(F\psi/RT). \qquad [1]$$

The potential, ψ, is the potential difference between that of the bulk aqueous solution (assumed as zero potential) and the mean potential at the location where the protons are bound to the humic substance. In reality, it is likely that protons are present at various locations (and potentials), and so the assumption that the distribution of potentials can be represented by an equivalent mean value is clearly only an approximation. The interpretation of Eq. [1] as applied to humics is discussed in detail by Bartschat *et al.* (1992). For humic substances, $p[H_s]$ is normally lower than p[H] since the residual negative charge leads to a negative potential and hence an accumulation of protons in the vicinity of the binding sites.

The charging curves measured at various background concentrations of an indifferent salt should merge into one curve if the chosen double-layer model and associated parameters are an appropriate model to account for the salt effect. The physical reality of the model can then be judged by comparing the value of the fitted model parameter(s) with estimates of the same parameter(s) based on independent measurements. Several authors have used the spherical or cylindrical double-layer model approach for humic substances (de Wit *et al.*, 1990, 1993a; Barak and Chen, 1992; Bartschat *et al.,* 1992). The relatively small salt effect seen with the charging curves of humic substances can best be explained by assuming that the adsorbing surface is strongly curved. It is also possible to consider the humics as worm-like chains with essentially a cylindrical geometry. In this case, the radius of the cylinder is the single adjustable parameter (de Wit *et al.,* 1993b). Either the (average) radius of the particles is treated as a single adjustable parameter or the humics are assumed to contain a mixture of particles with different radii. In the latter case, the "random" assumption for the potential no longer holds since the calculated potential depends on the assumed radius of the particle and so will differ for different-sized particles (Bartschat *et al.,* 1992). This can be seen as one source of the observed heterogeneity which applies to small particles (<5 nm radius) since larger particles approach the flat plate limit. Again, practical considerations mean that a decision has to be taken as to whether to put most of the observed heterogeneity into the size heterogeneity "bin" or into the site heterogeneity "bin." We favor the site heterogeneity approach because of the observed strong functional group heterogeneity of humics and its relative simplicity.

In practice it is usually possible to get a reasonable master curve by using a single fitted radius for a particular humic acid. This suggests that the radius is more or less independent of the ionic strength and that some form of average radius is sufficient to describe the distribution of surface potentials. This approach has been tested for a series of humic and fulvic acids (de Wit *et al.,* 1993a). For the spherical model, the inferred particle radii, which ranged from 0.6 to 4 nm (median, 0.85 nm), were in reasonable agreement with the "dry" radii that were estimated from the molecular weights of the various humic substances. De Wit *et al.* (1993a) also showed that the salt effect usually increased with the molecular weight of the humic substance, as expected. Therefore fulvic acids, which have a relatively low molecular weight, also show a relatively small salt effect compared with humic acids. In general, master curves of similar quality can be obtained if a cylindrical double-layer model is used. The resulting fitted radius (median, 0.32 nm) of the cylinder is on the order of what is expected for a chain of aliphatic and aromatic groups and is in general about half of that based on the spherical model.

Figure 1 shows the charging curves of a purified peat humic acid (PPHA) at various ionic strengths and the fitted curves based on the spherical and the cylindrical double-layer models in combination with a bimodal Langmuir–Freundlich (LF) isotherm. Figure 2 shows the calculated surface potential as a function of ionic strength and pH for the two models. The general shapes of the curves are similar and are in general highly non-Nernstian, especially at high ionic strengths and high p[H]. The negative surface potential for the spherical double-layer model is slightly greater than that for the cylindrical model. The resulting master curves are also similar for both model options (Fig. 3). The data for both models conform well to a single master curve over most of the $p[H_s]$ range; the greatest deviations occur at the extremes at both low and high $p[H_s]$.

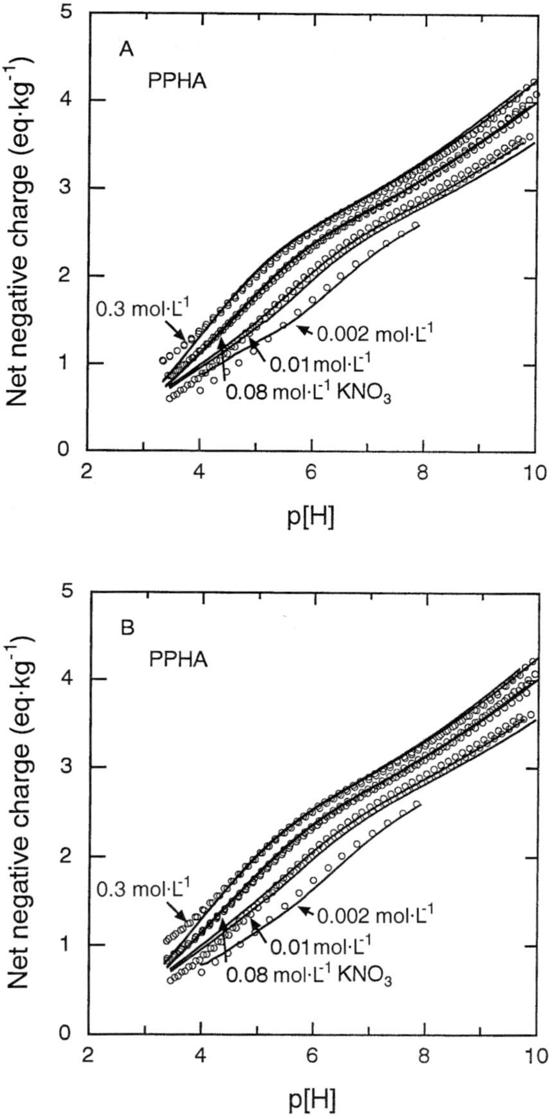

Figure 1 Experimentally measured charge development by a purified peat humic acid (PPHA) as a function of pH and ionic strength. Model fits (solid lines) are shown for a bimodal Langmuir–Freundlich isotherm combined with (A) a spherical and (B) a cylindrical double-layer model. "Duplicate" curves are shown for some of the ionic strengths. p[H] in the negative $\log_{10}$ of the proton *concentration.*

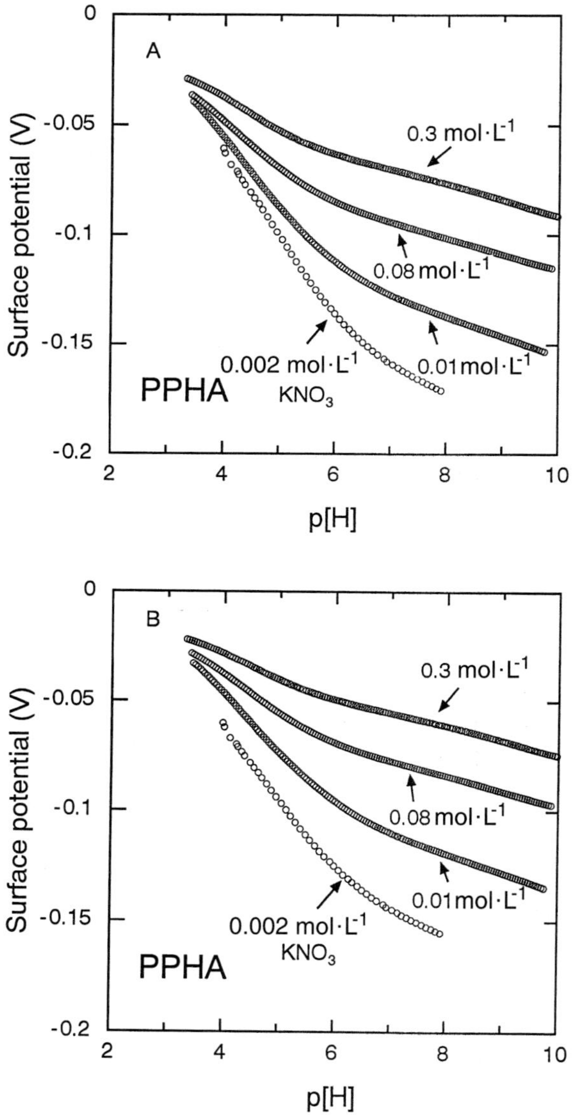

Figure 2 The estimated variation of surface potential of PPHA as a function of p[H] for the curves shown in Figure 1 according to (A) a spherical and (B) a cylindrical double-layer model.

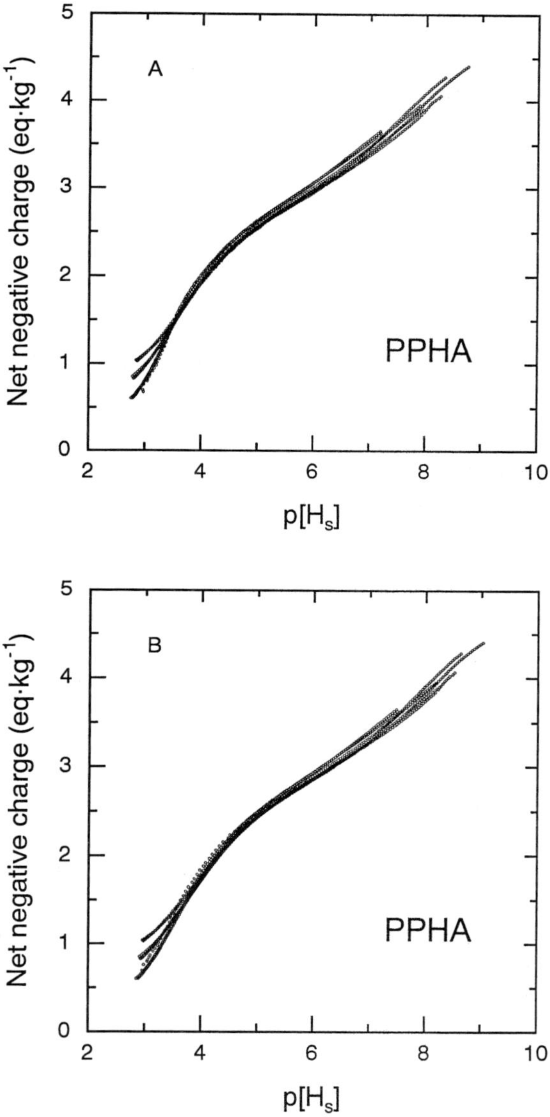

Figure 3 The "master curves" derived from the fits to the charging curves shown in Figure 1 according to (A) the spherical and (B) the cylindrical double-layer model.

An advantage of using a Poisson–Boltzmann double-layer model approach is that it takes into account the curvature of the surface and needs only one adjustable parameter in order to account for all of the electrostatic interactions. However, the calculations with such models are relatively computationally intensive since the Poisson–Boltzmann equation has to be solved numerically. This burden is especially significant when one also wants to calculate the composition of the double layer in mixed salt solutions. This could become a disadvantage if such models are intended to be used in chemical transport models.

The spherical model especially relies on the concept of a strongly hydrophobic nature for the humics. For weakly charged hydrophilic humics that may be strongly solvated random polyelectrolyte coils, the gel concept is more appropriate. The gel concept logically leads to the use of a type of Donnan model to account for the electrostatic interactions. An advantage of the Donnan approach is that the calculation of the composition of the double layer is mathematically very simple and thus requires relatively little computer time. The simple Donnan concept assumes that the electrical potential is constant throughout the gel volume. However, perhaps not unreasonably, it has been found that the simple Donnan concept does not work for humic substances if it is assumed that a particular humic substance can be characterized by a single specific gel volume which is independent of both ionic strength and charge density. Investigations of the variable charge behavior of bacterial cell walls have shown that the volume is more dependent on the ionic strength than on the charge density (Plette *et al.,* 1995). Assuming that this also holds for humics, this greatly simplifies the analysis. The variation of Donnan volume ($\mathbf{V}_D$) with ionic strength ($\boldsymbol{I}$) based on proton binding data for simple 1:1 electrolyte systems (Benedetti *et al.,* 1996a) is shown in Figure 4. There is an approximately linear relationship between log $\mathbf{V}_D$ and log $\boldsymbol{I}$. This relationship is not necessarily unique but appears to be in reasonable agreement with independent measurements based on viscometry. However, it is not expected to hold for all solution conditions, e.g., high concentrations of multivalent ions.

The fitted relationship between specific Donnan volume and ionic strength for this series of humic and fulvic acids (Fig. 4) can be described by the empirical relationship

$$\log \mathbf{V}_D = b(1 - \log \boldsymbol{I}) - 1, \qquad [2]$$

where $\mathbf{V}_D$ is the specific Donnan volume ($L \cdot kg^{-1}$), b is an empirical parameter reflecting the salt dependence of the Donnan volume, and $\boldsymbol{I}$ is the ionic strength ($mol \cdot L^{-1}$). Equation [2] describes a line that pivots around the point $\mathbf{V}_D = 0.1$ $L \cdot kg^{-1}$ for $\boldsymbol{I} = 10$ $mol \cdot L^{-1}$. The slope of this line appears to be similar for different humic acids but is quite different for fulvic acids (Benedetti *et al.,* 1996a). The variation of $\mathbf{V}_D$ with ionic strength indicates that the swelling and shrinking behavior of humic substances is an important factor with the Donnan concept. In the future, the physical reasonableness of the Donnan volumes needs to be tested

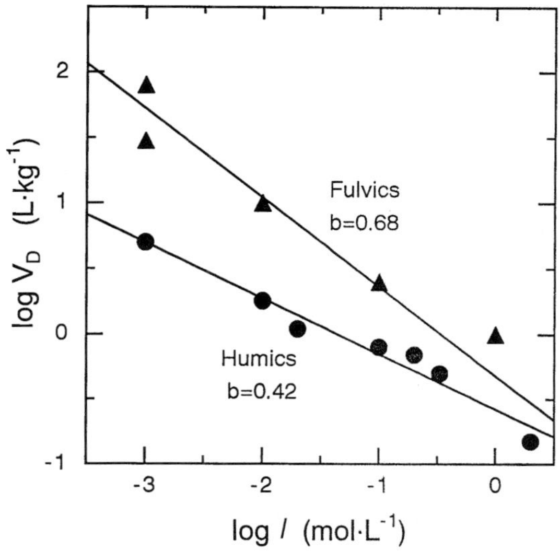

Figure 4 Variation of the Donnan volume with ionic strength derived from a variety of fulvic and humic acids (Benedetti *et al.*, 1996a). The lines shown are based on Eq. [2] with the b values indicated.

using well-characterized humic substances in which the gel volumes are measured independently as a function of pH and ionic strength.

Reasonable master curves can be obtained for humic and fulvic acids using this simple Donnan approach, but a unique solution does not exist (Benedetti *et al.*, 1996a). The specific volumes used by Benedetti *et al.* (1996a) for the humic acids seem to be somewhat low for hydrated humic acids, whereas the fitted volumes for the fulvic acids are relatively large considering the small size of the molecules anticipated from their molecular weights. The latter results may be physically justified if it is assumed that the gel does not consist of individual fulvic acid molecules but consists of micelle-like structures made up of many fulvic acid molecules.

As expected, the magnitude of the potentials calculated by the various models are considerably lower than the surface potentials that can develop on metal (hydr)oxide surfaces that often show a pseudo-Nernstian relation between pH and potential ($\Delta\psi/\Delta pH = 59$ mV/pH). The average calculated change of potential with pH for PPHA is less and in the region pH 4 to 8, varies from 50 mV/pH at low ionic strength to less than 10 mV/pH at high ionic strength.

Tipping and Hurley (1992) prefer a more empirical model to account for the effect of the electrostatic interactions on ion binding. In addition to estimating the interaction energy empirically, they use a semiempirical approach to calculate the composition of the double layer. This adopts the hard sphere concept for humics but uses a Donnan-type expression to simulate the effect of a diffuse double layer. The "Donnan volume" is estimated from the volume of the diffuse double layer surrounding the particle and can be related to the particle radius and the diffuse layer thickness using the Debye–Hückel characteristic distance ($1/\kappa$) (Tipping, 1993). With this approach, there is no direct link between the average (Donnan) potential implicit from the previous analysis of the diffuse double layer and the average electrostatic potential that can be inferred from the Boltzmann-type factor used for calculating ion binding to the specific binding sites. Unlike the Tipping and Hurley (1992) approach, the models already discussed do not require additional assumptions in order to calculate the composition of the double layer.

Other empirical approaches are possible but these have the disadvantage that they do not lead to much insight into the physical mechanisms influencing the interaction energy. Recently Westall *et al.* (1995) proposed that the salt dependency of the charging curves could be accounted for by assuming that all of the reactive sites of a humic substance have the same affinity for binding the indifferent salt cation. Such an approach involves a shifting of the charging curves with the background electrolyte concentration, but since it was only tested at two electrolyte concentrations, it was not sufficient to show whether or not such an approach can describe the salt dependency of a broader range of charging curves.

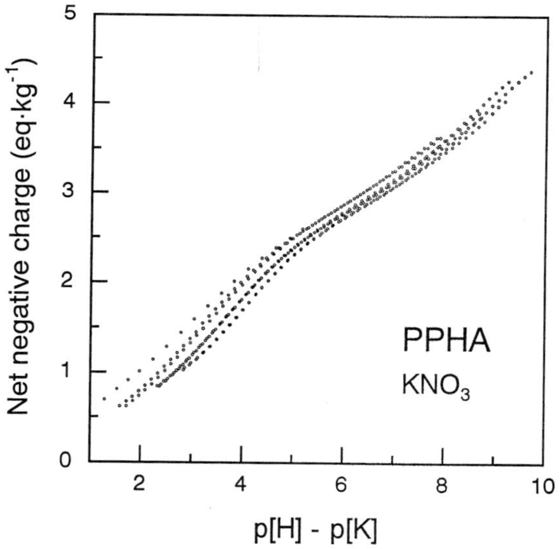

Figure 5 The "master curve" for the set of proton charging data shown in Figure 1 analyzed according to the simple ion exchange model proposed by Westall *et al.* (1995). The concentration of KNO_3 varies from 0.002 to 0.3 mol·L.

It is possible to combine the Westall *et al.* (1995) approach with the master curve concept if it is assumed that the sites are either protonated or occupied by a monovalent counterion, i.e., by assuming that there are no empty sites. This "ion exchange" approach should result in a master curve that is independent of the assumed binding constant of the monovalent ion, M, if one of the charging curves is arbitrarily chosen as a reference curve. The master curve should then result when the data are plotted as a function of p[H]–p[M] (Cabaniss *et al.,* 1989).

The results of such an approach are shown in Figure 5 for the PPHA data. The scatter in the master curve clearly shows that this approach cannot adequately account for the measured salt effects. Such an approach is equivalent to assuming a Donnan potential that is independent of salt concentration and charge density. One could argue that the results might have been somewhat better if dissociated sites, not associated with a "counterion," would have been included. However, unoccupied sites would lead to a charge that has to be compensated for by counterions in the double layer, and this should be accounted for in one way or another in a complete model. An improved master curve can be obtained if the independent variable is treated as p[H]–x(p[K]), where $x = 0.67$ in the case of the PPHA, but it is difficult to extend this more general approach to metal ion binding.

III. HETEROGENEITY ANALYSIS

A. Analyzing the Master Curve

The application of heterogeneity analysis (Nederlof *et al.,* 1993) to the master curve leads to an estimate of the intrinsic proton affinity distribution. Nederlof *et al.* (1993) have also discussed an objective way to find the charge–pH curves from measured titration curves. The conditional or apparent proton affinity distribution can be derived from the charge–pH curves as measured at a particular ionic strength by using one of the various techniques that are available from heterogeneity analysis. An overview of semianalytical methods that are based on a local isotherm approximation is given by Nederlof *et al.* (1990). An example of this type of analysis is shown in Figure 6 for the charging of the PPHA. The approximate intrinsic affinity distribution was calculated from the master curve by using the condensation approximation (CA) method discussed by Nederlof *et al.* (1990, 1993). The data for each ionic strength have been plotted separately in order to show the convergence. Two main peaks are visible. These are centered about $\log K_{Hs} = 3$ and $\log K_{Hs} \geq$ 8 to 10, but the results show that the data only give clear information on a part of the total distribution. No information is obtained for the very-high-affinity part of the distribution (high K_{Hs}) which is probed at high pH values or for the low-affinity part of the distribution which is probed at low pH values. Inevitable analytical errors make these parts of the distribution difficult to determine.

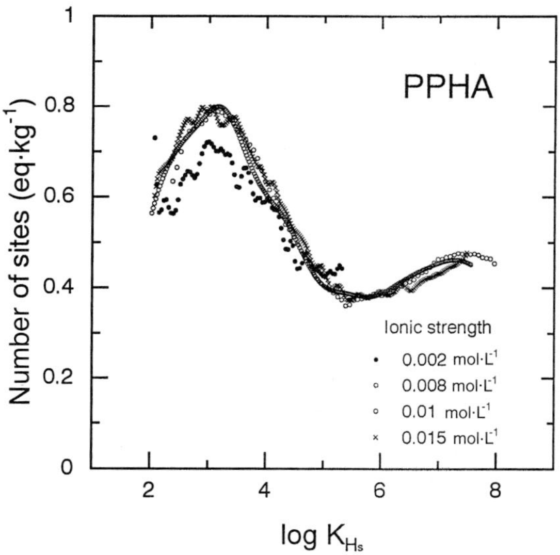

Figure 6 The intrinsic affinity distribution for the purified peat humic acid (PPHA) estimated from a heterogeneity analysis of the "master curve" according to the method of Nederlof *et al.* (1990) using the condensation approximation.

In general, intrinsic affinity distributions for humics show two distinct peaks which can broadly be identified with carboxylic-type groups (a log K_{Hs} of about 3 to 4) and phenolic-type groups (a log K_{Hs} of about 8 to 10.5). The overlap between these two peaks tends to be large, and so the attribution of sites titrating at near-neutral pH values is quite sensitive to the assumed shape of the component distributions and ultimately to the scope and quality of the data available for defining these distributions.

B. Discrete versus Continuous Distributions

In modeling the heterogeneous behavior shown in Figure 6, two schools of thought are apparent: the "discrete" school and the "continuous" school. The discrete school prefers to describe the observed behavior by assuming the presence of a certain number of distinct site types. Each type of site has a certain affinity and a certain number of sites per gram of humic material. In principle, the smooth and broad distribution observed requires a very large number of sites, leading to a very large number of model parameters (two per site type). In their Model V, Tipping and Hurley (1992) minimize the number of adjustable parameters by assuming the presence of eight types of site which are divided into two broad types of sites reflecting carboxylic- and phenolic-type sites. They further assume that there are twice as many carboxylic-type sites as phenolic-type sites and further subdivide each of these into four discrete sites which are defined by a mean log K and a spread, Δlog K, around this. The abundance of each of these sets of four sites is assumed to be the same, and the spacing between the affinity of the four sites on a log K scale is assumed to be constant. Characteristic parameters for a set of four sites are thus an average position on the log K axis, the total number of sites in the set, and the spread around the average log K. The required information is therefore similar to that required by a continuous distribution model. The shape of the distribution is constrained in Model V just as it is with continuous distribution models. Using eight sites without any *a priori* assumptions would have led to 16 adjustable parameters to describe the master curve rather than the 5 that are used in Model V. Therefore the approach adopted by Model V can be seen as a discretized version of a bimodal continuous affinity distribution.

Continuous distribution models assume a continuous distribution of affinities. These distributions are often assumed to be monomodal for simplicity but need not necessarily be so. For humics, bimodal distributions are often more appropriate because of the dominance of carboxylic- and phenolic-type functional groups. The corresponding distribution functions often have affinities ranging from $+\infty$ to $-\infty$, but the range can be truncated, e.g., the distribution derived from the generalized Freundlich isotherm (Kinniburgh *et al.,* 1983). In practice, the number of very-high- and very-low-affinity sites defined by these distributions tends to drop off rapidly at the extremes. Nevertheless, it is one of the strengths of continuous distribution models that a wide range of affinities is always considered—the sites are "assumed in" rather than "counted in."

An advantage of a continuous affinity distribution model over a completely discrete model without any *a priori* assumptions is that the continuous model can match the observed smooth charge–pH data observed with relatively few parameters. Invariably, a close inspection of the residuals resulting from a discrete site approach with a modest number of sites shows a characteristic sinusoidal pattern as the model attempts to fit itself to data in an optimal way. An example of this is shown for the fitting of a five-site Langmuir model to the 0.08 $mol \cdot L^{-1}$ KNO_3 PPHA data in Figure 7A. The log K values were fixed at log K = 3, 5, 7, 9, and 11. Altogether there were six adjustable parameters (the number of sites at each of the five log K values plus the initial charge on the humic acid). If the initial charge can be estimated independently, e.g., from the initial pH, then the number of adjustable parameters is reduced by one. Although the overall fit is excellent ($R^2 > 0.99$), the residuals clearly show a systematic error. This error is likely to be especially large if the model is used for extrapolation. Increasing the number of sites or allowing the position of the log K values to be optimized reduces the magnitude of the residuals, but the underlying systematic distribution of residuals remains. Fitting the same data to a bimodal LF isotherm gave an even better fit and there was no systematic distribution of residuals (Fig. 7B), giving credence to this method.

The previous analysis and the affinity distributions inferred from a heterogeneity analysis suggest that the combination of two LF equations should give an ex-

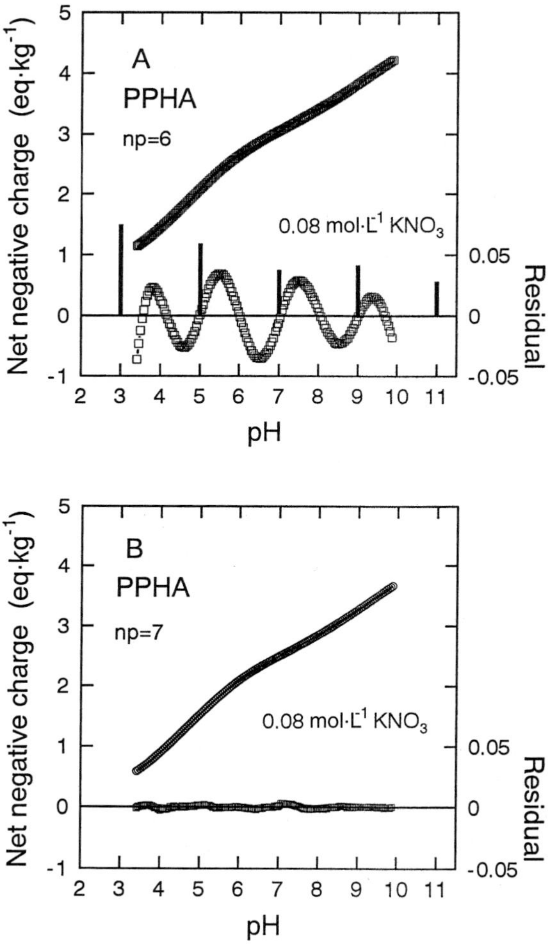

Figure 7 Charging curve and residuals resulting from fitting the charging data in 0.08 mol·L^{-1} KNO_3 to (A) a five-site Langmuir model with log K values fixed at 3, 5, 7, 9, and 11, and to (B) a bimodal Langmuir–Freundlich isotherm. np is the number of adjustable parameters. The vertical bars in (A) are the estimated number of sites at each log K value (refer to the left-hand axis).

cellent description of the data. The LF adsorption model is based on a semi-Gaussian symmetrical affinity distribution, the Sips distribution (an alternative way to arrive at the LF equation is to assume a block-type distribution). An advantage of the LF approach is that the Sips distribution leads to a simple analytical adsorption equation (Kinniburgh *et al.*, 1983). The Gaussian affinity distribution has also been used for describing ion binding by humic substances (Posner, 1966; Perdue and Lytle, 1983; Dzombak *et al.*, 1986; Dobbs *et al.*, 1989; Manunza *et al.*, 1995) but this requires numerical integration to give the overall isotherm. While the Gaussian distribution leads to a superficially similar distribution to the Sips distribution, the steeper decline of the Gaussian affinity distribution in the tail regions leads to a linear isotherm at low concentrations. The bimodal Langmuir-Freundlich isotherm (LF2) requires seven adjustable parameters: three parameters for each of the two distributions—a median log K_H locating the center of the distribution (log $\tilde{K}_H$), a parameter describing the width of the distribution (**m**), and the total number of sites (Q_{max})—plus the initial charge.

C. Ionic Strength Dependency

The preceding analysis was for a single ionic strength. In order to analyze data for a range of ionic strengths, it is necessary to use local proton concentrations rather than bulk solution proton concentrations. In this case, the corresponding bimodal LF model can be written as

$$Q_T = \frac{Q_{1,max}\,(\tilde{K}_1 H_s)^{\mathbf{m}_1}}{1 + (\tilde{K}_1 H_s)^{\mathbf{m}_1}} + \frac{Q_{2,max}\,(\tilde{K}_2 H_s)^{\mathbf{m}_2}}{1 + (\tilde{K}_1 H_s)^{\mathbf{m}_2}}, \qquad [3]$$

where Q_T is the total number of protons bound, $Q_{i,max}$ is the number of sites in distribution i (i = 1 or 2), $\tilde{K}_i$ is the median affinity of distribution i, $\mathbf{m}_i$ is the apparent heterogeneity parameter which reflects the width of distribution i, and H_s is the local (surface) concentration of protons close to the binding sites. The chosen double-layer model accounts for the relationship between H and H_s. This approach (Milne *et al.*, 1995b) leads to a reasonably good description of the salt-dependent charging curves of a purified peat humic acid (Fig. 1). The main discrepancy is at low p[H] where the model suggests convergence of the various charge curves, whereas the data show almost parallel behavior.

In principle, any double-layer approach that leads to a good description of the master curve can be used. Slightly different approaches for the description of the proton binding are possible as long as the strong heterogeneity in affinity is reproduced in a reasonable way.

IV. pH-DEPENDENT METAL ION BINDING

A. Basic Concepts

Before we start to analyze the various options for describing metal ion binding and to discuss the interpretation of the models and their results, we will first look at the data and discuss some basic concepts. The binding of Ca, Cd, and Cu to PPHA as a function of pH and metal ion concentration on a log–log plot is shown in Figure 8. The data were obtained using ion-selective electrodes and a comput-

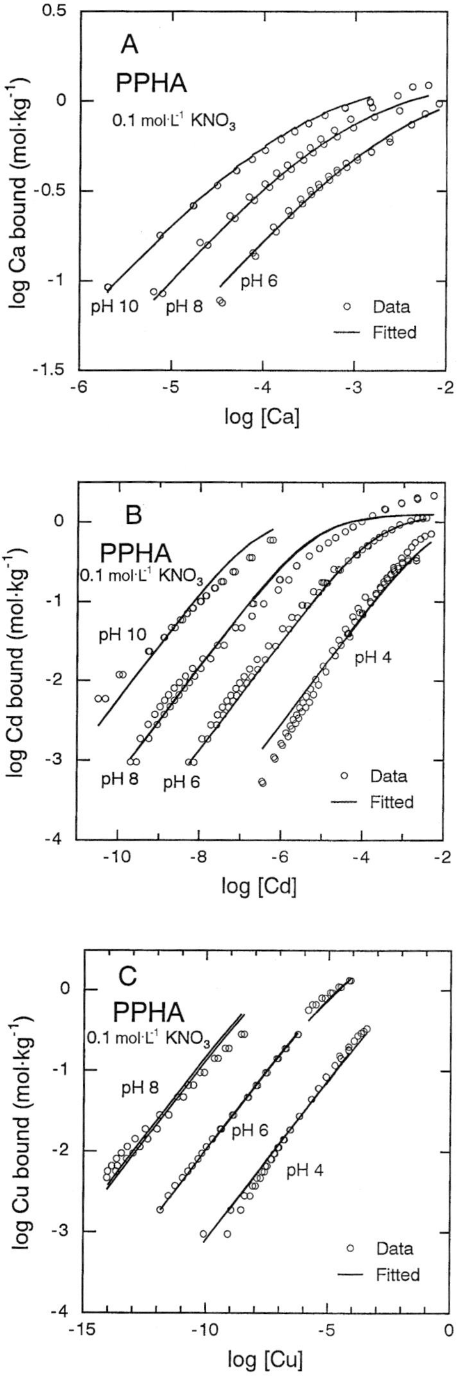

Figure 8 Isotherms for the binding of (A) Ca, (B) Cd, and (C) Cu to the purified peat humic acid (PPHA) at various pH values, all in 0.1 mol·L^{-1} KNO_3. The continuous lines are fits according to the model given in Eq. [8].

er-controlled titrator operating in pH stat mode (Kinniburgh *et al.*, 1995). The data show that the slope of the log–log plot for all metal ions is relatively constant over a large part of the pM range. At high metal ion concentrations, the slope gradually becomes lower, showing some signs of site saturation. The slope of the approximately linear part of the plot is metal ion specific and is invariably less than 1.

The data also show that the pH dependence is metal ion specific, as is the measured molar H^+/M^{2+} exchange ratio (Benedetti *et al.*, 1995) (Table I). In general, for a given ion, the exchange ratio appears to be remarkably constant, but often decreases somewhat with increasing metal loading and also sometimes with increasing pH. The exchange ratio was relatively constant for Ca at pH 6 and 8 and was about 0.3 (Benedetti *et al.*, 1995). The exchange ratio for Cd at pH 6 and 8 varied more than that for Ca and was around 0.6, but the ratio at low metal ion loadings was considerably higher. No reliable data were obtained at lower pH. The measured exchange ratio for Cu at pH 6 and 8 was greatest of all and was also the most variable. Most of the data fell in the range 1.3 to 1.7, a range generally confirmed in the work of Tipping *et al.* (1995) and Robertson (1996). The observed trend in the pH dependence of the various cations also follows the observed trend in the exchange ratio, i.e., the greater the pH dependence, the greater the exchange ratio.

Similar ion-specific nonstoichiometric exchange ratios have been measured for binding of divalent cations to metal (hydr)oxides which are also variable-charge, variable-potential colloids, although the ratios for oxides are generally higher. Kinniburgh *et al.* (1983) showed that the measured H^+/M^{2+} exchange ratios for ox-

Table I

Observed H^+/M^{2+} Exchange Ratios for Metal Ion Binding by Various Humic Acids

Cation	H^+/M^{2+} exchange ratio	pH	Comments	Reference
Ca	0.25 to 0.4	6 and 8	No significant trends with pH or Ca loading	Benedetti *et al.*, 1995
Cd	0.4 to 1.0	6 and 8	No significant trend with pH; generally decreases with increasing Cd loading	Benedetti *et al.*, 1995
Cu	1.1 to 1.4	4 and 4.5	Decreases with increasing pH and with increasing Cu loading	Tipping *et al.*, 1995
Cu	0.7 to 1.5	4, 5, and 6	Generally decreases with increasing pH, increasing Cu loading, and increasing ionic strength	Robertson, 1996
Cu	1.3 to 1.8	6 and 8	No significant trend with pH; generally decreases with increasing Cu loading	Benedetti *et al.*, 1995

ides were approximately equal to the horizontal shift of the adsorption isotherm when plotted in a log–log fashion. The exchange ratios were about 1.65 for Zn adsorption to ferrihydrite and 0.9 for Ca^{2+} adsorption. Perona and Leckie (1985) and Honeyman and Leckie (1986) have given a thermodynamic analysis of this relationship that confirms these observations. Such a thermodynamic analysis is based on what is called thermodynamic coinsistency (Cerník *et al.,* 1996; Borkovec *et al.,* 1998). An adsorption model should ideally be thermodynamically consistent.

The source of the pH dependency of metal ion binding to variable-charge, variable-potential colloids can be quite complicated to interpret. The pH dependency is in part due to electrostatic interactions. This effect can be very large for metal ion binding to oxides. The effect of the electrostatic interactions is clearly shown by Fokkink *et al.* (1990), who assume that the metal ions replace adsorbed water molecules and are adsorbed at the Stern plane some distance from the surface of the oxide. In their approach, direct competition between protons and metal ions for binding to the same type of reactive sites is not included; therefore the pH dependence is entirely ascribed to electrostatic interactions. The closer the metal ion is positioned to the surface, the stronger is the electrostatic interaction, the higher is the exchange ratio, and the greater is the pH dependence. The greatest pH dependence is obtained when the metal ions are placed in the same plane as the bound proton; this leads to an exchange ratio of 2 for divalent cations if it is assumed that the surface potential has a near-Nernstian response (59-mV change per pH unit change) (Fokkink, 1987). However, the Fokkink approach does not lead to a consistent description of the pH-dependent charging of oxides in the absence of metal ions, of the pH-dependent metal ion binding, or of the measured exchange ratio (Venema *et al.,* 1996). This is probably due to the fact that it neglects site competition.

None of the classical oxide models tested by Venema *et al.* (1996) gives a consistent description of all the available data. The exchange ratio is particularly difficult to reproduce accurately with most of the models and the same is expected to be true for humic models. The pH dependence and exchange ratio are not only dependent on electrostatic interactions, but also depend on the assumed surface species in the model. One can, for instance, assume that the basic proton and metal ion binding reactions can be described by the following monodentate reactions:

$$S^- + H^+ = SH \tag{4}$$

$$S^- + M^{z+} = SM^{(z-1)+}. \tag{5}$$

Competitive binding in this case leads, apart from electrostatic interactions, to a maximum exchange ratio of 1. The electrostatic contribution for humic substances is much less than that for oxides because of the much lower dependence of the potential on pH. The presence of bidentate complexes increases the maximum possible exchange ratio to 2. Another possibility is to assume that partially hydrolyzed metal ion species are preferentially adsorbed from solution. Adsorption of MOH^+ from solution under conditions where the degree of hydrolysis in solution is negligible leads to the release of one extra proton, leading to a maximum exchange ratio of 2 for monodentate binding and 3 for bidentate binding.

An exchange ratio greater than 2 for bivalent metal ion binding to a variable-charge colloid is extremely unlikely since it would imply that the surface becomes more negative upon binding a cation. This would lead to an increase in the affinity of the colloid for the metal ion as more metal ions are bound. It has been reported that the exchange ratio for zinc binding on metal (hydr)oxide is greater than 3 (Benjamin and Leckie, 1981), but this is probably an experimental artifact. The combination of chemical heterogeneity, electrostatic interactions, and competitive ion binding leads to a very complex situation. In the following sections, we will analyze various approaches to this complex problem in more detail.

B. Nonstoichiometric Exchange Models

The observation that the measured exchange ratio is not too variable for a particular metal ion can be used as a starting point for the development of a relatively simple equation that describes the pH-dependent metal ion binding at a constant ionic strength. The Freundlich-like behavior at constant pH and low metal ion concentration should also be accounted for by the model. The macroscopic formulation of the exchange process that these observations lead to is

$$SH_x + M^{z+} = SM^{(z-x)+} + xH^+ \qquad (K_{ex}) \qquad [6a]$$

or equivalently

$$S + M^{z+} + xH_2O = SM(OH)_x^{(z-x)+} + xH^+ \qquad (K^*_{ex}). \qquad [6b]$$

The sites are hypothetical sites and the origin of the released protons is not well defined. This follows from the mathematical equivalence of Eqs. [6a] and [6b]. In this approach, no use is made of the available information on the basic proton binding behavior previously discussed. The parameter x equals the proton–metal ion exchange stoichiometry and accounts for the pH dependence of the metal ion binding. Under the assumption of homogeneity, the expression for the amount of metal ion adsorbed corresponding to Eq. [6] is

$$Q_M = \frac{Q_{max}\,K\,(M/H^x)}{1 + K\,(M/H^x)}\,. \qquad [7]$$

Equation [7] does not lead to the observed Freundlich-like behavior. However, such behavior can be derived if it is assumed that the exchange constant is a distributed parameter. The combination of Eq. [7] with a Sips distribution function leads to

$$Q_M = \frac{Q_{max}\,\tilde{K}_M\,(M/H^x)^{\mathbf{m}}}{1 + \tilde{K}_M\,(M/H^x)^{\mathbf{m}}}, \qquad [8]$$

where M and H are the metal and proton activities (or concentrations), respectively. This equation was used for the description of the pH-dependent metal ion binding to oxides (Kinniburgh *et al.,* 1983) and for the binding of Ca and Cd to PPHA (Milne *et al.,* 1995a). The description for Ca, Cd, and Cu binding to PPHA with this model is shown in Figure 8. The parameter values are shown in Table II. The parameter x for the various cations corresponds reasonably well with the measured exchange ratio for the cations.

The $\tilde{K}$ values, one for each metal, represent the modal exchange constants for an $SH^x \leftrightarrow SM$ exchange reaction and therefore implicitly depend on the stoichiometry of the reaction, x. This makes them more difficult to interpret than affinity constants for 1:1 reactions. For example, the Cu ion, which has the highest affinity for PPHA, shows by far the lowest value of $\tilde{K}$. The values of the parameter **m,** which is a measure of the widths of the distributions of the metal–proton exchange constants, are governed by the slopes of the log–log plots at low metal ion concentrations. The lower the value of **m,** the wider the corresponding distribution and the lower the limiting slope.

The fitted adsorption maximum also varies somewhat for the various ions. However, the quality of the description is reasonably good for all three ions. Disadvantages of this approach are: (i) that it does not give much insight into the un-

Table II

Parameters for the H^x Model (Eq. [8]) for Ca, Cd, and Cu Binding to a Purified Peat Humic Acid

Metal ion	Parameter	Value
Ca	log $\tilde{K}_M$	0.85
	m	0.59
	x	0.29
	Q_{max} ($mol \cdot kg^{-1}$)	1.38
Cd	log $\tilde{K}$	−0.65
	m	0.71
	x	0.74
	Qmax ($mol \cdot kg^{-1}$)	1.24
Cu	log $\tilde{K}$	−4.69
	m	0.40
	x	1.43
	Q_{max} ($mol \cdot kg^{-1}$)	3.34

derlying adsorption mechanism; (ii) in its present form it is not possible to predict the effect of competition between two or three metal ions; and (iii) since no coupling is made with the basic proton binding, it is also not possible to estimate what effect a shift of the proton affinity distribution would have on the metal ion binding. An advantage is that it is a simple approach with relatively few parameters, and appears to be reasonably faithful in reflecting both the pH dependence of metal ion binding and the H^+/M^{z+} exchange ratio for a wide range of systems of environmental importance.

Far from the adsorption maximum, Eq. [8] can be simplified to what has been called the pH-dependent Freundlich isotherm, or two-component Freundlich isotherm (Boekhold *et al.*, 1993; Temminghoff *et al.*, 1994, 1995). This simplified equation has been used to describe the pH-dependent metal ion binding in various soil systems:

$$Q_M = KM^a H^b. \qquad [9]$$

The advantage of the derivation of this equation from Eq. [8] is that it shows clearly that the parameter K in Eq. [9] is proportional to the adsorption maximum which is a function of the organic matter content of the soil or sediment, providing that the organic matter is the dominant reactive component. However, one should be aware that metal (hydr)oxides or a combination of metal (hydr)oxides, organic matter, and clay can control the metal ion binding in a natural system. In this case, Eq. [9] may still fit the data (Altmann and Buffle, 1988). This can be seen as an advantage but means that great care must be taken over the interpretation of the parameters. The derivation also shows that $a = \mathbf{m}$ and $b = -x\mathbf{m}$, so that it is possible to obtain an estimate of the average exchange ratio from Eq. [9], i.e., $x = -b/a$.

The contributions of the solid and dissolved organic matter to metal ion binding in a field sample can be estimated if the distribution of the metal ion over the solid and solution phases is *a priori* estimated using a competitive binding model and if all the interactions are taken into account as well as possible (Goody *et al.*, 1995; Benedetti *et al.*, 1996b). In its present form, the preceding nonstoichiometric exchange approach is too limited an approach since it is at present not clear how the approach can be extended to include competition between various metal ions.

C. Competitive Adsorption: Correlated, Congruent Distribution Functions

All sites that can bind a proton can in principal also bind a bivalent metal ion. The affinity, however, can vary from that for protons. Let us assume for the time being that one has chosen to represent the proton binding behavior by using eight different sites. The number of sites per gram of humic substance has also already been determined from the interpretation of the acid–base behavior. In order to account for the metal ion binding, one first has to decide on the basic binding equations that are to be used. Even if one assumes that the ions bind only in a monodentate fashion as M^{2+}, one would still need to estimate an additional eight binding constants for each metal ion. Assuming a combination of mono- and bidentate binding and also considering the binding of partly hydrolyzed species leads to an even more complex situation. It is therefore useful to discuss the applicability of some simplifying assumptions.

One simplifying assumption could be to assume that the metal ions bind in monodentate fashion as M^{2+} and that the affinity distributions for the metal ion binding and the proton binding are fully correlated and congruent, resulting in a metal ion affinity distribution that has the same shape as the proton affinity distribution but is shifted along the affinity axis. This is essentially the model adopted by Dobbs *et al.* (1989) and Manunza *et al.* (1995), who assume a Gaussian distribution model which is solved numerically. Dobbs *et al.* (1989) assume a single Gaussian distribution whereas Manunza *et al.* (1995) assume two Gaussian distributions which are broadly identified with "carboxylic" and "phenolic" groups. When two such distributions are involved, there are two possibilities for the relationship between the proton and the metal binding—either a similar shift in log K ($=\log(K_M/K_H)$) is applied to both distributions or different shifts are applied to each distribution. We have called these two cases "congruent" distributions and "highly correlated" distributions, respectively (Milne *et al.,* 1995a). Apart from the computational simplicity, the advantage of the congruency assumption is that an analytical competitive LF equation can be derived for these conditions (van Riemsdijk *et al.,* 1987) which is similar to the Gaussian model but computationally much more straightforward. The competitive LF equation uses the competitive Langmuir equation as the adsorption equation for a homogeneous subset of sites, the so-called "local isotherm" equation.

The competitive Langmuir equation and the competitive LF equation (without an electrostatic component) have the desirable characteristic that the proton–metal ion exchange and the pH shift of the isotherm are directly coupled. However, without an electrostatic model this combination gives a relatively poor description of humic–metal ion binding data, as has been shown for Cd binding to fulvic acids at constant ionic strength (Koopal *et al.,* 1994). The model gives a linear isotherm at low metal ion binding and at low pH which is not in accord with the experimental data. The competitive LF equation can also be combined with a chosen double-layer model. The electrostatic Boltzmann term can then be seen as a correction term to account for some "nonideality." Since the valence of the proton differs from the valence of divalent metal ions, the overall electrochemical affinity is no longer fully coupled even though the intrinsic affinity distributions are. The competitive LF equation can be written as

$$\theta_{i,T} = \frac{Q_i}{Q_{max}} = \frac{\tilde{K}_i C_{s,i}}{\Sigma_j \tilde{K}_j C_{s,j}} \times \frac{\Sigma_j (\tilde{K}_j C_{s,j})^{\mathbf{m}}}{1 + \Sigma_j (\tilde{K}_j C_{s,j})^{\mathbf{m}}}, \qquad [10]$$

where $\theta_{i,T}$ is the fractional surface coverage of the bound ion i; Q_i is the amount of i bound; Q_{max} is the number of binding sites present and is equivalent to the maximum possible monodentate binding; $\tilde{K}_i$ is the modal binding constant for ion i and $C_{s,i}$ is its surface or local concentration; and **m** is related to the width of the underlying Sips distibution. It is clear from Eq. [10] that the shapes of the monocomponent distributions are all the same since, for a single component, the term on the left of the "times" symbol cancels and the remaining term has the same exponent, **m,** for each component. The summations apply to all species present, including the proton. It follows from Eq. [10] that the assumption of a full correlation between the distribution functions strongly suppresses the effect of the heterogeneity, leading to a linear isotherm at low metal ion concentrations.

The slope of the log–log isotherm plot is also affected by the electrostatics since the negative charge of the humic substances is considerably reduced on progressive binding of Ca and Cd because of the low exchange ratio. It has been shown elsewhere (Milne *et al.,* 1995a) that the coupling of a LF model with a cylindrical double-layer model can give a good description of the basic charging and pH-dependent binding of Ca and Cd to PPHA, providing that only the carboxylic-type sites were assumed to bind the metal ions. This means that only one metal-ion-specific parameter is needed to extend the proton binding model to the binding of Ca and Cd. The derived affinity constants are reasonable when compared with those for metal ion binding to the carboxylic groups of small organic molecules. The fact that only binding to the carboxylic-type groups needs to be invoked does not mean that the second ("phenolic") distribution does not affect the pH-dependent metal ion binding. The presence of these groups has an important effect on the pH dependence at high pH through their effect on the overall negative charge of the humic material and its consequent effect on the electrostatic Boltzmann factor.

At high pH, the carboxylic groups are almost completely dissociated and thus cannot give rise to a pH dependence! The use of an appropriate double-layer model can therefore alter the pH dependence of the model in a realistic way. Invoking an electrostatic model in this case does not lead to a complication but to an elegant and simple description which can also be physically interpreted. The approach is successful for the binding of protons, Ca, and Cd, but it fails completely for the description of the binding of Cu. The very low slopes of the measured Cu isotherms on a log–log plot and the strong pH dependence cannot be reproduced with this approach. All models which assume a full correlation between proton and metal ion binding (i.e., congruent distributions) are expected to predict an almost linear slope at low metal ion binding and will fail to represent the Cu binding to PPHA. This includes Model V (Tipping and Hurley, 1992), the Dobbs model (Dobbs *et al.,* 1989) as used in the MINTEQA2 speciation program (Allison *et al.,* 1991), and the correlated bimodal Gaussian model (Manunza *et al.,* 1995). Copper binding to other humics shows similar behavior to that of PPHA, and so this conclusion is expected to be quite general (Kinniburgh *et al.,* 1996). The slope of the log–log plot corresponds with the limiting slope that would have been expected from the exponent **m** that is found for both the carboxylic and the phenolic peaks (Milne *et al.,* 1995a) based on an analysis of basic charging behavior.

The suppression of the heterogeneity due to the assumption of correlated distributions which works satisfactorily for Ca and Cd seems not to occur for Cu binding. Further model development is therefore required in order to be able to obtain a satisfactory model that can account for all of the available data.

D. Competitive Adsorption: Uncorrelated, Congruent Distributions

Rudzinski *et al.* (1993) suggested that the following isotherm equation is related to a LF distribution, assuming that there is no correlation between the individual distributions:

$$\theta_i = \frac{(\tilde{K}_i C_{s,i})^{n_i}}{1 + (\Sigma_j \tilde{K}_j C_{s,j})^{n_j}} \, . \qquad [11]$$

The notation is similar to that for Eq. [10] except that n_i can be thought of as a nonideality factor. In Eq. [11], different values of n_i are present, which suggests that the monocomponent equations may have different widths. At present, there is no rigorous derivation of an analytical binding equation for the case in which the monocomponent distributions have different widths. Equation [11] in its general form is not thermodynamically consistent. We will come back to the general formulation of Eq. [11] in the next section. A special case of Eq. [11] is the situation where the n_i values are equal. In this case, Eq. [11] is thermodynamically consistent and Eq. [11] becomes Eq. [12], which can be compared with Eq. [10]:

$$\theta_i = \frac{(\tilde{K}_i C_{s,i})^{n}}{1 + (\Sigma_j \tilde{K}_j C_{s,j})^{n}} \, . \qquad [12]$$

Equations [10] and [12] differ in the way in which the competition is described. Equation [12] will result in Freundlich behavior at low metal ion concentrations whereas Eq. [10] becomes linear. Equation [12] is of little or no use for the description of metal ion binding to natural organic matter. The exponent for proton binding is on the order of 0.5 and so the binding of Ca and Cd at constant pH should give a similar slope in log–log plots at low concentrations and low pH. This is clearly not the case (Fig. 8). One might argue that it could be an appropriate equation for the description of Cu binding since the limiting slope for Cu binding is also about 0.5. However, this approach can give at most a theoretical proton/metal ion exchange ratio of 1. Since Eq. [12] is thermodynamically consistent, the exchange ratio and pH dependence are strictly coupled, which means that Eq. [12] will inevitably not be able to reproduce the observed large pH dependency of Cu binding.

E. The Nonideal Competitive Adsorption (NICA) Equation

The analysis so far shows that the heterogeneity is, to a greater or lesser extent, ion specific. Ion-specific heterogeneity may imply differences in the degree of correlation between the individual distributions, but also differences in the width of the individual distributions. Differences in the width of distributions for the same surface and different adsorbing components are well known in the field of gas adsorption. The modeling of competitive adsorption for such systems is complex and is the subject of ongoing research. Koopal *et al.* (1994) developed a new analytical competitive adsorption equation which is based upon the idea that part of the interaction is due to ion-specific heterogeneity or nonideality, and part is due to a generic heterogeneity which is considered to be congruent, fully correlated, and of the LF type. The starting point of their derivation is that they use Eq. [11] as the local isotherm, which gives

$$\theta_{i,L} = \frac{(K_i C_{s,i})^{n_i}}{1 + (\Sigma_j K_j C_{s,j})^{n_i}}. \qquad [13]$$

The ion-specific nonideality which this equation implies is also present to a certain extent in the use of Eq. [10] with an electrostatic model since the Boltzmann term differs for ions with differing charge. The extended Henderson–Hasselbalch equation (Katchalski and Spitnik, 1947), which is often used to describe ion binding to chemically homogeneous polyelectrolytes has the same form as Eq. [12] for the monocomponent case. The pH shift and the exchange ratio are only coupled in the way expected from a thermodynamic analysis if all of the n_i values are equal. The "nonideality" of Eq. [12] when n values are not all equal may be a theoretical disadvantage, but it can also have a practical advantage as we will show later.

The use of Eq. [13] as a local isotherm with the LF distribution function, and assuming full correlation between the individual distributions, leads to the NICA equation (Koopal *et al.*, 1994):

$$\theta_{i,T} = \frac{Q_i}{Q_{max}} = \frac{(\tilde{K}_i C_{s,i})^{n_i}}{\Sigma_j (\tilde{K}_j C_{s,j})^{n_j}} \times \frac{\{\Sigma_j \tilde{K}_j C_{s,j}\}^p}{1 + \{\Sigma_j (\tilde{K}_j C_{s,j})^{n_j}\}^p}. \qquad [14]$$

The monocomponent NICA equation simplifies to the classical LF equation, where the heterogeneity exponent **m** of the LF equation is now equal to the product of the ion-specific exponent n_i of the local isotherm and the parameter p which characterizes the width of the generic distribution. The ion-specific nonideality can therefore only be found from competitive binding studies. The NICA model was originally applied to Cd binding data for a water-derived fulvic acid (Koopal *et al.,* 1994). The data analyzed were from Saar and Weber (1979). The (monomodal) NICA model gives an excellent fit to the data. The fitted parameters are given in Table III.

It is interesting to note that the nonideality exponents are similar for the proton and the Cd, which thus leads to a thermodynamically consistent description of the data. The basic charging data that can be derived from this model are in very good agrement with the charging behavior that is expected for fulvic acids (de Wit *et al.,* 1993a). The average proton affinity constant and the monocomponent heterogeneity parameter for the proton are well within what is expected for fulvic acids. Unfortunately no charging data are available to test the prediction. The nonideality that is required to get a good description can thus either be obtained by the use of an electrostatic model as discussed in Section II.D, or can be incorporated into a nonideality exponent.

It is interesting to note that the nonideality exponent is, within experimental error, equal to the exponent for Cd and Ca that was obtained in the nonstoichiometric exchange approach for PPHA. In the NICA approach there is no need to introduce a fitted x parameter since the pH dependence and the exchange ratio are defined implicitly by the competitive binding model. The bimodal NICA model has been applied to the extensive data set of ion binding to PPHA both without (Benedetti *et al.,* 1995) and with an electrostatic model (Kinniburgh *et al.,* 1996).

The NICA–Donnan model gives an excellent fit to the proton charging data for

Table III

Fitted Parameters for the pH-Dependent Binding of Cd to a Water-Derived Fulvic Acid Based on the Monomodal NICA Model, Eq. [15] from Koopal *et al.* (1994)

Parameter	Value
$\log \tilde{K}_H$	3.1
n_H	0.66
$\log \tilde{K}_{Cd}$	0.48
n_{Cd}	0.62
p	0.56
Q_{max} ($mol \cdot kg^{-1}$)	5.31

PPHA (Kinniburgh *et al.,* 1996) and results in an excellent master curve for proton binding (Fig. 9), as well as a good fit to the pH-dependent binding of Ca, Cd, and Cu. An advantage of the use of an electrostatic submodel is that the effect of a change in ionic strength can in principle be described and can, for example, explain the ionic strength dependence of Cd and Cu binding (Kinniburgh *et al.,* 1996). Another advantage is that there is no need to assume binding of Ca to the "phenolic"-type sites in order to account for ion binding at high pH. The relevance of metal ion binding in the double layer can also be assessed. This tends to be most significant at high metal ion concentrations and low pH.

The results of the application of the NICA–Donnan model indicate that Ca binding is mainly due to binding to the carboxylic sites and in the double layer. Binding of Cd takes place mainly on carboxylic sites at Cd^{2+} concentrations above 10^{-8} mol·L^{-1} and on phenolic sites at lower concentrations. Only at low pH and high Cd concentrations does Cd binding in the double layer (in the presence of 0.1 mol·L^{-1} indifferent salt) become important. Larive *et al.* (1996) have recently shown with ^{113}Cd NMR that Cd binding to Suwannee River fulvic acid at millimolar Cd^{2+} concentrations is predominantly by monodentate binding to carboxylic groups. This confirms our speciation results. For Cu binding to PPHA, the "phenolic"-type sites dominate the binding at low free metal ion concentrations whereas the "carboxylic"-type sites become more important at higher free Cu con-

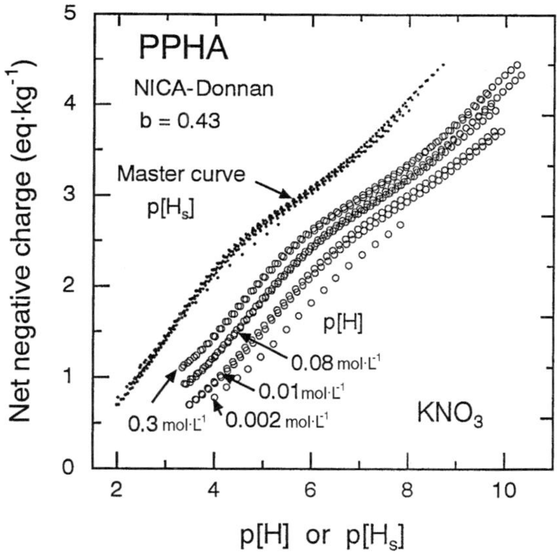

Figure 9 The experimentally measured charging curves for the purified peat humic acid (PPHA) at various ionic strengths and the estimated "master curve" based on the NICA–Donnan model (Kinniburgh *et al.,* 1996).

centrations (Kinniburgh *et al.,* 1996). The NICA–Donnan approach gives a good prediction of the competition between Ca and Cd, and between Ca and Cu at different pH values (Kinniburgh *et al.,* 1996). The model has also been shown to give good results when applied to soils, and to a freshwater lake (Benedetti *et al.,* 1996b).

It is obvious from the previous discussion that this approach cannot lead to a correct description of the exchange ratio of Cu. The fact that the model does indeed give a very good description of the pH dependence without giving the correct exchange ratio is caused by the thermodynamic inconsistency (which can also be interpreted as a "nonideality" of the model) when all n_i values are not equal.

F. Competitive Monodentate and Bidentate Metal Ion Binding Combined with Ion-Specific Nonideality: The CONICA Model

An exchange ratio greater than 1 at low pH clearly indicates that the mechanism of Cu binding cannot be due to monodentate binding, but must at least be partly due to bidentate binding. In contrast, the low exchange ratios for Ca and Cd are indicative of predominantly monodentate binding. Therefore a general purpose mechanistic model for the competitive binding of ions to heterogeneous surfaces should consider both monodentate and bidentate binding. The starting point in developing such a model can therefore be to consider protonation as occurring in two steps. As with the simple dibasic weak acid, this involves the release of the second proton only after the first proton has already been released.

The first attempt to combine such a two-step protonation concept with heterogeneity using an analytical competitive adsorption model was made by van Riemsdijk *et al.* (1986). The proton binding was assumed to occur in two consecutive steps:

$$S^{2-} + H^+ = SH^- \qquad (K_{SH}) \qquad [15a]$$

$$SH^- + H^+ = SH_2 \qquad (K_{SH_2}). \qquad [15b]$$

In the initial study, it was assumed that the parameter K_{SH} was a distributed parameter and that K_{SH_2} was independent of the heterogeneity. This assumption simplifies the derivation of the competitive equations, but a more recent heterogeneity analysis of the proton binding to PPHA humic acid has shown that this assumption is not physically realistic. Therefore van Riemsdijk *et al.* (1996) have derived equations in which both K_{SH} and K_{SH_2} are distributed parameters. Monodentate and bidentate metal ion binding are described by, respectively,

$$SH^- + M^{2+} = SHM^+ \qquad (K_{SHM}) \qquad [16a]$$

$$S^{2-} + M^{2+} = SM \qquad (K_{SM}). \qquad [16b]$$

The derivations of the competitive binding equations are calculated by taking ion-specific nonideality into account. The derivation and the resulting equations are more complex than the equations of the NICA model—four equations are needed to describe the overall proton and metal ion binding, one for each of the species SH (low-affinity proton), SH_2 (high-affinity proton), SHM (monodentate metal), and SM (bidentate metal) (van Riemsdijk *et al.,* 1996). The resulting model is called the CONICA model (COnsecutive NICA model) because there is a considerable similarity between the bimodal NICA model and the CONICA model. The CONICA model has one less parameter than the bimodal NICA model, and as with the NICA model, it can be combined with a Donnan model to give the binding of nonspecifically bound counterions. The basic charging data and the pH-dependent binding of Ca, Cd, and Cu can also be described very well with the CONICA–Donnan model, and the predicted exchange ratio for Ca and Cd is in reasonable agreement with the data, as expected. The predicted exchange ratio for Cu at pH 4 now equals 1.5, which is in reasonable agreement with the experimental measurements. The predicted exchange ratio at pH 8 is around 0.9, which is clearly below the experimentally measured value at this pH.

Bidentate binding significantly increases the predicted exchange ratio for pH values below pH 6. At pH 8, there are hardly any differences from the results obtained with the bimodal NICA–Donnan model. This observation is relatively easy to understand. At pH 8, most of the carboxylic-type protons are already desorbed and the formation of a bidentate bond will in general release only one proton at this pH. Two protons could be released at high pH if the metal ion reacts with two sites that have a high affinity for the proton. However, in the derivation of the CONICA model, it is assumed that the first proton that adsorbs (Eq. [15a]) has on average a high affinity ("phenolic") and that the affinity of the second step (Eq. [15b]) has on average a low affinity ("carboxylic"). The chance that a bidentate metal ion complex is formed with two high-affinity proton sites is therefore low in the CONICA model. The parameter values used to describe the binding of protons and the three metal ions are shown in Table IV. For details of the CONICA equations and the notation used, see van Riemsdijk *et al.* (1996). Limited data (pH 5) were also available for Pb (Kinniburgh *et al.,* 1996) and were also included. The parameters were estimated by simultaneously optimizing all of the parameters using a nonlinear least-squares approach. This involved combining the H^+, Ca, Cd, Cu, and Pb data into a single data file. Only one in three of the proton binding data points was used in order to balance the number of data points for each component. This and the simultaneous fitting explain the small differences in parameter values for the H and Cu binding compared with our earlier analysis (van Riemsdijk *et al.,* 1996). In general, the quality of the overall fits of the CONICA–Donnan model was very similar to that for the NICA–Donnan model (Kinniburgh *et al.,* 1996).

Table IV

Fitted Parameters for the Binding of H, Ca, Cd, and Pb to Purified Peat Humic Acid Based on the Two-Step Protonation (CONICA–Donnan) Model which Incorporates both Monodentate and Bidentate Binding

	N_s (mol sites·kg^{-1})	p_1	p_2	Donnan b, Eq. [2]
	2.67	0.60	0.61	0.34
	High-affinity proton		Low-affinity proton	
	$\log \tilde{K}_{SH}$	n_1	$\log \tilde{K}_{SH_2}$	n_2
	8.00	0.61	2.95	0.80
	Monodentate binding		Bidentate binding	
Metal	$\log \tilde{K}_{SHM}$	1_M	$\log \tilde{K}_{SM}$	ν_M
Ca	−2.93	0.50	−2.04	0.34
Cd	−0.08	0.79	2.04	0.53
Cu	0.24	0.48	5.93	0.34
Pb	1.11	0.65	2.82	1.00 (fixed)

The extent of competition between Ca and Cd and between Ca and Cu as predicted by the model and as measured can be seen in Figure 10. The extent of competition was predicted from the single-metal data (parameter values in Table IV)—no competition data were used during the parameter estimation. The predictions are generally good. Calcium competition is considerably greater for Cd than for Cu. There is a small underprediction of the extent of Ca competition for Cd at pH 10. The CONICA–Donnan model can describe most of the available data—the principal failing is the low predicted H^+/Cu^{2+} exchange ratio (0.8 to 1.5), especially at high pH (van Riemsdijk *et al.,* 1996). The high measured exchange ratio for Cu (1.3 to 1.8) suggests that bidentate formation is taking place during Cu binding. This is also supported by spectroscopic evidence, although the evidence is not unambiguous (Stevenson, 1994, p. 394). The high exchange ratio at high pH can in principle be explained if it is assumed that the bidentate binding predominantly involves sites where both proton steps have a high affinity. It is, however, physically unlikely that this is the only cause, considering the large amounts of Cu that can be bound at pH 8. The binding of partially hydrolyzed Cu at high pH is another quite likely explanation (Tipping *et al.,* 1995) since such hydrolysis releases an extra proton. It is not yet clear how frequently such a hydrolysis reaction will need to be invoked, but it appears that it may be needed to describe the pH de-

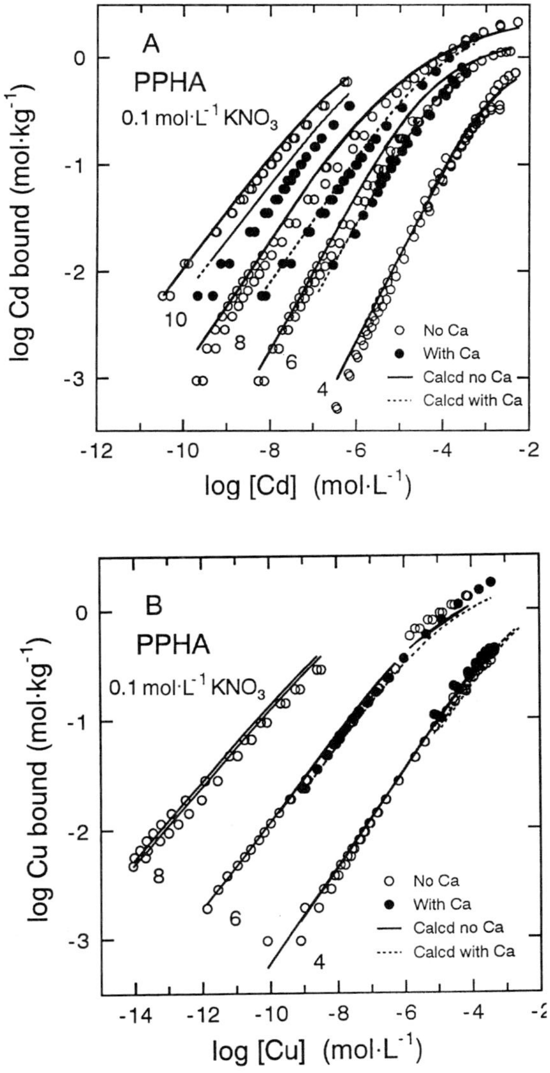

Figure 10 Effect of Ca on (A) Cd (pH 6, 8, and 10) and (B) Cu (pH 4 and 6) binding measured experimentally and calculated according to the CONICA–Donnan model. The free Ca concentration ranged from approximately $10^{-3.5}$ to $10^{-3.1}$ mol·L^{-1}. The Cd^{-}(pH 4) and Cu (pH 8) isotherms are also shown. The predictions of Ca competition (dashed lines) are based on the parameters derived from the single-metal data only.

pendence of binding of metal ions that hydrolyze strongly such as Cu^{2+}, Al^{3+}, and UO_2^{2+}.

V. CONCLUSIONS

Essential features that need to be considered by a complete model for describing ion binding to humic substances are: (1) the possibility of accounting for the large effect of chemical heterogeneity on competitive ion binding; (2) the bimodal character of proton binding, which is also of importance for the binding of Ca and Cd at high pH, and for Cu over the whole pH range; (3) bidentate ion binding and probably metal ion hydrolysis, which appear to be essential features in the case of Cu binding; and (4) that although the NICA/CONICA models can account in part for electrostatic effects via their nonideality exponents, it is an advantage to combine the ion binding models with a simple electrostatic model in order to account for effects arising from changes in the ionic strength, to account explicitly for electrostatic competitive interactions, and to account for the nonspecific binding of counterions in the double layer. This is particularly important for the macro ions (Ca, Na, K, and Al at low pH) and especially at high solid/solution ratios where the electrical double layer or Donnan volume makes up a significant part of the total volume of solution.

The bimodal NICA–Donnan model has been shown to be very successful in describing the pH-dependent metal ion binding of several metal ions over a very wide range of experimental conditions. The model also gives satisfactory predictions of the competition between various metal ions as far as the available data are able to show, and it has also been shown to be applicable to real environmental systems. The possibility of describing the pH dependence of Cu binding without taking into account bidentate metal ion binding stems from the nonideality of the model in the sense that it is thermodynamically inconsistent if the nonideality coefficients (n_i) are significantly different from each other. The inconsistency shows itself clearly in the incorrect prediction of the proton/metal exchange ratio for Cu. This weakness is often not very important in practice since the small release of protons in the environment at the low levels of trace metal binding usually found has little or no effect on the resulting pH of the system. Nevertheless, from a fundamental point of view, the weakness is a serious problem. This stimulated us to develop the CONICA model, which provides a considerable step forward by taking into account bidentate binding. However, the CONICA model remains thermodynamically inconsistent. Ongoing work suggests an alternative solution. We have found that it is possible to make the NICA model fully thermodynamically consistent by simply scaling the amount of each ion bound by its nonideality coefficient. This allows proton/metal exchange ratios to be greater than one ($n_i < n_H$) and appears to fit our existing data well. Fortunately, this consistent form of the NICA model requires no increase in the number of model parameters. We are continuing to explore its application.

Natural systems are characterized by the presence of a wide range of metal ions in different concentration ratios, at different pH values, and with different concentrations of heterogeneous binding ligands present. It is therefore important that speciation models can handle this complexity properly. This implies that speciation models should be able to handle multimetal ion competition in a reasonably rigorous way. This is not at all trivial and the predictions are likely to be highly model dependent. For example, the inclusion of nonideality and whether one uses a monomodal or bimodal approach are important decisions that can strongly affect the predicted metal–metal competition. The model development must be guided to a large extent by what we observe in nature. Therefore in order to test and develop such models further, more good data are required on multimetal ion systems.

ACKNOWLEDGMENT

D.G.K. publishes with the permission of the Director of the British Geological Survey (NERC).

REFERENCES

Allison, J. D., Brown, D. S., and Novo-Gradac, K. J. 1991. MINTEQA2/PRODEFA2: A geochemical assessment model for environmental systems, version 3.0 user's manual. EPA/600/3-91/021, U.S. Environmental Protection Agency, Athens, GA.

Altmann, R. S., and Buffle, J. 1988. The use of differential equilibrium functions for interpretation of metal binding in complex ligand systems: Its relation to site occupation and site affinity distributions. *Geochim. Cosmochim. Acta* 52:1505–1519.

Barak, P., and Chen, Y. 1992. Equivalent radii of humic macromolecules from acid-base titration. *Soil Sci* 154:184–195.

Bartschat, B. M., Cabaniss, S. E., and Morel, F. M. M. 1992. Oligoelectrolyte model for cation binding by humic substances. *Environ. Sci. Technol.* 26:284–294.

Benedetti, M. F., Milne, C. J., Kinniburgh, D. G., van Riemsdijk, W. H., and Koopal, L. K. 1995. Metal-ion binding to humic substances—Application of the nonideal competitive adsorption model. *Environ. Sci. Technol.* 29:446–457.

Benedetti, M. F., van Riemsdijk, W. H., and Koopal, L. K. 1996a. Humic substances considered as a heterogeneous Donnan gel phase. *Environ. Sci. Technol.* 30:1805–1813.

Benedetti, M. F., van Riemsdijk, W. H., Koopal, L. K., Kinniburgh, D. G., Gooddy, D. C., and Milne, C. J. 1996b. Metal ion binding by natural organic matter: From the model to the field. *Geochim. Cosmochim. Acta* 60:2503–2513.

Benjamin, M. M., and Leckie, J. O. 1981. Multiple-site adsorption of Cd, Cu, Zn, and Pb on amorphous iron oxyhydroxide. *J. Colloid Interface Sci.* 79:209–221.

Boekhold, A. E., Temminghoff, E. J. M., and van der Zee, S. E. A. T. M. 1993. Influence of electrolyte composition and pH on cadmium sorption by an acid sandy soil. *J. Soil Sci.* 44:85–96.

Borkovec, M., Rusch, U., and Westall, J. C. 1998. Modeling of competitive ion binding to heterogeneous materials with affinity distributions. *In* "Adsorption of Metals by Geomedia" (E. A. Jenne, Ed.). Academic Press, San Diego.

Cabaniss, S. E., Morel, F. M. M., and Marinsky, J. A. 1989. Comment on "A unified physicochemical description of the protonation and metal ion complexation equilibria of natural organic acids (humic and fulvic acids)." *Environ. Sci. Technol.* 23:746–747.

Cerník, M., Borkovec, M., and Westall, J. C. 1996. Affinity distribution description of competitive ion-binding to heterogeneous materials. *Langmuir,* 12:6127–6137.

De Wit, J. C. M., van Riemsdijk, W. H., and Koopal, L. K. 1993a. Proton binding to humic substances. 1. Electrostatic effects. *Environ. Sci. Technol.* 27:2005–2014.

De Wit, J. C. M., van Riemsdijk, W. H., and Koopal, L. K. 1993b. Proton binding to humic substances. 2. Chemical heterogeneity and adsorption models. *Environ. Sci. Technol.* 27:2015–2022.

De Wit, J. C. M., van Riemsdijk, W. H., Nederlof, M. M., Kinniburgh, D. G., and Koopal, L. K. 1990. Analysis of ion binding on humic substances and the determination of intrinsic affinity distributions. *Anal. Chim. Acta* 232:189–207.

Dobbs, J. C., Susetyo, W., Carreira, L. A., and Azarraga, L. V. 1989. Competitive binding of protons and metal ions in humic substances by lanthanide ion probe spectroscopy. *Anal. Chem.* 61:1519–1524.

Dzombak, D. A., Fish, W., and Morel, F. M. M. 1986. Metal-humate interactions. 1. Discrete ligand and continuous distribution models. *Environ. Sci. Technol.* 20:669–675.

Fokkink, L. G. J. 1987. Ion adsorption on oxides: Surface charge formation and cadmium binding on rutile and hematite. Ph.D. Thesis, Wageningen Agricultural University, Wageningen, The Netherlands.

Fokkink, L. G. J., Keizer, A. de, and Lyklema, J. 1990. Temperature dependence of cadmium adsorption on oxides. I. Experimental observation and model analysis. *J. Colloid Interface Sci.* 135:118–131.

Gooddy, D. C., Shand, P., Kinniburgh, D. G., and van Riemsdijk, W. H. 1995. Field-based partition coefficients for trace elements in soil solutions. *Eur. J. Soil Sci.* 42:265–285.

Hayes, M. H. B., MacCarthy, P., Malcolm, R. L., and Swift, R. S. 1989. "Humic Substances II: In Search of Structure." Wiley, Chichester, UK.

Honeyman, B. D., and Leckie, J. O. 1986. Macroscopic partitioning coefficients for metal ion adsorption. Proton stoichiometry at variable pH and adsorption density. *In* "Geochemical Processes at Mineral Surfaces" (J. A. Davis and K. F. Hayes, Eds.), Symposium Series Vol. 323, pp. 162–190. American Chemical Society, Washington, DC.

Katchalski, A., and Spitnik, P. 1947. Potentiometric titration of polymethacrylic acid. *J. Polymer Sci.* 2:432–446.

Kinniburgh, D. G., Barker, J. A., and Whitfield, M. 1983. A comparison of some simple adsorption isotherms for describing cation adsorption by ferrihydrite. *J. Colloid Interface Sci.* 95:370–383.

Kinniburgh, D. G., Milne, C. J., Benedetti, M. F., Pinheiro, J. P., Filius, J., Koopal, L. K., and van Riemsdijk, W. H. 1996. Metal ion binding by humic acid: Application of the NICA-Donnan model. *Environ. Sci. Technol.* 30:1687–1698.

Kinniburgh, D. G., Milne, C. J., and Venema, P. 1995. Design and construction of a personal-computer-based automatic titrator. *Soil Sci. Soc. Am. J.* 59:417–422.

Koopal, L. K., van Riemsdijk, W. H., de Wit, J. C. M., and Benedetti, M. F. 1994. Analytical isotherm equations for multicomponent adsorption to heterogeneous surfaces. *J. Colloid Interface Sci.* 166:51–60.

Larive, C. K., Rogers, A., Morton, M., and Carper, W. R. 1996. ^{113}Cd NMR binding studies of Cd-fulvic acid complexes: Evidence of fast exchange. *Environ. Sci. Technol.* 30:2828–2831.

Manunza, B., Deiana, S., Maddau, V., Gessa, C., and Seeber, R. 1995. Stability constants of metal-humate complexes: Titration data analyzed by bimodal Gausssian distribution. *Soil Sci. Soc. Am. J.* 59:1570–1574.

Milne, C. J., Kinniburgh, D. G., de Wit, J. C. M., van Riemsdijk, W. H., and Koopal, L. K. 1995a. Analysis of metal-ion binding by a peat humic acid using a simple electrostatic model. *J. Colloid Interface Sci.* 175:448–460.

Milne, C. J., Kinniburgh, D. G., de Wit, J. C. M., van Riemsdijk, W. H., and Koopal, L. K. 1995b. Analysis of proton binding by a peat humic acid using a simple electrostatic model. *Geochim. Cosmochim. Acta* 59:1101–1112.

Nederlof, M. M., de Wit, J. C. M., van Riemsdijk, W. H., and Koopal, L. K. 1993. Determination of proton affinity distributions for humic substances. *Environ. Sci. Technol.* 27:846–856.

Nederlof, M. M., van Riemsdijk, W. H., and Koopal, L. K. 1990. Determination of adsorption affinity distributions: A general framework for methods related to local isotherm approximations. *J. Colloid Interface Sci.* 135:410–426.

Perdue, E. M., and Lytle, C. R. 1983. Distribution model for binding protons and metal ions by humic substances. *Environ. Sci. Technol.* 17:654–660.

Perona, M. J., and Leckie, J. O. 1985. Proton stoichiometry for the adsorption of cations on oxide surfaces. *J. Colloid Interface Sci.* 106:64–69.

Plette, A. C. C., van Riemsdijk, W. H., Benedetti, M. F., and Wal, A. van der. 1995. pH dependent charging behavior of isolated cell walls of a gram-positive soil bacterium. *J. Colloid Interface Sci.* 173:354–363.

Posner, A. M. 1966. The humic acids extracted by various reagents from a soil. Part I. Yield, inorganic components, and titration curves. *J. Soil Sci.* 17:65–78.

Robertson, A. 1996. Goethite/humic acid interactions and their effects on Cu(II) binding. Ph.D. Thesis, Stanford University, CA.

Rudzinski, W., Charmas, R., Partyka, S., and Bottero, J. Y. 1993. On the nature of the energetic surface heterogeneity in ion adsorption at a water/oxide interface: Theoretical studies of some special features of ion adsorption at low ion concentrations. *Langmuir* 9:2641–2651.

Saar, R. A., and Weber, J. H. 1979. Complexation of cadmium(II) with water- and soil-derived fulvic acids: Effect of pH and fulvic acid concentration. *Can. J. Chem.* 57:1263–1268.

Stevenson, F. J. 1994. "Humus Chemistry: Genesis, Composition, Reactions," 2nd ed. Wiley, New York.

Temminghoff, E. J. M., van der Zee, S. E. A. T. M., and de Haan, F. A. M. 1995. Speciation and Ca competition effects on cadmium sorption by sandy soil at various pH values. *Eur. J. Soil Sci.* 46:649–655.

Temminghoff, E. J. M., van der Zee, S. E. A. T. M., and Keizer, M. 1994. The influence of pH on the desorption and speciation of Cu in a sandy soil. *Soil Sci.* 158:398–408.

Tipping, E. 1993. Modeling the competition between alkaline earth cations and trace metal species for binding by humic substances. *Environ. Sci. Technol.* 27:520–529.

Tipping, E., and Hurley, M. A. 1992. A unifying model of cation binding by humic substances. *Geochim. Cosmochim. Acta* 56:3627–3641.

Tipping, E., Fitch, A., and Stevenson, F. J. 1995. Proton and Cu binding by humic acid: Application of a discrete-site/electrostatic ion-binding model. *Eur. J. Soil Sci.* 46:95–101.

van Riemsdijk, W. H., Bolt, G. H., Koopal, L. K., and Blaakmeer, J. 1986. Electrolyte adsorption on heterogeneous surfaces: Adsorption models. *J. Colloid Interface Sci.* 109:219–228.

van Riemsdijk, W. H., Bolt, G. H., and Koopal, L. K. 1991. The electrified interface of the soil solid phase. *In* "Interactions at the Soil Colloid-Soil Solution Interface" (G. H. Bolt, Ed.), pp. 81–113. Kluwer Academic, Dordrecht, The Netherlands.

van Riemsdijk, W. H., de Wit, J. C. M., Koopal, L. K., and Bolt, G. H. 1987. Metal ion adsorption on heterogeneous surfaces: Adsorption models. *J. Colloid Interface Sci.* 116:511–522.

van Riemsdijk, W. H., de Wit, J. C. M., Mous, S. L. J., Koopal, L. K., and Kinniburgh, D. G. 1996. An analytical isotherm equation (CONICA) for nonideal mono- and bidentate competitive ion adsorption to heterogeneous surfaces. *J. Colloid Interface Sci.* 183:35–50.

Venema, P., Hiemstra, T., and van Riemsdijk, W. H. 1996. Comparison of different site binding models for cation sorption: Description of pH dependency, salt concentration and cation-proton exchange. *J. Colloid Interface Sci.* 181:45–59.

Westall, J. C., Jones, J. D., Turner, G. D., and Zachara, J. M. 1995. Models for association of metal-ions with heterogeneous environmental sorbents. 1. Complexation of Co(II) by leonardite humic-acid as a function of pH and NaClO4 concentration. *Environ. Sci. Technol.* 29:951–959.

Xue, H. B., and Sigg, L. 1993. Free cupric ion concentration and Cu(II) speciation in a eutrophic lake. *Limnol. Oceanogr.* 38:1200–1213.

Chapter 24

Predictive Double-Layer Modeling of Metal Sorption in Mine-Drainage Systems

Kathleen S. Smith,[1] James F. Ranville,[2] Geoffrey S. Plumlee,[1] and Donald L. Macalady[2]

[1]U.S. Geological Survey, Denver Federal Center, MS 973, Denver, Colorado;
[2]Department of Chemistry and Geochemistry, Colorado School of Mines, Golden, Colorado

We describe several applications of geochemical modeling to the study of metal sorption in a number of mine-drainage systems. As a validation exercise, the computer model MINTEQA2, coupled with the Generalized Two-Layer Model for sorption reactions, was successfully used to simulate results from batch pH-dependent sorption experiments with natural water and iron-rich sediment from a mine-drainage system. Based on this validation exercise, the model was then used to examine metal sorption onto suspended iron-rich particulates in several geologically and geochemically diverse mine-drainage systems. We conclude that metal partitioning in iron-rich mine-drainage systems can be approximated by assuming that iron-rich suspended sediment primarily controls sorption reactions and that sorption onto bed sediment plays a decidedly secondary role. Finally, we present a modeling approach that can determine the self-mitigating capacity of mine drainage and we demonstrate that sorption reactions in the water column can be exploited in the mitigation and remediation of some mine-drainage problems.

I. INTRODUCTION

Mine drainage can pose serious water-quality problems. In Colorado alone, more than 1300 miles of streams are impacted by metal input associated with mining activity. An understanding of the processes that control metal transport and attenuation in these waters is critical in assessments of the potential environmental impact of mine drainage.

Sorption reactions are known to control trace-metal concentrations in many natural systems; oxide minerals sorb metals to a greater degree than most other phases generally found in natural aquatic systems (Jenne, 1977). Sorption of cations and anions onto oxide minerals is strongly pH-dependent. A comprehensive review by Dzombak and Morel (1987) provides an overview of sorption onto oxide-mineral surfaces.

Hydrous iron oxides readily sorb trace metals as demonstrated by extensive work on metal sorption onto hydrous ferric oxide (HFO) in simple, synthetic laboratory systems (e.g., Kinniburgh *et al.,* 1976; Davis and Leckie, 1978; Swallow *et al.,* 1980; Benjamin and Leckie, 1981a,b, 1982; Benjamin, 1983). The pH-dependent sorption behavior observed in these synthetic systems has been observed in some natural iron-rich systems. For example, several researchers report pH-dependent partitioning of metals in mine drainage (e.g., Robinson, 1981, 1983; Chapman *et al.,* 1983; Johnson, 1986; Filipek *et al.,* 1987; Karlsson *et al.,* 1987; Davis *et al.,* 1991; Smith, 1991). Sorptive properties of hydrous iron oxides depend on the pH of the surrounding water, the chemical composition of the water, and the ratio of the amount of dissolved trace metals to the amount of hydrous iron oxides (Dzombak and Morel, 1987; Davis and Kent, 1990; Stumm, 1992).

Computer models have been developed to simulate sorption reactions (e.g., Davis and Leckie, 1978; Dzombak and Morel, 1990). These models can adequately predict trace-metal partitioning between water and sediment in iron-rich, metal-mine-drainage systems (Smith *et al.,* 1989, 1991, 1994; Davis *et al.,* 1991; Smith, 1991, 1994; Smith and Macalady, 1991), and in some other natural systems (e.g., Tessier *et al.,* 1996).

This chapter presents results from a series of metal-sorption studies in waters from mine-drainage systems. These studies are discussed in chronological order and build upon one another to eventuate in a predictive approach to determining metal mobility in mine-drainage systems. We first describe batch, pH-dependent metal-sorption experiments with stream water and streambed sediment from a stream affected by acidic, metal-rich mine drainage. Results from these experiments provide validation for a modeling approach that assumes that hydrous iron oxides dominate metal-sorption reactions in mine-drainage systems. This approach lays the foundation for further sorption-related studies and modeling at diverse mine-drainage sites.

II. SORPTION EXPERIMENTS AND MODELING WITH NATURAL MATERIALS

In this section we discuss batch, pH-dependent metal-sorption experiments conducted with stream water and iron-rich streambed sediment from St. Kevin Gulch, a small subalpine stream impacted by mine drainage near Leadville, Colorado. These experiments were done as a validation exercise to see if computer models can be used to approximate metal partitioning in iron-rich mine-drainage systems. We find general agreement between our pH-dependent sorption experiments using natural materials from St. Kevin Gulch and literature data from laboratory experiments of metal sorption onto synthetic HFO. We describe the general setting at St. Kevin Gulch; summarize our sampling, analytical, and experimental techniques; outline our computer-modeling approach; and discuss our findings.

A. Site Description

St. Kevin Gulch is located at the headwaters of the Arkansas River about 7 km northwest of Leadville, Colorado. St. Kevin Gulch receives acidic (pH 2.8), metal-rich discharge from an abandoned mine-dump/mill-tailings area. Downstream from the mine-dump inflow, the pH of St. Kevin Gulch varies between 3.3 and 4.6, and the stream water carries elevated concentrations of Fe, Al, Cu, Cd, Zn, and several other elements derived from the oxidative weathering of sulfides in the mining wastes. The suspended sediment load in St. Kevin Gulch is very small. Downstream from the inflow, the streambed material consists of rocks, cobbles, and pebbles coated with abundant red-to-yellow iron-rich hydrous oxides that precipitate as a result of mixing of the iron-rich mine-dump inflow with higher pH water in St. Kevin Gulch. These precipitates have been identified as schwertmannite ($Fe_8O_8(OH)_6SO_4$) (Bingham, 1994), ferrihydrite ($\approx Fe_5OH_8 \cdot 4H_2O$), and goethite ($\alpha$-FeOOH) (Smith, 1991).

Stream water and streambed sediment from two sites along St. Kevin Gulch were used for pH-dependent sorption experiments. The first site, referred to as the Tailings Site, is located a few meters downstream from the mine-dump inflow into St. Kevin Gulch. The reach of stream at this site is shallow, fairly level, and in full sun, and the streambed is extensively coated with hydrous iron oxide as a result of the mine-dump inflow. The second site, referred to as the Downstream Site, is about 1 km downstream from the mine-dump inflow. Between the mine-dump inflow and the Downstream Site is a fairly steep grade, a small tributary inflow (near-neutral pH), and a few small acidic groundwater seeps. The reach of stream at the Downstream Site is shallow, fairly level, and partly shaded. The Downstream Site generally has less iron-rich sediment coating the streambed and more algal growth

Table I

Chemical Composition of Unfiltered Stream Water and Iron-Rich Streambed Sediment from the Tailings and Downstream Sites[a] at St. Kevin Gulch, Colorado, 1989

Water constituents	July		October		Sediment constituents	July		October	
	Tailings	Downstream	Tailings	Downstream		Tailings	Downstream	Tailings	Downstream
Ca ($mg \cdot L^{-1}$)	18	12	23	15	Fe (wt %)	38.4	23.5	42.9	16.1
Zn ($mg \cdot L^{-1}$)	13	8.0	16	9.5	C (wt %)	3.7	8.8	3.2	8.5
Si ($mg \cdot L^{-1}$)	7.7	6.5	11	7.9	S (wt %)	3.1	1.1	3.3	0.60
Fe ($mg \cdot L^{-1}$)	7.5	1.2	6.0	1.8	Al (wt %)	0.67	4.5	0.61	5.0
Mn ($mg \cdot L^{-1}$)	7.4	3.6	10	5.2	Si (wt %)	0.5	2	2	>10
Mg ($mg \cdot L^{-1}$)	6.7	4.2	8.4	5.2	P (wt %)	0.31	0.30	0.35	0.20
Al ($mg \cdot L^{-1}$)	4.1	2.5	5.3	3.7	K (wt %)	0.23	1.4	0.20	1.9
Na ($mg \cdot L^{-1}$)	2.8	2.7	5.2	4.5	Mg (wt %)	0.05	0.30	0.04	0.32
K ($mg \cdot L^{-1}$)	0.5	0.5	0.9	0.9	Na (wt %)	0.04	0.33	0.04	0.52
Cu ($\mu g \cdot L^{-1}$)	300	160	259	154	Ca (wt %)	0.03	0.22	0.02	0.21
Cd ($\mu g \cdot L^{-1}$)	75	45	62	47	Ti (wt %)	0.02	0.10	0.02	0.14
P ($\mu g \cdot L^{-1}$)	70	120	80	120	Zn ($mg \cdot kg^{-1}$)	243	2,480	311	1,180
Ni ($\mu g \cdot L^{-1}$)	60	50	38	24	Pb ($mg \cdot kg^{-1}$)	132	560	124	558
Sr ($\mu g \cdot L^{-1}$)	40	30	60	40	Th ($mg \cdot kg^{-1}$)	130	156	85	115
Ba ($\mu g \cdot L^{-1}$)	30	30	40	30	Mn ($mg \cdot kg^{-1}$)	106	537	109	436
Pb ($\mu g \cdot L^{-1}$)	25	25	13	18	As ($mg \cdot kg^{-1}$)	72	59	92	40
Co ($\mu g \cdot L^{-1}$)	20	10	21	12	Cu ($mg \cdot kg^{-1}$)	57	201	65	130
SO_4^{2-} ($mg \cdot L^{-1}$)	179	75	224	121	BET surface area ($m^2 \cdot g^{-1}$)	183	76	167	47
pH (field)	3.6	4.0	3.3	3.7					
DOC ($mg\ C \cdot L^{-1}$)	0.4	0.7	0.4	0.4					

Note. Sediment was air-dried prior to analysis.

[a]See site description in Section II.A of text.

than does the Tailings Site. The streambed sediment also contains significant amounts of organic carbon and detrital materials. Table I lists the stream-water and streambed-sediment composition in St. Kevin Gulch at these two sites for samples collected during midsummer and fall, 1989.

B. Methods

1. Sampling Techniques

A composite sample of the iron-rich streambed sediment that coats streambed rocks and pebbles was obtained by placing the rocks and pebbles inside a 1-qt plastic container filled with stream water. The container was shaken, thus suspending the iron-rich sediment, and the sediment–stream-water mixture was decanted into a 1-gal polyethylene jug. This procedure was repeated several times at each sampling site until the 1-gal jug was filled with a streambed-sediment–stream-water mixture from that site. Unfiltered stream water was also collected at each site for pH-dependent sorption experiments in the absence of streambed sediment.

Stream water was collected for chemical analysis at the time of sediment collection. A 1-L polyethylene container was used to collect the water, and attempts were made to collect a composite sample from several different spots and different depths at each site. An untreated subsample was poured from the 1-L bottle into an acid-washed 100-ml polyethylene bottle and saved for metals analysis. The remaining water was filtered through a 0.1-μm nitrocellulose filter using new 60-cc plastic syringes fitted with acid-washed refillable 25-mm Millipore Swinnex disc filter holders and saved for various analyses. Samples for metals analysis were acidified to pH < 2 with ultrapure nitric acid. Samples for anion analysis were kept refrigerated. Temperature, pH, specific conductance, and dissolved oxygen were determined on site.

A portion of the sediment–stream-water mixture was allowed to settle. The overlying water was then decanted and the remaining solid material was air-dried and saved for analysis.

2. Analytical Techniques

Filtrates from the sorption-rate experiments and the July batch sorption experiments were analyzed for Cu and Cd content by flameless atomic absorption spectrophotometry (AAS). Filtrates from the October batch sorption experiments were analyzed by inductively coupled plasma–mass spectrometry (ICP-MS) to determine trace-element content. Both sets of sorption-study filtrates, all stream-water samples, and streambed sediment samples were analyzed by inductively coupled argon plasma–atomic emission spectrometry (ICP-AES) for major and minor elements. Dissolved Fe and Al were analyzed by AAS, and sulfate was analyzed by ion chromatography (IC). Carbon and sulfur were determined for streambed sediment by combustion techniques. Dissolved oxygen was determined on site using a commercially available kit.

3. Sorption-Rate Determination

Experiments were performed to determine the time period required for rapid-step sorption reactions to reach completion. This information was needed to design batch sorption experiments using stream water and streambed sediment from St. Kevin Gulch. For the sorption-rate experiment, 1 L of St. Kevin Gulch stream-water–streambed-sediment mixture was collected from the Tailings Site and the Downstream Site. Sufficient 0.5 $mol \cdot L^{-1}$ $NaHCO_3$ was added to these streambed-sediment–stream-water mixtures to raise the pH to 6. The mixtures were allowed to equilibrate at the stream-water temperature of about 14°C with occasional shaking. Ten-milliliter portions of the mixtures were periodically removed and filtered through 0.1-μm nitrocellulose filters over a 22-hr period. The filtrates were acidified with ultrapure HNO_3 and stored for Cu and Cd analysis. A stream-water–streambed-sediment mixture in which the pH remained unaltered was also run for comparison.

4. Batch Sorption Experiments

Batch, pH-dependent sorption experiments were conducted using St. Kevin Gulch stream-water–streambed-sediment mixtures. The mixtures were brought back to the lab and sorption experiments were performed within 24 hr of collection. Triplicate 10-ml portions were removed from each mixture, filtered through preweighed 0.1-μm filters, and dried in an oven at 30°C. The dried material was weighed. Each mixture was then adjusted to the same weight of solid material per unit volume (2.9 $g \cdot L^{-1}$) by addition of unfiltered stream water collected concurrently with the streambed-sediments–stream-water mixtures. This was done so that the sorption data could be directly compared on a weight-per-volume basis. These adjusted mixtures were used for the batch pH-dependent sorption experiments.

Forty-milliliter portions of the adjusted mixtures were dispensed into a series of polyethylene bottles. An initial potentiometric acid–base titration was conducted on a separate 40-ml portion of each adjusted stream-water–streambed-sediment mixture to determine the amount of 0.2 $mol \cdot L^{-1}$ NaOH necessary to bring the pH of the mixture to pH > 6. The pH was then incrementally raised (by addition of various volumes of 0.2 $mol \cdot L^{-1}$ NaOH) in the series of polyethylene bottles containing 40 ml of the mixtures. This was done to obtain a pH range from ambient to pH > 6. No other substances were added. The bottles were shielded from light and allowed to sit for approximately 6 hr at room temperature with occasional shaking. After the 6-hr period, part of each mixture was separated by filtration through 0.1-μm nitrocellulose filters. The filtrates were stored for chemical analysis, and final pH values were measured in the remaining mixtures. Determination of the solute concentrations remaining in solution at each pH was used to measure pH-dependent sorption reactions. Experiments were also carried out in unfiltered stream water in the absence of the bed sediment.

5. Computer-Model Simulations

The sorption model used in this work was the Generalized Two-Layer Model (GTLM). This model, a type of surface-complexation model (see Davis and Kent, 1990), and its accompanying database of surface-complexation constants for hydrous ferric oxide (HFO) were developed by Dzombak and Morel (1990). The HFO surface-complexation constants were determined using the GTLM. Consequently, the HFO surface-complexation constants and the GTLM were internally consistent, and a combination of the model with the constants provided a best-fit mathematical description of literature data on metal sorption onto HFO. The GTLM evokes a two-site model for cation sorption that includes a subset of high-affinity cation-binding sites. The model assumes that uncomplexed M^{2+} is the only metal species that sorbs onto HFO.

Computer-model simulations were conducted using the equilibrium speciation program MINTEQA2 (Allison *et al.*, 1991). This program can incorporate the GTLM and compute sorption reactions and solution equilibria simultaneously in a self-consistent manner. St. Kevin Gulch water chemistry (Table I) at each site was entered in MINTEQA2. The pH was fixed at desired values (incremental values ranging from 3 to 7 or 8), and the P_{CO_2} was fixed at ambient atmospheric levels ($10^{-3.5}$ atm). Neither trace-metal hydroxide and carbonate solid phases nor anglesite ($PbSO_4$) exceeded saturation in the computer-model simulations. Only two solid phases, ferrihydrite and amorphous aluminum hydroxide, were allowed to precipitate in the simulations because they were considered to be the most likely solids to control Fe and Al solubilities during the 6-hr time scale of the pH-dependent sorption experiments. The equilibrium formation constant used for ferrihydrite was log K = 4.891 (MINTEQA2 database), and that for amorphous aluminum hydroxide was log K = 10.8 (Nordstrom *et al.*, 1990; the MINTEQA2 database was modified to incorporate a more accurate constant for amorphous aluminum hydroxide).

HFO was the sole sorbent material in the computer-model simulations. All of the iron in the streambed sediments (Table I) was assumed to be present as HFO and calculations of binding-site concentrations and total HFO were based on this assumption. It was assumed that the sediment iron had the same properties (i.e., specific surface area of 600 $m^2 \cdot g^{-1}$) as those of HFO described by Dzombak and Morel (1990). Two site densities were used, 0.005 moles sites per mole Fe for the high-affinity site densities and 0.2 moles sites per mole Fe for the low-affinity site densities. As suggested by Dzombak and Morel (1990), for conversion from $g \cdot L^{-1}$ HFO to $mol \cdot L^{-1}$ Fe, the stoichiometry $Fe_2O_3 \cdot H_2O$ (89 g HFO per mole Fe) was assumed.

It should be emphasized that all of the computer-model input was derived either from chemical analyses of St. Kevin Gulch water and sediment or from information provided by Dzombak and Morel (1990). None of the model sorption parameters were modified and no fitting of parameters was involved in the model simulations. The simulations were performed as a validation exercise for the computer model using experimental data derived from natural materials. A similar modeling approach has been used by Loux *et al.* (1989).

C. Results and Discussion

1. Sorption Rate

Figure 1 illustrates results of sorption-rate experiments for Cu and Cd onto streambed sediment from the Tailings Site and the Downstream Site adjusted to pH 6. These data show very rapid initial sorption followed by a less well-defined, slower sorption process. Results from both sites exhibit a decrease in the amount of sorption between the initial rapid step and the slower step over a 1- to 2- or 4-hr period for the Tailings Site and the Downstream Site, respectively. The curves appear to level off after 4 to 6 hr; the reactions stabilize more quickly at the Tailings Site than at the Downstream Site. Concurrent experiments (not shown) in which the pH of the mixtures remained unaltered demonstrated that Cu and Cd are not desorbed from the streambed sediment at ambient pH.

Although it is likely that sorption rate varies with pH, a pH of 6 was chosen for sorption-rate experiments because it represents a pH at which Cu and Cd completely or nearly completely sorb. At lower pH values more typical of ambient stream conditions (i.e., pH 3.3 to 4.6) incomplete sorption of Cu and Cd takes place, which makes the sorption rate more difficult to determine. The results at pH 6 represent an approximate sorption rate that was used to define the duration of batch sorption experiments.

2. pH-Dependent Sorption and Sorption Modeling

Figure 2 shows results from batch pH-dependent sorption experiments using St. Kevin Gulch stream-water–streambed-sediment mixtures. Figure 2 expresses the amount sorbed as a percentage of the original metal concentration. To calculate micromoles of metal per gram sediment, convert the concentration of a given metal listed in Table I to molar units, multiply by the fraction sorbed (from Fig. 2),

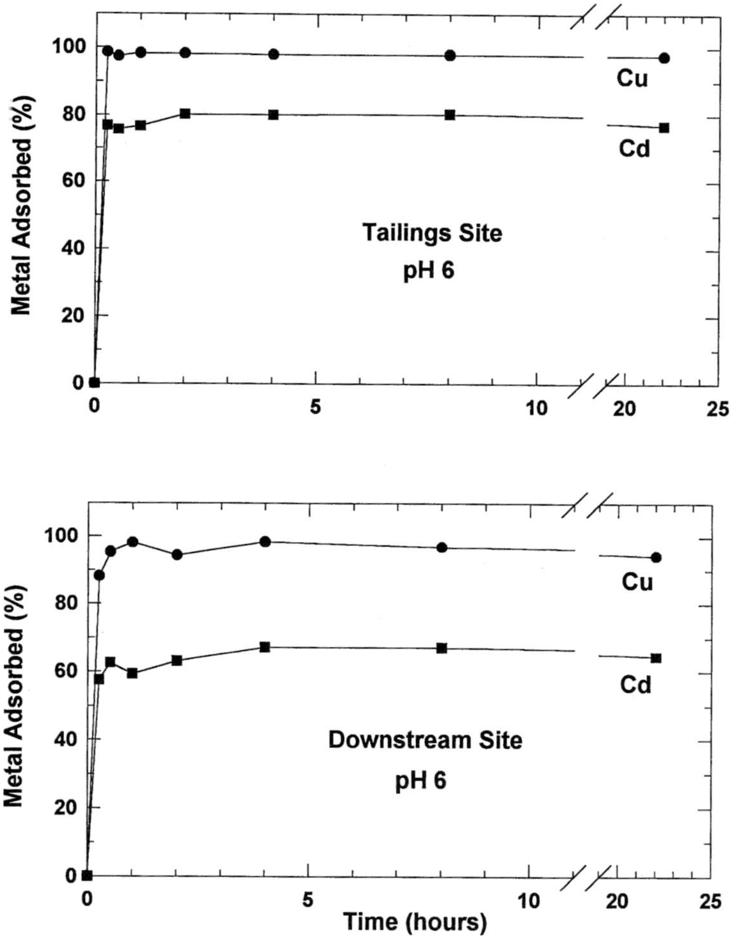

Figure 1 Rates of Cu and Cd sorption onto streambed sediment from St. Kevin Gulch, Colorado, adjusted to pH 6.

and divide by the sediment concentration of 2.9 $g \cdot L^{-1}$. There is an adsorption edge for all the metals examined in this study, and Pb and Cu are removed from solution at a much lower pH than are Zn, Cd, and Ni. The adsorption edge is a narrow pH region (generally about 2 pH units wide) in which partitioning changes from nearly all dissolved to nearly all sorbed, provided that sufficient sorption sites are available. The relative adsorption selectivity sequence for metals in this study is Pb > Cu > Zn > Ni = Cd. The adsorption edge for Pb lies between pH 3 and 4.5, for Cu between pH 4 and 5.5, and for Zn, Cd, and Ni between pH 5 and 7. This

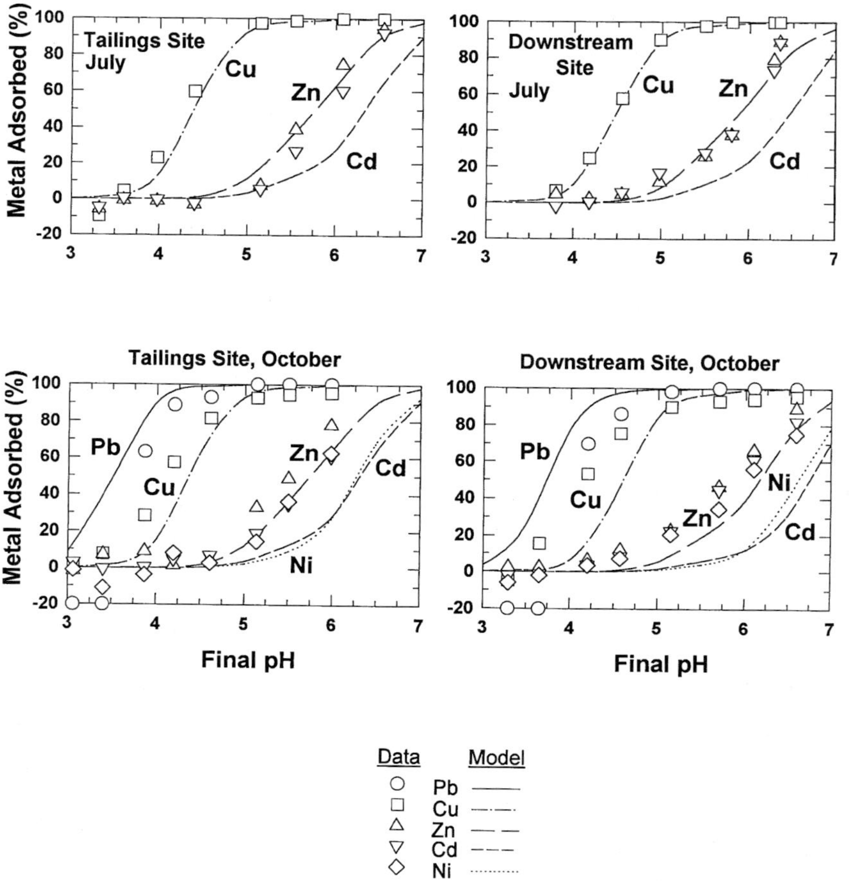

Figure 2 Comparison of experimental data (symbols) for Pb, Cu, Zn, Ni, and Cd sorption onto streambed sediment from St. Kevin Gulch, Colorado, with computer-model simulations (curves) for sorption onto hydrous ferric oxide. The concentration of the streambed sediment was 2.9 $g \cdot L^{-1}$.

adsorption sequence is in agreement with that from studies of metal sorption onto synthetic HFO reported in the literature (e.g., Kinniburgh *et al.,* 1976; Davis and Leckie, 1978; Swallow *et al.,* 1980; Benjamin and Leckie, 1981a,b, 1982; Benjamin, 1983). As a result, metal-partitioning behavior between stream water and bed sediment from St. Kevin Gulch can be approximated from general sorption trends reported in the literature.

Due to the fact that we were working with mixed systems as a consequence of using natural materials for our sorption experiments, we were not able to directly measure competition effects between metal and other components of the systems.

Benjamin and Leckie (1981a) found that there is little competition among Cd, Zn, Cu, and Pb for binding sites on amorphous HFO. This suggests that many of the high-energy metal-binding sites on HFO may be metal-specific. In this study we assumed that the hydrous iron oxides in St. Kevin Gulch streambed sediment were similar to HFO and we ignored possible competition effects among metals in our systems. As a result, our modeling efforts were greatly simplified.

Figure 2 provides a comparison between results from batch pH-dependent sorption experiments using St. Kevin Gulch stream-water–streambed-sediment mixtures and those from computer-model simulations of sorption at the Tailings and Downstream Sites in July and October. The model simulations are based on surface-complexation constants derived from well-defined, single-metal laboratory experiments using synthetic HFO. Therefore, the simulations represent a best-fit mathematical description of literature data for metal sorption onto synthetic HFO for the chemical conditions modeled.

The agreement between model simulations and experimental sorption data for Cu and Zn in the July experiments is generally good. In the October experiments computer-model simulations slightly underestimate Cu and Zn partitioning to the sediment. This underestimation is especially apparent for the Downstream Site where the least amount of iron and the greatest amount of carbon are present in the sediment. The carbon-rich material in the sediment may enhance Cu and Zn sorption at the Downstream Site.

Cadmium and Ni partitioning to the sediment is consistently underestimated by the model. Some insight from well-defined synthetic systems may be helpful in explaining this underestimation. However, there are many uncertainties when comparing results from well-defined laboratory experiments with results from this study, which used natural materials in a mixed system. The predominant ions in St. Kevin Gulch water are Ca and sulfate (see Table I). Benjamin and Leckie (1982) found similar underestimation of model predictions for Cd sorption onto HFO in synthetic systems containing sulfate. In their study, the ratio of sulfate to Cd was much greater than that in stream water from St. Kevin Gulch. However, based on their results, one possible explanation for enhanced Cd sorption in this study is binding of Cd–sulfato complexes to HFO. Another explanation may be the incorporation of Si into the sorbent HFO, which has been shown to enhance sorption of Cd. Anderson and Benjamin (1985) found that Cd sorption was continuously enhanced with increasing Si content of ferrihydrite sorbent, whereas the sorption behavior of Zn, Cu, and Co was not significantly affected by the Si content. Although we have no direct measurement of the Si content of hydrous iron-oxide minerals in our streambed sediment, there is abundant dissolved Si that could be incorporated upon precipitation of the hydrous iron-oxide minerals (6.5 to 11 $mg \cdot L^{-1}$ Si; see Table I). In contrast, Cowan *et al.* (1991) reported reduced sorption of Cd on HFO in the presence of high Ca concentrations. Based on the enhanced Cd sorption of our experimental data compared with model predictions, high aqueous concentrations of Ca in St. Kevin Gulch do not appear to reduce Cd sorption. The ratio of dissolved Cd and Ca in St. Kevin Gulch is within the ratio used by Cowan *et al.* (1991) but, according to model-calculated distributions of components, Ca does not successfully compete for sorption sites until the pH exceeds 7, which is after most of the Cd is already sorbed (see Fig. 2).

Lead was determined only in the October sorption experiments. There was release of Pb by the sediment at lower pH values in the sorption experiments. This may indicate either that some Pb was present as a mineral phase that was soluble at pH < 4, or that Pb was desorbed from the sediment at low pH during the course of the experiments. Desorption is likely because the beginning ambient pH in the sorption experiments was generally slightly lower than the measured field pH due to precipitation of hydrous iron oxides during transport and storage of the streambed-sediment–stream-water mixtures prior to conducting the batch sorption experiments. Because Pb is a cation, less Pb should be sorbed by the sediment at lower pH. In any case, this illustrates the point that sediment can serve both as a sink and as a source for metals. Lead was readily sorbed by the sediment at pH $>$ 4, and this sorption is approximated by the computer-model simulation.

There does not appear to be a large temporal variation in metal-partitioning trends over the 3-month time span between the sampling periods. However, the model–experimental agreement is far better for the July than the October data set. Active precipitation of hydrous iron oxides, observed in the stream during July, may have provided a larger proportion of fresh metal-binding sites than were available in October when active precipitation had greatly diminished.

Surface charge properties of streambed-sediment surfaces were investigated by measuring the electrophoretic mobility (EM) of sediment in the streambed-sediment–stream-water mixtures. Electrophoretic mobility was determined by means of a light-scattering technique (Ranville, 1992) for particulates from the July sorption experiment. Figure 3 illustrates the variation in EM of the streambed sediments with pH for the Tailing and Downstream Sites in July. Although computer-model simulations of metal sorption at these two sites performed equally well, it is clear that the electrical character and surface charge of sediment from the two sites are different. Streambed sediment from the Tailings Site has a weak, net positive surface charge whereas sediment from the Downstream Site has a net negative surface charge. Possible explanations for the more negative surface charge at the Downstream Site include one or more of the following: (1) increased presence of detrital minerals (e.g., clay minerals), (2) decreased amounts of positively charged surface complexes, (3) increased amounts of negatively charged surface complexes, and (4) the presence of sorbed or coprecipitated organic matter or biotic materials. Examination of streambed-sediment morphology by scanning electron microscopy (SEM) revealed that particulates from the two sites were of similar size, but the Downstream Site contained more detrital minerals in the bed sediment. The computer-model sorption simulations assume a pristine point of

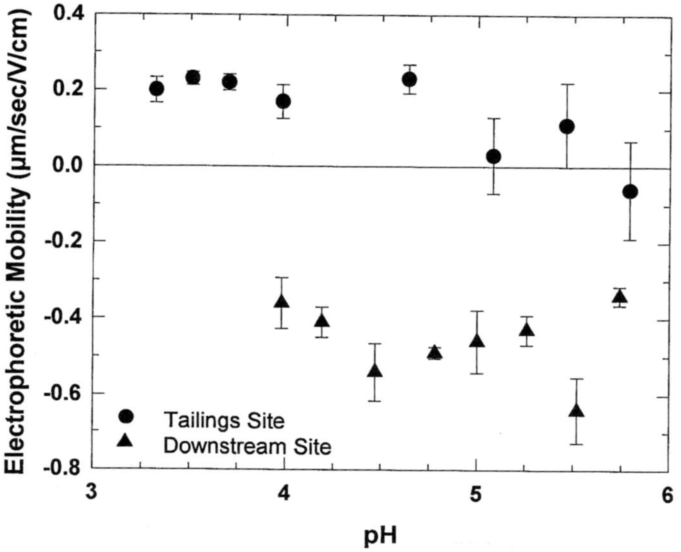

Figure 3 Electrophoretic mobility as a function of pH for pH-adjusted mixtures of streambed sediment and stream water collected from St. Kevin Gulch in July. Error bars represent the standard deviation of triplicate measurements.

zero charge of 8.1 for HFO. Comparison of the EM results with sorption experiments and modeling results indicated that the observed differences in surface charge did not appear to adversely affect our ability to simulate metal sorption. We did not attempt to alter the sorption model to fit our EM data.

Mineralogy of the streambed sediment was determined by X-ray diffraction. These data (not shown) indicate that there was not a distinct seasonal difference in sediment mineralogy for each site between July and October. However, there are some mineralogical differences between the Tailings and Downstream Sites. The sediments all consist of a mixture of poorly crystalline, nondetrital material and detrital material. The poorly crystalline, nondetrital component appears to be a mixture of schwertmannite ($Fe_8O_8(OH)_6SO_4$), ferrihydrite ($\approx Fe_5OH_8 \cdot 4H_2O$), and goethite ($\alpha$-FeOOH). The detrital component is most noticeable at the Downstream Site and primarily consists of quartz and clay minerals. Despite mineralogical differences both between the two sites and between streambed sediment from St. Kevin Gulch and synthetic HFO used in laboratory studies, computer-model simulations of metal sorption were able to approximate observed metal partitioning in experiments with water and streambed sediment from St. Kevin Gulch.

This work demonstrates that modeling can be used to approximate metal partitioning in iron-rich systems such as St. Kevin Gulch. In applying the model, we assume that metal sorption only takes place onto HFO in the streambed sediment; hence, we assume that sorption onto HFO controls metal sorption in these systems. This assumption appears to be adequate to simulate metal-sorption reactions between streambed sediment and stream water from St. Kevin Gulch. Although other mineral phases (e.g., aluminum oxides) also may sorb metals, restricting sorption to HFO appears to adequately simulate observed sorption behavior, especially for Pb, Cu, and Zn.

An instream pH-modification experiment was conducted at St. Kevin Gulch to examine the response of chemical species in the stream to pH changes and to observe instream sorption reactions (Kimball *et al.,* 1994). During the experiment, the pH of a reach of St. Kevin Gulch (normally about 3.5) was raised to 5.9. Run concurrently with the instream pH-modification experiment were batch pH-dependent sorption experiments using stream-water–streambed-sediment mixtures and stream water in the absence of sediment. Water and streambed sediment for these batch experiments were collected as a composite along the reach of stream used for the pH modification. The dissolved Cu concentrations as a function of pH measured during the instream experiment compare well with the batch sorption experiments conducted with unfiltered stream water in the absence of streambed sediment. These data indicate that processes controlling Cu attenuation with increasing pH in the stream occur primarily in the water column, and the streambed sediment appears to play only a secondary role in Cu attenuation. In both the instream and the batch experiments, particulates precipitated in the water column in response to increased pH. These water-column particulates appear to be responsible for the observed attenuation of Cu. Details of this study are discussed by Smith (1996).

3. Effect of Sulfate Sorption on Model Predictions

The GTLM assumes that uncomplexed M^{2+} is the only metal species that sorbs onto HFO. Two types of cation-binding sites are used, one for high-affinity sites and one for low-affinity sites. Typically, reactions for cation adsorption onto both high- and low-affinity sites are in the form

$$\equiv FeOH^0 + M^{2+} = \equiv FeOM^+ + H^+,$$

where $\equiv Fe$ represents surface-binding sites on HFO and M^{2+} is a divalent cation. Due to the high concentration of sulfate in these waters, we ran the model with and without sulfate sorption and with and without corrections for aqueous metal speciation. The GTLM describes anion adsorption with surface ligand-exchange reactions, such as the following reactions for sulfate:

$$\equiv FeOH^0 + SO_4^{2-} + H^+ = \equiv FeSO_4^- + H_2O$$

$$\equiv FeOH^0 + SO_4^{2-} = \equiv FeOHSO_4^{2-}.$$

Corrections for metal speciation were done by calculating the percentage of the free metal ion sorbed rather than the percentage of the total metal concentration sorbed. In the case of sulfate sorption with corrections for aqueous metal speciation, differences were too small to display. Figure 4 shows the results of these model simulations and the percent sulfate sorbed as a function of pH when sulfate sorption was allowed. As seen in Figure 4, there is generally fairly close agreement between the computer-model simulations without sulfate sorption and those results corrected for aqueous metal speciation, although the model predicts slightly more Cu, Zn, Cd, and Ni sorption when the data are corrected for aqueous metal speciation.

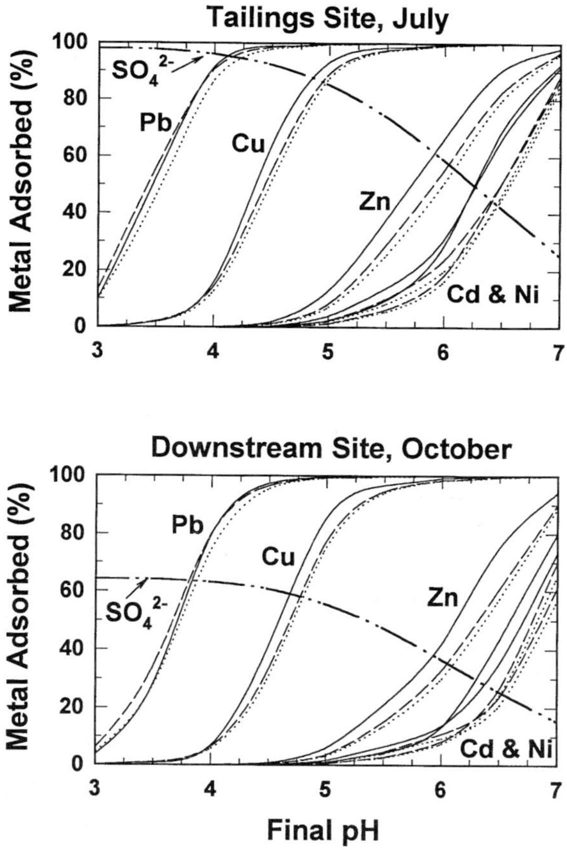

Figure 4 Computer-model simulations showing the percentage of sulfate sorption and the effect of sulfate sorption and metal speciation on model predictions of trace-metal sorption onto hydrous ferric oxide as a function of pH. (—) model prediction with sulfate sorption; (···) model prediction without sulfate sorption; (- - -) model prediction with speciation correction and without sulfate sorption.

Figure 4 shows some significant differences between computer-model simulations run with and without sulfate sorption. If one thinks of surface-complexation sites as analogous to a complexing ligand (e.g., sulfate), then one can consider competition between surface sites and sulfate for metal ions. There can also be competition between sulfate and metal ions for surface sites in the model. When sulfate sorbs in the model simulation it is less available for aqueous complexation with metals. Hence, a larger fraction of the aqueous metals is present as M^{2+}, resulting in a larger proportion of the metals that can sorb. This concept is illustrated in Figure 5 for Pb and Cd model simulations for the Tailings Site in July. Fig-

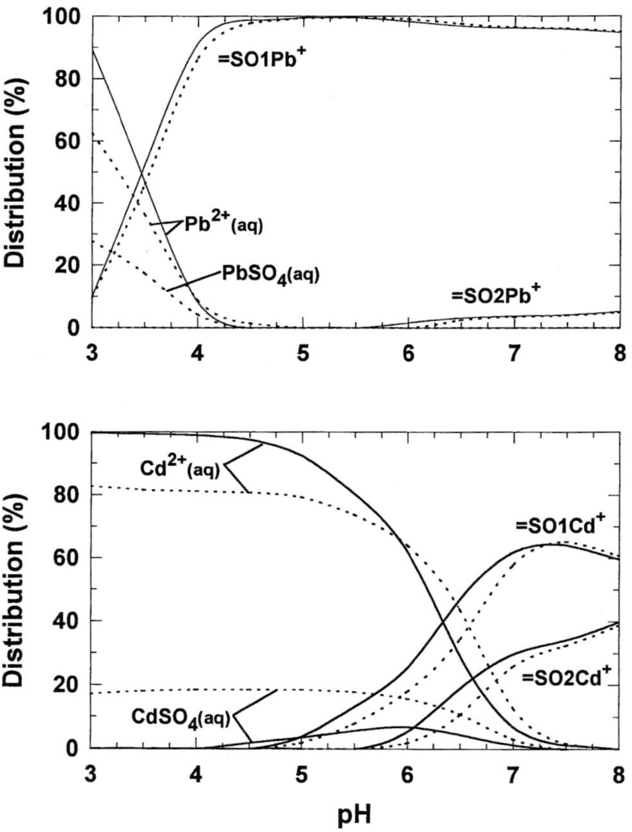

Figure 5 Distribution of model-calculated species for cases with and without sulfate sorption onto hydrous ferric oxide for conditions at the Tailings Site in July. $=SO1M^{+}$ represents high-affinity surface complexes and $=SO2M^{+}$ represents low-affinity surface complexes as defined by Dzombak and Morel (1990). (—) model species distribution with sulfate sorption; (- - -) model species distribution without sulfate sorption.

ure 5 shows the distribution of aqueous and sorbed species as a function of pH and illustrates the larger fraction of M^{2+} and the corresponding larger proportion of sorbed species in simulations where sulfate is allowed to sorb. Also shown in Figure 5 is the distribution between high-energy and low-energy surface complexes. Lead tends to sorb almost exclusively to the high-energy sites whereas Cd is more equally divided between the high- and low-energy sites.

It is clear that the treatment of sulfate in computer-model simulations can influence the outcome of metal-sorption results; this may be especially true when there are a limited number of surface-complexation sites in relation to sorbate metals. Consequently, it is important to try to set up the model to match the geochemical conditions to be simulated. In the case of St. Kevin Gulch, running the model without sulfate sorption more truly represents the geochemical conditions to be simulated since sulfate is already equilibrated with the sediment.

III. SORPTION MODELING AT DIVERSE MINE-DRAINAGE SITES

Our previous work at St. Kevin Gulch (see Section II) demonstrates that modeling can be used to approximate metal partitioning in iron-rich mine-drainage systems. Our next step was to expand this work to a variety of mine-drainage sites, so we collected water samples from mine-drainage sites throughout Colorado that represent a variety of geologic and chemical settings (Ficklin *et al.,* 1992; Plumlee *et al.,* 1992). Ten of these sites were chosen for studies of processes that influence metal transport and sorption reactions. These results are detailed in Smith *et al.* (1992) and are summarized here.

The emphasis of these studies was on metal sorption onto iron-rich suspended sediments rather than bed sediments. This emphasis was based on our observations that little metal attenuation takes place when metal-laden water flows over iron-rich streambed sediments. Most attenuation seems to take place at confluences where iron-rich suspended particulates are formed in the water column. This is similar to the results of the instream pH-modification study in St. Kevin Gulch (see Section II.C.2).

A. Site Descriptions

Ten mine-drainage sites were selected to examine metal transport and sorption processes. These sites all exhibited abundant hydrous iron oxide coatings in the mine-drainage channel. At a typical site, mine drainage flowed from a mine adit and continued for several tens to hundreds of meters in a confined channel until joining with a stream or percolating into porous material. The criteria used to select the ten sites included (1) the presence of iron-rich bottom sediment, (2) drainage in contact with the iron-rich bottom sediment, (3) presence of abundant dissolved oxygen, and (4) at least 0.5 $mg{\cdot}L^{-1}$ Fe present in the suspended particulate fraction. The pH at the 10 sites ranged from 3.8 to 7.7, and Fe as suspended particulates ranged from 0.5 to 10 $mg{\cdot}L^{-1}$.

B. Methods

1. Sampling and Analytical Techniques

Water samples were collected from 10 diverse mine-drainage sites in Colorado. The pH, temperature, specific conductivity, and dissolved oxygen were measured on site. A portion of each water sample was filtered through a 10,000-Da filter plate using a Millipore Minitan Ultrafiltration System. Both the filtered and unfiltered portions of the water samples were acidified with ultrapure nitric acid and saved for metals analysis.

Trace-metal concentrations were determined by ICP-MS, Fe and Al were determined by AAS, and major-element concentrations were determined by ICP-AES and IC. Suspended particulate iron in the water samples was calculated as the difference between Fe concentration in unfiltered water (total Fe) and Fe concentration in filtered water (dissolved Fe).

2. Sorption Modeling Approach

Sorption at the 10 mine-drainage sites was determined by observations of partitioning between filtered and unfiltered water collected at each site at the ambient pH. No pH-modification experiments were performed with waters from these 10 sites. Sorption of As, Cd, Cu, Ni, Pb, and Zn onto hydrous iron oxides present in suspended particulates was modeled using MINTEQA2 coupled with the GTLM as described in Section II.B.5. For modeling purposes, all suspended particulate iron determined in the water samples was assumed to be present as HFO, and calculations of binding-site concentrations and total HFO were based on this assumption as outlined in Section II.B.5. None of the model parameters were modified, and no fitting parameters were used in these computer-model simulations.

C. Results and Discussion

Metal attenuation was examined along the 10 mine-drainage channels coated with hydrous iron oxides. We decided to place our emphasis on metal sorption onto iron-rich suspended sediment in the mine-drainage systems. This is because little metal attenuation was observed when metal-laden water flowed over iron-rich streambed sediments. Most attenuation seems to take place at confluences where iron-rich suspended particulates are formed in the water column. This is similar to the results reported in Section II.C.2. Possible reasons why there is so little metal attenuation by the bed sediment compared with the suspended sediment may include (1) prior equilibration of the bed sediment, (2) hydrologic controls that prevent or retard contact between dissolved constituents and the bed sediment, (3) the presence of a stagnant liquid film at the water–bed-sediment interface that interferes with diffusion of dissolved constituents, (4) the physical proximity of suspended sediment to dissolved constituents, and (5) differences in sorption site-binding energies between the suspended particulates and the bed sediment.

Table II provides a comparison of model predictions with measured values at 3 of the 10 sites. Data from these three sites are representative of the other sites and show that there is good general agreement between the model predictions for pH-dependent sorption reactions and the measured metal partitioning in the waters. These data indicate that metal sorption on suspended iron-rich particulates is an important control of metal partitioning in mine-drainage waters. Table II shows that both the model predictions and the measured data indicate strong sorption of Pb and As, intermediate sorption of Cu, and little sorption of Zn, Cd, and Ni. Ranges given for some of the measured data incorporate analytical precision uncertainties of the data used to calculate the particulate fraction for that metal.

A similar approach was used to model metal attenuation in the Leviathan and Bryant Creek drainage system on the border of California and Nevada. MINTEQA2 coupled with the GTLM was used to calculate trace-metal sorption

Table II

Comparison of Model Predictions and Measured Values of Percent Metals Associated with the Suspended Particulate Fraction of Mine-Drainage Waters from Selected Sites

	Argo-3 (pH 5.6, HFO = 0.007 g·L^{-1})		Rawley-3 (pH 6.2, HFO = 0.011 g·L^{-1})		Leadville Drain (pH 7.2, HFO = 0.001 g·L^{-1})	
	Predicted (%)	Measured (%)	Predicted (%)	Measured (%)	Predicted (%)	Measured (%)
As	—	—	98	>78[a]	—	—
Pb	82	>71[a]	80	>93[a]	86	>71[a]
Cu	18	27	60	63	—	—
Zn	<1	0 to 8	<1	0 to 9	2	3
Ni	<1	<1	<1	1	—	—
Cd	<1	<1	<1	6	<1	<1

[a]Dissolved concentration was below the limit of determination. Value was calculated using the limit of detection for the dissolved concentration.

onto hydrous iron oxide over selected intervals of the Leviathan and Bryant Creek drainage. Because there were no data for unfiltered water, we were not able to directly determine the composition of suspended particulate phases. Instead, the amount of iron precipitated for a given reach of stream was calculated as the difference in mass flow of Fe over the stream reach divided by the upstream discharge. Upstream water chemistry and downstream pH were used in the model simulations, and metal attenuation in the Leviathan and Bryant Creek drainage was determined as the calculated percentage decrease in mass flow of the metal. Sorption onto suspended hydrous iron oxides appears to be the major mechanism regulating the concentrations of As, Cu, and Zn downstream from the mine site in the Leviathan and Bryant Creek drainage. Details of this study are discussed in Webster *et al.* (1994).

IV. PREDICTIVE SORPTION MODELING FOR MITIGATION AND REMEDIATION

Iron–rich particulates are commonly formed as a result of the oxidation and/or neutralization of mine-drainage waters. These iron-rich particulates have the capacity to sorb trace metals contained in the mine drainage. The sorption reactions are very sensitive to the pH of the mine-drainage water and also to the ratio of the concentration of suspended iron-rich particulates relative to the concentration of dissolved trace metals. A method is presented here that applies computer-model simulations to the prediction of trace-metal sorption by suspended iron-rich particulates and provides a measure of the "self-mitigating capacity" of a given mine-drainage water. The self-mitigating capacity refers to the amount of Fe contained in the mine-drainage water that can precipitate and attenuate dissolved metals. This method is discussed in Smith *et al.* (1993) and Smith (in press) and is summarized here.

A. Sorption-Modeling Approach

For each computer-model simulation, complete analytical information on water composition was entered into MINTEQA2, similar to the method described in Section II.B.5. The water had been filtered through a 0.1-μm filter to remove suspended particulates. For each simulation, pH was fixed at desired values (incremental values between 4 and 8) with the water in equilibrium with atmospheric carbon dioxide. Iron and Al were allowed to precipitate during the simulations in order to control their solubilities as described in Section II.B.5. The mechanism for removal of trace metals in the simulations was sorption onto HFO.

HFO was the sole sorbent material in the simulations. To calculate the amount of HFO present in each simulation, the dissolved Fe contained in each water was assumed to completely precipitate. The amount of HFO sorbent material was calculated from this original dissolved Fe content. This assumption was made in order to simulate the formation of suspended iron-rich particulates during progressive neutralization of mine-drainage water and to determine the self-mitigating capacity of the water.

It should be emphasized that all of the model input was obtained either from chemical analysis or from information given in Dzombak and Morel (1990). None of the model parameters were modified for these simulations, and no fitting parameters were used.

B. Examples

We use two mine-drainage sites to illustrate the self-mitigating capacity of the waters and to demonstrate the potential importance of suspended iron-rich particulates in the natural attenuation of metals in mine-drainage waters. Table III gives the pH, dissolved metal, and sulfate content in mine-drainage water for the two sites. The model-derived sorption results are shown in Figure 6 as a function of pH. The two cases presented here have been chosen to illustrate the differences in sorption reactions even when the pH is the same and both sites contain high concentrations of dissolved Fe (in excess of 100 $mg \cdot L^{-1}$).

The Argo Tunnel, in Idaho Springs, Colorado, drains mine workings that exploited pyrite-rich gold-quartz veins in the Central City mining district. This pH 2.9 water contains 120 $mg \cdot L^{-1}$ dissolved Fe. Figure 6 illustrates predictive sorp-

Table III

Chemical Composition of Filtered Water from the Argo Tunnel and the Reynolds Adit, Colorado

	Argo Tunnel	Reynolds Adit
As ($\mu g \cdot L^{-1}$)	90	400
Pb ($\mu g \cdot L^{-1}$)	40	320
Cu ($\mu g \cdot L^{-1}$)	4,500	120,000
Zn ($\mu g \cdot L^{-1}$)	30,000	20,000
Cd ($\mu g \cdot L^{-1}$)	130	200
Ni ($\mu g \cdot L^{-1}$)	120	800
SO_4^{2-} ($mg \cdot L^{-1}$)	2,100	1,900
Fe ($mg \cdot L^{-1}$)	120	310
pH (field)	2.9	2.9

Note. Samples were filtered through a 0.1-μm filter.

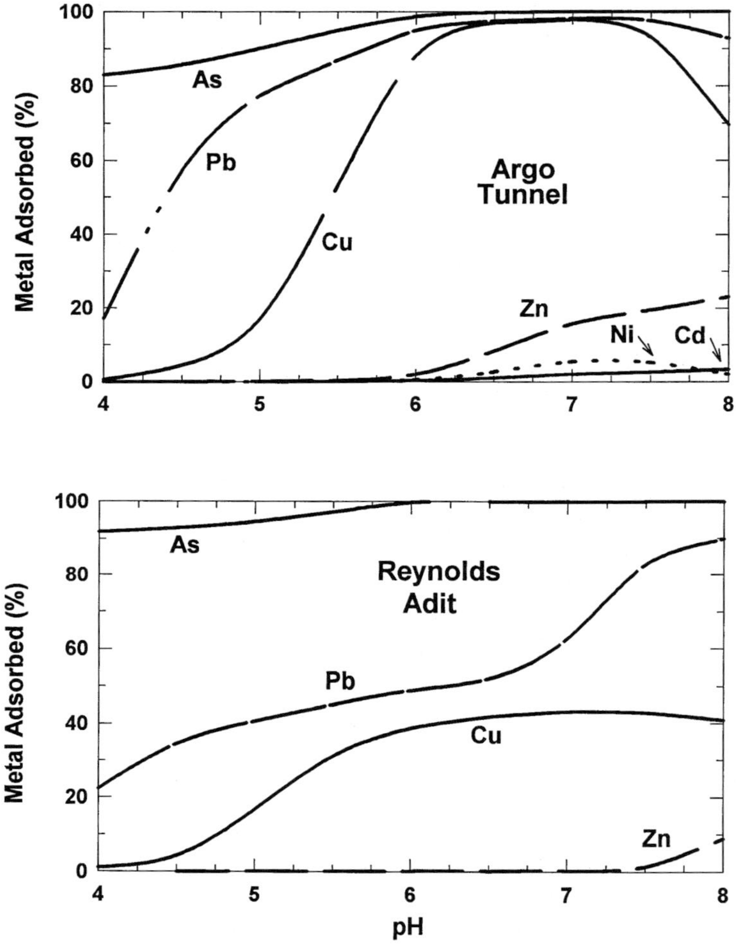

Figure 6 Computer-model simulations of trace-metal sorption onto hydrous ferric oxide as a function of pH for mine drainage from the Argo Tunnel (Idaho Springs, Colorado) and the Reynolds Adit (Summitville, Colorado).

tion-modeling results if this 120 $mg \cdot L^{-1}$ dissolved Fe were to precipitate as HFO and be available to sorb dissolved trace metals over the pH range 4 to 8. Modeling data indicate that there is a relatively low concentration of surface metal-binding sites compared to the high concentrations of dissolved trace metals, especially Zn and Cu. The concentration of high-affinity sites is exceeded by the concentration of dissolved trace metals. Hence, the number of surface-complexation sites available for sorption of trace metals is limited. Figure 6 shows that most of the As, Pb, and Cu and about 20% of the Zn are predicted to be sorbed by HFO at pH 8. The amount of sorbed Cu and Pb decreases between pH 7 and pH 8, however, due to metal complexation by aqueous ligands (mostly hydroxide and carbonate species). Trace-metal hydroxide and carbonate solid phases and anglesite ($PbSO_4$) did not exceed saturation in the computer-model simulations. Based on these model predictions, dissolved Zn, Cd, and Ni would be expected to be mobile at this site over the pH range of 3 to 8.

The Reynolds Adit, at Summitville, Colorado, drains acid-sulfate epithermal ores of the Summitville mining district. The Summitville ores are rich in pyrite and contain abundant copper-sulfide minerals and copper-arsenic sulfide minerals. This pH 2.9 water contains 310 $mg \cdot L^{-1}$ dissolved Fe. Figure 6 illustrates predictive modeling results if this 310 $mg \cdot L^{-1}$ dissolved Fe were to precipitate as HFO and be available to sorb dissolved trace metals over the pH range 4 to 8. Sorption of Ni over the pH range 4 to 8 was $<1\%$, and sorption of Cd was $<2\%$ (curves not shown in Figure 6). Even though a large amount of particulate iron is available to sorb trace metals, there are again a limited number of surface-complexation sites relative to the large amount of dissolved metals. There is saturation of surface-complexation sites on HFO as a result of the extremely high concentrations of dissolved trace metals, particularly Cu and Zn. Figure 6 shows that most of the As and Pb are predicted to be sorbed by HFO at pH 8. Less than 50% of the Cu, very little Zn, and virtually no Cd and Ni are predicted to be retained by HFO. Hence, dissolved Cu, Zn, Cd, and Ni would be expected to be mobile at this site over the pH range 3 to 8.

C. Implications for Mitigation and Remediation

The sorption simulations presented herein illustrate the potential predictive power of computer modeling and the potential self-mitigating capacity of some mine-drainage waters. This approach can provide guidance in mine-drainage remediation and planning efforts. Sorption of trace metals by suspended iron-rich particulates is an important aspect of mine-drainage remediation techniques that involve neutralization and/or oxidation of iron-rich waters. Mine-drainage pH is very important in determining these sorption reactions, as is the concentration of suspended iron-rich particulates relative to the concentration of dissolved trace metals. In the presence of a large excess of suspended iron-rich particulates, most trace-metal cations tend to be sorbed at neutral pH; however, when there is not a large excess of these particulates, differences in trace-metal sorption as a function of pH are apparent. These differences in the pH regime for sorption of a particular trace-metal ion can be used to estimate the pH necessary for removal of that trace metal and, conversely, to predict the degree of mobility of that trace metal. These differences could also be used to design a treatment system for the selective recovery of specific trace metals from water.

V. CONCLUSIONS

1. Some pH-dependent sorption data derived from experiments using natural materials from a stream impacted by acid mine drainage were simulated using a surface-complexation sorption model. The data were modeled with the assumption that metal sorption onto hydrous iron oxide dominates metal-sorption reactions in iron-rich mine-drainage systems. This work demonstrates that predictive modeling can be used for practical applications, and provides a validation test for the MINTEQA2 computer program.

2. Electrophoretic mobility measurements show distinct differences in surface charge in sediments used for sorption experiments. However, this difference in surface charge does not manifest itself in differences in sorption behavior.

3. Metals in mine-drainage waters appear to partition between the dissolved phase and iron-rich suspended particulates rather than with iron-rich bed sediment.

4. Studies at diverse mine-drainage sites indicate that generally Pb and As are largely associated with the suspended particulate fraction, Cu is partially associated with the suspended particulate fraction, and Zn, Cd, and Ni tend to remain in the dissolved fraction. This may help explain why Zn is the major base metal present in most mine drainages.

5. The prevalence of sorption reactions in the water column may be exploited in the mitigation and remediation of some mine drainages. Some mine-drainage waters have the potential to self-mitigate by sorption of dissolved metals onto iron-rich particulates that precipitate upon neutralization or oxidation of the waters.

ACKNOWLEDGMENTS

We thank T. T. Chao, M. Montour, and two anonymous reviewers for their helpful reviews of this paper. We gratefully acknowledge the help of C. Ball, K. Cabral, W. Ficklin, K. Hanula, B. Kimball, S. Kulinski, D. McKnight, and N. Loux (EPA). J. Herron, R. Kirkham, and J. Lake of the Division of Minerals and Geology, Colorado Department of Natural Resources, were an invaluable source of information and we appreciate their help in locating and gaining access to mines. Thanks also to the mine owners who allowed us to sample on their properties. Analytical assistance was provided by P. Briggs, J. Christie, D. Fey, W. Ficklin, M. Malcolm, J. McHugh, A. Meier, and G. Riddle. The project was funded by the U.S. Geological Survey Toxic Substances Hydrology Program and the Mineral Resources Program.

REFERENCES

Allison, J. D., Brown, D. S., and Novo-Gradac, K. J. 1991. MINTEQA2/PRODEFA2, a geochemical assessment model for environmental systems; version 3.0 user's manual. EPA/600/3-91/021, U.S. Environmental Protection Agency, Athens, GA.

Anderson, P. R., and Benjamin, M. M. 1985. Effects of silicon on the crystallization and adsorption properties of ferric oxides. *Environ. Sci. Technol.* 19:1048–1053.

Benjamin, M. M. 1983. Adsorption and surface precipitation of metals on amorphous iron oxyhydroxide. *Environ. Sci. Technol.* 17:686–692.
Benjamin, M. M., and Leckie, J. O. 1981a. Competitive adsorption of Cd, Cu, Zn, and Pb on amorphous iron oxyhydroxide. *J. Colloid Interface Sci.* 83:410–419.
Benjamin, M. M., and Leckie, J. O. 1981b. Multiple-site adsorption of Cd, Cu, Zn, and Pb on amorphous iron oxyhydroxide. *J. Colloid Interface Sci.* 79:209–221.
Benjamin, M. M., and Leckie, J. O. 1982. Effects of complexation by Cl, SO_4, and S_2O_3 on adsorption behavior of Cd on oxide surfaces. *Environ. Sci. Technol.* 16:162–170.
Bigham, J. M. 1994. Mineralogy of ochre deposits formed by sulfide oxidation. *In* "The Environmental Geochemistry of Sulfide Mine-Wastes" (D. W. Blowes and J. L. Jambor, Eds.), Short Course Handbook Vol. 22, pp. 103–132. Mineralogical Association of Canada. Nepean, Ontario.
Chapman, B. M., Jones, D. R., and Jung, R. F. 1983. Processes controlling metal ion attenuation in acid mine drainage streams. *Geochim. Cosmochim. Acta* 47:1957–1973.
Cowan, C. E., Zachara, J. M., and Resch, C. T. 1991. Cadmium adsorption on iron oxides in the presence of alkaline-earth elements. *Environ. Sci. Technol.* 25:437–446.
Davis, A., Olsen, R. L., and Walker, D. R. 1991. Distribution of metals between water and entrained sediment in streams impacted by acid mine discharge, Clear Creek, Colorado, U.S.A. *Appl. Geochem.* 6:333–348.
Davis, J. A., and Kent, D. B. 1990. Surface complexation modeling in aqueous geochemistry. *In* "Mineral-Water Interface Geochemistry" (M. F. Hochella and A. F. White, Eds.), Reviews in Mineralogy Vol. 23, pp. 177–260. Mineralogical Society of America. Washington, DC.
Davis, J. A., and Leckie, J. O. 1978. Surface ionization and complexation at the oxide/water interface. II. Surface properties of amorphous iron oxyhydroxide and adsorption of metal ions. *J. Colloid Interface Sci.* 67:90–107.
Dzombak, D. A., and Morel, F. M. M. 1987. Adsorption of inorganic pollutants in aquatic systems. *J. Hydraulic Eng.* 113:430–475.
Dzombak, D. A., and Morel, F. M. M. 1990. "Surface Complexation Modeling: Hydrous Ferric Oxide." Wiley, New York.
Ficklin, W. H., Plumlee, G. S., Smith, K. S., and McHugh, J. B. 1992. Geochemical classification of mine drainages and natural drainages in mineralized areas. *In* "Water-Rock Interaction" (Y. K. Kharaka and A. S. Maest, Eds.), pp. 381–384, "Proceedings, Seventh International Symposium on Water-Rock Interaction, Park City, Utah, July 13–18, 1992," Vol. 1. A.A. Balkema, Rotterdam.
Filipek, L. H., Nordstrom, D. K., and Ficklin, W. H. 1987. Interaction of acid mine drainage with waters and sediments of West Squaw Creek in the West Shasta Mining District, California. *Environ. Sci. Technol.* 21:388–396.
Jenne, E. A. 1977. Trace element sorption by sediments and soils—Sites and processes. *In* "Molybdenum in the Environment" (W. R. Chappell and K. K. Peterson, Eds.), Vol. 2, pp. 425–553. Marcel Dekker, Inc., New York.
Johnson, C. A. 1986. The regulation of trace element concentrations in river and estuarine waters contaminated with acid mine drainage: The adsorption of Cu and Zn on amorphous hydrous oxides-oxyhydroxides. *Geochim. Cosmochim. Acta* 50:2433–2438.
Karlsson, S., Sanden, P., and Allard, B. 1987. Environmental impacts of an old mine tailings deposit—Metal adsorption by particulate matter. *Nordic Hydrol.* 18:313–324.
Kimball, B. A., Broshears, R. E., McKnight, D. M., and Bencala, K. E. 1994. Effects of instream pH modification on transport of sulfide-oxidation products. *In* "Environmental Geochemistry of Sulfide Oxidation" (C. N. Alpers and D. W. Blowes, Eds.), ACS Am. Chem. Symp. Ser. 550, pp. 224–243. American Chemical Society, Washington, DC.
Kinniburgh, D. G., Jackson, M. L., and Syers, J. K. 1976. Adsorption of alkaline earth, transition, and heavy metal cations by hydrous oxide gels of iron and aluminum. *Soil Sci. Soc. Am. J.* 40:796–799.
Loux, N. T., Brown, D. S., Chafin, C. R., Allison, J. D., and Hassan, S. M. 1989. Chemical speciation

and competitive cationic partitioning on a sandy aquifer material. *Chem. Speciation Bioavailability* 1:111-125.

Nordstrom, D. K., Plummer, L. N., Langmuir, D., Busenberg, E., May, H. M., Jones, B. F., and Parkhurst, D. L. 1990. Revised chemical equilibrium data for major water-mineral reactions and their limitations. *In* "Chemical Modeling of Aqueous Systems II" (D. C. Melchior and R. L. Bassett, Eds.), Symposium Series 416, pp. 398–413. American Chemical Society, Washington, DC.

Plumlee, G. S., Smith, K. S., Ficklin, W. H., and Briggs, P. H. 1992. Geological and geochemical controls on the composition of mine drainages and natural drainages in mineralized areas. *In* "Water-Rock Interaction" (Y. K. Kharaka and A. S. Maest, Eds.), pp. 419–422, "Proceedings, Seventh International Symposium on Water-Rock Interaction, Park City, Utah, July 13–18, 1992," Vol. 1. A.A. Balkema, Rotterdam.

Ranville, J. F., 1992. "Factors Influencing the Electrophoretic Mobility of Suspended Sediments in Acid-Mine Drainage." Ph.D. Thesis. Colorado School of Mines, Golden, CO.

Robinson, G. D. 1981. Adsorption of Cu, Zn, and Pb near sulfide deposits by hydrous manganese-iron oxide coatings on stream alluvium. *Chem. Geol.* 33:65–79.

Robinson, G. D. 1983. Heavy metal adsorption by ferromanganese coatings on stream alluvium: Natural controls and implications for exploration. *Chem. Geol.* 38:157–174.

Smith, K. S. 1991. Factors Influencing Metal Sorption onto Iron-Rich Sediment in Acid-Mine Drainage. Ph.D. Thesis, Colorado School of Mines, Golden, CO.

Smith, K. S. 1994. Prediction of element dispersion in mine-drainage systems based upon geochemical principles. *In* "International Land Reclamation and Mine Drainage Conference and Third International Conference on the Abatement of Acidic Drainage, Proceedings of a Conference Held in Pittsburgh, Pennsylvania, April 24–29, 1994," Special Publication SP 06B-94, Vol. 2, pp. 425. Bureau of Mines, U.S. Department of the Interior.

Smith, K. S. 1996. Processes controlling dissolved copper concentrations during an instream pH-modification experiment. *In* "Proceedings of the U.S. Geological Survey Toxic Substances Hydrology Program—Technical Meeting, Colorado Springs, Colorado, September 20–24, 1993," Water-Resources Investigations Report 94-4015. Vol. 2. U.S. Geological Survey, (pp. 769–774). Reston, VA.

Smith, K. S. In press. Metal sorption on mineral surfaces: An overview with examples relating to mineral deposits. *In* "The Environmental Geochemistry of Mineral Deposits" (G. S. Plumlee and M. J. Logsdon, Eds.), Reviews in Economic Geology, Vol. 6A, Ch. 7. Society of Economic Geologists, Littleton, CO.

Smith, K. S., Ficklin, W. H., Plumlee, G. S., and Meier, A. L. 1992. Metal and arsenic partitioning between water and suspended sediment at mine-drainage sites in diverse geologic settings. *In* "Water-Rock Interaction" (Y. K. Kharaka and A. S. Maest, Eds.), pp. 443–447, "Proceedings, Seventh International Symposium on Water-Rock Interaction, Park City, Utah, July 13–18, 1992," Vol. 1. A.A. Balkema, Rotterdam.

Smith, K. S. Ficklin, W. H., Plumlee, G. S., and Meier, A. L. 1993. Computer simulations of the influence of suspended iron-rich particulates on trace-metal removal from mine-drainage waters. *In* "Proceedings of the Mined Land Reclamation Symposium, Billings, Montana, March 21–27, 1993," Vol. 2, pp. 107–115. Montana State University Reclaimation Unit, Bozeman, MT.

Smith, K. S., and Macalady, D. L. 1991. Water/sediment partitioning of trace elements in a stream receiving acid-mine drainage. *In* "Proceedings of the Second International Conference on the Abatement of Acidic Drainage, Montreal, Canada, September 16–18, 1991," Vol. 3, pp. 435–450. Mine Environmental Neutral Drainage Program, Ottawa, Ontario.

Smith, K. S., Macalady, D. L., and Briggs, P. H. 1989. Partitioning of metals between water and flocculated bed material in a stream contaminated by acid mine drainage near Leadville, Colorado. *In* "U.S. Geological Survey Toxic Substances Hydrology Program—Proceedings of the Technical Meeting, Phoenix, Arizona, September 26–30, 1988" (G. E. Mallard and S. E. Ragone, Eds.), Water-Resources Investigations Report 88-4220, pp. 101–109. U.S. Geological Survey, Reston, VA.

Smith, K. S., Plumlee, G. S., and Ficklin, W. H. 1994. Predicting water contamination from metal mines and mining wastes—Notes from a workshop presented at the International Land Reclamation and Mine Drainage Conference and the Third International Conference on the Abatement of Acidic Drainage, Pittsburgh, Pennsylvania, April 24, 1994. Open-File Report 94-264, U.S. Geological Survey, Reston VA.

Smith, K. S., Ranville, J. F., and Macalady, D. L. 1991. Predictive modeling of copper, cadmium, and zinc partitioning between streamwater and bed sediment from a stream receiving acid mine drainage, St. Kevin Gulch, Colorado. *In* "U.S. Geological Survey Toxic Substances Hydrology Program—Proceedings of the Technical Meeting, Monterey, California, March 11–15, 1991" (G. E. Mallard and D. A. Aronson, Eds.), Water-Resources Investigations Report 91-4034, pp. 380–386. U.S. Geological Survey.

Stumm, W. 1992. "Chemistry of the Solid-Water Interface." John Wiley & Sons, Inc., New York.

Swallow, K. C., Hume, D. N., and Morel, F. M. M. 1980. Sorption of copper and lead by hydrous ferric oxide. *Environ. Sci. Technol.* 14:1326–1332.

Tessier, A., Fortin, D., Belzile, N., DeVitre, R. R., and Leppard, G. G. 1996. Metal sorption to diagenetic iron and manganese oxyhydroxides and associated organic matter: Narrowing the gap between field and laboratory measurements. *Geochim. Cosmochim. Acta* 60:387–404.

Webster, J. G., Nordstrom, D. K., and Smith, K. S. 1994. Transport and natural attenuation of Cu, Zn, As, and Fe in the acid mine drainage of Leviathan and Bryant Creeks. *In* "Environmental Geochemistry of Sulfide Oxidation" (C. N. Alpers and D. W. Blowes, Eds.), ACS Symposium Series 550, pp. 244–260. American Chemical Society, Washington, DC.

Epilogue

Priorities for Future Metal Adsorption Research

Everett A. Jenne (Retired)
Battelle, Pacific Northwest National Laboratory, Richland, Washington

The proposed dual goals of future adsorption research are (1) to fully understand adsorption processes and mechanisms, and (2) as suggested by Kinniburgh *et al.* (1998, this volume), to further develop models with useful predictive capabilities utilizing a minimum number of parameters, each of which can be readily determined. With these goals in mind, there are several aspects of adsorption that merit special research attention.

I. INTEGRATION AND SYNTHESIS OF EXISTING DATA

Aspects of the adsorption of metals (especially the exchange of macrocations) by geomedia have been reviewed occasionally (Kelley, 1948; Thomas, 1977; Harter, 1986). The general topic of adsorption is dealt with in many geochemistry and environmental chemistry textbooks [recent ones include Appelo and Postma (1993) and Langmuir (1997)]. Distribution coefficients (K_d) have been tabulated and retabulated (see Jenne, 1998, this volume, for references), and the various aspects of adsorption have been dealt with in numerous symposia (Jenne, 1979; Anderson and Rubin, 1981; Tewari, 1981; Davis and Hayes, 1986; Kim and Warnecke, 1986; Sibley and Myttenaere, 1986; Melchior and Bassett, 1990) and in many scientific journals.

Given the hundreds of metal adsorption papers and reports published, there has been surprisingly little effort to integrate and evaluate these data or to reconcile apparent discrepancies in the published data, such as those illustrated by Tessier *et al.* (1989). Rare synthesis efforts include the (1) calculation of power function constants for a dozen or so published contaminant metal–macrocation and macrocation–macrocation exchange isotherms by Langmuir (1981); (2) calculation by Dzombak and Morel (1990) of double-layer (DL) constants, from original data, for the surface complexation (SC) model for the adsorption of metals and metalloids onto amorphic iron oxide [$Fe(OH)_3$(a)]; (3) recalculation of triple-layer (TL) constants for the SC model using average values for the first and second acidity constants (Smith and Jenne, 1989, 1991); (4) summarization of adsorption system information of upward of 50 literature data sets for metal adsorption by Al, Fe, Mn, and Ti oxides by Turner (1995) with tabulation of the point of zero charge (pH_{ZPC}) surface area (A_s), protonation and deprotonation constants (for CC, DL, and TL SC models), with plots of the actinide adsorption data; and (5) development of correlation methods for surface adsorption parameters (Sahai and Sverjensky, 1997).

Dzombak and Morel (1990), Smith and Jenne (1991), and Turner (1995) all included summary information on the papers which they used in their syntheses. In future synthesis studies, it is desirable that such summaries be included and that they specify (1) the method(s) of adsorbent preparation (and range in age of precipitated adsorbents used) and/or of adsorbent pretreatment, (2) characterization information (e.g., A_s, site concentration, pH_{zpc}, and particle or aggregate size range or distribution), (3) the experimental design [concentration range of the metal(s), adsorbent(s), competing metals, and macrocations; saturation indices of potential precipitates; temperature; container type; mixing parameters], (4) the rationale for selection of the reaction time used, (5) the final pH values (and initial values if different and any adjustment technique), (6) the method of phase separation, (7) the mode of blank correction, and (8) results of error analysis or the reviewer's estimate of overall error.

There is considerable current interest in the modeling of adsorption data. It has been nearly a decade since the calculation of DL constants from original data by Dzombak and Morel (1990) and recalculation of TL constants by Smith and Jenne (1989, 1991). The calculation of TL constants by Turner (1995) was restricted to actinides. Therefore, it is timely to search out all published DL and TL constants, recalculate them as necessary to achieve internal consistency, and calculate additional constants using available data, tabulating both by adsorbent and by metal. Further, it is desirable to rationalize appropriate data sets using models such as those of Borkovec *et al.* (1998, this volume) or Kinniburgh *et al.* (1998, this volume).

A note of caution is in order here; compilations per se *are not* called for. What is needed is careful evaluation (with due cognizance of potential adsorbent dissolution and precipitation) of the extant data, recalculation, and modeling of that data as appropriate with a maximum of sample and system characterization information (see above) attached.

II. SYSTEM CHARACTERIZATION

Kinniburgh *et al.* (1998, this volume) suggest that the objective of system characterization and modeling should be to allow laboratory data to be used in field transport applications. However, as the majority of current adsorption papers are process and mechanism oriented, another equally important objective is to produce quality data that can be used by other investigators in subsequent synthesis efforts. Campbell and Davies (1995) contend that with sufficient control of system variables in published studies, the issue of differing degrees of site heterogeneity among the samples could already have been addressed with the available data. Adequate system characterization is a key step in assuring quality of new adsorption data. Careful analysis of the aqueous phase can provide useful information on the stoichiometry of exchange adsorption, variation in preferential desorption as a function of pH, and concentration of cations dissolved from adsorbents. For example, careful investigations are increasingly finding that below pH 4, Fe and Al arising from the dissolution of adsorbents significantly affect metal adsorption by competing for adsorption sites or for aqueous ligands (Zachara *et al.,* 1995; Turner *et al.,* 1996; Redden *et al.,* 1998, this volume). Possible surface coprecipitates of the adsorbate metal and adsorbent dissolution products, which have recently been reported (Scheidegger *et al.,* 1996; Towle *et al.,* 1997; and H. Thompson, personal communication), need to be looked for.

Such characterization of the aqueous phase can often avoid the necessity for speculation as to the reasons for the results obtained. It needs to be recognized that artifact-free experiments are highly sample/system specific; for example, although 0.45-μm filter membranes are often inadequate (Horowitz *et al.,* 1996), Nagy *et al.* (1991) found no difference in the aqueous chemistry of a kaolinite suspension between filtrations with 0.05-μm and 0.45-μm pore size membranes.

Redden *et al.* (1998, this volume) suggest that experimental container walls need to be conceptualized as a competing adsorbent and container adsorption determined as a function of the major variables of the experiment, especially metal concentration and pH. Otherwise, blank corrections can increase the apparent analytical error rather than reduce it. In addition to reporting the magnitude of blank corrections, it is desirable for investigators to report errors for successive steps and report propagated error for the adsorption data and constants.

Bertetti *et al.* (1998, this volume) suggest that a slow mass transfer of CO_2 from laboratory air into experimental solutions that were CO_2 undersaturated may be one cause of slow equilibrium in metal adsorption experiments. The magnitude of this effect and the time required to reach equilibrium by diffusion through plastic containers need to be assessed experimentally.

III. PRINCIPAL ADSORBENTS

Iron oxides are nearly always an important adsorbent in natural systems and, except for reducing conditions or the case where the Fe oxide concentration is low or another adsorbent is relatively high, they are the dominant adsorbent (Jenne, 1968, 1998, this volume; Axe and Anderson, 1998, this volume; Goldberg et al.,

1998, this volume). For example, Smith *et al.* (1998, this volume) note that although the use of $Fe(OH)_3(a)$ as the only adsorbent gave good results at most of their sulfitic mine drainage sites, the fit of their data deteriorated at one site where the ratio of PROC to oxidic Fe was higher than those at the other sites. However, reliable metal adsorption modeling in oxic systems requires knowing the minimum particulate reactive organic carbon (PROC), Mn oxide, and/or carbonate concentrations, relative to that of Fe oxide, that requires their inclusion in the model or, alternatively, their being used in lieu of Fe oxides as the adsorbent(s). This may be particularly important for the suspended surface waters which may contain primary and secondary minerals with minimal digenetic coatings. It has been suggested that one effect of PROC is to block the access of metals to the pores of inorganic adsorbents such as $Fe(OH)_3(a)$ (A. Turner, personal communication). Indeed, pores in particles and aggregates may be a favored habitat for bacteria because of the protection and pores provide from grazing organisms (Holmen and Gschwend, 1997).

Because of the importance of Fe oxides as an adsorbent, there are two issues that merit special attention, First, given the fact that the surface area of $Fe(OH)_3(a)$ is so much larger than that of the crystalline oxides on a mass or molar basis and the increasingly common use of a reference site density for the amorphic fraction, a means is needed to estimate the additional sites contributed by the crystalline Fe oxides present in heterogeneous geomedia samples. Second, given that high aqueous concentrations of sulfate are reported to cause an increase in Cd adsorption on $Fe(OH)_3(a)$ (Benjamin and Leckie, 1982; Smith *et al.,* 1998, this volume) and that an increased Si content of $Fe(OH)_3(a)$ is reported to result in increased Cd adsorption (Anderson and Benjamin, 1985), there is a need for systematic studies to determine the effect of varying concentrations of SO_4^{2-}, Si, and perhaps other constituents (e.g., Al, Mn, and fulvic acid) upon the number of effective sites and on DL and TL SC model adsorption constants for $Fe(OH)_3(a)$. Mechanistic studies using surface spectroscopy methods will be required on $Fe(OH)_3(a)$ preparations containing a range of Si concentrations to understand why the presence of Si increases the adsorption of Cd but not of Cu or Zn.

IV. MULTIMETAL DATA ON INDIVIDUAL ADSORBENTS

Few adsorption data sets contain data for more than two metals on a well-characterized adsorbent, although exceptions exist (e.g., Payne *et al.,* 1998, this volume). The number and range of variables are generally insufficient to reliably evaluate interactions among variables. As pointed out by Kinniburgh *et al.* (1998, this volume, p. 517), it is necessary that "in order to test and develop . . . models further, more good data are required on multi-metal systems." Extensive data (multiple metals; multiple competitive cations; and a wide range in variables such as metal concentration, pH, and ionic strength) on a single adsorbent are more useful than less intensive data on multiple adsorbents. Because the number of metals that can be included in a given study, an alternative approach to obtaining multimetal adsorption data is for multiple investigators to use a common set of adsorbents, pretreatments, and/or common preparation methods for precipitates, and to use common ionic strengths and background electrolytes. Given the current interest both in the fundamental aspects of surface complex formation and in applying adsorption models to transport problems, the review and publication of a set of reference methods for adsorption studies and an interlaboratory comparison could be valuable.

V. LOADING

Benjamin and Leckie (1981) inferred that $Fe(OH)_3$(a) may contain multiple distinct binding sites, and further, based on competition experiments, that the preferred binding sites for one metal are not necessarily the preferred adsorption sites for other metals. However, this conclusion of multiple types of binding sites does not meld smoothly with the Dzombak and Morel (1986) hypothesis of one set of strong sites and one set of weak sites. These two views need to be reconciled.

The importance of loading on adsorption constants has been well demonstrated (Benjamin and Leckie, 1981; Smith and Jenne, 1991). However, no study has been found that attempts to predict the loading effect or that proposes a means of including the apparent decrease in the adsorption constant with increased loading in adsorption models. There is obviously a need for an experimental effort to ascertain how pH and cation competition varies as the loading of the metal of interest increases and vice versa. In studying the interactions of loading, pH, competing cations, etc., the constant aqueous concentration technique of Axe and Anderson (1998, this volume) may offer an advantage.

VI. SURFACE AREA, SITE DENSITY, AND METAL ADSORPTION DENSITY

Brady *et al.* (1998, this volume) note that the maximum metal adsorbed is commonly more than an order of magnitude less than the number of sites estimated by acid–base titration. Bottero *et al.* (1993) used ionic surfactants as probe molecules, obtaining A_s values of 163, 134, and 139 $m^2 \cdot g^{-1}$ for $Fe(OH)_3$(a) prepared with OH to Fe ratios of 0.0, 2.0, and 2.5, respectively, in the precipitation media. Using a continuous Ar adsorption technique, they obtained an A_s of 300 $m^2 \cdot g^{-1}$, compared to 139 $m^2 \cdot g^{-1}$ obtained by anionic surfactant adsorption. They suggest that hydration of $Fe(OH)_3$(a) aggregates in water may result in a rather different surface area than in aggregates in the dry state. Since virtually all surface area data are obtained on desiccated samples, current measurements may not reflect the areas encountered by adsorbates.

Site concentration and surface area are vital parameters in interpreting and modeling metal adsorption. They may become even more important as the loading effect is further evaluated. Thus, there is a need to reexamine surface area measurements and to exploit techniques that do not require dehydration of oxide and clay preparations because there may be only a poor correspondence between wet- and dry-surface areas. Further, in using SC models in conjunction with transport modeling in groundwater, means of obtaining adequate parameterization of the *in situ* surface area of aquifer material and of estimating the uncertainty in the surface area parameter need to be developed and evaluated.

VII. TIME DEPENDENCY

The data of Axe and Anderson (1998, this volume) showing that periods of a month and of 5 years are required for equilibrium in the Sr–$Fe(OH)_3$(a) system at pH 6 and 7, respectively, call into question the common practice of allowing a few hours for equilibration. The view of Axe and Anderson (1998, this volume) is that to reliably model the leaching of metals from contaminated soils or sediments, it is necessary to model diffusion and the adsorption process within oxide particles. A further evaluation of other oxide preparations in terms of diffusion rates and equilibration times is needed, as is the testing of means of quantifying the amount of adsorption associated with the fast (exterior sites) and slow (interior sites) processes, respectively. If further research supports the conclusions of Jenne (1995, 1996) and Axe and Anderson (1998, this volume) that virtually all metal adsorption by geomedia is diffusion limited, the validity of calculating reaction rates according to some assumed rate law needs to be reexamined.

With the possibility that continued metal adsorption over extended time may represent the formation of multinuclear complexes of such a size that they are *de facto* precipitates, there is a need to carry out adsorption rate experiments in conjunction with surface spectroscopic analysis to ascertain if the initial rate of precipitation via formation of multinuclear complexes is significantly different than the rate of metal diffusion to sites on the interior of $Fe(OH)_3$(a) aggregates.

VIII. ADDITIVITY OF MULTIPLE ADSORBENTS

Transport modeling approaches that calculate K_d values for individual flow path segments, weighting the K_d for individual minerals based on the amount of that mineral present (Vandergraaf *et al.,* 1993), implicitly assume additivity. However, Honeyman (1984) contended that significant changes in adsorptive properties occurred when certain adsorbents were present in mixtures; the largest effect was when thallium oxide, a quadravalent metal, was present in the mixture. In contrast, Davies-Colley *et al.* (1984) found no evidence of a reduction in adsorption effectiveness as a result of mixing selected solids [e.g., a $Fe(OH)_3$(a)–Wyoming bentonite–humic substance mixture]. Anderson and Benjamin (1990) found that decreased adsorption in $Fe(OH)_3$(a)–$Al(OH)_3$ mixtures was due to $Al(OH)_y^{x+}$ adsorption on $Fe(OH)_3$(a) surfaces. Thus there is an acute need to clarify whether nonadditivity laboratory data apply to natural materials and to determine the conditions under which additivity is a valid assumption.

IX. SOLIDS CONCENTRATION EFFECT

For several years there have been periodic reports that increasing the concentration of solids in an experimental system results in a progressive decrease in the K_d (e.g., Di Toro 1985; Honeyman and Santschi, 1988; Oscarson and Hume, 1998, this volume). Others have found that the decreased metal adsorption could often be attributed to experimental procedural artifacts or to data reduction and presentation artifacts (McKinley and Jenne, 1991). A critical examination of the available data is required to resolve this issue since, if the effect is real, the solids concentration effect would need to be taken into account in applying adsorption data obtained on dilute suspensions to field problems.

X. NATURE OF SURFACE COMPLEXES AND MECHANISTIC MODELING

There is a need for further collaborative studies to evaluate the extent to which surface complexes giving "best fits" in DL and TL SC models are confirmed by surface analysis techniques. Although SC models generally fit the data well, this could be a matter of sufficient adjustable variables (Payne *et al.*, 1998, this volume). Given the reports of multinuclear surface complexes (e.g., O'Day *et al.*, 1994; Payne *et al.*, 1998, this volume; Thompson *et al.*, 1998, this volume), there is a need to reevaluate whether or not it is possible to distinguish between adsorption and surface precipitation.

With regard to molecular dynamic modeling of adsorbed metal cations, Cygan *et al.* (1998, this volume) point out the desirability of including the electronic polarizability of the adsorbing ion and the electrostatics of several layers of surrounding water molecules in future modeling.

Redden *et al.* (1998, this volume) have found that at equimolar concentrations of U and citrate, each constituent was adsorbed by goethite in amounts similar to that of single-adsorbate systems. However, at higher citrate concentrations U adsorption increased markedly when both were present. In contrast, with gibbsite and kaolinite the increased citrate significantly decreased U adsorption. These authors raise the likelihood of a citrate–UO_2^{2+} bridging complex on goethite surfaces. This proposal calls for spectroscopic confirmation. Redden *et al.* (1998, this volume) note that the available SC models are not suited to dealing with the apparently variable dentate nature of the adsorption of trivalent ligands such as citrate. Similarly, there may be a need to develop the SC model's capability to deal with decreasing adsorption constants as loading with a particular metal increases.

XI. FIELD-SCALE APPLICATIONS

One of the goals in understanding and quantifying interfacial processes at solid surfaces is to assess reliably surface, ground, and soil water transport of reactive contaminants, such as metals. Significant increases in our mechanistic understanding of adsorption processes are occurring through the current widespread application of surface spectroscopy. The continued increase in computational power and speed of computers provides hope for the inclusion of increased adsorption modeling detail in transport codes. Therefore, consideration needs to be given to a means of utilizing mechanistic information in field-level transport problems. However, application of mechanistic adsorption information at the field level offers a considerable challenge given the notable chemical and physical spatial heterogeneity of sediments and soils (see for example Rosentreter *et al.,* 1998, this volume; Smith *et al.,* 1998, this volume).

Studies focused on translating laboratory adsorption measurements to field-scale applications are limited and need to be expanded. A fertile area for future research exists in developing appropriate methods for the scale up averaging of nonlinear adsorption process data obtained from laboratory measurements to field-scale applications. The applicability of adsorption parameters obtained in dilute suspension to aquifers and the vandose (i.e., unsaturated) zone of surfical sediments remains in question (Kaplan *et al.,* 1996; Oscarson and Hume, 1998, this volume). Field-scale investigations of the spatial variability of sediment adsorptive properties, relationships of adsorption properties to physical and geological characteristics of a site, and identification of cost-effective characterization strategies for field sites are required (e.g., Rosentreter *et al.,* 1998, this volume).

Evaluations of the trade-offs between adequate characterization of geomedia for the application of multiadsorbent surface complexation models and direct measurement of metal adsorption on a suite of site materials for use in semi-empirical models need to be made (see for example, Kinniburgh *et al.,* 1998, this volume). It appears that models are needed which bridge the gulf between mechanism-focused laboratory studies and processes-oriented field investigations, models that (1) capture the salient feature observed in laboratory experiments, (2) require a minimum but adequate number of property measurements, and (3) are computationally efficient. These models need to be tested in intermediate- and field-scale experiments.

XII. SUMMARY

In general, there is an acute need for synthesis of existing data. Particular issues meriting detailed reexamination and recalculation of existing data include additivity (i.e., the extent to which adsorbents in mixtures lose or retain their individual adsorptive properties), loading (i.e., the apparent decrease in the adsorption constant with increasing adsorption of a metal), and competition from macrocations and other metals (e.g., the pH and metal concentration ranges at which competition is significant for particular metal–adsorbent systems). Multimetal, multicompeting cation experiments as a function of pH and metal concentration are needed to increase our understanding of interaction effects which will allow differentiation of the suitability of competitive versions of surface complexation models. Additional adsorption data in conjunction with surface analyses are needed to further our understanding of (1) surface complexes, (2) the transition from adsorption to surface precipitation, and (3) the role of ligands in facilitating metal adsorption. Some aspects of adsorption models also need further development.

ACKNOWLEDGMENTS

A draft of this chapter was sent to a number of scientists for their comments and suggestions. The content of chapter has benefited from the suggestions of Drs. Patrick Brady, Michael Borkovec, Charles Kincade, Bobby Pabalan, Kathy Smith, Robert Smith, Hillary Thompson, Andrew Turner, and David Turner.

REFERENCES

Anderson, P. R., and Benjamin, M. M. 1985. Effects of silicon on the crystallization and adsorption properties of ferric oxides. *Environ. Sci. Technol.* 19:1048–1053.

Anderson, P. R., and Benjamin, M. M. 1990. Surface and bulk characteristics of binary oxide suspensions. *Environ. Sci. Technol.* 24:692–698.

Anderson, M. A., and Rubin, A. J. (Eds.). 1981. "Adsorption of Inorganics at Solid-Liquid Interfaces." Ann Arbor Science, Ann Arbor, MI.

Appelo, C. A. J., and Postma, D. 1993. "Geochemistry, Groundwater and Pollution." A. A. Balkema, Rotterdam.

Axe, L., and Anderson, P. R. 1998. Interparticle diffusion of metal contaminants in amorphous oxide minerals. *In* "Adsorption of Metals by Geomedia" (E. A. Jenne, Ed.). Academic Press, San Diego.

Benjamin, M. M., and Leckie, J. O. 1981. Multiple-site adsorption of Cd, Cu, Zn, and Pb on amorphous iron oxyhydroxide. *J. Colloid Interface Sci.* 79:209–221.

Benjamin, M. M., and Leckie, J. O. 1982. Effects of complexation by Cl, SO_4, and S_2O_3 on adsorption behavior of Cd on oxide surfaces. *Environ. Sci. Technol.* 16:162–170.

Bertetti, F. P., Pabalan, R. T., and Almendarez, M. G. 1998. Studies of neptunium^{V} sorption on quartz, clinoptilolite, montmorillonite, and α-alumina. *In* "Adsorption of Metals by Geomedia" (E. A. Jenne, Ed.). Academic Press, San Diego.

Borkovec, M., Rusch, U., and Westall, J. C. 1998. Modeling of competitive ion binding to heterogenous materials with affinity distributions. *In* "Adsorption of Metals by Geomedia" (E. A. Jenne, Ed.). Academic Press, San Diego.

Bottero, J. Y., Anaud, M., Villieras, F., Michot, L. J., De Donato, P., and Francois, M. 1993. Surface and textural heterogeneity of fresh hydrous ferric oxides in water and in the dry state. *J. Colloid Interface Sci.* 159:45–52.

Brady, P. V., Cygan, R. T., and Nagy, K. L. 1998. Surface charge and metal sorption on kaolinite. *In* "Adsorption of Metals by Geomedia" (E. A. Jenne, Ed.). Academic Press, San Diego.

Campbell, L. S., and Davies, B. E. 1995. Soil sorption of caesium modeled by the Langmuir and Freundlich isotherm equations. *Appl. Geochem.* 10:715–723.

Cygan, R. T., Nagy, K. L., and Brady, P. V. 1998. Molecular models of cesium sorption on kaolinite. *In* "Adsorption of Metals by Geomedia" (E. A. Jenne, Ed.). Academic Press, San Diego.

Davies-Colley, R. J., Nelson, P. O., and Williamson, K. J. 1984. Copper and cadmium uptake by estuarine sedimentary phases. *Environ. Sci. Technol.* 18:491–499.

Davis, J. A., and Hayes, K. F. (Eds.). 1986. "Geochemical Processes at Mineral Surfaces," Symp. Ser. No. 323, pp. 2–18. American Chemical Society, Washington, DC.

Di Toro, D. M. 1985. A particle interaction model of reversible organic chemical sorption. *Chemosphere* 14:1503–1538.

Dzombak, D. A., and Morel, F. M. M. 1986. Sorption of cadmium on hydrous ferric oxide at high sorbate/sorbent ratios: Equilibrium, kinetics, and modeling. *J. Colloid Interface Sci.* 112:588–598.

Dzombak, D. A., and Morel, F. M. M. 1990. "Surface Complexation Modeling: Hydrous Ferric Oxide." Wiley, New York.

Goldberg, S., Su, C., and Foster, H. S. 1998. Sorption of molybdenum of oxides, clay minerals, and soils. *In* "Adsorption of Metals by Geomedia" (E. A. Jenne, Ed.). Academic Press, San Diego.

Grenthe, I., and Puigdomenech, I. 1997. "Modeling in Aquatic Chemistry." Nuclear Energy Agency, Organization for Economic Co-operation for Economic Development, Paris.

Harter, R. D. (Ed.). 1986. "Adsorption Phenomena." Van Nostrand Reinhold, New York.

Holmen, B. A., and Gschwend, P. M. 1997. Estimating sorption rates of hydrophobic organic compounds in iron oxide- and aluminosilicate clay-coated aquifer sands. *Environ. Sci. Tech.* 31:105–113.

Honeyman, B. D. 1984. "Cation and Anion Adsorption at the Oxide/Solution Interface in Systems Containing Binary Mixtures of Adsorbents: An Investigation of the Concept of Adsorptive Additivity." Ph.D. Thesis, Stanford University, Stanford, CA.

Honeyman, B. D., and Santschi, P. H. 1988. Metals in aquatic systems. *Environ. Sci. Technol.* 22:862–871.

Horowitz, A. J., Lum, K. R., Garbarino, J. R., Hall, G. E. M., Lemieux, C., and Demas, C. R. 1996. Problems associated with using filtration to define dissolved trace element concentrations in natural water samples. *Environ. Sci. Technol.* 30:954–963.

Jenne, E. A. 1968. Controls on Mn, Fe, Co, Ni, Cu and Zn concentrations in soils and water—The significant role of hydrous Mn and Fe oxides. Adv. Chem. no. 73, pp. 337–387. American Chemical Society, Washington, DC.

Jenne, E. A. (Ed.). 1979. "Chemical Modeling in Aqueous Systems: Speciation, Sorption, Solubility, and Kinetics," Symp. Ser. No. 93. American Chemical Society, Washington, DC.

Jenne, E. A. 1995. Metal adsorption onto and desorption from sediments: 1. Rates. *In* "Metal Speciation and Contamination of Aquatic Sediments" (H. E. Allen, Ed.), pp. 81–112. Ann Arbor Press, Ann Arbor, MI.

Jenne, E. A. 1996. Metal adsorption onto and desorption from sediments: 2. Artifact effects. *Mar. Freshwater Res.* 64:1–18.

Jenne, E. A. 1998. Adsorption of metals by geomedia: Data analysis, models, controlling factors, and related issues. *In* "Adsorption of Metals by Geomedia" (E. A. Jenne, Ed.). Academic Press, San Diego.

Kaplan, D. I., Serne, R. J., Owen, A. T., Conca, J., Wietsma, T. W., and Gervais, T. L. 1996. "Radionuclide Adsorption Distribution Coefficients Measured in Hanford Sediments for the Low Level Waste Performance Assessment Project." PNNL-11385, Pacific Northwest National Laboratory, Richland, WA.

Kelley, W. P. 1948. "Cation Exchange in Soils," Am. Chem. Soc. Monograph No. 109. Reinhold, New York.

Kim, J.-I., and Warnecke, E. (Eds.). 1986. "Chemie und Migrationsverhalten der Aktioide und Spaltprodukte in Natürlichen Aquatischen Systemen." Berich SE-14, Vorträge des 66, PTB-Seminars, DE 88 750429, ISSN 0721-0892, ISBN 3-88-314-578-5.

Kinniburgh, D. G., van Riemskijk, W. H., Koopal, L. K., and Benedetti, M. F. 1998. Ion binding to humic substances: Measurements, models, mechanisms. *In* "Adsorption of Metals by Geomedia" (E. A. Jenne, Ed.). Academic Press, San Diego.

Langmuir, D. 1981. The power exchange function: A general model for metal adsorption onto geological materials. *In* "Adsorption from Aqueous Solutions" (P. H. Tewari, Ed.), pp. 1–17. Plenum Press, New York.

Langmuir, D. 1997. "Aqueous Environmental Geochemistry." Prentice Hall, Upper Saddle River, NJ.

McKinley, J. P., and Jenne, E. A. 1991. An experimental investigation and review of the "solids concentration" effect in adsorption studies. *Environ. Sci. Technol.* 25:2082–2087.

Melchior, D. C., and Bassett, R. L. 1990. "Chemical Modeling of Aqueous Systems II," Symp. Ser. 416. American Chemical Society, Washington, DC.

Nagy, K. L., Blum, A. E., and Lasaga, A. C. 1991. Dissolution and precipitation kinetics of kaolinite at 80°C and pH 3: The dependence on solution saturation state. *Am. J. Sci.* 291:649–686.

O'Day, P. A., Brown, G. E., Jr., and Parks, G. A. 1994. X-ray absorption spectroscopy of cobalt(II) multinuclear surface complexes and surface precipitates on kaolinite. *J. Colloid Interface Sci.* 165:269–289.

Oscarson, D. W., and Hume, H. B. 1998. Effect of the solid:liquid ratio on the sorption of Sr^{2+} and Cs^{+} on bentonite. In "Adsorption of Metals for Geomedia," (E. A. Jenne, Ed.). Academic Press, San Diego.

Payne, T. E., Lumpkin, G. R., and Waite, T. D. 1998. Uranium^{VI} adsorption on model minerals: Controlling factors and surface complexation modeling. *In* "Adsorption of Metals by Geomedia" (E. A. Jenne, Ed.). Academic Press, San Diego.

Redden, G. D., Li, J., and Leckie, J. O. 1998. Adsorption of U^{VI} and citric acid on goethite, gibbsite, and kaolinite: Comparing results for binary and ternary systems. *In* "Adsorption of Metals by Geomedia" (E. A. Jenne, Ed.). Academic Press, San Diego.

Sahi, N., and Sverjensky, D. A. 1997. Solvation and electrostatic model for specific electrolyte adsorption. *Geochim. Cosmochim. Acta.* 61:2827–2848.

Scheidegger, A. M., Lamble, G. M., and Sparks, D. L. 1996. Investigation of Ni sorption on pyrophyllite: An XAFS study. *Environ. Sci. Technol.* 30:548–554.

Sibley, T. H., and Myttenaere, C. (Eds.). 1986. "Application of Distribution Coefficients to Radiological Assessment Models." Elsevier, London.

Smith, K. S., Ranville, J. F., Plumlee, G. S., and Macalady, D. L. 1998. Predictive double-layer modeling of metal sorption in mine-drainage systems. *In* "Adsorption of Metals by Geomedia" (E. A. Jenne, Ed.). Academic Press, San Diego.

Smith, R. W., and Jenne, E. A. 1989. Compilation, evaluation, and prediction of triple-layer model constants for ions on Fe(III) and Mn(IV) hydrous oxides. PNL-6754, Pacific Northwest National Laboratory, Richland, WA.

Smith, R. W., and Jenne, E. A. 1991. Compilation, evaluation, and prediction of triple-layer model constants for ions on Fe(III) and Mn(IV) hydrous oxides. *Environ. Sci. Technol.* 25:525–531.

Stumm, W., and Morgan, J. 1996. "Aquatic Chemistry–Chemical Equilibria and Rates in Natural Waters," 3rd ed. Wiley, New York.

Tessier, A., Carignan, R., Dubreuil, B., and Rapin, F. 1989. Partitioning of zinc between the water column and the oxic sediments in lakes. *Geochim. Cosmochim. Acta* 53:1511–1522.

Tewari, P. H. (Ed.). 1981. "Adsorption from Aqueous Solutions." Plenum, New York.

Thomas, G. W. 1977. Historical developments in soil chemistry: Ion exchange. *Soil Sci. Soc. Am. J.* 41:230–238.

Thompson, H. A., Parks, G. A., and Brown, G. E., Jr. 1998. Structure and composition of uranium^{VI} sorption complexes at the kaolinite-water interface. *In* "Adsorption of Metals by Geomedia" (E. A. Jenne, Ed.). Academic Press, San Diego.

Towle, S. N., Barger, J. R., Brown, G. E., Jr., and Parks, G. A. 1997. Surface precipitation of Co(II)(aq) on Al_2O_3. *J. Colloid Interface Sci.* 187:62–82.

Turner, D. R. 1995. A uniform approach to surface complexation modeling of radionuclide sorption. CNWRA 95-001, Center for Nuclear Waste Regulatory Analyses, San Antonio, TX.

Turner, G. D., Zachara, J. M., McKinley, J. P., and Smith, S. C. 1996. Surface-charge properties and UO_2^{2+} adsorption of a subsurface smectite. *Geochim. Cosmochim. Acta* 60:3399–3414.

Vandergraaf, T. T., Ticknor, K. V., and Melnyk, T. W. 1993. The selection of a sorption database for the Geosphere model in the Canadian Nuclear Fuel Waste Management Program. *J. Contam. Hydrol.* 1:327–345.

Zachara, J. M., Smith, S. C., and Kuzel, L. S. 1995. Adsorption and dissociation of Co-EDTA complexes in iron oxide-containing subsurface sands. *Geochim. Cosmochim. Acta* 59:4825–4844.

Appendix

Abbreviations and Acronyms

Nonletter Marks

[ ]	signifies the number of a particular equations(s)
:	signifies the metal "S" is an adsorption site on solid surface

Greek Letters

α	alpha radiation
β	beta radiation
β	plane of outersphere adsorption in TLM
Γ	site density
ε_{ij}	Lennard-Jones well-depth potential
ε_b	bulk porosity
ε_o	permittivity of vacuum
ε_p	particle porosity
ηm	nanometer
ηmol	nanomole
$\theta_{i,T}$	fractional surface coverage of, I, the adsorbed ion
θ	angle of incident X-ray radiation
κ	$1/\kappa$ is Debye-Hückel characteristic distance
λ	equivalent ionic conductance
λ	distance between neutral adsorption sites
μ	mobility of dissolved metal
μ	chemical potential
μA	microamp
μCi	microcurie
μg	microgram

μl	microliter
μm	micrometer
μmol	micromole
μS	microSiemen
μs	microsecond
ν_{vib}	vibrational frequency
ν_4	frequency of infrared bending modes
ρ	bulk (dry) density
ρs	picosecond
Σ	summation
σ	standard deviation
σ	surface charge
σ_o	charge of outer triple layer
σ^2	Debye-Waller factor
σ^2	disorder parameter
τ	tortuosity
ϕ	electric potential
ϕ_d	bulk density of geomedia
ψ	potential (difference) between aqueous solution and the mean potential at the location where protons are bound at the surface
ψ_1	electrostatic potential
ω	Damkohler number

English Letters

Å	angstrom
$_{aq}$	aqueous
A_e	net (radio) activity at equilibrium
A_i	net initial (radio) activity
A_s	area of surface
A	amp
AAS	atomic absorption spectroscopy
AEM	analytical electron microscopy
AFM	atomic force spectrometry
AOD	ammonium oxalate extraction in darkness
ATR-FTIR	attenuated total reflectance-Fourier transform infrared
(a)	amorphous
$_{ads}$	adsorbed
a	activity of an ion
atm	atmosphere

BET	Brunauer, Emmett and Teller A_S measurement
Bi	Biot number
b	Donnan volume salt dependence
b_i	regression constant
C_a	concentration of acid
C_b	concentration of base
C_p	particle (solids) concentration
$C_{s,i}$	surface or local concentration of metal
$\boldsymbol{C}_1$	capacitance, inner Helmholtz layer
$\boldsymbol{C}_2$	capacitance, outer Helmholtz layer
C	capacitance
C	concentration of dissolved (aq) or adsorbed (ads) metal
C	charge
CA	condensation approximation
CEC	cation exchange capacity
ClP	chlorpyromorphite
CN	coordination number
CONICA	consecutive NICA model
CR	soil-plant concentration ratio
CVFF	consistent valence force field
CC	constant capacitance version, triple layer model
(c)	crystalline
c_i	concentration, initial
c_f	concentration, final
cc	cubic centimeters
cap	capacity
cmol	centimole
cpm	counts (i.e., radioactive disintegrations)/minute
D	dielectric constant of water
D_e	pore diffusion coefficient
D_s	surface diffusion coefficient within sorbent particle
D	diffusivity
DCB	dithionate-citrate-bicarbonate extractant for Fe oxides
DCP	direct current plasma
DDL, DL, DLM	diffuse double layer (model)
DF	degrees of freedom
DOE	(U.S.) Department of Energy
DOM	dissolved organic matter
DRIFT	diffuse reflectance infrared Fourier transform spectroscopy

$_d$	charge of diffuse layer in triple layer model
d	day
d	plane of diffuse ion swarm
d	diameter of sphere or of particle
den	density
dpm	disintegrations per minute
E_{ij}	potential energy of interaction between atoms I and J
E_0	energy of electron at which $k=0$
E	kinetic energy of electron
E	total energy
E	potential
EDL	electrical double layer term in SCM equation
EDS	energy dispersive spectrometry
EDTA	ethylene diamine tetraacetic acid
EPA	(U.S.) Environmental Protection Agency
EG	ethylene glycol
EGME	ethylene glycol monoethylene
Eqs.	equations
EM	electrophoretic mobility
EXAFS	extended X-ray absorption fine structure
EXD	energy dispersive x-ray
e	electronic charge
eq	equivalent
eV	electron volt
$_F$	signifies the constant is from Fruendlich equation
F	fluorescence
F	Faraday constant
F_{ML}	fluorescence of metal-ligand complex
Fe_s	iron adsorption sites, strong
Fe_w	iron adsorption sites, weak
FE-SEM	field emission-scanning electron microscopy
FEP	fluorinated ethylene propylene
FITEQL	model for optimizing fitting of triple layer model constants
[Flour]	flourophane concentration
FT	Fourier transform
FTIR	Fourier transform infrared
$_f$	final as in concentration or time
fsec	femtoseconds
GeV	gigaelectron volt
GTLM	generalized two layer (surface complexation) model

$_{(g)}$, (g)	gas
g	gram
g	gravity
H_s	concentration of protons close to binding sites
HA	hydroxylapatite
HDPE	high density polyethylene
HFO	hydrous ferric oxide
HLW	high level (nuclear) waste
HP	hydroxypyromorphite
HPLC	high pressure liquid chromatography
HYDRAQL	geochemical model
h	Planck's constant
hr	hour
I	ionic strength
IC	ion chromatograph
ICP	inductively coupled plasma
IMt	Montana illite
IOCS	iron oxide coated sand
IR	infrared
^{int}	superscripted constant is an intrinsic constant
i.d.	inside diameter
in	inch
is	subscript "is" refers to inner-sphere surface complexation
J	Joule
$\tilde{K}_i$	center of K_+ distribution for $SH^x \leftrightarrow SM$ reaction for i*th* ion
K′	conditional stability constant
K_d^A	activity-based distribution coefficient
^aK	exchange constant obtained using the activity of aqueous metal specie
^{int}K, K^{int}	intrinsic exchange constant, surface complexation model
K^{SC}	surface complexation constant, nonintrinsic
K_A	anion surface complexation constants, triple layer model
K_a	distribution coefficient, normalized to specific surface area
$K_{a'}$	distribution coefficient, normalized to "effective" surface area
K_+^{app}	surface acidity constant, apparent, protonated site
K_-^{app}	surface acidity constant, apparent, deprotonated site
K_{a1}	first acidity constant
K_{a2}	second acidity constant
K_d	distribution coefficient

$K_{d,a}$	distribution coefficient, surface area based
$K_{d,i}$	distribution coefficient of *ith* sample
$K_{d,i,5}$	distribution coefficient of *ith* sample at pH 5.0
K_p	Power function exchange constant
K_{Me}	metal-surface binding constant
$K_{L,aff}$	'affinity' constant in Langmuir equation
$K_{L,max}$	maximum adsorption constant in Langmuir equation
K_L	ligand-surface binding constant
K_{SH}	distribution parameters, monodentate adsorption equilibria
K_{SHM}	distribution parameters, bidentate adsorption equilibria
K	temperature, degrees Kelvin
KGa	kaolin, Georgia
k_1	rate constant, forward
k_2	rate constant, backward
k_f	mass transfer coefficient
k	photoelectron wave factor
k	rate constant, first order
k	factor in fluorescence equation
kcal	kilocalories
keV	kiloelectron volt
kJ	kilojoule
kV	kilovolt
L	signifies the constant is from Langmuir equation
L	length
L	ligand
L	liter
LDPE	low density polyethylene
LEA	linear adsorption with local equilibrium model
LF	Langmuir-Fruendlich bimodal isotherm
LM	ligand-metal complex
LSA	liquid scintillation analysis
M	sum of solid solution Pb plus Ca, hydroxypyromorphite
M	metal
M	mass
M	moles per liter
MA	mass action relationship
MD	molecular dynamics
MEP	molecular electrostatic potential
Mg	million grams
Mg-EG	magnesium-saturated and ethylene glycol treated

MINTEQ	geochemical model
MPa	milliPascal
MS	multiple scattering
MS	mass spectrometer
MVLRM	multiple variable linear regression model
MVNRM	multiple variable nonlinear regression model
$M_{j,k}$	undesignated metals
$\mathbf{m_i}$	apparent heterogenity parameter reflecting width of distribution i
m	exponent in LF equation related to width of Sips distribution
m_e	atomic mass
<u>m</u>	molecular mass
m	meter
mA	milliamp
me	mass of electron
mg	milligram
min	minute
ml	milliliter
mm	millimeter
mmol	millimole
mol	mole
N_A	Avogodro's number
N_i	ion specific exponent of local isotherm = non-ideality factor
N_s	surface site density
N	flux of a dissolved metal
N	normality of solution
N	coordination number
N	number (total) of atoms or ions
NAA	neutron activation analysis
NBS	National Bureau of Standards
NEA	Nuclear Energy Agency
NICA	non-ideal competitive adsorption equation
NIR	near infrared
NIST	National Institute of Standards and Technology
NLEA	nonlinear adsorption and local equilibrium model
NLNE	nonlinear adsorption and nonequilibrium model
NPL	National Priority List of the EPA
Nu	Nusselt number
n	number (equation term coefficient)
n_F	constant in Fruendlich equation
n_i	non-ideality factor

n_L	constant in Langmuir equation
nm	nanometer
O_{ax}	oxygen, axial position
O_{eq}	oxygen, equitorial position
O_S	atom, at solid surface
O	surface adsorption plane
OC	organic carbon
P	Peclet number
Pa	Pascal
PFBA	pentafluorobenzoate
$p[H_s]$	negative logarithim of proton activity at location of charged sites in humic acid phase
p[H]	negative logarithim of proton activity in bulk solution
pH_{pzc}	pH at point of zero charge
PIPES	piperazine-N,N′-bis(2-ethansulfonic acid)
POM	particulate organic matter
PPHA	humic acid, purified from peat
p	parameter characterizing width of the generic distribution in simplification of NICA equation to Langmuir-Fruendlick equation
pK	minus log of acidity, hydrolysis, or complex formation constant
ps	picosecond
Q_i	amount of i*th* ion adsorbed
$Q_{i,max}$	number of sites in distribution
Q_M	quantity of metal adsorbed
Q_T	total number of protons bound
Q	mass action equation, conditional constant
R^2	correlation coefficient, squared
R	vertex of polygonal surface
R_f	retardation factor for linear isotherm
R'_f	retardation factor for nonlinear isotherm
R	gas constant
R	bonding or interatomic distance
Re	Reynolds number
RCF	relative centrifical force
REE	rare earth elements
RM	reference material
RSD	relative standard deviation

RSF	radial structure function
RW	Ryan-Webster equations
R_f	retardation factor
r	radial position within sphere measured from center
$\mathbf{r}_i$	spatial distance between atoms
rpm	revolutions per minute
S′:	pH dependent plus pH independent adsorption sites
S:	structural metal atom exhibiting pH dependent charge
S:O	surface adsorption site, deprotonated
S:OH	surface adsorption site, neutral
$S{:}OH_2$	surface adsorption site, protonated
>S	surface adsorption site
S_A	specific surface area
S_c	central sulfur atom (in species or complex)
S_{EA}	effective surface area
S_o	other than central sulfur atom (in species or complex)
S_0	zero salinity
S_T	total site density or concentration
S_e	noncentral sulfur atom (in species or complex)
S‰	salinity
S	concentration of dissolved metal
Sc	Schmidt number
SAz	Arizona montmorillonite
SCM	surface complexation model
SEM	scanning electron microscope
SFM	scanning force microscopy
SGW	synthetic ground water
SH_2	high affinity adsorbed proton
SH	low affinity adsorbed proton
SHCa	California hectorite
SM	adsorbed metal, bidentate
SMM	adsorbed metal, monodentate
SOS	weighted sum of squares of residuals
SSRL	Stanford Synchrotron Radiation Laboratory
SV	Stern-Volmer equation
s	second
sec	second
$_T$	total amount of subscripted variable
T	temperature
TEM	transmission electron microscopy
TL, TLM	triple layer, triple layer model

TLS	total luminescence spectroscopy
TOC	total organic carbon
$t_{1/2}$	half-life
t_i	transference number
t	time
U	surface potential energy
USEPA	U.S. Environmental Protection Agency
UV	ultraviolet
V	volume of experimental system
V	electrostatic potential
V	volt
$\mathbf{V}_d$	Donnan volume
V_e	volume at reaction equilibrium
V_i	initial volume
V_o	height of energy barrier between adsorption sites
V_y	overall variance
v	bulk fluid velocity
v	volt
W	weight (mass)
wt %	weight percent
X	exchange site, cation *see also* Y and Z
XANES	X-ray absorption spectroscopy
XAS	X-ray absorption spectroscopy
XRD	x-ray diffraction
x	number of water molecules in reaction
x	stoichiometry in $S{:}H^x \leftrightarrow S{:}M^y$ reaction
x/m	amount of metal adsorbed per unit mass of geomedia
Y	exchange site, cation *see also* X and Z
yr	year
Z_i	charge on i*th* ion
Z	exchange site, cation *see also* X, Y
Z	atomic number
z	charge of ion
z	charge at solid surface

Index

B

C

D

E

F

G

H

L

M

N

O